EEG Signal Processing and Machine Learning

EEG Signal Processing and Machine Learning

Second Edition

Saeid Sanei
Imperial College London & Nottingham Trent University, UK

Jonathon A. Chambers
University of Leicester, UK

This second edition first published 2022

Edition History
John Wiley & Sons, Ltd. (1e, 2007)

Registered Offices
John Wiley & Sons, Inc., 111 River Street, Hoboken, NJ 07030, USA
John Wiley & Sons Ltd, The Atrium, Southern Gate, Chichester, West Sussex, PO19 8SQ, UK

Editorial Office
The Atrium, Southern Gate, Chichester, West Sussex, PO19 8SQ, UK

For details of our global editorial offices, customer services, and more information about Wiley products visit us at www.wiley.com.

Wiley also publishes its books in a variety of electronic formats and by print-on-demand. Some content that appears in standard print versions of this book may not be available in other formats.

Library of Congress Cataloging-in-Publication Data

Names: Sanei, Saeid, author. | Chambers, Jonathon A., author. | John Wiley & Sons, publisher
Title: EEG signal processing and machine learning / Saeid Sanei, Jonathon A. Chambers.
Description: Second edition. | Hoboken, NJ : Wiley, 2021. | Includes bibliographical references and index.
Identifiers: LCCN 2021003276 (print) | LCCN 2021003277 (ebook) | ISBN 9781119386940 (hardback) | ISBN 9781119386926 (adobe pdf) | ISBN 9781119386933 (epub)
Subjects: LCSH: Electroencephalography. | Signal processing. | Machine learning.
Classification: LCC RC386.6.E43 S252 2021 (print) | LCC RC386.6.E43 (ebook) | DDC 616.8/047547–dc23
LC record available at https://lccn.loc.gov/2021003276
LC ebook record available at https://lccn.loc.gov/2021003277

Cover Design: Wiley
Cover Image: © Andrea Danti/Shutterstock, imaginima/iStock/Getty Images, xijian/iStock/Getty Images, Marmaduke St. John/Alamy Stock Photo

Set in 9.5/12.5pt STIXTwoText by Straive, Pondicherry, India

10 9 8 7 6 5 4 3 2 1

Contents

Preface to the Second Edition

Brain research has reached a considerable level of maturity due, for example, to having access to: a wealth of recording and screening resources; availability of substantial data banks; advanced data processing algorithms; and emerging artificial intelligence (AI) for making more accurate clinical diagnosis. Neurotechnology is also now being exploited to design revolutionary interfaces to guide artificial prostheses for human rehabilitation. Moreover, the technology for brain repair, communications between live and AI-based body parts, mind reading, and intelligent recordings together with the use of virtual and augmented reality domains is advancing remarkably. The advances in brain research will soon make the Internet-of-brains feasible and enable fully monitoring the body for personal medicine purposes.

To progress this fast-growing technology, the demand for electroencephalography (EEG) data, as a widely accessible, informative, flexible, and expandable brain screening modality, together with suitable approaches in EEG processing, is rising dramatically.

Automatic clinical diagnosis requires signal processing and machine learning algorithms to bring more insight into interpretation of the data, devising a treatment plan, and defining the path for achieving personalized medicine which is the goal of future healthcare systems. EEG is of particular interest to researchers due to its very rich information content and its relation to the entire body function.

EEG signals represent three fundamental activities in the brain: firstly, they show the normal brain rhythms which exist in the EEGs of healthy subjects and indicate the human states such as awake and sleep; secondly, they demonstrate the brain responses to audio, visual, and somatosensory excitations, whose variations can represent the brain performance in the cases of mental fatigue, learning, and memory load; and thirdly, the communications between various brain zones which can change due to ageing, dementia, and many other factors. The study of these three aspects of EEG is the focus of this book.

Most of the concepts in single channel or multichannel EEG signal processing have their origin in distinct application areas such as communication, seismic, speech and music signal processing. EEG signals are generally slow-varying waveforms and therefore, similar to many other physiological signals, can be processed online without much computational effort.

This second edition of the book EEG Signal Processing, first published in 2007, highlights the major impact machine learning is now having on EEG analysis. This has been made possible by the recent developments in data analysis: firstly, due to the availability of supercomputers, powerful graphic cards, large volume computer clusters, and memory space within the public cloud, and secondly, due to introducing powerful classification algorithms such as deep neural networks (DNNs) which are suitable for numerous applications in brain–computer interfacing, mental task evaluation, brain disorder/disease recognition, and many others.

This edition is inclusive and comprehensive, encompassing almost all methodologies in EEG processing and learning together with their diverse applications. It is not only the result of the endeavours of our research teams, but also an encyclopaedia of the most recent works in EEG signal processing, machine learning, and their applications. Hence, this edition covers a wider, deeper, and richer content thereby alleviating the shortcomings in the first edition of this book. As such, this edition can be used as a reference by researchers in bioengineering, neuroscience, psychiatry, neuroimaging, and brain–computer interfacing. It can also be used for teaching bioengineering and neuroengineering at different university levels.

In this second edition, the number of chapters has increased from 7 to 18 by covering and extending the content in each chapter and adding many new topics for analysis of EEG signals including: (i) offering deeper understanding and insight into the generation of EEG signals and modelling the brain EEG generators in Chapters 2 and 3, (ii) being more inclusive in the domains of theoretical and practical aspects in EEG single- and multichannel signal processing including static and dynamic systems within multimodal and multiway mathematical models in Chapters 3–6, and (iii) providing a comprehensive and detailed approach to AI, particularly machine learning approaches, starting from traditional crisp classification to advanced deep feature learning approaches in Chapter 7.

Chapter 8 addresses brain coherency, synchrony, and connectivity. This chapter introduces a completely new topic of cooperative learning and adaptive filtering into the domain of brain connectivity and its applications. Chapter 9 introduces the brain response to audio, visual, and tactile events when they are regularly presented or targeted in an odd ball paradigm. The important topic of brain source localization, using both forward and inverse problems, is addressed in Chapter 10. A vast range of applications of these three chapters is given in the ensuing chapters.

From Chapter 11 onwards, more practical and clinically demanding approaches are discussed with the help and direct application of the theoretical developments in the previous chapters. Seizure and epileptic waveforms are studied comprehensively in Chapter 11. This chapter includes a very innovative approach using DNNs to model the pathways between the generators of epileptiform discharges to the scalp electrode recordings.

The fundamental objectives of new and advanced materials included in Chapters 12–14 are to assess the cortical brain waves, the coherency and connectivity within various brain zones, and the brain responses to different stimuli while the subject is in different states of awake, sleep, mentally tired, and under different emotions. Chapters 15 and 16 introduce the state-of the-art techniques in signal processing and machine learning for recognition of degenerative diseases and neurodevelopmental disorders respectively.

Brain–computer interfacing benefiting from a wealth of new neurotechnology together with applications of advanced AI systems for rehabilitation, computer gaming, and eventually brain communications purposes is comprehensively covered in Chapter 17 which concludes application of EEG signals and systems. Finally, Chapter 18 shows how EEG can be combined with other simultaneously recorded functional neuroimaging data. It introduces a number of applications where EEG can be combined with functional magnetic resonance images (fMRI) and functional near-infrared spectroscopy (fNIRS) images, to exploit their high spatial resolutions for enhancing the overall diagnostic performance.

In the treatment of various topics covered within this research monograph it is assumed that the reader has a background in the fundamentals of digital signal processing and machine learning and wishes to focus on EEG analysis. It is hoped that the concepts covered in each chapter provide a solid foundation for future research and development in the field.

As we concluded in the first edition, we do wish to stress that in this book there is no attempt to challenge previous clinical or diagnostic knowledge. Instead, the tools and algorithms described in this book can, we believe, potentially enhance the significant information within EEG signals and thereby aid physicians and ultimately provide more cost effective and efficient diagnostic tools.

Both authors wish to thank most sincerely our Research Associates and PhD students who have contributed so much to the materials in this work.

Saeid Sanei and Jonathon A. Chambers

Preface to the First Edition

There is ever-increasing global demand for more affordable and effective clinical and healthcare services. New techniques and equipment must therefore be developed to aid in the diagnosis, monitoring, and treatment of abnormalities and diseases of the human body. Biomedical signals (biosignals) in their manifold forms are rich information sources, which when appropriately processed have the potential to facilitate such advancements. In today's technology, such processing is very likely to be digital, as confirmed by the inclusion of digital signal processing concepts as core training in biomedical engineering degrees. Recent advancements in digital signal processing are expected to underpin key aspects of the future progress in biomedical research and technology, and it is the purpose of this research monograph to highlight this trend for the processing of measurements of brain activity, primarily electroencephalograms (EEGs).

Most of the concepts in multichannel EEG digital signal processing have their origin in distinct application areas such as communications engineering, seismics, speech and music signal processing, together with the processing of other physiological signals, such as electrocardiograms (ECGs) The particular topics in digital signal processing first explained in this research monograph include: definitions; illustrations; time domain, frequency domain, and time–frequency domain processing; signal conditioning; signal transforms; linear and nonlinear filtering; chaos definition, evaluation, and measurement; certain classification algorithms; adaptive systems; independent component analysis; and multivariate autoregressive modelling. In addition, motivated by research in the field over the last two decades, techniques specifically related to EEG processing such as brain source localization, detection and classification of event-related potentials, sleep signal analysis, seizure detection and prediction, together with brain–computer interfacing are comprehensively explained and, with the help of suitable graphs and (topographic) images, simulation results are provided to assess the efficacy of the methods.

Chapter 1 of this research monograph is a comprehensive biography of the history and generation of EEG signals, together with a discussion of their significance and diagnostic capability. Chapter 2 provides an in-depth introduction to the mathematical algorithms and tools commonly used in the processing of EEG signals. Most of these algorithms have only been recently developed by experts in the signal processing community and then applied to the analysis of EEG signals for various purposes. In Chapter 3, event-related potentials are explained and the schemes for their detection and classification are explored. Many neurological and psychiatric brain disorders are diagnosed and monitored using these techniques. Chapter 4 complements the previous chapter by specifically looking at the behaviour of EEG signals in patients suffering from epilepsy. Some very recent methods in seizure prediction are demonstrated. This chapter concludes by opening up a new methodology in joint, or bimodal, EEG–fMRI analysis of epileptic seizure signals. Localization of brain source signals is next covered in Chapter 5. Traditional dipole methods are described and some very recent processing techniques such as blind source separation are briefly reviewed. In Chapter 6, the concepts developed for the analysis and description of EEG sleep recordings are summarized and the important parameters and terminologies are explained. Finally, in Chapter 7, one of the most important applications of the developed mathematical tools for processing of EEG signals, namely brain–computer interfacing, is explored and recent advancements are briefly explained. Results of the application of these algorithms are described.

In the treatment of various topics covered within this research monograph it is assumed that the reader has a background in the fundamentals of digital signal processing and wishes to focus on processing of EEGs. It is hoped that the concepts covered in each chapter provide a foundation for future research and development in the field.

In conclusion, we do wish to stress that in this book there is no attempt to challenge previous clinical or diagnostic knowledge. Instead, the tools and algorithms described in this book can, we believe, potentially enhance the significant clinically related information within EEG signals and thereby aid physicians and ultimately provide more cost effective and efficient diagnostic tools.

Both authors wish to thank most sincerely our previous and current PhD students who have contributed so much to the material in this work and our understanding of the field. Special thanks to Min Jing, Tracey Lee, Kianoush Nazarpour, Leor Shoker, Loukianous Spyrou, and Wenwu Wang, who contributed to providing some of the illustrations. Finally, this book became truly possible due to spiritual support and encouragement of Maryam Zahabsaniei, Erfan Sanei, and Ideen Sanei.

Saeid Sanei
Jonathon A. Chambers
January 2007

List of Abbreviations

3D	Three-dimensional
AASM	American Academic of Sleep Medicine
ACC	Anterior cingulate cortex
ACE	Addenbrooke's cognitive examination
ACR	Accuracy of responses
ACT	Adaptive chirplet transform
AD	Alzheimer's disease
ADC	Analogue-to-digital converter
ADD	AD patients with mild dementia
ADD	Attention-deficit disorder
ADHD	Attention-deficit hyperactivity disorder
AE	Approximate entropy
AE	Autoencoder
AEP	Audio evoked potentials
AfC	Affective computing
Ag–AgCl	Silver–silver chloride
AI	Artificial intelligence
AIC	Akaike information criterion
ALE	Adaptive line enhancer
ALF	Adaptive standardized LORETA/FOCUSS
ALM	Augmented Lagrange multipliers method
ALS	Alternating least squares
ALS	Amyotrophic lateral sclerosis
AMDF	Average magnitude difference function
AMI	Average mutual information
AMM	Augmented mixing matrix
ANN	Artificial neural network
AOD	Auditory oddball
AP	Action potential
ApEn	Approximate entropy
APGARCH	Asymmetric power GARCH
AR	Autoregressive modelling
ARMA	Autoregressive moving average

ASCOT	Adaptive slope of wavelet coefficient counts over various thresholds
ASD	Autism spectrum disorder
ASDA	American Sleep Disorders Association
AsI	Asymmetry index
ASR	Automatic speaker recognition
ASS	Average artefact subtraction
AUC	Area under the curve
BAS	Behavioural activation system
BBCI	Berlin BCI
BBI	Brain-to-brain interface
BCG	Ballistocardiogram
BCI	Brain–computer interfacing
BDS	Brock, Dechert, and Scheinkman
BEM	Boundary-element method
BF	Beamformer
BGD	Bootstrapped geometric difference
BIC	Bayesian information criterion
BIS	Behavioural inhibition system
BIS	Bispectral index
BMI	Brain–machine interfacing
BOLD	Blood oxygenation level dependent
BP	Bereitschaftspotential
BP	Bipolar disorder
Brain/MINDS	Brain Mapping by Integrated Neurotechnologies for Disease Studies
BSE	Blind source extraction
BSR	Burst-suppression ratio
BSS	Blind source separation
bvFTD	Behaviour variant frontotemporal dementia
Ca	Calcium
CAE	Contractive autoencoder
CANDECOMP	Canonical decomposition
CBD	Corticobasal degeneration
CBF	Cerebral blood flow
CCA	Canonical correlation analysis
CEEMDAN	Complete ensemble EMD with adaptive noise
CF	Characteristic function
CF	Cognitive fluctuation
CFS	Chronic fatigue syndrome
Cl	Chloride
CDLSA	Coupled dictionary learning with sparse approximation
CDR	Current distributed-source reconstruction
CI	Covariance intersection

cICA	Constrained ICA
CIT	Concealed information test
CJD	Creutzfeldt–Jakob disease
CMA	Circumplex model of affects
CMTF	Coupled matrix and tensor factorizations
CMOS	Complementary metal oxide semiconductor
CNN	Convolutional neural network
CNS	Central nervous system
CORCONDIA	Core consistency diagnostic
CoSAMP	Compressive sampling matching pursuit
CPS	Cyber-physical systems
CRBPF	Constrained Rao-Blackwellised particle filter
CSA	Central sleep apnoea
CSD	Current source density
CSF	Cerebrospinal fluid
CSP	Common spatial patterns
CT	Computerized tomography
DAE	Denoising autoencoder
DARPA	Defence Advanced Research Projects Agency
DASM	Differential asymmetry
DBS	Deep brain stimulation
DC	Direct current
DCAU	Differential Causality
DCM	Dynamic causal modelling
DCT	Discrete cosine transform
dDTF	Direct directed transfer function
DE	Differential entropy
DeconvNet	Deconvolutional ANN
DFT	Discrete Fourier transform
DFV	Dominant frequency variability
DHT	Discrete Hermite transform
DL	Diagonal loading
DLE	Digitally linked ears
DM	Default mode
DMN	Default mode network
DNN	Deep neural network
DPF	Differential pathlength factor
DSM	Diagnostic and Statistical Manual
DSTCLN	Deep spatio-temporal convolutional bidirectional long short-term memory network
DT	Decision tree
DTF	Directed transfer function
DTI	Diffusion tensor imaging
DUET	Degenerate unmixing estimation technique

DWT	Discrete wavelet transform
ECD	Electric current dipole
ECD	Equivalent current dipole
ECG	Electrocardiogram
ECG	Electrocardiography
ECoG	Electrocorticogram
ECT	Electroconvulsive therapy
ED	Error distance
EEG	Electroencephalogram
EEG	Electroencephalography
EEMD	Ensemble empirical mode decomposition
EGARCH	Exponential GARCH
EGG	Electrogastrography
EKG	Electrocardiogram
EKG	Electrocardiography
EM	Expectation maximization
EMD	Empirical mode decomposition
EMG	Electromyogram
EMG	Electromyography
ENet	Efficient neural network
EOG	Electro-oculogram
EP	Evoked potential
EPN	Early posterior negativity
EPSP	Excitatory post-synaptic potential
ERBM	Entropy rate bound minimization
ERD	Event-related desynchronization
ERN	Error-related negativity
ERP	Event-related potential
ERS	Event-related synchronization
FA	Factor analysis
FC	Functional connectivity
FCM	Fuzzy *c*-means
FD	Fractal dimension
FDA	Food and Drug Administration
FDispEn	Fluctuation-based dispersion entropy
FDR	False detection rate
FEM	Finite element model
FFNN	Feed forward neural network
FET	Field-effect transistor
fICA	Fast independent component analysis
FIR	Finite impulse response
fMRI	Functional magnetic resonance imaging
FMS	Fibromyalgia syndrome
FN	False negative

fNIRS	Functional near-infrared spectroscopy
FO	Foramen ovale
FOBSS	First order blind source separation
FOCUSS	Focal underdetermined system solver
FOOBI	Fourth order cumulant based blind identification
FP	False positive
FRDA	Frontal rhythmic delta activity
FRN	Feedback related negativity
FSOR	Feature selection with orthogonal regression
FSP	Falsely detected source number (position)
FTD	Frontotemporal dementia
FuzEn	Fuzzy entropy
GA	Genetic algorithm
GAD	General anxiety disorder
GAN	Generative adversarial network
GARCH	Generalized autoregressive conditional heteroskedasticity
GARCH-M	GARCH-in-mean
GC	Granger causality
GCN	Graph convolutional network
GFNN	Global false nearest neighbours
GJR-GARCH	Glosten, Jagannathan, & Runkle GARCH
GLM	General linear model
GMM	Gaussian mixture model
GP	Gaussian process
GP-LR	Gaussian process logistic regression
GSCCA	Group sparse canonical correlation analysis
GSR	Galvanic skin response
GWN	Gaussian white noise
HBO/HbO	Oxyhaemoglobin
HBR/HbR	De-oxyhaemoglobin
HBT	Total haemoglobin
HCI	Human computer interaction
HD	Huntington's disease
HEOG	Horizontal electro-oculograph
HFD	Higuchi's fractal dimension
HHT	Hilbert–Huang transform
HMD	Head-mounted display
HMM	Hidden Markov model
HOPLS	Higher-order partial least squares
HOS	Higher-order statistics
HR	Hemodynamic response
HRF	Haemodynamic response function
HT	Hilbert transform
IAPS	International affective picture system

IBE	International Bureau for Epilepsy
IC	Independent component
ICA	Independent component analysis
iCOH	Imaginary part of coherency
IED	Interictal epileptiform discharge
iEEG	Intracranial electroencephalogram
ICA	Independent component analysis
IIR	Infinite impulse response
ILAE	International League Against Epilepsy
IMF	Intrinsic mode function
ImSCoh	Imaginary part of S-coherency
INDSCAL	INDividual Differences SCALing
IoB	Internet-of-brains
IPL	Inferior parietal lobule
IPSP	Inhibitory post-synaptic potential
IR	Impulse response
IRLS	Iterative recursive least squares
ISODATA	Iterative self-organizing data analysis technique algorithm
Isomap	Isometric mapping
ISSWT	Inverse synchro-squeezing wavelet transform
ITR	Information transfer rate
IVE	Immersive virtual environments
JAD	Joint approximate diagonalization
JADE	Joint approximate diagonalization of eigenmatrices
jICA	Joint ICA
K	Potassium
Kc	Kolmogorov complexity
KDT	Karolinska drowsiness test
KL	Kullback–Leibler
KLT	Karhunen–Loéve transform
KMI	Kinaesthetic motor imagery
KNN	k-nearest neighbour
KPCA	Kernel principal component analysis
KSS	Karolinska sleepiness scale
KT	Kuhn–Tucker
LBD	Lewy body dementia
LCMV	Linearly constrained minimum variance
LD	Linear discriminants
LD	Linearly distributed
LDA	Linear discriminant analysis
LDA	Long delta activity
LE	Lyapunov exponent
LEM	Local EEG model
LLE	Largest Lyapunov exponent

LMS	Least mean square
LORETA	Low-resolution electromagnetic tomography algorithm
LP	Lowpass
LPM	Letters per minute
LPP	Late positive potential
LRCN	Long-term recurrent convolutional network
LRT	Low-resolution tomography
LS	Least squares
LSE	Least-squares error
LSTM	Long short-term memory network
LVQ	Learning vector quantization
LWR	Levinson–Wiggins–Robinson
LZC	Lempel–Ziv complexity
M2M	Machine-to-machine
MA	Mental arithmetic
MA	Moving average
MAF	Multivariate ambiguity function
MAP	Maximum a posteriori
MCI	Mild cognitive impairment
MCMC	Markov chain Monte Carlo
MDI	Multidimensional directed information
MDP	Moving dipole
MEG	Magnetoencephalogram
MFDE	Multiscale fluctuation-based dispersion entropy
mHTT	Mutant huntingtin
MI	Mutual information
MIL	Matrix inversion lemma
ML	Maximum likelihood
MLE	Maximum likelihood estimation
MLE	Maximum Lyapunov exponent
MLP	Multilayer perceptron
MMN	Mismatch negativity
MMSE	Minimum mean squared error
MNI	Montreal Neurological Institute and Hospital
MNLS	Minimum norm least squares
MP	Matching pursuits
MRI	Magnetic resonance imaging
MRP	Movement-related potential
MSE	Multiple system atrophy
MSE	Mean squared error
MSE	Multiscale entropy
MS	Multiple sclerosis
MTLE	Mesial temporal lobe epilepsy
MUSIC	Multichannel signal classification

MVAR	Multivariate autoregressive
Na	Sodium
NC	Normal control
NCDF	Normal cumulative distribution function
NCSP	Nonparametric common spatial patterns
NDD	Neurodevelopmental disorder
NES	Nonepileptic seizure
NIH	National Institute of Health
NIR	Near infrared
NIRS	Near-infrared spectroscopy
NLMS	Normalized least mean square
NMF	Nonnegative matrix factorization
NMCSP	Nonparametric multiclass common spatial patterns
NMR	Nuclear magnetic resonance
NN	Neural network
NNQP	Nonnegative quadratic program
NP	Neural process
NREM	Non-rapid eye movement
NSI	Nonstationary index
OA	Ocular artefact
OBS	Optimal basis set
OBS	Organic brain syndrome
OFC	Orbital frontal cortex
OMP	Orthogonal matching pursuit
OP	Oddball paradigm
OSA	Obstructive sleep apnoea
OSAHS	Obstructive sleep apnoea hypopnea syndrome
PARAFAC	Parallel factor analysis
PCA	Principal component analysis
PCANet	Principal component analysis network
PCC	Pearson product correlation coefficient
PCC	Posterior cingulate cortices
PD	Parkinson's disease
PDC	Partial directed coherence
pdf	Probability density function
Pe	Error positivity
PerEn	Permutation entropy
PET	Positron emission tomography
PF	Particle filter
PFC	Prefrontal cortex
PIC	Power iteration clustering
PIF	Phase interaction function
PLED	Periodic literalized epileptiform discharges
PLI	Phase lag index

PLMD	Periodic limb movement disorder
PLS	Partial least squares
PMBR	Post-movement beta rebound
PMBS	Post-movement beta synchronization
PNRD	Nonrhythmic delta activity
POST	Positive occipital sharp transients
PPC	Phase–phase coupling
PPG	Photoplethysmography
PPM	Piecewise Prony method
PSD	Power spectrum density
PSDM	Phase-space dissimilarity measures
PSG	Polysomnography
PSI	Phase-slope index
PSP	Post-synaptic potential
PSP	Progressive supranuclear palsy
PSWC	Periodic sharp wave complexes
PTSD	Post-traumatic stress disorder
PWVD	Pseudo-Wigner–Ville distribution
QEEG	Quantitative EEG
QGARCH	Quadratic GARCH
QNN	Quantum neural networks
QP	Quadratic programming
R&K	Rechtschtschaffen and Kales
RAP	Recursively applied and projected
RASM	Rational asymmetry
RBD	REM sleep behaviour disorder
RBF	Radial basis function
RBPF	Rao-Blackwellised particle filter
RBR	Relative beta ratio
RCE	Recursive channel elimination
RE	Regional entropy
ReLU	Rectified linear unit
REM	Rapid eye movement
RF	Radio frequency
RFNN	Recurrent fuzzy neural network
RKHS	Reproducing kernel Hilbert spaces
RLS	Recursive least squares
RMBF	Robust minimum variance beamformer
RMS	Root mean square
RNN	Recurrent neural network
ROC	Receiver operating characteristic
RP	Readiness potential
RR	Respiratory rate
RT	Reaction time

rTMS	repetitive transcranial magnetic stimulation
RV	Residual variance
SAE	Stacked autoencoder
SampEn	Sample entropy
SAS	Sleep apnoea syndrome
SCA	Sparse component analysis
SCD	Sickle cell disease
SCP	Slow cortical potential
SCPS	Slow cortical potential shift
SCR	Skin conductance response
SCV	Spectral coherence value
SCWT	Stroop colour and word test
SDAE	Stacked denoising autoencoder
SDTF	Short-time DTF
SEM	Structural equation modelling
SFS	SyncFastSlow
SG	Sensory gating
SI	Synchronization index
SICA	Spatial ICA
SL	Synchronization likelihood
sLORETA	Standardized LORETA
SLTP	Short- and long-term prediction
SMI	Sample-matrix inversion
SMl	Sensorimotor left
SMOTE	Synthetic minority oversampling technique
sMRI	Structural MRI
SN	Salient network
SNN	Spike neural network
SNNAP	Simulator for Neural Networks and Action Potentials
SNR	Signal-to-noise ratio
SOBI	Second-order blind identification
SOBIUM	Second-order blind identification of underdetermined mixtures
SPET	Single photon emission tomography
SPM	Statistical parametric mapping
SPQ	Schizotypal personality questionnaire
SREDA	Subclinical rhythmic EEG discharges of adults
SRNN	Sleep EEG recognition neural network
SSA	Singular spectrum analysis
SSLOFO	Shrinking standard LORETA-FOCUSS
SSPE	Subacute sclerosing panencephalitis
SSVEP	Steady-state visual evoked potential
SSVER	Steady-state visual evoked response
SSWT	Synchro-squeezing wavelet transform

STF	Space–time–frequency
STFD	Spatial time–frequency distribution
STFT	Short time–frequency transform
STL	Short-term largest Lyapunov exponent
STS	Superior temporal sulcus
SV	Support vectors
SVD	Singular-value decomposition
SVM	Support vector machines
SVR	Support vector regression
SWA	Slow-wave activity
SWDA	Step-wise discriminant analysis
SW	Slow wave
SWP	Slow-wave power
SWS	Slow-wave sleep
TBI	Traumatic brain injury
TDNN	Time delay neural network
TDOA	Time difference of arrival
TDP-43	Transactive response DNA-binding protein 43 kDa
TENS	Transcutaneous electrical nerve stimulation
TF	Time–frequency
TGARCH	Threshold GARCH model
TICA	Temporal ICA
TIS	TMS induction simulator
TLE	Temporal lobe epilepsy
TMS	Transcranial magnetic stimulation
TN	True negative
TNM	Traditional nonlinear method
TOA	Time of arrival
TotHb	Total haemoglobin
TP	True positive
TR	Repeat time
TSSA	Tensor-based singular spectrum analysis
TTD	Thought translation device
UBI	Underdetermined blind identification
UOM	Underdetermined orthogonal model
USP	Undetected source number
USR	Underdetermined source recovery
VAE	Variational autoencoder
VEOG	Vertical electro-oculograph
VEP	Visual evoked potential
VLSI	Very large-scale integrated
vMPFC	Ventral medial prefrontal cortices
VPP	Vertex positive peak
VR	Virtual reality

WA	Wald tests on amplitudes
WC	Word chain
WCO	Weakly coupled oscillator
WDC	Weighted degree centrality
WL	Wald test on locations
WMN	Weighted minimum norm
WN	Wavelet network
WPE	Wavelet packet energy
wPLI	Weighted phase lag index
wSMI	Weighted symbolic mutual information
WT	Wavelet transform
WU	Weighted undersampling
WV	Wigner–Ville

1

Introduction to Electroencephalography

1.1 Introduction

The brain is the most amazing and complicated part of the human body and is responsible for controlling all other organs. The neural activity of the human brain starts between the seventeenth and twenty-third week of prenatal development. It is believed that from this early stage and throughout life electrical signals generated by the brain represent not only the brain function but also the status of the whole body. This assumption provides the motivation to study and understand the range of brain activities including normal brain rhythms, brain responses to stimuli, brain motor generators, and finally brain connectivity. One or more of these activities change in cases of brain disorder, disease, or abnormality. The brain status and often the entire body condition can then be recognized by applying advanced digital signal processing and machine learning methods to the electroencephalography (EEG) signals measured from the brain, and thereby underpin the later chapters of this book.

Although nowhere in this book do the authors attempt to comment on the physiological aspects of brain activities, there are several issues related to the nature of the original sources, their actual patterns, and the characteristics of the medium, that have to be addressed. The medium defines the path from the neurons, so-called signal sources, to the electrodes, which are the sensors where some form of mixtures of the sources (for the case of scalp electrodes) or individual sources (e.g. for subdural electrodes) are measured.

Understanding of neuronal functions and neurophysiological properties of the brain together with the mechanisms underlying the generation of signals and their recordings is however, vital for those who deal with these signals for detection, diagnosis, and treatment of brain disorders and the related diseases.

Examining brain activity or abnormality, however, is not limited to the use of EEG. For the abnormalities which affect the brain structure, different radiographical methods can be used. Also, for detecting the brain functional abnormalities, which is the main agenda for activity detection and functional monitoring, magnetoencephalography (MEG) and functional magnetic resonance imaging (fMRI) may be used. However, MEG is expensive and not readily accessible, and fMRI has very low temporal resolution and is not widely

EEG Signal Processing and Machine Learning, Second Edition. Saeid Sanei and Jonathon A. Chambers.

available. Therefore, EEG remains the main functional brain scanning modality as it is cheap, portable, and widely available.

We begin by providing a brief history of EEG measurements and looking at the journey from the time the brain function was initially recognized to the current time when new techniques in data processing, machine learning, and artificial intelligence have become popular research focuses.

1.2 History

The first understanding of the brain was in 1700 BCE when Imhotep lived in Egypt (in Edwin Smith Surgical Papyrus). At that time the hieroglyphic for 'brain' was presented as that in Figure 1.1. Then, during 460–379 BCE, Hippocrates discussed and introduced epilepsy as a disturbance of the brain. Since then many physicians, clinicians, and philosophers from around the world, particularly from Greece (Roman) and Iran (Persian), have encountered various brain diseases. In his Canon of Medicine (Al-Qanun fi al-Tibb), Avicenna (980–1037, also known as Abu Ali Sina), a Persian physician and philosopher, categorizes the causes of epilepsy into two main groups: those caused by brain diseases and those associated with the abnormalities and diseases of other organs.

However, the history of EEG, as an instrument to record the brain activity, goes back to when for the first time some activity of the brain was recorded or displayed. Carlo Matteucci (1811–1868, Pisa, Italy) and Emil Du Bois-Reymond (1818–1896, Berlin, Germany) were the first people who registered the electrical signals emitted from muscle nerves using a galvanometer and established the concept of neurophysiology [1, 2]. However, the concept of *action current* introduced by Hermann Von Helmholtz (1821–1894, Potsdam Germany) [3] clarified and confirmed the negative variations occurring during muscle contraction via measuring the speed of frog nerve impulses in 1849.

Richard Caton (British, 1842–1926) measured the brain activities of rabbits and monkeys from over the cortex in 1875. He discovered the electrical nature of the brain and laid the groundwork for Hans Berger to discover alpha wave activity in the human brain. He also placed two electrodes over the human scalp to record for the first time the brain activity in the form of electrical signals in 1875. Since then, the concepts of electro- (referring to registration of brain electrical activities) encephal- (referring to emitting the signals from head) and gram (or graphy), which means drawing or writing, were combined so that the term EEG was henceforth used to denote electrical neural activity of the brain.

Figure 1.1 Hieroglyphic symbol for the ancient Egyptian word for 'brain'.

Figure 1.2 Physiologists Adolf Beck (Polish, 1863–1942) on the left and Vladimir Pravdich-Neminsky (Ukrainian, 1879–1952) on the right who performed the first recording of brain activities from over the skull.

Fritsch (1838–1927) and Hitzig (1838–1907) discovered that the human cerebral can be electrically stimulated. Vasily Yakovlevich Danilevsky (1852–1939) followed Caton's work and finished his PhD thesis in the investigation of brain physiology in 1877 [4]. In this work he investigated the brain activity following electrical stimulation as well as spontaneous electrical activity in the brain of animals.

The cerebral electrical activity observed over the visual cortex of different species of animals was reported by Ernst Fleischl von Marxow (1845–1891). Napoleon Cybulski (1854–1919) provided EEG evidence of an epileptic seizure in a dog caused by electrical stimulation.

The idea of the association of epileptic attacks with abnormal electrical discharges was expressed by Kaufman [5].

Adolf Beck (Polish, 1863–1942) and Vladimir Pravdich-Neminsky (Ukrainian, 1879–1952) measured the EEG from over the skull of dogs. Therefore, these two scientists are indeed the pioneers in scalp EEG recording (Figure 1.2).

Pravdich-Neminsky recorded EEG from the brain, termed the dura, and the intact skull of a dog in 1912. He observed a 12–14 cycles s^{-1} rhythm under normal conditions which slowed under asphyxia and later called it the *electrocerebrogram*.

Although much research work on EEG principles and measurements has been performed by the above scientists, Hans Berger (1873–1941, Germany) was credited and named the first one for discovering and measuring human EEG signals. He began his study of human EEGs in 1920 [6]. Berger is well known by almost all electroencephalographers. He started working with a string galvanometer in 1910, then migrated to a smaller Edelmann model, and after 1924, to a larger Edelmann model. In 1926, Berger started to use the more powerful Siemens double coil galvanometer (attaining a sensitivity of 130 $\mu V\ cm^{-1}$) [7]. His first report of human EEG recordings of one to three minutes duration on photographic paper was in 1929. In this recording he only used a one channel bipolar method with fronto-occipital leads. Recording of the EEG became popular in 1924. The first report of 1929 by Berger included the alpha rhythm, as the major component of the EEG signals as described later in this chapter, and the alpha blocking response.

During the 1930s, the first EEG recording of sleep spindles was undertaken by Berger. He then reported the effect of hypoxia on the human brain, the nature of several diffuse and localized brain disorders, and gave an inkling of epileptic discharges [8]. During this time another group established in Berlin-Buch and led by Kornmüller, provided more precise recording of the EEG [9]. Berger was also interested in cerebral localization and particularly in the localization of brain tumours. He also found some correlation between mental activities and the changes in the EEG signals.

Toennies (1902–1970) from the group in Berlin built the first biological amplifier for the recording of brain potentials. A differential amplifier for recording of EEGs was later produced by the Rockefeller foundation in 1932.

The importance of multichannel recordings and using a large number of electrodes to cover a wider brain region was recognized by Kornmüller [10]. The first EEG work focusing on epileptic manifestation, and the first demonstration of epileptic spikes was presented by Fischer and Lowenbach [11, 12].

In England, W. Gray Walter became the pioneer of clinical EEG. He discovered the foci of slow brain activity (delta waves), which initiated enormous clinical interest in the diagnosis of brain abnormalities. In Brussels, Fredric Bremer (1892–1982) discovered the influence of afferent signals on the state of vigilance [13].

Research activities related to EEGs started in North America in around 1934. In this year, Hallowell Davis illustrated a good alpha rhythm for himself. A cathode ray oscilloscope was used around this date by the group in St. Louis University in Washington, in the study of peripheral nerve potentials. The work on human EEGs started at Harvard in Boston and the University of Iowa in the 1930s. The study of epileptic seizure developed by Fredric Gibbs was the major work on EEGs during these years, as the realm of epileptic seizure disorders was the domain of their greatest effectiveness. Epileptology may be divided historically into two periods [14]: before and after the advent of EEG. Gibbs and Lennox applied the idea of Fischer based on his studies about picrotoxin and its effect on the cortical EEG in animals to human epileptology. Berger [15] showed a few examples of paroxysmal EEG discharges in a case of presumed petit mal attacks and during a focal motor seizure in a patient with general paresis.

As the other great pioneers of EEG in North America, Hallowell Davis, Herbert H. Jasper, Frederic A. Gibbs, William Lennox, and Alfred L. Loomis were the earliest investigators of the nature of EEG during human sleep. Alfred L. Loomis, E. Newton Harvey, and Garret A. Hobart were the first who mathematically studied the human sleep EEG patterns and the stages of sleep. At McGill University, Herbert Jasper studied the related behavioural disorder before he found his niche in basic and clinical epileptology [16].

The American EEG Society was founded in 1947 and the first international EEG Congress was held in London, United Kingdom around this time. While the EEG studies in Germany were still limited to Berlin, Japan gained attention by the work of Motokawa, a researcher of EEG rhythms [17]. During these years the neurophysiologists demonstrated the thalamocortical relationship through anatomical methods. This leads to the development of the concept of centrencephalic epilepsy [18, 30].

Throughout the 1950s the work on EEGs expanded in many different places. During this time surgical operation for removing the epileptic foci became popular and the book entitled *Epilepsy and the Functional Anatomy of the Human Brain* (Jasper and Penfield) was published. During this time microelectrodes were invented. They were made of metals such as tungsten or glass, filled with electrolytes such as potassium chloride, with diameters of less than 3 μm.

Depth EEG of a human was first obtained with implanted intracerebral electrodes by Mayer and Hayne (1948). Invention of intracellular microelectrode technology revolutionized this method and was used in the spinal cord by Brock et al. in 1952, and in the cortex by Phillips in 1961.

Analysis of EEG signals started during the early days of EEG measurement. Berger assisted by Dietch (1932) applied Fourier analysis to EEG sequences which was rapidly developed during the 1950s. Analysis of sleep disorders with EEGs started its development in the 1950s through the work of Kleitman at the University of Chicago.

In the 1960s the analysis of EEGs of full-term and premature newborns began its development [19]. Investigation of evoked potentials (EPs), especially visual EPs, as commonly used for monitoring mental illnesses, progressed during the 1970s.

The history of EEG however has been a continuous process which started from the early 1300s and has brought daily development of clinical, experimental, and computational studies for discovery, recognition, diagnosis, and treatment of a vast number of neurological and physiological brain abnormalities as well as the rest of human central nervous system (CNS). At this time, EEGs are recorded invasively and noninvasively using fully computerized systems. The EEG machines are equipped with many signal processing tools, delicate and accurate measurement electrodes, and enough memory for very-long-term recordings of several hours. EEG or MEG machines may be integrated with other neuroimaging system such as fMRI. Very delicate needle-type electrodes can also be used for recording the EEGs from over the cortex (electrocorticogram), and thereby avoid the attenuation and nonlinearity effects induced by the skull. We next proceed to describe the nature of neural activities within the human brain.

1.3 Neural Activities

The CNS generally consists of nerve cells and glia cells, which are located between neurons. Each nerve cell consists of axons, dendrites, and cell bodies. Nerve cells respond to stimuli and transmit information over long distances. A nerve cell body has a single nucleus and contains most of the nerve cell metabolism especially that related to protein synthesis. The proteins created in the cell body are delivered to other parts of the nerve. An axon is a long cylinder, which transmits an electrical impulse and can be several metres long in vertebrates (giraffe axons go from the head to the tip of spine). In humans the length can be a percentage of a millimetre to more than a metre. An axonal transport system for delivering proteins to the ends of the cell exists and the transport system has 'molecular motors' which ride upon tubulin rails.

Dendrites are connected to either the axons or dendrites of other cells and receive impulses from other nerves or relay the signals to other nerves. In the human brain each nerve is connected to approximately 10 000 other nerves, mostly through dendritic connections.

The activities in the CNS are mainly related to the synaptic currents transferred between the junctions (called synapses) of axons and dendrites, or dendrites and dendrites of cells. A potential of 60–70 mV with negative polarity may be recorded under the membrane of the cell body. This potential changes with variations in synaptic activities. If an action potential (AP) travels along the fibre, which ends in an *excitatory* synapse, an excitatory post-synaptic potential (EPSP) occurs in the following neuron. If two APs travel along the same fibre over a short distance, there will be a summation of EPSPs producing an AP on the post-synaptic

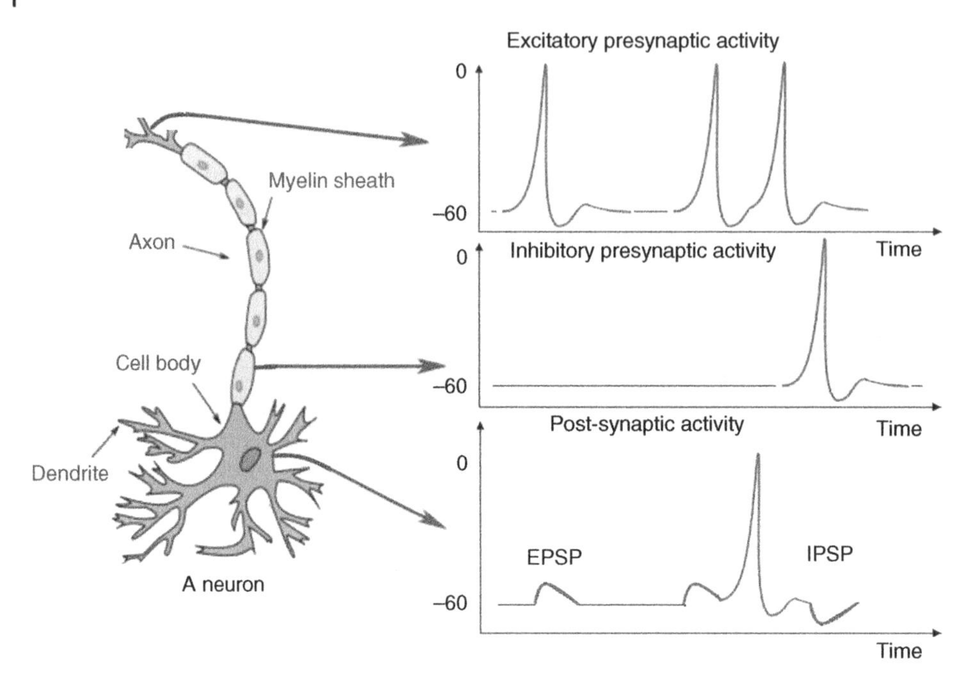

Figure 1.3 The neuron membrane potential changes and current flow during synaptic activation recorded by means of intracellular microelectrodes. APs in the excitatory and inhibitory presynaptic fibre respectively lead to EPSP and IPSP in the post-synaptic neuron.

neuron providing a certain threshold of membrane potential is reached. If the fibre ends in an *inhibitory* synapse, then hyperpolarization will occur, indicating an inhibitory post-synaptic potential (IPSP) [20, 21]. Figure 1.3 shows the above activities schematically.

Following the generation of an IPSP, there is an overflow of cations from the nerve cell or an inflow of anions into the nerve cell. This flow ultimately causes a change in potential along the nerve cell membrane. Primary transmembranous currents generate secondary ional currents along the cell membranes in the intracellular and extracellular space. The portion of these currents that flow through the extracellular space is directly responsible for the generation of field potentials. These field potentials, usually with less than 100 Hz frequency, are called EEGs when there are no changes in the signal average and called DC potential if there are slow drifts in the average signals, which may mask the actual EEG signals. A combination of EEG and DC potentials is often observed for some abnormalities in the brain such as seizure (induced by pentylenetetrazol), hypercapnia, and asphyxia [22]. We next focus on the nature of APs.

1.4 Action Potentials

AP is actually the information transmitted by a nerve. These potentials are caused by an exchange of ions across the neuron membrane and an AP is a temporary change in the membrane potential that is transmitted along the axon. It is usually initiated in the cell body and normally travels in one direction. The membrane potential depolarizes (becomes more

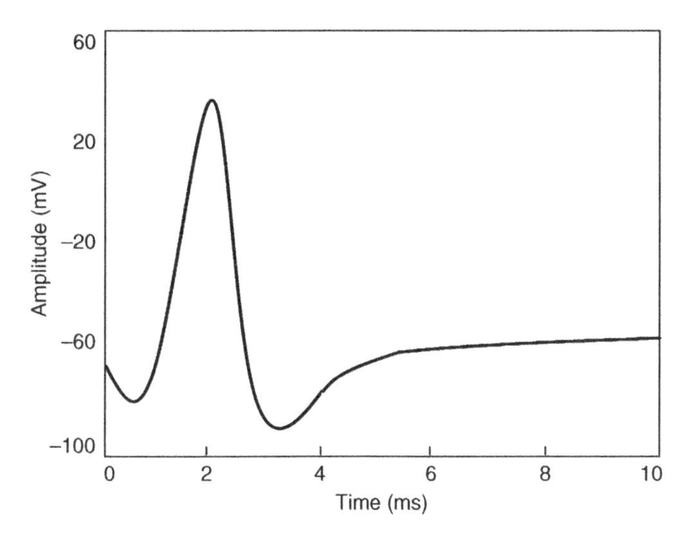

Figure 1.4 An example of an AP.

positive) producing a spike. After the peak of the spike the membrane repolarizes (becomes more negative). The potential becomes more negative than the resting potential and then returns to normal. The APs of most nerves last between 5 and 10 ms. Figure 1.4 shows an example of an AP.

The conduction velocity of APs lies between 1 and 100 m s^{-1}. APs are initiated by many different types of stimuli; sensory nerves respond to many types of stimuli, such as: chemical, light, electricity, pressure, touch, and stretching. Conversely, the nerves within the CNS (brain and spinal cord) are mostly stimulated by chemical activity at synapses.

A stimulus must be above a threshold level to set off an AP. Very weak stimuli cause a small local electrical disturbance, but do not produce a transmitted AP. As soon as the stimulus strength goes above the threshold, an AP appears and travels down the nerve.

The spike of the AP is mainly caused by opening of Na (sodium) channels. The Na pump produces gradients of both Na and K (potassium) ions, both are used to produce the AP; Na is high outside the cell and low inside. Excitable cells have special Na and K channels with gates that open and close in response to the membrane voltage (voltage-gated channels). Opening the gates of Na channels allows Na to rush into the cell, carrying a +ve charge. This makes the membrane potential positive (depolarization), producing the spike. Figure 1.5 shows the stages of the process during evolution of an AP for a giant squid.

For a human being the amplitude of the AP ranges between approximately −60 to 10 mV. During this process [23]:

I) When the dendrites of a nerve cell receive the stimulus the Na^+ channels will open.
II) If the opening is sufficient to drive the interior potential from −70 mV up to −55 mV, the process continues.
III) As soon as the action threshold is reached, additional Na^+ channels (sometimes called voltage-gated channels) open. The Na^+ influx drives the interior of the cell membrane up to about +30 mV. The process to this point is called depolarization.
IV) Then Na^+ channels close and the K^+ channels open. Since the K^+ channels are much slower to open, the depolarization has time to be completed. Having both Na^+ and K^+

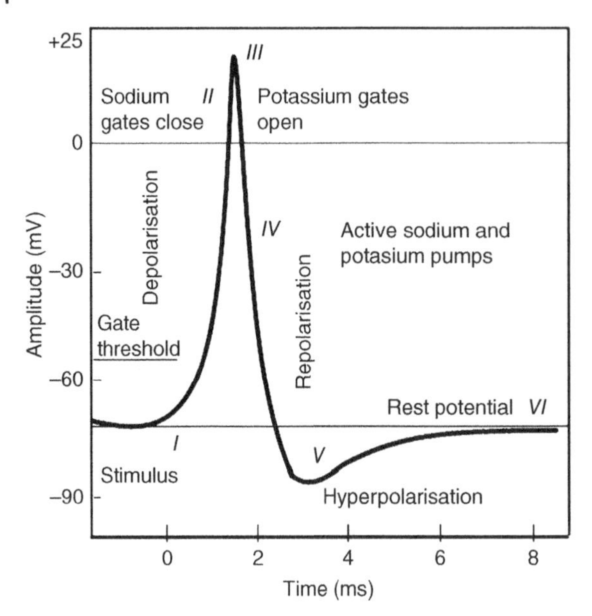

Figure 1.5 Changing the membrane potential for a giant squid by closing the Na channels and opening K channels. (Source: adapted from Ka Xiong Charand [23].)

channels open at the same time would drive the system towards neutrality and prevent the creation of the AP.

V) Having the K^+ channels open, the membrane begins to repolarize back towards its rest potential.

VI) The repolarization typically overshoots the rest potential to a level of approximately −90 mV. This is called hyperpolarization, and would seem to be counterproductive, but it is actually important in the transmission of information. Hyperpolarization prevents the neuron from receiving another stimulus during this time, or at least raises the threshold for any new stimulus. Part of the importance of hyperpolarization is in preventing any stimulus already sent up an axon from triggering another AP in the opposite direction. In other words, hyperpolarization assures that the signal is proceeding in one direction.

After hyperpolarization, the Na^+/K^+ pumps eventually bring the membrane back to its resting state of −70 mV.

The nerve requires approximately two milliseconds before another stimulus is presented. During this time no AP can be generated. This is called the refractory period. The generation of EEG signals is next described.

1.5 EEG Generation

An EEG signal is an indirect measurement of currents that flow during synaptic excitations of the dendrites of many pyramidal neurons in the cerebral cortex. When the brain cells (neurons) are activated, the synaptic currents are produced and propagate through the dendrites. This current generates a magnetic field measurable by EMG machines and a secondary electrical field over the scalp measurable by EEG systems.

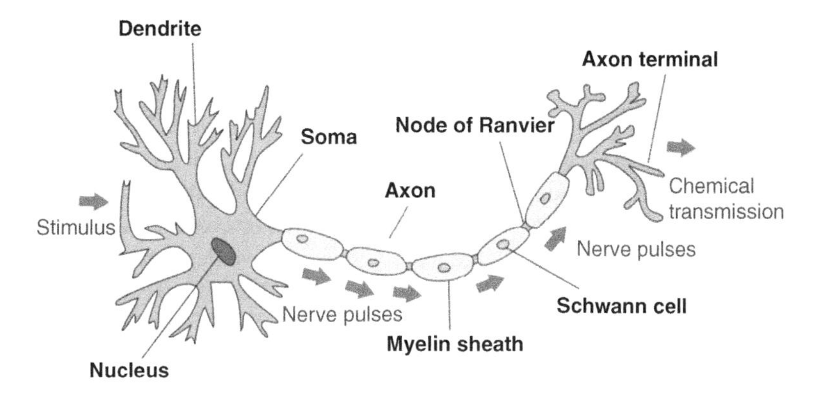

Figure 1.6 Structure of a neuron.

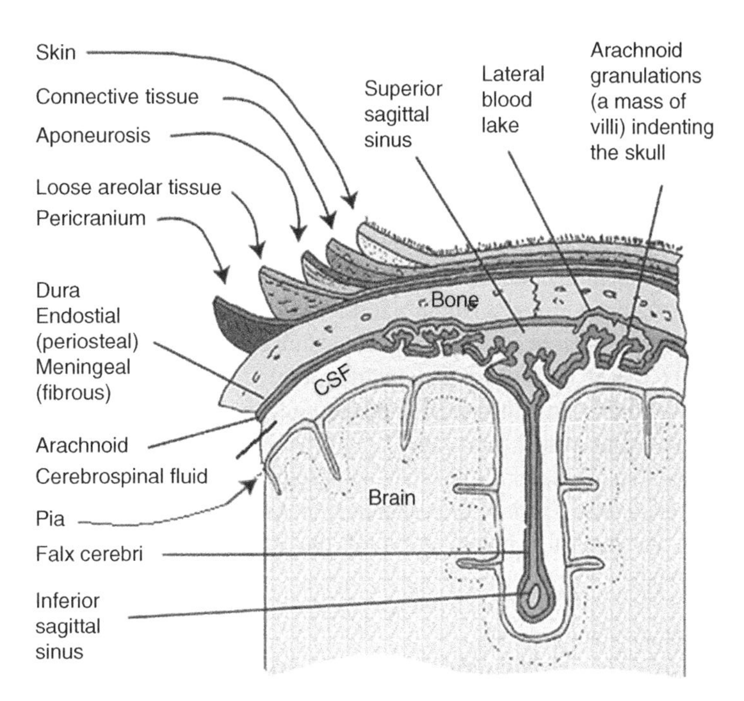

Figure 1.7 The head layers from brain to scalp.

Differences of electrical potentials are caused by summed post-synaptic graded potentials from pyramidal cells that create electrical dipoles between the soma (body of a neuron) and apical dendrites which branch from neurons (Figure 1.6). The current in the brain is generated mostly due to pumping the positive ions of sodium, Na^+, potassium, K^+, calcium, or Ca^{++}, and the negative ion of Cl^-, through the neuron membranes in the direction governed by the membrane potential [24].

The human head consists of three main layers of scalp, skull, brain (Figure 1.7) including many other thin layers in-between. In addition, the scalp consists of different layers such as skin, connective tissue, which is a thin layer of fat and fibrous tissue lying beneath the skin, the loose areolar connective tissue, and the pericranium, which is the periosteum of the skull bones and provides nutrition to bone and capacity for repair. Conversely, the brain is covered by a thin layer of cortex, which encompasses various brain tissues. The cortex includes arachnoid, meninges, dura, epidural, and subarachnoid space. The skull attenuates the signals approximately one hundred times more than the soft tissue. Conversely, most of the noise is generated either within the brain (internal noise) or over the scalp (system noise or external noise). Therefore, only large populations of active neurons can generate enough potential to be recordable using the scalp electrodes. These signals are later amplified greatly for display purposes. Approximately 10^{11} neurons are developed at birth when the CNS becomes complete and functional [25]. This makes an average of 10^4 neurons per cubic millimetre. Neurons are interconnected into neural nets through synapses. Adults have approximately 5.10^{14} synapses. The number of synapses per neuron increases with age, whereas the number of neurons decreases with age.

Given the diversity in electric and dielectric properties of the head layers, the distribution of attenuation of brain discharges including cortical, subcortical, and hippocampal activities is not uniform over the scalp and is subject to nonlinearity. Therefore, to model the neuronal pathways or localization of brain activity sources an accurate head electrical model should be available.

From an anatomical point of view the brain may be divided into three parts: the cerebrum, cerebellum, and brain stem (Figure 1.8). The cerebrum consists of both left and right lobes of the brain with highly convoluted surface layers called the cerebral cortex.

The cerebrum includes the regions for movement initiation, conscious awareness of sensation, complex analysis, and expression of emotions and behaviour. The cerebellum coordinates voluntary movements of muscles and balance maintaining.

The brain stem controls involuntary functions such as respiration, heart regulation, biorhythms, neurohormones, and hormone secretion [26].

The study of EEG paves the way for diagnosis of many neurological disorders and other abnormalities in the human body. The acquired EEG signals from a human (and also from animals) may for example be used for investigation of the following clinical problems [26, 27]:

- Monitoring alertness, coma, and brain death.
- Locating areas of damage following head injury, stroke, and tumour.
- Testing afferent pathways (by EPs).
- Monitoring cognitive engagement (alpha rhythm).
- Producing biofeedback situations.
- Controlling anaesthesia depth (servo anaesthesia).
- Investigating epilepsy and locating seizure origin.
- Testing epilepsy drug effects.
- Assisting in experimental cortical excision of epileptic focus.
- Monitoring the brain development.
- Testing drugs for convulsive effects.
- Investigating sleep disorders and physiology.
- Investigating mental disorders.

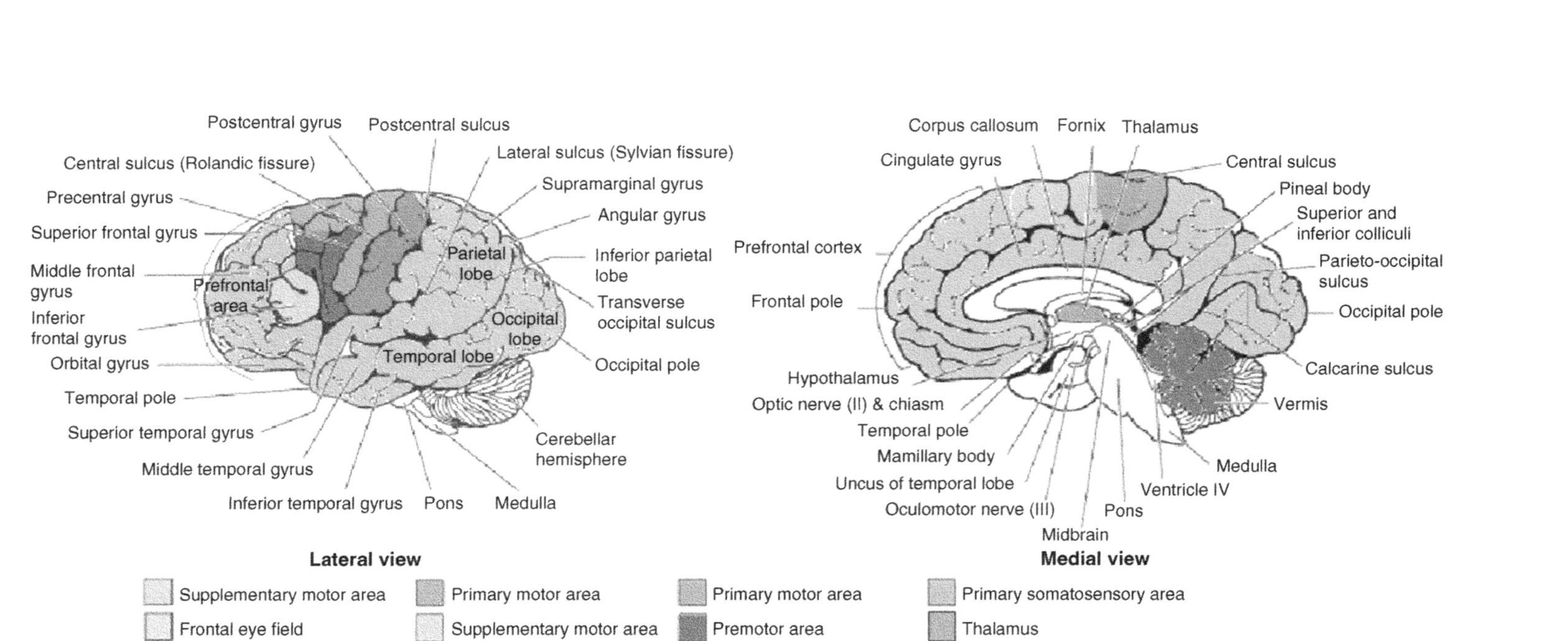

Figure 1.8 Diagrammatic representation of the major parts of the brain.

- Recognition of emotions for autistics.
- Monitoring mental fatigue for pilots and drivers.
- Providing a hybrid data recording system together with other imaging modalities.
- This list confirms the rich potential for EEG analysis and motivates the need for advanced signal processing techniques to aid the clinician in their interpretation. We next proceed to describe the brain rhythms, which are expected to be measured within EEG signals.
- Some of the mechanisms that generate the EEG signals are known at the cellular level and rest on a balance of excitatory and inhibitory interactions within and between populations of neurons. Although it is well established that the EEG (or MEG) signals result mainly from extracellular current flow, associated with summed post-synaptic potentials in synchronously activated and vertically oriented neurons, the exact neurophysiological mechanisms resulting in such synchronization to a given frequency band, remain obscure.

1.6 The Brain as a Network

Networks are already an integral part of human daily social, business, and intellectual life. Networking science is appealing to the field of neuroscience as the brain function stems from the communication and signalling between the neurons. The measured EEG amplitudes probed at each electrode have been the main parameter in evaluating brain function. Synchrony between left and right brain lobes gives further insight into detection of abnormalities such as mental fatigue and dementia and is associated with many other brain states such as emotions, as stated in the next chapter. The synchrony is often measured in the frequency domain where the variations in frequency and phase, corresponding to the time delay between the lobes, can be easily measured. More generally, recent developments in network science, however, have created a new direction in the study of brain normal and abnormal functions.

Although the fundamental concepts in network science originated from mathematics [28] and are used mostly in communications, a number of well established approaches, such as autoregressive modelling, have been used in characterizing the brain functional connectivity from the multichannel EEG. In addition, graph theory has become popular in designing effective classifiers which can segment the EEGs in time–space into the regions each encompassing a separate functionally connected brain region. In [29], a review of recent advances in neuroscience research in the specific area of brain connectivity as a potential biomarker of Alzheimer's disease with a focus on the application of graph theory can be studied.

Later in this book, we derive equations for the graphs applied to EEG in a similar way to those of brain connectivity estimators. We also observe that machine learning techniques such as deep neural networks can be directly applied to graphs for recognition of the brain state.

1.7 Summary

Following some details on EEG history, this chapter overviews the neuronal level analysis of the brain function. It also provides some information about the head anatomy. Generation of EEG signals as the result of signalling at the dendrite-dendrite or axon-dendrite synapses and production of APs, is an important and fundamental concept covered in this chapter. It is also highlighted that normal brain rhythms, brain evoked responses, and brain connectivity are the outcomes of neuronal activities and should be treated differently to recognize the brain normal and abnormal states.

References

1 Caton, R. (1875). The electric currents of the brain. *British Medical Journal* **2**: 278.
2 Walter, W.G. (1964). Slow potential waves in the human brain associated with expectancy, attention and decision. *Archiv für Psychiatrie und Nervenkrankheiten* **206**: 309–322.
3 Cobb, M. (2002). Exorcizing the animal spirits: Jan Swammerdam on nerve function. *Neuroscience* **3**: 395–400.
4 Danilevsky, V.D. (1877). *Investigation into the physiology of the brain.* Doctoral thesis. University of Charkov, Quoted after Brazier.
5 Brazier, M.A.B. (1961). *A history of the electrical activity of the brain; the first half-century.* New York: Macmillan.
6 Massimo, A. (July 2004). In memoriam Pierre Gloor (1923–2003): an appreciation. *Epilepsia* **45** (7): 882.
7 Grass, A.M. and Gibbs, F.A. (1938). A Fourier transform of the electroencephalogram. *Journal of Neurophysiology* **1**: 521–526.
8 Haas, L.F. (2003). Hans Berger (1873–1941), Richard Caton (1842–1926), and electroencephalography. *Journal of Neurology, Neurosurgery, and Psychiatry* **74**: 9.
9 Spear, J.H. (2004). Cumulative change in scientific production: research technologies and the structuring of new knowledge. *Perspectives on Science* **12** (1): 55–85.
10 Kornmüller, A.E. (1935). Der Mechanismus des epileptischen anfalles auf grund bioelektrischer untersuchungen am zentralnervensystem. *Fortschritte der Neurologie-Psychiatrie* **7**: 391–400; 414–432.
11 Fischer, M.H. (1933). Elektrobiologische auswirkungen von krampfgiften am zentralnervensystem. *Medizinische Klinik* **29**: 15–19.
12 Fischer, M.H. and Lowenbach, H. (1934). Aktionsstrome des zentralnervensystems unter der einwirkung von krampfgiften, 1. Mitteilung Strychnin und Pikrotoxin. *Naunyn-Schmiedebergs Archiv für experimentelle Pathologie und Pharmakologie* **174**: 357–382.
13 Bremer, F. (1935). Cerveau isole' et physiologie du sommeil. *Compte Rendu de la Sociéte de Biologie (Paris)* **118**: 1235–1241.
14 Niedermeyer, E. (1999). Chapter 1, Electroencephalography, basic principles, clinical applications, and related fields. In: *Historical Aspects*, 4e (eds. E. Niedermeyer and F.L. da Silva), 1–14. Lippincott Williams & Wilkins.
15 Berger, H. (1929). Über das Elektrenkephalogramm des Menschen. 7th report. *Archiv für Psychiatrie, Nervenkr*, **100**: 301–320.
16 Avoli, M. (1969). *Jasper's Basic Mechanisms of the Epilepsies [Internet].* 4e, Bethesda, MD: NCBI. https://www.ncbi.nlm.nih.gov/books/NBK98150 (accessed 9 September 2020).
17 Motokawa, K. (1949). Electroencephalogram of man in the generalization and differentiation of condition reflexes. *The Tohoku Journal of Experimental Medicine* **50**: 225.

18 Niedermeyer, E. (1973). Common generalized epilepsy. The so-called idiopathic or centrencephalic epilepsy. *European Neurology* **9** (3): 133–156.

19 Aserinsky, E. and Kleitman, N. (1953). Regularly occurring periods of eye motility, and concomitant phenomena, during sleep. *Science* **118**: 273–274.

20 Speckmann, E.-J. and Elger, C.E. (1999). Introduction to the neurophysiological basis of the EEG and DC potentials. In: *Electroencephalography*, 4e (eds. E. Niedermeyer and F. Da Silva), 15–34. Lippincott Williams and Wilkins.

21 Shepherd, G.M. (1974). *The Synaptic Organization of the Brain*. London: Oxford University Press.

22 Caspers, H., Speckmann, E.-J., and Lehmenkühler, A. (1986). DC potentials of the cerebral cortex, seizure activity and changes in gas pressures. *Reviews of Physiology, Biochemistry and Pharmacology* **106**: 127–176.

23 Ka Xiong Charand. (2011). Action potentials. http://hyperphysics.phy-astr.gsu.edu/hbase/biology/actpot.html (accessed 19 August 2021).

24 Attwood, H.L. and MacKay, W.A. (1989). *Essentials of Neurophysiology*. Hamilton, Canada: B. C. Decker.

25 Nunez, P.L. (1995). *Neocortical Dynamics and Human EEG Rhythms*. New York: Oxford University Press.

26 Teplan, M. (2002). Fundamentals of EEG measurements. *Measurement Science Review* **2** (Sec. 2): 1–11.

27 Bickford, R.D. (1987). Electroencephalography. In: *Encyclopedia of Neuroscience* (ed. G. Adelman), 371–373. Cambridge (USA): Birkhauser.

28 Sporns, O. (2011). *Networks of the Brain*. MIT Press.

29 del Etoile, J. and Adeli, H. (2017). Graph theory and brain connectivity in Alzheimer's disease. *The Neuroscientist* **23** (6): 616–626.

30 Shipton, H.W. (1975). EEG analysis: a history and prospectus. *Annual Reviews*, University of Iowa, USA: 1–15.

2

EEG Waveforms

2.1 Brain Rhythms

Traditionally, many brain disorders are diagnosed by visual inspection of EEG signals. The clinical experts in the field are familiar with manifestation of brain rhythms in the EEGs. In healthy adults, the amplitudes and frequencies of such signals change from one state of a human to another such as wakefulness to sleep and vice versa. The characteristics of the waves also change with age. There are five major brain waves distinguished by their different frequency ranges. These frequency bands from low to high frequencies respectively are called alpha (α), theta (θ), beta (β), delta (δ) and gamma (γ). The alpha and beta waves were introduced by Berger in 1929. Jasper and Andrews (1938) used the term 'gamma' to refer to the waves of above 30 Hz. The delta rhythm was introduced by Walter (1936) to designate all frequencies below the alpha range. He also introduced theta waves as those having frequencies within the range 4–7.5 Hz. The notion of a theta wave was introduced by Walter and Dovey in 1944 [1].

Delta waves lie within the range of 0.5–4 Hz. These waves are primarily associated with deep sleep and may be present in the waking state. It is very easy to confuse artefact signals caused by the large muscles of the neck and jaw with the genuine delta response. This is because the muscles are near the surface of the skin and produce large signals, whereas the signal, which is of interest, originates from deep within the brain and is severely attenuated in passing through the skull. Nevertheless, by applying simple signal analysis methods to the EEG, it is very easy to see when the response is caused by excessive movement.

Theta waves lie within the range of 4–7.5 Hz. The term theta might be chosen to allude to its presumed thalamic origin. Theta waves appear as consciousness slips towards drowsiness. Theta waves have been associated with access to unconscious material, creative inspiration and deep meditation. A theta wave is often accompanied by other frequencies and seems to be related to level of arousal. We know that healers and experienced mediators have an alpha wave which gradually lowers in frequency over long periods of time. The theta wave plays an important role in infancy and childhood. Larger contingents of theta wave activity in the waking adult are abnormal and are caused by various pathological problems. The changes in the rhythm of theta waves are examined for maturational and emotional studies [2].

The alpha waves appear in the posterior half of the head and are usually found over the occipital region of the brain, and can be detected in all parts of posterior lobes of the brain.

EEG Signal Processing and Machine Learning, Second Edition. Saeid Sanei and Jonathon A. Chambers.

For alpha waves the frequency lies within the range 8–13 Hz, and commonly appears as a round or sinusoidal shape signal. However, in rare cases it may manifest itself as sharp waves. In such cases, the negative component appears to be sharp and the positive component appears to be rounded, similar to the wave morphology of the rolandic mu (μ) rhythm. Alpha waves have been thought to indicate both a relaxed awareness without any attention or concentration. The alpha wave is the most prominent rhythm in the whole realm of brain activity and possibly covers a greater range than has been previously accepted. You can regularly see a peak in the beta wave range in frequencies even up to 20 Hz, which has the characteristics of an alpha wave state rather than one for a beta wave. Again, we very often see a response at 75 Hz which appears in an alpha' setting. Most subjects produce some alpha waves with their eyes closed and this is why it has been claimed that it is nothing but a waiting or scanning pattern produced by the visual regions of the brain. It is reduced or eliminated by opening the eyes, by hearing unfamiliar sounds, by anxiety or mental concentration or attention. Albert Einstein could solve complex mathematical problems while remaining in the alpha state; although generally, beta and theta waves are also present. An alpha wave has a higher amplitude over the occipital areas and has an amplitude of normally less than 50 μV. The origin and physiological significance of an alpha wave is still unknown and yet more research has to be undertaken to understand how this phenomenon originates from cortical cells [3].

A beta wave is the electrical activity of the brain varying within the range of 14–26 Hz (though in some literature no upper bound is given). A beta wave is the usual waking rhythm of the brain associated with active thinking, active attention, focus on the outside world or solving concrete problems, and is found in normal adults. A high-level beta wave may be acquired when a human is in a panic state. Rhythmical beta activity is encountered chiefly over the frontal and central regions. Importantly, a central beta rhythm is related to the rolandic mu rhythm and can be blocked by motor activity or tactile stimulation. The amplitude of beta rhythm is normally under 30 μV. Similar to the mu rhythm the beta wave may also be enhanced because of a bone defect [1] and also around tumoural regions.

The frequencies above 30 Hz (mainly up to 45 Hz) correspond to the gamma range (sometimes called as the fast beta wave). Although the amplitudes of these rhythms are very low and their occurrence is rare, detection of these rhythms can be used for confirmation of certain brain diseases. The regions of high EEG frequencies and highest levels of cerebral blood flow (as well as oxygen and glucose uptake) are located in the frontocentral area. The gamma wave band has also been proved to be a good indication of event-related synchronization (ERS) of the brain and can be used to demonstrate the locus for right and left index finger movement, right toes and the rather broad and bilateral area for tongue movement [4].

Waves in frequencies much higher than the normal activity range of EEG, mostly in the range of 200–300 Hz have been found in cerebellar structures of animals, but they have not played any role in clinical neurophysiology [5, 6].

Figure 2.1 shows the typical normal brain rhythms with their usual amplitude levels. In general, the EEG signals are the projection of neural activities which are attenuated by leptomeninges, cerebrospinal fluid, dura matter, bone, galea, and the scalp. Cortiographic discharges show amplitudes of 0.5–1.5 mV in range and up to several millivolts for spikes. However, on the scalp the amplitudes commonly lie within 10–100 μV.

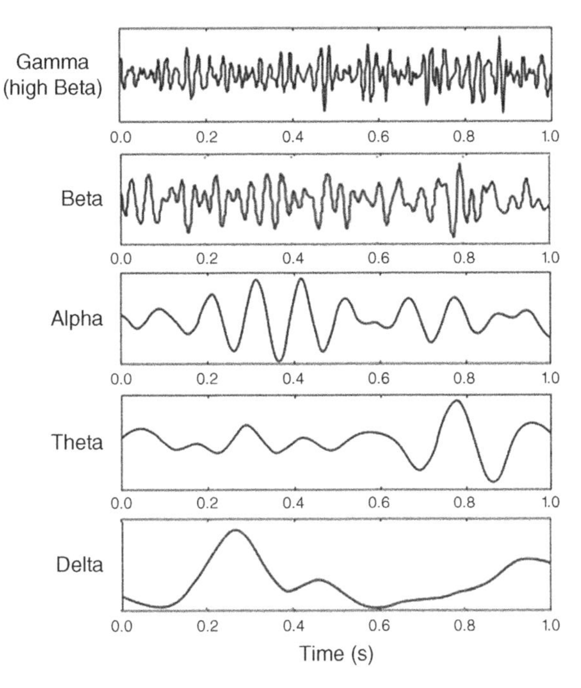

Figure 2.1 Five (can be categorized as four) typical dominant brain normal rhythms, from high to low frequencies. The delta wave is observed in infants and sleeping adults, the theta wave in children and sleeping adults, the alpha wave is detected in the occipital brain region when there is no attention, the beta wave appears frontally and parietally with low amplitude during attention and concentration, and gamma for stressed brain under heavy workload.

The above rhythms may last if the state of the subject does not change and therefore they are approximately cyclic in nature. Conversely, there are other brain waveforms, which may:

a) Have a wide frequency range or appear as spiky type signals such as K-complexes, vertex waves (which happen during sleep), or a breach rhythm, which is an alpha-type rhythm due to cranial bone defect [7], which does not respond to movement, and is found mainly over the midtemporal region (under electrodes T3 or T4), and some seizure signals.
b) Be a transient such as an event-related potential (ERP) and contain positive occipital sharp transient (POST) signals (also called rho [ρ]) waves.
c) Originate from the defected regions of the brain such as tumoural brain lesions.
d) Be spatially localized and considered as cyclic in nature, but can be easily blocked by physical movement such as mu rhythm. Mu denotes motor and is strongly related to the motor cortex. Rolandic (central) mu is related to posterior alpha in terms of amplitude and frequency. However, the topography and physiological significance are quite different. From the mu rhythm one can investigate the cortical functioning and the changes in brain (mostly bilateral) activities subject to physical and imaginary movements. The mu rhythm has also been used in feedback training for several purposes such as treatment of epileptic seizure disorder [1].

Also, there are other rhythms introduced by researchers such as:

e) Phi (φ) rhythm (less than 4 Hz) occurring within two seconds of eye closure. The phi rhythm was introduced by Daly [3].
f) The kappa (κ) rhythm, which is an anterior temporal alpha-like rhythm and it is believed to be the result of discrete lateral oscillations of the eyeballs and is considered to be an artefact signal.
g) The sleep spindles (also called sigma [σ] activity) within the 11–15 Hz frequency range.
h) Tau (τ) rhythm which represents the alpha activity in the temporal region.
i) Eyelid flutter with closed eyes which gives rise to frontal artefacts in the alpha band.
j) Chi rhythm is a mu-like activity believed to be a specific rolandic pattern of 11–17 Hz. This wave has been observed during the course of Hatha Yoga exercises [8].
k) Lambda (λ) waves are most prominent in waking patients, although they are not very common. They are sharp transients occurring over the occipital region of the head of walking subjects during visual exploration. They are positive and time-locked to saccadic eye movement with varying amplitude, generally below 90 μV [9].

The chart in Figure 2.2 shows all possible waveforms which may appear in a scalp EEG. The waveforms can be normal or abnormal rhythms during awake or sleep as well as various artefacts.

Often it is difficult to understand and detect the brain rhythms from the scalp EEGs even with trained eyes. Application of advanced signal processing tools, however, should enable separation and analysis of the desired waveforms from within the EEGs. Therefore, definition of foreground and background EEG is very subjective and entirely depends on the abnormalities and applications. We next consider the development in the recording and measurement of EEG signals.

An early model for the generation of brain rhythms is that of Jansen and Rit [10]. This model uses a set of parameters to produce alpha activity through an interaction between inhibitory and excitatory signal generation mechanisms in a single area. The basic idea behind these models is to make excitatory and inhibitory populations interact such that oscillations emerge. This model was later modified and extended to generate and emulate the other main brain rhythms, i.e. delta, theta, beta, and gamma, too [11]. The assumptions and mathematics involved in building the Jansen model and its extension are explained in this chapter. Application of such models in generation of post-synaptic potentials and using them as the template to detect, separate, or extract ERPs is of great importance. In Chapter 3 of this book, we can see the use of such templates in the extraction of the ERPs.

2.2 EEG Recording and Measurement

Acquiring signals and images from the human body has become vital for early diagnosis of a variety of diseases. Such data can be in the form of electrobiological signals such as electrocardiogram (ECG) from the heart, electromyography (EMG) from muscles, electroencephalography (EEG) from the brain, magnetoencepalogram (MEG) from the brain, electrogastrography (EGG) from the stomach, and electro-oculography (electro-optigraphy,

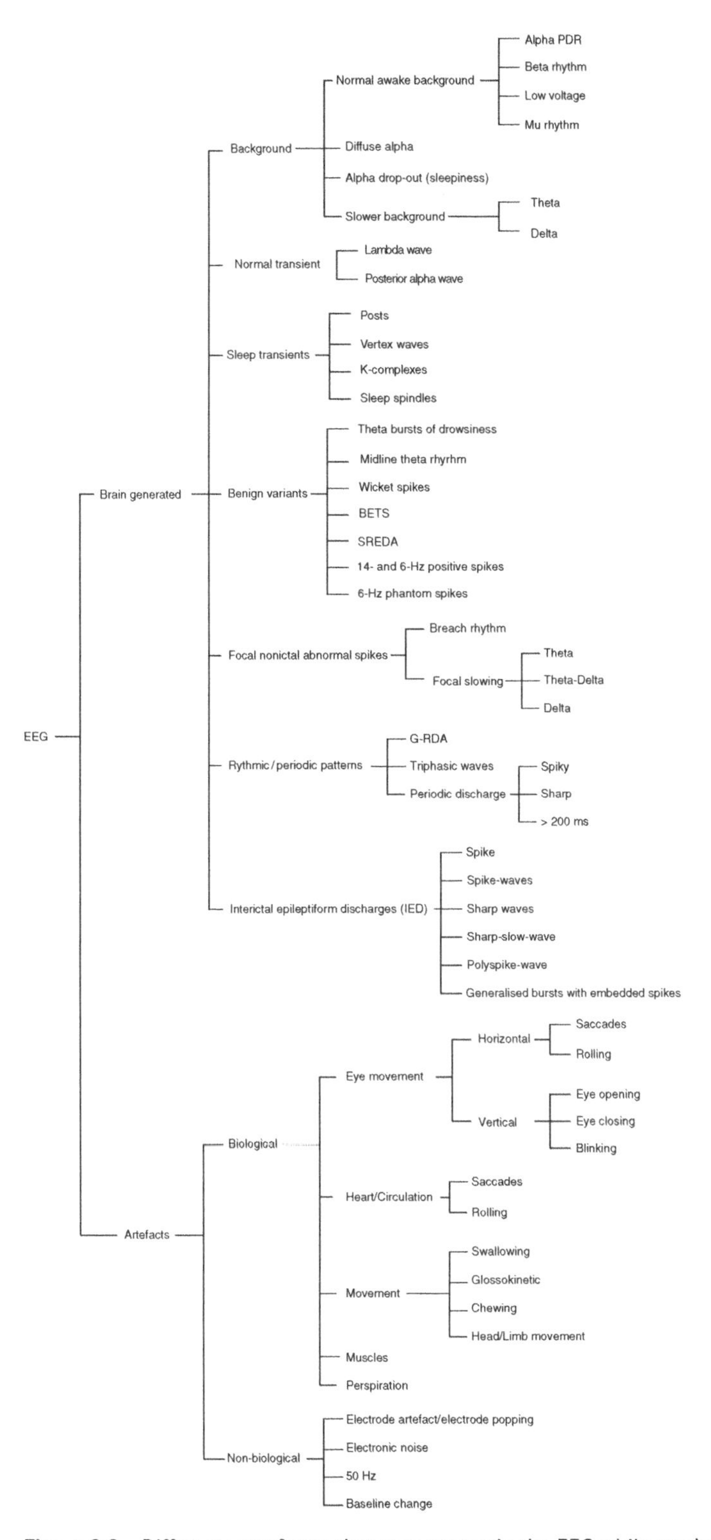

Figure 2.2 Different waveforms that may appear in the EEG while awake or during sleep periods.

EOG) from eye nerves. Measurements can also have the form of one type of ultrasound or radiograph such as sonograph (or ultrasound image), computerized tomography (CT), magnetic resonance imaging (MRI) or functional MRI (fMRI), positron emission tomography (PET), and single photon emission tomography (SPET).

Functional and physiological changes within the brain may be registered by either EEG, MEG, or fMRI. Application of fMRI is however very limited in comparison with EEG or MEG due to a number of important reasons:

a) The time resolution of fMRI image sequences is very low (for example approximately two frames per second), whereas complete EEG bandwidth can be viewed using EEG or MEG signals.
b) Many types of mental activities, brain disorders, and mal functions of the brain cannot be registered using fMRI since their effect on the level of oxygenated blood is low.
c) The accessibility to fMRI (and currently to MEG) systems is limited and costly.
d) The spatial resolution of EEG however, is limited to the number of recording electrodes (or number of coils for MEG).

The first electrical neural activities were registered using simple galvanometers. In order to magnify very fine variations of the pointer a mirror was used to reflect the light projected to the galvanometer on the wall. The d'Arsonval galvanometer later featured a mirror mounted on a movable coil and the light focused on the mirror was reflected when a current passed the coil. The capillary electrometer was introduced by Marey and Lippmann [12]. The string galvanometer, as a very sensitive and more accurate measuring instrument, was introduced by Einthoven in 1903. This became a standard instrument for a few decades and enabled photographic recording.

More recent EEG systems consist of a number of delicate electrodes, a set of differential amplifiers (one for each channel) followed by filters [9], and needle (pen) type registers. The multichannel EEGs could be plotted on plane paper or paper with a grid. Soon after this system came to the market, researchers started looking for a computerized system, which could digitize and store the signals. Therefore, to analyze EEG signals it was soon understood that the signals must be in digital form. This required sampling, quantization, and encoding of the signals. As the number of electrodes grows the data volume, in terms of the number of bits, increases. The computerized systems allow variable settings, stimulations, and sampling frequency, and some are equipped with simple or advanced signal processing tools for processing the signals.

The conversion from analogue-to-digital EEG is performed by means of multichannel analogue-to-digital converters (ADCs). Fortunately, the effective bandwidth for EEG signals is limited to approximately 100 Hz. For many applications this bandwidth may be considered even half of this value. Therefore, a minimum frequency of 200 Hz (to satisfy the Nyquist criterion) is often enough for sampling the EEG signals. In some applications where a higher resolution is required for representation of brain activities in the frequency domain, sampling frequencies of up to 2000 samples per second may be used.

In order to maintain the diagnostic information the quantization of EEG signals is normally very fine. Representation of each signal sample with up to 16 bits is very popular for the EEG recording systems. This makes the necessary memory volume for archiving the signals massive, especially for sleep EEG and epileptic seizure monitoring records.

However, in general, the memory size for archiving the images is often much larger than that used for archiving the EEG signals.

A simple calculation shows that for a one hour recording from 128-electrode EEG signals sampled at 500 samples per second a memory size of $128 \times 60 \times 60 \times 500 \times 16 \approx 3.68$ Gbits $\approx$ 0.45 Gbyte is required. Therefore, for longer recordings of a large number of patients there should be enough storage facilities such as in today's technology Zip disks, CDs, large removable hard drives, and optical disks.

Although the format of reading the EEG data may be different for different EEG machines, these formats are easily convertible to spreadsheets readable by most signal processing software packages such as MATLAB.

The EEG recording electrodes and their proper function are crucial for acquiring high quality data. There are different types of electrodes often used in the EEG recording systems as:

- disposable (gel-less, and pre-gelled types)
- reusable disc electrodes (gold, silver, stainless steel, or tin)
- headbands and electrode caps
- saline-based electrodes
- needle electrodes.

For multichannel recordings with a large number of electrodes, electrode caps are often used. Commonly used scalp electrodes consist of Ag–AgCl discs, less than 3 mm in diameter, with long flexible leads that can be plugged into an amplifier. Needle electrodes are those which have to be implanted under the skull with minimal invasive operations. High impedance between the cortex and the electrodes as well as the electrodes with high impedances can lead to distortion, which can even mask the actual EEG signals. Commercial EEG recording systems are often equipped with impedance monitors. To enable a satisfactory recording the electrode impedances should read less than 5 kΩ and be balanced to within 1 kΩ of each other. For more accurate measurement the impedances are checked after each trial.

Due to the layered and spiral structure of the brain, however, distribution of the potentials over the scalp (or cortex) is not uniform [13]. This may affect some of the results of source localization using the EEG signals.

2.2.1 Conventional Electrode Positioning

The International Federation of Societies for Electroencephalography and Clinical neurophysiology has recommended the conventional electrode setting (also called 10–20) for 21 electrodes (excluding the earlobe electrodes) as depicted in Figure 2.3 [14]. Often, the earlobe electrodes called A1 and A2, connected respectively to the left and right earlobes, are used as the reference electrodes. The 10–20 system avoids both eyeball placement and considers some constant distances by using specific anatomic landmarks from which the measurement would be made and then uses 10 or 20% of that specified distance as the electrode interval. The odd electrodes are on the left and the even ones on the right.

For setting a larger number of electrodes using the above conventional system, the rest of the electrodes are placed in between the above electrodes with equidistance between them.

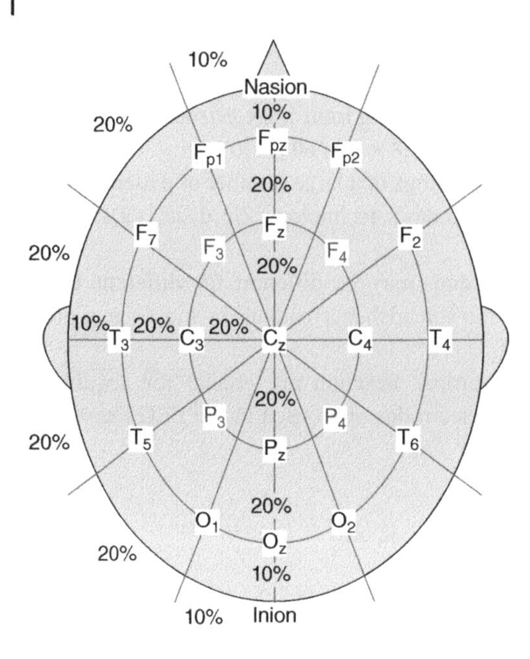

Figure 2.3 Conventional 10–20 EEG electrode positions for the placement of 21 electrodes.

For example, C_1 is placed between C_3 and C_z. Figure 2.4 represents a larger setting for 75 electrodes including the reference electrodes based on the guidelines by the American EEG Society. Extra electrodes are sometimes used for the measurement of EOC, ECG, and EMG of the eyelid and eye surrounding muscles. In some applications such as ERP analysis and brain–computer interfacing a single channel may be used. In such applications, however, the position of the corresponding electrode has to be well determined. For example, C_3 and C_4 can be used to record, respectively, the right and left finger movement-related signals for BCI applications. Also, F_3, F_4, P_3, and P_4 can be used for recordings of the ERP P300 signals.

Two different modes of recordings namely differential and referential, are used. In the differential mode the two inputs to each differential amplifier are from two electrodes. In referential mode, conversely, one or two reference electrodes are used. Several different reference electrode placements can be found in the literature. Physical references can be used as vertex (C_z), linked ears, linked mastoids, ipsilateral ear, contralateral ear, C_7, bipolar references, and tip of the nose [15]. There are also reference-free recording techniques which actually use a common average reference. The choice of reference may produce topographic distortion if the reference is not relatively neutral. In modern instrumentation, however, the choice of a reference does not play an important role in the measurement [16]. In such systems other references such as FP_z, hand, or leg electrodes may be used [17]. The overall setting includes the active electrodes and the references.

In another similar setting called the Maudsley electrode positioning system the conventional 10–20 system has been modified to better capture the signals from epileptic foci in epileptic seizure recordings. The only difference between this system and the 10–20 conventional system is that the outer electrodes are slightly lowered to enable better capturing of the required signals. The advantage of this system over the conventional one is that it provides a more extensive coverage of the lower part of the cerebral convexity, increasing the

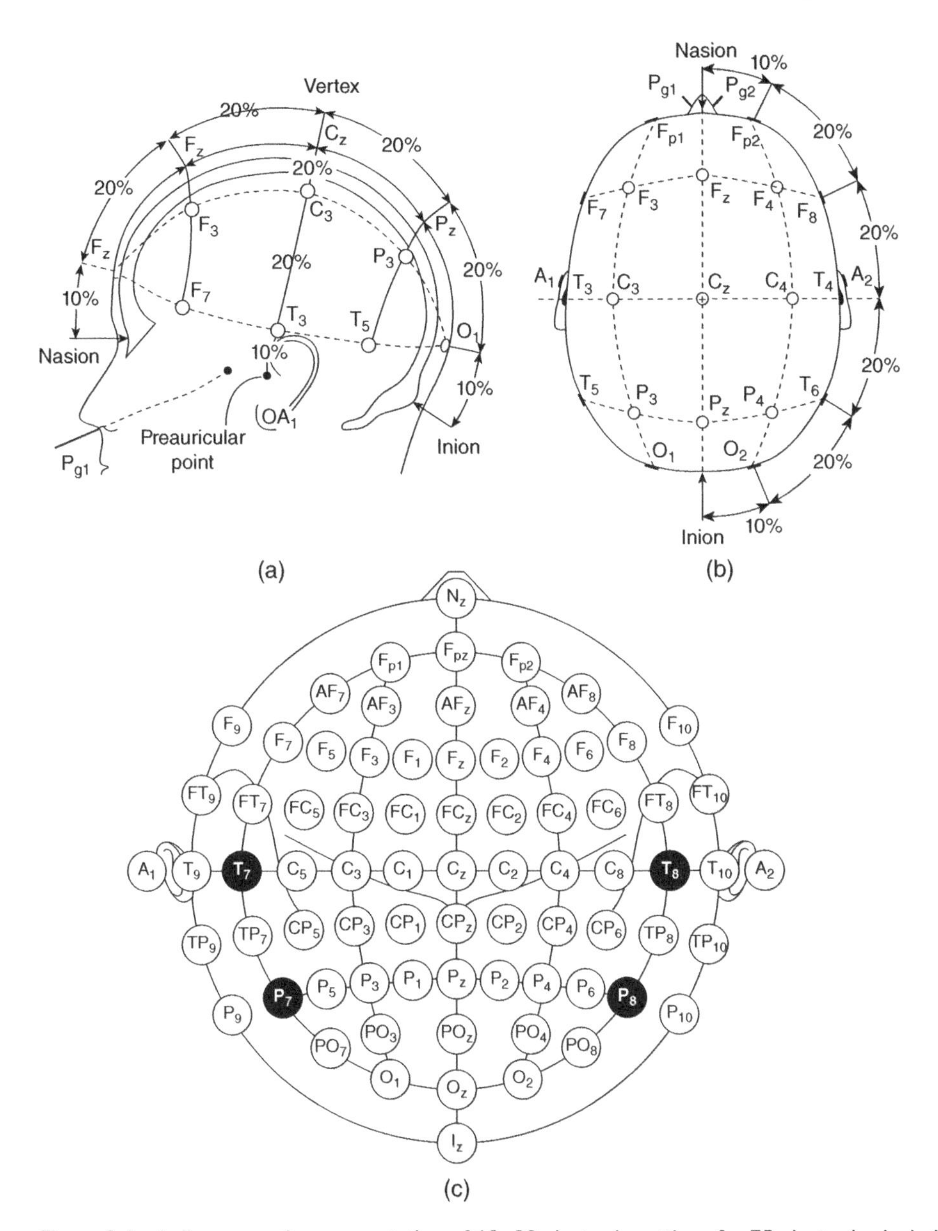

Figure 2.4 A diagrammatic representation of 10–20 electrode settings for 75 electrodes including the reference electrodes: (a and b) represent the three-dimensional measures and (c) indicates a two-dimensional view of the electrode setup configuration.

sensitivity for the recording from basal sub-temporal structures [18]. Other deviations from the international 10–20 system as used by researchers are found in [19, 20].

In many applications such as brain–computer interfacing (BCI) and study of mental activity, often a small number of electrodes around the movement-related regions are selected and used from the 10–20 setting system.

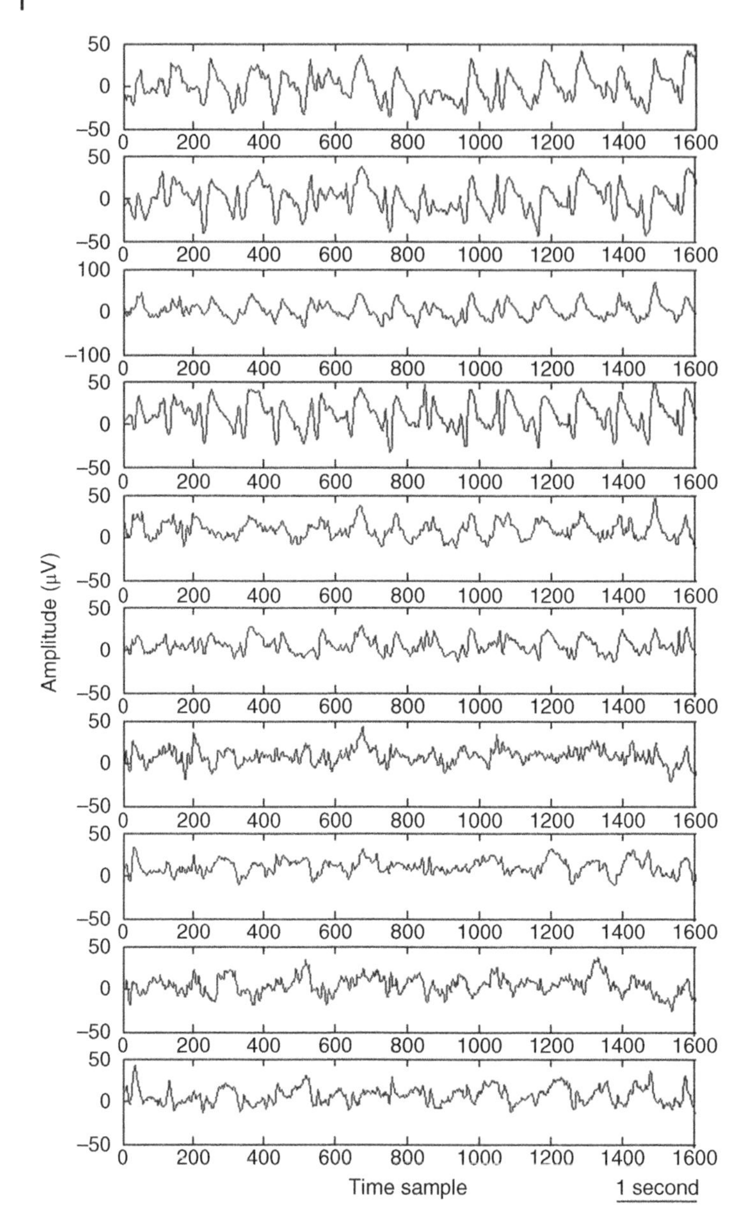

Figure 2.5 A typical set of EEG signals during approximately seven seconds of normal adult brain activity.

Figure 2.5 illustrates a typical set of EEG signals during approximately seven seconds of normal adult brain activity.

2.2.2 Unconventional and Special Purpose EEG Recording Systems

Despite many small-electrode portable EEG systems such as 5-channel or 14-channel EEG Emotive systems, advances in wearable technology have led to special purpose and miniaturized designs for EEG recordings. GlassEEG, produced by Mobile Health Technologies for real-time detection of epileptic seizures, EarEEG, designed mainly for SSVEP-based brain–computer interface (BCI), and tattooed EEG, very recently developed by Graz University in Austria for BCI are among such systems (Figure 2.6).

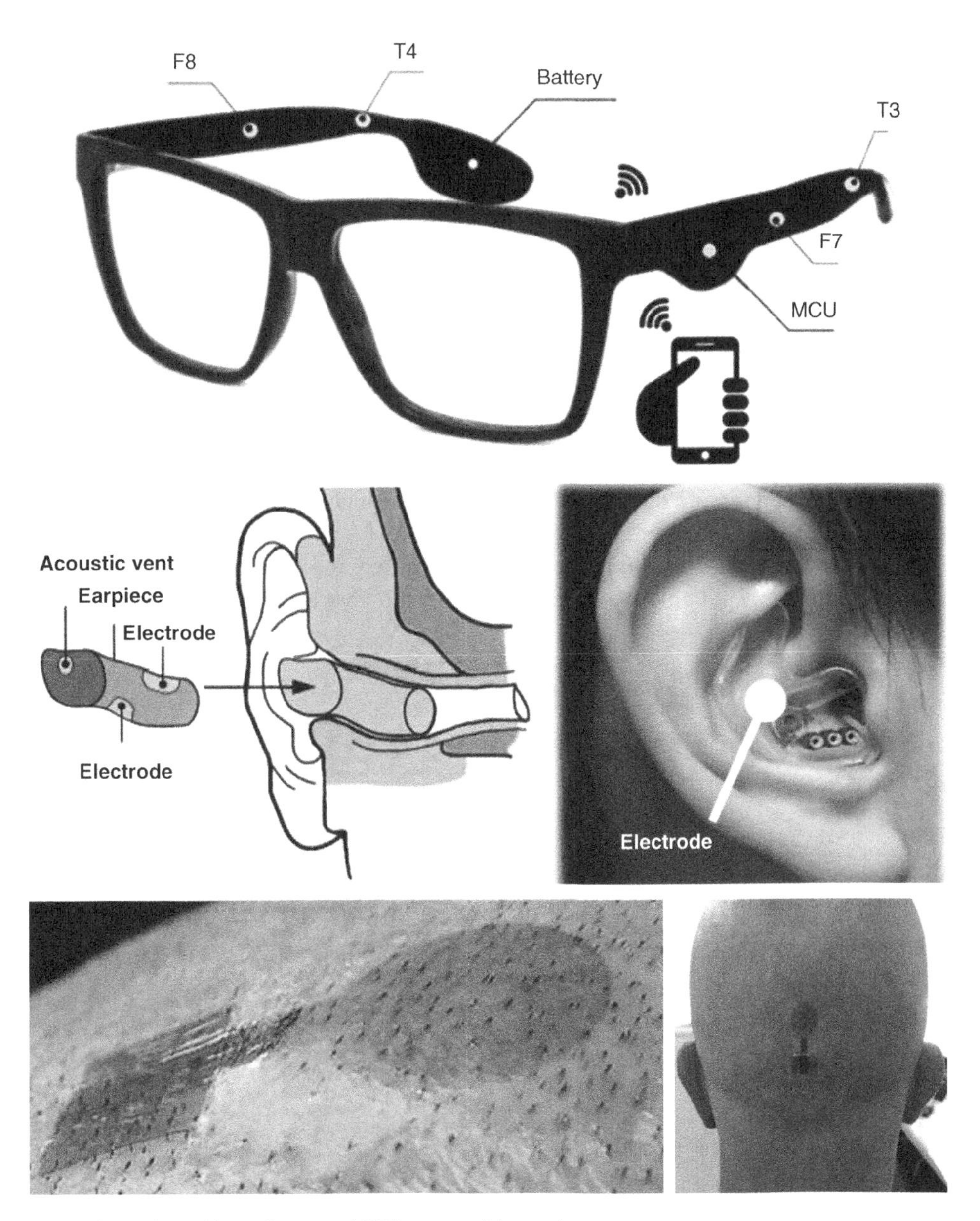

Figure 2.6 Wearable and tattooed EEG systems/electrodes.

2.2.3 Invasive Recording of Brain Potentials

Invasive brain screening is mainly to get closer to the neurons and thereby gain information about timing of neuron firing and the exact locations of defected neurons (such as for seizure) for various purposes such as deep brain stimulation, surgical operation, or rehabilitative implants.

The recording directly from over the cortex, called electrocorticography (ECoG), or intracranial electroencephalography (iEEG), is a type of electrophysiological brain monitoring that uses a grid of electrodes placed directly on the exposed surface of the brain to record electrical activity from the cerebral cortex. Figure 2.7 shows an example.

A popular application of ECoG is the localization of the central sulcus by means of phase reversal of somatosensory evoked potentials (EPs) for BCI-based rehabilitation.

Another technique for iEEG recording is by using strings of electrodes inserted into foramen ovale holes (Figure 2.8) deep into the hippocampus mainly to record the neuron activities during seizure and localization of infected neurons for surgical purposes. A set of these electrodes can be seen in Figure 2.9.

Later in Chapter 11 of this book we will show the application of this recording modality in detection of interictal epileptiform discharges (IEDs) from the hippocampus.

With the advances in microelectronics and microsensor designs, a new type of microsensor array insertable into the brain blood vessels within the motor cortex without any open surgery has been designed. The sensor array, called Stentrode™ is a small metallic mesh tube (stent), with not more than 4 mm diameter and with electrode contacts (small metal discs) within the stent structure.

The Stentrode™ (Figure 2.10) is expected to accurately record the neuronal activities within motor area and therefore help patients with loss of motor function due to

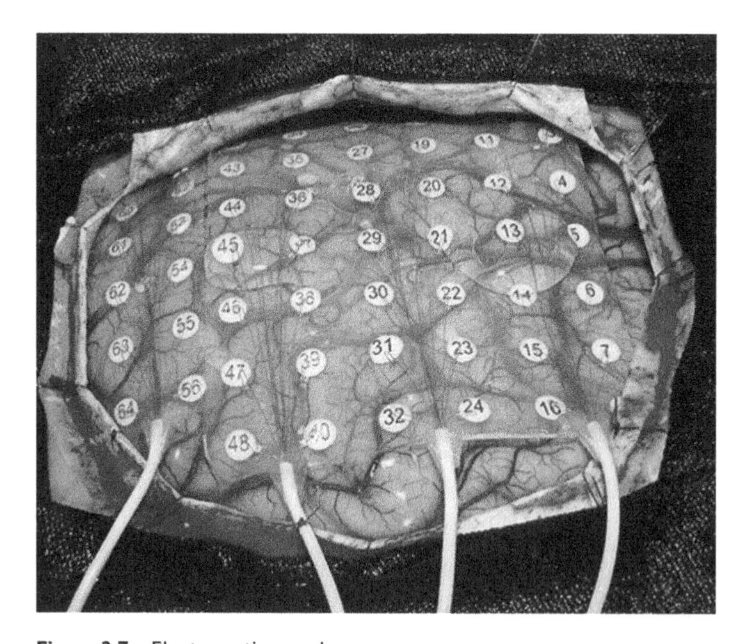

Figure 2.7 Electrocorticography.

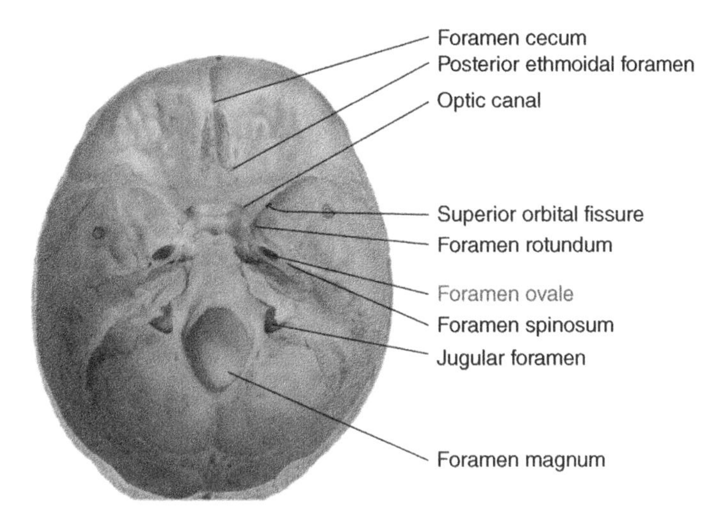

Figure 2.8 Foramen ovale holes within facial skeleton.

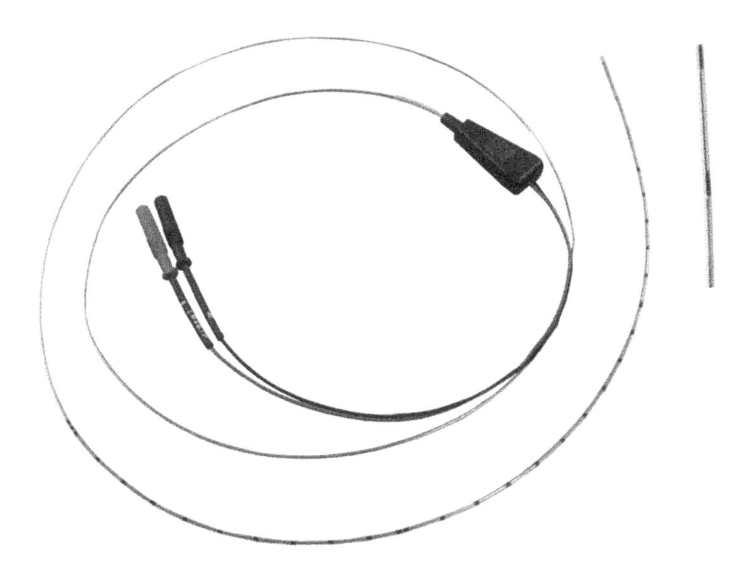

Figure 2.9 Foramen ovale electrodes.

paralysis from spinal cord injury, motor neuron disease, stroke, muscular dystrophy, or loss of limbs.

2.2.4 Conditioning the Signals

The raw EEG signals have amplitudes of the order of μV and contain frequency components of up to 300 Hz. To retain the effective information the signals have to be amplified before the ADC and filtered, either before or after the ADC, to reduce the noise and make the

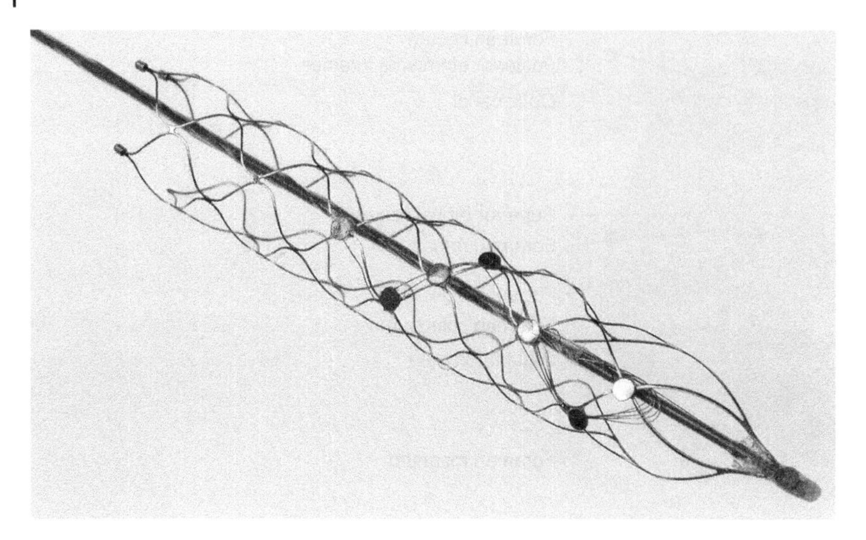

Figure 2.10 A 4-mm diameter Stentrode with electrode contacts within the stent structure.

signals suitable for processing and visualization. The filters are designed in such a way not to introduce any change or distortion to the signals. Highpass filters with cut-off frequency of usually less than 0.5 Hz are used to remove the disturbing very low frequency components such as those of breathing. Conversely, high-frequency noise is mitigated using lowpass filters with cut-off frequency of approximately 50–70 Hz. Notch filters with the null frequency of 50 Hz are often necessary to ensure perfect rejection of the strong 50 Hz power supply. In this case the sampling frequency can be as low as twice the bandwidth as commonly used by most EEG systems. The commonly used sampling frequencies for EEG recordings are 100, 250, 500, 1000, and 2000 samples per second. The main artefacts can be divided into patient related (physiological) and system artefacts. The patient related or internal artefacts are body movement-related, EMG, ECG (and pulsation), EOG, ballistocardiogram, and sweating. The system artefacts are 50/60 Hz power supply interference, impedance fluctuation, cable defects, electrical noise from the electronic components, and unbalanced impedances of the electrodes. Often in the preprocessing stage these artefacts are highly mitigated and the informative information is restored. Some methods for removing the EEG artefacts will be discussed in the related chapters of this book. Figure 2.11 shows a set of normal EEG signals affected by eye-blinking artefact. Similarly, Figure 2.12 represents a multichannel EEG set with clear appearance of ECG signals over the electrodes in the occipital region.

In the following sections we highlight the most popular changes in EEG measurements which correlate with physiological and mental abnormalities in the brain.

2.3 Sleep

Brain waves change during the various stages of normal sleep. When sleep starts the high-frequency waveforms gradually die away and continuous slow waves (in the delta band) followed by spindles and K-complexes appear from stage two of sleep and

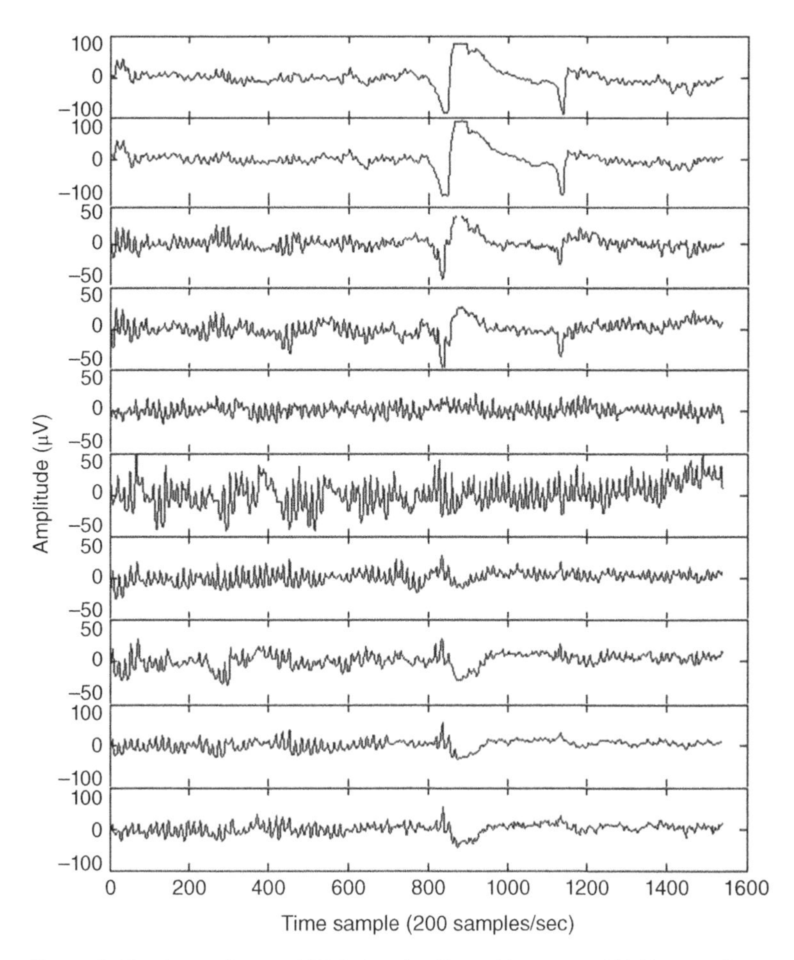

Figure 2.11 A set of normal EEG signals affected by an eye-blinking artefact.

extend towards stage four. During the last stages of sleep rapid eye movement (REM) starts and finishes before thc pcrson wakes up. All these stages can be observed and accurately detected and scored automatically by analyzing the EEG signals during normal sleep.

Conversely, there are cases where the subject has sleep disorder. These include obstructive sleep apnoea (OSA), insomnia (including parainsomnia and hyperinsomnia), REM sleep behaviour disorder, circadian rhythm sleep disorders, non-24-hour sleep–wake disorder, periodic limb movement disorder (PLMD), shift work sleep disorder, narcolepsy. Sleep disorders need to be diagnosed and treated.

Such disorders are often detected, monitored, or diagnosed through other recording modalities such as heartrate and respiration. Nevertheless, the use of sleep EEG in detection or monitoring of abnormal sleep has been a topic of research recently. Some methods are discussed in the related chapter of this book.

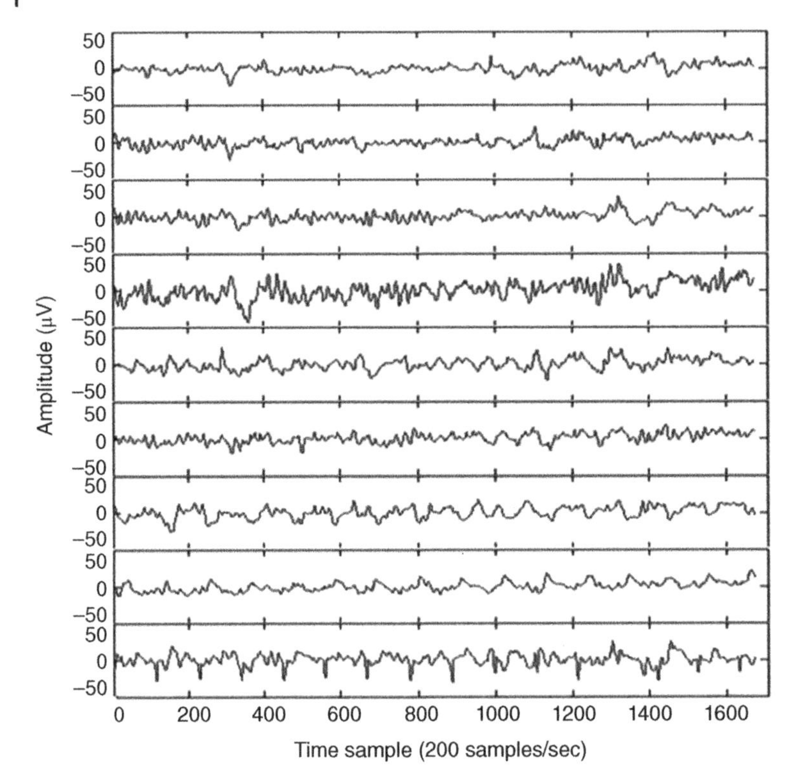

Figure 2.12 A multichannel EEG set with the clear appearance of ECG signals over the electrodes in the occipital region.

2.4 Mental Fatigue

Mental fatigue is often the result of prolonged cognitive load and can negatively affect people over a short or long time and can cause failure of the brain in performing its regular tasks. Mental fatigue mainly affects the function of the brain network rather than changing the normal brain rhythms significantly [21, 22] In these studies it has been shown that there is less synchrony between the left and right brain lobes and less connectivity between various brain lobes when the brain is under mental fatigue.

Mental fatigue also deteriorates the brain response to audio, visual, and other stimulations by attenuating and time shifting the ERPs. It has been shown that the amplitude of ERP (particularly P3b which is a subcomponent of P300,) reduces and the ERP latency, particularly for P3b, increases with the increase in mental fatigue [23–25].

A chapter of this book is devoted to detailed analysis of the EEG for under fatigue brains.

2.5 Emotions

The brain right hemisphere is associated with emotions. Lateralized electrophysiological parameters measured during emotionally charged states found relative activation (as measured by decreased alpha power) using EEG during the recollection of past events associated with anger or relaxation. Similar findings were obtained during self-reported emotional reactions to visual material [26]. Other experimental studies have verified the relative right-hemisphere activation when participants were asked to generate emotional imagery [27], and during hypnotically induced depression [28]. More recent quantitative EEG research has demonstrated reliable relationships between the magnitude of cerebral activation and the intensity of emotional arousal [29]. In related research, the ages of emotional memories correlated with the magnitude of activation using quantitative EEG [30]. Most of these studies conclude from the asymmetry in the activities of lateral brain lobes [31].

Perhaps the strongest evidence in support of the valence hypothesis was derived from a plethora of EEG studies that have associated relative increased left-hemisphere activity with positive emotional states and relative increased right-hemisphere activity with negative emotional states such as in [32]. Frontal EEG shows relative left-hemisphere activity during a positive emotional response, whereas the opposite pattern is displayed during a negative emotional response.

Some recent machine learning applications [33] have shown that the ratio between the power of EEG in the beta band and the theta band and their asymmetry is associated with change in emotions.

In [34] the authors have shown that six features namely, power spectral density (PSD), differential entropy (DE), differential asymmetry (DASM), rational asymmetry (RASM), asymmetry (ASM), and differential causality (DCAU) features from EEG, are associated with emotions and can be classified for emotion recognition.

Further unpublished research has focused on brain connectivity for emotion recognition. The patterns of interdependency between different brain regions for emotional and non-emotional film stimuli from EEGs have been analyzed and the emotion-related differences evaluated. A simple measure of synchronization index (SI) has then been used to detect interdependencies in EEG signals mainly for happiness and sadness. The SI significantly changes/increases during emotional stimulation and, in particular, during sadness, yielding an enhanced connectivity among frontal channels. Conversely, happiness is associated with a wider synchronization among frontal and occipital sites, although happiness itself was less synchronized [35].

2.6 Neurodevelopmental Disorders

Neurodevelopmental disorders are a group of disorders that affect the development of the nervous system, leading to abnormal brain function which may affect emotion, learning ability, self-control, and memory [36]. Such disorders, often starting from childhood, though their effects vary over time, usually remain with the subject and affect the person throughout their lifetime.

Examples of neurodevelopmental disorders in children include attention deficit hyperactivity disorder (ADHD), autism, also called autism spectrum disorder (ASD), learning disabilities, intellectual disability (also known as mental retardation), conduct disorders, cerebral palsy, and impairments in vision and hearing. One may add depression to this category of disorders.

ADHD, ASD, and depression have been under EEG studies in recent years and therefore some examples are provided in the later chapters of this book.

2.7 Abnormal EEG Patterns

Variations in the EEG patterns for certain states of the subject indicate abnormality. This may be due to distortion and disappearance of abnormal patterns, appearance and increase of abnormal patterns, or disappearance of all patterns. Sharbrough [37] divided the nonspecific abnormalities in the EEGs into three categories: (i) widespread intermittent slow-wave abnormalities often in the delta wave range and associated with brain dysfunction, (ii) bilateral persistent EEG usually associated with impaired conscious cerebral reactions, and (iii) focal persistent EEG usually associated with focal cerebral disturbance.

The first category is a burst type signal, which is attenuated by alerting the individual and eye opening, and accentuated with eye closure, hyperventilation, or drowsiness. The peak amplitude in adults is usually localized in the frontal region and influenced by age. In children, however, it appears over the occipital or posterior head region. Early findings showed that this abnormal pattern frequently appears with an increased intracranial pressure with tumour or aqueductal stenosis. Also, it correlates with grey matter disease, both in cortical and subcortical locations. However, it can be seen in association with a wide variety of pathological processes varying from systemic toxic or metabolic disturbances to focal intracranial lesions.

Regarding the second category, i.e. bilateral persistent EEG, the phenomenon in different stages of impaired, conscious, purposeful responsiveness are etiologically nonspecific and the mechanisms responsible for their generation are only partially understood. However, the findings in connection with other information concerning aetiology and chronicity may be helpful in arriving more quickly at an accurate prognosis concerning the patient's chance of recovering his previous conscious life.

As for the third category, i.e. focal persistent EEG, these abnormalities may be in the form of distortion and disappearance of normal patterns, appearance and increase of abnormal patterns, or disappearance of all patterns, but such changes are seldom seen at the cerebral cortex. The focal distortion of normal rhythms may produce an asymmetry of amplitude, frequency, or reactivity of the rhythm. The unilateral loss of reactivity of a physiological rhythm, such as the loss of reactivity of the alpha rhythm to eye opening [38] or to mental alerting [39], may reliably identify the focal side of abnormality. A focal lesion may also distort or eliminate the normal activity of sleep-inducing spindles and vertex waves.

Focal persistent nonrhythmic delta activity (PNRD) may be produced by focal abnormalities. This is one of the most reliable findings of a focal cerebral disturbance. The more persistent, the less reactive, and the more nonrhythmic and polymorphic is such focal slowing, the more reliable an indicator it becomes for the appearance of a focal cerebral disturbance [40–42]. There are other cases such as focal inflammation, trauma, vascular disease, brain tumour, or almost any other cause of focal cortical disturbance, including an asymmetrical onset of CNS degenerative diseases that may result in similar abnormalities in the brain signal patterns.

The scalp EEG amplitude from cerebral cortical generators underlying a skull defect is also likely to increase unless acute or chronic injury has resulted in significant depression of underlying generator activity. The distortions in cerebral activities are because focal abnormalities may alter the interconnections, number, frequency, synchronicity, voltage output, and access orientation of individual neuron generators, as well as the location and amplitude of the source signal itself.

With regards to the three categories of abnormal EEGs, their identification and classification requires a dynamic tool for various neurological conditions and any other available information. A precise characterization of the abnormal patterns leads to a clearer insight into some specific pathophysiologic reactions, such as epilepsy, or specific disease processes, such as subacute sclerosing panencephalitis (SSPE) or Creutzfeldt–Jakob disease (CJD) [37].

Over and above the reasons mentioned above there are many other causes for abnormal EEG patterns. The most common abnormalities are briefly described in the following sections.

2.8 Ageing

The ageing process affects the normal cerebral activity in awake and sleep human, and changes the response of the brain to stimuli. The changes stem from reducing the number of neurons and due to a general change in the brain pathology. This pathology indicates that the frontal and temporal lobes of the brain are more affected than the parietal lobes, resulting in shrinkage of large neurons and increasing the number of small neurons and glia [43]. A diminished cortical volume indicates that there is age related neuronal loss. A general cause for ageing of the brain may be the decrease in cerebral blood flow [43].

A reduction of the alpha frequency is probably the most frequent abnormality in EEG. This often introduces a greater anterior spread to frontal regions in the elderly and reduces the alpha wave blocking response and reactivity. The diminished mental function is somehow related to the degree of bilateral slowing in the theta and delta waves [43].

Although the changes in high-frequency brain rhythms have not been well established, some researchers have reported an increase in beta wave activity. This change in beta wave activity may be considered as an early indication of intellectual loss [43].

As for the sleep EEG pattern, older adults enter into drowsiness with a more gradual decrease in EEG amplitude. Over the age of 60, the frontocentral waves become slower, the frequency of the temporal rhythms also decreases, and frequency lowering with slow eye movements become more prominent, and spindles appear in the wave pattern after the dropout of the alpha rhythm. The amplitudes of both phasic and tonic NREM sleep EEG [43] reduce with age. There is also significant change in REM sleep organization with age; the REM duration decreases during the night and there is significant increase in the sleep disruption [43].

Dementia is the most frequent mental disorder that occurs predominantly in the elderly. Therefore, the prevalence of dementia increases dramatically with ageing of the society. Generally, EEGs are a valuable diagnostic tool in differentiation between organic brain syndromes (OBSs) and functional psychiatric disorders [43], and together with EPs play an important role in the assessment of normal and pathological ageing. Ageing is expected to change most neurophysiological parameters. However, the variability of these parameters must exceed the normal degree of spontaneous variability to become a diagnostic factor in acute and chronic disease conditions. Automatic analysis of the EEG during sleep and wakefulness may provide a better contrast in the data and enable a robust diagnostic tool. We next describe particular and very common mental disorders whose early onset may be diagnosed with EEG measurements.

2.9 Mental Disorders

2.9.1 Dementia

Dementia is a syndrome that consists of a decline in intellectual and cognitive abilities. This consequently affects the normal social activities, mode, and the relationship and interaction with other people [44]. EEG is often used to study the effect of dementia. In most cases such as in primary degenerative dementia, e.g. Alzheimer's, and psychiatric disorder, e.g. depression with cognitive impairment, the EEG can be used to detect the abnormality [45].

In [45] dementia is classified into cortical and subcortical forms. The most important cortical dementia is Alzheimer's disease (AD), which accounts for approximately 50% of the cases. Other known cortical abnormalities are Pick's disease and CJD. They are characterized clinically by findings such as aphasia, apraxia, and agnosia. CJD can often be diagnosed using EEG signals. Figure 2.13 shows a set of EEG signals from a patient with CJD. Conversely, the most common subcortical diseases are Parkinson's disease, Huntington's disease, lacunar state, normal pressure hydrocephalus, and progressive supranuclear palsy. These diseases are characterized by forgetfulness, slowing of thought processes, apathy, and depression. Generally, subcortical dementias introduce less abnormality to the EEG patterns than the cortical ones.

In AD, the EEG posterior rhythm (alpha rhythm) slows down and the delta and theta wave activities increase. Conversely, beta wave activity may decrease. In the severe cases epileptiform discharges and triphasic waves can appear. In such cases, cognitive impairment often results. The spectral power also changes; the power increases in delta and theta bands and decreases in beta and alpha bands and also in mean frequency.

The EEG wave morphology is almost the same for AD and Pick's disease. Pick's disease involves the frontal and temporal lobes. An accurate analysis followed by an efficient classification of the cases may discriminate these two diseases. CJD is a mixed cortical and subcortical dementia. This causes slowing of the delta and theta wave activities and, after approximately three months of the onset of the disease, periodic sharp wave complexes are generated which occur almost every second, together with decrease in the background activity [45]. Parkinson's disease is a subcortical dementia, which causes slowing down of the background activity and an increase of the theta and delta wave activities. Some works have been undertaken using spectral analysis to confirm the above changes [46]. Some other disorders such as depression have lesser effect on the EEGs and more accurate analysis of the EEGs has to be performed to detect the signal abnormalities for these brain disorders.

Generally, EEG is usually used in the diagnosis and evaluation of many cortical and subcortical dementias. Often it can help to differentiate between a degenerative disorder such as AD, and pseudodementia due to psychiatric illness [45]. The EEG may also show whether the process is focal or diffuse (i.e. involves the background delta and theta wave activities).

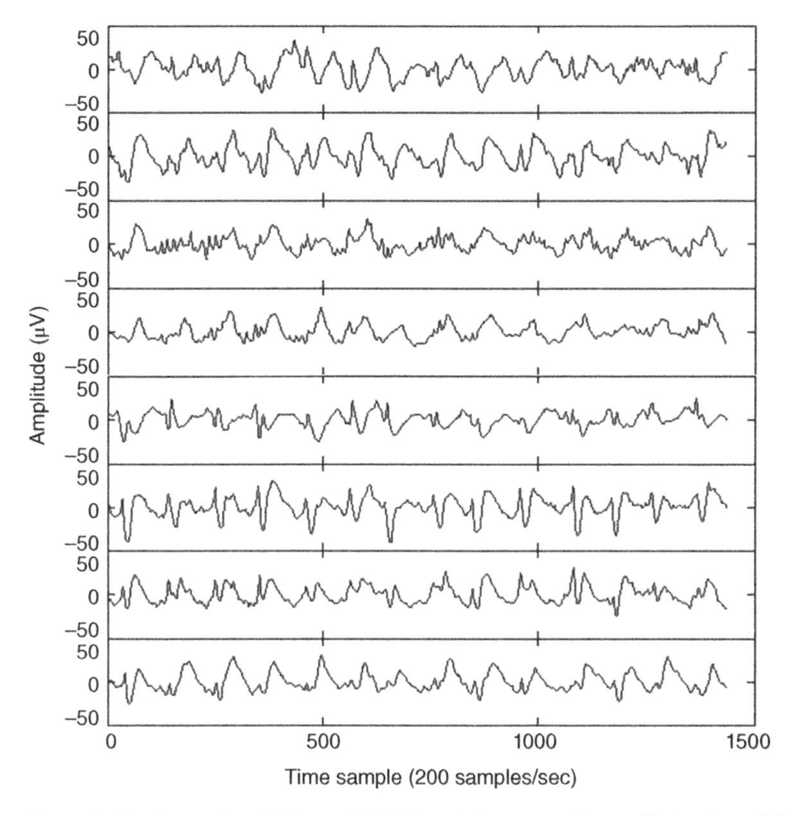

Figure 2.13 A set of multichannel EEG signals from a patient suffering from CJD.

The EEG may also reveal the early CJD-related abnormalities. However, more advanced signal processing and quantitative techniques may be implemented to achieve robust diagnostic and monitoring performance.

2.9.2 Epileptic Seizure and Nonepileptic Attacks

Often the onset of a clinical seizure is characterized by a sudden change of frequency in the EEG measurement. It is normally within the alpha wave frequency band with slow decrease in frequency (but increase in amplitude) during the seizure period. It may or may not be spiky in shape. Sudden desynchronization of electrical activity is found in electrodecremental seizures. The transition from preictal to ictal state, for a focal epileptic seizure, consists of gradual change from chaotic to ordered waveforms. The amplitude of the spikes does not necessarily represent the severity of the seizure. Rolandic spikes in a child of 4–10 years old for example, are very prominent, however the seizure disorder is usually quite benign or there may not be clinical seizure [47].

In terms of spatial distribution, in childhood the occipital spikes are very common. Rolandic central–midtemporal–parietal spikes are normally benign, whereas frontal spikes or multifocal spikes are more epileptogenic. The morphology of the spikes varies significantly with age. However, the spikes may occur in any level of awareness including wakefulness and deep sleep.

Distinction of seizure from common artefacts is not difficult. Seizure artefacts within an EEG measurement have prominent spiky but repetitive (rhythmical) nature, whereas the majority of other artefacts are transients or noise-like in shape. For the case of the ECG, the frequency of occurrence of the QRS waveforms is approximately 1 Hz. These waveforms have a certain shape which is very different from that of seizure signals.

The morphology of an epileptic seizure signal slightly changes from one type to another. The seizure may appear in different frequency ranges. For example, a petit mal discharge often has a slow spike around 3 Hz, lasting for approximately 70 ms, and normally has its maximum amplitude around the frontal midline. Conversely, higher frequency spike wave complexes occur for patients over 15 years old. Complexes at 4 and 6 Hz may appear in the frontal region of the brain of epileptic patients. As for the 6 Hz complex (also called benign EEG variants and patterns), patients with anterior 6 Hz spike waves are more likely to have epileptic seizures and those with posterior discharges tend to have neuro-autonomic disturbances [48]. The experiments do not always result in the same conclusion [47]. It was also found that the occipital 6 Hz spikes can be seen and are often drug related (due to hypoanalgetics or barbiturates) and withdrawal [49].

Among nonepileptics, the discharges may occur in patients with cerebrovascular disorder, syncopal attacks, and psychiatric problems [47]. Fast and needle-like spike discharges may be seen over the occipital region in most congenitally blind children. These spikes are unrelated to epilepsy and normally disappear in older age patients.

Bursts of 13–16 Hz or 5–7 Hz, as shown in Figure 2.14, (also called 14 and 6 Hz waves) with amplitudes less than 75 μV and arch shapes may be seen over the posterior temporal and the nearby regions of the head during sleep. These waves are positive with respect to the background waves. The 6 and 14 Hz waves may appear independently and be found respectively in younger and older children. These waves may be confined to the regions lying beneath a skull defect. Despite the 6 Hz wave, there are rhythmical theta bursts of wave activities relating to drowsiness around the midtemporal region with a morphology very similar to ictal patterns. In old age patients other similar patterns such as *subclinical rhythmic EEG discharges of adults* (SREDA) over the 4–7 Hz frequency band around the centroparietal region, and a wide frequency range (2–120 Hz) *temporal minor sharp transient* and *wicket spikes* over anterior temporal and midtemporal lobes of the brain may occur. These waves are also nonepileptic but with seizure-type waveform [47].

The epileptic seizure patterns, called ictal wave patterns, appear during the onset of epilepsy. Although the next chapters of this book focus on analysis of these waveforms from signal processing and machine learning points of view, here a brief explanation of morphology of these waveforms is given. Researchers in signal processing may exploit these concepts in the development of their algorithms. Although these waveform patterns are often highly obscured by the muscle movements, they normally maintain certain key characteristics.

Tonic–clonic seizure (also called grand mal) is the most common type of epileptic seizure. It appears in all electrodes but more towards the frontal electrodes (Figure 2.15). It has a rhythmic but spiky pattern in the EEG and occurs within the frequency range of 6–12 Hz. Petit mal is another interictal paroxysmal seizure pattern which occurs at approximately 3 Hz with a generalized synchronous spike wave complex of prolonged bursts.

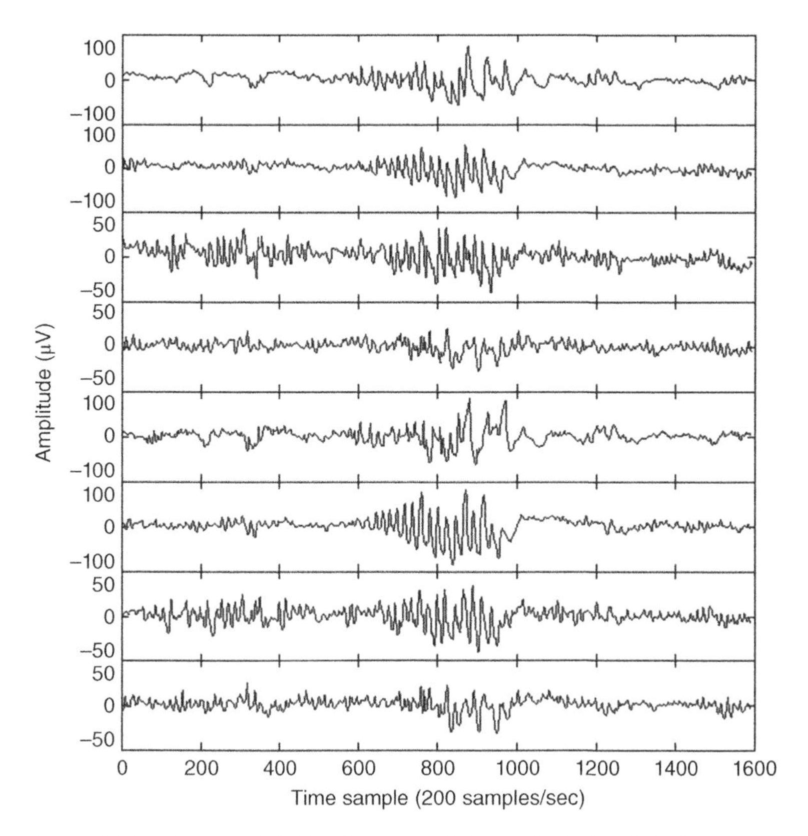

Figure 2.14 Bursts of 3–7 Hz seizure activity in a set of adult EEG signals.

A temporal lobe seizure (also called a psychomotor seizure or complex partial seizure) is presented by bursts of serrated slow waves with relatively high amplitude of above 60 μV and frequencies of 4–6 Hz. Cortical (focal) seizures have contralateral distribution with rising amplitude and diminishing frequency during the ictal period. The attack is usually initiated by local desynchronization, i.e. very fast and very low voltage spiky activity, which gradually rises in amplitude with diminishing frequency. Myoclonic seizures have concomitant polyspikes seen clearly in the EEG signals. They can have generalized or bilateral spatial distribution more dominant in the frontal region [50]. Tonic seizures occur in patients with Lennox–Gastaut syndrome [51] and have spikes which repeat with frequency approximately 10 Hz. Atonic seizures may appear in the form of a few seconds drop attack or be inhibitory, lasting for a few minutes. They show a few polyspike waves or spike waves with generalized spatial distribution of approximately 10 Hz followed by large slow waves of 1.5–2 Hz [52]. Akinetic seizures are rare and characterized by arrest of all motion, which, however, is not caused by sudden loss of tone as in atonic seizure and the patient is in an absent-like state. They are rhythmic with frequency of 1–2 Hz. Jackknife seizures also called salaam attacks, are common in children with hypsarrhythmia (infantile spasms, West syndrome) are either in the form of sudden generalized flattening desynchronization or have rapid spike discharges [51].

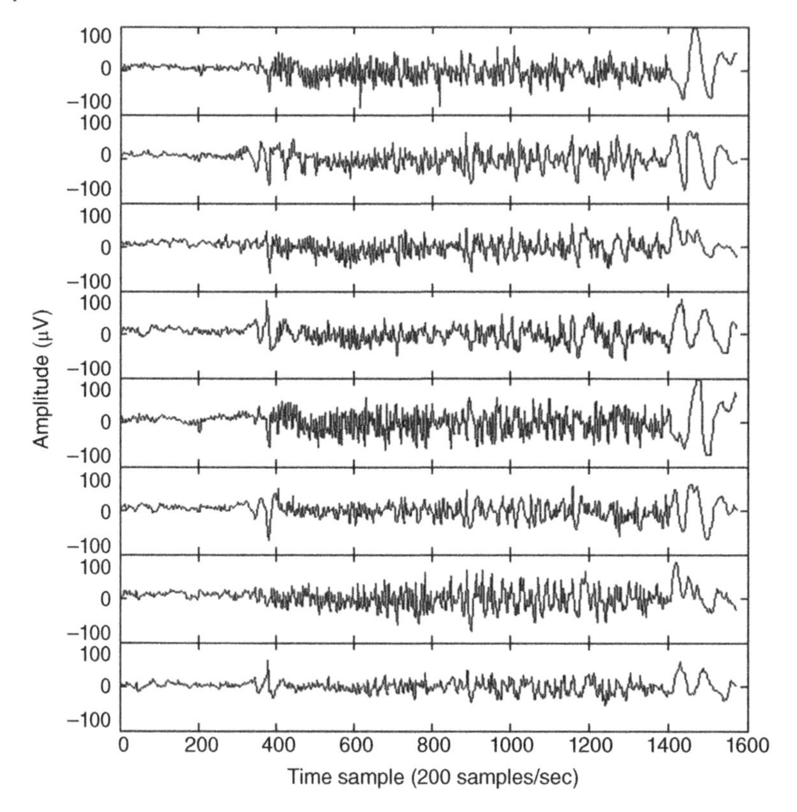

Figure 2.15 Generalized tonic–clonic (grand mal) seizure. The seizure appears in almost all the electrodes.

There are generally several varieties of recurring or quasi-recurring discharges, which may or may not be related to epileptic seizure. These abnormalities may be due to psychogenic changes, variation in body metabolism, circulatory insufficiency (which appears often as acute cerebral ischaemia). Of these, the most important ones are: periodic or quasiperiodic discharges related to severe CNS diseases; periodic complexes in subacute sclerosing panencephalitis (SSPE); periodic complexes in herpes simplex encephalitis; syncopal attacks; breath holding attacks; hypoglycaemia and hyperventilation syndrome due to sudden changes in blood chemistry [53]; and periodic discharges in CJD (mad cow disease) [54, 55]. The waveforms for this latter abnormality consist of a sharp wave or a sharp triphasic transient signal of 100–300 ms duration, with a frequency of 0.5–2 Hz. The periodic activity usually shows a maximum over the anterior region except for the Heidenhain form, which has a posterior maximum [47]. Other epileptic waveforms include periodic literalized epileptiform discharges (PLED), periodic discharges in acute cerebral anoxia, and periodic discharges of other etiologies.

Despite the above epileptiform signals there are spikes and other paroxysmal discharges in healthy nonepileptic persons. These discharges may be found in healthy individuals without any other symptoms of diseases. However, they are often signs of certain cerebral dysfunctions that may or may not develop into an abnormality. They may appear during periods of particular mental challenge on individuals, such as soldiers in the war front line, pilots, and prisoners.

Generation of epileptiform brain discharges from deeper brain layers such as the hippocampus during pre-ictal or interictal periods is an indication of upcoming seizure. These discharges which are spike-type and have particular morphology which can be seen by inserting electrodes such as multichannel foramen ovale electrodes deep into the hippocampus. More than 90% of these discharges cannot be seen over the scalp due to their attenuation and smearing. A comprehensive overview of epileptic seizure disorders and nonepileptic attacks can be found in many books and publications such as [53, 56]. In this book a chapter is dedicated to the methods for analyzing intracranial and scalp EEGs.

2.9.3 Psychiatric Disorders

Not only can functional and certain anatomical brain abnormalities be investigated using EEG signals, pathophysiological brain disorders can also be studied by analyzing such signals. According to the 'Diagnostic and Statistical Manual (DSM) of Mental Disorders' of the American Psychiatric Association, changes in psychiatric education have evolved considerably since the 1970s. These changes have mainly resulted from physical and neurological laboratory studies based upon EEG signals [57].

There have been evidences from EEG coherence measures suggesting differential patterns of maturation between normal and learning disabled children [58]. This finding can lead to the establishment of some methodology in monitoring learning disorders.

Several psychiatric disorders are diagnosed by analysis of EPs achieved by simply averaging a number of consecutive trails having the same stimuli.

Some pervasive mental disorders such as: dyslexia which is a developmental reading disorder; autistic disorder which is related to abnormal social interaction, communication, and restricted interests and activities, and starts appearing from the age of three; Rett's disorder, characterized by the development of multiple deficits following a period of normal postnatal functioning; and Asperger's disorder which leads to severe and sustained impairments in social interaction and restricted repetitive patterns of behaviour, interests, and activities; cause significant losses in multiple functioning areas [57].

ADHD and attention-deficit disorder (ADD), conduct disorder, oppositional defiant disorder, and disruptive behaviour disorder have also been under investigation and considered within the DSM. Most of these abnormalities appear during childhood and often prevent children from learning and socializing well. The associated EEG features have been rarely analytically investigated, but the EEG observations are often reported in the literature [59–63]. However, most of such abnormalities tend to disappear with advancing age.

EEG has also been analyzed recently for the study of delirium [64, 65], dementia [66, 67], and many other cognitive disorders [68]. In EEGs, characteristics of delirium include slowing or dropout of the posterior dominant rhythm, generalized theta or delta slow-wave activity, poor organization of the background rhythm, and loss of reactivity of the EEG to eye opening and closing. In parallel with that, the quantitative EEG (QEEG) shows increased absolute and relative slow-wave (theta and delta) power, reduced ratio of fast-to-slow band power, reduced mean frequency, and reduced occipital peak frequency [65].

Dementia includes a group of neurodegenerative diseases that cause acquired cognitive and behavioural impairment of sufficient severity to interfere significantly with social and occupational functioning. Alzheimer's disease is the most common of the diseases that cause dementia. At present, the disorder afflicts approximately 5 million people in the United States and more than 30 million people worldwide. Larger numbers of individuals have lesser levels of cognitive impairment, which frequently evolves into full-blown dementia. Prevalence of dementia is expected to nearly triple by 2050, since the disorder preferentially affects the elderly, who constitute the fastest-growing age bracket in many countries, especially in industrialized nations [67].

Among other psychiatric and mental disorders, amnestic disorder (or amnesia), mental disorder due to general medical condition, substance-related disorder, schizophrenia, mood disorder, anxiety disorder, somatoform disorder, dissociative disorder, sexual and gender identity disorder, eating disorders, sleep disorders, impulse-controlled disorder, and personality disorders have often been addressed in the literature [57]. However, the corresponding EEGs have been seldom analyzed by means of advanced signal processing tools.

2.9.4 External Effects

EEG signal patterns may significantly change when using drugs for the treatment and suppression of various mental and CNS abnormalities. Variation in the EEG patterns may also rise by just looking at the TV screen or listening to music without any attention. However, among the external effects the most significant ones are the pharmacological and drug effects. Therefore, it is important to know the effects of these drugs on the changes of EEG waveforms due to chronic overdosage, and the patterns of overt intoxication [69].

The effect of administration of drugs for anaesthesia on EEGs is of interest to clinicians. The related studies attempt to find the correlation between the EEG changes and the stages of anaesthesia. It has been shown that in the initial stage of anaesthesia a fast frontal activity appears. In deep anaesthesia this activity become slower with higher amplitude. In the last stage, a burst-suppression pattern indicates the involvement of brainstem functions, including respiration and finally the EEG activity ceases [69]. In the cases of acute intoxication, the EEG patterns are similar to those of anaesthesia [69].

Barbiturate is commonly used as an anticonvulsant and antiepileptic drug. With small dosage of barbiturate the activities within the 25–35 Hz frequency band around the frontal cortex increases. This changes to 15–25 Hz and spreads to the parietal and occipital regions. Dependence and addiction to barbiturates are common. Therefore, after a long-term ingestion of barbiturates, its abrupt withdrawal leads to paroxysmal abnormalities. The major complications are myoclonic jerks, generalized tonic–clonic seizures, and delirium [69].

Many other drugs are used in addition to barbiturates as sleeping pills such as melatonin, and bromides. Very pronounced EEG slowing is found in chronic bromide encephalopathies [69]. Antipsychotic drugs also influence the EEG patterns. For example, neuroleptics increase the alpha wave activity but reduce the duration of beta wave bursts and their average frequency. As another example, clozapine increases the delta, theta, and above 21 Hz beta wave activities. As another antipsychotic drug, tricyclic antidepressants such as imipramine, amitriptyline, doxepin, desipramine, nortriptyline, and protriptyline increase the amount of slow and fast activity along with instability of frequency and voltage, and also slow down the alpha wave rhythm. After administration of tricyclic antidepressants the seizure frequency in chronic epileptic patients may increase. With high dosage, this may further lead to single or multiple seizures occurring in nonepileptic patients [69].

During acute intoxication, a widespread, poorly reactive, irregular 8–10 Hz activity and paroxysmal abnormalities including spikes, as well as unspecific coma patterns, are observed in the EEGs [69]. Lithium is often used in the prophylactic treatment of bipolar mood disorder. The related changes in the EEG pattern consist of slowing of the beta rhythm and of paroxysmal generalized slowing, occasionally accompanied by spikes. Focal slowing also occurs, which is not necessarily a sign of a focal brain lesion. Therefore, the changes in the EEG are markedly abnormal with lithium administration [69]. The beta wave activity is highly activated by using benzodiazepines, as an anxiolytic drug. These activities persist in the EEG as long as two weeks after ingestion. Benzodiazepine leads to a decrease in an alpha wave activity and its amplitude, and slightly increases the 4–7 Hz frequency band activity. In acute intoxication the EEG shows prominent fast activity with no response to stimuli [69]. The psychotogenic drugs such as lysergic acid diethylamide and mescaline decrease the amplitude and possibly depress the slow waves [69].

The CNS stimulants increase the alpha and beta wave activities and reduce the amplitude and the amount of slow waves and background EEGs [69].

The effect of many other drugs especially antiepileptic drugs is investigated and new achievements are published frequently. One of the significant changes of the EEG of epileptic patients with valproic acid consists of reduction or even disappearance of generalized spikes along with seizure reduction. Lamotrigine is another antiepileptic agent that blocks voltage-gated sodium channels thereby preventing excitatory transmitter glutamate release. With the intake of lamotrigine a widespread EEG attenuation occurs [69].

Penicillin if administered in high dosage may produce jerks, generalized seizures, or even status epilepticus [69].

2.10 Summary

In this chapter the formation of EEG signals have been briefly explained. The conventional measurement setups for EEG recording and the brain rhythms present in normal or abnormal EEGs have also been described. In addition, the effects of popular brain abnormalities such as mental diseases, ageing, and epileptic and nonepileptic attacks have been pointed out. Despite the known neurological, physiological, pathological, and mental abnormalities of the brain mentioned in this chapter, there are many other brain disorders and dysfunctions which may or may not manifest some kinds of abnormalities in the related EEG signals.

Sleep, fatigue, ageing, emotions and many other states of the human body can directly or indirectly manifest themselves in the EEG patterns. Neurodevelopmental disorders, particularly those with human behaviour have become attractive areas of research as they are associated with child personality development.

Degenerative disorders of the CNS [70, 71] such as a variety of lysosomal disorders, several peroxisomal disorders, a number of mitochondrial disorders, inborn disturbances of the urea cycle, many aminoacidurias, and other metabolic and degenerative diseases as well as chromosomal aberrations have to be evaluated and their symptoms correlated with the changes in the EEG patterns. The similarities and differences within the EEGs of these diseases have to be well understood. Conversely, the developed mathematical algorithms need to take the clinical observations and findings into account to further enhance the outcome of such processing. Although a number of technical methods have been well established for the processing of the EEGs with relation to the above abnormalities, there is still a long way to go and many questions to be answered.

The following chapters of this book introduce new digital signal processing and machine learning techniques employed mainly for analysis of EEG signals followed by a number of examples in the applications of such methods.

References

1 Walter, W.G. and Dovey, V.J. (1944). Electro-encephalography in cases of sub-cortical tumour. *Journal of Neurology, Neurosurgery, and Psychiatry* **7** (3–4): 57–65. https://doi.org/10.1136/jnnp.7.3-4.57.

2 Ashwal, S. and Rust, R. (2003). Child neurology in the 20th century. *Pediatric Research* **53**: 345-361.

3 Niedermeyer, E. (1999). The normal EEG of the waking adult, Chapter 10. In: *Electroencephalography, Basic Principles, Clinical Applications, and Related Fields*, 4e (eds. E. Niedermeyer and F.L. Da Silva), 174–188. Lippincott Williams & Wilkins.

4 Pfurtscheller, G., Flotzinger, D., and Neuper, C. (1994). Differentiation between finger, toe and tongue movement in man based on 40 Hz EEG. *Electroencephalography and Clinical Neurophysiology* **90**: 456–460.

5 Adrian, E.D. and Mattews, B.H.C. (1934). The Berger rhythm, potential changes from the occipital lob in man. *Brain* **57**: 345–359.

6 Trabka, J. (1963). High frequency components in brain waves. *Electroencephalography and Clinical Neurophysiology* **14**: 453–464.

7 Cobb, W.A., Guiloff, R.J., and Cast, J. (1979). Breach rhythm: the EEG related to skull defects. *Electroencephalography and Clinical Neurophysiology* **47**: 251–271.

8 Roldan, E., Lepicovska, V., Dostalek, C., and Hrudova, L. (1981). Mu-like EEG rhythm generation in the course of hatha-yogi exercises. *Electroencephalography and Clinical Neurophysiology* **52**: 13.

9 IFSECN (1974). A glossary of terms commonly used by clinical electroencephalographers. *Electroencephalography and Clinical Neurophysiology* **37**: 538–548.

10 Jansen, B.H. and Rit, V.G. (1995). Electroencephalogram and visual evoked potential generation in a mathematical model of coupled cortical columns. *Biological Cybernetics* **73**: 357–366.

11 David, O. and Friston, K.J. (2003). A neural mass model for MEG/EEG coupling and neuronal dynamics. *NeuroImage* **20**: 1743–1755.

12 Marey, E.J. and Lippmann, G. (1876). Des variations electriques des muscles du coeur en particulier etudiees au moyen de l'electrometre di. M. Lippmann. *Comptes Rendus* **82**: 975–977.

13 Gotman, J., Ives, J.R., and Gloor, R. (1979). Automatic recognition of interictal epileptic activity in prolonged EEG recordings. *Electroencephalography and Clinical Neurophysiology* **46**: 510–520.

14 Jasper, H. (1958). Report of committee on methods of clinical exam in EEG. *Electroencephalography and Clinical Neurophysiology* **10**: 370–375.

15 Bickford, R.D. (1987). Electroencephalography. In: *Encyclopedia of Neuroscience* (ed. G. Adelman), 371–373. Cambridge (USA): Birkhauser.

16 Montoya-Martínez, J., Vanthornhout, J., Bertrand, A., and Francart, T. (2021). Effect of number and placement of EEG electrodes on measurement of neural tracking of speech. *PLoS ONE* **16** (2): e0246769.

17 Collura, T. (1998). A guide to electrode selection, location, and application for EEG Biofeedback, Ohio. *Proceedings of the 6th Annual Conference on Brain Function/EEG, Modification and Training*, Palm Springs, CA (21–25 February).

18 Nayak, D., Valentin, A., Alarcon, G. et al. (2004). Characteristics of scalp electrical fields associated with deep medial temporal epileptiform discharges. *Clinical Neurophysiology* **115**: 1423–1435.

19 Barrett, G., Blumhardt, L., Halliday, L. et al. (1976). A paradox in the lateralization of the visual evoked responses. *Nature* **261**: 253–255.

20 Halliday, A.M. (1978). *Commentary: Evoked Potentials in Neurological Disorders, Chapter: Event-Related Brain Potentials in Man* (eds. E. Calloway, P. Tueting and S.H. Coslow), 197–210. Academic Press.

21 Jarchi, D. and Sanei, S. (2010). Mental fatigue analysis by measuring synchronization of brain rhythms incorporating enhanced empirical mode decomposition. *Proceedings of the 2nd International Workshop on Cognitive Information Processing (CIP)*. Elba, Italy.

22 Jarchi, D. and Sanei, S. (2010). A novel method for analysis of mental fatigue from normal brain rhythms. *Proceedings of the 17th European Signal Processing Conference, EUSIPCO*. Denmark.

23 Jarchi, D., Sanei, S., and Lorist, M.M. (2011). Coupled particle filtering: a new approach for P300-based analysis of mental fatigue. *Journal of Biomedical Signal Processing and Control* **6** (2): 175–185.

24 Jarchi, D., Makkiabadi, B., and Sanei, S. (2009). Estimation of trial to trial variability of P300 subcomponents by coupled Rao-Blackwellised particle filtering. *Proceedings of the IEEE Workshop on Statistical Signal Processing, SSP2009*. Cardiff, UK.

25 Jarchi, D., Makkiabadi, B., and Sanci, S. (2009). Separating and tracking ERP subcomponents using constrained particle filter. *Proceedings of the 16th International Conference on Digital Signal Processing, DSP2009*, Greece.

26 Davidson, R.J. and Henriques, J.B. (2000). Regional brain function in sadness and depression. In: *The Neuropsychology of Emotion* (ed. J.C. Borod), 269–297. New York: Oxford Press.

27 Karlin, R., Weinapple, M., Rochford, J., and Goldstein, L. (1979). Quantitated EEG features of negative affective states: report of some hypnotic studies. *Research Communications in Psychology, Psychiatry, and Behavior* **4**: 397–413.

28 Tucker, D.M., Stenslie, C.E., Roth, R.S., and Shearer, S.L. (1981). Right frontal lobe activation and right hemisphere performance decrement during a depressed mood. *Archives of General Psychiatry* **38** (2): 169–174.

29 Foster, P.S. and Harrison, D.W. The relationship between magnitudes of cerebral activation and intensity of emotional arousal. *International Journal of Neuroscience* **112**: 1463–1477, 2002.

30 Foster, P.S. and Harrison, D.W. (2004). Cerebral correlates of varying ages of emotional memories. *Cognitive and Behavioral Neurology* **17** (2): 85–92.

31 Demaree, H.A., Everhart, D.E., Youngstrom, E.A., and Harrison, D.W. (2005). Brain lateralization of emotional processing: historical roots and a future incorporating dominance. *Behavioral and Cognitive Neuroscience Reviews* **4** (1): 3–20. https://doi.org/10.1177/1534582305276837.

32 Lee, G.P., Meador, K.J., Loring, D.W. et al. (2004). Neural substrates of emotion as revealed by functional magnetic resonance imaging. *Cognitive and Behavioral Neurology* **17** (1): 9–17.

33 Xu, X., Wei, F., Zhu, Z. et al. (2020). EEG feature selection using orthogonal regression: application to emotion recognition. *ICASSP 2020–2020 IEEE International Conference on Acoustics, Speech and Signal Processing (ICASSP)*, 1239–1243. Barcelona, Spain. https://doi.org/10.1109/ICASSP40776.2020.9054457.

34 Zheng, W.-L., Zhu, J.-Y., and Lu, B.-L. (2019). Identifying stable patterns over time for emotion recognition from EEG. *IEEE Transactions on Affective Computing* **10** (3): 417–429.

35 Costa, T., Rognoni, E., and Galati, D. (2006). EEG phase synchronization during emotional response to positive and negative film stimuli. *Neuroscience Letters* **406**: 159–164.

36 Mullin, A.P., Gokhale, A., Moreno-De-Luca, A. et al. (2013). Neurodevelopmental disorders: mechanisms and boundary definitions from genomes, interactomes and proteomes. *Translational Psychiatry* **3**: e329. https://doi.org/10.1038/tp.2013.108.

37 Sharbrough, F.W. (1999). Nonspecific abnormal EEG patterns, Chapter 12. In: *Electroencephalography, Basic Principles, Clinical Applications, and Related Fields*, 4e (eds. E. Niedermeyer and F.L. Da Silva). Lippincott Williams & Wilkins.

38 Bancaud, J., Hecaen, H., and Lairy, G.C. (1955). Modification de la reactivite E.E.G., troubles des functions symboliques et troubles con fusionels dans les lesions hemispherigues localisees. *Electroencephalography and Clinical Neurophysiology* **7**: 179.

39 Westmoreland, B. and Klass, D. (1971). Asymetrical attention of alpha activity with arithmetical attention. *Electroencephalography and Clinical Neurophysiology* **31**: 634–635.

40 Cobb, W. (1976). EEG interpretation in clinical medicine. In: *Part B, Handbook of Electroencephalography and Clinical Neurophysiology*, vol. **11** (ed. A. Remond), B1–B6. Elsevier.

41 Hess, R. (1975). Brain tumors and other space occupying processing. In: *Part C, Handbook of Electroencephalography and Clinical Neurophysiology*, vol. **14** (ed. A. Remond), C1–C6. Elsevier.

42 Klass, D. and Daly, D. (eds.) (1979). *Current Practice of Clinical Electroencephalography*, 1e. Raven Press.

43 Van Sweden, B., Wauquier, A., and Niedermeyer, E. (1999). Normal aging and transient cognitive disorders in the elderly, Chapter 18. In: *Electroencephalography, Basic Principles, Clinical Applications, and Related Fields*, 4e (eds. E. Niedermeyer and F.L. Da Silva), 340–348. Lippincott Williams & Wilkins.

44 America Psychiatric Association (1994). *Committee on Nomenclature and Statistics, Diagnostic and Statistical Manual of Mental Disorder: DSM-IV*, 4e. Washington DC: American Psychiatric Association.

45 Brenner, R.P. (1999). EEG and dementia, Chapter 19. In: *Electroencephalography, Basic Principles, Clinical Applications, and Related Fields*, 4e (eds. E. Niedermeyer and F.L. Da Silva), 349–359. Lippincott Williams & Wilkins.

46 Neufeld, M.Y., Bluman, S., Aitkin, I. et al. (1994). EEG frequency analysis in demented and nondemented parkinsonian patients. *Dementia* **5**: 23–28.

47 Niedermeyer, E. (1999). Abnormal EEG patterns: epileptic and paroxysmal, Chapter 13. In: *Electroencephalography, Basic Principles, Clinical Applications, and Related Fields*, 4e (eds. E. Niedermeyer and F.L. Da Silva), 235–260. Lippincott Williams & Wilkins.

48 Hughes, J.R. and Gruener, G.T. (1984). Small sharp spikes revisited: further data on this controversial pattern. *Electroencephalography and Clinical Neurophysiology* **15**: 208–213.

49 Hecker, A., Kocher, R., Ladewig, D., and Scollo-Lavizzari, G. Das Minature-spike-wave. *Das EEG Labor* **1**: 51–56.

50 Geiger, L.R. and Harner, R.N. (1978). EEG patterns at the time of focal seizure onset. *Archives of Neurology* **35**: 276–286.

51 Gastaut, H. and Broughton, R. (1972). *Epileptic Seizure*. Springfield, IL: Charles C. Thomas.

52 Oller-Daurella, L. and Oller-Ferrer-Vidal, L. (1977). *Atlas de Crisis Epilepticas*. Geigy Division Farmaceut.

53 Niedermeyer, E. (1999). Nonepileptic attacks, Chapter 28. In: *Electroencephalography, Basic Principles, Clinical Applications, and Related Fields*, 4e (eds. E. Niedermeyer and F.L. Da Silva), 586–594. Lippincott Williams & Wilkins.

54 Creutzfeldt, H.G. (1968). Uber eine eigenartige herdformige erkrankung des zentralnervensystems. *Zeitschrift für die gesamte Neurologie und Psychiatrie* **57**: 1–18, Quoted after W. R. Kirschbaum, 1920.

55 Jakob, A. (1968). Uber eigenartige erkrankung des zentralnervensystems mit bemerkenswerten anatomischen befunden (spastistische pseudosklerose, encephalomyelopathie mit disseminerten degenerationsbeschwerden). *Deutsche Zeitschrift für Nervenheilkunde* **70**: 132, Quoted after W. R. Kirschbaum, 1921.

56 Niedermeyer, E. (1999). Epileptic seizure disorders, Chapter 27. In: *Electroencephalography, Basic Principles, Clinical Applications, and Related Fields*, 4e (eds. E. Niedermeyer and F.L. Da Silva), 476–585. Lippincott Williams & Wilkins.

57 Small, J.G. (1999). Psychiatric disorders and EEG, Chapter 30. In: *Electroencephalography, Basic Principles, Clinical Applications, and Related Fields*, 4e (eds. E. Niedermeyer and F.L. Da Silva), 235–260. Lippincott Williams & Wilkins.

58 Marosi, E., Harmony, T., Sanchez, L. et al. (1992). Maturation of the coherence of EEG activity in normal and learning disabled children. *Electroencephalography and Clinical Neurophysiology* **83**: 350–357.

59 Linden, M., Habib, T., and Radojevic, V. (1996). A controlled study of the effects of EEG biofeedback on cognition and behavior of children with attention deficit disorder and learning disabilities. *Biofeedback and Self-Regulation* **21** (1): 35–49.

60 Hermens, D.F., Soei, E.X., Clarke, S.D. et al. (2005). Resting EEG theta activity predicts cognitive performance in attention-deficit hyperactivity disorder. *Pediatric Neurology* **32** (4): 248–256.

61 Swartwood, J.N., Swartwood, M.O., Lubar, J.F., and Timmermann, D.L. (2003). EEG differences in ADHD-combined type during baseline and cognitive tasks. *Pediatric Neurology* **28** (3): 199–204.

62 Clarke, A.R., Barry, R.J., McCarthy, R., and Selikowitz, M. (2002). EEG analysis of children with attention-deficit/hyperactivity disorder and comorbid reading disabilities. *Journal of Learning Disabilities* **35** (3): 276–285.
63 Yordanova, J., Heinrich, H., Kolev, V., and Rothenberger, A. (2006). Increased event-related theta activity as a psychophysiological marker of comorbidity in children with tics and attention-deficit/hyperactivity disorders. *NeuroImage* **32** (2): 940–955.
64 Jacobson, S. and Jerrier, H. (2000). EEG in delirium. *Seminars in Clinical Neuropsychiatry* **5** (2): 86–92.
65 Onoe, S. and Nishigaki, T. (2004). EEG spectral analysis in children with febrile delirium. *Brain & Development* **26** (8): 513–518.
66 Brunovsky, M., Matousek, M., Edman, A. et al. (2003). Objective assessment of the degree of dementia by means of EEG. *Neuropsychobiology* **48** (1): 19–26.
67 Koenig, T., Prichep, L., Dierks, T. et al. (2005). Decreased EEG synchronization in Alzheimer's disease and mild cognitive impairment. *Neurobiology of Aging* **26** (2): 165–171.
68 Babiloni, C., Binetti, G., Cassetta, E. et al. (2006). Sources of cortical rhythms change as a function of cognitive impairment in pathological aging: a multicenter study. *Clinical Neurophysiology* **117** (2): 252–268.
69 Bauer, G. and Bauer, R. (1999). EEG, drug effects, and central nervous system poisoning, Chapter 35. In: *Electroencephalography, Basic Principles, Clinical Applications, and Related Fields*, 4e (eds. E. Niedermeyer and F.L. Da Silva), 671–691. Lippincott Williams & Wilkins.
70 Beck, E. and Daniel, P.M. (1969). Degenerative diseases of the central nervous system transmissible to experimental animals. *Postgraduate Medical Journal* **45** (524): 361–370.
71 Naidu, S. and Niedermeyer, E. (1999). Digenerative disorders of the central nervous system, Chapter 20. In: *Electroencephalography, Basic Principles, Clinical Applications, and Related Fields*, 4e (eds. E. Niedermeyer and F.L. Da Silva), 360–382. Lippincott Williams & Wilkins.

3

EEG Signal Modelling

3.1 Introduction

Generation of electrical potentials or magnetic fields, measurable from the brain, is due to a nonlinear sum/distribution of electrochemical active potentials within all the neurons involved in cognitive or movement-related processes. An accurate model that can link the chemical processes within corresponding neurons generating the active potentials is hard to achieve due to the involvement of various chemicals and chemical processes. Some models, however, have been introduced since 1950s.

Neurons may be considered as signal converters. The brain is a complicated network of a tremendous number of neurons. Figure 3.1a shows a small network of three neurons. Each axon is extended from a soma which is the main neuron body. Neurons transmit and exchange electric signals called action potentials (APs) or spikes, among each other. Neurons receive the spikes at a synapse. Then, the electric signals or information is transmitted in the direction from a dendrite to an axon. Figure 3.1b illustrates the waveform of APs which have an amplitude of approximately 100 mV in the human brain.

Typical neurons do not generate any spikes without input signals which often come from other neurons. A sufficiently large input pulse causes a neuron to generate an output spike whereas no output spike is generated by a small input. Therefore, a neuron possesses a *threshold* or *all-or-none* characteristic. There is a special period or timing called the *refractory period* (the timing of the downstroke of the AP) in which the neuron cannot produce any output spike even though a sufficient amount of inputs is applied to the neuron. Hence, a neuron may be considered as a device which transforms or converts the input spike train to another spike train where each output spike (AP), as we will see later, is the integral of a number of input spikes.

3.2 Physiological Modelling of EEG Generation

The popular physiological models aim to best simulate the coupling between two or more neurons. In [1] three models for generation of brain potentials have been introduced and compared.

EEG Signal Processing and Machine Learning, Second Edition. Saeid Sanei and Jonathon A. Chambers.

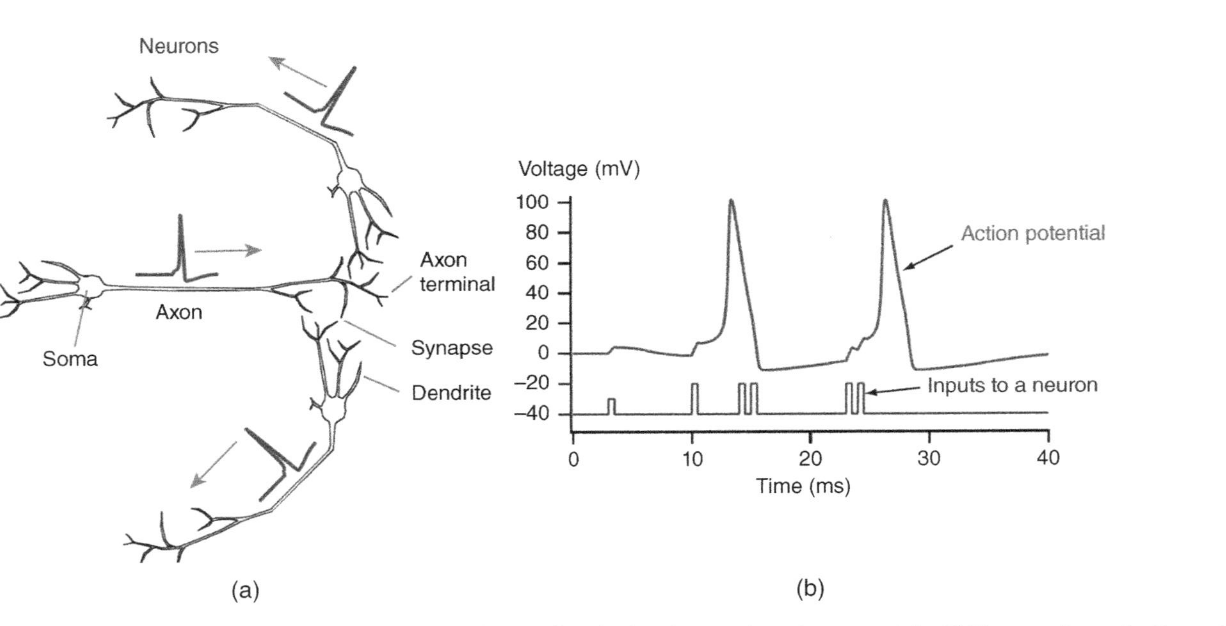

Figure 3.1 (a) A network of three neurons that exchange electric signals, namely action potentials. (b) The waveform of action potentials including the input to a neuron.

3.2.1 Integrate-and-Fire Models

For coupling two neurons two integrate-and-fire neurons with mutual excitatory or inhibitory coupling have been described in [1]. The neurons with activation variables x_i for $i = 1$, 2, satisfy:

$$\frac{dx_i}{dt} = \xi - x_i + E_i(t) \tag{3.1}$$

where $\xi > 1$ is a constant, $0 < x_i < 1$, and $E_i(t)$ is the synaptic input to neuron i. Neuron i fires when $x_i = 1$ and hen resets x_i to 0. If cell $j \neq i$ fires at time t_j the function E_i is augmented to $E_i(t) + E_s(t\text{-}t_j)$, where E_s is the contribution coming from one spike [1]. In an example in [1] this function is selected as:

$$E_s(t) = g\alpha^2 te^{-\alpha t} \tag{3.2}$$

where g and α are parameters determining the strength and speed of the synapse respectively and the factor of α^2 in (3.2) normalizes the integral of E_s over time to the value g. In the considered cases the two neurons continue firing periodically when they are coupled together. Assuming that neuron 1 fires at times $t = nT$, where T is the period and n is an integer, while neuron 2 fires at $t = (n - \varphi)T$. Therefore, both neurons are firing at the same frequency but are separated by a phase φ. We wish to determine possible values of the phase difference φ and conditions under which they arise.

3.2.2 Phase-Coupled Models

Neuronal synchronization processes, measured with brain imaging data, can be described using weakly coupled oscillator (WCO) models. Dynamic causal modelling (DCM) is used to fit the WCOs to brain imaging data and so make inferences about the structure of neuronal interactions [2]. The complex behaviours are mediated by the interaction of particular brain regions. Recent studies agreed that such interactions may be instantiated by the transient synchronization of oscillatory neuronal ensembles [3]. For example, contour detection is accompanied by gamma band synchronization in distant parts of visual cortex, multimodal object processing by parieto-temporal synchronization in the beta band [4] and spatial memory processes by hippocampal–prefrontal synchronization in the theta band [2, 5]. DCM allows for different model structures to be compared using Bayesian model selection [6]. In [2] DCM has been extended to the study of phase coupling. One direction is based on the WCO models in which the rate of change of phase of one oscillator is related to the phase differences between itself and other oscillators [7].

The WCO theory applies to system dynamics close to limit cycles. By assuming that weak coupling leads to only small perturbations away from these cycles, one can reduce a high-dimensional system of differential equations to one based solely on the phases of the oscillators, and pairwise interactions between them [2].

Dynamics on the limit cycle are given by [2]:

$$\dot{X}_0 = F(X_0)$$
$$X_0(t + T) = X_0(t)$$

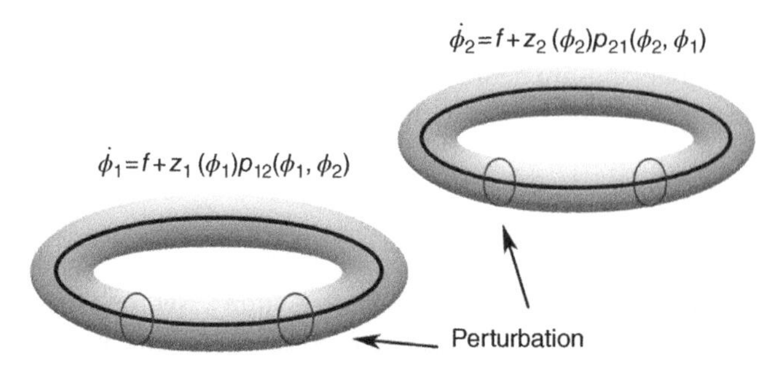

Figure 3.2 A pair of oscillators weakly coupled via the perturbation function $p(\varphi_1, \varphi_2)$.

$$\dot{\phi}(X_0) = f(X_0) \tag{3.3}$$

Then, for any perturbation of phase $p(\phi)$ this changes to [2]:

$$\dot{X} = F(X) + P(X) \tag{3.4}$$

$$\dot{\phi}(X_0) = f(X_0) + z(\phi)p(\phi) \tag{3.5}$$

where:

$$z(\phi) = \frac{d\phi(X_0)}{dX_0} \tag{3.6}$$

For a pair of oscillators of phases φ_1 and φ_2, using the same analysis (see Figure 3.2, assuming $p_{12} = p_{21}$):

$$\dot{\phi}_1 = f + z_1(\phi_1)p_{12}(\phi_1, \phi_2) \tag{3.7}$$

$$\dot{\phi}_2 = f + z_2(\phi_2)p_{21}(\phi_2, \phi_1) \tag{3.8}$$

If it is further assumed that the phase difference $\varphi_2 - \varphi_1 = \varphi$ changes slowly, then [2]:

$$\dot{\phi}_1 = f + \Gamma_{12}(\phi_1 - \phi_2) \tag{3.9}$$

$$\dot{\phi}_2 = f + \Gamma_{21}(\phi_2 - \phi_1) \tag{3.10}$$

$$\Gamma_{ij}(\phi) = \frac{1}{2\pi}\int_0^{2\pi} z_i(\psi)p_{ij}(\psi, \psi + \phi)d\psi \tag{3.11}$$

$\Gamma_{ij}(\phi)$ is called phase interaction function (PIF). Similarly, for N_R regions the rate of change of phase of the ith oscillator is given by:

$$\dot{\phi}_i = f_i + \sum\nolimits_{j=1}^{N_R} \Gamma_{ij}\left(\left(\phi_i - \phi_j\right) - d_{ij}\right. \tag{3.12}$$

where f_i is the intrinsic frequency of the ith oscillator. In these formulations there are two key assumptions: the first one is that the perturbations are sufficiently small that the differentiations can equivalently be evaluated at X_0 rather than X. The second assumption is that the relative changes in the oscillator phase are sufficiently slow with respect to the oscillation frequency, that the phase offset term can be replaced by a time average.

In [2] an extension of the above WCO model is used to describe the dynamic phase changes in a network of oscillators. The use of Bayesian model comparison allows one to infer the mechanisms underlying synchronization processes in the brain. This has been applied to synthetic bimanual finger movement data from physiological models and on magnetoencephalogram (MEG) data from a study of visual working memory. The WCO approach accommodates signal nonstationarity by using differential equations which describe how the changes in phase are driven by pairwise differences in instantaneous phase.

3.2.3 Hodgkin–Huxley Model

Most probably the earliest physical model is based on the Hodgkin and Huxley's Nobel Prize winning mathematical model for a squid axon published in 1952 [8–10]. The Hodgkin and Huxley equations are important not only because they represent the most successful mathematical model in quantitatively describing the related biological phenomena but also due to the fact that deriving the model of a squid is directly applicable to many kinds of neurons and other excitable cells. According to this model, a nerve axon may be stimulated and the activated sodium (Na^+) and potassium (K^+) channels produced in the vicinity of the cell membrane can lead to the electrical excitation of the nerve axon. The excitation arises from the effect of the membrane potential on the movement of ions, and from interactions of the membrane potential with the opening and closing of voltage activated membrane channels. The membrane potential increases when the membrane is polarized with a net negative charge lining the inner surface and an equal but opposite net positive charge on the outer surface. This potential may be simply related to the amount of electrical charge Q, using:

$$E = Q/C_m, \tag{3.13}$$

where Q is in terms of Coulombs cm^{-2}, C_m is the measure of the capacity of the membrane and has units farads cm^{-2} and E has units of volts. In practise, in order to model the APs the amount of charge Q^+ on the inner surface (and Q^- on the outer surface) of the cell membrane has to be mathematically related to the stimulating current I_{stim} flowing into the cell through the stimulating electrodes. Figure 3.1b illustrates how the neuron excitation results in generation of APs by acting as a signal converter [11].

The electrical potential (often called electrical force) E is then calculated using Eq. (3.13). The Hodgkin and Huxley's model is illustrated in Figure 3.3. In this figure I_{memb} is the result of positive charges flowing out of the cell. This current consists of mainly three currents namely, Na, K, and leak currents. The leak current is due to the fact that the inner and outer Na and K ions are not exactly equal.

Hodgkin and Huxley estimated the activation and inactivation functions for the Na and K currents and derived a mathematical model to describe an AP similar to that of a giant squid. The model is a neuron model that uses voltage-gated channels. The space-clamped version of the Hodgkin–Huxley model may be well described using four ordinary differential equations [12]. This model describes the change in the membrane potential (E) with respect to time and is described in [13]. The overall membrane current is the sum of capacity current and ionic current as:

$$I_{memb} = C_m \frac{dE}{dt} + I_i \tag{3.14}$$

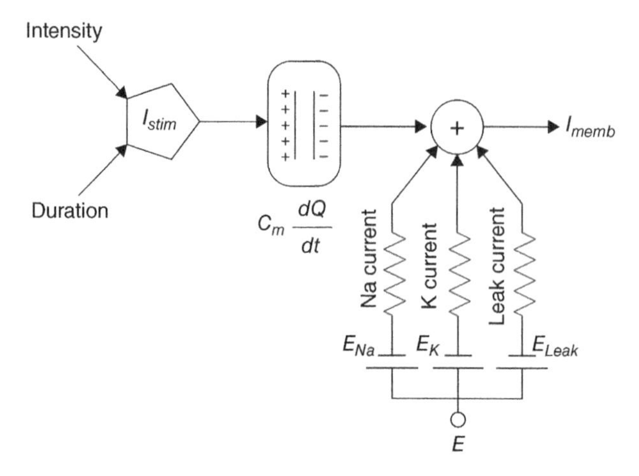

Figure 3.3 The Hodgkin–Huxley excitation model.

where I_i is the ionic current and as indicated in Figure 3.1 which can be considered as the sum of three individual components: Na, K, and leak currents:

$$I_i = I_{Na} + I_K + I_{leak} \tag{3.15}$$

I_{Na} can be related to the maximal conductance $\bar{g}_{Na}$, activation variable a_{Na}, inactivation variable h_{Na}, and a driving force $(E - E_{Na})$ through:

$$I_{Na} = g_{Na} a_{Na}^3 h_{Na} (E - E_{Na}) \tag{3.16}$$

Similarly I_k can be related to the maximal conductance $\bar{g}_K$, activation variable a_K, and a driving force $(E - E_K)$ as:

$$I_K = g_K a_K (E - E_K) \tag{3.17}$$

and I_{leak} is related to the maximal conductance $\bar{g}_l$ and a driving force $(E - E_l)$ as:

$$I_l = g_l (E - E_l) \tag{3.18}$$

The changes in the variables a_{Na}, a_k, *and* h_{Na} vary from 0 to 1 according to the following equations:

$$\frac{da_{Na}}{dt} = \lambda_t [\alpha_{Na}(E)(1 - a_{Na}) - \beta_{Na}(E) a_{Na}] \tag{3.19}$$

$$\frac{dh_{Na}}{dt} = \lambda_t [\alpha_h(E)(1 - h_{Na}) - \beta_h(E) h_{Na}] \tag{3.20}$$

$$\frac{da_K}{dt} = \lambda_t [\alpha_K(E)(1 - a_K) - \beta_K(E) a_K] \tag{3.21}$$

where $\alpha(E)$ and $\beta(E)$ are respectively forward and backward rate functions and λ_t is a temperature-dependent factor. The forward and backward parameters depend on voltage and were empirically estimated by Hodgkin and Huxley as:

$$\alpha_{Na}(E) = \frac{3.5 + 0.1E}{1 - e^{-(3.5 + 0.1E)}} \tag{3.22}$$

$$\beta_{Na}(E) = 4e^{-(E+60)/18} \tag{3.23}$$

$$\alpha_h(E) = 0.07e^{-(E+60)/20} \tag{3.24}$$

$$\beta_h(E) = \frac{1}{1 + e^{-(3+0.1E)}} \tag{3.25}$$

$$\alpha_K(E) = \frac{0.5 + 0.01E}{1 - e^{-(5+0.1E)}} \tag{3.26}$$

$$\beta_K(E) = 0.125e^{-(E+60)/80} \tag{3.27}$$

As stated in the Simulator for Neural Networks and Action Potentials (SNNAP) literature [12], the $\alpha(E)$ and $\beta(E)$ parameters have been converted from the original Hodgkin–Huxley version to agree with the present physiological practice where depolarization of the membrane is taken to be positive. In addition, the resting potential has been shifted to −60 mV (from the original 0 mV). These equations are used in the model described in the SNNAP. In Figure 3.4 an AP has been simulated. For this model the parameters are set to $C_m = 1.1$ μF cm^{-2}, $\bar{g}_{Na} = 100$ ms cm^{-2}, $\bar{g}_K = 35$ mS cm^{-2}, $\bar{g}_l = 0.35$ mS cm^{-2}, and $E_{Na} = 60$ mV.

The simulation can run to generate a series of APs as practically happens in the case of event-related potential (ERP) signals. If the maximal ionic conductance of the potassium current $\bar{g}_K$, is reduced the model will show a higher resting potential. Also, for $\bar{g}_K = 16$ mS cm^{-2}, the model will begin to exhibit oscillatory behaviour. Figure 3.5 shows the result of a Hodgkin–Huxley oscillatory model with reduced maximal potassium conductance.

The SNNAP can also model bursting neurons and central pattern generators. This stems from the fact that many neurons show cyclic spiky activities followed by a period of

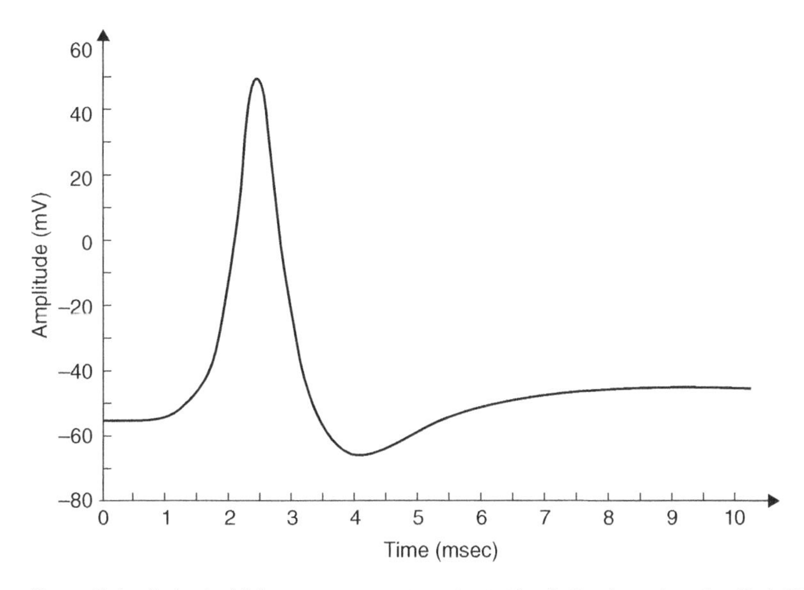

Figure 3.4 A single AP in response to a transient stimulation based on the Hodgkin–Huxley model. The initiated time is at t = 0.4 ms and the injected current is 80 μA cm^{-2} for a duration of 0.1 ms. The selected parameters are $C_m = 1.2$ μF cm^{-2}, $\bar{g}_{Na} = 100$ mS cm^{-2}, $\bar{g}_K = 35$ mS cm^{-2}, $\bar{g}_l = 0.35$ mS cm^{-2}, and $E_{Na} = 60$ mV.

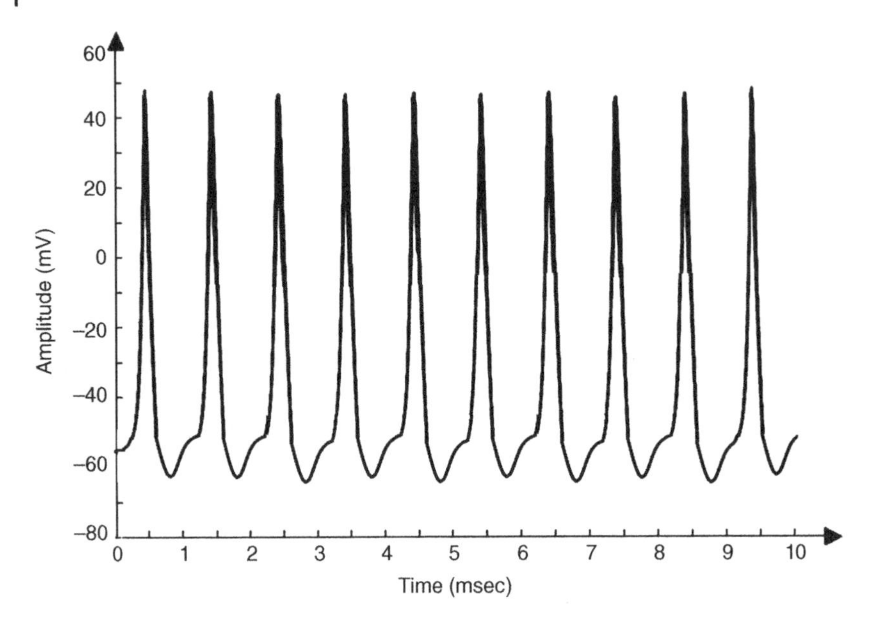

Figure 3.5 The AP from a Hodgkin–Huxley oscillatory model with reduced maximal potassium conductance.

inactivity. Several invertebrate as well as mammalian neurons are bursting cells and exhibit alternating periods of high-frequency spiking behaviour followed by a period of no spiking activity.

3.2.4 Morris–Lecar Model

A simpler model than that of Hodgkin–Huxley for simulating spiking neurons is the Morris–Lecar model [14]. This model is a minimal biophysical model, which generally exhibits single AP. This model considers that the oscillation of a slow calcium wave depolarizing the membrane leads to a bursting state. The Morris–Lecar model was initially developed to describe the behaviour of barnacle muscle cells. The governing equations relating the membrane potential (E) and potassium activation w_k to the activation parameters are given as:

$$C\frac{dE}{dt} = I_i - g_{ca}a_{Ca}(E)(E - E_{ca}) - g_k w_k(E - E_k) - g_l(E - E_l) \tag{3.28}$$

$$\frac{dw_k}{dt} = \lambda_t\left(\frac{w_\infty(E) - w_k}{\tau_k(E)}\right) \tag{3.29}$$

where I_i is the combination of three ionic currents, calcium (Ca), potassium (K) and leak (l) and similar to the Hodgkin–Huxley model, are products of a maximal conductance $\overline{g}$, activation components (in such as a_{Ca}, w_k), and the driving force E. The changes in the potassium activation variable w_k is proportional to a steady-state activation function $w_\infty(E)$ (a sigmoid curve) and a time-constant function $\tau_k(E)$ (a bell-shaped curve). These functions are respectively defined as:

$$w_{\infty}(E) = \frac{1}{1 + e^{-(E - h_w)/S_w}} \tag{3.30}$$

$$\tau_k(E) = \frac{1}{e^{(E - h_w)/2S_w} + e^{-(E - h_w)/2S_w}} \tag{3.31}$$

The steady-state activation function $a_{ca}(E)$, involved in calculation of the calcium current, is defined as:

$$a_{ca}(E) = \frac{1}{1 + e^{-(E - h_{Ca})/S_m}} \tag{3.32}$$

Similar to the sodium current in the Hodgkin–Huxley model, the calcium current is an inward current. Since the calcium activation current is a fast process in comparison with the potassium current, it is modelled as an instantaneous function. This means that for each voltage E, the steady-state function $a_{Ca}(E)$ is calculated. The calcium current does not incorporate any inactivation process. The activation variable w_k here is similar to a_k in the Hodgkin–Huxley model, and finally the leak currents for both models are the same [12]. A simulation of the Morris–Lecar model is presented in Figure 3.6.

Calcium-dependent potassium channels are activated by intracellular calcium, the higher the calcium concentration the higher the channel activation [12]. For the Morris–Lecar model to exhibit bursting behaviour, the two parameters of maximal time constant and the input current have to be changed [12]. Figure 3.7 shows the bursting behaviour of the Morris–Lecar model. The basic characteristics of a bursting neuron are the duration

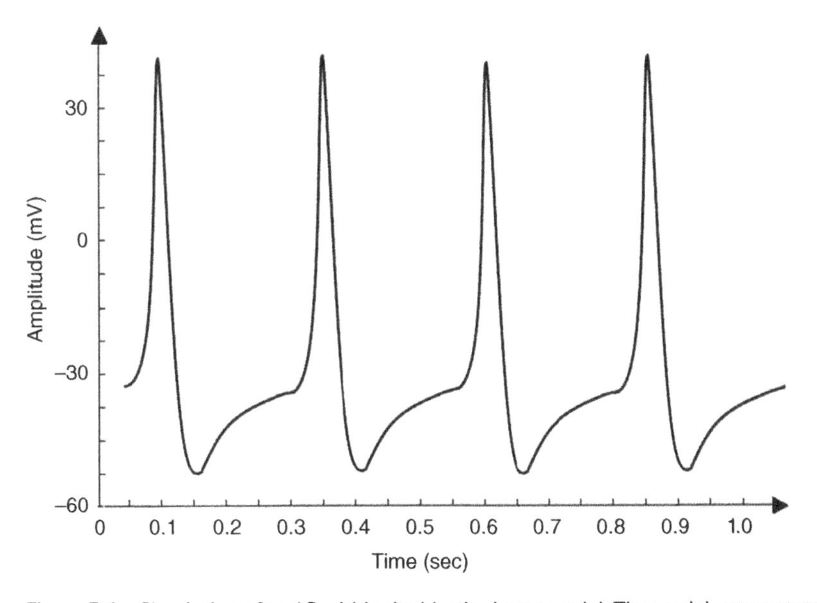

Figure 3.6 Simulation of an AP within the Morris–Lecar model. The model parameters are: C_m = 22 μF cm^{-2}, $\bar{g}_{Ca}$=3.8 mS cm^{-2}, $\bar{g}_k$=8.0 mS cm^{-2}, $\bar{g}_l$=1.6 mS cm^{-2}, E_{Ca} = 125 mV, E_k = −80 mV, E_l = −60 mV, λ_t = 0.06, h_{Ca} = −1.2, S_m = 8.8.

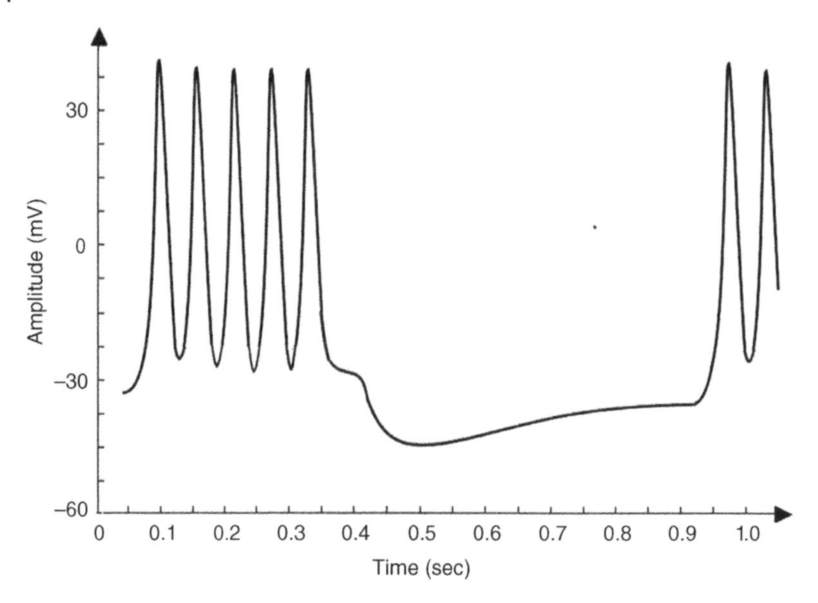

Figure 3.7 An illustration of the bursting behaviour that can be generated by the Morris–Lecar model.

of the spiky activity, the frequency of the APs during a burst, and the duration of the quiescence period. The period of an entire bursting event is the sum of both active and quiescence duration [12].

Neurons communicate with each other across synapses through axon-dendrites or dendrites-dendrites connections, which can be excitatory, inhibitory, or electric [12]. By combining a number of the above models, a neuronal network can be constructed. The network exhibits oscillatory behaviour due to the synaptic connection between the neurons. It is commonly assumed that excitatory synaptic coupling tends to synchronize neural firing, while inhibitory coupling pushes neurons towards anti-synchrony. Such behaviour has been seen in models of neuronal circuits [1].

A synaptic current is produced as soon as a neuron fires an AP. This current stimulates the connected neuron and may be modelled by an alpha function multiplied by a maximal conductance and a driving force as:

$$I_{syn} = g_{syn} \cdot g_{syn}(t)\left(E(t) - E_{syn}\right) \tag{3.33}$$

where:

$$g_{syn}(t) = t.e^{(-t/u)} \tag{3.34}$$

and t is the latency or time since the trigger of the synaptic current, u is the time to reach to the peak amplitude, E_{syn} is the synaptic reversal potential, and $\bar{g}_{syn}$ is the maximal synaptic conductance. The parameter u alters the duration of the current while $\bar{g}_{syn}$ changes the strength of the current. This concludes the treatment of the modelling of APs.

As the nature of the EEG sources cannot be determined from the electrode signals directly, many researchers have tried to model these processes on the basis of information extracted using signal processing techniques. The method of linear prediction (LP) described in the later sections of this chapter is frequently used to extract a parametric description.

A tutorial on realistic neural modelling using the Hodgkin–Huxley excitation model by David Beeman has been documented at the first annual meeting of the World Association of Modelers (WAM) Biologically Accurate Modeling Meeting (BAMM) in 2005 in Texas, USA. This can be viewed at http://www.brains-minds-media.org/archive/218.

3.3 Generating EEG Signals Based on Modelling the Neuronal Activities

The objective in this section is to introduce some established models for generating normal and some abnormal EEGs. These models are generally nonlinear and some have been proposed [15] for modelling a normal EEG signal and some others for the abnormal EEGs.

A simple distributed model consisting of a set of simulated neurons, thalamocortical relay cells, and interneurons was proposed [16, 17] that incorporates the limited physiological and histological data available at that time. The basic assumptions were sufficient to explain the generation of the alpha rhythm, i.e. the EEGs within the frequency range of 8–13 Hz.

A general nonlinear lumped model may take the form shown in Figure 3.8. Although the model is analogue in nature all the blocks are implemented in discrete form. This model can take into account the major characteristics of a distributed model and it is easy to investigate

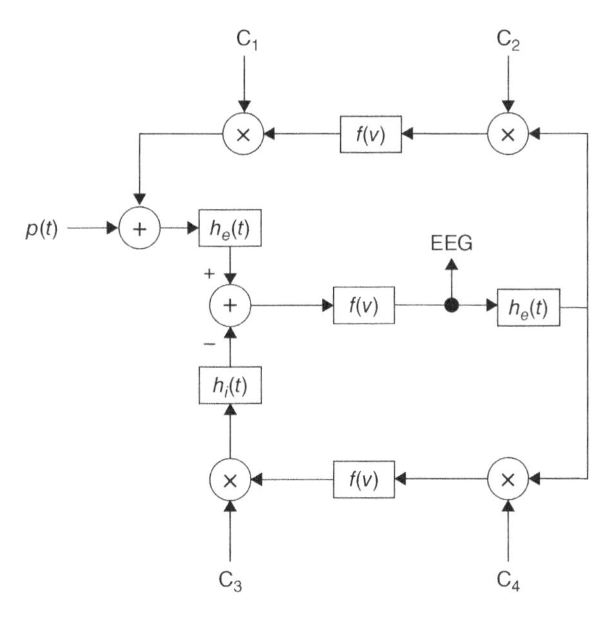

Figure 3.8 A nonlinear lumped model for generating the rhythmic activity of the EEG signals; $h_e(t)$ and $h_i(t)$ are the excitatory and inhibitory post synaptic potentials, $f(v)$ is normally a simplified nonlinear function, and the C_is are respectively the interaction parameters representing the interneurons and thalamocortical neurons [16].

the result of changing the range of excitatory and inhibitory influences of thalamocortical relay cells and interneurons.

In this model [16] there is a feedback loop including the inhibitory post-synaptic potentials, the nonlinear function, and the interaction parameters C_3 and C_4. The other feedback includes mainly the excitatory potentials, nonlinear function, and the interaction parameters C_1 and C_2. The role of the excitatory neurons is to excite one or two inhibitory neurons. The latter, in turn, serve to inhibit a collection of excitatory neurons. Thus, the neural circuit forms a feedback system. The input $p(t)$ is considered as a white noise signal. This is a general model; more assumptions are often needed to enable generation of the EEGs for the abnormal cases. Therefore, the function $f(v)$ may change to generate the EEG signals for different brain abnormalities. Accordingly, the C_i coefficients can be varied. In addition, the output is subject to environment and measurement noise. In some models, such as the local EEG model (LEM) [16] the noise has been considered as an additive component in the output.

Figure 3.9 shows the LEM model. This model uses the formulation by Wilson and Cowan [18] who provided a set of equations to describe the overall activity (not specifically the EGG) in a cartel of excitatory and inhibitory neurons having a large number of interconnections [19]. Similarly, in the LEM the EEG it is assumed that the rhythms are generated by distinct neuronal populations, which possess frequency selective properties. These populations are formed by the interconnection of the individual neurons and are assumed to be driven by a random input. The model characteristics, such as the neural interconnectivity, synapse pulse response, and threshold of excitation are presented by the LEM parameters. The changes in these parameters produce the relevant EEG rhythms.

The input $p(t)$ is assumed to result from the summation of a randomly distributed series of random potentials which drive the excitatory cells of the circuit, producing the ongoing background EEG signal. Such signals originate from other deeper brain sources within the thalamus and brain stem and constitute part of the ongoing or spontaneous firing of the central nerve system (CNS). In the model, the average number of inputs to an inhibitory

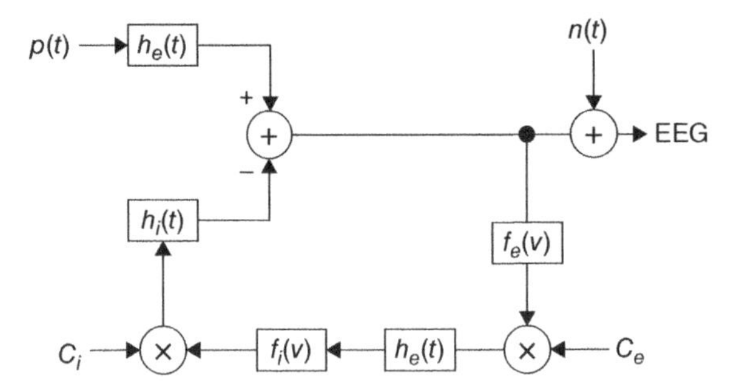

Figure 3.9 The local EEG model (LEM). The thalamocortical relay neurons are represented by two linear systems having impulse responses $h_e(t)$, on the upper branch, and the inhibitory post-synaptic potential by $h_i(t)$. The nonlinearity of this system is denoted by $f_e(v)$ representing the spike generating process. The interneuron activity is represented by another linear filter $h_e(t)$ in the lower branch, which generally can be different from the first linear system, and a nonlinearity function $f_i(v)$. C_e and C_i represent respectively, the number of interneuron cells and the thalamocortical neurons.

neuron from the excitatory neurons is designated by C_e and the corresponding average number from inhibitory neurons to each individual excitatory neuron is C_i. The difference of two decaying exponentials are used for modelling each post-synaptic potential h_e or h_i:

$$h_e(t) = A[exp(-a_1 t) - exp(-a_2 t)] \tag{3.35}$$

$$h_i(t) = B[exp(-b_1 t) - exp(-b_2 t)] \tag{3.36}$$

where A, B, a_k, and b_k are constant parameters, which control the shape of the pulse waveforms. The membrane potentials are related to the axonal pulse densities via the static threshold functions f_e and f_i. These functions are generally nonlinear; however, to ease the manipulations they are considered linear for each short time interval. Using this model, the normal brain rhythms such as alpha wave is considered as filtered noise.

A more simplified model is the one presented by Jansen and Rit [20]. This model is shown in Figure 3.10. The model is a neurophysiologically inspired model simulating electrical brain activity (including EEG or evoked potentials (EPs)). A previously developed lumped-parameter model [16] of a single cortical column has been implemented in their work. The model could produce a large variety of EEG-like waveforms and rhythms. Coupling two models, with delays in the interconnections to simulate the synaptic connections within and between cortical areas, made it possible to replicate the spatial distribution of alpha and beta activity.

EPs were simulated by presenting pulses to the input of the coupled models. In general, the responses were more realistic than those produced using a single model. The proposed model is based on a nonlinear model of a cortical column described by Jansen and Rit [20] and also based upon Lopes da Silva's lumped-parameter model [16, 18]. The cortical column is modelled by a population of 'feed forward' pyramidal cells, receiving inhibitory and excitatory feedback from local interneurons (i.e. other pyramidal, stellate or basket cells residing in the same column) and excitatory input from neighbouring or more distant columns. The input can be a pulse, arbitrary function, or noise.

Each of the neuron populations is modelled by two blocks. The first block transforms the average pulse density of APs coming to the population of neurons into an average post-synaptic membrane potential which can either be excitatory or inhibitory. This block is

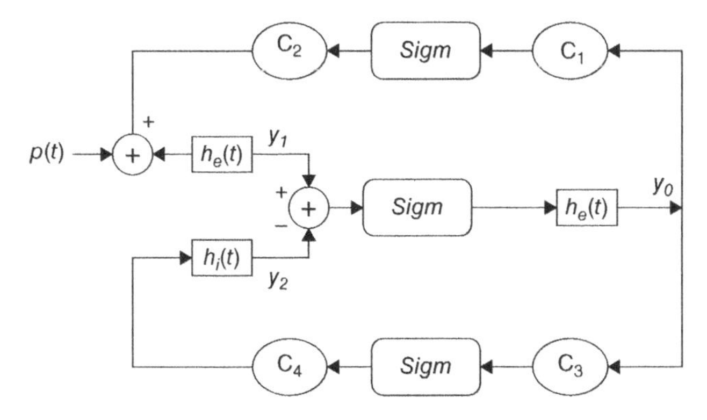

Figure 3.10 Simplified model for brain cortical alpha generation. The input is a pulse-shaped waveform.

referred to as the post-synaptic potential (PSP) block and represents a linear transformation with an impulse response given by [20]:

$$h_e(t) = \begin{cases} Aate^{-at} \; t \geq 0 \\ 0 \; t < 0 \end{cases} \tag{3.37}$$

for the excitatory case, and the following for the inhibitory case:

$$h_i(t) = \begin{cases} Bbte^{-bt} \; t \geq 0 \\ 0 \; t < 0 \end{cases} \tag{3.38}$$

A and B determine the maximum amplitudes of the excitatory and inhibitory PSPs respectively, and a and b are the lumped representation of the sum of reciprocal of the time constant of passive membrane and all other spatially distributed delays in the neuron dendric network. The *Sigm* function is defined as [21]:

$$Sigm(v) = \frac{2e_0}{1 + e^{r(v_0 - v)}} \tag{3.39}$$

where e_0 indicates the maximum firing rate of the neural population, v_0 the PSP for which a 50% firing rate is achieved, and r the steepness of sigmoidal transformation. The connectivity constants $C_1 - C_4$ in the figure characterize the interaction between the pyramidal cells and the excitatory and inhibitory interneurons which account for the total number of synapses established between the neurons.

As an empirical value, considering $C = C_1 = 135$, typical values suggested for the rest of the parameters are $C_2 = 0.8C$, $C_3 = 0.25C$, $C_4 = 0.25C$, $A = 3.25$, $B = 22$, $v_0 = 6$, $a = 100\,s^{-1}$, $b = 50\,s^{-1}$ and $a_d = 30$.

The above system has been extended to modelling the visual evoked potentials (VEPs) based on the fact that VEPs are the results of interaction of two or more of so-called cortical columns. A proposed two-column model for VEP can be seen in Figure 3.11 where each block represents the model in Figure 3.10.

The main problem with such a model is due to the fact that only a single channel EEG is generated and unlike the phase-coupling model explained in subsection 3.2.2, there is no modelling of interchannel relationships and the inherent connectivity of the brain zones. Therefore, a more accurate model has to be defined to enable simulation of a multichannel EEG generation system. This is still an open question and remains an area of research.

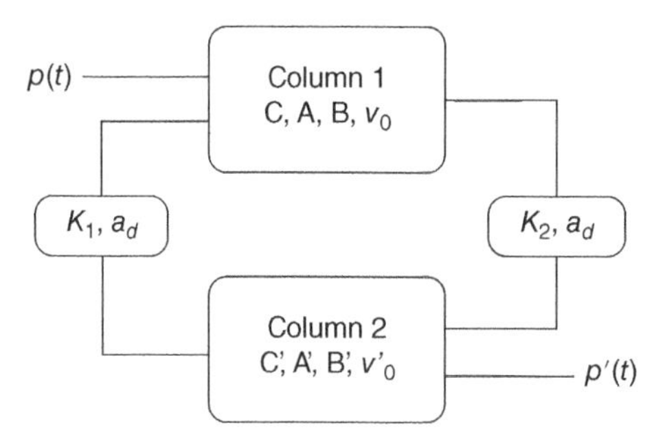

Figure 3.11 A two-column model for generation of VEP. Two connectivity constants K_1 and K_2 attenuate the output of a column before it is fed to the other.

3.4 Mathematical Models Derived Directly from the EEG Signals

These models are basically the description of single and multichannel signals in terms of a limited number of statistical parameters which not only represent the morphology of the waveforms but also their temporal or spatial sample correlations. In these models the internal and external noise is often considered as an uncorrelated temporal signal independent of the brain generated signals.

3.4.1 Linear Models

3.4.1.1 Prediction Method

The main objective of using prediction methods is to find a set of model parameters which best describe the signal generation system. Such models generally require a noise type input. In autoregressive (AR) modelling of signals each sample of a single channel EEG measurement is defined to be linearly related with respect to a number of its previous samples, i.e.:

$$y(n) = -\sum_{k=1}^{p} a_k y(n-k) + x(n) \tag{3.40}$$

where a_k, $k = 1,2,\ldots,p$, are the linear parameters, n denotes the discrete sample time normalized to unity, p is called the model or prediction order and $x(n)$ is the noise input. In an autoregressive moving average (ARMA) linear predictive model each sample is obtained based on a number of its previous input and output sample values, i.e.:

$$y(n) = -\sum_{k=1}^{p} a_k y(n-k) + \sum_{k=0}^{q} b_k x(n-k) \tag{3.41}$$

where b_k, $k = 1, 2,\ldots, q$ are the additional linear parameters. The parameters p and q are the model orders. The Akaike criterion can be used to determine the order of the appropriate model of a measurement signal by maximizing the log-likelihood equation [22] with respect to the model order:

$$AIC(p,q) = ln\left(\sigma_{pq}^2\right) + \frac{2(p+q)}{N} \tag{3.42}$$

where p and q represent respectively, the assumed AR and MA model prediction orders, N is the number of signal samples, and σ_{pq}^2 is the noise power of the ARMA model at the pth and qth stage. Later in this chapter we will see how the model parameters are estimated either directly or by employing some iterative optimization techniques.

In a multivariate AR (MVAR) approach a multichannel scheme is considered. Therefore, each signal sample is defined versus both its previous samples and the previous samples of the signals from other channels, i.e. for channel i we have:

$$y_i(n) = -\sum_{j=1}^{m} \sum_{k=1}^{p} a_{jk} y_j(n-k) + x_i(n) \tag{3.43}$$

where m represents the number of channels and $x_i(n)$ represents the noise input to channel i. Similarly, the model parameters can be calculated iteratively in order to minimize the error between the actual and predicted values [23].

Figure 3.12 A linear model for the generation of EEG signals.

There are numerous applications for linear models. These applications are discussed in other chapters of this book. Different algorithms have been developed to find efficiently the model coefficients. In the maximum likelihood estimation (MLE) method [23–25] the likelihood function is maximized over the system parameters formulated from the assumed real, Gaussian distributed, and sufficiently long input signals of approximately 10–20 seconds (consider a sampling frequency of $f_s = 250$ Hz as often used for EEG recordings). Using Akaike's method the gradient of the squared error is minimized using the Newton–Raphson approach applied to the resultant nonlinear equations [23, 26]. This is considered as an approximation to the MLE approach. In the Durbin method [27] the Yule–Walker equations, which relate the model coefficients to the autocorrelation of the signals, are iteratively solved. The approach and the results are equivalent to those using a least-squared-based scheme [28]. The MVAR coefficients are often calculated using the Levinson–Wiggins–Robinson (LWR) algorithm [29]. The MVAR model and its application in representation of what is called a direct transfer function (DTF), and its use in quantification of signal propagation within the brain, will be explained in detail in Chapter 8 of this book. After the parameters are estimated the synthesis filter can be excited with wide sense stationary noise to generate the EEG signal samples. Figure 3.12 illustrates the simplified system.

The prediction models can be easily extended to multichannel data. This leads to estimation of matrices of prediction coefficients. These parameters stem from both the temporal and interchannel correlations.

3.4.1.2 Prony's Method

Prony's method has been previously used to model EPs [30, 31]. Based on this model an EP, which is obtained by applying a short audio or visual brain stimulus to the brain, can be considered as the impulse response (IR) of a linear infinite impulse response (IIR) system. The original attempt in this area was to fit an exponentially damped sinusoidal model to the data [32]. This method was later modified to model sinusoidal signals [33]. Prony's method is used to calculate the LP parameters. The angles of the poles in the z-plane of the constructed LP filter are then referred to the frequencies of the damped sinusoids of the exponential terms used for modelling the data. Consequently, both the amplitude of the exponentials and the initial phase can be obtained following the methods used for an AR model, as follows.

Based on the original method we can consider the output of an AR system with zero excitation to be related to its IR as:

$$y(n) = \sum_{k=1}^{p} a_k y(n-k) = \sum_{j=1}^{p} w_j \sum_{k=1}^{p} a_k r_j^{n-k-1} \tag{3.44}$$

where $y(n)$ represents the exponential data samples, p is the prediction order, $w_j = A_j e^{j\theta_j}$, $r_k = exp\left((\alpha_k + j2\pi f_k)T_s\right)$, T_s is the sampling period normalized to 1, A_k is the amplitude of the exponential, α_k is the damping factor, f_k is the discrete-time sinusoidal frequency in samples per second, and θ_j is the initial phase in radians.

Therefore, the model coefficients are first calculated using one of the methods previously mentioned in this section, i.e. $\boldsymbol{a} = -\boldsymbol{Y}^{-1}\breve{\boldsymbol{y}}$, where:

$$\boldsymbol{a} = \begin{bmatrix} a_0 \\ a_1 \\ . \\ . \\ a_p \end{bmatrix}, \boldsymbol{Y} = \begin{bmatrix} y(p)...y(1) \\ y(p-1)...y(2) \\ . \\ . \\ y(2p-1)...y(p) \end{bmatrix}, \text{and } \breve{\boldsymbol{y}} = \begin{bmatrix} y(p+1) \\ y(p+2) \\ . \\ . \\ y(2p) \end{bmatrix} \tag{3.45}$$

and $a_0 = 1$. On the basis of (3.39), $y(n)$ is calculated as the weighted sum of its p past values. $y(n)$ is then constructed and the parameters f_k and r_k are estimated. Hence, the damping factors are obtained as

$$\alpha_k = ln\,|r_k| \tag{3.46}$$

and the resonance frequencies as

$$f_k = \frac{1}{2\pi} tan^{-1}\left(\frac{Im(r_k)}{Re\,(r_k)}\right) \tag{3.47}$$

where Re(.) and Im(.) denote respectively the real and imaginary parts of a complex quantity. The w_k parameters are calculated using the fact that $y(n) = \sum_{k=1}^{p} w_k r_k^{n-1}$ or

$$\begin{bmatrix} r_1^0 & r_2^0, & \dots & r_p^0 \\ r_1^1 & r_2^1, & \dots & r_p^1 \\ . & . & & . \\ . & . & & . \\ r_1^{p-1} & r_2^{p-1} & \dots & r_p^{p-1} \end{bmatrix} \begin{bmatrix} w_1 \\ w_2 \\ . \\ . \\ w_p \end{bmatrix} = \begin{bmatrix} y(1) \\ y(2) \\ . \\ . \\ y(p) \end{bmatrix} \tag{3.48}$$

In vector form this can be illustrated as $\boldsymbol{Rw} = \boldsymbol{y}$, where $[\boldsymbol{R}]_{k,l} = r_l^k, k = 0, 1, \cdots, p-1, l = 1, \cdots, p$ denoting the elements of the matrix in the above equation. Therefore, $\boldsymbol{w} = \boldsymbol{R}^{-1}\boldsymbol{y}$, assuming R is a full-rank matrix, i.e. there are no repeated poles. Often, this is simply carried out by implementing the Cholesky decomposition algorithm [34]. Finally, using w_k, the amplitude and initial phases of the exponential terms are calculated as follows:

$$A_k = |w_k| \tag{3.49}$$

and

$$\theta_k = tan^{-1}\left(\frac{Im(w_k)}{Re\,(w_k)}\right). \tag{3.50}$$

In the above solution we considered that the number of data samples N is equal to $N = 2p$, where p is the prediction order. For the cases where $N > 2p$ a least-squares (LS) solution for **w** can be obtained as:

$$\boldsymbol{w} = \left(\boldsymbol{R}^H\boldsymbol{R}\right)^{-1}\boldsymbol{R}^H\boldsymbol{y} \tag{3.51}$$

where $(.)^H$ denotes conjugate transpose. This equation can also be solved using the Cholesky decomposition method. For real data such as EEG signals this equation changes to $\boldsymbol{w} = (\boldsymbol{R}^T\boldsymbol{R})^{-1}\boldsymbol{R}^T\boldsymbol{y}$, where $(.)^T$ represents the transpose operation. A similar result can be achieved using principal component analysis (PCA) [25].

In cases for which the data are contaminated with white noise, the performance of Prony's method is reasonable. However, for non-white noise, the noise information is not easily separable from the data and therefore the method may not be sufficiently successful.

As we will see in a later chapter of this book, Prony's algorithm has been used in modelling and analysis of audio and visual EPs (AEP and VEP) [31, 35].

3.4.2 Nonlinear Modelling

An approach similar to AR or MVAR modelling in which the output samples are nonlinearly related to the previous samples, may be followed based on the methods developed for forecasting financial growth in economical studies.

In the generalized autoregressive conditional heteroskedasticity (GARCH) method [36], each sample relates to its previous samples through a nonlinear (or sum of nonlinear) function(s). This model was originally introduced for time-varying volatility (honoured with the Nobel Prize in Economic sciences in 2003).

Nonlinearities in the time series are declared with the aid of the McLeod and Li [37] and BDS (Brock, Dechert, and Scheinkman) tests [38]. However, both tests lack the ability to reveal the actual kind of nonlinear dependency.

Generally, it is not possible to discern whether the nonlinearity is deterministic or stochastic in nature, nor can we distinguish between multiplicative and additive dependencies. The type of stochastic nonlinearity may be determined based on Hsieh test [39]. The additive and multiplicative dependencies can be discriminated by using this test. However, the test itself is not used to obtain the model parameters.

Considering the input to a nonlinear system to be $u(n)$ and the generated signal as the output of such a system to be $x(n)$, a restricted class of nonlinear models suitable for the analysis of such process is given by:

$$x(n) = g(u(n-1), u(n-2), \cdots) + u_n.h(u(n-1), u(n-2), \cdots) \tag{3.52}$$

Multiplicative dependence means nonlinearity in the variance, which requires the function $h(.)$ to be nonlinear; additive dependence, conversely, means nonlinearity in the mean, which holds if the function $g(.)$ is nonlinear. The conditional statistical mean and variance are respectively defined as:

$$EE(x(n) \mid \chi_{n-1}) = g(u(n-1), u(n-2), \cdots) \tag{3.53}$$

and

$$Var(x(n) \mid \chi_{n-1}) = h^2(u(n-1), u(n-2), \cdots) \tag{3.54}$$

where $\chi_{n\text{-}1}$ contains all the past information up to time n-1. The original GARCH(p,q) model, where p and q are the prediction orders, considers a zero mean case, i.e. $g(.) = 0$.

If $e(n)$ represents the residual (error) signal using the above nonlinear prediction system, we have:

$$Var(e(n) \mid \chi_{n-1}) = \sigma^2(n) = \alpha_0 + \sum_{j=1}^{q} \alpha_j \sigma^2(n-j) + \sum_{j=1}^{p} \beta_j \sigma^2(n-j) \qquad (3.55)$$

where α_j and β_j are the nonlinear model coefficients. The second term (first sum) in the right side corresponds to a qth order moving average (MA) dynamical noise term and the third term (second sum) corresponds to an AR model of order p. It is seen that the current conditional variance of the residual at time sample n depends on both its previous sample values and previous variances.

Although in many practical applications such as forecasting of stock prices the orders p and q are set to small fixed values such as $(p,q) = (1,1)$; for a more accurate modelling of natural signals such as EEGs the orders have to be determined mathematically. The prediction coefficients for various GARCH models or even the nonlinear functions g and h are estimated iteratively as for the linear ARMA models [36, 37].

Such simple GARCH models are only suitable for multiplicative nonlinear dependence. In addition, additive dependencies can be captured by extending the modelling approach to the class of GARCH-M models [40].

Another limitation of the above simple GARCH model is failing to accommodate sign asymmetries. This is because the squared residual is used in the update equations. Moreover, the model cannot cope with rapid transitions such as spikes. Considering these shortcomings, numerous extensions to the GARCH model have been proposed. For example, the model has been extended and refined to include the asymmetric effects of positive and negative jumps such as the exponential GARCH model EGARCH [41], the GJR-GARCH model [42], the threshold GARCH model (TGARCH) [43], the asymmetric power GARCH model APGARCH [44], and quadratic GARCH model QGARCH [45].

In these models different functions for $g(.)$ and $h(.)$ in (3.48) and (3.49) are defined. For example, in the EGARCH model proposed by Glosten et al. [41] $h(n)$ is iteratively computed as:

$$h_n = b + \alpha_1 u^2(n-1)(1-\eta_{t-1}) + \alpha_2 u^2(n-1)\eta_{t-1} + \kappa h_{n-1} \qquad (3.56)$$

where b, α_1, α_2, and κ are constants and η_n is an indicator function that is zero when u_n is negative and one otherwise.

Despite modelling the signals, the GARCH approach has many other applications. In some recent works [46] the concept of GARCH modelling of covariance is combined with Kalman filtering to provide a more flexible model with respect to space and time for solving the inverse problem. There are several alternatives for solution to the inverse problem. Many approaches fall into the category of constrained least-squares methods employing Tikhonov regularization [47]. Among numerous possible choices for the GARCH dynamics, the EGARCH model [41] has been used to estimate the variance parameter of the Kalman filter iteratively.

Nonlinear models have not been used for EEG processing. To enable use of these models the parameters and even the order should be adapted to the EEG properties. Also, such a model should incorporate the changes in the brain signals due to abnormalities and onset of diseases.

3.4.3 Gaussian Mixture Model

In this very popular modelling approach the signals are characterized using the parameters of their distributions. The distributions in terms of probability density functions are sum of a number of Gaussian functions with different variances which are weighted and delayed differently [48]. The overall distribution subject to a set of K Gaussian components is defined as:

$$p(x \mid \theta_k) = \sum_{k=1}^{K} w_k p(x \mid \mu_k, \sigma_k) \tag{3.57}$$

The vector of unknown parameters $\theta_k = [w_k, \mu_k, \sigma_k]$ for $k = 1,2, \ldots, K$. w_k is equivalent to the probability (weighting) that the data sample is generated by the kth mixture component density subject to:

$$\sum_{k=1}^{K} w_k = 1 \tag{3.58}$$

μ_k, and σ_k are mean and variances of the kth Gaussian distribution and $p(x| \mu_k, \sigma_k)$ is a Gaussian function of x with parameters μ_k, and σ_k. Expectation maximization (EM) [49] is often used to estimate the above parameters by maximizing the log-likelihood of the mixture of Gaussian (MOG) for an N-sample data defined as:

$$L(\theta_k) = \sum_{i=1}^{N} log\left(\sum_{k=1}^{K} w_k p(x(i) \mid \mu_k, \sigma_k\right) \tag{3.59}$$

The EM algorithm alternates between updating the posterior probabilities used for generating each data sample by the kth mixture component (in a so-called E-step) as:

$$h_i^k = \frac{w_k p(x_i \mid \mu_k, \sigma_k)}{\sum_{j=1}^{K} w_j p\left(x_i \mid \mu_j, \sigma_j\right)} \tag{3.60}$$

and weighted maximum likelihood updates of the parameters of each mixture component (in a so-called M-step) as:

$$w_k \leftarrow \frac{1}{N} \sum_{i=1}^{N} h_i^k \tag{3.61}$$

$$\mu_k \leftarrow \frac{\sum_{i=1}^{N} h_i^k x_i}{\sum_{j=1}^{N} h_j^k} \tag{3.62}$$

$$\sigma_k \leftarrow \sqrt{\frac{\sum_{i=1}^{N} h_i^k \left(x_i - \mu_k^*\right)^2}{\sum_{j=1}^{N} h_j^k}} \tag{3.63}$$

The EM algorithm (especially in cases of high-dimensional multivariate Gaussian mixtures) may converge to spurious solutions when there are singularities in the log-likelihood function due to small sample sizes, outliers, repeated data points or rank deficiencies leading to 'variance collapse'. Some solutions to these shortcomings have been provided by many researchers such as those in [50–52]. Figure 3.13 demonstrates how an unknown multimodal distribution can be estimated using weighted sum of Gaussians with different mean and variances:

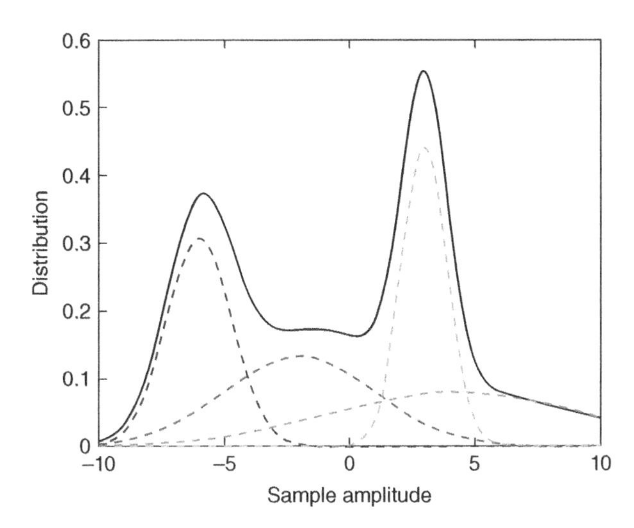

Figure 3.13 Mixture of Gaussian (dotted curves) models of a multimodal unknown distribution (bold curve).

Similar to prediction-based models, the model order can be estimated using Akaike information criterion (AIC) or by iteratively minimizing the error for best model order. Also, mixture of exponential distributions can also be used instead of MOGs where there are sharp transients within the data. In [53] it has been shown that these models can be used for modelling variety of physiological data such as EEG, EOG, EMG, and ECG. In another work [54], the MOG model has been used for segmentation of magnetic resonance brain images. As a variant of the Gaussian mixture model (GMM), Bayesian GMMs have been used for partial amplitude synchronization detection in brain EEG signals [55]. This work introduces a method to detect subsets of synchronized channels that do not consider any baseline information. It is based on a Bayesian GMM applied at each location of a time-frequency map of the EEGs.

3.5 Electronic Models

These models describe the cell as an independent unit. The well established models such as the Hodgkin–Huxley model have been implemented using electronic circuits. Although, for accurate models large numbers of components are required, in practise it has been shown that a good approximation of such models can be achieved using simple circuits [56].

3.5.1 Models Describing the Function of the Membrane

Most of the models describing the excitation mechanism of the membrane are electronic realizations of the theoretical membrane model of Hodgkin and Huxley. In the following sections, two of these realizations are discussed.

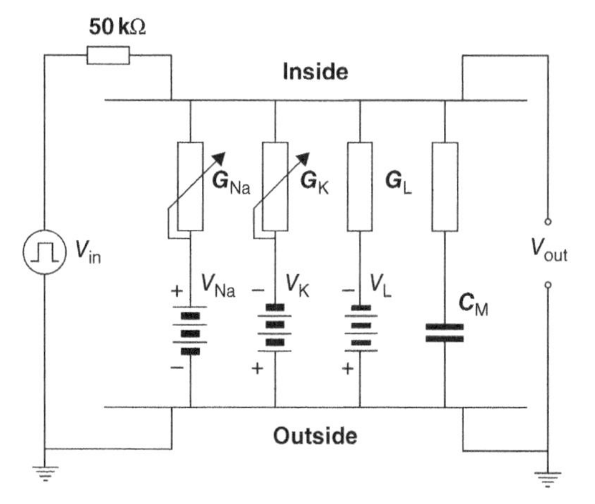

Figure 3.14 The Lewis membrane model [57].

3.5.1.1 Lewis Membrane Model

Lewis electronic membrane models are based on the Hodgkin–Huxley equations. All the components are parallel circuits connected between nodes representing the inside and outside of the membrane. He realized the sodium and potassium conductances using electronic hardware in the form of active filters. Since the model output is the transmembrane voltage V_m, the potassium current can be evaluated by multiplying the voltage corresponding to G_K by $(V_m - V_K)$. Figure 3.14 is consequently an accurate physical analogue to the Hodgkin–Huxley expressions, and the behaviour of the output voltage V_m corresponds to that predicted by the Hodgkin–Huxley equations. The electronic circuits in the Lewis neuromime had provision for inserting (and varying) not only such constants as $G_{K\ max}$, $G_{Na\ max}$, V_K, V_{Na}, V_{Cl}, which enter the Hodgkin–Huxley formulation, but also τ_h, τ_m, τ_n, which allow modifications from the Hodgkin–Huxley equations. In this realization the voltages of the biological membrane are multiplied by 100 to fit the electronic circuit. In other quantities, the original values of the biological membrane have been used.

3.5.1.2 Roy Membrane Model

As in Figure 3.15, Roy also introduced a model based on the Hodgkin–Huxley model [58]. He used field-effect transistors (FETs) to simulate the sodium and potassium conductances in the membrane model [58].

In the Roy model the conductance is controlled by a circuit including an operational amplifier, capacitors, and resistors. This circuit is designed to make the conductance behave according to the Hodgkin–Huxley model. Roy's main goal was to achieve a very simple model rather than to simulate accurately the Hodgkin–Huxley model.

3.5.2 Models Describing the Function of a Neuron

3.5.2.1 Lewis Neuron Model

The Lewis model is based on the Hodgkin–Huxley membrane model and the theories of Eccles on synaptic transmission [59]. The model circuit is illustrated in Figure 3.16. This neuron model is divided into two sections: the synaptic section and the section generating the

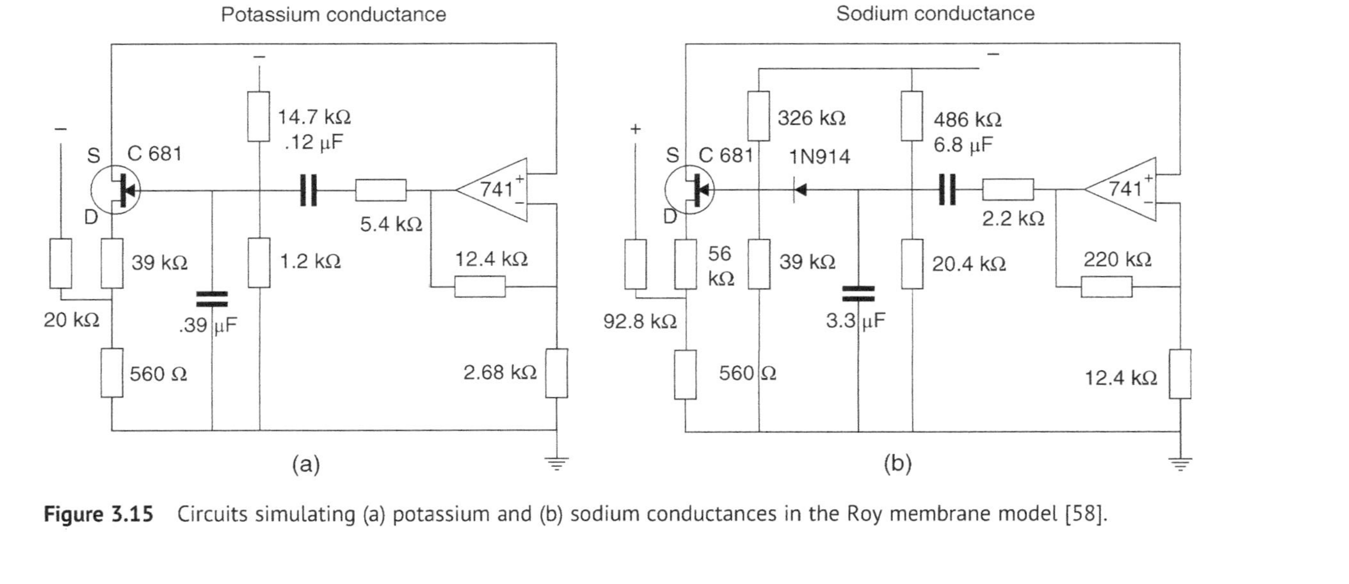

Figure 3.15 Circuits simulating (a) potassium and (b) sodium conductances in the Roy membrane model [58].

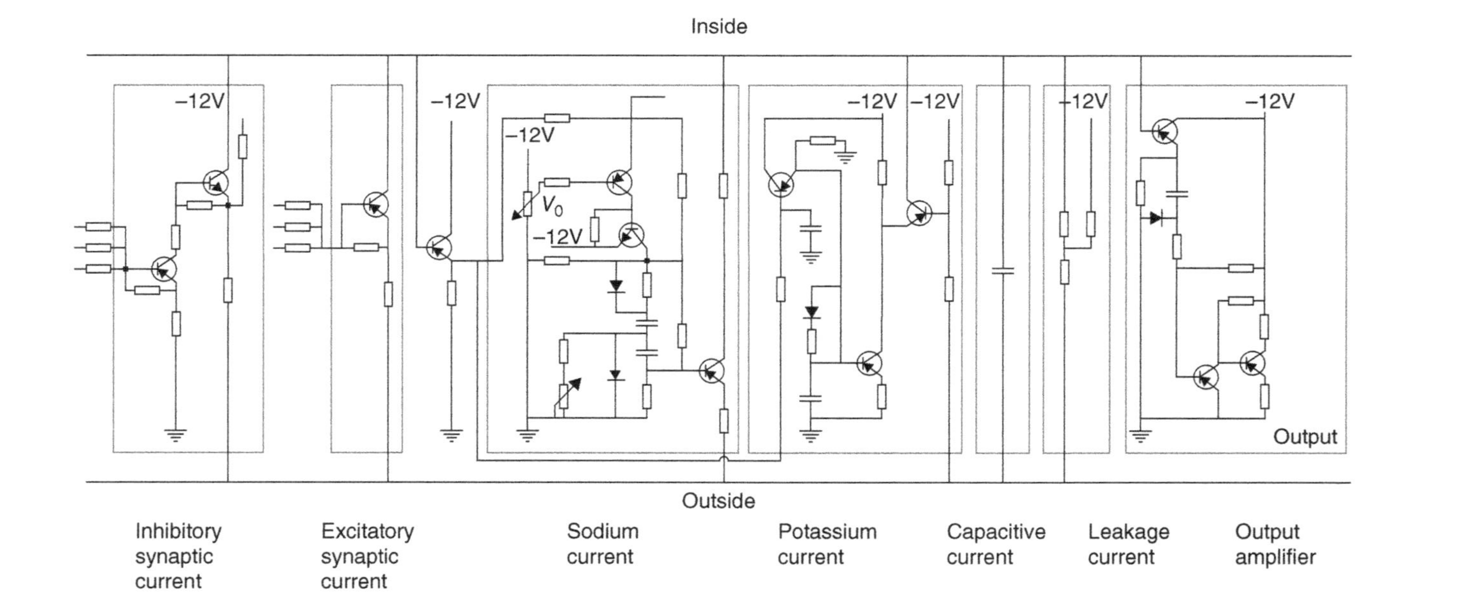

Figure 3.16 The Lewis neuron model from 1968 [57].

action pulse. Both sections consist of parallel circuits connected to the nodes representing the intracellular and extracellular sites of the membrane.

The section representing the synaptic junction is divided into two components: the inhibitory junction and the excitatory junction. The sensitivity of the section generating the action pulse to a stimulus introduced at the excitatory synaptic junction is reduced by the voltage introduced at the inhibitory junction. The section generating the action pulse is based on the Hodgkin–Huxley model which consists of the normal circuits simulating the sodium and potassium conductances, the leakage conductance, and the membrane capacitance. The circuit also includes an amplifier for the output signal. This model may be used in research on neural networks. However, it is actually a simplified version of Lewis's 46-transistor network having the same form. The purpose of this simplified Lewis model is to simulate the form of the action pulse with moderate accuracy following simple models.

3.5.2.2 The Harmon Neuron Model

The electronic realizations of the Hodgkin–Huxley model are very accurate in simulating the function of a single neuron. However, these circuits are often very complicated. Leon D. Harmon managed to develop a neuron model having a very simple circuit [60]. A circuit of the Harmon neuron model is given in Figure 3.17. The model is equipped with five excitatory inputs which can be adjusted. These include diode circuits representing various synaptic functions. The signal introduced at excitatory inputs charges the 0.02 μF capacitor which, after reaching a voltage of approximately 1.5 V, allows the monostable multivibrator, formed by transistors T1 and T2, to generate the action pulse. This impulse is amplified by transistors T3 and T4. The output of one neuron model may drive the inputs of a large number of the neighbouring neuron models.

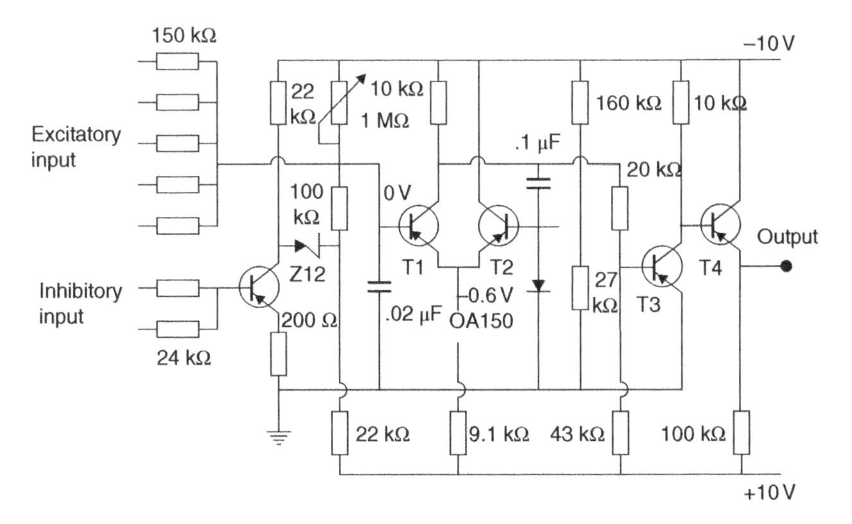

Figure 3.17 The Harmon neuron model [60].

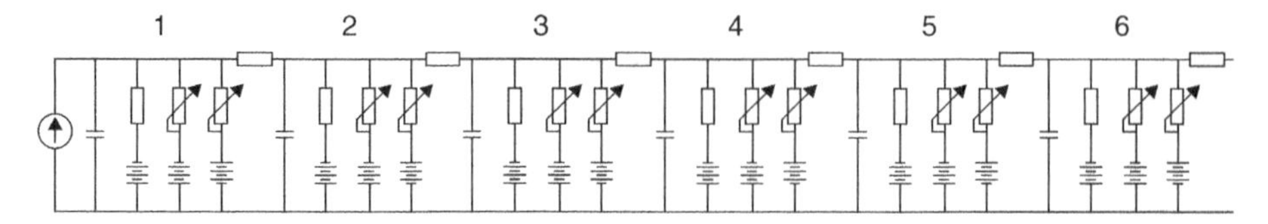

Figure 3.18 The Lewis model for simulation of the propagation of the action pulse [57].

3.5.3 A Model Describing the Propagation of the Action Pulse in an Axon

Lewis simulated the propagation of an action pulse in a uniform axon and obtained interesting results [57]. The model structure, illustrated in Figure 3.18, can be seen to include a network of membrane elements as well as axial resistors representing the intracellular resistance. A total of six membrane elements are depicted in the figure. The model is an electronic realization of the linear core-conductor model with active membrane elements. An approximation of an AP is generated in the output of each membrane element.

3.5.4 Integrated Circuit Realizations

Mahowald et al. [61] used electronic neuron models or neuron-like circuits as processing elements for electronic computers, called neurocomputers. Two examples of these models are as follows. The electronic neuron model developed by [62] is realized with integrated circuit technology. The circuit includes one neuron with eight synapses. The chip area of the integrated circuit is $4.5 \times 5\ \text{mm}^2$. The array contains about 200 NPN and 100 PNP transistors, and about 200 of them are used. In 1991, Mahowald and Douglas [63] published an integrated circuit realization of electronic neuron model. The model was realized with complementary metal oxide semiconductor (CMOS) circuits using very large-scale integrated (VLSI) technology. Their model accurately simulates the spikes of a neocortical neuron. The power dissipation of the circuit is 60 μW, and it occupies less than $0.1\ \text{mm}^2$. It is estimated that 100–200 such neurons could be fabricated on a $1\ \text{cm} \times 1\ \text{cm}$ die.

3.6 Dynamic Modelling of Neuron Action Potential Threshold

Neuron threshold is the transmembrane voltage level at which any further depolarization will activate a sufficient number of sodium channels to enter a self-generative positive feedback phase [64]. This threshold is often considered constant. Any alteration to the threshold influences the neuron spike train temporal transformation. There are however evidences that the threshold is nonlinearly affected by the AP firing history [65–67]. In [64] a method for dynamically varying the threshold for intercellular activity has been proposed. The method is suitable for systems with spikes in both their inputs and outputs.

3.7 Summary

Most of the models generated mathematically are very primitive representations of EEG generators. More complicated models will be necessary to represent the brain EPs and the abnormal EEGs recorded under various brain diseases and disorders. The model based on phase coupling explained here introduces the WCOs which can be used to model the interactions between the neurons. Hodgkin and Huxley model conversely provides a detailed and accurate model for generation of active potentials. Linear and nonlinear prediction filers can be used in modelling the neuro generators. Gaussian mixtures are capable in modelling the EEG particularly event-related, evoked, and movement-related potentials. Finally, circuit models have been introduced to combine the excitatory and inhibitory postsynaptic potentials for generation of an EEG signal. The synaptic currents have also been modelled using electronic circuits. These circuits can be expanded to have very accurate model of either a single neuron or a membrane. They can also be used to model a large number of neurons for generation of particular brain waveforms. The electronic models have potential to be used in brain morphing technology.

References

1 van Vreeswijk, C., Abbott, L.F., and Ermentrout, G.B. (1994). When inhibition not excitation synchronizes neural firing. *Journal of Computational Neuroscience* **1**: 313–321.

2 Penny, W.D., Litvak, V., Fuentemilla, L. et al. (2009). Dynamic causal models for phase coupling. *Journal of Neuroscience Methods* **183** (1): 19–30.

3 Ward, L. (2003). Synchronous neural oscillations and cognitive processes. *Trends in Cognitive Sciences* **7** (12): 553–559.

4 von Stein, A., Rappelsberger, P., Sarnthein, J., and Petsche, H. (1999). Synchronization between temporal and parietal cortex during multimodal object processing in man. *Cerebral Cortex* **9** (2): 137–150.

5 Jones, M. and Wilson, M. (2005). Theta rhythms coordinate hippocampal–prefrontal interactions in a spatial memory task. *PLoS Biology* **3** (12): e402.

6 Penny, W.D., Stephan, K.E., Mechelli, A., and Friston, K.J. (2004). Comparing dynamic causal models. *NeuroImage* **22** (3): 1157–1172.

7 Hoppensteadt, F. and Izhikevich, E. (1997). *Weakly Connected Neural Networks*. New York, USA: Springer-Verlag.

8 Benedek, G. and Villars, F. (2000). *Physics, with Illustrative Examples from Medicine and Biology*. New York: Springer-Verlag.

9 Hille, B. (1992). *Ionic Channels of Excitable Membranes*. Sunderland, MA: Sinauer.

10 Hodgkin, A. and Huxley, A. (1952). A quantitative description of membrane current and its application to conduction and excitation in nerve. *Journal of Physiology (London)* **117**: 500–544.

11 Doi, S., Inoue, J., Pan, Z., and Tsumoto, K. (2010). *Computational Electrophysiology*. Springer.

12 Simulator for Neural Networks and Action Potentials (SNNAP) (2003). Tutorial, The University of Texas-Houston Medical School. https://med.uth.edu/nba/wp-content/uploads/sites/29/2019/09/snnap_8_tutorial_v2019.pdf (accessed 19 August 2021).

13 Ziv, I., Baxter, D.A., and Byrne, J.H. (1994). Simulator for neural networks and action potentials: description and application. *Journal of Neurophysiology* **71**: 294–308.

14 Gerstner, W. and Kistler, W.M. (2002). *Spiking Neuron Models*, 1e. Cambridge University Press.

15 Lagerlund, T.D., Sharbrough, F.W., and Busacker, N.E. (1997). Spatial filtering of multichannel electroencephalographic recordings through principal component analysis by singular value decomposition. *Journal of Clinical Neurophysiology* **14** (1): 73–82.

16 Da Silva, F.H., Hoeks, A., Smits, H., and Zetterberg, L.H. (1974). Model of brain rhythmic activity: the alpha-rhythm of the thalamus. *Kybernetic* **15**: 27–37.

17 da Silva, F.H.L., van Rotterdam, A., Barts, P. et al. (1976). Models of neuronal populations: the basic mechanisms of rhythmicity. In: *Perspective of Brain Research*, Prog. Brain Res, vol. **45** (eds. M.A. Corner and D.F. Swaab), 281–308.

18 Wilson, H.R. and Cowan, J.D. (1972). Excitatory and inhibitory interaction in localized populations of model neurons. *Biophysical Journal* **12**: 1–23.

19 Zetterberg, L.H. (1973). *Stochastic activity in a population of neurons – a system analysis approach*. Report No. 2.3.153/1, 23, Issue no. 197418, p. 32. Utrecht: Institute of Medical Physics TNO, Utrecht.

20 Jansen, B.H. and Rit, V.G. (1995). Electroencephalogram and visual evoked potential generation in a mathematical model of coupled cortical columns. *Biological Cybernetics* **73**: 357–366.

21 Jansen, B.H., Zouridakis, G., and Brandt, M.E. (1993). A neurophysiologically based mathematical model of flash visual evoked potentials. *Biological Cybernetics* **68**: 275–283.

22 Akaike, H. (1974). A new look at statistical model order identification. *IEEE Transactions on Automatic Control* **19**: 716–723.

23 Kay, S.M. (1988). *Modern Spectral Estimation: Theory and Application*. Prentice Hall.

24 Guegen, C. and Scharf, L. (1980). Exact maximum likelihood identification of ARMA models: a signal processing perspective. In: *Signal Processing Theory Applications* (eds. M. Kunt and F. de Coulon), 759–769. Amsterdam: North Holland Publishing Co.

25 Akay, M. (2001). *Biomedical Signal Processing*. Academic Press.

26 Cavanaugh, J.E. and Neath, A.A. (2019). The Akaike information criterion: background, derivation, properties, application, interpretation, and refinements. *WIREs Computational Statistics* **11** (3): e1460.

27 Durbin, J. (1959). Efficient estimation of parameters in moving average models. *Biometrika* **46**: 306–316.

28 Trench, W.F. (1964). An algorithm for the inversion of finite Toelpitz matrices. *Journal of the Society for Industrial and Applied Mathematics* **12**: 515–522.

29 Morf, M., Vieria, A., Lee, D., and Kailath, T. (1978). Recursive multichannel maximum entropy spectral estimation. *IEEE Transactions on Geoscience Electronics* **16**: 85–94.

30 Spreckelesen, M. and Bromm, B. (1988). Estimation of single-evoked cerebral potentials by means of parametric modelling and Kalman filtering. *IEEE Transactions on Biomedical Engineering* **33**: 691–700.

31 Demiralp, T. and Ademoğlu, A. (1992). Modeling of evoked potentials as decaying sinusoidal oscillations by PRONY-method. *1992 14th Annual International Conference of the IEEE Engineering in Medicine and Biology Society*, Paris, 2452–2453. https://doi.org/10.1109/IEMBS.1992.5761536.

32 De Prony, B.G.R. (1795). Essai experimental et analytique: sur les lois de la dilatabilite de fluids elastiques et sur celles de la force expansive de la vapeur de l'eau et de la vapeur de l'alkool, a differentes temperatures. *Journal of Engineering Polytechnique* **1** (2): 24–76.

33 Marple, S.L. (1987). *Digital Spectral Analysis with Applications*. Prentice-Hall.

34 Lawson, C.L. and Hanson, R.J. (1974). *Solving Least Squares Problems*. Englewood Cliffs, NJ: Prentice Hall.

35 Bouattoura, D., Gaillard, P., Villon, P. et al. (1996). Multilead evoked potentials modelling based on the Prony's method. *Proceedings of Digital Processing Applications (TENCON '96)*, Perth, WA, Australia, pp. 565–568.

36 Dacorogna, M., Muller, U., Olsen, R.B., and Pictet, O. (1998). Modelling short-term volatility with GARCH and HARCH models. In: *Nonlinear Modelling of High Frequency Financial Time Series, Econometrics* (eds. L.C. Dunis and B. Zhou). Elsevier, 17 pp.

37 McLeod, A.J. and Li, W.K. (1983). Diagnostics checking ARMA time series models using squared residual autocorrelations. *Journal of Time Series Analysis* **4**: 269–273.

38 Ray, W.D. (1993). Nonlinear dynamics, chaos and instability statistical theory and economic evidence. *Journal of the Operational Research Society* **44** (2): 202–203.

39 Hsieh, D.A. (1989). Testing for nonlinear dependence in daily foreign exchange rates. *Journal of Business* **62**: 339–368.

40 Engle, R.F., Lilien, D.M., and Robin, R.P. Estimating time-varying risk premia in the term structure: the ARCH-M model. *Econometrica* **55**: 391–407.

41 Nelson, D.B. (1990). Stationarity and persistence in the GARCH(1,1) model. *Journal of Econometrics* **45**: 7–35.

42 Glosten, L.R., Jagannathan, R., and Runkle, D. (1995). On the relation between the expected value and the volatility of the nominal excess return on stocks. *The Journal of Finance* **2**: 225–251.

43 Zakoian, J.M. (1994). Threshold heteroskedastic models. *Journal of Economic Dynamics & Control* **18**: 931–955.

44 Ding, Z., Engle, R.F., and Granger, C.W.J. (1993). A long memory property of stock market returns and a new model. *Journal of Empirical Finance* **1**: 83–106.

45 Sentana, E. (1991). *Quadratic ARCH Models: A Potential Reinterpretation of ARCH Models as Second-Order Taylor Approximations*. London School of Economics.

46 Galka, A., Yamashita, O., and Ozaki, T. (2004). GARCH modelling of covariance in dynamical estimation of inverse solutions. *Physics Letters A* **333**: 261–268.

47 Tikhonov, A. (1992). *Ill-Posed Problems in Natural Sciences*. Coronet.

48 Roweis, S. and Ghahramani, Z. (1999). A unifying review of linear Gaussian models. *Neural Computation* **11**: 305–345.

49 Dempster, A.P., Laird, N.M., and Rubin, D.B. (1977). Maximum likelihood from incomplete data via the EM algorithm. *Journal of the Royal Statistical Society B* **39**: 1–38.

50 Redner, R.A. and Walker, H.F. (1984). Mixture densities, maximum likelihood and the EM algorithm. *SIAM Review* **26**: 195–239.

51 Ormoneit, D. and Tresp, V. (1998). Averaging, maximum penalized likelihood and Bayesian estimation for improving Gaussian mixture probability density estimates. *IEEE Transactions on Neural Networks* **9** (4): 639–650.

52 Archambeau, C., Lee, J.A., and Verleysen, M. (2003). On convergence problems of the EM algorithm for finite Gaussian mixtures. *Proceedings of the European Symposium on Artificial Neural Networks (ESANN 2003)*, Bruges, Belgium (23–25 April 2003), 99–106.

53 Hesse, C.W., Holtackers, D., and Heskes, T. (2006). On the use of mixtures of gaussians and mixtures of generalized exponentials for modelling and classification of biomedical signals. *Belgian Day on Biomedical Engineering, IEEE Benelux EMBS Symposium* (7–8 December).

54 Greenspan, H., Ruf, A., and Goldberger, J. (2006). Constrained Gaussian mixture model framework for automatic segmentation of mr brain images. *IEEE Transactions on Medical Imaging* **25** (9): 1233–1245.

55 Rio, M., Hutt, A., and Loria, B.G. (2010). Partial amplitude synchronization detection in brain signals using Bayesian Gaussian mixture models. *Cinquième conférence plénière française de Neurosciences Computationnelles, "Neurocomp'10"*, Lyon, France (October 2010), 109–113.

56 Malmivuo, J. and Plonsey, R. (1995). *Bioelectromagnetism: Principles and Applications of Bioelectric and Biomagnetic Fields*. Oxford University Press.

57 Lewis, E.R. (1964). An electronic model of the neuron based on the dynamics of potassium and sodium ion fluxes. In: *Proceedings of the 1962 Ojai Symposium on Neural Theory and Modelling* (eds. R.F. Reiss, H.J. Hamilton, L.D. Harmon, et al.), 427. Stanford: Stanford University Press.

58 Roy, G. (1972). "A simple electronic analog of the squid axon membrane," The Neurofet. *IEEE Transactions on Biomedical Engineering* **19** (1): 60–63.

59 Eccles, J.C. (1964). *The Physiology of Synapses*. Berlin: Springer-Verlag.

60 Harmon, L.D. (1961). Studies with artificial neurons, I: properties and functions of an artificial neuron. *Kybernetik Heft* **3** (Dez.):: 89–101.

61 Mahowald, M.A., Douglas, R.J., LeMoncheck, J.E., and Mead, C.A. (1992). An introduction to silicon neural analogs. *Seminars in Neuroscience* **4**: 83–92.

62 Prange, S. (1990). Emulation of biology-oriented neural networks. In: *Proceedings of the International Conference on Parallel Processing in Neural Systems and Computers (ICNC)* (ed. M. Eckmiller). Düsseldorf.

63 Mahowald, M.A. and Douglas, R.J. (1991). A silicon neuron. *Nature* **354**: 515–518.

64 Lu, U., Roach, S.M., Song, D., and Berger, T.W. (2012). Nonlinear dynamic modelling of neuron action potential threshold during synaptically driven broadband intercellular activity. *IEEE Transactions on Biomedical Engineering* **59** (3): 706–716.

65 Azouz, R. and Gray, C.M. (1999). Cellular mechanisms contributing to response variability of cortical neurons *in Vivo*. *Journal of Neuroscience* **19** (6): 2209–2223.

66 Henze, D.A. and Buzsáki, G. (2001). Action potential threshold of hippocampal pyramidal cells in vivo is increased by recent spiking activity. *Neuroscience* **105** (1): 121–130.

67 Chacron, M.J., Lindner, B., and Longtin, A. (2007). Threshold fatigue and information transfer. *Journal of Computational Neuroscience* **23** (3): 301–311.

4

Fundamentals of EEG Signal Processing

4.1 Introduction

Electroencephalography (EEG) signals are the signatures of neural activities and generally are the integrals of active potentials which elicit from the brain with different latencies and populations around each time instant. They are captured by multiple-electrode EEG machines either from inside the brain, over the cortex under the skull, or in the majority of applications, certain locations over the scalp. The EEG file formats are different for different recording machines but nowadays they can be easily read or converted by conventional software. The signals are normally presented in the time domain, however, many new EEG machines are capable of applying simple signal processing tools such as the Fourier transform to perform frequency analysis and equipped with some imaging tools to visualize EEG topographies (maps of the brain activities in the spatial domain).

There have been many algorithms developed so far for processing EEG signals. The operations include, but are not limited to, time-domain analysis, frequency-domain analysis, spatial-domain analysis, and multiway processing. Also, several algorithms have been developed to visualize the brain activity from images reconstructed from only the EEGs namely topographs. Separation of the desired sources from the multisensor EEGs has been another research area. This can later lead to the detection of brain abnormalities such as epilepsy and the sources related to various physical and mental activities. In Chapter 17 of this book we will also see that the recent works in brain–computer interfacing (BCI) [1] have been focused upon the development of advanced signal processing as well as the vast advances in cooperative networking and deep learning tools and algorithms.

Modelling of neural activities is probably more difficult than modelling the function of any other organ. However, some simple models for generating EEG signals have been proposed. Some of these models have also been extended to include generation of abnormal EEG signals.

Localization of brain signal sources is another very important field of research [2]. In order to provide a reliable algorithm for localization of the sources within the brain sufficient knowledge about both propagation of electromagnetic waves, and how the information from the measured signals can be exploited in separation and localization of the sources within the brain is required. The sources might be considered as magnetic dipoles for which the well known inverse problem has to be solved, or they can be considered as distributed current sources.

EEG Signal Processing and Machine Learning, Second Edition. Saeid Sanei and Jonathon A. Chambers.

Patient monitoring and sleep monitoring require real-time processing of (up to few days) long EEG sequences. The EEG provides important and unique information about the sleeping brain. Major brain activities during sleep can be captured using the developed algorithms such as the method of matching pursuit (MP) [3] discussed later in this chapter.

Epilepsy monitoring, detection, and prediction have also attracted many researchers. Dynamical analysis of time series together with application of blind separation of the signal sources has enabled prediction of focal epilepsies from the scalp EEGs. Conversely, application of time–frequency (TF) domain analysis for detection of the seizure in neonates has paved the way for further research in this area. An important recent contribution in the field of epilepsy is how to model the pathways between intracranial epileptiform discharges (mainly during the interictal period) and the scalp electrodes. Recent developments in this area have been made by introducing some new methods such as the multiview approach and dictionary learning [4–6] as well as deep neural network structures [7, 8] which enable capturing a considerable percentage of so-called interictal epileptiform discharges (IEDs) from over the scalp. This technique is described in Chapters 7 and 11 of this book.

In the following sections most of the tools and algorithms for the above objectives are explained and the mathematical foundations discussed. The application of these algorithms to analysis of the normal and abnormal EEGs however will follow in the later chapters of this book. The reader should also be aware of the required concepts and definitions borrowed from linear algebra. Further details of which can be found in [9]. Throughout this chapter and the reminder of this book continuous time is denoted by 't' and discrete time, with normalized sampling period $T = 1$, with 'n'. In some illustrations however the actual timings in seconds are presented.

4.2 Nonlinearity of the Medium

The head as a mixing medium combines EEG signals which are locally generated within the brain at the sensor positions. As a system, the head may be more or less susceptible to such sources in different situations. Generally, an EEG signal can be considered as the output of a nonlinear system, which may be characterized deterministically. This concept represents the fundamental difference between EEG and magnetoencephalography (MEG) signals. Unlike in EEG, for MEG the resistivity (or conductivity) of the head has significantly less effect on the magnetic field although the slight nonlinearity in MEG stems from minor differences in head tissue permittivity values.

Electromagnetic source characterization and localization requires accurate volume conductor models representing head geometry and the electrical conductivity field. It has been shown that, despite extensive research, the measurements of head tissue parameters are inconsistent and vary in different experiments [10]. There is more discussion on this concept in Chapter 10 of this book where we introduce brain source localization.

The changes in brain metabolism as a result of biological and physiological phenomena in the human body can change the process of combining the EEG source signals. Some of these changes are influenced by the activity of the brain itself. These effects make the system nonlinear. Analysis of such a system is very complicated and up to now nobody has fully modelled the system to aid in the analysis of brain signals.

Conversely, some measures borrowed from chaos theory and analysis of the dynamics of time series such as dissimilarity, attractor dimension, and largest Lyapunov exponents (LLE) can characterize the nonlinear behaviour of EEG signals. These concepts are discussed in Section 4.7 and some of their applications given in Chapter 11.

4.3 Nonstationarity

Nonstationarity of the signals can be quantified by evaluating the changes in signal distribution over time. In a strict-sense stationary process the signal distribution remains the same for different time intervals. However, often wide-sense stationarity is not required and therefore, it is sufficient to have various statistics such as mean and variance fixed (or without significant change) over time.

Although in many applications the multichannel EEG distribution is considered as multivariate Gaussian, the mean and covariance properties generally change from segment to segment. As such EEGs are considered stationary only within short intervals and are generally quasi-stationary. Although this interval can change due to the rapid changes in the brain state such as going from closed eye to open eye, sleep to wakefulness, normal to seizure, change in alertness, brain responding to a stimulation in the form of event-related potential (ERP) and evoked potential (EP) signals, eye blinking, and emotion change, in practise, a 10 seconds window of EEG is considered stationary.

The change in the distribution of the signal segments can be measured in terms of both the parameters of a Gaussian process and the deviation of the distribution from Gaussian. The non-Gaussianity of the signals can be checked by measuring or estimating some higher-order moments such as skewness, kurtosis, negentropy, and Kullback–Leibler (KL) distance (aka KL divergence).

Skewness is a measure of asymmetry, or more precisely, the lack of symmetry in distribution. A distribution is symmetric if it looks the same to the left and right of its midline or mean point. The skewness is defined for a real signal as

$$Skewness = \frac{E\left[(x(n)-\mu)^3\right]}{\sigma^3} \tag{4.1}$$

where μ and σ are respectively, the mean and variance and E denotes statistical expectation. If the distribution is more to the right of the mean point the skewness is negative, and vice versa. For a symmetric distribution such as Gaussian, the skewness is zero.

Kurtosis is a measure for showing how peaked or flat a distribution is relative to a normal distribution. That is, datasets with high kurtosis tend to have a distinct peak near the mean, decline rather rapidly, and have heavy tails. Datasets with low kurtosis tend to have a flat top near the mean rather than a sharp peak. A uniform distribution would be the extreme case. The kurtosis for a signal $x(n)$ is defined as:

$$kurt = \frac{m_4(x(n))}{m_2^2(x(n))} \tag{4.2}$$

where $m_i(x(n))$ is the ith central moment of the signal $x(n)$, i.e. $m_i(x(n)) = E[(x(n) - \mu)^i]$. The kurtosis for signals with normal distributions is three. Therefore, an excess or normalized kurtosis is often used and defined as

$$Ex\ kurt = \frac{m_4(x(n))}{m_2^2(x(n))} - 3 \tag{4.3}$$

which is zero for Gaussian distributed signals. Often the signals are considered ergodic, hence the statistical averages can be assumed identical to time averages so that they can be estimated with time averages.

The negentropy of a signal $x(n)$ [11] is defined as:

$$J_{neg}(x(n)) = H(x_{Gauss}(n)) - H(x(n)) \tag{4.4}$$

where $x_{Gauss}(n)$ is a Gaussian random signal with the same covariance as $x(n)$ and H(.) is the differential entropy [12] defined as:

$$H(x(n)) = \int_{-\infty}^{\infty} p(x(n))\, log \frac{1}{p(x(n))} dx(n) \tag{4.5}$$

and $p(x(n))$ is the signal distribution. Negentropy is always nonnegative.

Entropy, by itself, is an important measure of EEG behaviour particularly in the cases in which the brain synchronization changes such as when brain waves become gradually more synchronized when the brain approaches the seizure onset. It is also a valuable indicator of other neurological disorders presented in psychiatric diseases.

By replacing the probability density function (pdf) with joint or conditional pdfs in Eq. (4.5), joint or conditional entropy is defined respectively. In addition, there are new definitions of entropy catering for neurological applications such as the multiscale fluctuation-based dispersion entropy defined in [13], which is briefly explained in this chapter, and those in the references herein.

The KL distance between two distributions p_1 and p_2 is defined as:

$$KL = \int_{-\infty}^{\infty} p_1(z)\, log \frac{p_1(z)}{p_2(z)} dz \tag{4.6}$$

It is clear that the *KL* distance is generally asymmetric, therefore by changing the position of p_1 and p_2 in this Eq. (4.6) the *KL* distance changes. The minimum of the *KL* distance occurs when $p_1(z) = p_2(z)$.

4.4 Signal Segmentation

Often it is necessary to label the EEG signals into the segments of similar characteristics particularly meaningful to clinicians and for the assessment by neurophysiologists. Within each segment, the signals are considered statistically stationary usually with similar time and frequency statistics. As an example, an EEG recorded from an epileptic patient may be divided into three segments of preictal, ictal, and postictal segments. Each may have a different duration. Figure 4.1 represents an EEG sequence including all the above segments.

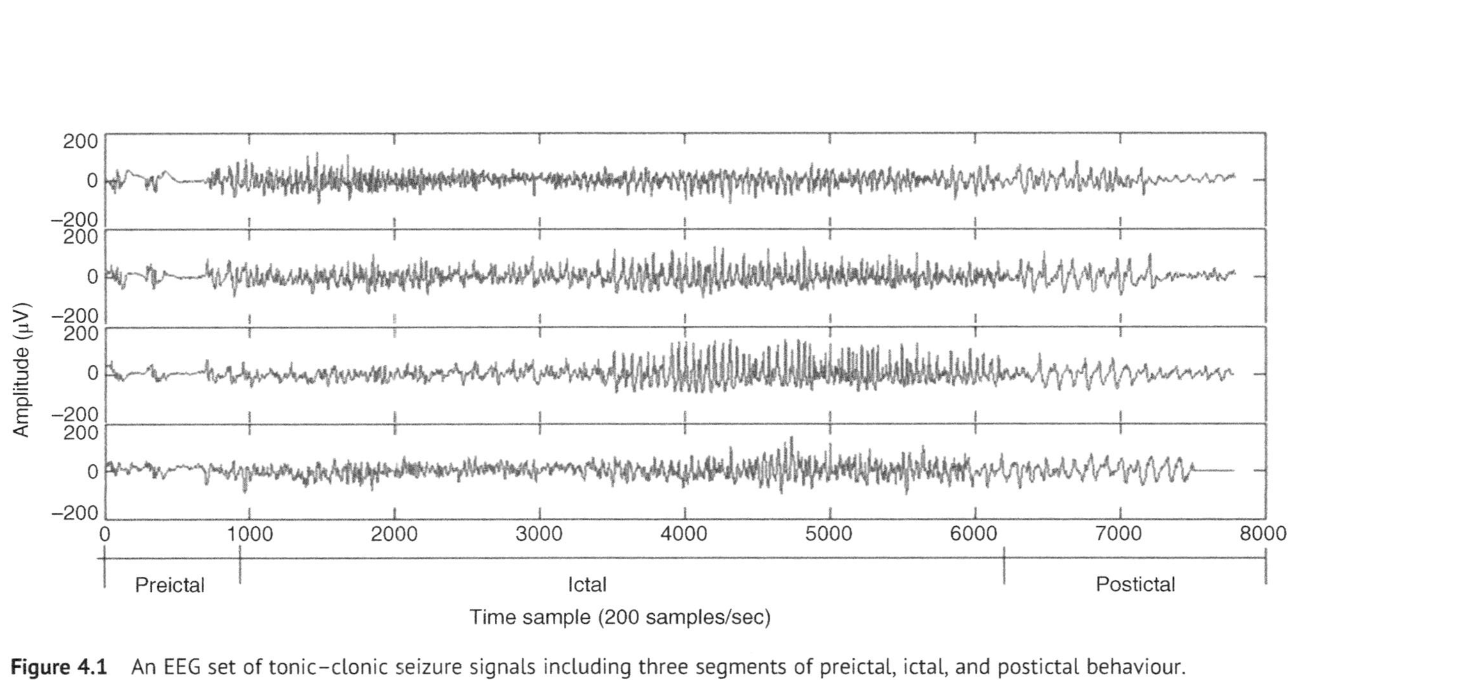

Figure 4.1 An EEG set of tonic–clonic seizure signals including three segments of preictal, ictal, and postictal behaviour.

In segmentation of EEGs time or frequency properties of the signals may be exploited. This eventually leads to a dissimilarity measurement denoted as $d(m)$ between the adjacent EEG frames where m is an integer value indexing the frame and the difference is calculated between the m and $(m-1)$th (consecutive) signal frames. The boundary of the two different segments is then defined as the boundary between the m and $(m-1)$th frames provided $d(m) > \eta_T$, and η_T is an empirical threshold level. An efficient segmentation is possible by highlighting and effectively exploiting the diagnostic information within the signals with the help of expert clinicians. However, access to such experts is not always possible and therefore, there are needs for algorithmic methods.

A number of different dissimilarity measures may be defined based on the fundamentals of signal processing. One criterion is based on the autocorrelations for segment m defined as:

$$r_x(k,m) = E[x(n,m)x(n+k,m)] \tag{4.7}$$

The autocorrelation function of the mth length N frame for an assumed time interval $n, n+1, \dots, n+(N-1)$, can be approximated as:

$$\hat{r}_x(k,m) = \begin{cases} \frac{1}{N}\sum_{l=0}^{N-1-k} x(l+m+k)x(l+m), & k = 0,\dots,N-1 \\ 0 & k = N, N+1,\dots \end{cases} \tag{4.8}$$

Then the criterion is set to:

$$d_1(m) = \frac{\sum_{k=-\infty}^{\infty} (\hat{r}_x(k,m) - \hat{r}_x(k,m-1))^2}{\hat{r}_x(0,m)\hat{r}_x(0,m-1)} \tag{4.9}$$

A second criterion can be based on higher-order statistics. The signals with more uniform distributions such as normal brain rhythms have a low kurtosis, whereas seizure signals or ERP signals often have high kurtosis values. Kurtosis is defined as the fourth-order cumulant at zero time lags and related to the second-order and fourth-order moments as given in Eqs. (4.1)–(4.3). A second level discriminant $d_2(m)$ is then defined as:

$$d_2(m) = kurt_x(m) - kurt_x(m-1) \tag{4.10}$$

where m refers to the mth frame of the EEG signal $x(n)$. A third criterion is defined from the spectral error measure of the periodogram. A periodogram of the mth frame is obtained by Fourier transforming of the correlation function of the EEG signal:

$$S_x(\omega,m) = \sum_{k=-\infty}^{\infty} \hat{r}_x(k,m)e^{-j\omega k} \quad \omega \in [-\pi,\pi] \tag{4.11}$$

where $\hat{r}_x(.,m)$ is the autocorrelation function for the mth frame as defined above. The criterion is then defined based on the normalized periodogram as:

$$d_3(m) = \frac{\int_{-\pi}^{\pi} (S_x(\omega,m) - S_x(\omega,m-1))^2 d\omega}{\int_{-\pi}^{\pi} S_x(\omega,m)d\omega \int_{-\pi}^{\pi} S_x(\omega,m-1)d\omega} \tag{4.12}$$

The test window sample autocorrelation for the measurement of both $d_1(m)$ and $d_3(m)$ can be updated through the following recursive equation (Eq. 4.13) over the test windows of size N:

$$\hat{r}_x(k,m) = \hat{r}_x(k,m-1) + \frac{1}{N}(x(m-1+N)x(m-1+N-k) - x(m-1+k)x(m-1)) \tag{4.13}$$

and thereby computational complexity can be reduced in practise. A fourth criterion corresponds to the error energy in autoregressive (AR) -based modelling of the signals. The prediction error in the AR model of the mth frame is simply defined as:

$$e(n,m) = x(n,m) - \sum_{k=1}^{p} a_k(m)x(n-k,m) \tag{4.14}$$

where p is the prediction order and $a_k(m)$, $k = 1, 2, \ldots, p$, are the prediction coefficients. For certain p the coefficients can be found directly (for example by using Durbin's method) in such a way to minimize the error (residual) between the actual and predicted signal energy. In this approach it is assumed that the frames of length N overlap by one sample. The prediction coefficients estimated for the $(m-1)$th frame are then used to predict the first sample in the mth frame, which we denote it as $\hat{e}(1,m)$. If this error is small, it is likely that the statistics of the mth frame are similar to those of the $(m-1)$th frame. Conversely, a large value is likely to indicate a change. An indicator for the fourth criterion can then be the differentiation of this signal (difference of digital signal) with respect to time, which gives a peak at the segment boundary, i.e.:

$$d_4(m) = max(\nabla_n e(n,m)) \tag{4.15}$$

where $\nabla_n(.)$ denotes the gradient with respect to n, approximated by a first-order difference operation. Figure 4.2 shows the residual and the gradient defined in Eq. (4.15), wherein the prediction order has been selected as 12.

Finally, a fifth criterion $d_5(m)$ may be defined by using the AR-based spectrum of the signals in the same way as short-time Fourier transform (STFT) for $d_3(m)$. The above AR model is a univariate model, i.e. it models a single-channel EEG. A similar criterion may be defined when multichannel EEGs are considered [14]. In such cases a multivariate AR (MVAR) model is analyzed. The MVAR can also be used for characterization and quantification of the signal propagation within the brain and is discussed in Chapter 8 of this book.

Although the above criteria can be effectively used for segmentation of EEG signals, better systems may be defined for the detection of certain abnormalities. In order to do that, the features, which best describe the behaviour of the signals, need to be identified and used. Therefore, the segmentation problem becomes a classification problem for which different classifiers can be used.

4.5 Signal Transforms and Joint Time-Frequency Analysis

If the signals are statistically stationary it is straightforward to characterize them in either time- or frequency-domains. The frequency-domain representation of a finite-length signal can be found by using linear transforms such as the (discrete) Fourier transform (DFT),

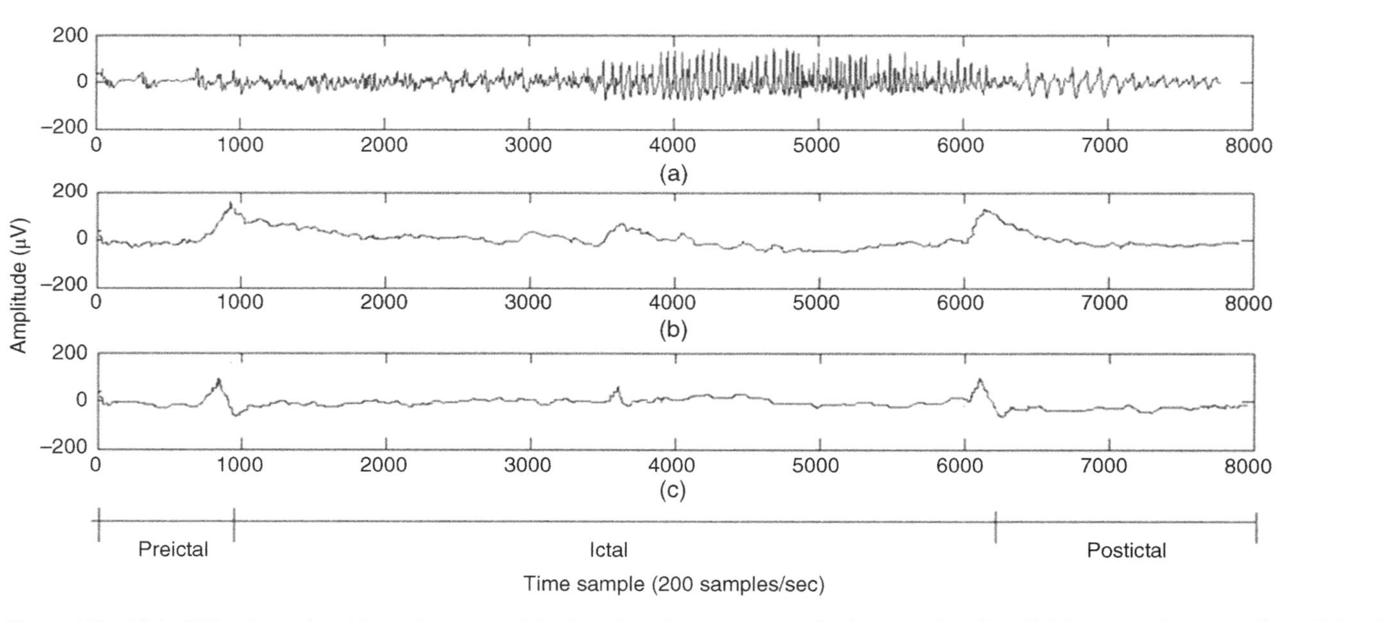

Figure 4.2 (a) An EEG seizure signal including preictal, ictal, and postictal segments, (b) the error signal and (c) the approximate gradient of the signal, which exhibits a peak at the boundary between the segments. The number of prediction coefficients $p = 12$.

cosine transform (DCT) or other semi-optimal transform, which have kernels independent of the signal. However, the results of these transforms can be degraded by spectral smearing due to the short-term time-domain windowing of the signals and fixed transform kernels. An optimal transform such as Karhunen–Loéve transform (KLT) requires complete statistical information, which may not be available in practise.

Parametric spectrum estimation methods such as those based on AR or autoregressive moving average (ARMA) modelling can outperform the DFT in accurately representing the frequency-domain characteristics of a signal, but they may suffer from poor estimation of the model parameters mainly due to the limited length of the measured signals. For example, in order to model the EEGs using an AR model, accurate values for the prediction order and coefficients are necessary. A high prediction order may result in splitting the true peaks in the frequency spectrum and low prediction order results in combining peaks in close proximity in the frequency domain.

For an AR model of the signal $x(n)$ the error or driving signal is considered to be zero-mean white noise. Therefore, by applying z-transform to Eq. (4.14), dropping the block index m, and replacing z by $e^{j\omega}$ we have:

$$\frac{X_p(\omega)}{E(\omega)} = \frac{1}{1 - \sum_{k=1}^{p} a_k e^{-jk\omega}} \tag{4.16}$$

where, $E(\omega) = K_\omega$ (Constant), is the power spectrum of the white noise and $X_p(\omega)$ is used to denote the signal power spectrum. Hence,

$$X_p(\omega) = \frac{K_\omega}{1 - \sum_{k=1}^{p} a_k e^{-jk\omega}} \tag{4.17}$$

and the parameters K_ω, a_k, $k = 1, \ldots, p$, are the exact values. In practical AR modelling these would be estimated from the finite-length measurement, thereby degrading the estimate of the spectrum. Figure 4.3 provides a comparison of the spectrum of an EEG segment of approximately 1550 samples of a single-channel EEG using both DFT analysis and AR modelling.

The fluctuations in the signal DFT as shown in Figure 4.3b is a consequence of the statistical inconsistency of periodogram-like power spectral estimation techniques. The result from the AR technique (Figure 4.3c) overcomes this problem provided the model fits the actual data by selecting a proper prediction order for the AR estimates. EEG signals are often statistically nonstationary, particularly where there is an abnormal event captured within the signals. In these cases the frequency-domain components are integrated over the observation interval and do not show the characteristics of the signals accurately. A TF approach is the solution to the problem.

In the case of multichannel EEGs, where the geometrical positions of the electrodes reflect the spatial dimension, space–time–frequency (STF) analysis through multiway processing methods has also become popular [15]. The main concepts in this area together with the parallel factor analysis (PARAFAC) algorithm will be reviewed in Chapter 17 aimed at detection of movement-related potentials or eye-blinking artefacts from the EEG signals.

The STFT is defined as the discrete-time Fourier transform evaluated over a sliding window. The STFT can be performed as:

$$X(n,\omega) = \sum_{-\infty}^{\infty} x(\tau) w(n-\tau) e^{-j\omega\tau} \tag{4.18}$$

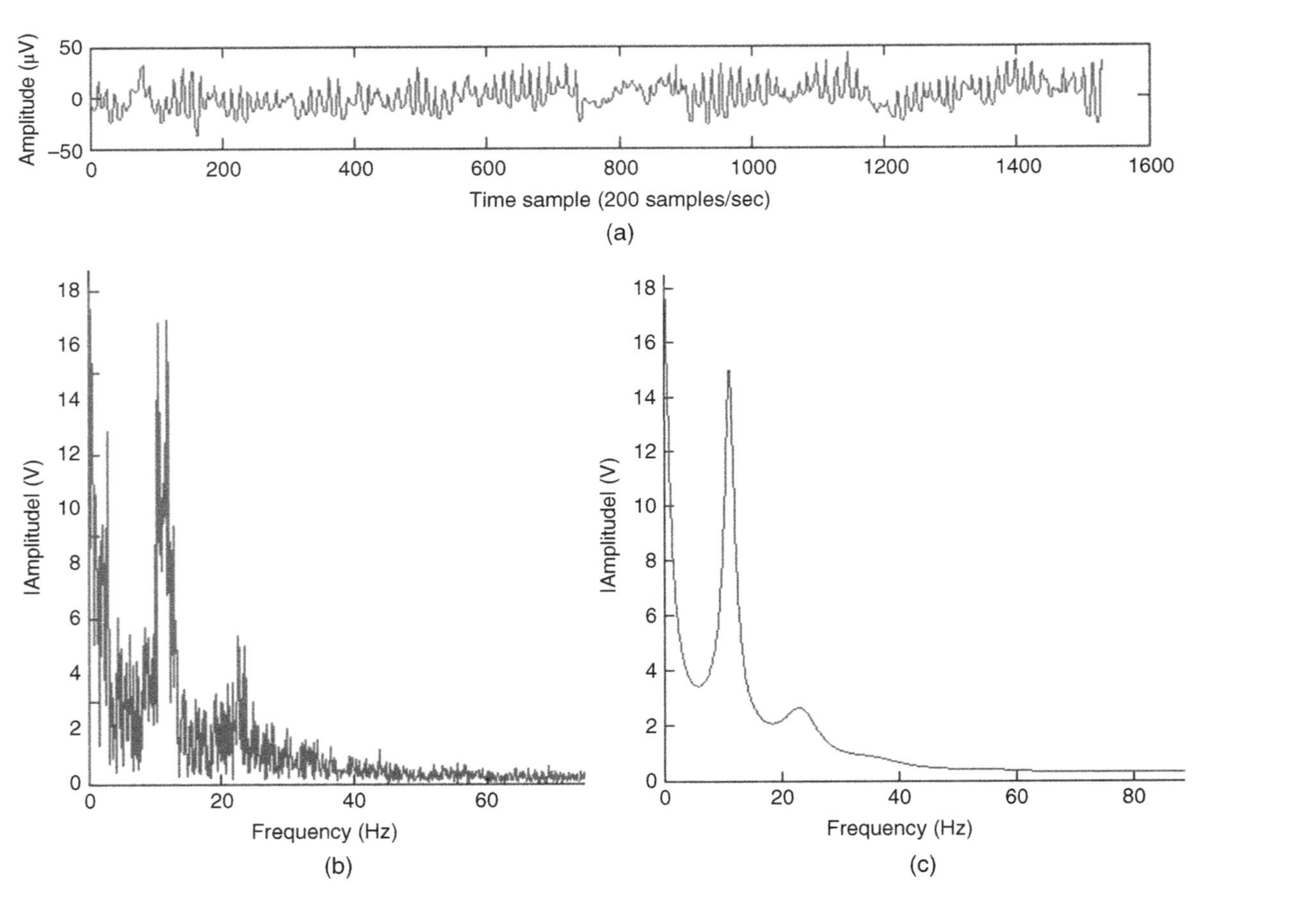

Figure 4.3 Single-channel EEG spectrum. (a) A segment of an EEG signal with a dominant alpha rhythm. (b) Spectrum of the signal in (a) using DFT. (c) Spectrum of the signal in (a) using a 12-order AR model.

where the discrete-time index n refers to the position of the window $w(n)$. Analogous with the periodogram a spectrogram is defined as:

$$S_x(n,\omega) = |X(n,\omega)|^2 \tag{4.19}$$

Based on the uncertainty principle, i.e. $\sigma_t^2\sigma_\omega^2 \geq \frac{1}{4}$, where σ_t^2 and σ_∞^2 are respectively time and frequency-domain variances, perfect resolution cannot be achieved in both time- and frequency-domains. Windows are typically chosen to eliminate discontinuities at block edges and to retain positivity in the power spectrum estimate. The choice also impacts upon the spectral resolution of the resulting technique, which, put simply, corresponds to the minimum frequency separation required to resolve two equal amplitude-frequency components.

Figure 4.4 shows the TF representation of an EEG segment during the evolution from preictal to ictal and to postictal stages. In Figure 4.4 the effect of time resolution has been illustrated using a Hanning windows of different durations of one and two seconds. Importantly, in Figure 4.4 the drift in frequency during the ictal period is observed clearly.

4.5.1 Wavelet Transform

The wavelet transform (WT) is another alternative for TF analysis. There is already a well established literature detailing the WT such as [16, 17]. Unlike the STFT, the TF kernel for the WT-based method can better localize the signal components in TF space. This efficiently exploits the dependency between time and frequency components. Therefore, the main objective of introducing the WT by Morlet [16] was likely to have a coherence time proportional to the sampling period. To proceed, consider the context of a continuous time signal.

4.5.1.1 Continuous Wavelet Transform

The Morlet–Grossmann definition of the continuous WT for a 1D signal $f(t)$ is:

$$W(a,b) = \frac{1}{\sqrt{a}}\int_{-\infty}^{\infty} f(t)\psi^*\left(\frac{t-b}{a}\right)dt \tag{4.20}$$

where $(.)^*$ denotes the complex conjugate, $\psi(t)$ is the analyzing wavelet, a (>0) is the scale parameter (inversely proportional to frequency) and b is the position parameter. The transform is linear and is invariant under translations and dilations, i.e.:

$$\text{If } f(t) \rightarrow W(a,b) \text{ then } f(t-\tau) \rightarrow W(a,b-\tau) \tag{4.21}$$

and

$$f(\sigma t) \rightarrow \frac{1}{\sqrt{\sigma}}W(\sigma a, \sigma b) \tag{4.22}$$

The last property makes the WT very suitable for analyzing hierarchical structures. It is similar to a mathematical microscope with properties that do not depend on the

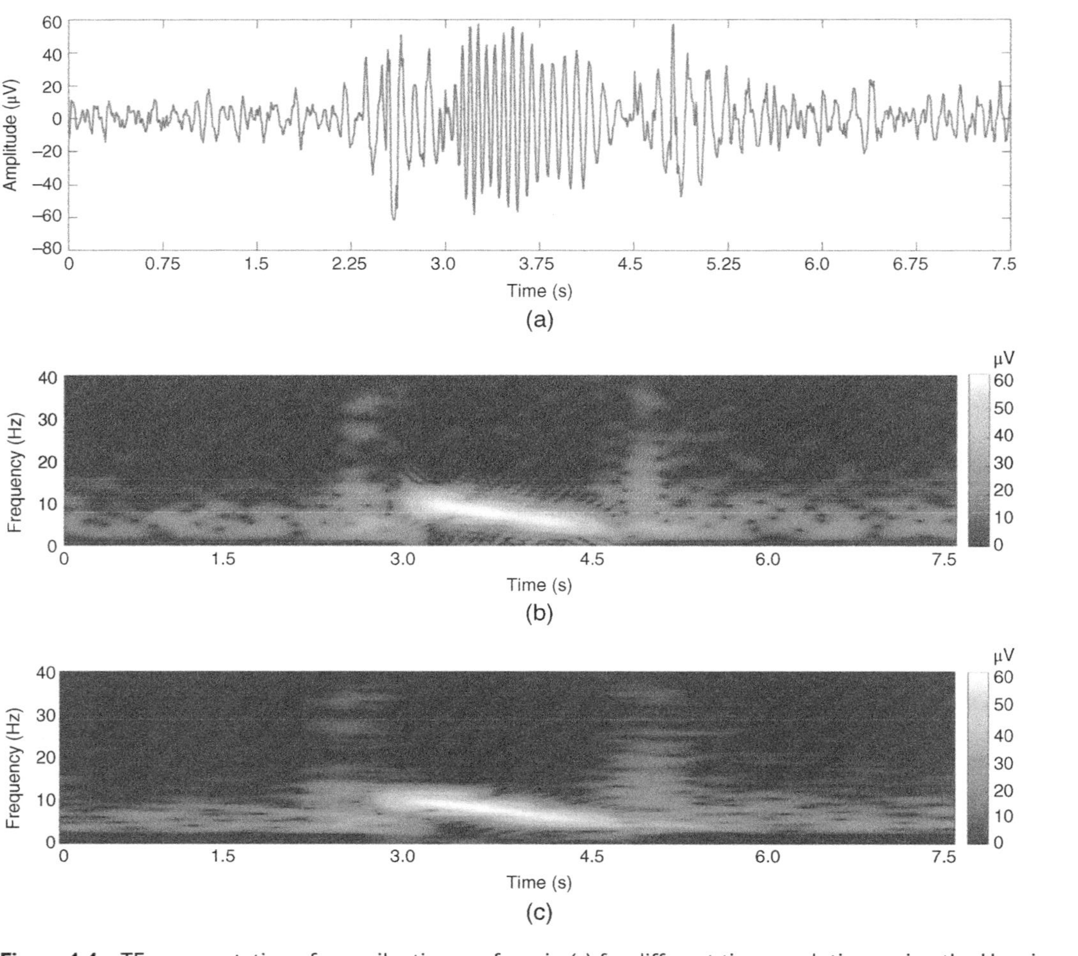

Figure 4.4 TF representation of an epileptic waveform in (a) for different time resolutions using the Hanning window of (b) 1 ms, and (c) 2 ms duration.

magnification. Consider a function $W(a,b)$ which is the WT of a given function $f(t)$. It has been shown [18, 19] that $f(t)$ can be recovered according to:

$$f(t) = \frac{1}{C_\phi}\int_0^{\infty}\int_{-\infty}^{\infty}\frac{1}{\sqrt{a}}W(a,b)\phi\left(\frac{t-b}{a}\right)\frac{dadb}{a^2} \tag{4.23}$$

where

$$C_\phi = \int_0^{\infty}\frac{\hat{\psi}^*(\nu)\hat{\phi}(\nu)}{\nu}d\nu = \int_{-\infty}^{0}\frac{\hat{\psi}^*(\nu)\hat{\phi}(\nu)}{\nu}d\nu \tag{4.24}$$

Although often it is considered that $\psi(t) = \phi(t)$, other alternatives for $\phi(t)$ may enhance certain features for some specific applications [20]. The reconstruction of $f(t)$ is subject to having C_ϕ defined (admissibility condition). The case $\psi(t) = \phi(t)$ implies $\hat{\psi}(0) = 0$, i.e. the mean of the wavelet function is zero.

4.5.1.2 Examples of Continuous Wavelets

Different waveforms/wavelets/kernels have been defined for the continuous WTs. The most popular ones are given below.

Morlet's wavelet is a complex waveform defined as:

$$\psi(t) = \frac{1}{\sqrt{2\pi}}e^{-\frac{t^2}{2}+j2\pi b_0 t} \tag{4.25}$$

This wavelet may be decomposed into its constituent real and imaginary parts as:

$$\psi_r(t) = \frac{1}{\sqrt{2\pi}}e^{-\frac{t^2}{2}}\cos(2\pi b_0 t) \tag{4.26}$$

$$\psi_i(t) = \frac{1}{\sqrt{2\pi}}e^{-\frac{t^2}{2}}\sin(2\pi b_0 t) \tag{4.27}$$

where b_0 is a constant, and it is considered that $b_0 > 0$ to satisfy the admissibility condition. Figure 4.5 shows respectively the real and imaginary parts.

The Mexican hat defined by Murenzi [17] is:

$$\psi(t) = \left(1-t^2\right)e^{-0.5t^2} \tag{4.28}$$

which is the second derivative of a Gaussian waveform (see Figure 4.6).

4.5.1.3 Discrete-Time Wavelet Transform

In order to process digital signals a discrete approximation of the wavelet coefficients is required. The discrete wavelet transform (DWT) can be derived in accordance with the sampling theorem if we process a frequency band-limited signal.

The continuous form of the WT may be discretized with some simple considerations on the modification of the wavelet pattern by dilation. Since generally the wavelet function $\psi(t)$ is not band limited, it is necessary to suppress the values outside the frequency components above half the sampling frequency to avoid aliasing (overlapping in frequency) effects.

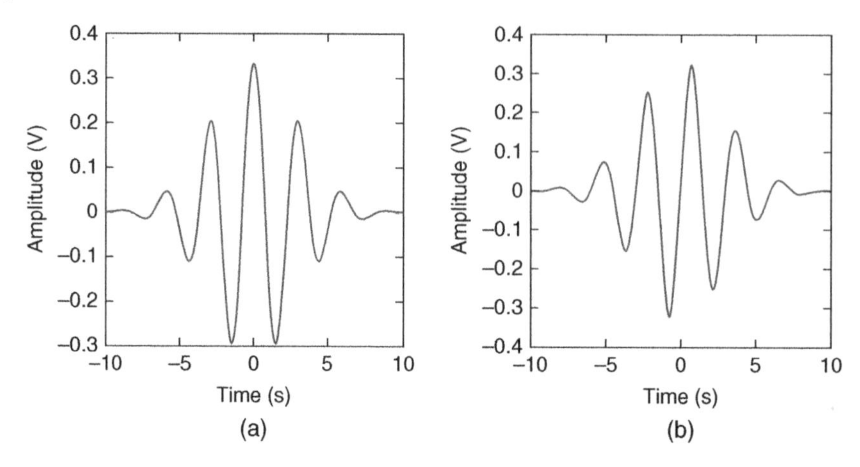

Figure 4.5 Morlet's wavelet: real and imaginary parts shown respectively in (a) and (b).

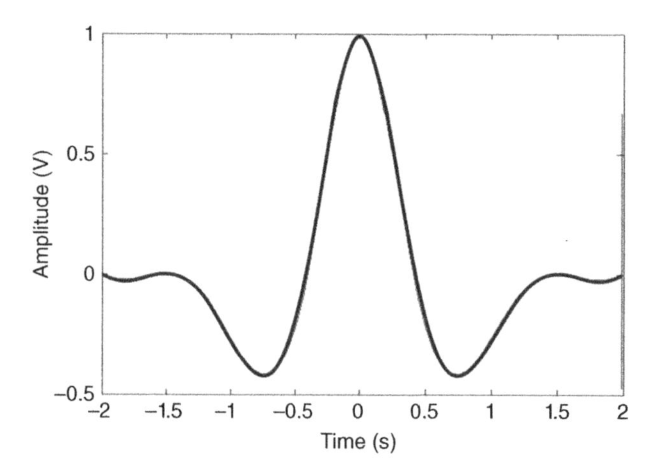

Figure 4.6 Mexican hat wavelet.

A Fourier space may be used to compute the transform scale-by-scale. The number of elements for a scale can be reduced if the frequency bandwidth is also reduced. This requires a band-limited wavelet. The decomposition proposed by Littlewood and Paley [21] provides a very nice illustration of the reduction of elements scale-by-scale. This decomposition is based on an iterative dichotomy of the frequency band. The associated wavelet is well localized in Fourier space where it allows a reasonable analysis to be made although not in the original space. The search for a discrete transform, which is well localized in both spaces leads to multiresolution analysis.

4.5.1.4 Multiresolution Analysis

Multiresolution analysis results from the embedded subsets generated by the interpolations (or down-sampling and filtering) of the signal at different scales. A function $f(t)$ is projected at each step j onto the subset V_j. This projection is defined by the scalar product $c_j(k)$ of $f(t)$ with the scaling function $\varphi(t)$, which is dilated and translated:

$$C_j(k) = \langle f(t), 2^{-j}\varphi(2^{-j}t - k)\rangle \tag{4.29}$$

where $\langle\cdot,\cdot\rangle$ denotes an inner product and $\varphi(t)$ has the property:

$$\frac{1}{2}\varphi\left(\frac{t}{2}\right) = \sum_{n=-\infty}^{\infty} h(n)\varphi(t-n) \tag{4.30}$$

where the right side is convolution of h and ϕ. By taking the Fourier transform of both sides:

$$\Phi(2\omega) = H(\omega)\Phi(\omega) \tag{4.31}$$

where $H(\omega)$ and $\Phi(\omega)$ are the Fourier transforms of $h(t)$ and $\phi(t)$ respectively. For a discrete frequency space (i.e. using the DFT) the above equation (Eq. 4.31) permits the computation of the wavelet coefficient $C_{j+1}(k)$ from $C_j(k)$ directly. If we start from $C_0(k)$ we compute all $C_j(k)$, with $j > 0$, without directly computing any other scalar product:

$$C_{j+1}(k) = \sum_n C_j(n)h(n-2k) \tag{4.32}$$

where k is the discrete frequency index.

At each step, the number of scalar products is divided by two and consequently the signal is smoothed. Using this procedure, the first part of a filter bank is built up. In order to restore the original data, Mallat uses the properties of orthogonal wavelets, but the theory has been generalized to a large class of filters by introducing two other filters $\tilde{h}$ and $\tilde{g}$, also called conjugate filters. The restoration is performed with:

$$C_j(k) = 2\sum_l\left[C_{j+1}(l)\tilde{h}(k+2l) + w_{j+1}(l)\tilde{g}(k+2l)\right] \tag{4.33}$$

where $w_{j+1}(\bullet)$ are the wavelet coefficients at the scale $j+1$ defined later in this section. For an exact restoration, two conditions have to be satisfied for the conjugate filters:

Anti-aliasing condition:

$$H\left(\omega + \frac{1}{2}\right)\tilde{H}(\omega) + G\left(\omega + \frac{1}{2}\right)\tilde{G}(\omega) = 0 \tag{4.34}$$

Exact restoration:

$$H(\omega)\tilde{H}(\omega) + G(\omega)\tilde{G}(\omega) = 1 \tag{4.35}$$

In the decomposition, the input is successively convolved with the two filters H (low frequencies) and G (high frequencies). Each resulting function is decimated by suppression of one sample out of two. The high frequency signal is left, and we iterate with the low frequency signal (left side of Figure 4.7). In the reconstruction, we restore the sampling by inserting a zero between each sample, then we convolve with the conjugate filters $\tilde{H}$ and $\tilde{G}$, we add the resulting outputs and we multiply the result by 2. We iterate up to the smallest scale (right side of Figure 4.7). Orthogonal wavelets correspond to the restricted case where:

$$G(\omega) = e^{-2\pi\omega}H^*\left(\omega + \frac{1}{2}\right) \tag{4.36}$$

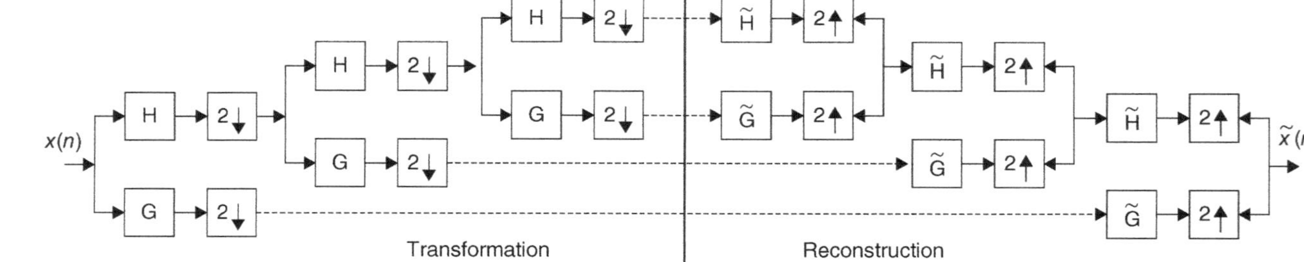

Figure 4.7 The filter bank associated with the multiresolution analysis.

$$\widetilde{H}(\omega) = H^*(\omega) \tag{4.37}$$

$$\widetilde{G}(\omega) = G^*(\omega) \tag{4.38}$$

and

$$|H(\omega)|^2 + \left|H\left(\omega + \frac{1}{2}\right)\right|^2 = 1 \quad \forall\omega \tag{4.39}$$

We can easily see that this set satisfies the two basic relations (4.72) and (4.73). Among various wavelets, Daubechies wavelets are the only compact solutions to satisfy the above conditions. For bi-orthogonal wavelets we have the relations:

$$G(\omega) = e^{-2\pi\omega}\widetilde{H}^*\left(\omega + \frac{1}{2}\right) \tag{4.40}$$

$$\widetilde{G}(\omega) = e^{2\pi\omega}H^*\left(\omega + \frac{1}{2}\right) \tag{4.41}$$

and

$$H(\omega)\widetilde{H}(\omega) + H^*\left(\omega + \frac{1}{2}\right)\widetilde{H}^*\left(\omega + \frac{1}{2}\right) = 1 \tag{4.42}$$

The relations (4.34) and (4.35) have to be also satisfied. A large class of compact wavelet functions can be used. Many sets of filters were proposed, especially for coding [22]. It was shown that the choice of these filters must be guided by the regularity of the scaling and the wavelet functions. The complexity is proportional to N. The algorithm provides a pyramid of N elements.

4.5.1.5 Wavelet Transform Using Fourier Transform

Consider the scalar products $c_0(k) = \langle f(t).\ \varphi(t-k)\rangle$ for continuous wavelets. If $\varphi(t)$ is band limited to half of the sampling frequency, the data are correctly sampled. The data at the resolution $j = 1$ are:

$$c_1(k) = \left\langle f(t).\frac{1}{2}\varphi\left(\frac{t}{2} - k\right)\right\rangle \tag{4.43}$$

and we can compute the set $c_1(k)$ from $c_0(k)$ with a discrete-time filter with frequency response $H(\omega)$:

$$H(\omega) = \begin{cases} \dfrac{\Phi(2\omega)}{\Phi(\omega)} & \text{if } |\omega| < \omega_c \\ 0 & \text{if } \omega_c \le |\omega| < \dfrac{1}{2} \end{cases} \tag{4.44}$$

and for $\forall\omega$ and $\forall$ integer m

$$H(\omega + m) = H(\omega) \tag{4.45}$$

Therefore, an estimate of the coefficients is:

$$C_{j+1}(\omega) = C_j(\omega)H(2^j\omega) \tag{4.46}$$

The cut-off frequency is reduced by a factor 2 at each step, allowing a reduction of the number of samples by this factor. The wavelet coefficients at the scale $j + 1$ are:

$$w_{j+1} = \left\langle f(t), 2^{-(j+1)}\psi\left(2^{-(j+1)}t - k\right)\right\rangle \tag{4.47}$$

and they can be computed directly from C_j by:

$$W_{j+1}(\omega) = C_j(\omega)G(2^j\omega) \tag{4.48}$$

where G is the following discrete-time filter:

$$G(\omega) = \begin{cases} \dfrac{\Psi(2\omega)}{\Phi(\omega)} & \text{if } |\omega| < \omega_c \\ 0 & \text{if } \omega_c \leq |\omega| < \dfrac{1}{2} \end{cases} \tag{4.49}$$

and for $\forall\omega$ and $\forall$integer m:

$$G(\omega + m) = G(\omega) \tag{4.50}$$

The frequency band is also reduced by a factor of two at each step. These relationships are also valid for DWT following Section 4.5.1.4.

4.5.1.6 Reconstruction

The reconstruction of the data from its wavelet coefficients can be performed step-by-step, starting from the lowest resolution. At each scale, we compute:

$$C_{j+1} = H(2^j\omega)C_j(\omega) \tag{4.51}$$

$$W_{j+1} = G(2^j\omega)C_j(\omega) \tag{4.52}$$

we look for C_j knowing C_{j+1}, W_{j+1}, h, and g. Then $C_j(\omega)$ is restored by minimizing:

$$P_h(2^j\omega)\left|C_{j+1}(\omega) - H(2^j\omega)C_j(\omega)\right|^2 + P_g(2^j\omega)\left|W_{j+1}(\omega) - G(2^j\omega)C_j(\omega)\right|^2 \tag{4.53}$$

using a least minimum squares estimator. $P_h(\omega)$ and $P_g(\omega)$ are weight functions which permit a general solution to the restoration of $C_j(\omega)$. The relationship from of $C_j(\omega)$ is in the form of:

$$C_j(\omega) = C_{j+1}(\omega)\widetilde{H}(2^j\omega) + W_{j+1}(\omega)\widetilde{G}(2^j\omega) \tag{4.54}$$

where the conjugate filters have the expressions:

$$\widetilde{H}(\omega) = \frac{P_h(\omega)H^*(\omega)}{P_h(\omega)|H(\omega)|^2 + P_g(\omega)|G(\omega)|^2} \tag{4.55}$$

$$\widetilde{H}(\omega) = \frac{P_g(\omega)G^*(\omega)}{P_h(\omega)|H(\omega)|^2 + P_g(\omega)|G(\omega)|^2} \tag{4.56}$$

It is straightforward to see that these filters satisfy the exact reconstruction condition given in Eq. (4.35). In fact, Eqs. (4.55) and (4.56) give the general solutions to this equation. In this analysis, the Shannon sampling condition is always respected. No aliasing exists, so that the anti-aliasing condition (4.76) is not necessary.

The denominator is reduced if we choose:

$$G(\omega) = \sqrt{1 - |H(\omega)|^2} \tag{4.57}$$

This corresponds to the case where the wavelet is the difference between the squares of two resolutions:

$$|\Psi(2\omega)|^2 = |\Phi(\omega)|^2 - |\Phi(2\omega)|^2 \tag{4.58}$$

The reconstruction algorithm then carries out the following steps:

1) Compute the fast Fourier transform (FFT) of the signal at the low resolution.
2) Set j to n_p (number of WT resolutions); perform the following iteration steps:
3) Compute the FFT of the wavelet coefficients at the scale j.
4) Multiply the wavelet coefficients W_j by $\widetilde{G}$.
5) Multiply the signal coefficients at the lower resolution C_j by $\widetilde{H}$.
6) The inverse Fourier transform of $W_j\widetilde{G} + C_j\widetilde{H}$ gives the coefficients C_{j-1}.
7) $j = j - 1$ and return to step 3.

The use of a band-limited scaling function allows a reduction of sampling at each scale and limits the computation complexity.

The WT has been widely used in EEG signal analysis. Its application to seizure detection, especially for neonates, modelling of the neuron potentials, and the detection of EP and ERPs will be discussed in the corresponding chapters of this book.

4.5.2 Synchro-Squeezed Wavelet Transform

The synchro-squeezing wavelet transform (SSWT) has been introduced as a post-processing technique to enhance the TF spectrum obtained by applying the WT [23]. Assuming that the input $f(t)$ is a pure harmonic signal ($f(t) = A\cos(\omega t)$), using Plancherel's theorem, the following equations are derived from (4.20) [23]:

$$\begin{aligned} W(a,b) &= \int_{-\infty}^{\infty} f(t)a^{-\frac{1}{2}}\overline{\psi}\left(\frac{t-b}{a}\right)dt = \frac{1}{2\pi}\int_{-\infty}^{\infty} \hat{f}(\varepsilon)a^{\frac{1}{2}}\overline{\hat{\psi}}(a\varepsilon)e^{ib\varepsilon}d\varepsilon \\ &= \frac{A}{4\pi}\int_{0}^{\infty} [\delta(\varepsilon-\omega) + \delta(\varepsilon+\omega)]a^{\frac{1}{2}}\overline{\hat{\psi}}(a\varepsilon)e^{ib\varepsilon}d\varepsilon = \frac{A}{4\pi}a^{\frac{1}{2}}\overline{\hat{\psi}}(a\omega)e^{ib\omega} \end{aligned} \tag{4.59}$$

One assumption in the above equation (Eq. 4.59) is that the selected mother wavelet is concentrated within the positive energy range, which means $\hat{\psi}(\varepsilon) = 0$, for $\varepsilon < 0$. If $\hat{\psi}(\varepsilon)$ is concentrated around $\varepsilon = \omega_0$, then, $W(a, b)$ is concentrated around $a = \omega_0/\omega$. This is spread out over a region of the horizontal line (e.g. $a = \omega_0/\omega$). In the situation where ω is almost but not exactly similar to the actual instantaneous frequency (IF) of the input signal, then, $W(a, b)$ has non-zero energy. By synchro-squeezing, the idea is to move this energy away from ω. The proposed method in the SSWT aims to reassign the frequency locations which are closer to the actual IF to obtain an enhanced spectrum. To do that, first, the candidate IFs are calculated. For these IFs, $W(a, b) \neq 0$:

$$\omega(a,b) = -i(W(a,b))^{-1}\frac{\partial}{\partial b}W(a,b) \tag{4.60}$$

Considering the selected pure harmonic signal $f(t) = A\cos(\omega t)$, it is simple to observe that $\omega(a, b) = \omega$. The candidate IFs are exploited to recover the actual frequencies. Therefore, a reallocation technique has been used to map the time domain into TF domain using $(b, a) \Rightarrow (b, \omega(a, b))$. Based on this, each value of $W(a, b)$ (computed at discrete values of a_k) is re-allocated into $T_f(\omega_l, b)$ as provided in the following equation (Eq. 4.61):

$$T_f(\omega_l, b) = (\Delta\omega)^{-1} \sum_{a_k:|\omega(a_k,b)-\omega_l| \le \frac{\Delta\omega}{2}} W(a_k, b)a_k^{-\frac{3}{2}}(\Delta a)_k \tag{4.61}$$

where ω_l is the nearest frequency to the original point $\omega(a, b)$, $\Delta\omega$ is the width of the frequency bins $\left[\omega_l - \frac{1}{2}\Delta\omega, \omega_l + \frac{1}{2}\Delta\omega\right]$,$\Delta\omega = \omega_l - \omega_{l-1}$, and $(\Delta a)_k = a_k - a_{k-1}$. $T_f(\omega_l, b)$ represents the synchro-squeezed transform at the centres ω_l of consecutive frequency bins. For each fixed time point b, the reassigned frequencies should be estimated for all scales using Eq. (4.103). For each desired IF of ω_l, $T_f(\omega_l, b)$ is calculated using summation of all $W(a_k, b)$ considering that the distance between the reassigned frequency $\omega(a_k, b)$ and ω_l must be within the specified frequency bin width ($\Delta\omega$). It has been shown that the original signal can be reconstructed after the synchro-squeezing process [23].

4.5.3 Ambiguity Function and the Wigner–Ville Distribution

The ambiguity function for a continuous time signal is defined as:

$$A_x(\tau, \nu) = \int_{-\infty}^{\infty} x^*\left(t - \frac{\tau}{2}\right)x\left(t + \frac{\tau}{2}\right)e^{j\nu t}dt \tag{4.62}$$

This function has its maximum value at the origin as

$$A_x(0, 0) = \int_{-\infty}^{\infty} |x(t)|^2 dt \tag{4.63}$$

As an example, if we consider a continuous time signal consisting of two modulated signals with different carrier frequencies such as

$$x(t) = x_1(t) + x_2(t)$$

$$= s_1(t)e^{j\omega_1 t} + s_2(t)e^{j\omega_2 t} \tag{4.64}$$

The ambiguity function $A_x(\tau,\nu)$ will be in the form of:

$$A_x(\tau, \nu) = A_{x_1}(\tau, \nu) + A_{x_2}(\tau, \nu) + \text{cross terms} \tag{4.65}$$

This concept is very important in separation of signals using the TF domain. This will be addressed in the context of blind source separation (BSS) later in this chapter. Figure 4.8 demonstrates this concept.

The Wigner–Ville frequency distribution of a signal $x(t)$ is then defined as the two-dimensional Fourier transform of the ambiguity function:

$$\begin{aligned} X_{WV}(t,\omega) &= \frac{1}{2\pi}\int_{-\infty}^{\infty}\int_{-\infty}^{\infty} A_x(\tau,\nu)e^{-j\nu t}e^{-j\omega t}d\nu d\tau \\ &= \frac{1}{2\pi}\int_{-\infty}^{\infty}\int_{-\infty}^{\infty}\int_{-\infty}^{\infty} x^*\left(\beta-\frac{\tau}{2}\right)x\left(\beta+\frac{\tau}{2}\right)e^{-j\nu(t-\beta)}e^{-j\omega\tau}d\beta d\nu d\tau \end{aligned} \tag{4.66}$$

which changes to the dual form of the ambiguity function as:

$$X_{WV}(t,\omega) = \int_{-\infty}^{\infty} x^*\left(t-\frac{\tau}{2}\right)x\left(t+\frac{\tau}{2}\right)e^{-j\omega\tau}d\tau \tag{4.67}$$

A quadratic form for the TF representation with the Wigner–Ville distribution can also be obtained using the signal in the frequency domain as:

$$X_{WV}(t,\omega) = \int_{-\infty}^{\infty} X^*\left(\omega-\frac{\nu}{2}\right)X\left(\omega+\frac{\nu}{2}\right)e^{-j\nu t}d\nu \tag{4.68}$$

The Wigner–Ville distribution is real and has very good resolution in both the time- and frequency-domains. Also it has time and frequency support properties, i.e. if $x(t) = 0$ for $|t| > t_0$, then $X_{WV}(t, \omega) = 0$ for $|t| > t_0$, and if $X(\omega) = 0$ for $|\omega| > \omega_0$, then $X_{WV}(t, \omega) = 0$ for $|\omega| > \omega_0$. It has also both time-marginal and frequency-marginal conditions of the form:

$$\frac{1}{2\pi}\int_{-\infty}^{\infty} X_{WV}(t,\omega)dt = |X(t)|^2 \tag{4.69}$$

and

$$\int_{-\infty}^{\infty} X_{WV}(t,\omega)dt = |X(\omega)|^2 \tag{4.70}$$

If $x(t)$ is the sum of two signals $x_1(t)$ and $x_2(t)$, i.e. $x(t) = x_1(t) + x_2(t)$, the Wigner–Ville distribution of $x(t)$ with respect to the distributions of $x_1(t)$ and $x_2(t)$ will be:

$$X_{WV}(t,\omega) = X_{1WV}(t,\omega) + X_{2WV}(t,\omega) + 2\,Re\{X_{12WV}(t,\omega)\} \tag{4.71}$$

where Re{.} denotes the real part of a complex value and

$$X_{12WV}(\tau,\omega) = \int_{-\infty}^{\infty} x_1^*\left(t-\frac{\tau}{2}\right)x_2\left(t+\frac{\tau}{2}\right)e^{-j\omega\tau}d\tau \tag{4.72}$$

It is seen that the distribution is related to the spectra of both auto- and cross-correlations. A pseudo-Wigner–Ville distribution (PWVD) is defined by applying a window function, $w(\tau)$, centred at $\tau = 0$ to the time-based correlations, i.e.:

$$\breve{X}_{WV}(t,\omega) = \int_{-\infty}^{\infty} x^*\left(t-\frac{\tau}{2}\right)x\left(t+\frac{\tau}{2}\right)w(\tau)e^{-j\omega\tau}d\tau \tag{4.73}$$

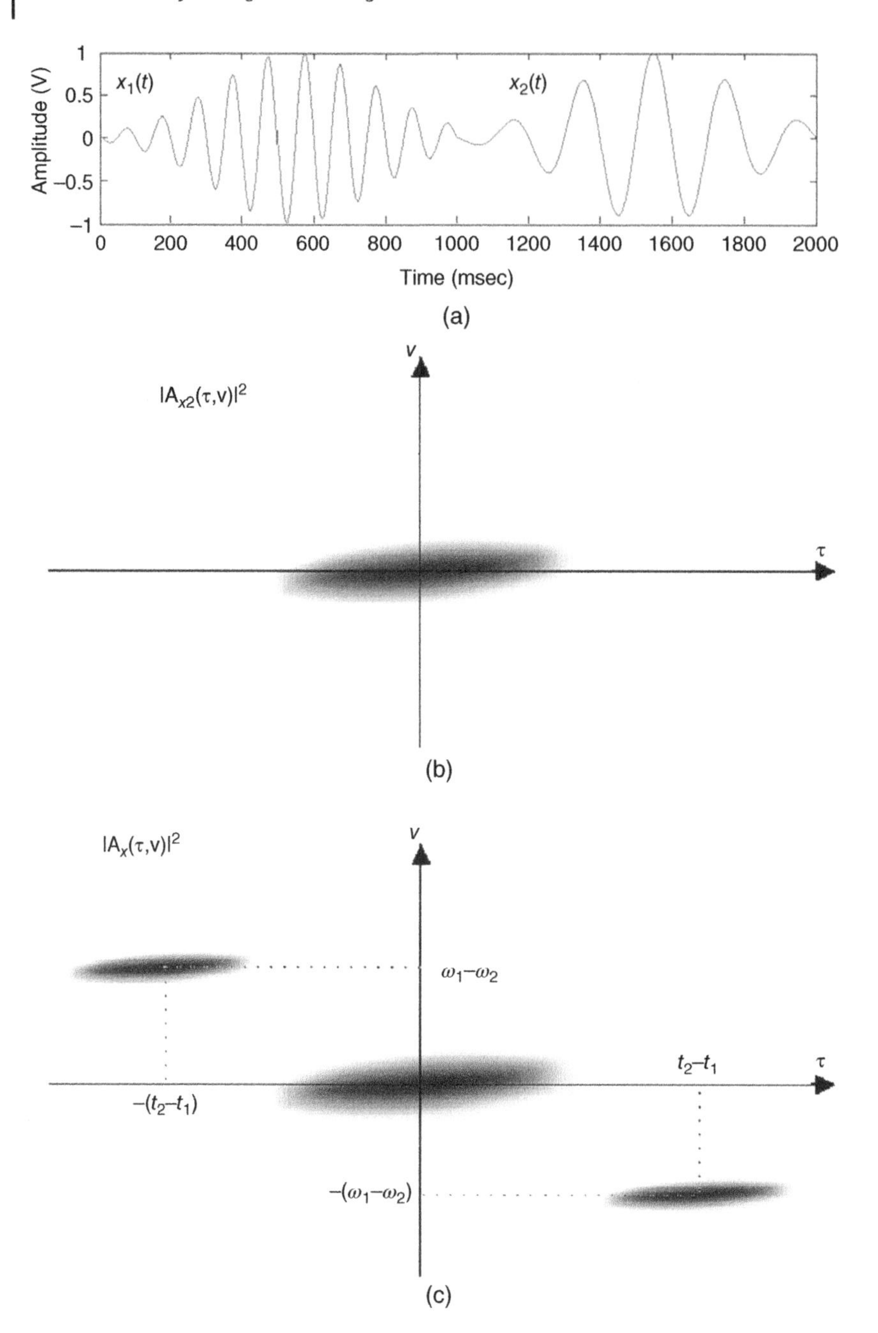

Figure 4.8 (a) A segment of a signal consisting of two modulated components, (b) ambiguity function for $x_1(t)$ only, and (c) the ambiguity function for $x(t) = x_1(t) + x_2(t)$.

In order to suppress the undesired cross-terms the two-dimensional Wigner–Ville (WV) distribution may be convolved with a TF-domain window. The window is a two-dimensional lowpass filter, which satisfies the time and frequency-marginal (uncertainty) conditions, as described earlier. This can be performed as:

$$C_x(t,\omega) = \frac{1}{2\pi}\int_{-\infty}^{\infty}\int_{-\infty}^{\infty} X_{WV}(t',\omega')\Phi(t-t',\omega-\omega')dt'd\omega' \tag{4.74}$$

where

$$\Phi(t,\omega) = \frac{1}{2\pi}\int_{-\infty}^{\infty}\int_{-\infty}^{\infty} \varphi(\tau,\nu)e^{-j\nu t}e^{-j\omega\tau}d\nu d\tau \tag{4.75}$$

and $\varphi(\tau,\nu)$ is often selected from a set of well known waveforms called *Cohen's class*. The most popular member of the Cohen's class of functions is the bell-shaped function defined as:

$$\varphi(\tau,\nu) = e^{-\frac{\nu^2\tau^2}{4\pi^2\sigma}}, \quad \sigma > 0 \tag{4.76}$$

A graphical illustration of such a function can be seen in Figure 4.9. In this case the distribution is referred to as a Choi–Williams distribution.

The application of a discrete-time form of the Wigner–Ville distribution to BSS will be discussed later in this chapter and its application to seizure detection has been explained

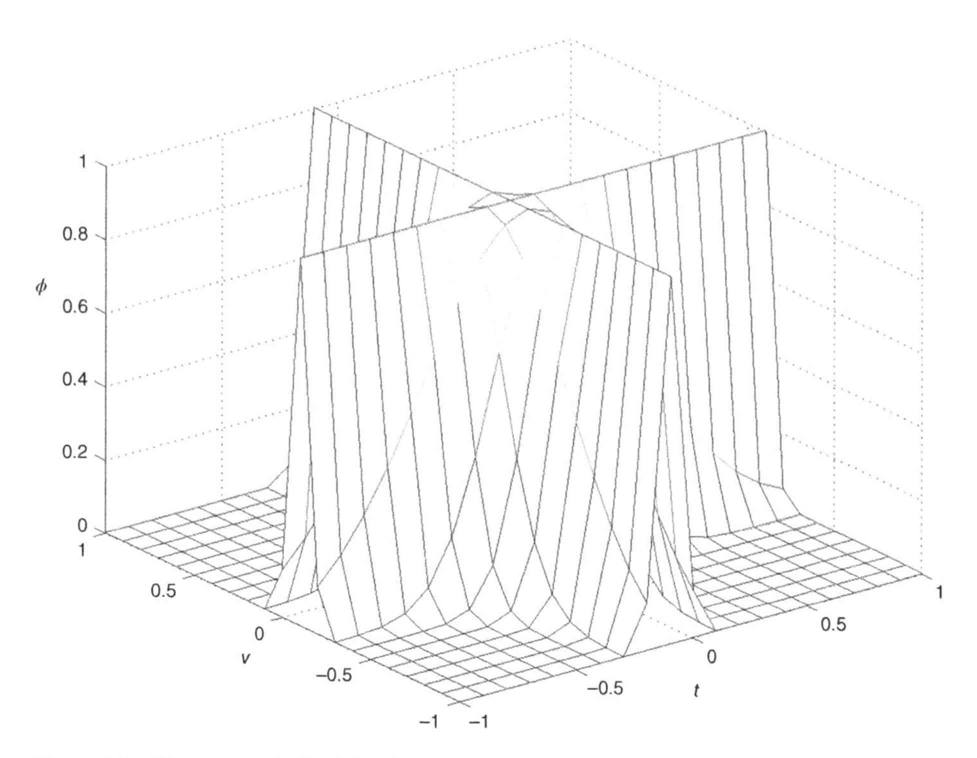

Figure 4.9 Illustration of $\phi(\tau, v)$ for the Choi–Williams distribution.

in the chapter devoted to seizure. To improve the distribution a signal-dependent kernel may also be used [24].

4.6 Empirical Mode Decomposition

Empirical mode decomposition (EMD) may be considered as a multiresolution signal decomposition technique. EMD is an adaptive time–space analysis method suitable for processing nonstationary and nonlinear time series. EMD partitions a series into 'modes' namely, intrinsic mode functions (IMFs) in the time domain. Like Fourier and WTs EMD does not follow any physical concept. However, the modes may provide insight into various signals contained within the data and have distinct frequency bands/components.

EMD is based on The Hilbert–Huang transform (HHT) which is a way to decompose a signal into IMFs along with a trend and obtain IF data. Its difference from other common transforms like the Fourier transform, is that the HHT is more like an algorithm (an empirical approach) that can be applied to a data set, rather than a theoretical tool.

IMF represents a simple oscillatory mode as a counterpart to the simple harmonic function, but instead of constant amplitude and frequency in a simple harmonic component, an IMF can have variable amplitude and frequency along the time axis.

The procedure of extracting an IMF is called sifting. The sifting process is as follows [25, 26]:

1) Identify all the local extrema in the test data.
2) Connect all the local maxima by a cubic spline line as the upper envelope.
3) Repeat the procedure for the local minima to produce the lower envelope.

The upper and lower envelopes should cover all the data between them. Their mean is m_1. The difference between the data and m_1 is the first component d_1:

$$x(t) - m_1 = d_1 \tag{4.77}$$

Ideally, d_1 should satisfy the definition of an IMF, since the construction of d_1 described above should have made it symmetric and having all maxima positive and all minima negative. After the first round of sifting, a crest may become a local maximum. New extrema generated in this way actually reveal the proper modes lost in the initial examination. In the subsequent sifting process, d_1 can only be treated as a proto-IMF. In the next step, d_1 is treated as data:

$$d_1 - m_{11} = d_{11} \tag{4.78}$$

After repeated sifting up to k times, d_1 becomes an IMF, i.e.:

$$d_{1(k-1)} - m_{1(k-1)} = d_{1k} \tag{4.79}$$

$C_1 = d_{1k}$ is considered as the first IMF of the signal $x(t)$.

The iteration above can be stopped in different ways such as when the power (standard deviation) of the difference (between current and previous iteration) signal becomes less than a predefined threshold, or when the number of iterations reaches a reasonable number [27].

For calculation of the other IMFs:

$$r_1 = x(t) - C_1 \tag{4.80}$$

The residue r_1 is then treated as the new signal and the same processing is applied to that. Therefore

$$r_n = r_{n-1} - C_n \tag{4.81}$$

The sifting process finally stops when the residue, r_n, becomes a monotonic function from which no more IMFs can be extracted. From the above equations, it is induced that:

$$\boldsymbol{x}(\boldsymbol{t}) = \sum_{j=1}^{n} C_j + \boldsymbol{r}_n \tag{4.82}$$

This results in decomposition of the data into n-empirical modes [25, 26].

Ensemble EMD (EEMD) is a noise assisted data analysis method. EEMD consists of 'sifting' an ensemble of white noise-added signal. EEMD can separate scales naturally without any a priori subjective criterion selection as in the intermittence test for the original EMD algorithm. Complete ensemble EMD with adaptive noise (CEEMDAN) is a variation of the EEMD algorithm that provides an exact reconstruction of the original signal and a better spectral separation of the IMFs.

4.7 Coherency, Multivariate Autoregressive Modelling, and Directed Transfer Function

In some applications such as in detection and classification of finger movement, it is very useful to find out how the associated movement signals propagate within the neural network of the brain. As will be shown in Chapter 16, there is a consistent movement of the source signals from the occipital to temporal regions. It is also clear that during the mental tasks different regions within the brain communicate with each other. The interaction and cross-talk among the EEG channels may be the only clue to understanding this process. This requires recognition of the transient periods of synchrony between various regions in the brain. These phenomena are not easy to observe by visual inspection of the EEGs. Therefore, some signal processing techniques have to be used in order to infer such causal relationships. One time series is said to be causal to another if the information contained in that time series enables the prediction of the other time series.

The spatial statistics of scalp EEG are usually presented as coherence in individual frequency bands, these coherences result both from correlations among neocortical sources and volume conduction through the tissues of the head, i.e. brain, cerebrospinal fluid, skull, and scalp. Therefore, spectral coherence [28] is a common method for determining the synchrony in EEG activity. Coherency is given as:

$$Coh_{ij}^2(\omega) = \frac{E\left\{\left|C_{ij}(\omega)\right|^2\right\}}{E\{C_{ii}(\omega)\}E\left\{C_{jj}(\omega)\right\}} \tag{4.83}$$

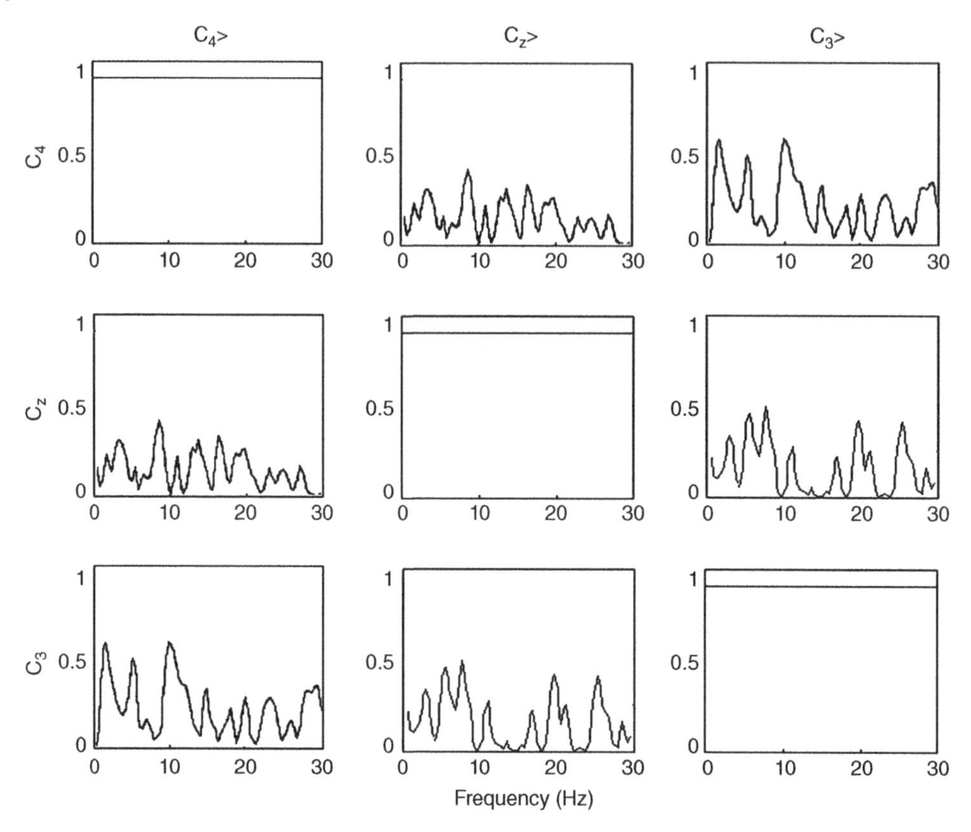

Figure 4.10 Cross-spectral coherence for a set of three electrode EEGs, one second before the right-finger movement. Each block refers to one electrode. By careful inspection of the figure, it is observed that the same waveform is transferred from C_z to C_3.

where $C_{ij}(\omega) = X_i(\omega)X_j^*(\omega)$ is the Fourier transform of the cross-correlation coefficients between channel i and channel j of the EEGs. Figure 4.10 shows an example of the cross-spectral coherence around one second prior to finger movement. A measure of this coherency, such as an average over a frequency band, is capable of detecting zero time-lag synchronization and fixed time non-zero time-lag synchronization, which may occur when there is a significant delay between the two neuronal population sites [29]. However, it does not provide any information on the directionality of the coupling between the two recording sites.

Granger causality (also called as Wiener–Granger causality) [30] is another measure, which attempts to extract and quantify the directionality from EEGs. Granger causality is based on bivariate AR estimates of the data. In a multichannel environment this causality is calculated from pair-wise combinations of electrodes. This method has been used to evaluate the directionality of the source movement from the local field potential in the visual system of cats [31].

For multivariate data in a multichannel recording, however, application of the Granger causality is not computationally efficient [31, 32]. The directed transfer function (DTF) [33], as an extension of Granger causality, is obtained from multichannel data and can be used to detect and quantify the coupling directions. The advantage of the DTF over spectral coherence is that it can determine the directionality in the coupling when the frequency spectra of the two brain regions have overlapping spectra. The DTF has been adopted by some researchers for determining the directionality in the coupling [34, 35] since it has been demonstrated that [36] there is a directed flow of information or cross-talk between the sensors around the sensory motor area before finger movement. The DTF is based on fitting the EEGs to an MVAR model. Assuming that $\mathbf{x}(n)$ is an M-channel EEG signal, it can be modelled in vector form as:

$$\mathbf{x}(n) = -\sum_{k=1}^{p} \mathbf{L}_k \mathbf{x}(n-k) + \mathbf{v}(n) \tag{4.84}$$

where n is the discrete-time index, p is the prediction order, $\mathbf{v}(n)$ is zero-mean noise, and $\mathbf{L}_k$ is generally an $M \times p$ matrix of prediction coefficients. A similar method to the Durbin algorithm for single-channel signals, namely the Levinson–Wiggins–Robinson (LWR) algorithm is used to calculate the MVAR coefficients [14]. The Akaike information criterion (AIC) [37] is also used for the estimation of prediction order p. By multiplying both sides of the above equation (Eq. 4.84) by $\mathbf{x}^T(n-k)$ and performing the statistical expectation the A set of Yule–Walker equation is obtained as [38]:

$$\sum_{k=0}^{p} \mathbf{L}_k \mathbf{R}(-k+p) = 0; \quad \mathbf{L}_0 = 1 \tag{4.85}$$

where $\mathbf{R}(q) = E[\mathbf{x}(n)\mathbf{x}^T(n+q)]$ is the covariance matrix of $\mathbf{x}(n)$, and the cross-correlations of the signal and noise are zero since they are assumed uncorrelated. Similarly, the noise autocorrelation is zero for non-zero shift since the noise samples are uncorrelated. The data segment is considered short enough for the signal to remain statistically stationary within that interval and long enough to enable accurate measurement of the prediction coefficients. Given the MVAR model coefficients, a multivariate spectrum can be achieved. Here it is assumed that the residual signal, $\mathbf{v}(n)$, is white noise. Therefore,

$$\boldsymbol{L}_f(\omega)\boldsymbol{X}(\omega) = \boldsymbol{V}(\omega) \tag{4.86}$$

where

$$\boldsymbol{L}_f(\omega) = \sum_{m=0}^{p} \boldsymbol{L}_m e^{-j\omega m} \tag{4.87}$$

and $\mathbf{L}(0) = \mathbf{I}$. Rearranging the above equation (Eq. 4.87) and replacing noise by $\sigma_v^2\mathbf{I}$ yields

$$\boldsymbol{X}(\omega) = \boldsymbol{L}_f^{-1}(\omega) \times \sigma_{\mathbf{v}}^2 \mathbf{I} = \boldsymbol{H}(\omega) \tag{4.88}$$

which represents the model spectrum of the signals or the transfer matrix of the MVAR system. The DTF or causal relationship between channel i and channel j can be defined directly from the transform coefficients [32] given by:

$$\Theta_{ij}^2(\omega) = |H_{ij}(\omega)|^2 \tag{4.89}$$

Electrode i is causal to j at frequency f if:

$$\Theta_{ij}^2(\omega) > 0 \tag{4.90}$$

A time-varying DTF can also be generated (mainly to track the source signals) by calculating the DTF over short windows to achieve the short-time DTF (SDTF) [32].

As an important feature in classification of left and right-finger movements, or tracking the mental task related sources, SDTF plays an important role. Some results of using SDTF for detection and classification of finger movement have been given in the context of BCI.

4.8 Filtering and Denoising

The EEG signals are subject to noise and artefacts. Electrocardiograms (ECGs), electro-oculograms (EOG) or eye blinks affect the EEG signals. Any multimodal recording such as EEG–functional magnetic resonance imaging (fMRI) significantly disturbs the EEG signals because of both magnetic fields and the change in the blood oxygen level and sensitivity of oxygen molecule to the magnetic field (balisto-cardiogram). Artefact removal from the EEGs will be explained in the related chapters. The noise in the EEGs, however, may be estimated and mitigated using adaptive and non-adaptive filtering techniques.

The EEG signals contain neuronal information below 100 Hz (in many applications the information lies below 30 Hz). Any frequency component above these frequencies can be simply removed by using lowpass filters. In the cases where the EEG data acquisition system is unable to cancel out the 50 Hz line frequency (due to a fault in grounding or imperfect balancing of the inputs to the differential amplifiers associated with the EEG system) a notch filter is used to remove it.

The nonlinearities in the recording system related to the frequency response of the amplifiers, if known, are compensated by using equalizing filters. However, the characteristics of the internal and external noises affecting the EEG signals are often unknown. The noise may be characterized if the signal and noise subspaces can be accurately separated. Using principal component analysis (PCA) or independent component analysis (ICA) we are able to decompose the multichannel EEG observations to their constituent components such as the neural activities and noise. Combining these two together, the estimated noise components can be extracted, characterized, and separated from the actual EEGs. These concepts are explained in the following sections and their applications to the artefact and noise removal will be brought in the later chapters.

Adaptive noise cancellers used in communications, signal processing, and biomedical signal analysis can also be used for removing noise and artefacts from the EEG signals. An effective adaptive noise canceller however requires a reference signal. Figure 4.11 shows a general block diagram of an adaptive filter for noise cancellation. The reference signal carries significant information about the noise or artefact and its statistical properties. For example, in the removal of eye-blinking artefacts (discussed in Chapter 16) a signature of the eye-blink signal can be captured from the FP1 and FP2 EEG electrodes. In detection of the ERP signals, as another example, the reference signal can be obtained by averaging a number of ERP segments. There are many other examples such as ECG cancellation from EEGs and the removal of fMRI scanner artefacts from EEG-fMRI simultaneous recordings where the reference signals can be provided.

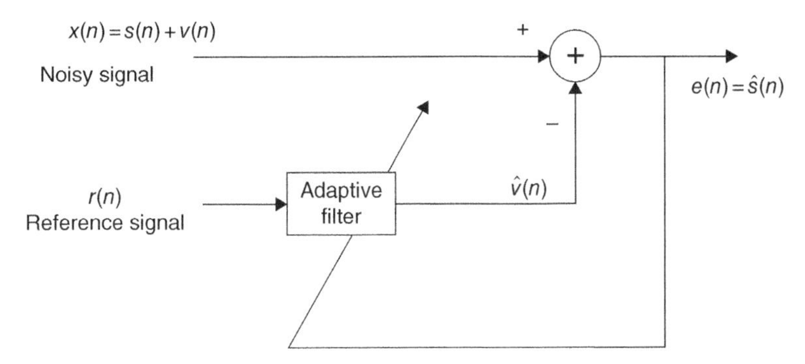

Figure 4.11 An adaptive noise canceller.

Adaptive Wiener filters are probably the most fundamental type of adaptive filters. In Figure 4.11 the optimal weights for the filter, $\mathbf{w}(n)$, are calculated such that $\hat{s}(n)$ is the best estimate of the actual signal $\mathbf{s}(n)$ in the mean square sense. The Wiener filter minimizes the mean square value of the error defined as:

$$e(n) = x(n) - \hat{v}(n) = x(n) - \mathbf{w}^T r(n) \tag{4.91}$$

where $\mathbf{w}$ is the Wiener filter coefficient vector. Using the orthogonality principle [39] the final form of the mean squared error will be:

$$E\left[e(n)^2\right] = E\left[x(n)^2\right] - 2\mathbf{p}^T\mathbf{w} + \mathbf{w}^T\mathbf{R}\mathbf{w} \tag{4.92}$$

where $E(.)$ represents statistical expectation:

$$\mathbf{p} = E[x(n)\mathbf{r}(n)], \tag{4.93}$$

and

$$\mathbf{R} = E\left[\mathbf{r}(n)\mathbf{r}^T(n)\right]. \tag{4.94}$$

By taking the gradient with respect to $\mathbf{w}$ and equating it to zero we have:

$$\mathbf{w} = \mathbf{R}^{-1}\mathbf{p} \tag{4.95}$$

As $\mathbf{R}$ and $\mathbf{p}$ are usually unknown the above minimization is performed iteratively by substituting time averages for statistical averages. The adaptive filter in this case, decorrelates the output signals. The general update equation is in the form of:

$$\mathbf{w}(n+1) = \mathbf{w}(n) + \Delta\mathbf{w}(n) \tag{4.96}$$

where n is the iteration number which typically corresponds to discrete-time index. $\Delta\mathbf{w}(n)$ has to be computed such that $E[\mathbf{e}(n)]^2$ reaches to a reasonable minimum. The simplest and most common way of calculating $\Delta\mathbf{w}(n)$ is by using gradient descent or steepest descent algorithm [39]. In both cases, a criterion is defined as a function of the squared error (often called a performance index) such as $\eta\,(e(n)^2)$, such that it monotonically decreases after each iteration and converges to a global minimum. This requires:

$$\eta(\mathbf{w} + \Delta\mathbf{w}) \le \eta(\mathbf{w}) = \eta\left(\mathbf{e}(n)^2\right) \tag{4.97}$$

Assuming ΔW is very small, it is concluded that:

$$\eta(\mathbf{w}) + \Delta\mathbf{w}^T\nabla_{\mathbf{w}}(\eta(\mathbf{w})) \leq \eta(\mathbf{w}) \tag{4.98}$$

where, $\nabla_{\mathbf{w}}(.)$represents gradient with respect to **w**. This means that the above equation (Eq. 4.98) is satisfied by setting $\Delta\mathbf{w} = -\mu\nabla_{\mathbf{w}}(.)$, where μ is the learning rate or convergence parameter. Hence, the general update equation takes the form:

$$\mathbf{w(n)} = \mathbf{w}T - \eta\nabla_{\mathbf{w}}(\eta(\mathbf{w}(n))) \tag{4.99}$$

Using the least mean square (LMS) approach, $\nabla_{\mathbf{w}}(\eta(\mathbf{w}))$ is replaced by an instantaneous gradient of the squared error signal, i.e.:

$$\nabla_{\mathbf{W}}(\eta(\mathbf{w}(n))) \cong -2e(n)\mathbf{r}(n) \tag{4.100}$$

Therefore, the LMS-based update equation is

$$\mathbf{w}(n+1) = \mathbf{w}(n) + 2\mu e(n)\mathbf{r}(n) \tag{4.101}$$

Also, the convergence parameter, μ, must be positive and should satisfy the following:

$$0 < \mu < \frac{1}{\lambda_{max}} \tag{4.102}$$

where $\lambda_{\max}$ represents the maximum eigenvalue of the autocorrelation matrix $\boldsymbol{R}$. The LMS algorithm is the most simple and computationally efficient algorithm. However, the speed of convergence can be slow especially for correlated signals. The recursive least-squares (RLS) algorithm attempts to provide a high speed stable filter, but it is numerically unstable for real-time applications [40, 41]. Defining the performance index as:

$$\eta(\mathbf{w}) = \sum_{i=0}^{n}\gamma^{n-i}e^2(i) \tag{4.103}$$

Then, by taking the derivative with respect to **w** we obtain

$$\nabla_{\mathbf{W}}\eta(\mathbf{w}) = -2\sum_{i=0}^{n}\gamma^{n-i}e(i)\mathbf{r}(i) \tag{4.104}$$

where $0 < \gamma \leq 1$ is the forgetting factor [40, 41]. Replacing for $e(n)$ in the above equation (Eq. 4.104) and writing it in vector form gives:

$$\mathbf{R}(n)\mathbf{w}(n) = \mathbf{p}(n) \tag{4.105}$$

where

$$\mathbf{R}(n) = \sum_{i=0}^{n}\lambda^{n-i}\mathbf{r}(i)\mathbf{r}^T(i) \tag{4.106}$$

and

$$\mathbf{p}(n) = \sum_{i=0}^{n}\lambda^{n-i}x(i)\mathbf{r}(i) \tag{4.107}$$

From this equation:

$$\mathbf{w}(n) = \mathbf{R}^{-1}(n)\mathbf{p}(n) \tag{4.108}$$

The RLS algorithm performs the above operation recursively such that **P** and **R** are estimated at the current time n as:

$$\mathbf{p}(n) = \lambda\mathbf{p}(n-1) + x(n)\mathbf{r}(n) \tag{4.109}$$

$$\mathbf{R}(n) = \lambda\mathbf{R}(n-1) + \mathbf{r}(n)\mathbf{r}^T(n) \tag{4.110}$$

In this case

$$\mathbf{r}(n) = \begin{bmatrix} r(n) \\ r(n-1) \\ . \\ . \\ r(n-M) \end{bmatrix} \tag{4.111}$$

where M represents the finite impulse response (FIR) filter order. Conversely:

$$\mathbf{R}^{-1}(n) = \left[\lambda\mathbf{R}^{-1}(n-1) + \mathbf{r}(n)\mathbf{r}^T(n)\right]^{-1} \tag{4.112}$$

which can be simplified using the matrix inversion lemma [42]:

$$\mathbf{R}^{-1}(n) = \frac{1}{\lambda}\left[\mathbf{R}^{-1}(n-1) - \frac{\mathbf{R}^{-1}(n-1)\mathbf{r}(n)\mathbf{r}^T(n)\mathbf{R}^{-1}(n-1)}{\lambda + \mathbf{r}^T(n)\mathbf{R}^{-1}(n-1)\mathbf{r}(n)}\right] \tag{4.113}$$

and finally, the update equation can be written as:

$$\mathbf{w}(n) = \mathbf{w}(n-1) + \mathbf{R}^{-1}(n)\mathbf{r}(n)\mathbf{g}(n) \tag{4.114}$$

where

$$\mathbf{g}(n) = x(n) - \mathbf{w}^T(n-1)\mathbf{r}(n) \tag{4.115}$$

and the error $e(n)$ after each iteration is recalculated as:

$$e(n) = x(n) - \mathbf{w}^T(n)\mathbf{r}(n) \tag{4.116}$$

The second term in the right-hand side of the above equation is $\hat{\mathbf{v}}(n)$. Presence of $\mathbf{R}^{-1}(n)$ in Eq. (4.115) is the major difference between RLS and LMS, but the RLS approach increases computation complexity by an order of magnitude.

4.9 Principal Component Analysis

All suboptimal transforms such as the DFT and DCT decompose the signals into a set of coefficients, which do not necessarily represent the constituent components of the signals. Moreover, the transform kernel is independent of the data hence they are not efficient in terms of both decorrelation of the samples and energy compaction. Therefore, separation of the signal and noise components is generally not achievable using these suboptimal transforms.

Expansion of the data into a set of orthogonal components certainly achieves maximum decorrelation of the signals. This enables separation of the data into the signal and noise subspaces.

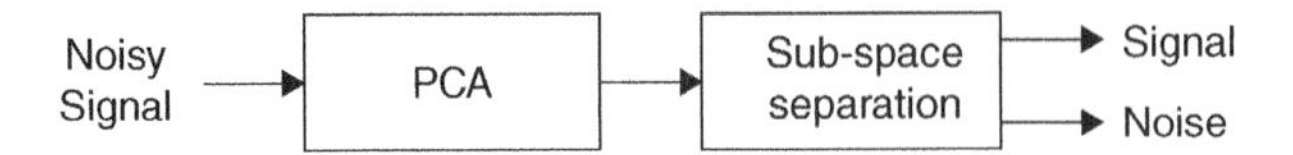

Figure 4.12 The general application of PCA.

For a single-channel EEG the Karhunen–Loéve transform is used to decompose the ith channel signal into a set of weighted orthogonal basis functions:

$$x_i(n) = \sum_{k=1}^{N} w_{i,k}\phi_k(n) \text{ or } \mathbf{x}_i = \mathbf{\Phi}\mathbf{w}_i \tag{4.117}$$

where $\mathbf{\Phi} = \{\phi_k\}$ is the set of orthogonal basis functions. The weights $w_{i,\ k}$ are then calculated as:

$$\mathbf{w}_i = \mathbf{\Phi}^{-1}\mathbf{x}_i \text{ or } w_{i,k} = \sum_{n=0}^{N-1} \phi_k^{-1}(n)x_i(n) \tag{4.118}$$

Often noise is added to the signal, i.e. $x_i(n) = s_i(n) + v_i(n)$, where $v_i(n)$ is additive noise. This degrades the decorrelation process. The weights are then estimated in order to minimize a function of the error between the signal and its expansion by the orthogonal basis, i.e. $\boldsymbol{e}_i = \mathbf{x}_i - \mathbf{\Phi}\mathbf{w}_i$. Minimization of the error in this case is generally carried out by solving the least-squares problem. In a typical application of PCA as depicted in Figure 4.12, the signal and noise subspaces are separated by means of some classification procedure.

4.9.1 Singular Value Decomposition

Singular value decomposition (SVD) is often used for solving the least-squares (LS) problem. This is performed by decomposition of the $M \times M$ square autocorrelation matrix $\mathbf{R}$ into its eigenvalue matrix $\mathbf{\Lambda} = diag(\lambda_1, \lambda_2, \ldots \lambda_M)$ and an $M \times M$ orthogonal matrix of eigenvectors $\mathbf{V}$, i.e. $\mathbf{R} = \mathbf{V}\mathbf{\Lambda}\mathbf{V}^H$, where $(.)^H$ denotes Hermitian (conjugate transpose) operation. Moreover, if $\mathbf{A}$ is an $M \times M$ data matrix such that $\mathbf{R} = \mathbf{A}^H\mathbf{A}$ then there exist an $M \times M$ orthogonal matrix $\mathbf{U}$, an $M \times M$ orthogonal matrix $\mathbf{V}$, and an $M \times M$ diagonal matrix $\mathbf{\Sigma}$ with diagonal elements equal to $\lambda_i^{1/2}$, such that:

$$\mathbf{A} = \mathbf{U}\mathbf{\Sigma}\mathbf{V}^H \tag{4.119}$$

Hence $\mathbf{\Sigma}^2 = \mathbf{\Lambda}$. The columns of $\mathbf{U}$ are called left singular vectors and the rows of $\mathbf{V}^H$ are called right singular vectors. If $\mathbf{A}$ is rectangular $N \times M$ matrix of rank k then $\mathbf{U}$ will be $N \times N$ and $\mathbf{\Sigma}$ will be:

$$\mathbf{\Sigma} = \begin{bmatrix} \mathbf{S} & 0 \\ 0 & 0 \end{bmatrix} \tag{4.120}$$

where $\mathbf{S} = diag(\sigma_1, \sigma_2, \ldots \sigma_k)$, where $\sigma_i = \lambda_i^{1/2}$. For such a matrix the Moore–Penrose pseudo-inverse is defined as an $M \times N$ matrix $\mathrm{A}^\dagger$ defined as:

$$\mathbf{A}^\dagger = \mathbf{U}\mathbf{\Sigma}^\dagger\mathbf{V}^H \tag{4.121}$$

where $\mathbf{\Sigma}^{\dagger}$ is an $M \times N$ matrix defined as:

$$\mathbf{\Sigma}^{\dagger} = \begin{bmatrix} \mathbf{S}^{-1} & 0 \\ 0 & 0 \end{bmatrix} \quad (4.122)$$

$\mathbf{A}^{\dagger}$ has a major role in the solutions of least-squares problems, and $\mathbf{S}^{-1}$ is a $k \times k$ diagonal matrix with elements equal to the reciprocals of the singular values of $\mathbf{A}$, i.e.

$$\mathbf{S}^{-1} = diag\left(\frac{1}{\sigma_1}, \frac{1}{\sigma_2}, \ldots, \frac{1}{\sigma_k}\right) \quad (4.123)$$

In order to see the application of the SVD in solving the LS problem consider the error vector $\mathbf{e}$ defined as:

$$\mathbf{e} = \mathbf{d} - \mathbf{Ah} \quad (4.124)$$

where $\mathbf{d}$ is the desired signal vector and $\mathbf{Ah}$ is the estimate $\hat{\mathbf{d}}$. To find $\mathbf{h}$ we replace $\mathbf{A}$ with its SVD in the above equation (Eq. 4.124) and find $\mathbf{h}$, which thereby minimizes the squared Euclidean norm of the error vector, $\|\mathbf{e}^2\|$. By using the SVD we obtain:

$$\mathbf{e} = \mathbf{d} - \mathbf{U}\sum \mathbf{V}^H\mathbf{h} \quad (4.125)$$

or equivalently

$$\mathbf{U}^H\mathbf{e} = \mathbf{U}^H\mathbf{d} - \Sigma\mathbf{V}^H\mathbf{h} \quad (4.126)$$

Since $\mathbf{U}$ is a unitary matrix, $\|\mathbf{e}^2\| = \|\mathbf{U}^H\mathbf{e}\|^2$. Hence, the vector $\mathbf{h}$ that minimizes $\|\mathbf{e}^2\|$ also minimizes $\|\mathbf{U}^H\mathbf{e}\|^2$. Finally, the unique solution as an optimum $\mathbf{h}$ (coefficient vector) may be expressed as [43]:

$$\boldsymbol{h} = \sum_{i=1}^{k} \frac{\boldsymbol{u}_i^H\boldsymbol{d}}{\sigma_i}\boldsymbol{v}_i \quad (4.127)$$

where k is the rank of $\mathbf{A}$. Alternatively, as the optimum least-squares coefficient vector:

$$\mathbf{h} = (\mathbf{A}^H\mathbf{A})^{-1}\mathbf{A}^H\mathbf{d} \quad (4.128)$$

Performing PCA is equivalent to performing an SVD on the covariance matrix. PCA uses the same concept as SVD and orthogonalization to decompose the data into its constituent uncorrelated orthogonal components such that the autocorrelation matrix is diagonalized. Each eigenvector represents a principal component and the individual eigenvalues are numerically related to the variance they capture in the direction of the principal components. In this case the mean squared error (MSE) is simply the sum of the N-K eigenvalues, i.e.:

$$MSE = \sum_{k=N-K}^{N} \phi_k^T R_x \phi_k = \sum_{k=N-K}^{N} \phi_k^T (\lambda_k \phi_k) = \sum_{k=N-K}^{N} \lambda_k \quad (4.129)$$

PCA is widely used in data decomposition, classification, filtering, and whitening. In filtering applications, the signal and noise subspaces are separated and the data are reconstructed from only the eigenvalues and eigenvectors of the actual signals. PCA is also used for BSS of correlated mixtures if the original sources can be considered statistically uncorrelated.

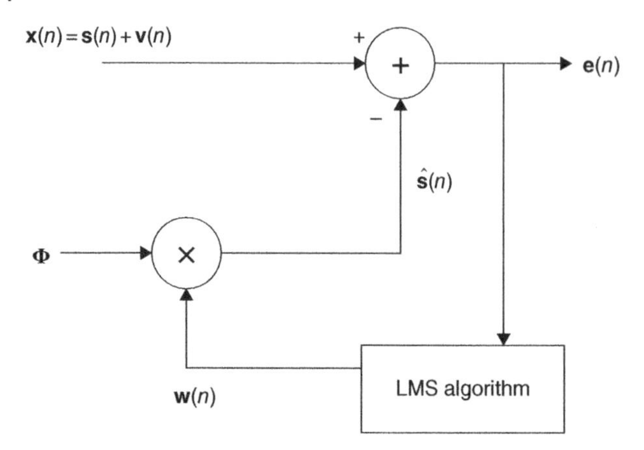

Figure 4.13 Adaptive estimation of the weight vector **w**(*n*).

The PCA problem is then summarized as how to find the weights **w** in order to minimize the error given the observations only. The LMS algorithm is used here to iteratively minimize the MSE as:

$$J_n = E\left[\left(\mathbf{x}(n) - \mathbf{\Phi}^T(n)\mathbf{w}(n)\right)^2\right] \tag{4.130}$$

The update rule for the weights is then:

$$\mathbf{w}(n+1) = \mathbf{w}(n) + \mu\mathbf{e}(n)\mathbf{\Phi}(n) \tag{4.131}$$

where the error signal $\mathbf{e}(n) = \mathbf{x}(n) - \mathbf{\Phi}^T(n)\mathbf{w}(n)$, $\mathbf{x}(n)$ is the noisy input and n is the iteration index. The step size μ may be selected empirically or adaptively. These weights are then used to reconstruct the sources from the set of orthogonal basis functions. Figure 4.13 shows the overall system for adaptive estimation of the weight vector **w** using the LMS algorithm.

4.10 Summary

In this chapter some basic signal processing tools and algorithms applicable to EEG signals have been reviewed. These fundamental techniques can be applied to reveal the inherent structure and the major characteristics of the signals based on which the state of the brain can be determined. TF-domain analysis is indeed a good EEG descriptive for both normal and abnormal cases. The change in entropy Conversely, may describe the transitions between preictal to ictal states for epileptic patients. These concepts will be exploited in the following chapters of this book.

References

1 Lebedev, M.A. and Nicolelis, M.A. (2006). Brain-machine interfaces: past, present and future. *Trends in Neurosciences* **29**: 536–546.

2 Lopes Da Silva, F. (2004). Functional localization of brain sources using EEG and/or MEG data: volume conductor and source models. *Journal of Magnetic Resonance Imaging* **22** (10): 1533–1538.

3 Durka, P.J., Dobieslaw, I., and Blinowska, K.J. (2001). Stochastic time–frequency dictionaries for matching pursuit. *IEEE Transactions on Signal Processing* **49** (3).
4 Spyrou, L., Lopez, D.M., Alarcon, G. et al. (2016). Detection of intracranial signatures of interictal epileptiform discharges from concurrent scalp EEG. *International Journal of Neural Systems* **26** (4): 1650016.
5 Spyrou, L., Kouchaki, S., and Sanei, S. (2016). Multiview classification and dimensionality reduction of EEG data through tensor factorisation. *Journal of Signal Processing Systems* **90**: 273–284.
6 Spyrou, L. and Sanei, S. (2016). Coupled dictionary learning for multimodal data: an application to concurrent intracranial and scalp EEG. *2016 IEEE International Conference on Acoustics, Speech and Signal Processing (ICASSP)*, 2349–2353. Shanghai, China.
7 Antoniades, A., Spyrou, L., Martin-Lopez, D. et al. (2018). Deep neural architectures for mapping scalp to intracranial EEG. *International Journal of Neural Systems* **28** (8): 1850009. https://doi.org/10.1142/S0129065718500090.
8 Antoniades, A., Spyrou, L., Martin-Lopez, D. et al. (2017). Detection of interictal discharges using convolutional neural networks from multichannel intracranial EEG. *IEEE Transactions on Neural Systems and Rehabilitation Engineering* **25** (12): 2285–2294.
9 Strang, G. (1998). *Linear Algebra and its Applications*, 3e. Thomson Learning.
10 McCann, H., Pisano, G., and Beltrachini, L. (2019). Variation in reported human head tissue electrical conductivity values. *Brain Topography* **32**: 825–858.
11 Hyvarinen, A., Kahunen, J., and Oja, E. (2001). *Independent Component Analysis*. Wiley.
12 Cover, T.M. and Thomas, J.A. (2001). *Elements of Information Theory*. Wiley.
13 Azami, H., Arnold, S.E., Sanei, S. et al. (2019). Multiscale fluctuation-based dispersion entropy and its applications to neurological diseases. *IEEE Access* **7** (1): 68718–68733, ISSN: 2169-3536. https://doi.org/10.1109/ACCESS.2019.2918560.
14 Morf, M., Vieria, A., Lee, D., and Kailath, T. (1978). Recursive multichannel maximum entropy spectral estimation. *IEEE Transactions on Geoscience Electronics* **16**: 85–94.
15 Bro, R. (1998). Multi-way analysis in the food industry: models, algorithms, and applications. PhD thesis. University of Amsterdam (NL) and Royal Veterinary and Agricultural University, MATLAB toolbox. http://www.models.kvl.dk/users/rasmus (accessed 19 August 2021).
16 Franaszczuk, P.J., Bergey, G.K., and Durka, P.J. (1996). Time–frequency analysis of mesial temporal lobe seizures using the matching pursuit algorithm. *Society for Neuroscience – Abstracts* **22**: 184.
17 Murenzi, R., Combes, J.M., Grossman, A., and Tchmitchian, P. (eds.) (1988). *Wavelets*. Heidelberg, New York: Springer Berlin.
18 Vaidyanathan, P.P. (1993). *Multirate Systems and Filter Banks*. Prentice Hall.
19 Holschneider, M., Kronland-Martinet, R., Morlet, R.J., and Tchamitchian, P. (1989). A real-time algorithm for signal analysis with the help of the wavelet transform. In: *Wavelets: Time-Frequency Methods and Phase Space* (eds. J.M. Combes, A. Grossman and P. Tchamitchian), 286–297. Berlin: Springer-Verlag.
20 Chui, C.K. (1992). *An Introduction to Wavelets*. Academic Press.
21 Stein, E.M. (1958). On the functions of Littlewood-Paley, Lusin and Marcinkiewicz. *Transactions of the American Mathematical Society* **88**: 430–466.
22 Vetterli, M. and Kovačevic, J. (1995). *Wavelets and Subband Coding*. Prentice Hall.

23 Daubechies, I., Lu, J., and Wu, H. (2011). Synchro-squeezed wavelet transforms: an empirical mode decomposition-like tool. *Applied and Computational Harmonic Analysis* **30** (2): 243–261.

24 Glassman, E.L. (2005). A wavelet-like filter based on neuron action potentials for analysis of human scalp electroencephalographs. *IEEE Transactions on Biomedical Engineering* **52** (11): 1851–1862.

25 Chen, Y. and Ma, J. (2014). Random noise attenuation by f-x empirical-mode decomposition predictive filtering. *Geophysics* **79** (3): V81–V91.

26 Chen, Y., Zhou, C., Yuan, J., and Jin, Z. (2014). Application of empirical mode decomposition in random noise attenuation of seismic data. *Journal of Seismic Exploration* **23**: 481–495.

27 Huang, Y.X., Schmitt, F.G., Lu, Z.M., and Liu, Y.L. (2008). An amplitude-frequency study of turbulent scaling intermittency using Hilbert spectral analysis. *Europhysics Letters* **84**: 40010.

28 Gerloff, G., Richard, J., Hadley, J. et al. (1998). Functional coupling and regional activation of human cortical motor areas during simple, internally paced and externally paced finger movements. *Brain* **121** (8): 1513–1531.

29 Sharott, A., Magill, P.J., Bolam, J.P., and Brown, P. (2005). Directional analysis of coherent oscillatory field potentials in cerebral cortex and basal ganglia of the rat. *Journal of Physiology* **562** (3): 951–963.

30 Granger, C.W.J. (1969). Investigating causal relations in econometric models and cross-spectral methods. *Econometrica* **37**: 424–438.

31 Bernosconi, C. and König, P. (1999). On the directionality of cortical interactions studied by spectral analysis of electrophysiological recordings. *Biological Cybernetics* **81** (3): 199–210.

32 Kaminski, M., Ding, M., Truccolo, W., and Bressler, S. (2001). Evaluating causal relations in neural systems: Granger causality, directed transfer function, and statistical assessment of significance. *Biological Cybernetics* **85**: 145–157.

33 Kaminski, M. and Blinowska, K. (1991). A new method of the description of information flow in the brain structures. *Biological Cybernetics* **65**: 203–210.

34 Jing, H. and Takigawa, M. (2000). Observation of EEG coherence after repetitive transcranial magnetic stimulation. *Clinical Neurophysiology* **111**: 1620–1631.

35 Kuś, R., Kaminski, M., and Blinowska, K. (2004). Determination of EEG activity propagation: pair-wise versus multichannel estimate. *IEEE Transactions on Biomedical Engineering* **51** (9): 1501–1510.

36 Ginter, J. Jr., Kaminski, M.J., Blinowska, K.J., and Durka, P. (2001). Phase and amplitude analysis in time–frequency–space; application to voluntary finger movement. *Journal of Neuroscience Methods* **110**: 113–124.

37 Akaike, H. (1974). A new look at statistical model order identification. *IEEE Transactions on Automatic Control* **19**: 716–723.

38 Ding, M., Bressler, S.L., Yang, W., and Liang, H. (2000). Short-window spectral analysis of cortical event-related potentials by adaptive multivariate autoregressive modelling: data preprocessing, model validation, and variability assessment. *Biological Cybernetics* **83**: 35–45.

39 Widrow, B., Glover, J.R., McCool, J. Jr. et al. (1975). Adaptive noise cancelling principles and applications. *Proceedings of the IEEE* **63** (12): 1692–1716.

40 Satorius, E.H. and Shensa, M.J. (1980). Recursive lattice filters: a brief overview. *Proceedings of the 19th IEEE Conference on Decision Control*, 955–959.

41 Lee, D., Morf, M., and Friedlander, B. (1981). Recursive square-root ladder estimation algorithms. *IEEE Transactions on Accoustic, Speech, Signal Processing* **29**: 627–641.

42 Lawsen, C.L. and Hansen, R.J. (1974). *Solving Least-Squares Problem*. Prentice Hall.

43 Proakis, J.G., Rader, C.M., Ling, F. et al. (2001). *Algorithms for Statistical Signal Processing*. Prentice Hall.

5

EEG Signal Decomposition

5.1 Introduction

Electroencephalography (EEG) signals record the combination of many signals originated from inside the brain. These sources can be cortical, deep sources, or artefacts. Since there is not any considerable differences between the time lags from the source to sensors, the combinations are considered as linear mixtures.

Although in the majority of cases EEG signals are processed as multichannel data there are cases where only a single EEG channel is available. Examples include some nonintrusive brain–computer interfacing (BCI) applications, monitoring driver fatigue, or human machine interfacing mainly for computer games. This makes single-channel EEG processing, and in some cases decomposition, an agenda for research.

Single-channel source decomposition is not always effective and not very accurate. Nevertheless, it is useful for analysis of many single-channel biological signals including simple and nonintrusive EEG recordings as stated above.

Another limitation in decomposition of multichannel EEGs is the relationship between the number of channels and the number of possible sources. In such cases, regularization methods such as sparse component analysis (SCA) may be applied.

In the following sections of this chapter, we start with single-channel source separation, then move to multichannel cases which can be performed by independent component analysis (ICA), matrix factorization, or tensor factorization. Then we will focus on SCA for two important scenarios: first when only one source is active at each time sample, and second when the number of active sources can be up to one less than the number of electrodes at each time sample or signal segment.

5.2 Singular Spectrum Analysis

One of the single-channel source decomposition methods is singular spectrum analysis (SSA).

The basic SSA method consists of two complementary stages: decomposition and reconstruction [148]; each stage includes two separate steps. In the first step the time series is decomposed and in the second stage the original time series is reconstructed and used

EEG Signal Processing and Machine Learning, Second Edition. Saeid Sanei and Jonathon A. Chambers.

for further analysis. The main concept in studying the properties of SSA is *separability*, which characterizes how well different components can be separated from each other. The absence of approximate separability is often observed in series with complex structure. For these time series and time series with special structure, there are different ways of modifying SSA leading to different versions such as SSA with single and double centring, Toeplitz SSA, and sequential SSA [148].

SSA is becoming an effective and powerful tool for time series analysis in meteorology, hydrology, geophysics, climatology, economics, biology, physics, medicine and other sciences where short and long, one-dimensional and multidimensional, stationary and non-stationary, almost deterministic and noisy time series are to be analyzed [149]. A brief description of the SSA (decomposition and reconstruction) stages is given in the following subsections.

5.2.1 Decomposition

This stage includes an embedding operation followed by singular value decomposition (SVD). The embedding operation maps a one-dimensional time series **f** into a $l \times k$ matrix as [148]:

$$\mathbf{X} = \{x_{ij}\} = [\mathbf{x}_1, \mathbf{x}_2, \ldots, \mathbf{x}_k] = \begin{bmatrix} f_0 & f_1 & f_2 & \cdots & f_{k-1} \\ f_1 & f_2 & f_3 & \cdots & f_k \\ f_2 & f_3 & f_4 & \cdots & f_{k+1} \\ \vdots & \vdots & & \ddots & \vdots \\ f_{l-1} & f_l & f_{l+1} & \cdots & f_{r-1} \end{bmatrix} \tag{5.1}$$

where $r = k + l - 1$ is the window length $(1 \leq l \leq r)$, and superscript T denotes the transpose of a vector. Vectors $\mathbf{x}_i$ are called *l-lagged vectors* (or, simply, *lagged vectors*). The window length l should be sufficiently large. Note that the trajectory matrix **X** is a Hankel matrix, which means that all the elements along the diagonal $i + j = const.$ are equal.

In the SVD stage the SVD of the trajectory matrix is computed and represented as sum of rank one bi-orthogonal elementary matrices. Consider the eigenvalues of the covariance matrix $\mathbf{C}_x = \mathbf{X}\mathbf{X}^T$ are denoted by $\lambda_1, \ldots, \lambda_l$ in decreasing order of magnitude $(\lambda_1 \geq \ldots \geq \lambda_l \geq 0)$ and the corresponding orthogonal eigenvectors by $\mathbf{u}_1, \mathbf{u}_2, \ldots, \mathbf{u}_l$. Set $d = \max(i;$ such that $\lambda_i > 0) = rank(\mathbf{X})$: If we denote $\mathbf{v}_i = \mathbf{X}^T\mathbf{u}_i/\sqrt{\lambda_i}$, then the SVD of the trajectory matrix can be written as:

$$\mathbf{X} = \sum_{i=1}^{d} \mathbf{X}_i = \sum_{i=1}^{d} \sqrt{\lambda_i}\mathbf{u}_i\mathbf{v}_i^T \tag{5.2}$$

The constituting components are separated through a constrained procedure of configuring **X** and the SVD operation. Since the desired signal might have different energy levels, in general there is no assumption about the relationship between the order of the eigenvalues and the physiologically dominant component (note that if $\lambda_1 >> \lambda_2$, the information energy will be mostly concentrated in the most dominant matrix $\mathbf{X}_1 = \sqrt{\lambda_1}\mathbf{u}_1\mathbf{v}_1^T$ associated with the first singular values).

In general, ICA can be used instead of SVD. This changes the subspace decomposition problem into the ICA problem which can be solved by employing a suitable ICA technique. ICA is discussed later in this chapter.

5.2.2 Reconstruction

During reconstruction the elementary matrices are first split into several groups (depending on the number of components in the time series – in this application electromyogram (EMG) and electrocardiograms (ECG) signals, and the matrices within each group are added together). The size of the group or in other words, the length of each subspace may be specified based on some a priori information. In the case of ECG removal from recorded EMGs this feature and the desired component are jointly (and automatically) identified based on a constraint on the statistical properties of the signal to be extracted. Let $I = \{i_1,\ldots, i_p\}$ be the indices corresponding to the p eigenvalues of the desired component. Then the matrix $\hat{\mathbf{X}}_I$ corresponding to the group I is defined as $\hat{\mathbf{X}}_I = \sum_{j=i_1}^{i_p} \mathbf{X}_j$. In splitting the set of indices $J = \{1, \ldots, d\}$ into disjoint subsets I_1 to I_m we always have:

$$\mathbf{Z} = \sum_{j=I_1}^{I_m} \hat{\mathbf{X}}_j \tag{5.3}$$

The procedure of choosing the sets $I_1, \ldots I_m$ is called eigentriple grouping. For a given group I the contribution of the component $\mathbf{X}_I$ in expansion (5.3) is measured by the contribution of the corresponding eigenvalues: $\sum_{i\in I}\lambda_i / \sum_{i=1}^{d}\lambda_i$. In the next step the obtained matrix is transformed to the form of a Hankel matrix which can be subsequently converted to a time series. If z_{ij} stands for an element of a matrix $\mathbf{Z}$, then the k-th term of the resulting series is obtained by averaging z_{ij} over all i,j such that $i + j = k + 1$. This procedure is called diagonal averaging or Hankelization of matrix $\mathbf{Z}$. The Hankelized $m \times n$ matrix $\mathbf{X}_h$ is defined as:

$$\mathbf{X}_h = \begin{bmatrix} \hat{x}_1\, \hat{x}_2 \cdots \hat{x}_i \\ \hat{x}_2\, \hat{x}_3 \cdots \hat{x}_{i+1} \\ \vdots\ \vdots\ \ddots\ \vdots \\ \hat{x}_j\, \hat{x}_{j+1} \cdots \hat{x}_{i+j-1} \end{bmatrix} \tag{5.4}$$

where

$$\hat{x}_k = \frac{1}{num(D_k)} \sum_{i,j\in D_k} z_{i,j} \text{ for } D_k = \{(p,q) : 1 \le p \le i, 1 \le q \le j, p + q = k + 1\} \tag{5.5}$$

where *num* abbreviates number.

A recursive SSA algorithm has been developed recently [150]. Such a method is advantageous for real-time processing of the data. The main problems in solving SSA problem are selection of appropriate embedding dimension (i.e. the window lengths) and establishing an effective criterion for selection of the desired subspace of eigentriples.

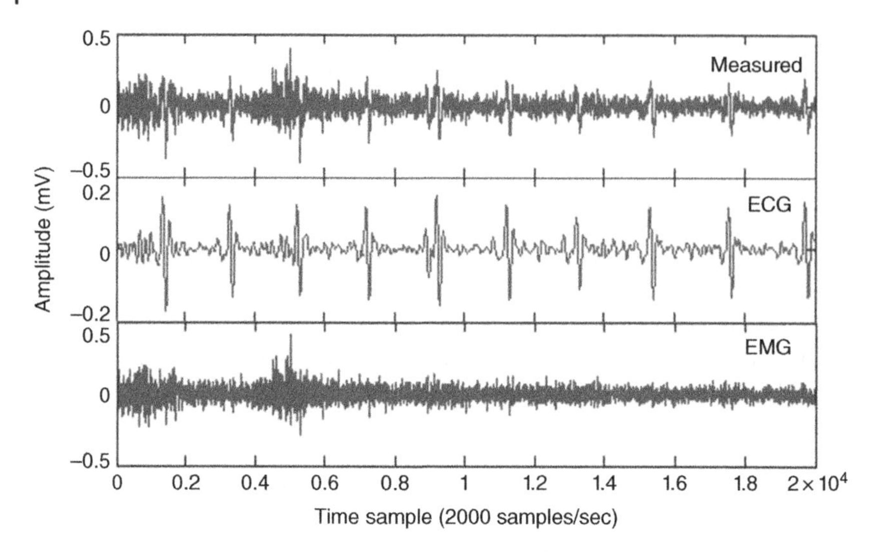

Figure 5.1 Separation of EMG and ECG using the SSA technique; top signal is the measurement, middle signal is the separated ECG, and the EMG signal can be viewed at the bottom.

As an example, Figure Error! Reference source not found. shows separation of ECG signals from an EMG measurement by carefully selecting the above parameters. Figure 5.1 clearly demonstrates the high potential of SSA in single-channel source decomposition.

5.3 Multichannel EEG Decomposition

EEG signals are inherently multichannel in their normal recording formats. However, the number of channels can be as low as 16 or as high as 256 in conventional clinical EEG systems.

5.3.1 Independent Component Analysis

In order to effectively separate the underlying sources from the electrode signals we should have more electrodes than source signals over any segment of time.

As a very popular approach for decomposition of multichannel EEGs, the concept of ICA lies in the fact that the constituent brain sources are independent from each other. In places where the combined source signals can be assumed independent from each other this concept plays a crucial role in separation and denoising the signals.

A measure of independency may be easily described to evaluate the independence of the decomposed components. Generally, considering the multichannel signal as **y**(n) and the constituent signal components as $y_i(n)$, the $y_i(n)$ are independent if:

$$p_Y(\mathbf{y}(n)) = \prod_{i=1}^{m} p_y(y_i(n)) \forall n \tag{5.6}$$

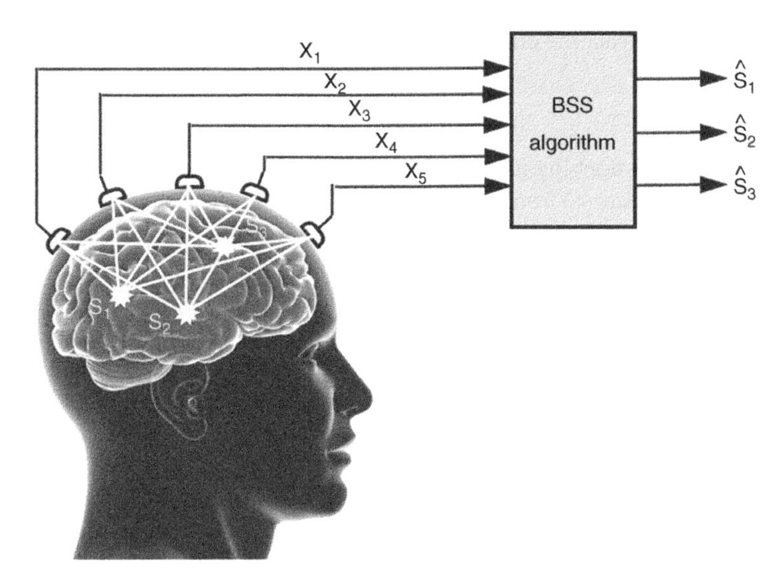

Figure 5.2 BSS concept; mixing and blind separation of the EEG signals.

where $p(\mathbf{Y})$ is the joint probability distribution, $p_y(y_i(n))$ are the marginal distributions and m is the number of independent components.

An important application of ICA is in blind source separation (BSS). BSS is an approach to estimate and recover the independent source signals using only the information of their mixtures observed at the recording channels. Due to its variety of applications BSS has attracted much attention recently. BSS of acoustic signals is often referred to as the Cocktail Party Problem [104], which means separation of individual sounds from a number of recordings in an uncontrolled environment such as a cocktail party. Figure 5.2 illustrates the BSS concept. As expected, ICA can be useful if the original sources are independent, i.e. $\boldsymbol{p}(\boldsymbol{s}(\boldsymbol{n})) = \prod_{i=1}^{m} \boldsymbol{p}_i(\boldsymbol{s}_i(\boldsymbol{n}))$.

A perfect separation of the signals requires taking into account the structure of the mixing process. In a real-life application, however, this process is unknown, but some assumptions may be made about the source statistics.

Generally, the BSS algorithms do not make realistic assumptions about the environment in order to make the problem more tractable. There are typically three assumptions about the mixing medium. The simplest but very popular and applicable BSS type is the instantaneous case, where the source signals arrive at the sensors at the same time. This has been considered for separation of biological signals such as the EEG where the signals have narrow bandwidths and the sampling frequency is normally low. The BSS model in this case can be easily formulated as:

$$\mathbf{x}(n) = \mathbf{H}\mathbf{s}(n) + \mathbf{v}(n) \tag{5.7}$$

where $m \times 1$ $\mathbf{s}(n)$, $n_e \times 1$ $\mathbf{x}(n)$, and $n_e \times 1$ $\mathbf{v}(n)$ denote respectively the vector of source signals, observed signals, and noise at discrete time n. $\mathbf{H}$ is the mixing matrix of size $n_e \times m$. The separation is performed by means of a separating $m \times n_e$ matrix, $\mathbf{W}$, which uses only the information about $\mathbf{x}(n)$ to reconstruct the original source signals (or the independent components) as:

$$\mathbf{y}(n) = \mathbf{W}\mathbf{x}(n) \tag{5.8}$$

In the context of EEG signal processing n_e denotes the number of electrodes. The early approaches in instantaneous BSS started from the work by Herault and Jutten [6] in 1986. In their approach, they considered non-Gaussian sources with similar number of independent sources and mixtures. They proposed a solution based on a recurrent artificial neural network (ANN) for separation of the sources.

In acoustic applications, however, there are usually time lags between the arrival of the signals at the sensors. The signals also may arrive through multiple paths. This type of mixing model is called a convolutive model. One example is in places where the acoustic properties of the environment vary, such as a room environment surrounded by walls. Based on these assumptions the convolutive mixing model can be classified into two more types: anechoic and echoic. In both cases the vector representations of mixing and separating processes are changed to $\mathbf{x}(n) = \mathbf{H}(n) * \mathbf{s}(n) + \mathbf{v}(n)$ and $\mathbf{y}(n) = \mathbf{W}(n) * \mathbf{x}(n)$ respectively, where $*$ denotes the convolution operation.

In an anechoic model, however, the expansion of the mixing process may be given as:

$$x_i(n) = \sum_{j=1}^{M} h_{ij} s_j(n - \delta_{ij}) + v_i(n), \quad \text{for } i = 1, \cdots, N \tag{5.9}$$

where the attenuation, h_{ij}, and delay, δ_{ij}, of source j to sensor i would be determined by the physical position of the source relative to the sensors. Then the unmixing process will be given as:

$$y_j(m) = \sum_{i=1}^{N} w_{ji} x_i(m - \delta_{ji}), \quad \text{for } j = 1, \cdots, M \tag{5.10}$$

where the w_{ji}s are the elements of W. In an echoic mixing environment it is expected that the signals from the same sources reach to the sensors through multiple paths. The reverberation time changes and can continue infinitely. However, in practical acoustic models, the number of paths in a multipath scenario as well as the overall reverberation time are considered limited. Therefore, the expansion of the mixing and separating models will be changed to:

$$x_i(n) = \sum_{j=1}^{M} \sum_{k=1}^{K} h_{ij}^k s_j\left(n - \delta_{ij}^k\right) + v_i(n), \quad \text{for } i = 1, \cdots, N \tag{5.11}$$

where K denotes the number of paths and $v_i(n)$ is the accumulated noise at sensor i. The unmixing process will be formulated similarly to the anechoic one. Obviously, for a known number of sources an accurate result may be expected if the number of paths is known.

The aim of BSS using ICA is to estimate an unmixing matrix $\mathbf{W}$ such that $\mathbf{Y} = \mathbf{WX}$ best approximates the independent sources S, where $\mathbf{Y}$ and $\mathbf{X}$ are respectively matrices with columns $\mathbf{y}(n) = [y_1(n), y_2(n), \cdots y_m(n)]^T$ and $\mathbf{x}(n) = [x_1(n), x_2(n), \cdots x_{ne}(n)]^T$.

In any case, the unmixing matrix for the instantaneous case is expected to be equal to the inverse of the mixing matrix, i.e. $\mathbf{W} = \mathbf{H}^{-1}$. However, since in all ICA algorithms are based upon restoring independence, the separation is subject to permutation and scaling ambiguities in the output independent components, i.e. $\mathbf{W} = \mathbf{PDH}^{-1}$, where $\mathbf{P}$ and $\mathbf{D}$ are respectively the permutation and scaling matrices.

There are three major approaches in using ICA for BSS:

1) Factorizing the joint pdf of the reconstructed signals into its marginal pdfs. Under the assumption that the source signals are stationary and non-Gaussian, the independence of the reconstructed signals can be measured by a statistical distance between the joint distribution and the product of its marginal pdfs. Kullback–Leibler (KL) divergence (distance) is an example. For nonstationary cases and for the short-length data, there will be poor estimation of the pdfs. Therefore, in such cases, this approach may not lead to good results. Conversely, such methods are not robust for noisy data since in this situation the pdf of the signal will be distorted.
2) Decorrelating the reconstructed signals through time, that is, diagonalizing the covariance matrices at every time instant. If the signals are mutually independent, the off-diagonal elements of the covariance matrix vanish. Although the reverse of this statement is not always true. If the signals are nonstationary, we can utilize the time-varying covariance structure to estimate the unmixing matrix. An advantage of this method is that it only uses second-order statistics (SOS), which implies that it is likely to perform better in noisy and short data length conditions than higher-order statistics (HOS).
3) Eliminating the temporal cross-correlation functions of the reconstructed signals as much as possible. In order to perform this, the correlation matrix of observations can be diagonalized at different time lags simultaneously. Here, SOS are also normally used. As another advantage, it can be applied in the presence of white noise since such noise can be avoided by using the cross-correlation only for $\tau \neq 0$. Such a method is appropriate for stationary and weakly stationary sources (i.e. when the stationarity condition holds within a short segment of data).

It has been shown [64] that mutual information (MI) is a measure of independence and that maximizing the non-Gaussianity of the source signals is equivalent to minimizing the MI between them.

In the majority of cases the number of sources is known. This assumption avoids any ambiguity caused by false estimation of the number of sources. In exactly-determined cases the number of sources is equal to the number of mixtures. In over-determined situations however, the number of mixtures is more than the number of sources.

There have been many attempts to apply BSS to EEG signals [72–82] for separation of normal brain rhythms, event-related signals, or mental or physical movement-related sources.

If the number of sources is unknown, a criterion has to be established to estimate the number of sources beforehand. This process is a difficult task especially when noise is involved.

In those cases where the number of sources is more than the number of mixtures (known as underdetermined systems), the above BSS schemes can not be applied simply because the unmixing matrix will not be invertible, and generally the original sources cannot be extracted. However, when the signals are sparse other methods based on clustering may be utilized.

A signal is said to be sparse when it has many zero or at least approximately zero samples. Separation of the mixtures of such signals is potentially possible in the situation where at each sample instant the number of non-zero sources is not more than the number of sensors. The mixtures of sparse signals can also be instantaneous or convolutive. However, as we will briefly describe later, the solution for only simple case of a small number of idealized sources has been given in the literature.

In the context of EEG analysis, although the number of signals mixed at the electrodes seems to be limited, the number of sources corresponding to the neurons firing at a time can be enormous. However, if the objective is to study a certain rhythm in the brain the problem can be transformed to the time–frequency (TF) domain or even to the space–time–frequency (STF) domain. In such domains the sources may be considered disjoint and generally sparse. Also it is said that in the brain neurons encode data in a sparse way if their firing pattern is characterized by a long period of inactivity [65, 66].

5.3.2 Instantaneous BSS

This is the most commonly used scheme for processing of the EEGs. The early work by Herault and Jutten led to a simple but fundamental adaptive algorithm [67]. Linsker [68] proposed unsupervised learning rules based on information theory that maximize the average MI between the inputs and outputs of an ANN. Comon [64] performed minimization of MI to make the outputs independent. The Infomax algorithm [69] in spirit is similar to the Linsker method. It uses an elegant stochastic gradient learning rule previously proposed by Amari et al. [70]. Non-Gaussianity of the sources was first exploited by Hyvarinen and Oja [71] in developing their fast ICA (fICA) algorithm. fICA is actually a blind source extraction algorithm, which extracts the sources one by one based on their kurtosis; the signals with transient peaks have high kurtosis. Later it was demonstrated that the Infomax algorithm and maximum likelihood estimation are in fact equivalent [1, 28].

Based on the Infomax algorithm [69] for signals with positive kurtosis such as simultaneous EEG-fMRI and speech signals, minimizing the MI between the source estimates and maximizing the entropy of the source estimates are equivalent. Therefore, a stochastic gradient ascent algorithm can be used to iteratively find the unmixing matrix by maximization of the entropy. The Infomax finds a **w** which minimizes the following cost function:

$$J(\mathbf{w}) = I(\mathbf{z},\mathbf{x}) = H(\mathbf{z}) - H(\mathbf{z} \mid \mathbf{x}) \tag{5.12}$$

where $H(\mathbf{z})$ is the entropy of the output, $H(\mathbf{z}|\mathbf{x})$ is the entropy of the output subject to a known input, and $\mathbf{z} = f(\mathbf{y})$ is a nonlinear activation function applied element-wise to $\mathbf{y}$, the estimated sources. $I(\mathbf{z},\mathbf{x})$ is the MI between the input and output of the constructed ANN. $H(\mathbf{z}|\mathbf{x})$ is independent of $\mathbf{W}$, therefore, the gradient of J is only proportional to the gradient of $H(\mathbf{z})$. Correspondingly, the natural gradient [70] of J devoted as $\nabla_{\mathbf{W}}J$ will be:

$$\nabla_{\mathbf{W}}J = \nabla_{\mathbf{W}}\mathrm{I}(\mathbf{z},\mathbf{x})\mathbf{W}^T\mathbf{W} = \nabla_{\mathbf{W}}\mathrm{I}(\mathbf{z},\mathbf{x})\mathbf{W}^T\mathbf{W} \tag{5.13}$$

in which the time index n is dropped for convenience of presentation. Then, the sequential adaptation rule for the unmixing matrix $\mathbf{W}$ becomes:

$$\mathbf{W}(n+1) = \mathbf{W}(n) + \mu\left(\boldsymbol{I} + (1 - 2f(\boldsymbol{y}(n)))\boldsymbol{y}^T(n)\right)\mathbf{W}(n) \tag{5.14}$$

where $f(\boldsymbol{y}(n)) = (1 + exp\,(-\boldsymbol{y}(n)))^{-1}$, assuming the outputs are super-Gaussian, and μ is the learning rate, which is either a small constant or gradually changes following the speed of convergence.

Joint diagonalization of eigen matrices (JADE) is another well known BSS algorithm [2] based on HOS. The JADE algorithm effectively diagonalizes the fourth-order cumulant of the estimated sources. This procedure uses certain matrices $Q_z(M)$ formed by the inner product of the fourth-order cumulant tensor of the outputs with an arbitrary matrix **M**, i.e.:

$$\{\mathbf{Q}_z(\mathbf{M})\}_{ij} = \sum_{k=1}^{n_e} \sum_{l=1}^{n_e} Cum\left(z_i, z_j^*, z_k, z_l^*\right) m_{lk} \tag{5.15}$$

where the (l,k)th component of the matrix M is written as m_{lk}, $\mathbf{Z} = \mathbf{CY}$, and $*$ denotes complex conjugate. The matrix $\mathbf{Q}_z(\mathbf{M})$ has the important property that it is diagonalized by the correct rotation matrix **U**, i.e. $\boldsymbol{U}^H\boldsymbol{QU} = \boldsymbol{\Lambda}_M$, and $\boldsymbol{\Lambda}_M$ is a diagonal matrix whose diagonal elements depend on the particular matrix M as well as **Z**. By using Eq. (5.15), for a set of different matrices **M**, a set of cumulant matrices $\mathbf{Q}_z(\mathbf{M})$ can be calculated. The desired rotation matrix **U** then, jointly diagonalizes these matrices. In practise, only approximate joint diagonalization is possible [2], i.e. the problem can be stated as minimization of $J(\mathbf{u}) = \sum_{j=1}^{n_e}\sum_{i=1}^{n_e} off\left\{\mathbf{u}^H Q_{ij}\mathbf{u}\right\}$ where:

$$off(\mathbf{M}) = \sum_{i \neq j} \left|m_{ij}\right|^2 \tag{5.16}$$

EEG signals are however nonstationary. Nonstationarity of the signals has been exploited in developing an effective BSS algorithm based on SOS called SOBI (second-order blind identification) [102]. In this algorithm separation is performed at a number of discrete time lags simultaneously. At each lag the algorithm unitarily diagonalizes the whitened data covariance matrix. It also mitigates the effect of noise on the observation by using a whitening matrix calculation, which can improve robustness to noise. Unitary diagonalization can be explained as follows: If **V** is a whitening matrix and **X** is the observation matrix, the covariance matrix of the whitened observation is $\mathbf{C_X} = \mathrm{E}[\mathbf{VXX}^H\mathbf{V}^H] = \mathbf{VR_xV}^H = \mathbf{VHR_sH}^H\mathbf{V}^H = \mathbf{I}$, where $\mathbf{R}_x$ and $\mathbf{R}_s$ denote respectively the covariance matrices of the observed data and the original sources. It is assumed that $\mathbf{R}_s = \mathbf{I}$, i.e. the sources have unit variance and they are uncorrelated, so **VH** is a unitary matrix. Therefore, **H** can be factored as $\mathbf{H} = \mathbf{V}^{-1}\mathbf{U}$, where $\mathbf{U} = \mathbf{VH}$. The joint approximate diagonalization for a number of time lags can be obtained efficiently using a generalization of the Jacobi technique for the exact diagonalization of a single Hermitian matrix. The SOBI algorithm is implemented through thc following steps as given in [102]:

1) The sample covariance matrix $\hat{\mathbf{R}}(0)$ is estimated from T data samples. The m largest eigenvalues and their corresponding eigenvectors of $\hat{\mathbf{R}}(0)$ are denoted as $\lambda_1, \lambda_2, ..., \lambda_m$ and $\mathbf{h}_1, \mathbf{h}_2, ..., \mathbf{h}_m$ respectively.
2) Under the white noise assumption, an estimate $\hat{\sigma}^2$ of the noise variance is the average of the n_e-m smallest eigenvalues of $\hat{\mathbf{R}}(0)$. The whitened signals are $\mathbf{z}(n) = [z_1(n), z_2(n), ..., z_{ne}(n)]^T$, computed by $z_i(n) = \left(\lambda_i - \hat{\sigma}^2\right)^{-\frac{1}{2}}\mathbf{h}_i^H\mathbf{x}(n)$ for $1 \leq i \leq n_e$. This is equivalent to forming a whitening matrix as $\hat{W} = \left[\left(\lambda_1 - \hat{\sigma}^2\right)^{-\frac{1}{2}}\mathbf{h}_1, \cdots, \left(\lambda_{n_e} - \hat{\sigma}^2\right)^{-\frac{1}{2}}\mathbf{h}_{n_e}\right]^H$.
3) Form sample estimates $\hat{\mathbf{R}}(\tau)$ by computing the sample covariance matrices of $\mathbf{z}(t)$ for a fixed set of time lags $\bar{\tau} \in \{\tau_j \,|\, j = 1, ..., K\}$.

4) A unitary matrix $\hat{\mathbf{U}}$ is then obtained as a joint diagonalizer of the set $\left\{\hat{R}(\tau_j) \mid j = 1, \dots, K\right\}$.
5) The source signals are estimated as $\hat{\mathbf{s}}(t) = \hat{\mathbf{U}}^H \hat{\mathbf{W}}\mathbf{x}(t)$ or the mixing matrix $\mathbf{A}$ is estimated as $\hat{\mathbf{A}} = \hat{\mathbf{W}}^{\#} \hat{\mathbf{U}}$, where superscript $^{\#}$ denotes the Moore–Penrose pseudo-inverse.

The Fast Intelligent Crawling Algorithm (FICA) [71] is another very popular BSS technique, which extracts the signals one by one based on their kurtosis. In fact, the algorithm uses an independence criterion that exploits non-Gaussianity of the estimated sources. In some places where the objective is to remove the spiky artefacts, such as the removal of the fMRI artefact from the simultaneous EEG-fMRI recordings, application of an iterative fICA followed by deflation of the artefact component gives excellent results [83]. A typical signal of this type is given in Figure 5.3.

Practically, fICA maximizes the negentropy, which represents the distance between a distribution and a Gaussian distribution having the same mean and variance, i.e.:

$$Neg(y) \propto \{E[f(y)] - E[f(y_{Gaussian})]\}^2 \tag{5.17}$$

where, *f* is a score function [3] and *Neg* stands for negentropy. This, as mentioned previously is equivalent to maximizing the kurtosis. Therefore, the cost function can be simply defined as:

$$J(\mathbf{w}) = -\frac{1}{4}|k_4(\mathbf{y})| = -\frac{\beta}{4}k_4(\mathbf{y}) \tag{5.18}$$

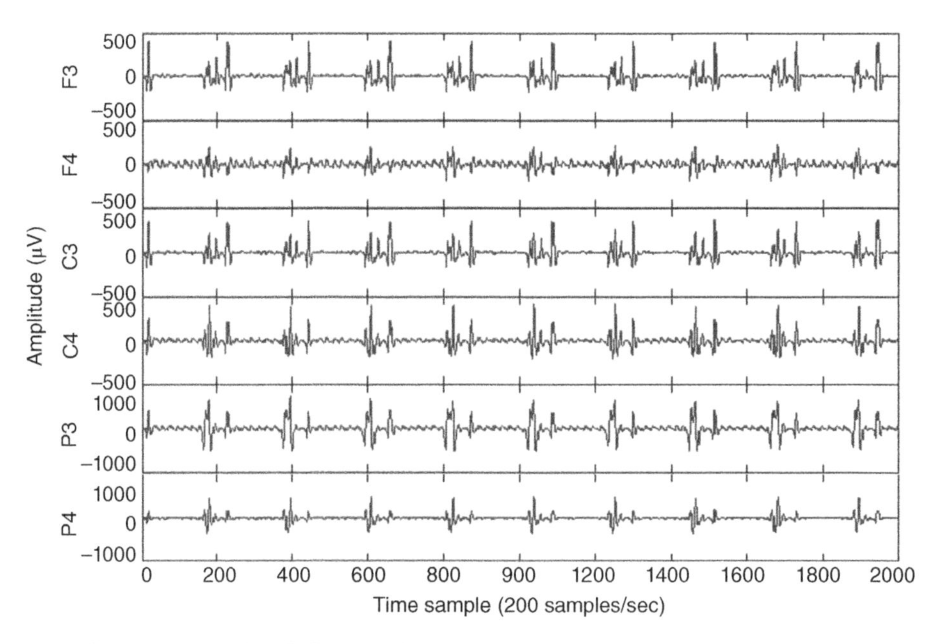

Figure 5.3 A sample of an EEG signal simultaneously recorded with fMRI.

where $k_4(\mathbf{y})$ is the kurtosis, β is the sign of the kurtosis. Applying the standard gradient decent approach to minimize the cost function one can obtain:

$$\mathbf{w}(n+1) = \mathbf{w}(n) - \mu \frac{\partial J(\mathbf{w})}{\partial \mathbf{w}}\bigg|_{\mathbf{w}=\mathbf{w}(n)} \tag{5.19}$$

where

$$-\mu \frac{\partial J(\mathbf{w})}{\partial \mathbf{w}}\bigg|_{\mathbf{w}=\mathbf{w}(n)} = \mu(n)\phi(\mathbf{y}(n))\mathbf{x}(n) \tag{5.20}$$

Here $\mu(n)$ is a learning rate:

$$\phi(y_i) = \beta \frac{\hat{m}_4(y_i)}{\hat{m}_2^3(y_i)} \left[\frac{\hat{m}_2(y_i)}{\hat{m}_4(y_i)} y_i^{\,3} - y_i \right] \tag{5.21}$$

and $\hat{m}_q(y_i) = E[y_i^{\,q}(n)]$, which is an estimate of the qth -order moment of the actual sources. Since fICA extracts the sources one by one a deflation process is followed to exclude the extracted source from the mixtures. The process reconstructs the mixtures iteratively by:

$$\mathbf{x}_{j+1} = \mathbf{x}_j - \widetilde{\mathbf{w}}_j \mathbf{y}_j, \quad j = 1, 2, \cdots \tag{5.22}$$

where $\widetilde{\boldsymbol{w}}_j$ is estimated by minimization of the following cost function:

$$J(\widetilde{\mathbf{w}}_j) = \frac{1}{2} E\left[\sum\nolimits_{p=1}^{n_r} \mathbf{x}_{j+1,p}^2\right] \tag{5.23}$$

where n_r is the number of remaining mixtures.

Figure 5.4 shows the results after application of fICA to remove the scanner artefact from the EEGs.

In a TF approach which assume the sources are approximately cyclostationary and nonstationary, the auto-terms and cross-terms of the covariance matrix of the mixtures are

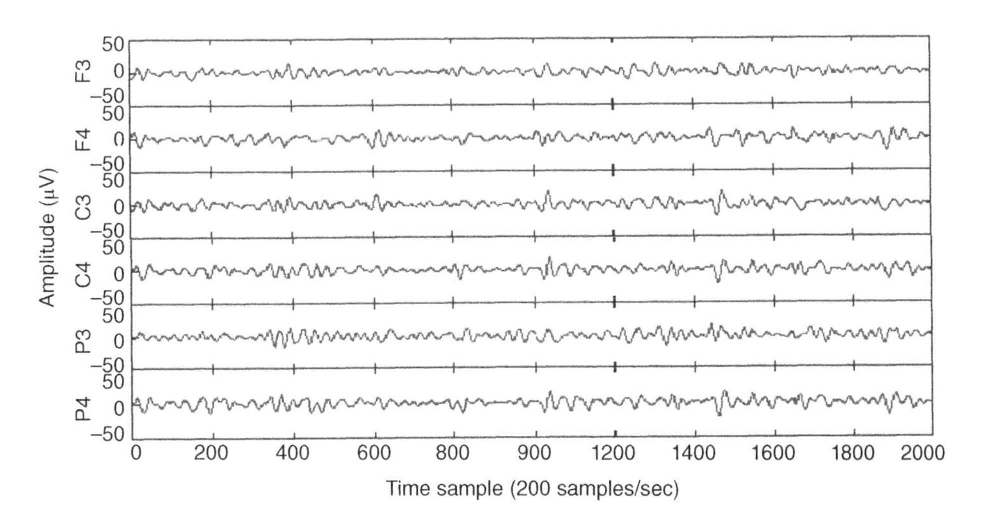

Figure 5.4 The EEG signals after removal of the scanner artefact.

first separated and BSS is applied to both terms [103, 105]. In this approach, the spatial time–frequency distribution (STFD) of the mixed signals is defined as

$$\mathbf{D}_{xx}(n,\omega) = \frac{1}{2\pi}\sum_{u=\frac{\tau}{2}}^{N-\frac{\tau}{2}}\sum_{\tau=0}^{\frac{N}{2}-1}\varphi(n-u,\tau)e^{-i\omega\tau}E\left[\mathbf{x}\left(u+\frac{\tau}{2}\right)\mathbf{x}\left(u-\frac{\tau}{2}\right)\right] \tag{5.24}$$

where $\varphi(.)$ is the discretized kernel function defining a distribution from Cohen's class of TF distributions [4] and $\mathbf{x}(.)$ is an N sample observation of the signals, which is normally contaminated by noise. Assuming $\mathbf{x}(t) = \mathbf{A}\mathbf{s}(t) + \mathbf{v}(t)$, using the above equation we find

$$\mathbf{D}_{xx}(n,\omega) = \mathbf{A}\mathbf{D}_{ss}(n,\omega)\mathbf{A}^H + \sigma^2\mathbf{I} \tag{5.25}$$

where $\mathbf{D}_{ss}(\bullet,\bullet)$ is the STFD of the source signals, σ^2 is the noise variance and depends on both noise power and the kernel function. From this equation it is clear that both $\mathbf{D}_{xx}$ and $\mathbf{D}_{ss}$ exhibit the same eigen-structure. The covariance matrix of the source signals is then replaced by the source STFD matrix composed of auto-source and cross-source time–frequency distributions (TFDs) respectively, on the diagonal and off-diagonal entries.

Defining a whitening matrix $\mathbf{W}$ such that $\mathbf{U} = \mathbf{WA}$ is unitary, a whitened and noise compensated STFD matrix is defined as:

$$\begin{aligned}\widetilde{\mathbf{D}}_{xx}(n,\omega) &= \mathbf{W}\left(\mathbf{D}_{xx}(n,\omega) - \sigma^2\mathbf{I}\right)\mathbf{W}^H \\ &= \mathbf{U}\mathbf{D}_{ss}(n,\omega)\mathbf{U}^H\end{aligned} \tag{5.26}$$

$\mathbf{W}$ and σ^2 can be estimated from the sample covariance matrix and $\mathbf{D}_{xx}$ is estimated based on the discrete-time formulation of the TFDs. From Eq. (5.24) it is known that the mixture STFD matrix exhibits the same eigen-structure as the data covariance matrix commonly used for cyclic data [105]. The covariance matrix of the source signals is replaced by a source STFD matrix composed of the auto- and cross-source TFDs on the diagonal and off-diagonal entries respectively. The peaks occur in mutually exclusive locations on the TF plane. The kernel function can be defined in such a way as to maximize disjointedness of the points in the TF plane. By estimation of the STFD in (5.24) at appropriate TFD points, one may recover the source signals by estimating a unitary transformation $\hat{\boldsymbol{U}}$, via optimization of a joint diagonal and off-diagonal criterion, to have:

$$\hat{\mathbf{s}}(n) = \hat{\mathbf{U}}^H\mathbf{W}\mathbf{x}(n) \text{ for } n = 1,\ldots,N-1 \tag{5.27}$$

In order to define and extract the peaks of the $\mathbf{D}_{xx}$ a suitable clustering approach has to be followed. This algorithm has potential application for estimating the EEG sources since in the most normal cases the sources are cyclic or quasi-cyclic.

5.3.3 Convolutive BSS

In many practical situations the signals reach the sensors with different time delays. The corresponding delay between source j and sensor i, in terms of number of samples, is directly proportional to the sampling frequency and conversely to the speed of sound, i.e. $\delta_{ij} \propto d_{ij} \times f_s/c$, where d_{ij}, f_s, and c are respectively, the distance between source j and sensor i, the sampling frequency, and the speed of sound. For speech and music in the air as an example we may have d_{ij} in terms of metres, f_s between 8 and 44 KHz, and c=330 m/s. Also, in an acoustic environment the sound signals can reach the sensors through multipaths after reflections by obstacles (such as walls). The above cases have been addressed respectively, as anechoic and echoic BSS models and formulated at the beginning of this section. The solution to echoic cases is obviously more difficult and it normally involves some approximations to the actual system. As an example, in the previously mentioned Cocktail Party Problem the source signals propagate through a dynamic medium with many parasitic effects such as multiple echoes and reverberation. So, the received signals are a weighted sum of mixed and delayed components. In other words, the received signals at each microphone are the convolutive mixtures of speech signals.

In the case of spatial ICA, convolutive BSS should be applied where the source locations change within a segment. A good example is separation of video signals. Spatial ICA can also be applied to EEGs in order to separate the movement-related cortical sources when distributed sources are considered.

Unfortunately, most of the proposed BSS approaches to instantaneous mixtures fail or are limited in separation of convolutive mixtures mainly because either there are delays involved for propagation of the sources or a number of copies of the sources, delayed and weighted differently, reach to the sensors.

5.3.3.1 General Applications

Convolutive BSS has been a focus of research in the acoustic signal processing community. Two major approaches have been followed for both anechoic and echoic cases; the first approach is to solve the problem in the time domain. In such methods in order to have accurate results both the weights of the unmixing matrix and the delays have to be estimated. However, in the second approach, the problem is transformed into the frequency domain as $\boldsymbol{h}(n)*\boldsymbol{s}(n) \xrightarrow{F} \boldsymbol{H}(\omega)\cdot\boldsymbol{S}(\omega)$ and instantaneous BSS applied to each frequency bin mixed signal. The separated signals at different frequency bins are then combined and transformed to the time domain to reconstruct the estimated sources. The short-term Fourier transform is often used for this purpose. However, the inherent permutation problem of BSS severely deteriorates the results since the order of the separated sources in different frequency bins can vary.

An early work in convolutive BSS by Platt and Faggin [5] who applied the adaptive noise cancellation network to the BSS model of Herault and Jutten [6], which has delays in the feedback path was based on the minimum output power principle. This scheme exploits the fact that the signal corrupted by noise has more power than the clean signal. The feedback path cancels out the interferences as the result of delayed versions of the other sources. This circuit was also used later to extend the Infomax BSS to convolutive cases [7]. The combined network maximizes the entropy at the output of the network with respect to the weights and delays. Torkkola [8] extended this algorithm to the echoic cases too. In order to achieve a reasonable convergence, some prior knowledge of the recording situation is necessary.

In another work an extension of SOBI has been used for anechoic BSS [9]; the problem has been transformed to the frequency domain and joint diagonalization of spectral matrices has been utilized to estimate the mixing coefficients as well as the delays [10]. In attempts by Parra et al. [11], Ikram and Morgan [12], and Cherkani and Deville [13] SOS have been used to ensure that the estimated sources, $\mathbf{Y}(\omega,m)$, are uncorrelated at each frequency bin. $\mathbf{W}(\omega)$ is estimated in such a way that it diagonalizes the covariance matrices $\mathbf{R}_Y(\omega, k)$ simultaneously for all time blocks k; $k = 0, 1, \ldots, K\text{-}1$, i.e.:

$$\begin{aligned}\mathbf{R}_Y(\omega,k) &= \mathbf{W}(\omega)\mathbf{R}_X(\omega,k)\mathbf{W}^H(\omega)\\ &= \mathbf{W}(\omega)\mathbf{H}(\omega)\boldsymbol{\Lambda}_S(\omega,k)\mathbf{H}^H(\omega)\mathbf{W}^H(\omega)\\ &= \boldsymbol{\Lambda}_c(\omega,k)\end{aligned} \tag{5.28}$$

where $\Lambda_S(\omega, k)$ is the covariance matrix of the source signals, which changes with k, $\Lambda_c(\omega, k)$ is an arbitrary diagonal matrix, and $\boldsymbol{R}_X(\omega, k)$ is the covariance matrix of $\mathbf{X}(\omega)$.

The unmixing filter $\mathbf{W}(\omega)$ for each frequency bin ω that simultaneously satisfies the K decorrelation equations can be obtained using an over-determined least-squares solution. Since the output covariance matrix $\mathbf{R}_Y(\omega, k)$ has to be diagonalized the update equation for estimation of the unmixing matrix $\mathbf{W}$ can be found by minimizing the off-diagonal elements of $\mathbf{R}_Y(\omega, k)$, which leads to:

$$\mathbf{W}_{\rho+1}(\omega) = \mathbf{W}_\rho(\omega) - \mu(\omega)\cdot\frac{\partial}{\partial\mathbf{W}_\rho^H(\omega)}\left\{\|\mathbf{V}_\rho(\omega,k)\|^2\right\} \tag{5.29}$$

where ρ is the iteration index $\|\bullet\|^2$ is the squared Frobenius norm:

$$\mu(\omega) = \frac{\alpha}{\sum_k\|\mathbf{R}_X(\omega,k)\|^2} \tag{5.30}$$

and

$$\mathbf{V}(\omega,k) = \mathbf{W}(\omega)\mathbf{R}_X(\omega,k)\mathbf{W}^H(\omega) - diag\left[\mathbf{W}(\omega)\boldsymbol{R}_X(\omega,k)\mathbf{W}^H(\omega)\right] \tag{5.31}$$

where α is a constant which is set empirically.

In these methods a number of solutions for mitigating the permutation ambiguity have been suggested. Smaragdis [14] reformulated the Infomax algorithm for the complex domain and used it to solve BSS in the frequency domain. Murata et al. [15] also formulated the problem of BSS in each frequency bin using a simultaneous diagonalization method similar to the SOBI method. To mitigate the permutation problem a method based on the temporal structure of signals, which exploits the nonstationarity of speech was introduced. The method exploits the correlations between the frequency bins of the spectrum of the signals. Detecting and exploiting the silence periods to either localize the speakers individually or to detect the source statistics such as pitch period for each speaker, are other options to alleviate the permutation problem.

5.3.3.2 Application of Convolutive BSS to EEG

For the EEG mixing model the f_s is normally low (since the bandwidth <100 Hz) and the propagation velocity is equivalent to that of electromagnetic waves (300 000 km s^{-1}). Therefore, the delay is almost zero and we can always consider the mixing model for temporal BSS as instantaneous. For EEGs the signal sources are considered stationary (fixed in position). This important assumption which in many cases, such as separation of normal brain rhythms of eye-blink artefact, is acceptable, enables effective application of linear BSS. However, in some other cases such as prolonged physical movements, the sources are considered as the current source, i.e. the location can change. To extract such sources the linear BSS solution is no longer valid. Spatial BSS of EEGs, where the mixed signals are in sample per channel (i.e. alternating samples in the electrode space), therefore, this can be considered convolutive. Spatial shift, analogous to time delay, is mainly due to source movement within one spatial signal segment. This idea has been developed in [100] in a spatiotemporal analysis of EEG using maximum likelihood convolutive ICA.

The main drawbacks for the application of BSS to separation of EEG signals is due to the:

- unknown physiological and measurement noise statistics
- unknown and varying number of sources
- nonstationarity of the sources mainly due to evoked and movement-related brain responses
- existence of distributed moving sources related to synaptic currents.

Although many attempts have been made to solve the above problems more efforts are required to provide robust solutions for different applications.

5.4 Sparse Component Analysis

Solution to underdetermined system problems where the number of sources are more than the number of mixtures, is often very challenging and cannot be carried out using the conventional BSS systems. This means in such cases more a priori information is required to be integrated within the decomposition algorithm. In nature, there are many events and sources with sparse nature. Some examples include interictal discharges within the hippocampus, QRS signals within ECGs, eye blinks within EEG, and mutations within genomic sequence. In places where the sources are sparse, i.e. at each time instant only one of the sources has significant non-zero value, the columns of the mixing matrix may be calculated individually, which makes the solution to the underdetermined case possible. The problem can be stated as a clustering problem since the lines in the scatter plot can be separated based on their directionalities by means of clustering [16, 17]. The same idea has been followed more comprehensively by Li et al. [18]. In his method, however, the separation has been performed in two different stages. First, the unknown mixing matrix is estimated using the k-means clustering method. Then, the source matrix is estimated using a standard linear programming algorithm. The line orientation of a data set may be thought of as the direction of its greatest variance. One way is to perform eigenvector decomposition on the covariance matrix of the data, the resultant principal eigenvector, i.e. the eigenvector with the largest eigenvalue, indicates the direction of the data. There are many cases for which the sources are disjoint in other domains rather than the time domain. In these cases the SCA can be performed in those domains more efficiently. One such approach called degenerate unmixing estimation technique (DUET) [19] transforms the anechoic convolutive observations into the time–frequency domain using a short-time Fourier transform and the relative attenuation and delay values between the two observations are calculated from the ratio of corresponding TF points. The regions of significant amplitudes (atoms) are then considered to be the source components in the TF domain.

For instantaneous cases, in separation of sparse sources the common approach used by most researchers is to attempt to maximize the sparsity of the extracted signals in the output of the separator. The columns of the mixing matrix A assign each observed data point to only one source based on some measure of proximity to those columns [20], i.e. at each instant only one source is considered active. Therefore, the mixing system can be presented as:

$$x_i(n) = \sum_{j=1}^{M} a_{ji} s_j(n) \quad i = 1, \cdots, N \tag{5.32}$$

where in an ideal case $a_{ji} = 0$ for $i \neq j$. Minimization of the l_0-norm (i.e. the number of non-zero samples), which in most of the cases is approximated by the l_1-norm (sum of the absolute values of the non-zero samples), is one of the most logical methods for estimation of the sources. l_1-norm minimization is a piecewise linear operation that partially assigns the energy of $\mathbf{x}(n)$ to the M columns of $\mathbf{A}$ that form a cone around $\mathbf{x}(n)$ in $\Re^M$ space. The remaining N-M columns are assigned zero coefficients. Therefore, the l_1-norm minimization can be manifested as:

$$\min_{\mathbf{s}} \|\mathbf{s}(n)\|_1 \text{ subject to } \mathbf{A}\mathbf{s}(n) = \mathbf{x}(n) \tag{5.33}$$

A detailed discussion of signal recovery using l_1-norm minimization is presented by Takigawa et al. [21].

A number of methods such as basis pursuit, smoothed l_0 [161], focal underdetermined system solver (FOCUSS), and orthogonal matching pursuit (OMP) are used to approximate the solution to $\min_{\mathbf{s}} \|\mathbf{s}(n)\|_0$ **subject to** $\mathbf{A}\mathbf{s}(n) = \mathbf{x}(n)$. In the following subsection a brief explanation of the most popular methods is provided.

5.4.1 Standard Algorithms for Sparse Source Recovery

Among the popular approaches are family of the greedy methods which focus on the support, and then, the non-zero values of **s** are obtained using a simple least-squares method. The greedy algorithms are mostly discrete due to the discrete nature of the support. In contrast, a different family of the algorithm, namely the relaxation family, attempts to approximate the sparse signal with a continuous one, and adopts a continuous optimization procedure.

5.4.1.1 Greedy-Based Solution

Mallat and Zhang [159] introduced a greedy method for sparse decomposition called matching pursuit (MP). They showed that when the dictionary (mixing matrix) is orthogonal and the observed signal is composed of $k << n$ atoms, the algorithm recovers the sources exactly after n steps [163]. In fact, MP is based on a suboptimal forward search through the mixing matrix. Succeeding greedy algorithms such as OMP [164] and compressive sampling matching pursuit (CoSaMP) [165] is based on the MP concept. The OMP method chooses the column of maximal correlation with residue, in fact, it is also the one having the steepest decline in the residue, which implies that OMP is greedy.

Recalling $\mathbf{x} = \mathbf{A}\mathbf{s}$, where $\mathbf{s}$ is a k-sparse signal. Since the number of non-zero coefficients of a k-sparse signal is equal to k, if the locations of these coefficients are determined, the non-zero values could be found by inverting or pseudo-inverting the corresponding $m \times k$ sub-matrices in **A**. This is the most complicated part of the algorithm. In the OMP algorithm, the non-zero places are found one by one. First, the highest degree of belonging (relating) a row of **x** to a column of **A** provides the most likelihood of that column of **A** is the support. To find the other members of the support, the recovered signal (and the related column) is deflated. This process is repeated until the other members of the support are estimated.

5.4.1.2 Relaxation-Based Solution

The relaxation methods smooth the l_0-norm by approximating that with other functions or norms.

Given $\rho_\alpha(s) \approx \|s\|_0$, then, as examples, $\rho_\alpha(s)$ can be $\rho_\alpha(s) = 1 - \exp\left(-\frac{s^2}{\alpha}\right)$, $\rho_\alpha(s) = \frac{s^2}{s^2+\alpha}$ or $\rho_\alpha(s) = |s|^\alpha$. By α approaching zero, these functions get closer to the l_0-norm. It is worth mentioning that, these functions are applied to the source vector elements and the final norm is calculated according to the following equation (Eq. 5.34):

$$f(\mathbf{s}) = \sum_{q=1}^{n} \rho_\alpha(s_q) \tag{5.34}$$

where s_q and $f(\mathbf{s})$ refer to the qth element and smoothed norm of **s**, respectively. A popular algorithm of this family is the FOCUSS [171]. FOCUSS employs an optimization method called iterative recursive least squares (IRLS). Based on this concept the smoothed l_0-norm ($\rho_\alpha(s)$) is posed as a weighted l_2-norm as:

$$\sum_{q=1}^{n} \rho_\alpha(s_q) \rightarrow \sum_{q=1}^{n} \frac{\rho_\alpha(s_q)}{s_q^2} s_q^2 = \sum_{q=1}^{n} w_q s_q^2 \tag{5.35}$$

IRLS iterates between a solution of the l_2-norm and update of the weights **w**. Similar approaches have been followed by many other researchers. The l_1-norm, i.e. $\|\mathbf{s}\|_1$ is the one more commonly used in the literature. The l_1-norm is presented by an absolute value of **s**, leading to an l_1-norm penalty as presented in Eq. (5.33). This convex relaxation technique is called basis pursuit [160, 161] and is solved via linear programming.

5.4.2 k-Sparse Mixtures

Most SCA-based methods for source separation and underdetermined blind identification (UBI) problems [18, 152–156] use the information about the signal intervals which include only one dominant source active in particular instants. These methods fail when there are not enough intervals including one single dominant source to enable estimation of all columns of **A**.

Some more recent approaches have extended the SCA to the cases where more than one source, say k_a, where $k_a < N_x$, are active in an instant or interval. Their methods are able to use the information about these intervals to estimate the columns of mixing matrix **A** [22–24]. These methods are called k-SCA methods [25] and under some mild conditions (i.e. i. having enough time samples to have enough planes/subspaces to cover all columns of **A** and ii. having $k_a < N_x$) [25, 154], are able to estimate **A** and the sources using some clustering methods.

In [22] it is assumed that at each instant of time there are on average $k_a < N_x/2$ sources active (non-zero). The main idea in this paper is to convert the problem of multiple dominant sources to a series of single dominant source problems, which may be solved by the well known methods. Each column of **A** is then separately estimated by solving these single dominant problems. In this approach the number of sources has to be known. This latter problem has been addressed in [23] and a solution has been proposed. The adaptivity to changing environment and the robustness against noise and outliers (i.e. the intervals which do not satisfy sparseness conditions) have been improved by the algorithm proposed in [24]. Conversely, in this method k_a satisfies $N_x - 1 > k_a > N_x/2$. SCA has been applied for EEG source estimation for ERP detection and source localization [157, 158].

The separation performance of current underdetermined source recovery (USR) solutions, including the relaxation and greedy families, reduces with decreasing the mixing system dimension and increasing the sparsity level k. Hence, in [151] the authors consider the applications where the sources are densely sparse, i.e. the number of active sources is high and very close to the number of sensors. In their work they developed a k-SCA-based algorithm that is suitable for USR in low-dimensional mixing systems. Assuming the sources are at most $(m - 1)$ sparse where m is the number of mixtures. Their method is capable of recovering the sources after estimating the mixing matrix using a subspace detection approach.

The k-SCA algorithms for solving the underdetermined blind source separation (UBSS) problem often focus on identifying the mixing matrix through UBI such as in [162] and utilize conventional source recovery algorithms, i.e. greedy methods, such as OMP [159], relaxation methods, e.g. basis pursuit [160] or the smoothed l_0 algorithm [161].

It has been proved that [153], as long as $k \leq m - 1$, where k is the number of active sources at each column of **S**, **A** can be identified and **S** estimated. One recent approach for solving this problem is based on the above proof and the definitions of disjoint subspaces [151]. These definitions state that if there is no common mixing vector between the subspaces, then they are called disjoint subspaces and if there is f common mixing vector between the subspaces, then they are called f-nondisjoint subspaces. In that case, the number of f-nondisjoint subspaces is $n_f = C_{m-f-1}^{n-f}$ where $f = 1, 2, \ldots\ m - 1$.

Then, in order to find the closest subspace to each observed vector, we first find the set of all subsets of $\{1,\ldots, n\}$ containing $k \leq m - 1$ elements called I. I_i is the ith member of I representing the indices of k columns of **A** where $i = 1 : C_k^n$. Next, we should identify an orthogonal complement space which is orthogonal to the observed data point and k columns of **A**. Then, the observed data point lies in that subspace spanned by those k columns of **A**, i.e. the observed data point will be the linear combination of those k vectors. This requires identification of the subspaces of each data point.

For this, the columns of the mixing matrix are normalized with respect to their l_2-norms. Then, for a given data point $\mathbf{x}_t$ and all members of I, the eigenvalues and eigenvectors of the covariance matrix, are calculated; set $\mathbf{F}^i = [\mathbf{x}_t\mathbf{A}(:, \mathrm{I}_i)]$ and then calculate $E\left(\mathbf{F}^i\mathbf{F}^{i^T}\right)$, which is called the augmented mixing matrix (AMM). Then, the eigenvector corresponding to the minimum eigenvalue which is most orthogonal to $\mathbf{F}_i$ is calculated. For $k = m - 1$ the number of disjoint subspaces is 1. The smaller $k < m - 1$ becomes the higher number of disjoint subspaces become available in which a sample $\mathbf{x}_i$ can be assigned. This means

$n_k = C^{n-k}_{m-k-1} > 1$. Then, the k basis vectors that have the maximum redundancy between n_k members of I are estimated. These indices are inserted in a new vector $\mathbf{h}$. These k basis vectors span the subspace where $\mathbf{x}_t$ is assigned to. In this way the generating mixing vectors of each data point, i.e. the columns of $\mathbf{W}$ are found. Finally, each source point is recovered by pseudo-inverse minimum mean squared error (MMSE) [151]:

$$\hat{\mathbf{s}}^{t^T} = \mathbf{W}^{t^T}\left(\mathbf{W}^t\mathbf{W}^{t^T} + \sigma_N^2\mathbf{I}_m\right)^{-1}\mathbf{x}_t \tag{5.36}$$

where $t = 1, \ldots T$, the number of data points, $\mathbf{W}^t = \mathbf{A}(:, \mathbf{h}^t)$, $\hat{\mathbf{s}}^t = \hat{\mathbf{S}}(\mathbf{h}^t, t)$, $\mathbf{I}_m$ is an $m \times m$ identity matrix and σ_N^2 is the noise variance estimated iteratively. To find $\mathbf{h}_t$, for each time instant we can have $\mathbf{h}_t \leftarrow C^{n-k}_{m-k-1}$ the indices from I corresponding to the first lower eigenvalues of the AMM.

It has been shown that [151] this method outperforms the state-of-the-art USR algorithms such as basis pursuit, minimizing l_1 – norm based constrained system, smoothed l_0 [161], FOCUSS, and OMP.

5.5 Nonlinear BSS

Consider the cases where the parameters of the mixing system change because of the changes in the mixing environment or the changes in the source statistics. For example, if the images of both sides of a semi-transparent paper are photocopied the results will be two mixtures of the original sources. However, since the minimum observable grey level is black (or zero) and the maximum is white (say 1), the sum of the grey levels cannot go beyond these limits. This represents a nonlinear mixing system. As another example, think of the joint sounds heard from surface electrodes from over the skin. The mixing medium involves acoustic parameters of the body tissues. Since the tissues are not rigid, in such cases if the tissues vibrate due to the sound energy then the mixing system will be a nonlinear. The mixing and unmixing can be generally modelled respectively as:

$$\mathbf{x}(n) = f(\mathbf{A}\mathbf{s}(n) + \mathbf{n}(n)) \tag{5.37}$$

$$\mathbf{y}(n) = g(\mathbf{W}\mathbf{x}(n)) \tag{5.38}$$

where $f(\bullet)$ and $g(\bullet)$ represent respectively the nonlinearities in the mixing and unmixing processes. There have been some attempts in solving nonlinear BSS problems especially for separation of image mixtures [94, 95]. In the attempt in [95] the mixing system has been modelled as a radial basis function (RBF) neural network. The parameters of this network are then computed iteratively. However, in these methods often an assumption about the mixing model is made [101].

An interesting example of a nonlinear mixing system is combination of heart and lung sounds mixing environment. Both heart and lung vary in shape during both heart beating and inhaling and exhaling processes. In order to solve the BSS for such a system, both heart and lung acoustics should be accurately modelled. Such a model has not been revealed so far and unfortunately, none of the nonlinear methods currently give satisfactory results.

5.6 Constrained BSS

The optimization problem underlying the solution to BSS problem may be subject to fulfilment of a number of conditions. These may be based on a priori knowledge of the sources or the mixing system. Any constraint on the estimated sources or the mixing system (or unmixing system) can lead to a more accurate estimation of the sources. Statistical [1] as well as geometrical constraints [2] have been very recently used in developing new BSS algorithms. In most of the cases the constrained problem is converted to an unconstrained one by means of a regularization parameter such as a Lagrange multiplier or more generally a nonlinear penalty function as used in [1].

Incorporating nonlinear penalty functions [13] into a joint diagonalization problem not only exploits nonstationarity of the signals but also ensures fast convergence of the update equation. A general formulation for the cost function of such a system can be in the form of:

$$J(\mathbf{W}) = J_m(\mathbf{W}) + \kappa\phi(J_c(\mathbf{W})) \tag{5.39}$$

where $J_m(\mathbf{W})$ and $J_c(\mathbf{W})$ are respectively the main and the constraint cost functions, $\phi(\cdot)$ is the nonlinear penalty function, and κ is the penalty parameter.

Constrained BSS has a very high potential in incorporating clinical and physiological information into the main optimization formulation.

As a new application of constrained BSS, an effective algorithm has been developed for removing the eye-blinking artefacts from EEGs. A similar method to the joint diagonalization of correlation matrices by using gradient methods [26] has been developed [27], which exploits the temporal structure of the underlying EEG sources. The algorithm is an extension of SOBI with the aim of iteratively performing the joint diagonalization of multiple time lagged covariance matrices of the estimated sources and exploiting the statistical structure of the eye-blink signal as a constraint. The estimated source covariance matrix is given by

$$\mathbf{R_Y}(k) = \mathbf{W}\mathbf{R_X}(k)\mathbf{W}^T \tag{5.40}$$

where $\boldsymbol{R}_{\mathbf{X}}(k) = E\{\boldsymbol{x}(n)\boldsymbol{x}^T(n-k)\}$ is the covariance matrix of the electrode signal. Following the same procedure as in [28], the least-squares (LS) estimate of W is found from:

$$J_m(\mathbf{W}) = arg\min_{\mathbf{W}} \sum\nolimits_{k=1}^{T_B} \|E(k)\|_F^2 \tag{5.41}$$

where $\|.\|_F^2$ is the squared Frobenius norm and $E(k)$ is the error to be minimized between the covariances of the source signals, $\mathbf{R_s}(k)$ and the estimated sources, $\mathbf{R_Y}(k)$. The corresponding cost function has been defined based on minimizing the off-diagonal elements for each time block, that is:

$$J(\mathbf{W}) = J_m(\mathbf{W}) + \Lambda J_c(\mathbf{W}) \tag{5.42}$$

where

$$J_m(\boldsymbol{W}) = \sum\nolimits_{k=1}^{T_B} \|\boldsymbol{R}_Y(k) - diag(\boldsymbol{R}_Y(k)\|_F^2 \tag{5.43}$$

and

$$J_c(\mathbf{W}) = \xi\left(E\left\{\mathbf{g}(n)\mathbf{y}^T(n)\right\}\right) \tag{5.44}$$

is a second-order constraint term. $\xi(\bullet)$ is a nonlinear function approximating the cumulative density function (CDF) of the data and $\mathbf{\Lambda} = \{\lambda_{ij}\}$ $(i = 1, ..., N)$ is the weighted factor which is governed by the correlation (matrix) between the EOG and EEG signals ($\mathbf{R}_{GY}$), defined as $\Lambda = \kappa \text{diag}(R_{GY})$, where κ is an adjustable constant. Then a gradient approach [4] is followed to minimize the cost function. The incremental update equation is:

$$\mathbf{W}(n+1) = \mathbf{W}(n) - \mu \frac{\partial J(\mathbf{W})}{\partial \mathbf{W}} \tag{5.45}$$

which concludes the algorithm.

BSS has been widely used for processing of EEG signals some referred or presented in this book for different applications. Although, the main assumptions about the source signals such as uncorrelatedness or independency of such signals have not been verified yet, the empirical results illustrate the effectiveness of such methods. EEG signals are noisy and nonstationary and are normally affected by one or more types of internal artefacts. The most efficient approaches are those which consider all different domain statistics of the signals and take the nature of the artefacts into account. In addition, a major challenge is in how to incorporate and exploit the physiological properties of the signals and characteristics of the actual sources into the BSS algorithm. Some examples have been given here; some more will be presented in the other chapters of this book.

In the case of brain signals the independency or uncorrelatedness conditions for the sources may not be satisfied. This however may be acceptable for abnormal sources, movement-related sources, or ERP signals. Transforming the problem into a domain such as STF-domain, where the sources can be considered disjoint, may be a good solution.

5.7 Application of Constrained BSS; Example

In practise, natural signals such as EEG source signals are not always independent. A topographic ICA proposed in [29] incorporates the dependency among the nearby sources in not only grouping the independent components related to nearby sources but also separating the sources originating from different regions of the brain. In this ICA model it is proposed that the residual dependency structure of the independent components (ICs), defined as dependencies that cannot be cancelled by ICA, could be used to establish a topographic order between the components. Based on this model, if the topography is defined by a lattice or grid, the dependency of the components is a function of the distance of the components on that grid. Therefore, the generative model that implies correlation of energies for components that are close in the topographic grid is defined. The main assumption is then, the nearby sources are correlated and those far from each other are independent.

To develop such an algorithm a neighbourhood relation is initially defined as:

$$h(i,j) = \begin{cases} 1, & \text{if } |i-j| \leq m \\ 0, & \text{otherwise} \end{cases} \tag{5.46}$$

where the constant m specifies the width of the neighbourhood. Such a function is therefore, a matrix of hyperparameters. This function can be incorporated into the main cost function of BSS. The update rule is then given as [29]:

$$\mathbf{w}_i \propto E\{\mathbf{z}(\mathbf{w}_i^T\mathbf{z})\mathrm{r}_i\} \tag{5.47}$$

where $\mathbf{z}_i$ is the whitened mixed signals and

$$r_i = \sum\nolimits_{k=1}^{n} h(i,k)g\left(\sum\nolimits_{j=1}^{n} h(k,j)\left(\mathbf{w}_j^T\mathbf{z}\right)^2\right) \tag{5.48}$$

The function g is the derivative of a nonlinear function such as those defined in [29]. It is seen that the vectors $\mathbf{w}_i$ are constrained to some topographic boundary defined by $h(i, j)$. Finally, the orthogonalization and normalization can be accomplished, for example, by the classical method involving matrix square roots:

$$\mathbf{W} \leftarrow \left(\mathbf{W}\mathbf{W}^T\right)^{-\frac{1}{2}}\mathbf{W} \tag{5.49}$$

where $\mathbf{W}$ is the matrix of the vectors $\mathbf{w}_i$, i.e. $\mathbf{W} = [\mathbf{w}_1, \mathbf{w}_2, ..., \mathbf{w}_n]^T$. The original mixing matrix $\mathbf{A}$ can be computed by inverting the whitening process as $\mathbf{A} = (\mathbf{W}\mathbf{V})^{-1}$, where $\mathbf{V}$ is the whitening matrix.

This algorithm has been modified for separation of seizure signals, by (*i*) iteratively finding the best neighbourhood m, and (*ii*) constraining the desired estimated source to be within a specific frequency band originating from certain brain zones (confirmed clinically). Figure 5.5 illustrates the independent components of a set of EEG signals from an epileptic patient, using the above constrained topographic ICA method. In Figure 5.6 the corresponding topographic maps for all independent components (i.e. backprojection of each IC to the scalp using the inverse of the estimated unmixing matrix,) are shown. From these figures the sixth IC from the top clearly shows the seizure component. Consequently, the corresponding topograph shows the location of a seizure over the left temporal electrodes. The high quality of the results is due to two important facts: first, the sources including those elicited from the brain and noise are independent and second, the mixtures are linear combination of the sources. The two properties are the main requirement for a linear ICA system.

5.8 Multiway EEG Decompositions

The multichannel methods based on ICA exploit the time–space (channel) diversity in the source decomposition. In these cases, despite independency of the sources they are considered stationary (within the segments being processed).

As emphasized at the beginning of this chapter, most of physiological signals have nonstationary behaviour, the existing artefacts are nonstationary or the mixing system may be nonlinear. Conversely, the additive noise may not be Gaussian. As an example, consider the state of the brain when transferring from relax to attention, from preictal to ictal for epileptic patients, or from before to after hand movement (when motor area is engaged). In these and many other cases the nonstationarity stems from the changes in the signal trends or

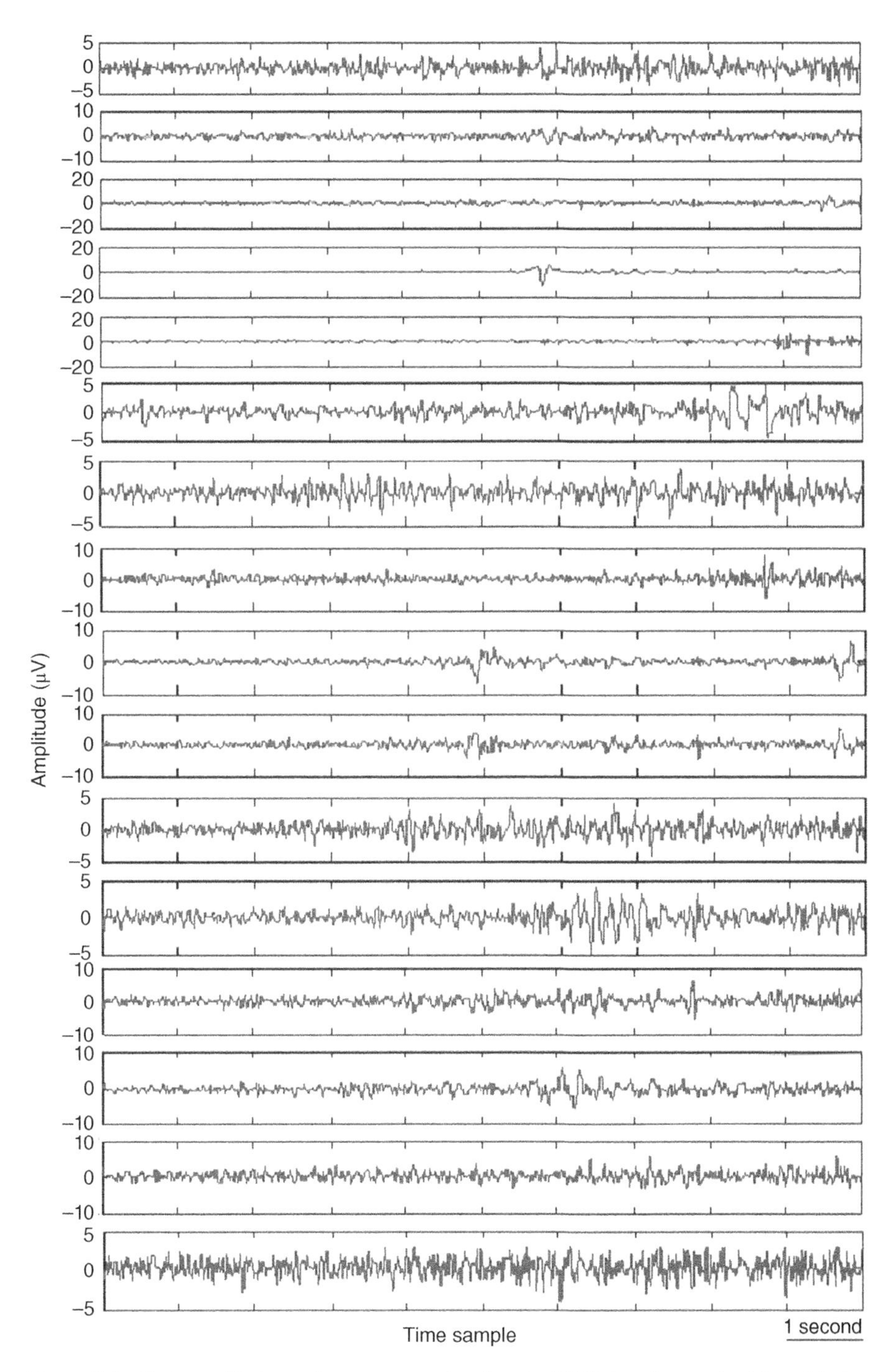

Figure 5.5 Estimated independent components of a set of EEG signals, acquired from 16 electrodes, using constrained topographic ICA.

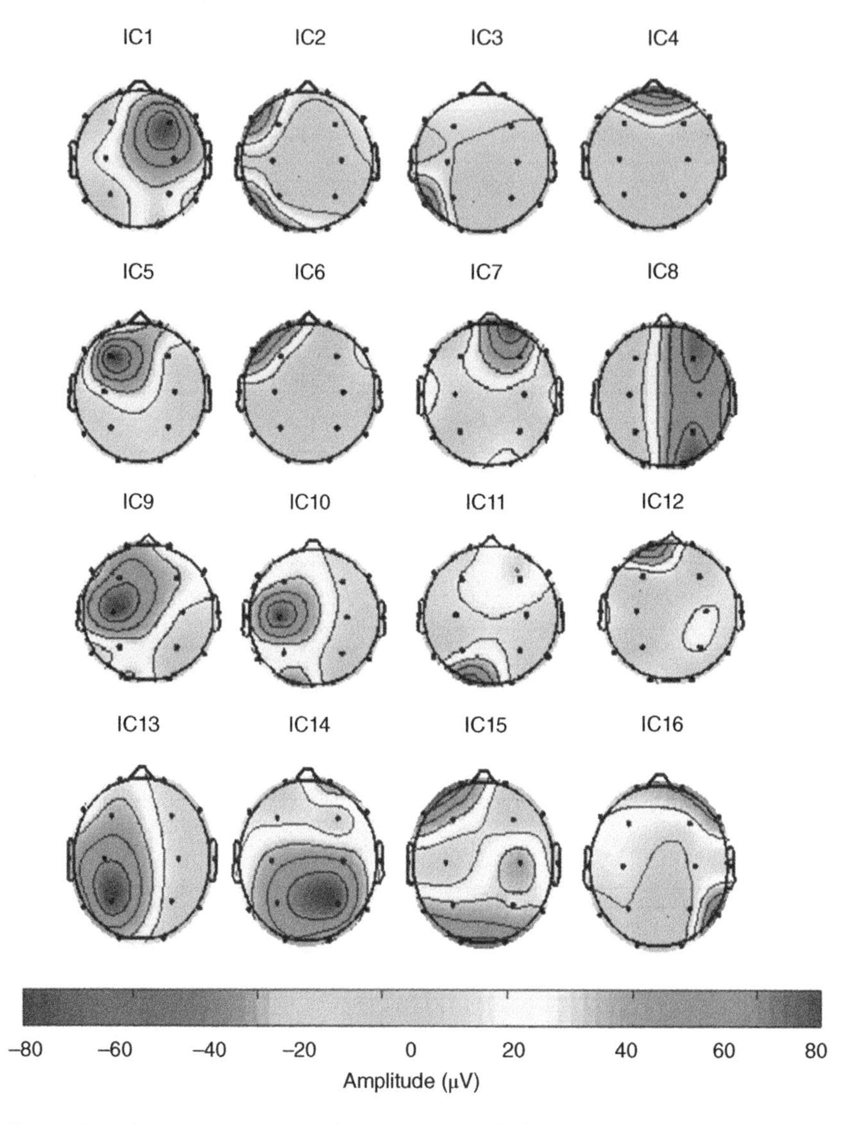

Figure 5.6 Topographic maps, each illustrating an IC. It is clear that the sources are geometrically localized.

statistics. These shortcomings have been taken into consideration and in some cases nicely exploited in development of a new class of BSS systems called nonstationary BSS. Such algorithms not only exploit the mutual statistical behaviour of the source signals but also exploit the variations in the statistical properties of those sources across smaller signal segments.

In addition, often the components of interest are not sufficiently distinct in the time–space domain. Therefore, algorithms which can extend the analysis into other domains such as frequency, segment (to exploit nonstationarity), trials and subjects become very favourable.

To accommodate this, a new mathematical representation of the signals and systems and their corresponding mixing model are needed to best exploit these variations as new dimensions for the data. The multiway approaches mainly based on tensor factorization is a solution to these problems.

Therefore, in the following subsections a number of these approaches are discussed. Before we move forward a new concept which has been recently used in source separation, channel identification, and source localization called tensor factorization, is described.

5.8.1 Tensor Factorization for BSS

With no doubt it is favourable to represent the data in a domain where the variation of such data in all aspects can be assessed and tracked. The so-called *multiway* signal processing is an effective step in this direction. The multiway technique incorporates the concept of tensors and tensor factorization to achieve that. Tensor decompositions were first introduced by Hitchcock in 1927 [30, 31], and the idea of a multiway model is attributed to Cattell in 1944 [32, 33]. These concepts received scant attention until the work of Tucker in the 1960s [34–36] and Carroll and Chang [37] and Harshman [38] in 1970, all of which appeared in psychometrics. Appellof and Davidson [39] are generally credited as being the first to use tensor decompositions (in 1981) in chemometrics, and tensors have since become extremely popular in that field [40].

A tensor is a multidimensional array. More formally, an N-way or Nth-order tensor is an element of the tensor product of N vector spaces, each of which has its own coordinate system. A first-order tensor is a vector, a second-order tensor is a matrix, and a third-order tensor has three indices. Tensors of order three or higher are called higher-order tensors. The order of a tensor is the number of dimensions also known as way or mode. A tensor decomposition has been widely used in signal processing such as in [41–50] and neuroscience such as in [51–60, 96]. Figure 5.7 illustrates the concept.

Decomposition of higher-order tensors has applications in chemometrics, signal processing, communications, linear algebra, computer vision, data mining, neuroscience, and graph analysis. For detection, separation, localization, noise removal, and tracking. Traditionally, two particular tensor decompositions can be considered to be higher-order extensions of the matrix SVD, namely Canonical decomposition (CANDECOMP)/PARAFAC (CP) tensor decomposition.

From a mathematical viewpoint, there are two general tensor factorization approaches. The first one is the Tucker model which tries to factorize a three-way tensor to a smaller tensor (core tensor) and three other factors/matrices. This model is not unique since there are various ways the core tensor can be organized. The second one is the PARAFAC tensor model with a superdiagonal core tensor. Since there is no other variation of such a core tensor the PARAFAC factorization is unique which means for orthogonal or independent sources, the mixtures can be decomposed into their constituent sources without any ambiguity. Figure 5.8 represents the above models. Also, PARAFAC decomposes a tensor into the sum of rank one tensors and the Tucker decomposition is a higher-order form of principal component analysis (PCA).

There are many other tensor decompositions, including INDSCAL, PARAFAC2, CANDELINC, DEDICOM, and PARATUCK2 as well as nonnegative variants of all of

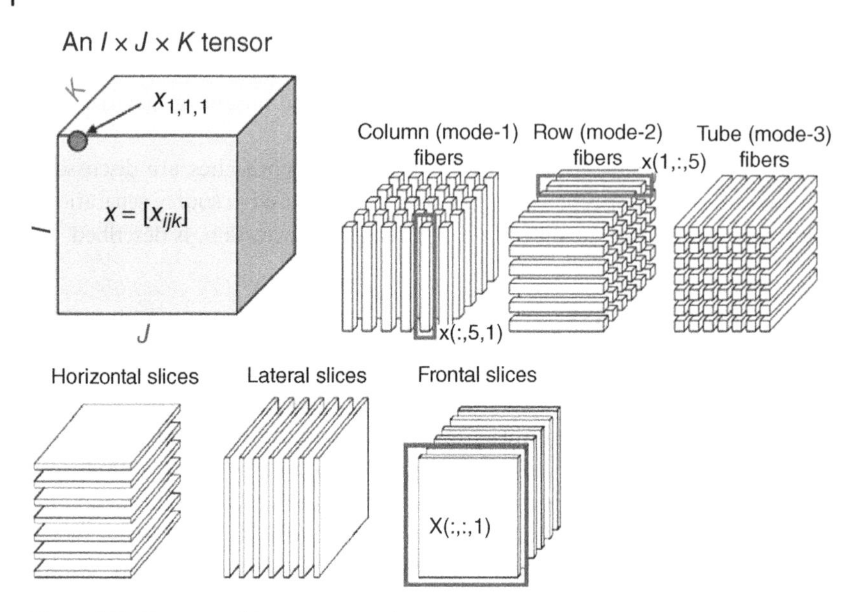

Figure 5.7 Tensor and its various modes.

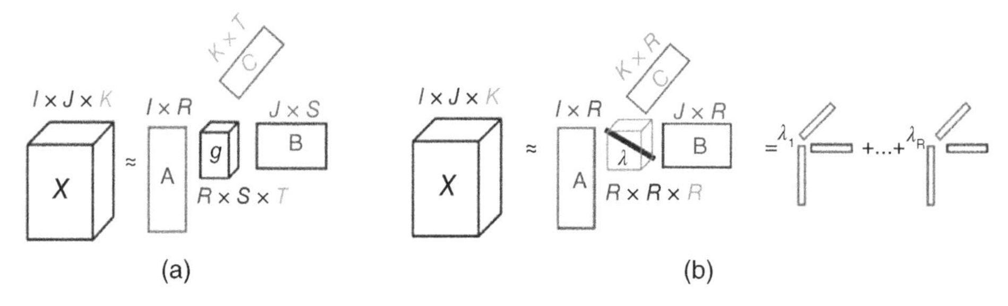

Figure 5.8 Tensor factorization using: (a) Tucker and (b) PARAFAC models.

them [61]. The N-way Toolbox, Tensor Toolbox, and Multilinear Engine are examples of software packages for working with tensors.

Recently, tensor factorization has been widely used for BSS of both instantaneous and convolutive mixtures. To be able to benefit from tensor modelling, we need to build up a tensor from our two-dimensional multichannel measurements. As an example, a time–space-frequency tensor can be built by initially transforming the data into the TF domain and then incorporating the space (geometrical locations of the sensors) dimension. For analysis of EEG or magnetoencephalography (MEG) this can be illustrated as in Figure 5.9.

In general, any other dimension in which the data variation is significant and desired can be selected. For example, to exploit the nonstationarity, one of the dimensions can be represented by the signal segment number. This allows the changes in statistics of the data to be tracked effectively within the tensor model.

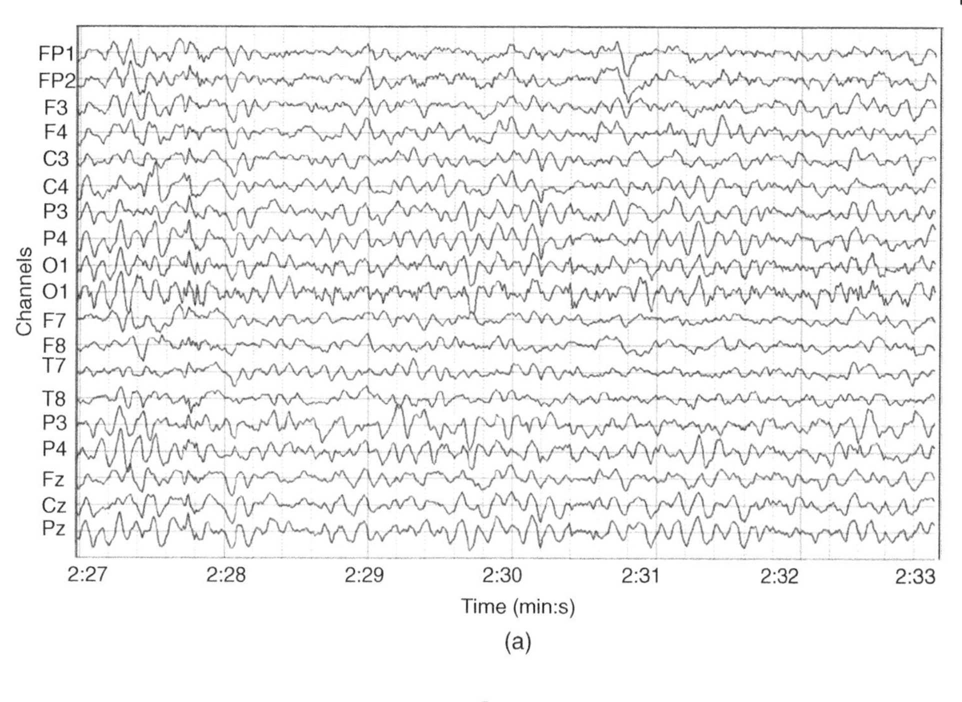

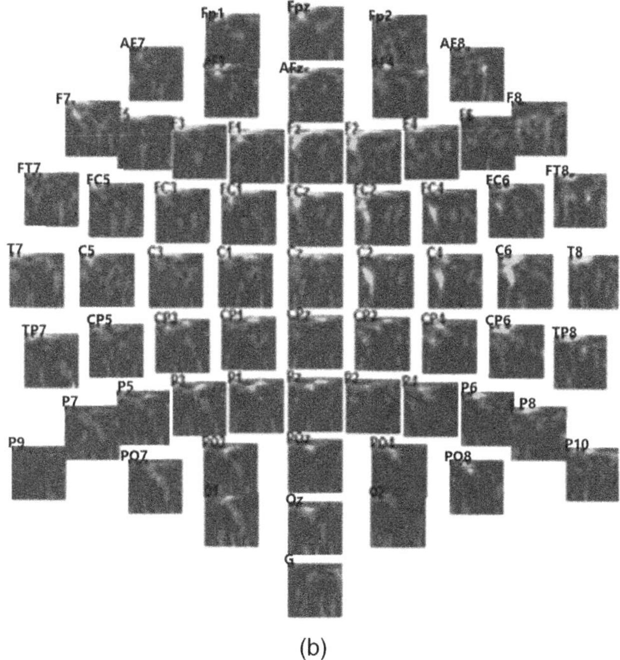

Figure 5.9 A tensor representation of a set of multichannel EEG signals. (a) Time-domain representation. (b) Time–frequency representation of each electrode signal. (c) The constructed tensor.

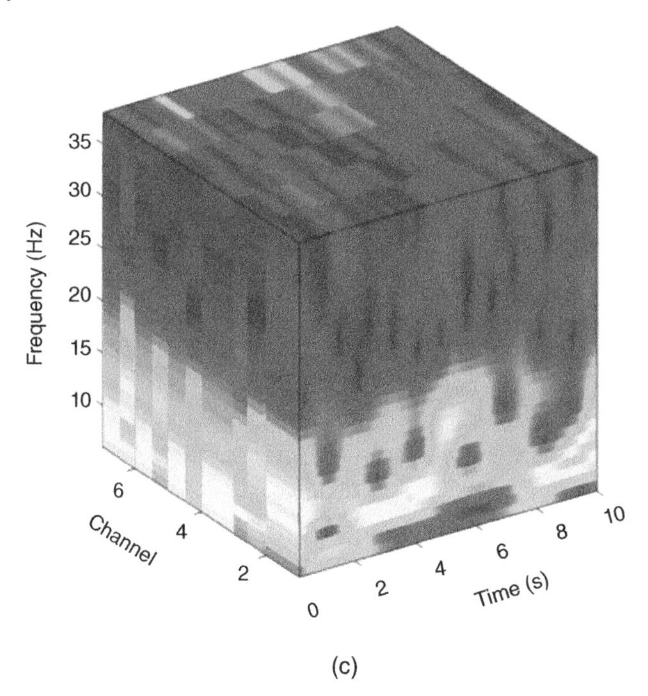

Figure 5.9 (Continued)

In [60] an approach for the removal of eye-blink artefacts from EEG signals based on a novel STF model of EEGs and the robust minimum variance beamformer (RMVB) has been proposed. In this method, in order to remove the artefact, the RMVB has been provided with a priori information, namely, an estimation of the steering vector corresponding to the point source of the eye blink. The artefact-removed EEGs are subsequently reconstructed by deflation. The a priori knowledge, the vector corresponding to the spatial distribution of the eye-blink factor, is identified using the STF model of EEGs, provided by the PARAFAC method. In order to reduce the computational complexity present in the estimation of the STF model using the three-way PARAFAC, the time domain is sub-divided into a number of segments and a four-way array is then set to estimate the space–time–frequency-time/segment (STF-TS) model of the data using the four-way PARAFAC. The correct number of the factors of the STF model has been effectively estimated by using a novel core consistency diagnostic (CORCONDIA)-based measure.

Subsequently, the STF-TS model is shown to approximate closely the classic STF model, with significantly lower computational cost. The results demonstrate that such an algorithm effectively identifies and removes the eye-blink (EB) artefact from raw EEG measurements. Using this approach, Figure 5.10 shows the space, time, and frequency signatures of an eye blink as the result of applying PARAFAC. Figure 5.11 shows the restored EEGs after applying the STF-TS method. In this method the nonstationarity of the data is also exploited through segmentation as discussed in Section 5.8.2.

The above approach shows that PARAFAC can be combined with another modality mainly to intelligently select and extract the atom (source) of interest.

More applications of PARAFAC such as in [59] related to BCI are explained in other chapters of this book.

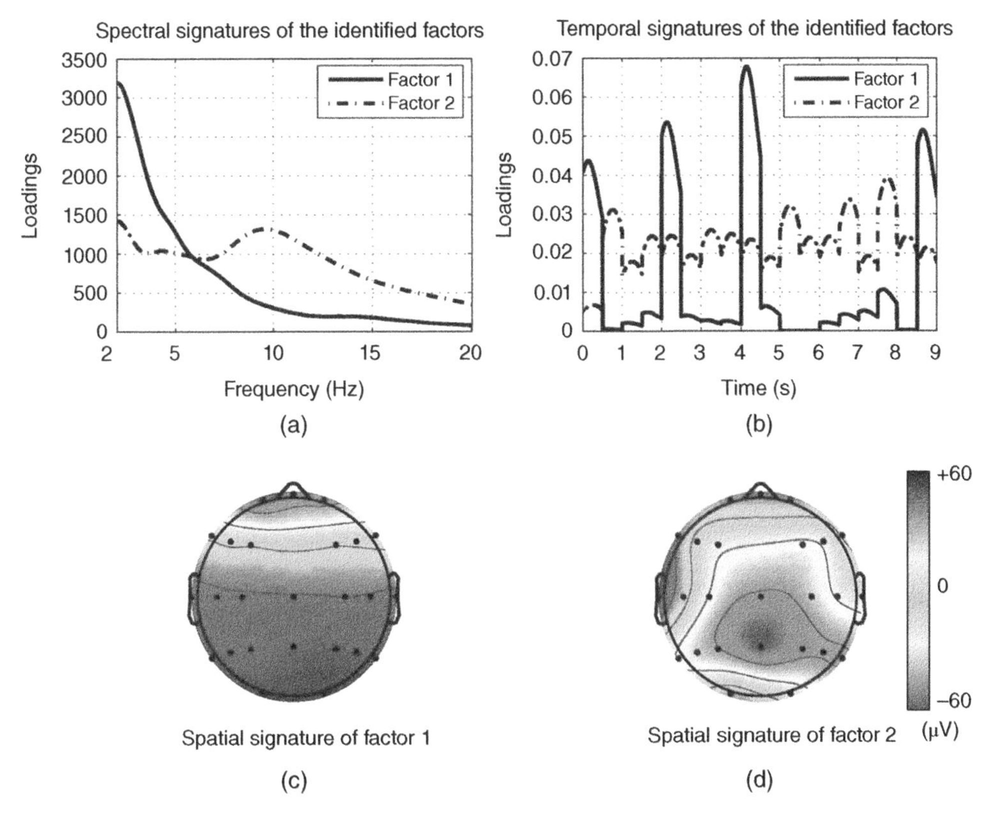

Figure 5.10 The extracted factors using STF–TS modelling. (a) and (b) illustrate respectively the spectral and temporal signatures of the extracted factors. (c) and (d) represents the spatial distributions of those extracted factors [60].

5.8.2 Solving BSS of Nonstationary Sources Using Tensor Factorization

Using a simple PARAFAC model may provide some insightful information about various components in the signal mixtures. As an example, Figure 5.12 demonstrates the results of PARAFAC operation for detecting two components (atoms) of an EEG segment recording during left finger movement. The main component in Figure 5.12b is in the right-hand side (contralateral) and the second component in Figure 5.12a is in the same side (ipsilateral) of the brain motor cortex.

In this approach, in order to localize seizure signals (within the scalp through an analysis of EEG), a tensor with three modes, i.e. time samples, scales, and electrodes, has been constructed through wavelet analysis of multichannel EEG. Next, they demonstrate that PARAFAC provides promising results in modelling the complex structure of epileptic seizure, localizing a seizure origin and extracting the artefacts.

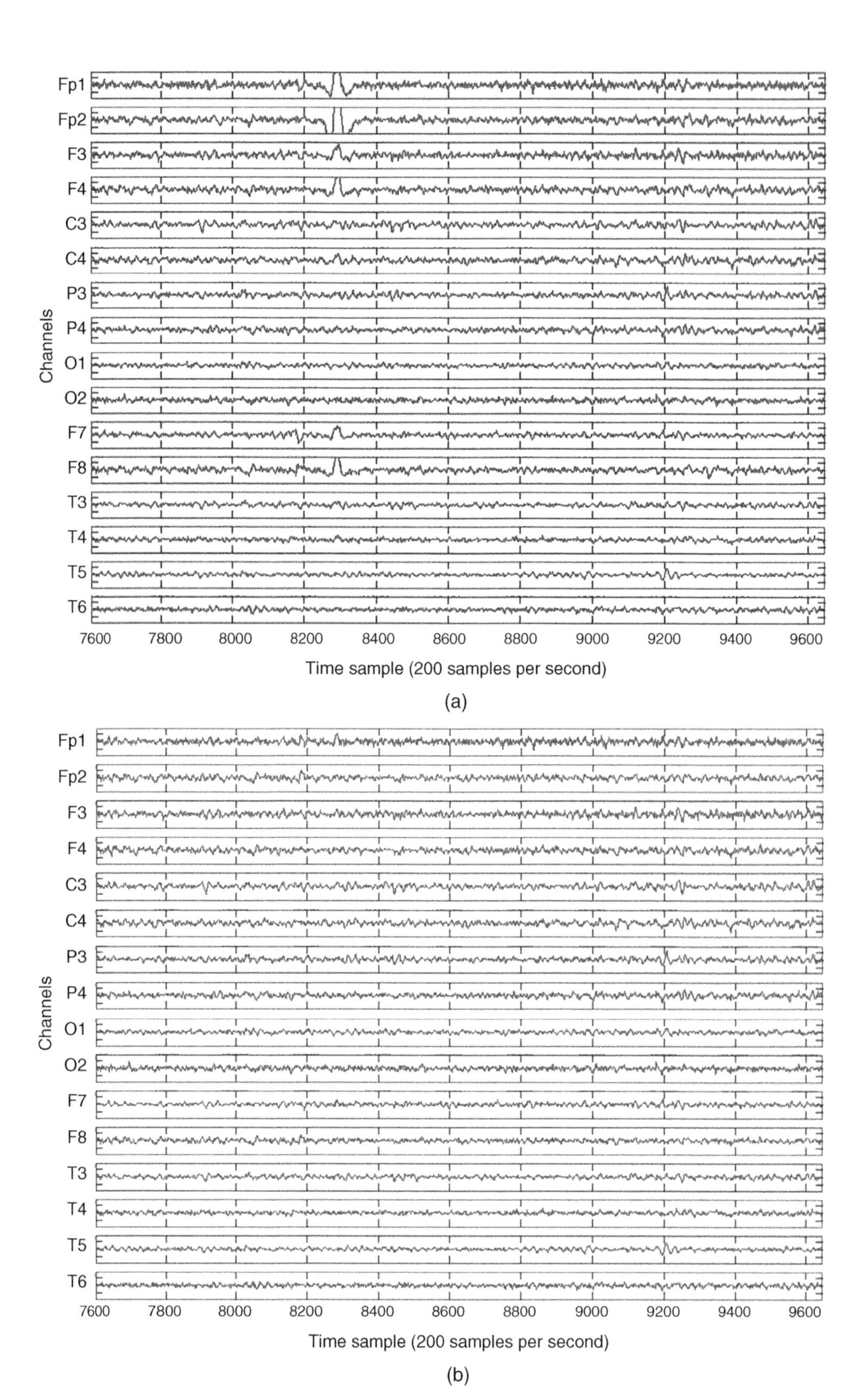

Figure 5.11 Restoration of the EEG signals in (a) from multiple eye blinks after applying the STF–TS method [60]; the results are shown in (b).

The fundamental expression of the PARAFAC model, which is used to describe decomposition of trilinear data sets, is given as:

$$x_{ijk} = \sum_{r=1}^{R} a_{ir} b_{jr} c_{kr} + e_{ijk} \tag{5.50}$$

In matrix form this can be described as:

$$\mathbf{X}_k = \mathbf{B}\mathbf{D}_k\mathbf{A}^T + \mathbf{E}_k \text{ for } k = 1, 2, \ldots, K \tag{5.51}$$

where $\mathbf{X}_k$ represents the kth frontal slice of tensor $\underline{\boldsymbol{X}}$ built from multichannel data $\mathbf{X}$, and $\mathbf{A}$ and $\mathbf{B}$ are the component matrices in the first and second modes, respectively. $\mathbf{D}_k$ is a diagonal matrix, whose diagonal elements correspond to the kth row of the third component matrix $\mathbf{C}$ denoted as $\mathbf{D}_k = diag(\mathbf{C}_{k,:})$. Finally, $\mathbf{E}_k$ contains the error terms corresponding to the entries in the kth frontal slice.

In order to estimate the factors an alternating least-squares (ALS) approach is applied. Ignoring the noise term, this leads to performing the following alternating operations until convergence:

$$\mathbf{A} = \mathbf{X}_{(1)}\left((\mathbf{C}\circ\mathbf{B})^T\right)^{\dagger} \tag{5.52}$$

$$\mathbf{B} = \mathbf{X}_{(2)}\left((\mathbf{C}\circ\mathbf{A})^T\right)^{\dagger} \tag{5.53}$$

$$\mathbf{C} = \mathbf{X}_{(3)}\left((\mathbf{B}\circ\mathbf{A})^T\right)^{\dagger} \tag{5.54}$$

where $(.)^{\dagger}$ denotes the pseudo-inverse operation, $\mathbf{X}_{(i)}$ represents the unfolded version of tensor $\mathbf{X}$ in the ith dimension and $\circ$ denotes the Khatri–Rao product.

The results of PARAFAC decomposition is unique if $k_A + k_B + k_C \geq 2R + 2$, where $k_{\mathbf{A}}$ refers to the Kruskal rank of $\mathbf{A}$. For an N rank tensor, this however has been modified to [62]:

$$\sum_{n=1}^{N} k_n \geq 2R + (N-1) \tag{5.55}$$

Although most of the properties of the sources can be evaluated using these so-called 'signatures' of the data [63] and they are useful for diagnostic applications but unfortunately, the original sources (in the time domain) cannot be recovered. Also, it is seen that the representing signatures do not vary for different time segments. Hence, in this model the signals are considered stationary. Therefore, such a model cannot be utilized for BSS. So, the main question here is that how tensor factorization can be used for source separation. Moreover, how this can be extended to separation of nonstationary data.

As an extension to PARAFAC, PARAFAC2 supports variation in one mode of the tensor. Mathematically, PARAFAC can be expressed in the following form:

$$\mathbf{X}_k = \mathbf{A}_k\mathbf{D}_k\mathbf{B}^T + \mathbf{E}_k \tag{5.56}$$

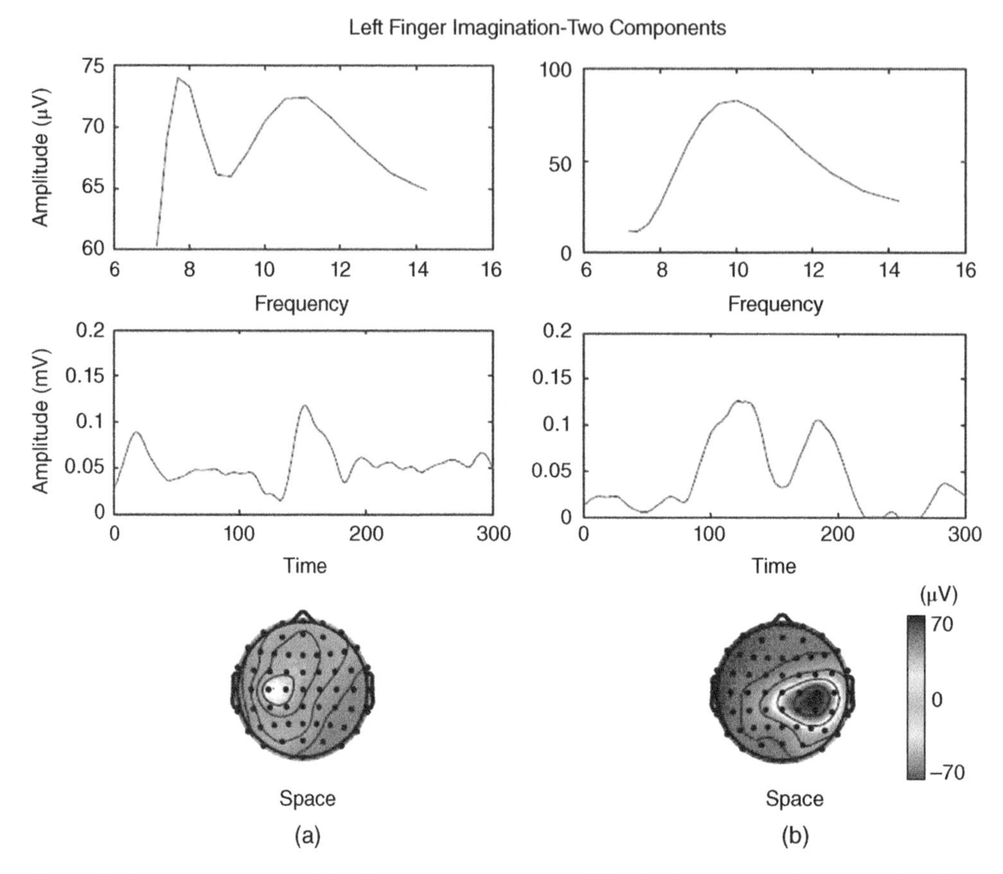

Figure 5.12 Representation of the first two components (a, b) in the time–space–frequency domain for a segment of EEG signal during left finger movement.

Compared with PARAFAC, PARAFAC2 is designed to deal with non-trilinear data sets, while keeping uniqueness of the solutions as for PARAFAC. To do so, PARAFAC2 allows some freedom in the shape of the k slabs ($\mathbf{X}_k$) in the variable mode. To maintain the uniqueness of the solutions $\mathbf{X}_k\mathbf{X}_k^T$ are forced to have the same structure for all k, i.e. $\mathbf{X}_k\mathbf{X}_k^T = \mathbf{A}\mathbf{D}_k\boldsymbol{\Phi}\mathbf{D}_k\mathbf{A}^T$. In this case the above equation is subject to $\mathbf{A}_k^T\mathbf{A}_k = \boldsymbol{\Phi}$. Consequently, $\mathbf{A}_k$ can be modelled as $\mathbf{A}_k = \mathbf{P}_k\,\mathbf{A}$, where $\mathbf{P}_k\mathbf{P}_k^T = \mathbf{I}$. Therefore, the *direct* form for PARAFAC2 can be in the following form:

$$\mathbf{X}_k = \mathbf{P}_k\mathbf{H}\mathbf{D}_k\mathbf{A}^T + \boldsymbol{E}_k \text{ subject to } \mathbf{P}_k\mathbf{P}_k^T = \mathbf{I} \tag{5.57}$$

ALS optimization is proposed to estimate the direct model parameters. Unlike for indirect PARAFAC2 [84], for the direct PARAFAC it is possible to apply a priori conditions, impose constraints, and generalize it to n-way models. The steps in estimating the factors are as follows [85]:

Considering $\mathbf{X}_k$ as an $R \times J \times K$ three-way array with frontal planes $\mathbf{P}_k^T\mathbf{X}_k$, $k = 1,..., K$.

Step 0. If $J < n_k$, where n_k is the number of columns in the kth data matrix, replace $\mathbf{X}_k$ by $\mathbf{H}_k$, e.g. from the Cholesky decomposition $\mathbf{X}_k\mathbf{X}_k^T = \mathbf{H}_k\mathbf{H}_k^T$.

Step 1. Initialize $\mathbf{B}$ as the loading matrix by applying PCA to $\sum_{k=1}^{K}\mathbf{X}_k\mathbf{X}_k^T$ and initialize $\mathbf{A}$ and $\mathbf{D}_1, \cdots, \mathbf{D}_k$ as $\mathbf{I}_R$.

Step 1a. Compute the SVD $\mathbf{A}\mathbf{D}_k\mathbf{B}^T\mathbf{X}_k^T = \mathbf{U}_k\mathbf{\Lambda}_k\mathbf{V}_k^T$ and update $\mathbf{P}_k$ as $\mathbf{U}_k\mathbf{V}_k^T$, $k = 1,\ldots,K$.

Step 1b. Update $\mathbf{A}$, $\mathbf{B}$ and $\mathbf{D}_1,\ldots, \mathbf{D}_K$ by one cycle of a PARAFAC algorithm applied to the $R \times J \times K$ three-way array with frontal planes $\mathbf{P}_k^T\mathbf{X}_k$, $k = 1,\ldots, K$.

Step 1c. Evaluate $\sigma(\mathbf{P}_1, \cdots, \mathbf{P}_k, \mathbf{A}, \mathbf{D}_1, \cdots, \mathbf{D}_k) = \sum_{k=1}^{K}\left\|\mathbf{X}_k - \mathbf{P}_k\mathbf{A}\mathbf{D}_k\mathbf{B}^T\right\|^2$. If $\sigma_{old} - \sigma_{new} > \varepsilon\sigma_{old}$ for some small value ε, repeat Step 1; else go to Step 2.

Step 2. Given that $\mathbf{X}_k$ has been replaced by $\mathbf{H}_k$ in Step 0, now replace $\mathbf{H}_k$ by $\mathbf{X}_k$ again and compute $\mathbf{P}_k$ according to Step 1a, $k = 1,\ldots, K$.

As all the factors are estimated through the above steps, the sources can be easily estimated as $\mathbf{S}_k = \mathbf{P}_k\mathbf{H}\mathbf{D}_k$.

In the case of nonstationary sources it is logical to have k as the number of segments. This allows to track and exploit the variation in signal statistics across the segments and therefore, the nonstationarities can be tolerated.

Some nonstationary BSS approaches by using tensor factorization have been developed recently. SOBI [3] was probably the first BSS approach tried to solve the problem of nonstationarity in BSS by applying diagonalization of covariance matrices in multiple lags. Although for nonstationary signal sources the performance of the SOBI algorithm is better than those of many other BSS approaches, for highly nonstationary cases not only the performance is not acceptable but also the statistical fluctuation within the source signals is ignored. The fundamental approach followed by the researchers recently involves dividing the input into small segments and converting the entire input signal matrices into tensors which can be processed using the corresponding mathematical techniques as described above. A number of approaches have been followed recently such as those in [86–88].

In the work by Hyvarinen [86] inspired by Matsukas et al.'s research [89], the nonstationarity is interpreted by smooth variation in variance and it is shown that the cross-cumulants for nonstationary sources are positive. Consequently, a method similar to the one for maximizing the kurtosis for BSS has been developed to maximize the cumulants which measure the nonstationarity of the separated source. Since the defined positive cross-cumulants do not require the marginal distribution of the source signals to be Gaussian, it can be used for nonstationary Gaussian sources too.

In another attempt [87] the algorithm utilizes the principle of maximum likelihood and minimum MI. In both cases the optimization amounts to approximate joint diagonalization of a set of L matrices. Intuitively, nonstationarity allows blind identification in the Gaussian case. This is because a single covariance matrix does not provide enough constraints to uniquely determine $\mathbf{A}$. A collection of several covariance matrices estimated over different time periods does determine $\mathbf{A}$, provided the source distributions change enough over the whole observation period. In his algorithm it is considered that the profile of changing variance of each estimated source is simply a smoothed version of the square of that source. Therefore, the variances are calculated over smaller segments of the data.

A similar concept has also been exploited in [88]. Similarly, this method can be used for both Gaussian and non-Gaussian sources. Also, in this approach it is assumed that the additive noise signals are temporally uncorrelated and mutually uncorrelated with the source signals. Unlike general BSS methods using SOS or HOS concepts, a first-order blind source separation (FOBSS) has been proposed by defining a first-order model for the mixtures. This system was then applied to separation of seizure components of a set of EEG signals. Using the FOBSS method is based on the following assumptions:

- A1: The columns of all $\mathbf{S}_k$ are mutually uncorrelated, particularly, $\mathbf{S}_k$ matrices are all orthogonal.
- A2: The source columns of all $\mathbf{S}_k$ can be temporally correlated/uncorrelated.
- A3: The source envelopes (as in [88]) change independently, i.e. each source has varying variance over its segments k and the variation for different sources is independent.
- A4: The mixing channel A is full rank and remains unchanged for all time segments k.
- A5: The additive noise signals are temporally uncorrelated and mutually uncorrelated with the sources.

Based on A1 all the source covariance matrices, i.e. $\mathbf{S}_k\mathbf{S}_k^T$, are diagonal with nonnegative diagonal values for all $k = 1,\ldots, K$. So, the basic model after temporal segmentation is simply stated as:

$$\mathbf{X}_k = \mathbf{S}_k\mathbf{A} + \mathbf{E}_k \text{ subject to } \mathbf{S}_k\mathbf{S}_k^T = \mathbf{D}_k^2; k = 1, \ldots, K \tag{5.58}$$

where $\mathbf{D}_k^2$ is a diagonal matrix. $\mathbf{A}$ is fixed across the segments. Each $\mathbf{S}_k$ is decomposed into one row-wise orthonormal matrix $\mathbf{P}_k$ and one diagonal matrix $\mathbf{D}_k$ which absorbs the norm of different columns of $\mathbf{S}_k$ at each segment k as $\mathbf{S}_k = \mathbf{P}_k\mathbf{D}_k$. This satisfies orthogonality of the segmented sources, i.e. $\mathbf{S}_k\mathbf{S}_k^T = \mathbf{D}_k^2$. Similar to the previous model, this can change the above model to:

$$\mathbf{X}_k = \mathbf{P}_k\mathbf{D}_k\mathbf{A}^T + \mathbf{E}_k \text{ subject to } \mathbf{P}_k^T\mathbf{P}_k = \mathbf{I}_{N_s}; k = 1, \ldots, K \tag{5.59}$$

The parameters (factors) of the above model using the FOBSS approach in [89] can be estimated by going through the following steps:

Step 1: Initialize all the model parameters randomly.

Step 2: Estimation of $\mathbf{P}_k$ using $\mathbf{P}_k = \mathbf{U}_k\mathbf{V}_k^T$ for all k = 1, ..., K, where $\mathbf{U}_k$ and $\mathbf{V}_k$ are the eigenvectors.

Step 3: $\mathbf{A}$ is estimated using $\mathbf{A} = \sum_{k=1}^{K}\mathbf{D}_k\mathbf{P}_k^T\mathbf{X}_k$.

Step 4: To calculate $\mathbf{D}_k$ the following relations (as the result of solving the above constrained optimization problem) are used:

$$\sum_{k=1}^{K}\mathbf{A}\mathbf{P}_k^T\mathbf{X}_k = \sum_{k=1}^{K}\mathbf{D}_k\Lambda \tag{5.60}$$

where $\mathbf{\Lambda}$ is a diagonal matrix defined as $\mathbf{\Lambda} = Ddiag(\mathbf{A}^T\mathbf{A})$ which sets all the non-diagonal elements of $\mathbf{A}^T\mathbf{A}$ to zero. Then, for all k we use:

$$\sum_{m=1}^{K} \mathbf{Q}_{km}\mathbf{D}_m = \mathbf{D}_k\mathbf{\Lambda} \tag{5.61}$$

where $\mathbf{Q}_{km} = Ddiag\left(\mathbf{P}_k^T\mathbf{X}_k\mathbf{X}_m^T\mathbf{P}_m\right)$ are diagonal matrices for $k,m = 1, \ldots, K$. Finally, for rth source using $\mathbf{G}_r\mathbf{d}_r = \lambda_r\mathbf{d}_r$, where $\mathbf{d}_r$ is the result of stacking all the rth diagonal elements of all $\mathbf{D}_k$. Solving the above problem is equal to finding the nearest eigenvalue of $\mathbf{G}_r$ to $\mathbf{a}_r^T\mathbf{a}_r$ where $\mathbf{a}_r$ is the rth column of mixing matrix $\mathbf{A}$, and respectively its corresponding eigenvectors $\mathbf{d}_r$.

Step 5: If $\|\mathbf{X}_k - \mathbf{P}_k\mathbf{D}_k\mathbf{A}^T\| \geq \sigma^2$, where σ^2 is the expected noise variance, go to **Step 2**, otherwise, stop.

Figure 5.13 shows the results of the FOBSS algorithm in separation of seizure component from the scalp EEG. The data are just the last segment of a long data recorded from the scalp of an epileptic patient in the Clinical Neuroscience Department, King's College London simultaneously with a set of intracranial signals using electrodes implanted within the brain. In Figure 5.14 the signals from three intracranial source signals have been illustrated. From Figure 5.14 the instant of ictal onset is very clear.

5.9 Tensor Factorization for Underdetermined Source Separation

The SCA methods in Section 5.4 exploit the sparsity of the constituent sources in grouping and separation of the signals. Moreover, the main assumption in these methods is the fact that only one signal has non-zero amplitude at each instant of time. Although in some applications with sparse sources identification of the mixing channels is achieved successfully, in many cases reconstruction of the original source signals are either difficult or less accurate.

In tensor factorization approaches recently developed for underdetermined or sparse source separation, such as in [41, 90–93], two main concepts have been exploited. The first property is sparsity of the sources as for the SCA approaches, and the second one, which represents one of the main advantages of tensor factorization to other separation methods, is the fact that tensors can have the ranks higher than those of related matrices. Therefore, it is expected that the maximum number of separable sources increases with increase in the rank of the tensor. Moreover, in some applications such as in [91] the idea of sparse events has been presented. This means that at each period of time up to a certain number of sources can be active. Such an approach can also be considered as generalization of the SCA methods described in Section 7.4 to the situation where we can have more than one sources active and also, the duration of activity of a source can be more than a sample period.

Therefore, tensor factorization techniques, especially PARAFAC [44, 92], have been used to solve the UBI problem even when the sources are non-sparse or fully active (e.g. $k_a \leq N_s$) [88]. Later, second and HOS based methods have been developed to solve the UBI problem using joint/simultaneous diagonalization of a series of symmetric square matrices such as second-order blind identification of underdetermined mixtures (SOBIUM) [90] and fourth-order cumulant-based blind identification of underdetermined mixtures

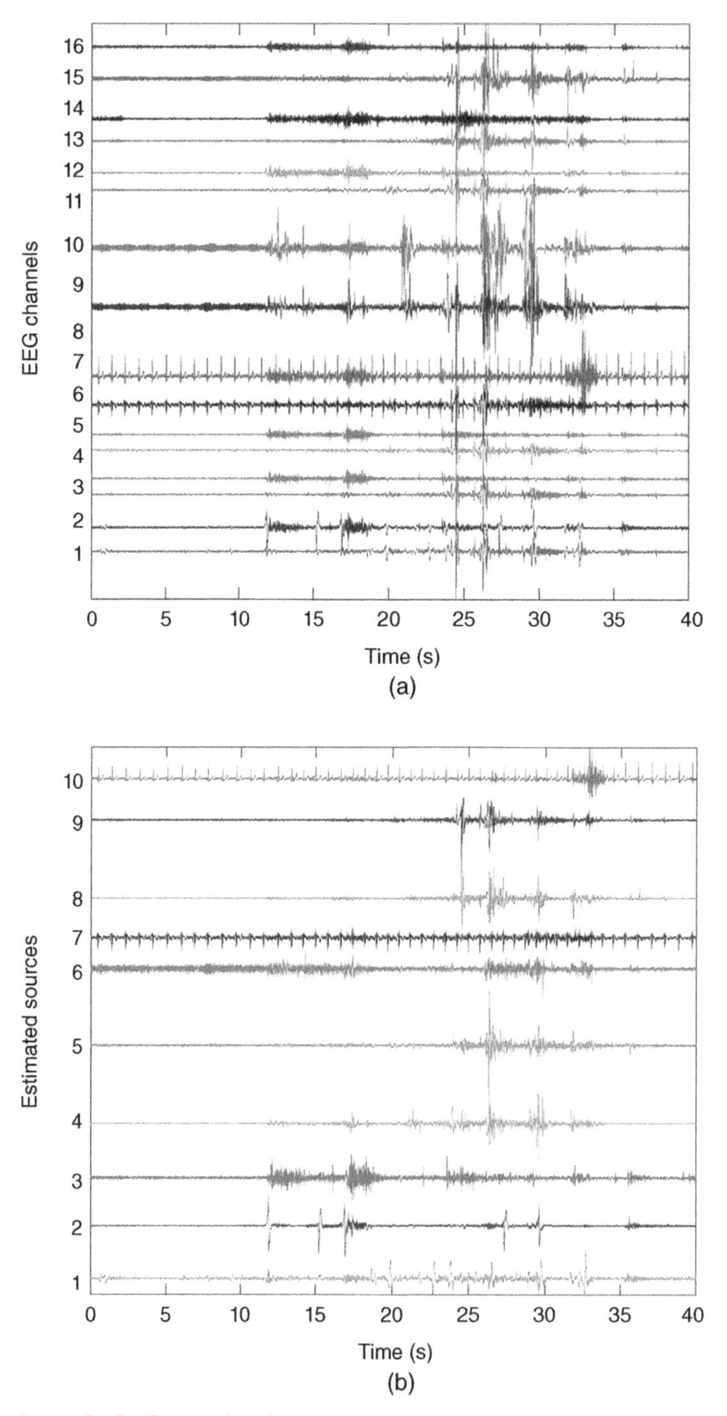

Figure 5.13 The results of application of the FOBSS algorithm to a set of scalp EEGs in (a), (b) the separated sources, and (c) the corresponding topographs respectively.

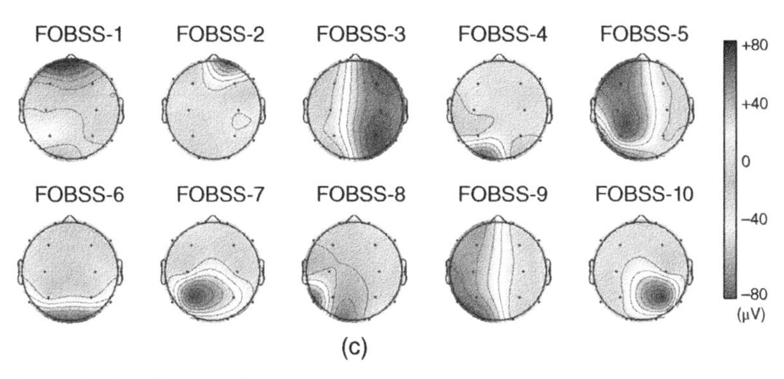

Figure 5.13 (Continued)

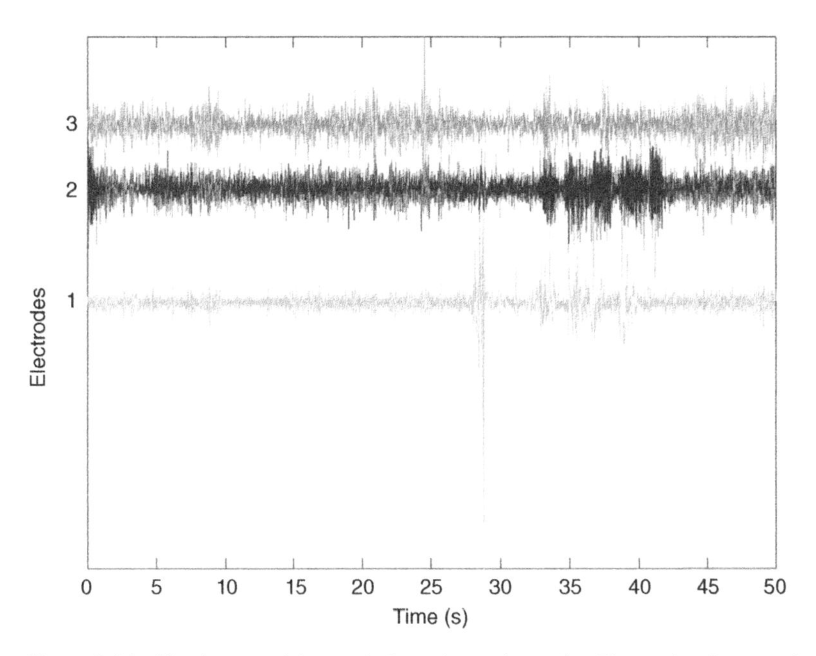

Figure 5.14 The intracranial records from three electrodes. These signals were simultaneously recorded together with the scalp EEGs in Figure 5.13a.

(so-called FOOBI) [41]. The SOBIUM method deals with the covariance matrices of different segments and tries to convert the UBI problem to a PARAFAC problem, with two identical factors called INdividuals Differences in SCALing (INDSCAL) [43], to joint block diagonalization of some exact-determined Hermitian matrices. Table 5.1 shows the upper bounds for SOBIUM and FOOBI in terms of maximum number of separable sources.

In [91] a tensor factorization based approach to k_a-SCA (called k-SCA in [28]) has been developed to solve UBSS and especially UBI problems. k_a sources have been considered active in each signal segment. Similar to k_a-SCA methods, introduced in Section 7.4, k_a must

Table 5.1 Maximum number of separable sources using different tensor-based underdetermined source separation methods.

N_s \ N_x	2	3	4	5	6	7
SOBIUM (real)	2	4	6	10	15	20
SOBIUM (complex)	2	4	9	14	21	30
FOOBI (real)	2	4	8	13	20	29
FOOBI (complex)	2	5	12	22	36	55

satisfy $k_a \leq N_x - 1$. However, in this approach instead of sparse signals (with limited signals active in each instant) sparse events (for which limited sources are active within some signal segments of particular duration) have been introduced and considered. Therefore, initially, the number of active sources in each signal segment has been estimated. A necessary condition for such a method to work is to have sufficient number of segments to have each source active in some of the segments. The UBI is then performed using the segments of sparsely active sources.

The algorithm, so-called UOM-BSS (underdetermined orthogonal model BSS) for solving a similar problem as in (7.51) has the following steps [91]:

Step 1: Initialize all the model parameters randomly; $iter = 1$.
Step 2: Estimate $\mathbf{P}_k$ using $\mathbf{P}_k = \mathbf{U}_k\mathbf{V}_k^T$ for all $k = 1, \ldots, K$.
Step 3: Estimate $\mathbf{A}$ using:

$$\mathbf{A} = \left(\sum\nolimits_{k=1}^{K} \mathbf{X}_k^T \mathbf{P}_k \mathbf{D}_k\right)\left(\sum\nolimits_{k=1}^{K} \mathbf{D}_k^2\right)^{\dagger}$$

Step 4: Estimate $\mathbf{D}_k$ for all $k = 1, \ldots, K$ using

$$\mathbf{D}_k \leftarrow \lambda\breve{\mathbf{D}}_k + (1-\lambda)\mathbf{D}_k$$

where $\lambda = \lambda(1 - e^{-0.01(iter-1)})_{max}$; $\lambda_{\max} = 0.2$ as an empirical choice, and $\breve{\mathbf{D}}_k = diag\left(\breve{\mathbf{d}}_k\right)$, a diagonal matrix with $\breve{\boldsymbol{d}}_k$ as the diagonal elements, computed separately as given later, $iter = iter + 1$:

Step 5: If $\|\mathbf{X}_k - \mathbf{P}_k\mathbf{D}_k\mathbf{A}^T\| \geq \sigma^2$, where σ^2 is the expected noise variance, go to **Step 2**, otherwise, stop. $\breve{\boldsymbol{d}}_k$ is also estimated for each segment k separately using the following steps [91]:
Step 1: $n = 0$, $\boldsymbol{y} = vec\left(\mathbf{P}_k^T\mathbf{X}_k\right) = vec(\mathbf{D}_k\mathbf{A})$, $\mathbf{\Phi} = \mathbf{A}\circ\mathbf{I}_{N_s}$, $\mathbf{s} = []$
Step 2: Find the index number of maximum element of $|\mathbf{\Phi}^T\mathbf{e}|$ as m. Update the selected active list $\mathbf{s}$ by $\boldsymbol{s} \leftarrow [\boldsymbol{s}\ m]$
Step 3: Build $\breve{\mathbf{\Phi}}$ by selecting columns of $\boldsymbol{\Phi}$ indexed in $\mathbf{s}$.

$$\mathbf{e} = \mathbf{y} - \breve{\mathbf{\Phi}}\left(\breve{\mathbf{\Phi}}^{\dagger}\mathbf{y}\right)$$

$n \leftarrow n+1$

Step 4: **if** $\|\mathbf{e}\| > \varepsilon$ OR $n < N_x - 1$

go to **Step 2**

else

converged: $\breve{\mathbf{d}}_{\mathbf{k}}(\mathbf{s}) = \breve{\boldsymbol{\Phi}}^{\dagger}\mathbf{y}$

end if.

This approach improves the general upper bound in terms of maximum possible number of sources that can be achieved by the SOBIUM method [90]. Using some synthetically mixed signals it has been shown that the separation error is less for the above UOM-BSS method compared to those of both the SOBIUM and FOOBI methods.

5.10 Tensor Factorization for Separation of Convolutive Mixtures in the Time Domain

Due to flexibility in definition of multiway systems, convolutive BSS in the time domain can be modelled using tensors. In [48, 49, 97, 98] the anechoic convolutive BSS problem has been formulated as a tensor factorization problem. In a noiseless environment the problem has been stated as:

$$\mathbf{X}_k = \sum\nolimits_{\tau=1}^{M} \Xi_\tau \mathbf{P}_k \mathbf{D}_k \mathbf{A}_\tau^T \text{ subject to } \mathbf{P}_k^T \mathbf{P}_k = \mathbf{I}_{N_s} \tag{5.62}$$

where Ξ_τ accounts for the delays and $\mathbf{A}_\tau$ accommodates all the weights for the delayed terms. An effective method for estimation of the parameters/factors has been proposed particularly for estimation of $\mathbf{A}_\tau$ [98].

In this approach it has been demonstrated that the impulse response of the channels can be more accurately estimated if the locations of the sources and the sensors are known. This geometrical information can be nicely fused into the separation process.

5.11 Separation of Correlated Sources via Tensor Factorization

Unfortunately, many physiological and biological signal sources are somehow correlated mainly because of having the same origin or cause for their generation. Consider for example the actual components of a QRS signal within an ECG signal. Such components cannot be separated using conventional BSS which relies on independency of the sources. Another example is separation of P3a and P3b, subcomponents of a P300 signal, from EEGs. These components are related to the same event or stimulus and partially overlap in time and space (and albeit in frequency).

In [99] this problem has been addressed using tensor factorization. Promising outcomes have been obtained using synthetically mixed components. However, assessment of the application of this method to real signals is still a subject of research.

Exploitation of matrix and tensor factorization together with new optimization techniques have become central to EEG signal processing. An application of such methods to joint scalp-intracranial EEGs namely, group component analysis for separation of common and individual components [166] aiming at detection of interictal epileptiform discharges (IED) [167] is explored in Chapter 11 of this book.

5.12 Common Component Analysis

Common component analysis can be posed as a tensor decomposition problem, unlike PCA, standard approaches to tensor decomposition have two drawbacks: (i) they are generally iterative and rely on the initialization. A bad initialization may lead to poor local optima and (ii) for a given level of approximation error, selection of suitable low dimensionality is not always possible.

Common component analysis can be formulated as follows. Assume a set of high-dimensional covariance matrices $\mathbf{X}_t \in \Re^{n \times n}$, $1 \leq t \leq T$. The key hypothesis driving the common component analysis concept is that the high-dimensional covariance matrices are indeed a linearly transformed version of a set of low-dimensional covariance matrices $\mathbf{Y}_t \in \Re^{r \times r}$, $1 \leq t \leq T$, i.e.:

$$\mathbf{X}_t = \mathbf{U}\mathbf{Y}_t\mathbf{U}^T + \mathbf{E}_t \tag{5.63}$$

where $\mathbf{U} \in \Re^{n \times r}$ and $\mathbf{E}_t$ is the residual matrix. In the above equation both $\mathbf{U}$ and $\mathbf{Y}_t$ are unknown given that $\mathbf{U}$ is orthonormal. In order to estimate them, the following optimization is carried out:

$$\min_{\mathbf{U},\mathbf{Y}_t} \sum_{t=1}^{T} \left\| \mathbf{X}_t - \mathbf{U}\mathbf{Y}_t\mathbf{U}^T \right\|_F^2 \text{ subject to} : \mathbf{U}\mathbf{U}^T = \mathbf{I}_r \tag{5.64}$$

This is called common component analysis because $\mathbf{U}$ determines a common subspace for all the covariance matrices. If there is only one covariance matrix $\mathbf{X}_1$, then the model reduces to standard PCA. For more than one covariance matrices, the problem can be solved using tensor decompositions approaches.

5.13 Canonical Correlation Analysis

Another popular method for EEG analysis is called canonical correlation analysis (CCA). CCA is used to identify and measure the associations among two sets of variables. CCA determines a set of canonical variates, orthogonal linear combinations of the variables within each set that best explain the variability both within and between sets. In this decorrelation process, very similar to common spatial patterns (CSP) the sources are estimated from their mixtures in such a way that they are temporally maximally correlated and mutually uncorrelated. The main purposes for CCA are: (i) Data reduction: this explains the covariation between two sets of variables using small number of linear combinations and (ii) Data interpretation: this is to find features (i.e. canonical variates) that are important for explaining covariation between sets of variables.

Let $\mathbf{X} = (\mathbf{x}_1, \mathbf{x}_2, \ldots, \mathbf{x}_p)'$ and $\mathbf{Y} = (\mathbf{y}_1, \mathbf{y}_2, \ldots, \mathbf{y}_q)'$ with covariance matrices Σ_x and Σ_y respectively and cross-covariance of Σ_{xy}. Considering $\mathbf{U} = \mathbf{W}_x\mathbf{X}$ and $\mathbf{V} = \mathbf{W}_y\mathbf{Y}$, the objective is to find the optimum $\mathbf{W}_x$ and $\mathbf{W}_y$ to maximize the correlation between $\mathbf{U}$ and $\mathbf{V}$ denoted as $\rho(\mathbf{U}, \mathbf{V})$:

$$\rho(\mathbf{U},\mathbf{V}) = \frac{\mathbf{W}_x\Sigma_{xy}\mathbf{W}_y{}^T}{\sqrt{\mathbf{W}_x\Sigma_x\mathbf{W}_x{}^T}\sqrt{\mathbf{W}_y\Sigma_y\mathbf{W}_y{}^T}} \tag{5.65}$$

CCA is used for feature detection [168] or source separation [169]. There have been many applications of CCA to analysis of EEG signals such as in [170] for separation of muscle artefacts from EEG signals.

5.14 Summary

During the past two decades a large number of BSS methods have been proposed. A BSS problem becomes more difficult to solve when the signals are convolutively mixed, the number of sources is more than the number of sensors, the signals are nonstationary, or the system is nonlinear. In this chapter the BSS problem for different scenarios have been explained. Recently introduced tensor factorization approaches can highly alleviate these problems by relying on their three fundamental properties: first, they can map the signals into a multidimensional space where the sources can become disjoint; second, their ranks are often higher than any of the corresponding two-dimensional representations of the data; and third, any related constraint can be easily incorporated to ensure the desired unique solution. The concepts in this chapter can be a good foundation for understanding many other approaches in BSS. The new techniques in underdetermined source separation allow k-sparsity for which the number of active sources at each time can be very close to the number of mixtures. Representation of multichannel data, which vary from segment to segment and from subject to subject, in tensor form, efficiently exploits the data diversity for separation and classification purposes.

References

1 Cardoso, J.-F. (1997). Infomax and maximum likelihood for blind source separation. *IEEE Signal Processing Letters* 4 (4): 112–114.

2 Cardoso, J. (1989). Source separation using higher order moments. *Proceedings of the International Conference on Acoustics, Speech and Signal Processing (ICASSP)*, 2109–2112.

3 Mathis, H. and Douglas, S.C. (2004). On the existence of universal nonlinearities for blind source separation. *IEEE Transactions on Signal Processing* 50: 1007–1016.

4 Cohen, L. (1995). *Time-Frequency Analysis*. Prentice Hall.

5 Platt, C. and Fagin, F. (1992). Networks for the separation of sources that are superimposed and delayed. In: *Advances in Neural Information Processing Systems*, vol. 4, 730–737. Morgan Kaufmann.
6 Herault, J. and Jutten, C. (1986). Space or time adaptive signal processing by neural models. *Proceedings of the AIP Conference on Neural Network for Computing*, 206–211. American Institute of Physics.
7 Torkkola, K. (1996). Blind separation of delayed sources based on information maximization. *Neural Networks for Signal Processing VI. Proceedings of the 1996 IEEE Signal Processing Society Workshop*, Kyoto, Japan, 423–432, 3509–3512. Atlanta, Georgia.
8 Torkkola, K. (1996). Blind separation of convolved sources based on information maximization. *Neural Networks for Signal Processing VI. Proceedings of the 1996 IEEE Signal Processing Society Workshop*, Kyoto, Japan, 315–323, 423–432.
9 Wang, W., Sanei, S., and Chambers, J.A. (2005). Penalty function based joint diagonalization approach for convolutive blind separation of nonstationary sources. *IEEE Transactions on Signal Processing* 53 (5): 1654–1669.
10 Yeredor, A. (2001). Blind source separation with pure delay mixtures. *Independent Component Analysis and Blind Source Separation, ICA 2001*. San Diego.
11 Parra, L., Spence, C., Sajda, P. et al. (2000). Unmixing hyperspectral data. *Advances in Neural Information Processing Systems* 13: 942–948.
12 Ikram, M. and Morgan, D. (2000). Exploring permutation inconstancy in blind separation of speech signals in a reverberant environment. *Proceedings of ICASSP*. Turkey.
13 Cherkani, N. and Deville, Y. (1999). Self adaptive separation of convolutively mixed signals with a recursive structure, part 1: stability analysis and optimisation of asymptotic behaviour. *Signal Processing* 73 (3): 225–254.
14 Smaragdis, P. (1998). Blind separation of convolved mixtures in the frequency domain. *Neurocomputing* 22: 21–34.
15 Murata, N., Ikeda, S., and Ziehe, A. (2001). An approach to blind source separation based on temporal structure of speech signals. *Neurocomputing* 41: 1–4.
16 Zibulevsky, M. (2002). Relative Newton Method for Quasi-ML Blind Source Separation. http://ie.technion.ac.il/~mcib/newt_ica_jmlr1.ps.gz (accessed 19 August 2021).
17 Luo, Y., Chambers, J., Lambotharan, S., and Proudler, I. (2006). Exploitation of source non-stationarity in underdetermined blind source separation with advanced clustering techniques. *IEEE Transactions on Signal Processing* 54 (6): 2198–2212.
18 Li, Y., Amari, S., Cichocki, A. et al. (2006). Underdetermined blind source separation based on sparse representation. *IEEE Transactions on Signal Processing* 54 (2): 423–437.
19 Jurjine, A., Rickard, S., and Yilmaz, O. (2000). Blind separation of disjoint orthogonal signals: demixing N sources from 2 mixtures. *Proceedings of the IEEE Conference on Acoustic, Speech, and Signal Processing (ICASSP 2000)*, 5: 2985–2988.
20 Vielva, L., Erdogmus, D., Pantaleon, C. et al. (2002). Underdetermined blind source separation in a time-varying environment. *Proceedings of the IEEE Conference on Acoustic, Speech, and Signal Processing (ICASSP 2000)*, 3: 3049–3052.
21 Takigawa, I., Kudo, M., Nakamura, A., and Toyama, J. (2004). On the minimum l1-norm signal recovery in underdetermined source separation. *Proceedings of the 5th International Conference on Independent Component Analysis, ICA 2004*, 22–24. Granada, Spain.

22 Noorshams, N., Babaie-Zadeh, M., and Jutten, C. (2007). Estimating the mixing matrix in sparse component analysis based on converting a multiple dominant to a single dominant problem. *ICA'07 Proceedings of the 7th International Conference on Independent Component Analysis and Signal Separation*, 397–405.

23 Movahedi Naini, F., Mohimani, G.H., Babaie-Zadeh, M., and Jutten, C. (2008). Estimating the mixing matrix in sparse component analysis (SCA) based on partial k-dimensional subspace clustering. *Neurocomputing* 71 (10–12): 2330–2343.

24 Washizawa, Y. and Cichocki, A. (2006). On-line K-PLANE clustering learning algorithm for sparse comopnent analysis. *2006 IEEE International Conference on Acoustics Speech and Signal Processing Proceedings*, V–V. Toulouse.

25 Gao, F., Sun, G., Xiao, M., and Lv, J. (2010). Matrix Estimation Based on Normal Vector of Hyperplane in Sparse Component Analysis. In: *Advances in Swarm Intelligence. ICSI 2010*, Lecture Notes in Computer Science, vol. 6146 (eds. Y. Tan, Y. Shi and K.C. Tan). Berlin, Heidelberg: Springer.

26 Joho, M. and Mathis, H. (2002). Joint diagonalization of correlation matrices by using gradient methods with application to blind signal processing. *Proceedings of the 1st Annual Conference in Sensor Array and Multichannel Signal Processing (SAM2002)*, 273–277.

27 Shoker, L., Sanei, S., and Chambers, J. (2005). Artefact removal from electroencephalograms using a hybrid BSS-SVM algorithm. *IEEE Signal Processing Letters* 12 (10): 721–724.

28 Parra, L. and Spence, C. (2000). Convolutive blind separation of non-stationary sources. *IEEE Transactions on Speech and Audio Processing* 8: 320–327.

29 Hyvärinen, A., Hoyer, P.O., and Inkl, M. (2001). Topographic independent component analysis. *Neural Computation* 13: 1527–1558.

30 Hitchcock, F.L. (1927). The expression of a tensor or a polyadic as a sum of products. *Journal of Mathematics and Physics* 6: 164–189.

31 Hitchcock, F.L. (1927). Multiple invariants and generalized rank of a p-way matrix or tensor. *Journal of Mathematics and Physics* 7: 39–79.

32 Cattell, R.B. (1944). Parallel proportional profiles and other principles for determining the choice of factors by rotation. *Psychometrika* 9: 267–283.

33 Cattell, R.B. (1952). The three basic factor-analytic research designs | their interrelations and derivatives. *Psychological Bulletin* 49: 499–452.

34 Tucker, L.R. (1963). Implications of factor analysis of three-way matrices for measurement of change. In: *Problems in Measuring Change* (ed. C.W. Harris), 122–137. University of Wisconsin Press.

35 Tucker, L.R. and Tucker, L. (1964, 110–127). The extension of factor analysis to three-dimensional matrices. In: *Contributions to Mathematical Psychology* (eds. H. Gulliksen and N. Frederiksen). New York: Holt, Rinehardt, & Winston.

36 Tucker, L.R. (1966). Some mathematical notes on three-mode factor analysis. *Psychometrika* 31: 279–311.

37 Carroll, J.D. and Chang, J.J. (1970). Analysis of individual differences in multidimensional scaling via an N-way generalization of `Eckart-Young' decomposition. *Psychometrika* 35: 283–319.

38 Harshman, R.A. (1970). Foundations of the PARAFAC procedure: Models and conditions for an "explanatory" multi-modal factor analysis. *UCLA Working Papers in Phonetics* 16: 1–84.

39 Appellof, C.J. and Davidson, E.R. (1981). Strategies for analyzing data from video uorometric monitoring of liquid chromatographic effluents. *Analytical Chemistry* 53: 2053–2056.

40 Smilde, A., Bro, R., and Geladi, P. (2004). *Multi-Way Analysis: Applications in the Chemical Sciences*. West Sussex, England: Wiley.

41 De Lathauwer, L., Castaing, J., and Cardoso, J.-F. (2007). Fourth-order cumulant based blind identification of underdetermined mixtures. *IEEE Transactions on Signal Processing* 55: 2965–2973.

42 De Lathauwer, L. and de Baynast, A. (2007). Blind deconvolution of DS-CDMA signals by means of decomposition in rank-(1; l; l) terms. *IEEE Transactions on Signal Processing* 56 (4): 1562–1571.

43 De Lathauwer, L. and De Moor, B. (1998). From matrix to tensor: multilinear algebra and signal processing. In: *Mathematics in Signal Processing IV* (eds. J. McWhirter and I. Proudler), 1–15. Oxford: Clarendon Press.

44 De Lathauwer, L. and Vandewalle, J. (2004). Dimensionality reduction in higher-order signal processing and rank-(R1;R2; : : : ;RN) reduction in multilinear algebra. *Linear Algebra and Its Applications* 391: 31–55.

45 Chen, B., Petropulu, A., and De Lathauwer, L. (2002). Blind identification of convolutive MIMO systems with 3 sources and 2 sensors. *EURASIP Journal on Applied Signal Processing (Special Issue on Space-Time Coding and Its Applications, Part II)* 5: 487–496.

46 Comon, P. (2001). Tensor decompositions: state of the art and applications. In: *Mathematics in Signal Processing V* (eds. J.G. McWhirter and I.K. Proudler), 1–24. Oxford University Press.

47 Muti, D. and Bourennane, S. (2005). Multidimensional filtering based on a tensor approach. *Signal Processing* 85: 2338–2353.

48 Sanei, S. and Makkiabadi, B. (2009). Tensor Factorization with Application to Convolutive Blind Source Separation of Speech. In: *Machine Audition: Principles, Algorithms and Systems* (ed. W. Wang). pp. 186-206, IGI-Global Pub.

49 Makkiabadi, B. and Sanei, S. (2012). A new time domain convolutive BSS of heart and lung sounds. *Proceedings of the 2012 IEEE International Conference on Acoustics, Speech and Signal Processing (ICASSP)*, 605–608. Kyoto, Japan.

50 Makkiabadi, B., Sarrafzadeh, A., Jarchi, D. et al. (2009). Semi-blind signal separation and channel estimation in MIMO communication systems by tensor factorization. *2009 IEEE/SP 15th Workshop on Statistical Signal Processing*, 305–308. Cardiff, UK.

51 Beckmann, C. and Smith, S. (2005). Tensorial extensions of independent component analysis for multisubject FMRI analysis. *NeuroImage* 25: 294–311.

52 Acar, E., Bingol, C.A., Bingol, H. et al. (2007). Multiway analysis of epilepsy tensors. *Bioinformatics* 23: i10–i18.

53 De Vos, M., De Lathauwer, L., Vanrumste, B. et al. (2007). Canonical decomposition of ictal scalp EEG and accurate source localisation: principles and simulation study. *Computational Intelligence and Neuroscience* 2007: 1–8.

54 De Vos, M., Vergult, A., De Lathauwer, L. et al. (2007). Canonical decomposition of ictal scalp EEG reliably detects the seizure onset zone. *NeuroImage* 37: 844–854.

55 Martinez-Montes, E., Valdés-Sosa, P.A., Miwakeichi, F. et al. (2004). Concurrent EEG/fMRI analysis by multiway partial least squares. *NeuroImage* 22: 1023–1034.

56 Miwakeichi, F., Martinez-Montes, E., Valds-Sosa, P.A. et al. (2004). Decomposing EEG data into space-time-frequency components using parallel factor analysis. *NeuroImage* 22: 1035–1045.

57 Ocks, J.M. (1988). Topographic components model for event-related potentials and some biophysical considerations. *IEEE Transactions on Biomedical Engineering* 35: 482–484.

58 Morup, M., Hansen, L.K., and Arnfred, S.M. (2007). ERPWAVELAB a toolbox for multichannel analysis of time-frequency transformed event related potentials. *Journal of Neuroscience Methods* 161: 361–368.

59 Nazarpour, K., Praamstra, P., Miall, R.C., and Sanei, S. (2009). Steady-state movement related potentials for brain computer interfacing. *IEEE Transactions on Biomedical Engineering* 56 (8): 2104–2113.

60 Nazarpour, K., Wangsawat, Y., Sanei, S. et al. (2008). Removal of the eye-blink artefacts from EEGs via STF-TS modeling and robust minimum variance beamforming. *IEEE Transactions on Biomedical Engineering* 55 (9): 2221–2231.

61 Kolda, T.G. and Bader, B.W. (2008). Tensor decompositions and applications. *SIAM Review* 51 (3): 455–500.

62 Stegeman, A. and Sidiropoulos, N.D. (2007). On Kruskal's uniqueness condition for the candecomp/parafac decomposition. *Linear Algebra and Its Applications* 420 (2–3): 540–552.

63 Rong, Y., Vorobyov, S.A., Gershman, A.B., and Sidiropoulos, N.D. (2005). Blind spatial signature estimation via time-varying user power loading and parallel factor analysis. *IEEE Transactions on Signal Processing*: 1697–1710.

64 Comon, P. (1994). Independent component analysis: a new concept. *Signal Processing* 36: 287–314.

65 Földiák, P. and Young, M. (1995). Sparse coding in the primate cortex. In: *The Handbook of Brain Theory and Neural Networks* (ed. M.A. Arbib), 895–898. MIT Press.

66 Einhauser, W., Kayser, C., Konig, P., and Kording, K.P. (2002). Learning the invariance properties of complex cells from their responses to natural stimuli. *European Journal of Neuroscience* 15: 475–486.

67 Jutten, C. and Herault, J. (1991). Blind separation of sources, part I: an adaptive algorithm based on neuromimetic architecture. *Signal Processing* 24: 1–10.

68 Linsker, R. (1989). An application of the principle of maximum information preservation to linear systems. In: *Advances in Neural Information Processing Systems*, vol. 1 (ed. D.S. Touretzky), 186–194. Morgan Kaufmann.

69 Bell, A.J. and Sejnowski, T.J. (1995). An information-maximization approach to blind separation, and blind deconvolution. *Neural Computation* 7 (6): 1129–1159.

70 Amari, S., Cichocki, A., and Yang, H.H. (1996). A new learning algorithm for blind signal separation. In: *Advances in Neural Information Processing Systems*, vol. 8, 757–763. MIT Press.

71 Hyvarinen, A. and Oja, E. (1997). A fast-fixed point algorithm for independent component analysis. *Neural Computation* 9 (7): 1483–1492.

72 Jung, T.P., Makeig, S., Westereld, M. et al. (1999). Analyzing and visualizing single-trial event-related potentials. In: *Advances in Neural Information Processing Systems*, vol. 11, 118–127. MIT Press.

73 Jung, T.P., Makeig, S., Humphries, C. et al. (2000). Removing electroencephalographic artifacts by blind source separation. *Psychophysiology* 37: 163–178.

74 Shoker, L., Sanei, S., and Chambers, J. (2005). Artifact removal from electroencephalograms using a hybrid BSS-SVM algorithm. *IEEE Signal Processing Letters* 12 (10): 721–724.

75 Shoker, L., Sanei, S., and Chambers, J. (2005). A hybrid algorithm for the removal of eye blinking artifacts from electroencephalograms. *Proceedings of the IEEE Statistical Signal Processing Workshop, SSP2005* (February 2005), France.

76 Corsini, J., Shoker, L., Sanei, S., and Alarcon, G. (2006). Epileptic seizure prediction from scalp EEG incorporating BSS. *IEEE Transactions on Biomedical Engineering* 53 (5): 790–799.

77 Latif, M.A., Sanei, S., and Chambers, J. (2006). Localization of abnormal EEG sources blind source separation partially constrained by the locations of known sources. *IEEE Signal Processing Letters* 13 (3): 117–120.

78 Spyrou, L., Jing, M., Sanei, S., and Sumich, A. (2006). Separation and localisation of P300 sources and the subcomponents using constrained blind source separation. *EURASIP Journal on Advances in Signal Processing* 2007: 82912.

79 Sanei, S. (2004). Texture segmentation using semi-supervised support vector machines. *International Journal of Computational Intelligence and Applications* 4 (2): 131–142.

80 Tang, A.C., Sutherland, T., and Wang, Y. (2006). Contrasting single trial ERPs between experimental manipulations: improving differentiability by blind source separation. *NeuroImage* 29: 335–346.

81 Makeig, S., Bell, A.J., Jung, T., and Sejnowski, T.J. (1996). Independent component analysis of electroencephalographic data. In: *Advances in Neural Information Processing Systems*, vol. 8, 145–151. MIT Press.

82 Makeig, S., Jung, T.P., Bell, A.J. et al. (1997). Blind separation of auditory event-related brain responses into independent components. *Proceedings of the National Academy of Sciences. United States of America*, 94: 10979–10984.

83 Jing, M. and Sanei, S. (2006). Scanner artifact removal in simultaneous EEG–fMRI for epileptic seizure prediction. *IEEE 18th International Conference on Pattern Recognition, ICPR* 3: 722–725.

84 Harshman, R.A. and Lundy, M.E. (1994). PARAFAC: parallel factor analysis. *Computational Statistics and Data Analysis* 8: 32–79.

85 Kiers, H.A.L., Ten Berge, J.M.F., and Bro, R. (1999). PARAFAC2 – part I. A direct fitting algorithm for the PARAFAC2 model. *Journal of Chemometrics* 13: 275–294.

86 Hyvarinen, A. (Nov 2001). Blind source separation by nonstationarity of variance: a cumulant-based approach. *IEEE Transactions on Neural Networks* 12 (6): 1471–1474.

87 Matsuoka, K. and Kawamoto, M. (1994). A neural net for blind separation of nonstationary signal sources. *IEEE World Congress on Computational Intelligence*, 1: 221–232.

88 Pham, D.T. and Cardoso, J.F. (2001). Blind separation of instantaneous mixtures of nonstationary sources. *IEEE Transactions on Signal Processing* 49 (9): 1837–1848.

89 Makkiabadi, B. (2011). Advances in factorization based blind source separation. PhD Thesis. University of Surrey.

90 De Lathauwer, L. and Castaing, J. (2008). Blind identification of underdetermined mixtures by simultaneous matrix diagonalization. *IEEE Transactions on Signal Processing* 56 (3): 1096–1105.

91 Makkiabadi, B., Sanei, S., and Marshall, D. (2010). A k-subspace based tensor factorization approach for under-determined blind identification. *Proceedings of the IEEE Asilomar Conference on Signals, Systems, and Computers,* 18–22. CA, USA.

92 Bro, R. (1997). PARAFAC. Tutorial and applications. *Chemometrics and Intelligent Laboratory Systems* 38 (2): 149–171.
93 Shoker, L., Sanei, S., Wang, W., and Chambers, J. (2004). Removal of eye blinking artefact from EEG incorporating a new constrained BSS algorithm. *IEEE Journal of Medical and Biological Engineering and Computing*: 290–295.
94 Almeida, L. (2005). Nonlinear source separation. In: *Synthesis Lectures on Signal Processing*, vol. 1, 1–114. Morgan&Claypool Publishers.
95 Jutten, C. and Karhunen, J. (2004). Advances in blind source separation (BSS) and independent component analysis (ICA) for nonlinear mixtures. *International Journal of Neural Systems* 14 (5): 267–292.
96 Nazarpour, K., Mohseni, H.R., Hesse, C. et al. (2008). A novel semi-blind signal extraction approach incorporating PARAFAC for the removal of eye-blink artefact from EEGs. *EURASIP Journal on Advances in Signal Processing* 2008: 857459. https://doi.org/10.1155/2008/857459.
97 Kouchaki, S. and Sanei, S. (2015). A new tensor factorisation approach for convolutive separation of complex signals. *Proceedings of 12th International Conference on Latent Variable Analysis and Signal Separation (LVA/ICA 2015) (25–28 August).* Liberec: Czech Republic.
98 Makkiabadi, B., Jarchi, D., Abolghasemi, V., and Sanei, S. (2011). A time domain geometrically constrained multimodal approach for convolutive blind source separation. *Proceedings of EUSIPCO 2011 (19th European Signal Processing Conference 2011).* Barcelona, Spain.
99 Makkiabadi, B., Jarchi, D., and Sanei, S. (2011). Blind separation and localization of correlated P300 subcomponents from single trial recordings using extended PARAFAC2 tensor model. *Proceedings of IEEE EMBC 2011*, 6955–6958. Boston, USA.
100 Dyrholm, M., Makeig, S., and Hansen, L.K. (2007). Model selection for convolutive ICA with an application to spatio-temporal analysis of EEG. *Neural Computation* 19 (4): 934–955.
101 Wang, L. and Ohtsuki, T. (2018). Nonlinear blind source separation unifying vanishing component analysis and temporal structure. *IEEE Access* 6: 42837–42850.
102 Belouchrani, A., Abed-Meraim, K., Cardoso, J.-F., and Moulines, E. (1997). A blind source separation technique using second order statistics. *IEEE Transactions on Signal Processing* 45 (2): 434–444.
103 Cirillo, L. and Zoubir, A. (2005). On blind separation of nonstationary signals. *Proceedings of the 8th Symposium on Signal Processing and its Applications (ISSPA)*, Sydney, Australia.
104 Cherry, C.E. (1953). Some experiments in the recognition of speech, with one and two ears. *Journal of the Acoustical Society of America* 25: 975–979.
105 Belouchrani, A., Abed-Mariam, K., Amin, M.G., and Zoubir, A.M. (2004). Blind source separation of nonstationary signals. *IEEE Signal Processing Letters* 11 (7): 605–608.
106 Golyandina, N., Nekrutkin, V., and Zhigljavsky, A. (2001). *Analysis of Time Series Structure: SSA and Related Techniques.* New York: Chapman & Hall/CRC.
107 Sanei, S. and Hassani, H. (2015). *Singular Spectrum Analysis of Biomedical Signals.* CRC.
108 Haavisto, O. (2010). Detection and analysis of oscillations in a mineral flotation circuit. *Journal of Control Engineering Practice* 18 (1): 23–30.

109 Eqlimi, E., Makkiabadi, B., Samadzadehaghdam, N. et al. A novel underdetermined source recovery algorithm based on k-sparse component analysis. *Circuits, Systems, and Signal Processing* 38: 1264–1286. https://doi.org/10.1007/s00034-018-0910-9.

110 Chichocki, A., Li, Y., Georgiev, P., and Amari, S. (2004). Beyond ICA: robust sparse signal representations. *Proceedings of the 2004 International Symposium on Circuits and Systems, ISCAS'04*, 5: V–V.

111 Georgiev, P., Theis, F., and Cichocki, A. (2004). Blind source separation and sparse component analysis of overcomplete mixtures. *Proceedings of the IEEE International Conference on Acoustics, Speech, and Signal Processing ICASSP'04*, 5: V–493.

112 Georgiev, P., Theis, F., and Cichocki, A. (2005). Sparse component analysis and blind source separation of underdetermined mixtures. *IEEE Transactions on Neural Networks* 16 (4): 992–996. https://doi.org/10.1109/TNN.2005.849840.

113 Georgiev, P., Theis, F., Cichocki, A., and Bakardjian, H. (2007). Sparse component analysis: a new tool for data mining. In: *Data Mining in Biomedicine. Springer Optimization and Its Applications*, vol. 7 (eds. P.M. Pardalos, V.L. Boginski and A. Vazacopoulos), 91–116. Boston, MA: Springer.

114 Zibulevsky, M. and Pearlmutter, B.A. (2001). Blind source separation by sparse decomposition in a signal dictionary. *Neural Computation* 13 (4): 863–882.

115 Li, Y., Cichocki, A., and Amari, S.I. (2006). Blind estimation of channel parameters and source components for EEG signals: a sparse factorization approach. *IEEE Transactions on Neural Networks* 17 (2): 419–431.

116 Li, Y., Yu, Z.L., Bi, N. et al. (2014). Sparse representation for brain signal processing: a tutorial on methods and applications. *IEEE Signal Processing Magazine* 31 (3): 96–106.

117 Mallat, S.G. and Zhang, Z. (1993). Matching pursuits with time-frequency dictionaries. *IEEE Transactions on Signal Processing* 41 (12): 3397–3415.

118 Donoho, D.L., Elad, M., and Temlyakov, V.N. (2006). Stable recovery of sparse overcomplete representations in the presence of noise. *IEEE Transactions on Information Theory* 52 (1): 6–18.

119 Mohimani, H., Babaie-Zadeh, M., and Jutten, C. (2009). A fast approach for overcomplete sparse decomposition based on smoothed l_0-norm. *IEEE Transactions on Signal Processing* 57 (1): 289–301.

120 He, Z., Cichocki, A., Li, Y. et al. (2009). K-hyperline clustering learning for sparse component analysis. *Signal Processing* 89 (6): 1011–1022.

121 Marvasti, F., Amini, A., Haddadi, F. et al. (2012). A unified approach to sparse signal processing. *EURASIP Journal on Advances in Signal Processing* 2012 (1): 44.

122 Tropp, J.A. (2004). Greed is good: algorithmic results for sparse approximation. *IEEE Transactions on Information Theory* 50 (10): 2231–2242.

123 Needell, D. and Tropp, J.A. (2009). CoSaMP: iterative signal recovery from incomplete and inaccurate samples. *Applied and Computational Harmonic Analysis* 26 (3): 301–321.

124 Zhou, G., Cichocki, A., Zhang, Y., and Mandic, D.P. (2015). Group component analysis for multiblock data: common and individual feature extraction. *IEEE Transactions on Neural Networks and Learning Systems* 27 (11): 2426–2439.

125 Abdi-Sargezeh, B., Valentin, A., Alarcon, G. et al. (2020). Sparse common feature analysis for detection of interictal epileptiform discharges from concurrent intracranial and scalp

EEGs. *IEEE Transactions on Biomedical Engineering* https://doi.org/10.1142/s0129065721500192.

126 Kaya, H., Eyben, F., Ali Salah, A., and Schuller, B. (2014). CCA based feature selection with application to continuous depression recognition from acoustic speech features. *IEEE International Conference on Acoustics, Speech and Signal Processing (ICASSP).* Florence, Italy.

127 Wu, X.-j., Hu, Y.-a., Li, M. et al. (2017). An Improved Group BSS-CCA Method for Blind Source Separation of Functional MRI Scans of the Human Brain. In: *2017 IEEE 2nd International Conference on Big Data Analysis (ICBDA), Beijing*, 758–761.

128 De Clercq, W., Vergult, A., Vanrumste, B. et al. (2006). Canonical correlation analysis applied to remove muscle artifacts from the electroencephalogram. *IEEE Transactions on Biomedical Engineering* 53 (12): 2583–2587.

129 Gorodnitsky, I.F. and Rao, B.D. (1997). Sparse signal reconstruction from limited data using FOCUSS: a re-weighted minimum norm algorithm. *IEEE Transactions on Signal Processing* 45 (3): 600–616.

6

Chaos and Dynamical Analysis

6.1 Introduction to Chaos and Dynamical Systems

Brain is a complex system whose characteristics change dynamically through time. Borrowed the concepts from time series analysis and forecasting, often used by mathematicians, astronomers, weather broadcasters, and researchers in economy, signal trends are not limited to only random and those with well defined trends (with correlated samples, stationary, cyclo-stationary, etc.). The trends and signals can follow particular dynamics when they are generated by well defined linear or nonlinear systems. In physiological systems there is no clear model for the system and therefore, only some approximations for the system model can be reached using the generated signals. Hence, characterization and evaluation of chaotic behaviour of such signals and system become an important problem.

The change in brain as a dynamic system can be clearly seen in many cases; when human goes to sleep the brain signal trends change significantly during the four stages of sleep. In the case of epileptic seizure, the signals from the neurogenerators around the epileptic loci gradually change from chaotic to ordered before the seizure onset. The same may happen with many biological signals. As another example, an intention to move a body part results in desynchronizing alpha rhythm and as the movement starts synchronizing beta rhythm in the brain.

As an effective tool for prediction and characterization of the signals, *deterministic chaos* plays an important role. Although the electroencephalography (EEG) signals are considered chaotic, there are rules, which do not in themselves involve any element of change, which can be used in their characterization [1]. Mathematical research about chaos started before 1890 when certain people such as Andrey Kolmogorov or Henri Poincaré tried to establish whether planets would indefinitely remain in their orbits. In the 1960s Stephan Smale formulated a plan to classify all the typical kinds of dynamic behaviour. Many chaos-generating mechanisms have been created and used to identify the behaviour of the dynamics of the system. The Rössler system was designed to model a strange attractor using a simple stretch and fold mechanism. This was however inspired by the Lorenz attractor introduced more than a decade earlier [1].

To evaluate how chaotic a dynamical system is different measures can be taken into account. A straightforward parameter is the attractor dimension. Different multidimensional attractors have been defined by a number of mathematicians. In many cases it is difficult to find the attractor dimension unless the parameters of the system can be approximated. However, later in this section we will show that the attraction dimension [2] can be simply achieved using the Lyapunov exponents.

6.2 Entropy

Entropy is a measure of uncertainty. The level of chaos may also be measured using entropy of the system. Higher entropy represents higher uncertainty and a more chaotic system. Entropy is given as:

$$Entropy\ of\ the\ Signal\, x(n) = \int_{min\,(x)}^{max\,(x)} p_x \log\left(1/p_x\right)dx \tag{6.1}$$

where p_x is the probability density function (pdf) of signal $x(n)$. Generally, the distribution can be a joint pdf when the EEG channels are jointly processed. Conversely, the pdf can be replaced by conditional pdf in places where the occurrence of the event is subject to another event. In this case, the entropy is called conditional entropy. Entropy is very sensitive to noise. Noise increases the uncertainty and noisy signals have higher entropy even if the original signal is ordered.

6.3 Kolmogorov Entropy

Also known as metric entropy, is an effective measure of chaos. To find the Kolmogorov entropy the phase space is divided into multidimensional hypercubes. Phase space is the space in which all possible states of a system are represented, with each corresponding to one unique point in the phase space. In phase space, every degree of freedom or parameter of the system is represented as an axis of a multidimensional space. A phase space may contain many dimensions. The hypercube is a generalization of a 3-cube to n-dimensions, also called an n-cube or measure polytope. It is a regular polytope with mutually perpendicular sides and is therefore an orthotope. Now, let $P_{i_0,\ldots,i_n}$ be the probability that a trajectory falls inside the hypercube; i_0 at $t = 0$, i_1 at $t = T$, i_2 at $t = 2\,T$,... Then define:

$$K_n = -\sum_{i_0,\ldots i_n} P_{i_0,\ldots i_n} \ln P_{i_0,\ldots i_n} \tag{6.2}$$

where $K_{n+1} - K_n$ is the information needed to predict which hypercube the trajectory will be in at $(n+1)T$ given trajectories up to nT. The Kolmogorov entropy is then defined as:

$$K = \lim_{N \to \infty} \frac{1}{NT} \sum_{n=0}^{N-1} \left(K_{n+1} - K_n\right) \tag{6.3}$$

However, estimation of the above joint probabilities for large dimensional data is computationally costly. Conversely, in practise, long data sequences are normally required to perform a precise estimation of the Kolmogorov entropy.

6.4 Multiscale Fluctuation-Based Dispersion Entropy

Entropy is not only a measure of uncertainty but also represents the minimum amount of information we can use to encode a binary code. However, if the objective is to evaluate the uncertainty or irregularities in the signals at one temporal scale, other measures such as sample entropy (SampEn), fuzzy entropy (FuzEn), and permutation entropy (PerEn) can be used.

Nevertheless, these approaches fail to account for the inherent multiple time scales of physiological signals such as EEG [3, 4]. To deal with this limitation, multiscale SampEn was proposed [5] and it has become a prevalent algorithm to quantify the complexity of univariate time series, especially physiological recordings [6].

A multiscale fluctuation-based dispersion entropy (MFDE) is based on a coarse-graining process [3] and fluctuation-based dispersion entropy (FDispEn) [7].

Consider a signal $u(t)$ of length L, it is first divided into non-overlapping segments x_j of length l, where $j = 1, 2..., L/l$. the normal cumulative distribution function (NCDF) maps x_j into y_j in the [0 1] interval as [6]:

$$y_j = \frac{1}{\sigma\sqrt{2\pi}}\int_{-\infty}^{x_j} e^{\frac{-(t-\mu)^2}{2\sigma^2}}dt \tag{6.4}$$

where μ and σ represent the centre and width of the mapping waveform. This value is then linearly quantized into C levels as:

$$z_j^C = round\left(C \cdot y_j + 0.5\right) \tag{6.5}$$

Then we define $z_i^{m,C} = \left\{z_i^C, z_{i+d}^C, ..., z_{i+(m-1)d}^C\right\}$ for $i = 1, 2, ..., N-(m-1)d$. Then a fluctuation-based dispersion pattern $\pi_{v_0 v_1 ... v_{m-1}}$ is associated with each $z_i^{m,C}$ where $v_{m-1} = z_{i+(m-1)d}^C$. The number of possible fluctuation-based dispersion patterns that can be assigned to each time series $z_i^{m,C}$ is equal to $(2C-1)^{m-1}$. Then we define:

$$p(\pi_{v_0 v_1 ... v_{m-1}}) = \frac{\#\{i \mid i \leq N-(m-1)d\}}{N-(m-1)d} \tag{6.6}$$

where # refers to cardinality. This, in fact, shows the number of dispersion patterns of $\pi_{v_0 v_1 ... v_{m-1}}$ assigned to $z_i^{m,C}$. The MFDE is then defined as the related entropy, i.e.:

$$\text{FDispEn}(x, m, C, d) = -\sum_{\pi=1}^{(2C-1)^{m-1}} p(\pi_{v_0 v_1 ... v_{m-1}}) \cdot \ln\left(p(\pi_{v_0 v_1 ... v_{m-1}}\right) \tag{6.7}$$

Some applications of MFDE to physiological signals including EEG can be found in [6].

6.5 Lyapunov Exponents

A chaotic model can be generated by a simple feedback system. Consider a quadratic iterator of the form $x(n) \rightarrow \alpha x(n)(1 - x(n))$ with an initial value of x_0. This generates a time series such as that in Figure 6.1 (for $\alpha = 3.8$).

Although in the first 20 samples the time series seems to be random noise its semi-ordered alterations (cyclic behaviour) later show that some rules govern its chaotic behaviour. This time series is subject to two major parameters α and x_0.

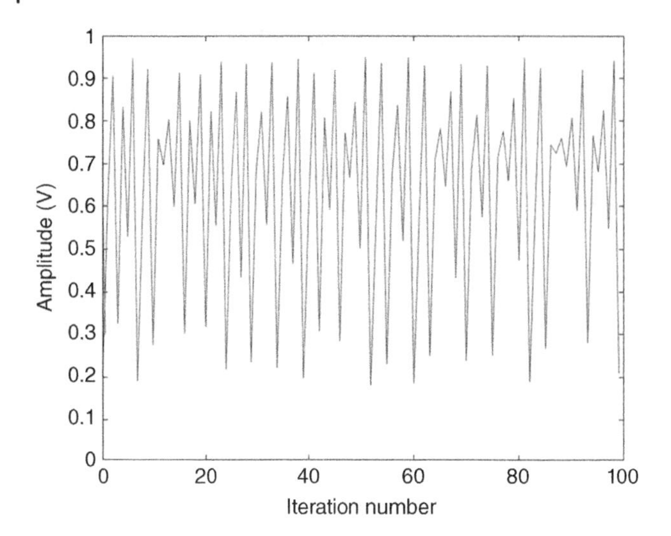

Figure 6.1 Generated chaotic signal using the model $x(n) \rightarrow \alpha x(n)(1 - x(n))$ using $\alpha = 3.8$ and $x_0 = 0.2$.

Now to adopt this model within a chaotic system a different initial value may be selected. Perturbation of an initial value generates an error E_0, which propagates during the signal evolution. After n samples the error changes to E_n. E_n/E_0 is a measure of how fast the error grows. The average growth of infinitesimally small errors in the initial point x_0 is quantified by Ljapunov (Lyapunov) exponents $\lambda(x_0)$. The total error amplification factor $|E_n/E_0|$, can be written in terms of sample error amplifications as:

$$\left|\frac{E_n}{E_0}\right| = \left|\frac{E_n}{E_{n-1}}\right| \cdot \left|\frac{E_{n-1}}{E_{n-2}}\right| \cdots \left|\frac{E_1}{E_0}\right| \tag{6.8}$$

The average logarithm of this becomes:

$$\frac{1}{n} \ln \left|\frac{E_n}{E_0}\right| = \frac{1}{n} \sum_{k=1}^{n} \ln \left|\frac{E_k}{E_{k-1}}\right| \tag{6.9}$$

Obviously, the problem is how to measure $|E_k/E_{k-1}|$. For the iterator $f(x(n))$ ($f(x(n)) = \alpha x(n)(1 - x(n))$ in the above example) having a small perturbation ε at the initial point, the term in the above equation (Eq. 6.9) may be approximated as:

$$\frac{1}{n} \ln \left|\frac{E_n}{E_0}\right| = \frac{1}{n} \sum_{k=1}^{n} \ln \left|\frac{\widetilde{E}_k}{\varepsilon}\right| \tag{6.10}$$

where $\widetilde{E}_k = f(x_{k-1} + \varepsilon) - f(x_{k-1})$. By replacing this in the above equation the Lyapunov exponent is approximated as:

$$\lambda(x_0) = \lim_{n \to \infty} \frac{1}{n} \sum_{k=1}^{n} \ln |f'(x_{k-1})| \tag{6.11}$$

This measure is very significant in separating unstable, unpredictable, or chaotic behaviour from predictable, stable, or ordered ones. If λ is positive the system is chaotic whereas it is negative for ordered systems.

Kaplan and Yorke [8] empirically concluded that it is possible to predict the dimension of a strange attractor from the knowledge of the Lyapunov exponents of the corresponding transformation. This is termed the Kaplan–Yorke conjecture, and has been investigated by many other researchers [9]. This is a very important conclusion since in many dynamical systems the various dimensions of the attractors are hard to compute, while the Lyapunov exponents are relatively easy to compute. This conjecture also claims that generally the information dimension D_I and Lyapunov dimension D_L, respectively, are defined as [1]:

$$D_I = \lim_{s \to 0} \frac{I(s)}{\log_2 1/s} \tag{6.12}$$

where s is the size of a segment of the attractor and $I(s)$ is the entropy of s, and

$$D_L = m + \frac{1}{|\lambda_{m+1}|} \sum_{k=1}^{m} \lambda_k \tag{6.13}$$

where m is the maximum integer with $\gamma(m) = \lambda_1 + \ldots + \lambda_m \geq 0$ given that $\lambda_1 > \lambda_2 > \ldots > \lambda_m$ (for $\lambda_1 < 0$ we set $D_L = 0$) are equivalent.

6.6 Plotting the Attractor Dimensions from Time Series

Very often it is necessary to visualize a phase space attractor and decide about the stability, chaosity, or randomness of a signal (time series). The attractors can be multidimensional. For a three-dimensional attractor we can choose a time delay T (a multiple of τ) and construct the following sequence of vectors:

$$\begin{array}{l}
[x(0) \quad x(T) \quad x(2T)] \\
[x(\tau) \quad x(\tau + T) \quad x(\tau + 2T)] \\
[x(2\tau) \quad x(2\tau + T) \quad x(2\tau + 2T)] \\
\ldots \\
\ldots \\
\ldots \\
[x(k\tau) \quad x(k\tau + T) \quad x(k\tau + 2T)]
\end{array}$$

By plotting these points in a three-dimensional coordinate space and linking the points together successively we can observe the attractor. Figure 6.2 shows the attractors for a sinusoidal and the above chaotic time series.

Although the attractors can be defined for a higher dimensional space, visualization of the attractors is not possible when the number of dimensions increases above three.

6.7 Estimation of Lyapunov Exponents from Time Series

Calculation of the Lyapunov exponents from time series was first proposed by Wolf et al. [10]. In their method, initially a finite embedding sequence is constructed from the finite time series of $2N + 1$ components as:

$$x(0), x(\tau), x(2\tau), \ldots$$

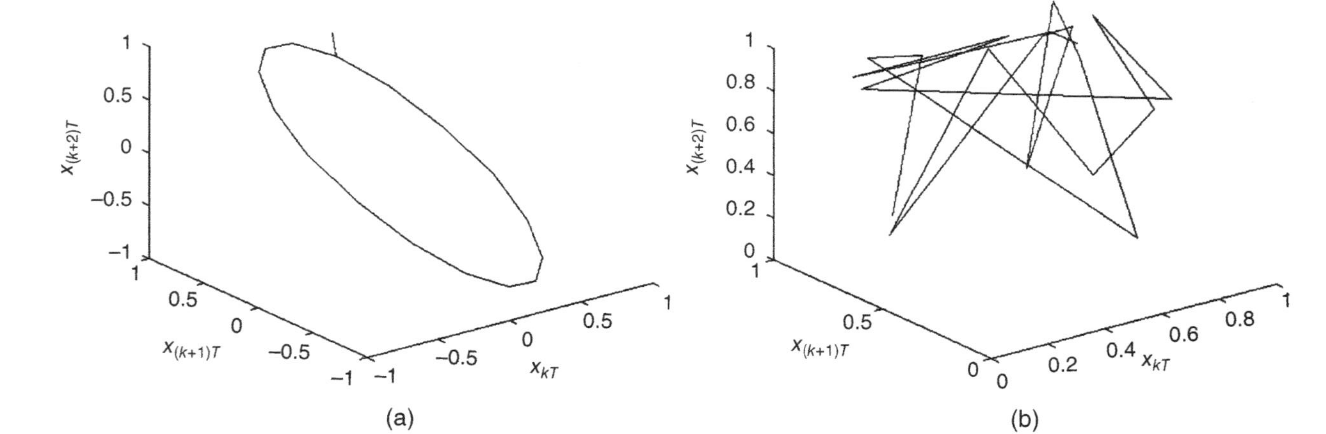

Figure 6.2 The attractors for (a) a sinusoid and (b) the above chaotic time sequence, both started from the same initial point.

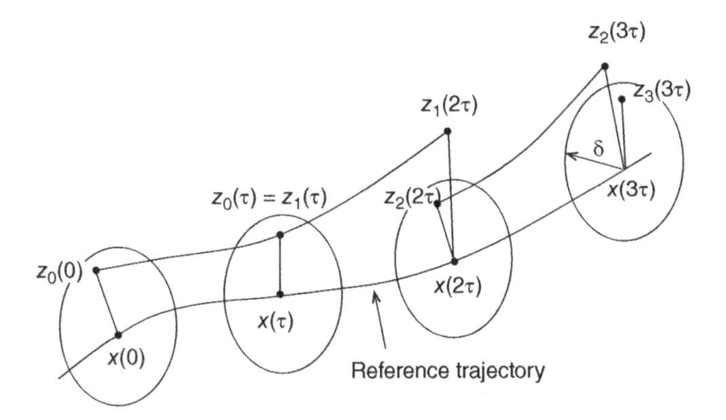

Figure 6.3 The reference and the model trajectories, evolution of the error, and start and end of the model trajectory segments. The model trajectory ends when its deviation from the reference trajectory is more than a threshold.

These are the basic data (often called reference trajectory or reference orbit) upon which the model builds. Generally, we do not have the start point since there is no explicit governing equation that would generate the trajectory. From this sequence we may choose a point $x(k_0\tau)$, which approximates the desired initial point $z(0)$. Considering Figure 6.3, these approximations should satisfy:

$$|x(k_0\tau) - x(0)| < \delta \tag{6.14}$$

where δ is an a priori chosen tolerance. We may rename this point as

$$z_0(0) = x(k_0\tau) \tag{6.15}$$

The successors of this point are known as:

$$z_0(r\tau) = x((k_0 + r)\tau), \quad r = 1, 2, 3, \ldots \tag{6.16}$$

Now there are two trajectories to compare. The logarithmic error amplification factor for the first time interval becomes:

$$l_0 = \frac{1}{\tau} \log \frac{|z_0(\tau) - z_0(0)|}{|x(\tau) - x(0)|} \tag{6.17}$$

This procedure is repeated for the next point $x(\tau)$ of the reference trajectory. For that point we need to find another point $z_1(\tau)$ from the trajectory, which represents an error with a direction close to the one obtained from $z_0(\tau)$ relative to $x(\tau)$. In the case that the previous trajectory is still close to the reference trajectory we may simply continue with that, thus setting $z_1(\tau) = z_0(\tau)$. This yields an error amplification factor l_1. Other factors, l_1, l_2, … , l_{m-1}, can also be found by following the same procedure until the segment of the time series is exhausted. An approximation to the largest Lyapunov exponent for the current segment of the time series is obtained by averaging the logarithmic amplification factors over the whole reference trajectory:

$$\lambda = \frac{1}{m} \sum_{j=0}^{m-1} l_j \tag{6.18}$$

Instead of the above average, the maximum value of the error amplification factor may also be considered as the largest Lyapunov exponent. It is necessary to investigate the effect of noise here. The data usually stem from a physical measurement and therefore contain noise. Hence, the perturbed points, $z_k(k\tau)$ should not be taken very close to each other, because then the noise would dominate the stretching effect on the chaotic attractor. Conversely, we should not allow the error to become too large in order to avoid nonlinear effects. Thus, in practise we prescribe some minimal error, δ_1, and a maximal error, δ_2, and require:

$$\delta_1 < |x(k\tau) - z_k(k\tau)| < \delta_2 \tag{6.19}$$

6.7.1 Optimum Time Delay

In the above calculation it is important to find the *optimum time delay* τ. Very small time delays may result in near-linear reconstructions with high correlations between consecutive phase space points and very large delays might ignore any deterministic structure of the sequence. In an early proposal [11] the autocorrelation function is used to estimate the time delay. In this method τ is equivalent to the duration after which the autocorrelation reaches to a minimum or drops to a small fraction of its initial value. In another attempt [12, 13] it has been verified that the values of τ at which the mutual information (MI) has a local minimum are equivalent to the values of τ at which the logarithm of the correlation sum has a local minimum.

6.7.2 Optimum Embedding Dimension

To further optimize the measurement of Lyapunov exponents we need to specify the optimum value for *m*, named the *embedding dimension*. Before doing that some definitions have to be given as follows.

Fractal dimension is another statistic related to the dynamical measurement. The strange attractors [14] are fractals and their fractal dimension D_f is simply related to the minimum number of dynamical variables needed to model the dynamics of the attractor. Conceptually, a simple way to measure D_f is to measure the *Kolmogorov capacity*. In this measurement a set is covered with small cells, depending on the dimensionality (i.e. squares for sets embedded in two dimensions, cubes for sets embedded in three dimensions, and so on), of size ε. If $M(\varepsilon)$ denotes the number of such cells within a set, the fractal dimension is defined as:

$$D_f = \lim_{\varepsilon \to 0} \frac{\log(M(\varepsilon))}{\log\left(\frac{1}{\varepsilon}\right)} \tag{6.20}$$

for a set of single points $D_f = 0$, for a straight line $D_f = 1$, and for a plane area it is $D_f = 2$. The fractal dimension, however, may not be an integer.

Correlation dimension is defined as:

$$D_r = \lim_{r \to 0} \frac{\log C(r)}{\log r} \tag{6.21}$$

where

$$C(r) = \sum_{i=1}^{M(r)} p_i^2 \tag{6.22}$$

is the correlation sum.

Optimal embedding dimension, m, as required for accurate estimation of the Lyapunov exponents, has to satisfy $m \geq 2D_f + 1$. D_f is however, not often known a priori. The Grassberger–Procaccia algorithm can nonetheless be employed to measure the correlation dimension, C_r. The minimum embedding dimension of the attractor is $m + 1$, where m is the embedding dimension above which the measured value of the correlation dimension C_r remains constant.

As another very important conclusion:

$$D_f = D_L = 1 + \frac{\lambda_1}{|\lambda_2|} \tag{6.23}$$

i.e. the fractal dimension D_f is equivalent to the Lyapunov dimension [1].

Chaos has been used as a measure in analysis of many types of signals and data. Its application to epileptic seizure prediction will be shown in Chapter 4.

6.8 Approximate Entropy

Approximate entropy (AE) is a statistic which can be estimated from the discrete-time sequences especially for real-time applications [15, 16]. This measure can quantify the complexity or irregularity of the system. AE is less sensitive to noise and can be used for short-length data. In addition, it is resistant to short strong transient interferences (outliers) such as spikes [16].

Given the embedding dimension m, the m-vector $\mathbf{x}(i)$ is defined as:

$$\boldsymbol{x}(i) = [x(i), x(i+1), \dots, x(i+m-1)], \quad i = 1, \dots, N-m+1 \tag{6.24}$$

where N is the number of data points. The distance between any two of the above vectors, $\mathbf{x}$ (i) and $\mathbf{x}$ (j), is defined as:

$$d[\boldsymbol{x}(i), \boldsymbol{x}(j)] = \max_k |x(i+k) - x(j+k)| \tag{6.25}$$

where |.| denotes the absolute value. Considering a threshold level of β, we find the number of times, $M^m(i)$, that the above distance satisfies $d[\boldsymbol{x}(i), \boldsymbol{x}(j)] \leq \beta$. This is performed for all i. For the embedding dimension m we form:

$$\xi_\beta^m(i) = \frac{M^m(i)}{N-m+1}, \quad \text{for } i = 1, \dots, N-m+1 \tag{6.26}$$

Then, the average natural logarithm of $\xi_\beta^m(i)$ is found as:

$$\psi_\beta^m = \frac{1}{N-m+1} \sum_{i=1}^{N-m+1} \ln \xi_\beta^m(i) \tag{6.27}$$

By repeating the same method for an embedding dimension of $m+1$, the AE will be given as:

$$AE(m,\beta) = \lim_{N\to\infty}\left[\psi_\beta^m - \psi_\beta^{m+1}\right] \tag{6.28}$$

In practise, however, N is limited and therefore the AE is calculated for N data samples. In this case the AE depends on m, β, and N, i.e.:

$$AE(m,\beta,N) = \psi_\beta^m - \psi_\beta^{m+1} \tag{6.29}$$

The embedding dimension can be found as previously mentioned. However, the threshold value has to be set correctly. In some applications the threshold value is taken as a value between 0.1 and 0.25 times the data standard deviation [15].

6.9 Using Prediction Order

It is apparent that for signals with highly correlated time samples the prediction order of an autoregressive (AR) or autoregressive–moving-average (ARMA) model is low and for noise type signals where the correlation among the samples is low the order is high. This means

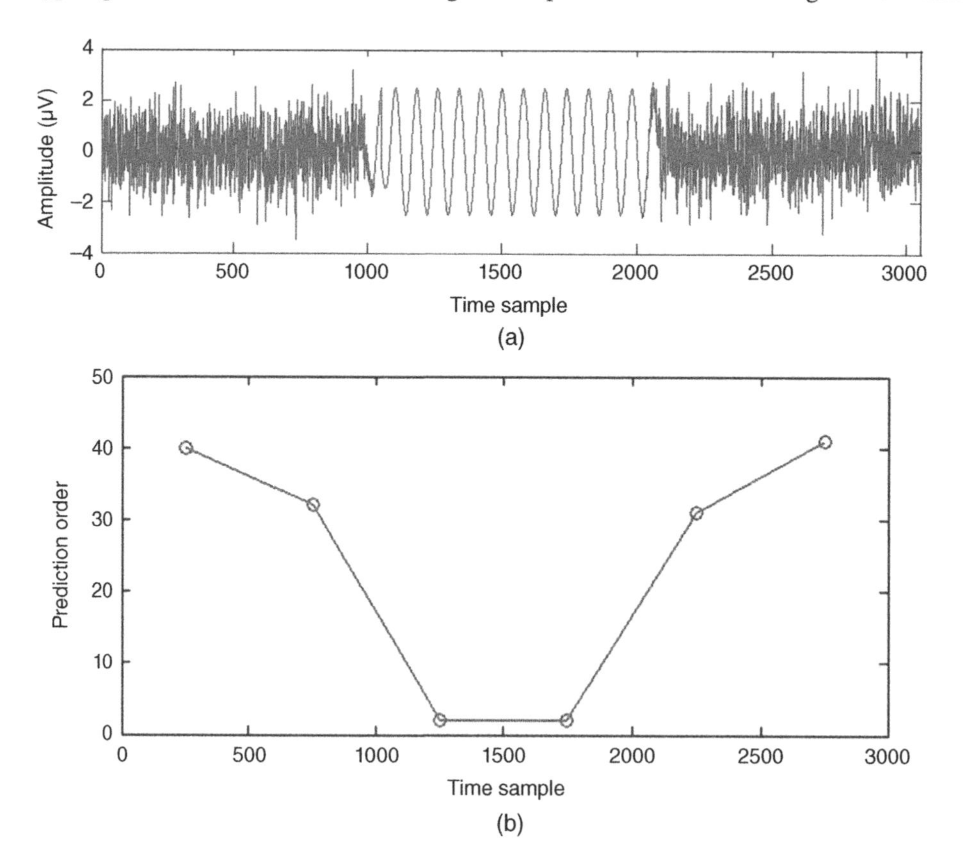

Figure 6.4 (a) The signal and (b) prediction order measured for overlapping segments of the signal.

for the latter case a large number of previous samples are required for prediction of the current sample. A different criterion such as the Akaike information criterion (AIC) may be employed to find the prediction order from the time series. Figure 6.4 shows the prediction order automatically computed for overlapping segments of three-sections of a time series in which the middle section is a sinusoidal and the first and third sections are random noise signals.

6.10 Summary

Brain is a complex system with nonlinear behaviour. It is generally very difficult to accurately characterize and express the dynamics of a complex system and the behaviour of a signal trend generated by a nonlinear system unless the nonlinearity is well defined and mathematically formulated. In order to more accurately define chaos in a system often other parameters such as optimum time delay and attractor dimension have to be estimated. These parameters inherently change in time and therefore, a more complex evaluation is required. Without correctly estimating these parameters the above measures are only rough chaos estimators. Conversely, most of the measures introduced in this chapter, particularly different entropy estimators, cannot distinguish between chaos and noise in a system.

Another interesting conclusion of this chapter is that in a multichannel case, an ensemble measure of entropy corresponds to dependency between the channels and therefore can represent the MI.

References

1 Peitgen, H.-O., Lurgens, H., and Saupe, D. (1992). *Chaos and Fractals*. New York: Springer-Verlag.
2 Grassberger, P. and Procaccia, I. (1983). Characterization of strange attractors. *Physical Review Letters* 50: 346–349.
3 Costa, M., Goldberger, A.L., and Peng, C.-K. (2005). Multiscale entropy analysis of biological signals. *Physical Review E* 71 (2): 021906.
4 Azami, H., Fernández, A., and Escudero, J. (2017). Refined multiscale fuzzy entropy based on standard deviation for biomedical signal analysis. *Medical & Biological Engineering & Computing* 55 (11): 2037–2052.
5 Costa, M., Goldberger, A.L., and Peng, C.-K. (2002). Multiscale entropy analysis of complex physiologic time series. *Physical Review Letters* 89 (6): 068102.
6 Azami, H., Arnold, S.E., Sanei, S. et al. (2019). Multiscale fluctuation-based dispersion entropy and its applications to neurological diseases. *IEEE Access* 7 (1): 68718–68733, Print ISSN: 2169-3536. https://doi.org/10.1109/ACCESS.2019.2918560.
7 Azami, H. and Escudero, J. (2018). Amplitude- and fluctuation-based dispersion entropy. *Entropy* 20 (3): 210.

8 Kaplan, L. and Yorke, J.A. (1979). Chaotic behaviour of multidimensional difference equations. In: *Functional Differential Equations and Approximation of Fixed Points* (ed. H.-O. Walther). Springer-Verlag.

9 Russell, D.A., Hanson, J.D., and Ott, E. (1980). Dimension of strange attractors. *Physical Review Letters* 45: 1175–1179.

10 Wolf, A., Swift, J.B., Swinny, H.L., and Vastano, J.A. (1985). Determining Lyapunov exponents from a time series. *Physica* 16D: 285–317.

11 Albano, A.M., Muench, J., Schwartz, C. et al. (1988). Singular value decomposition and the Grassberger–Procaccia algorithm. *Physical Review A* 38: 3017.

12 King, G.P., Jones, R., and Broomhead, D.S. (1987). Phase portraits from a time series: a singular system approach. *Nuclear Physics B* 2: 379.

13 Fraser, A.M. and Swinney, H. (1986). Independent coordinates for strange attractors from mutual information. *Physical Review A* 33: 1134–1139.

14 Grassberger, P. and Procaccia, I. (1983). Characterization of strange attractors. *Physical Review Letters* 50: 346–349.

15 Pincus, S.M. (1991). Approximate entropy as a measure of system complexity. *Proceedings of the National Academy of Sciences of the United States of America* 88: 2297–2301.

16 Fusheng, Y., Bo, H., and Qingyu, T. (2001). Approximate entropy and its application in biosignal analysis. In: *Nonlinear Biomedical Signal Processing*, vol. II (ed. M. Akay), 72–91. IEEE Press.

7

Machine Learning for EEG Analysis

7.1 Introduction

Machine learning, often including training and testing, refers to a combination of learning from data and data clustering or classification/prediction. In the majority of applications, it is combined with some signal processing techniques to extract the best discriminating data features prior to taking learning and classification steps. The emerging machine learning techniques, such as deep neural networks (DNNs), are also capable of feature learning.

The features may be extracted from a particular EEG channel, such as C3 or C4, to best capture the brain motor activity. To better identify the state of the human body, multimodal data such as EEG together with polysomnography [1] or functional magnetic resonance imaging (fMRI) may also be taken into account.

Newly emerging sensor technology is beginning to closely mimic the ultimate sensing machine, i.e. the human being. Sensor fusion technology leverages a microcontroller (which mimics the human brain) to fuse the individual data collected by multielectrodes. Multichannel recording techniques allow for a more accurate and reliable realization of the data than one would acquire from a single electrode. In the case of EEG, this becomes more obvious when multiple simultaneous brain activities are to be realized.

Availability of multichannel EEG recordings enables both data/sensor fusion and data factorization when presented in the form of multiway data (tensors). Advances in sensor fusion for remote emotive computing (emotion sensing and processing) could also lead to exciting new applications in the future, including smart healthcare. This approach motivates personalized healthcare providers to fine tune and customize systems which best suit individual's need.

Conversely, the brain zones communicate to each other and behave like the nodes of a communication network. Surely, such a network can be better characterized if the recordings were from over the cortex. The topology of such a network in terms of number of nodes and strength of the links between them changes with respect to time and the particular brain activity. At the same time, the new sensor networking technology is shifting towards consensus or diffusion communication networks whereby the network nodes can communicate with each other and decide on their next action without communicating to any hub or master node. This decentralized decision-making approach brings a new direction in machine learning called *cooperative learning* which exploits the information from a number

EEG Signal Processing and Machine Learning, Second Edition. Saeid Sanei and Jonathon A. Chambers.

of electrodes in a neighbourhood to decide on the action of each particular one within that neighbourhood.

Traditionally, there are mainly two types of machine learning algorithms for classification/prediction of data; supervised and unsupervised. In unsupervised learning the classifier clusters the data into the groups having farthest distances from each other. A popular example for these algorithms is the *k*-means algorithm. Conversely, for supervised classifiers, the target is known during the training phase and the classifier is trained to minimize a difference between the actual output and the target values. In both cases, effective estimation of data features often helps reduce the number of input samples thus enhancing the overall algorithm speed. The more efficiently the features are estimated, the more accurate the result of clustering or classification will be.

Feature estimation or detection requires the use of single or multichannel signal processing techniques. These algorithms are often capable of changing the dimensionality of the signals to make them more separable. Numerous statistical measures can be derived from the data with known distributions. Various statistics such as mean, variance, skewness, and kurtosis are very popular statistical measures to describe the data in scalar forms. The vector forms may become necessary for multichannel data.

In addition to the above two main classes of learning systems, semi-supervised techniques benefit from partially known information during the training or classification/prediction steps. This group of learning systems exploit various constraints on the input, output, or the classifier in order to best optimize the system for each particular application. One useful application of such systems is for rejection of anomalies in the datasets during the classifier training [2]. Conversely, machine learning is considered as the subset of artificial intelligence (AI). The AI learning systems may be categorized into supervised learning, unsupervised learning, semi-supervise learning, reinforcement learning (learning by trial and error), deep learning, and ensemble learning (by using multiple models and learning algorithms).

In the clustering methods there is no need to label the classes in advance and only the number of clusters maybe identified and fed into the clustering algorithm by the user. Classification of data is similar to clustering except, the classifier is trained using a set of labelled data before being able to classify new data. In practise, the objective of classification is to draw a boundary between two or more classes and to label them based on their measured features. In a multidimensional feature space this boundary takes the form of a separating hyperplane. The objective here is to find the best hyperplane, maximizing the distance from all the classes (inter-class members) while minimizing proximities within members of each class (intra-class members).

Clustering and classification become ambiguous when deciding which features to use and how to extract or enhance those features. Principal component analysis (PCA) and independent component analysis (ICA) are two very common approaches to extract the dominant data features. Traditionally, a wide range of features (such as various orders of statistics) are measured or estimated from the raw data and used by the machine learning algorithms for clustering or classification.

In the context of human-related signals, classification of the data in feature space is often required. For example, the power of alpha and beta waves in the EEG signals [3] may be classified to not only detect brain abnormalities but also determine the stage of the disease.

In sleep studies, one can look at the power of alpha and theta waves as well as appearance of spindles and k-complexes to classify the stages of sleep. As another example, in brain–computer interfacing (BCI) systems for left and right-finger movement detection one needs to classify the time, frequency, and spatial features.

Traditional methods based on adaptive filters using mean squared error (MSE) or least-squares error (LSE) minimizations are still in use. Popular examples of such classifiers are linear discriminant analysis (LDA), hidden Markov model (HMM), Gaussian mixture model (GMM), decision tree [4], random forest, and support vector machines (SVM). *k*-means clustering and fuzzy logic have also been widely used for clustering, pre-classification, and other AI-based applications [5]. These techniques have been developed and well explained in the literature [6]. DNNs have become even more popular recently mainly due to their feature learning/extracting capability. For DNNs the number of layers and neurons highly depends on the data complexity. Detailed explanation of all these methods is beyond the objective of this chapter. Instead, here we provide a summary of the most popular machine learning methods which have been widely used for EEG analysis.

Unlike many mathematical problems in which some forms of explicit formulae based on a number of inputs result in an output, in data classification there is no model or formulation of this kind. Instead, the system should be generally trained (using labelled data) to be able to recognize a new (test) input.

In the case of data with considerable noise or artefacts, successful classifiers are those which can minimize the effect of outliers. Outliers are the data samples or features very different from the class centres and therefore, they do not belong to any class. In many machine learning algorithms, outliers have destructive effect during the classifier training and increase the cross-validation error. Preprocessing (including smoothing and denoising) can significantly reduce the outliers or their impact. More advanced classifiers are able to reject the outliers during the training phase. Therefore, the right choice of classifiers and their associated feature detection systems can also reduce the influence of outliers.

Machine learning is important and becoming increasingly popular for EEG applications since:

1) Where EEG is used for BCI or integrated within a rehabilitation system, the learning block usually monitors dynamic environments which rapidly change over time. It is therefore desirable to develop systems that can adapt and operate efficiently in such environments.
2) The increase in the data captured from healthy and diseased persons enables the training of machine learning systems and the detection of new abnormal cases. By looking at distinguishing features of EEGs for different brain states and abnormalities, the cases such as seizure and Alzheimer disease can be detected easily subject to availability of moderate amount of EEG records.
3) EEG patterns for different brain abnormalities and various brain states are significantly different. In the following chapters some of these patterns are demonstrated. One popular example is sleep EEG which shows an obvious shift (from high to low) in normal brain rhythms and the rise of distinguishing waveforms such as spindles and k-complexes. The differences between such diverse patterns can be captured by machine learning systems.

4) New applications, such as in-patient status monitoring, cyber-physical systems (CPS), machine-to-machine (M2M) communications, and Internet of Things (IoT) technologies, have been introduced with a motivation to support more intelligent decision making and autonomous control [7]. Here, machine learning is important to extract the different levels of abstractions necessary to perform the AI tasks with limited human intervention [8].
5) EEG artefact waveforms have different morphology compared to normal brain rhythms and therefore the corrupted EEG segments can be detected and removed by means of classifiers [9, 10].

Many machine learning algorithms do not perform efficiently when:

- the number of features or input samples is high
- there is a limited time to perform the classification
- there is a non-uniform weighting amongst features
- there is a nonlinear map between inputs and outputs
- the distribution of data is not known
- there are significant number of outliers
- the algorithm is not convex (monotonic), so it may fall into a local minima.

In addition, in using machine learning for EEG analysis:

- Online learning or testing within a fully AI framework is feasible mainly due to a low sampling frequency of the signals. The sampling rate for EEG processing is often selected between 100 to 1000 Hz.
- As we deal with a multichannel system, time, frequency, and spatial features can be estimated and incorporated into the machine learning algorithms. This benefits from full signal diversity which introduces more efficiency in the overall learning algorithm and inclusiveness of the feature spaces.

For the data with unknown distributions often nonparametric techniques are applied. In these techniques a subset of labelled data is used to estimate or model the data distributions through learning before application of any feature selection or classification algorithm.

To assess the performance quality of the classifiers a multifold cross-validation is performed. For this, only the training data are considered. In each fold a percentage of the labelled data is used to train the classifier and the rest used for testing. The cross-validation error is then calculated as the amount of misclassified data over multiple folds, say k-fold.

One important requirement for machine learning is *dimensionality reduction* mainly for high-dimensionality datasets, where the number of available features is large. Reducing dimensionality often reduces the effect of outliers and noise and therefore enhances the classifier performance. PCA is the most popular dimensionality reduction technique. PCA projects the data into a lower dimensional space using singular value decomposition (SVD), $\mathbf{X} = \mathbf{U\Sigma V}^{\mathrm{T}}$ where $\mathbf{U}$ and $\mathbf{V}$ are respectively left and right singular matrices and $\mathbf{\Sigma}$ is the diagonal matrix of eigenvalues. Input to SVD can have mixed signs. Nonnegative matrix factorization (NMF), as another approach, restricts factors to be nonnegative and can be used when the input data are nonnegative. Images and power spectrum are examples of such data. Hence, it works based on imposing a non-negativity constraint on the extracted components, $\mathbf{X}_+ = \mathbf{W}_+\mathbf{H}_+$ where $(.)_+$ represents the non-negativity of all input data components. Binary matrix factorization is another extension of NMF for binary data by constraining the components to be binary, $\mathbf{X}_{0\text{–}1} = \mathbf{W}_{0\text{–}1}\mathbf{H}_{0\text{–}1}$ where $(.)_{0\text{–}1}$ represents the binary elements. Sparsity constraints (by constraining an appropriate norm of the factorized components) can be added to the PCA and NMF optimization to enhance the interoperability and stability of components for spatial principal components analysis (sPCA) and spatial nonnegative matrix factorization (sNMF). Sparsity constraints are particularly important where the data are naturally sparse [11].

7.2 Clustering Approaches

The overarching objective of clustering algorithms is to separate the members of a dataset (or feature set) into distinct groups so that the groups have maximum distance (inter-class distance) between them (their centres) and the members of each group or class (intra-class) have minimum average distance between them. In places where the classes are not separable, nonlinear kernels are often used to map the features into a higher dimensionality space where the data become separable.

7.2.1 *k*-Means Clustering Algorithm

The k-means algorithm [12] is an effective and generally a simple clustering tool that has been widely used for many applications such as discovering spike patterns in neural responses [13]. It is employed either for clustering only or as a method to find the target values (cluster centres) for a supervised classification in a later stage. This algorithm divides a set of features (such as points in Figure 7.1) into k clusters.

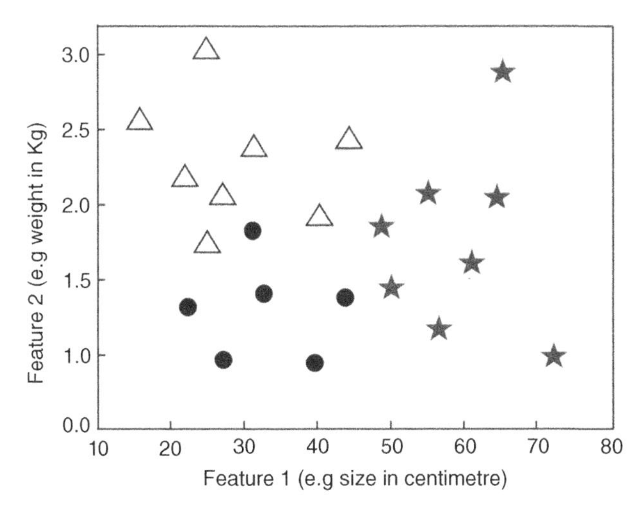

Figure 7.1 A two-dimensional feature space with three clusters, each with members of different shapes (circle, triangle, and asterisk).

The algorithm is initialized by setting C as the desired or expected number of clusters. Then, the centre for each cluster k is identified by selecting k representative data points. The next step is to assign the remaining data points to the closest cluster centre. Mathematically, this means that each data point needs to be compared with every existing cluster centre and the minimum distance between the point and the cluster centre found. Most often, this is performed in the form of error checking (which will be discussed shortly). However, prior to this, the new cluster centres are calculated. This is essentially the remaining step in k-means clustering: once clusters have been established (i.e. each data point is assigned to its closest cluster centre), the geometric centre of each cluster is recalculated.

The Euclidean distance between each data point within a cluster to the cluster centre can be calculated. This can be repeated for all other clusters, whose resulting sums can themselves be summed together. The final sum is known as the *sum of within-cluster sum of squares*. Consider the within-cluster variation (*sum of squares* for cluster c) error as ε_c:

$$\varepsilon_c = \sum_{i=1}^{n_c} \left\| \mathbf{x}_i^c - \overline{\mathbf{x}}_c \right\|_2^2 \quad \forall c \tag{7.1}$$

where $\|.\|_2^2$ is the squared Euclidean distance between its arguments, $\overline{\mathbf{x}}_c$ is the cluster centre, n_c is the total number of data points (features) in cluster c, and x_i^c is an individual data point in cluster c. The cluster centre (mean of data points in cluster c) can be defined as:

$$\overline{\mathbf{x}}_c = \frac{1}{n_c} \sum_{i=1}^{n_c} \mathbf{x}_i^c \tag{7.2}$$

and the total error is

$$E_C = \sum_{c=1}^{C} \varepsilon_c \tag{7.3}$$

The overall k-means algorithm may be summarized as:

1) Initialization
   a) Define the number of clusters (C).
   b) Designate a cluster centre (a vector quantity that is of the same dimensionality of the data) for each cluster, typically chosen from the available data points.
2) Assign each remaining data point to the closest cluster centre. That data point is now a member of that cluster.
3) Calculate the new cluster centre (the geometric average of all the members of a certain cluster).
4) Calculate the sum of within-cluster sum of squares. If this value has not significantly changed over a certain number of iterations, stop the iterations. Otherwise, go back to Step ii.

Therefore, an optimum clustering procedure depends on an accurate estimation of the number of clusters. A common problem in k-means partitioning is that if the initial partitions are not chosen carefully enough the computation will run the chance of converging to a *local* minimum rather than the *global* minimum solution. The initialization step is therefore very important. The following two sections address the problem of unknown number of clusters.

7.2.2 Iterative Self-Organizing Data Analysis Technique

To estimate the number of clusters, one way is to run the algorithm several times with different initializations and with iteratively increasing the number of clusters from the lowest number, e.g. 2. If the results converge to the same partitions, then it is likely that a global minimum has been reached. However, this is very time consuming and computationally expensive. Another solution is to dynamically change the number of partitions (i.e. number of clusters) as the iterations progress. The iterative self-organizing data analysis technique algorithm (ISODATA) is an improvement on the original k-means algorithm that does exactly this. ISODATA involves a number of additional parameters into the algorithm allowing it to progressively check within- and between-cluster similarities so that the clusters can dynamically split and merge.

7.2.3 Gap Statistics

Another approach for solving this problem is gap statistics [14]. In this approach the number of clusters is iteratively estimated. The steps of this algorithm are:

1) For a varying number of clusters $k = 1, 2, \ldots, C$, compute the error E_c using Eq. (7.3).
2) Generate B *number* of reference datasets. Cluster each one with the k-means algorithm and compute the dispersion measures, $\widetilde{E}_{kb}$, $b = 1, 2, \ldots B$. The gap statistics are then estimated using

$$G_k = \frac{1}{B}\sum_{b=1}^{B} \log\left(\widetilde{E}_{kb}\right) - \log\left(E_k\right) \tag{7.4}$$

   where the dispersion measure $\widetilde{E}_{kb}$ is the E_k of the reference dataset B.
3) To account for the sample error in approximating an ensemble average with B reference distributions, compute the standard deviation S_k as:

$$S_k = \left[\frac{1}{B}\sum_{b=1}^{B}\left(\log\left(\widetilde{E}_{kb}\right) - \overline{E}_b\right)^2\right]^{1/2} \tag{7.5}$$

   where

$$\overline{E}_b = \frac{1}{B}\sum_{b=1}^{B} \log\left(\widetilde{E}_{kb}\right) \tag{7.6}$$

4) By defining $\widetilde{S}_k = S_k\left(1 + \frac{1}{B}\right)^{1/2}$, estimate the number of clusters as the smallest k such that $G_k \geq G_{k+1} - \widetilde{S}_{k+1}$,
5) With the number of clusters identified, utilize the k-means algorithm to partition the feature space into k subsets (clusters).

The above clustering method has several advantages over k-means since it can estimate the number of clusters within the feature space. It is also a multiclass clustering system and unlike SVM can provide the boundary between the clusters.

7.2.4 Density-Based Clustering

The density-based clustering approach clusters the point features within surrounding noise based on their spatial distribution. It works by detecting areas where the points are concentrated and are separated by areas that are empty or sparse. Points that are not part of a cluster are labelled as noise. In this clustering method unsupervised machine learning clustering algorithms are used, which automatically detect patterns based purely on spatial location and the distance to a specified number of neighbours. These algorithms are considered unsupervised as they do not need any training on what it means to be a cluster [15].

Density-based spatial clustering of applications with noise (DBSCAN) is a popular density-based clustering algorithm with the aim of discovering clusters from approximate density distribution of the corresponding data points. DBSCAN does not need the number of clusters, instead, has two parameters to be set: an *epsilon* that indicates the closeness of the points in each and *minPts*, the minimum neighbourhood size a point should fall into to be considered as the member of that cluster. The routine is initialized randomly. The neighbourhood of this point is then retrieved and if it consists of an acceptable number of elements, a cluster is formed, otherwise the element is considered as noise. Hence, DBSCAN may result in some samples which are not clustered.

Usually, DBSCAN parameters are not known in advance and there are several ways to select their values. One way is to calculate the distance of each point to its closest nearest neighbour and use the histogram of distances to select epsilon. Then, a histogram can be obtained of the average number of neighbours for each point using the epsilon. Some of the samples do not have enough neighbouring points and are counted as noise samples. Implementation of the parameter selection is available in spark dbscan (https://github.com/alitouka/spark dbscan).

DBSCAN can find arbitrary-shaped clusters and is robust to outliers. Nevertheless, it may not identify clusters of various densities or may fail if the data are very sparse. It is also sensitive to the selection of its parameters and the distance measure (usually Euclidean distance), which affects other clustering techniques too.

7.2.5 Affinity-Based Clustering

This combines k-means clustering with a spectral based clustering method where the algorithms cluster the points using eigenvectors of matrices derived from the data. It uses k eigenvectors simultaneously and identifies the best setting under which the algorithm can perform favourably [16].

7.2.6 Deep Clustering

Deep clustering is another modification to clustering which combines k-means with a neural network (NN) to jointly learn the parameters of an NN and the cluster assignments of the resulting features. It iteratively groups the features using k-means and uses the

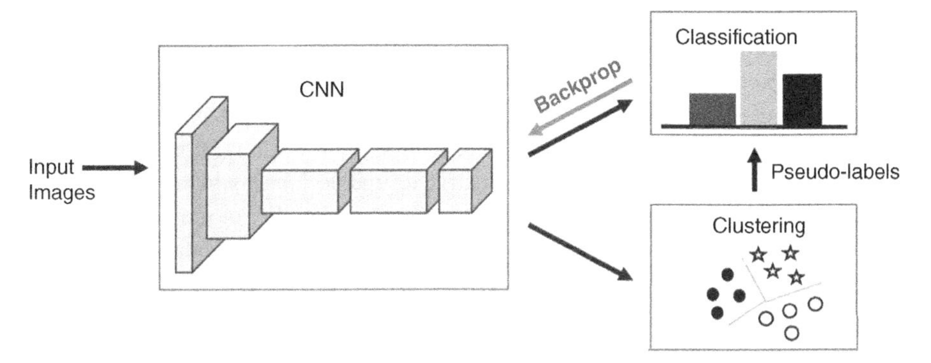

Figure 7.2 Schematic diagram of deep clustering [16]. The deep features are iteratively clustered and the cluster assignments are used as pseudolabels to learn the parameters of a CNN. In Figure 7.2, Backprop refers to the backpropagation algorithm often used to estimate the NN parameters.

subsequent assignments as supervision to update the weights of the network. This organizes an unsupervised training of convolutional neural networks (CNNs). This approach, demonstrated in Figure 7.2, is similar to the standard supervised training of a CNN and it integrates the CNN in its structure [17]. The CNN is explained later in this chapter.

In addition to the above clustering approaches, there are other clustering methods such as power iteration clustering (PIC) [18], which use similar concepts. PIC finds a very low-dimensional embedding of a dataset using truncated power iteration on a normalized pair-wise similarity matrix of the data. It has been shown that it is relatively fast for processing large datasets.

7.2.7 Semi-Supervised Clustering

Conventional clustering methods are unsupervised, meaning that there is no available label (target) or anything known about the relationship between the observations in the dataset. In many situations, however, information about the clusters is available in addition to the feature values. For example, the cluster labels of some observations may be known, or certain observations may be known to belong to the same cluster. In other cases, one may wish to identify clusters that are associated with a particular outcome (or target).

The main purposes for so-called semi-supervised methods are accurate determination of number of clusters and minimization of the clustering error. Very similar to the concept in Section 7.2.6, there are semi-supervised clustering techniques where supervised and unsupervised learnings are combined to iteratively improve the performance of clustering methods in terms of number of clusters and clustering error. Among these algorithms deep networks are often used for supervised clustering [16, 19, 20]. The algorithm in [19] simultaneously minimizes the sum of supervised and unsupervised cost functions by backpropagation.

7.2.7.1 Basic Semi-Supervised Techniques

Two basic semi-supervised techniques are label spreading (LS) and label propagation (LP). LS is based on considering samples (isolates) as nodes in a graph in which their relations are defined by edge weights, e.g. $w_{ij} = \exp(-||x_i - x_j||^2/2\sigma^2)$ if $i \neq j$ and $w_{ij} = 0$. The weight matrix is symmetrically normalized for better convergence. Each node receives information from its neighbouring points and its label is selected based on the class with most received information. A regularization term was also introduced for a better label assignment based on having a smooth classifier (not changing between similar points). LP is based on an iterative technique considering a transition matrix to update labels which starts with a random initialization. The transition matrix refers to the probability of moving from one node to another by propagating in high density areas of the unlabelled data.

7.2.7.2 Deep Semi-Supervised Techniques

As described in the later Section 7.3.9.3, an autoencoder (AE) is a DNN which has two main stages; an encoder that maps the input space to a lower dimension (latent space) and a decoder that reconstructs the input space from the latent representation. Hence, the network is based on unsupervised learning. AE can be stacked to form a deep stacked autoencoder (SAE) network that adds more power to the network and provides nonlinear latent representations for each hidden layer. Moreover, to improve generalization and increase the possibility of extracting more interesting patterns in the data, various noise levels can be added to the input of each layer, denoted as stacked denoising autoencoder (SDAE). The optimization for SDAE is based on minimizing the reconstruction error between the original data and the predicted network output.

Although AE, SAE, and SDAE are based on unsupervised learning (using the reconstruction error), they have been used for semi-supervised learning by adding a layer to the learned encoders and fine tuning the network based on the labelled data for the supervised task (supervised SDAE). The main disadvantage of this technique is to have two stages for supervised and unsupervised learning. Consequently, the first stage may extract information that is not useful for the final supervised task. Due to this, the fine-tuning stage cannot extensively change a fully learned network [21].

Ladder is an extension of SAE/SDAE considering a joint objective function of reconstruction error for unsupervised learning and likelihood of estimating the correct labels for supervised learning. It has two noiseless and noisy encoder paths and a denoising decoder [19].

7.2.8 Fuzzy Clustering

In this clustering method each element has a set of membership coefficients corresponding to the degree of being in a given cluster [22]. This is different from k-means, where each object belongs exactly to one cluster. For this reason, unlike k-means, which is known as hard or non-fuzzy clustering, this is termed as soft clustering method.

In fuzzy clustering, points close to the centre of a cluster, may be in the cluster to a higher degree than points at the edge of a cluster. The degree, to which an element belongs to a given cluster, is a numerical value varying from 0 to 1.

However, fuzzy c-means (FCM), the most widely used fuzzy clustering algorithms, is very similar to k-means. In FCM the centroid of a cluster is calculated as the mean of all points, weighted by their degree of belonging to the cluster. The aim of c-means is to cluster the points into k cluster by minimizing the objective function defined as:

$$\sum_{j=1}^{k} \sum_{x_i \in c_j} u_{ij}^{m} \left(\mathbf{x}_i - \mu_j \right)^2 \tag{7.7}$$

where μ_j is the centre of the cluster j, m the fuzzifier, often selected manually, and u_{ij}^m is the degree to which an observation $\mathbf{x}_i$ belongs to a cluster $\mathbf{c}_j$, defined as:

$$u_{ij}^{m} = \frac{1}{\sum_{l=1}^{k} \left(\frac{|\mathbf{x}_i - \mathbf{c}_j|}{|\mathbf{x}_i - \mathbf{c}_l|} \right)^{\frac{2}{m-1}}} \tag{7.8}$$

The degree of belonging, u_{ij}^m, is linked inversely to the distance from $\mathbf{x}$ to the cluster centre.

The parameter m is a real number within $1.0 < m \leq \infty$ and defines the level of cluster fuzziness. For m close to 1 the solution becomes very similar to hard clustering such as k-means; whereas a value of m close to infinity leads to complete fuzziness. The centroid of a cluster, $\mathbf{c}_j$, is the mean of all points, weighted by their degree of belonging to the cluster:

$$\mathbf{c}_j = \frac{\sum_{x \in c_j} u_{ij}^{m} \mathbf{x}}{\sum_{x \in c_j} u_{ij}^{m}} \tag{7.9}$$

The algorithm of fuzzy clustering can be summarized as follows:

1) Set the number of clusters k.
2) Assign randomly to each point coefficients for being in the clusters.
3) Repeat until the maximum number of iterations is reached, or when the algorithm has converged for a predefined error:

   a) Compute the centroid for each cluster using Eq. (9.9).
   b) For each point, use (9.8) to compute its coefficients of being in (degree of belonging to) the clusters.

This concludes the well established clustering approaches. Most of these methods are sensitive to outliers and the objective of new weighted or regularized techniques is to reduce this sensitivity. Other clustering methods such as k-medoid are very similar to the above methods [23] but are more robust to noise and outliers.

In addition to fuzzy classification systems, very often a combination of fuzzy systems and NNs, namely neuro-fuzzy learning systems, are used [24].

7.3 Classification Algorithms

Training using a large number of data ensamples and testing a new ensample are the two main stages of classification algorithms. In the training stage, the separability of data plays a crucial role in minimizing the so-called cross-validation error. Although by using kernels the data feature space can be mapped to a higher dimensional space where the features become separable, identification of suitable and accurate kernels is not easy to achieve for all types of data. In the following sections more popular classifiers used for classification of biomedical signals such as EEG are briefly reviewed.

7.3.1 Decision Trees

Decision trees (DTs) are popular, simple, powerful, and generally nonlinear tools for classification and prediction. They represent *rules* which can be deciphered by human being and utilized in knowledge-based systems.

The main steps in a decision tree algorithm are:

1) Selecting the *best* attribute(s) to split the remaining instances and make that (those) attribute(s) a decision node (decision nodes)
2) Repeating this process recursively for each child
3) Stop when:
   a) All the instances have the same target attribute value
   b) There are no more attributes
   c) There are no more instances.

One simple DT example for a humidity sensor can be:

> If the time is 9:00am and there is no rain what would be the humidity level (out of low, medium, and severe)?

Depending on the number of attributes, a decision tree may look like Figure 7.3. DTs can be perfectly fitted to any training data, with zero bias and high variance. Different algorithms are used to determine the 'best' split at a node.

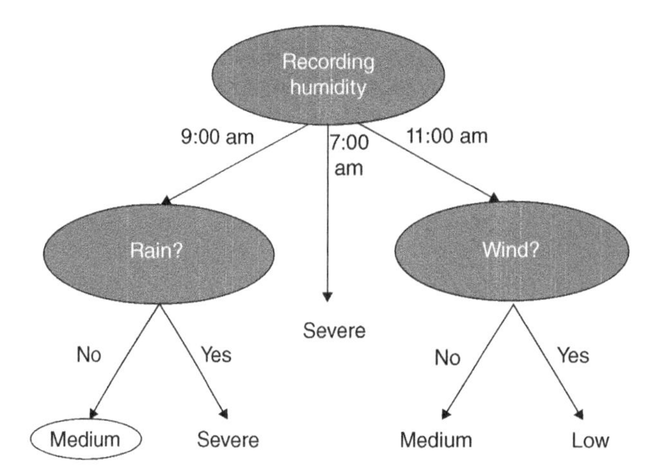

Figure 7.3 An example of a decision tree to show the humidity level at 9:00 a.m. when there is no rain.

DTs are often used to predict data labels by iterating the input data through a learning tree [25]. During this process, the feature properties are compared relative to the decision conditions to reach a specific category. A DT works only with linearly separable data and the process of building optimal learning trees is NP-complete [26].

In application of DT to EEG signals we need to consider that the input signals are naturally nonstationary, have poor signal to noise ratio, depend on physical or mental tasks, and are contaminated with various artefacts such as external electromagnetic waves, electromyogram and electrooculogram.

In an example [27] the authors proposed a simple DT structure to classify the EEG data related to up/down/right/left computer cursor movement imagery EEG data. The mean and variance of continuous Morlet wavelet coefficients, autoregressive (AR) model parameters, average derivative, and skewness are calculated from the EEG of three subjects and used as the input features to the DT system.

In another application [28] a classification framework and a data reduction method have been proposed to distinguish between multiclass motor imagery EEG for BCI based on the manifold of covariance matrices using Riemannian geometry [29]. Riemannian geometry is useful when the solution iterations follow a manifold (rather than for example Euclidean space) due to the imposed constraints. In the context of BCI for example, such constraints are spatial constraints related to the different zones within the motor cortex each belonging to the movement of a particular body part. The function of interest should be differentiable over the Riemannian manifold. In the BCI context, therefore, a spatial covariance matrix is used which is a symmetric and positive definite matrix [27, 28].

To perform the classification a DT classifier is designed to identify motor imagery tasks and reduce the classification error for the non-separable cases. Later this has been combined with other classifiers to enable semi-supervised classification.

The main idea is to use spatial covariance matrices as EEG signal descriptors and to rely on Riemannian geometry to directly classify these matrices using the topology of the manifold of symmetric and positive definite matrices. This allows to extract the spatial information contained in EEG signals without using spatial filtering. Based on the spatial covariance matrices, the distance between each two trials is estimated (and used in classification) using two different methods namely, (i) filter geodesic minimum distance to the Riemannian mean and (ii) tangent space mapping, both over a Riemannian manifold [28].

A simple Emotive EEG system has been used in this work and three types of shoulder joint movements, i.e. flexion, extension, and abduction have been classified. It has been shown that for some subjects, accuracies of over 90% are achievable.

7.3.2 Random Forest

Random forest (or random forests) classifier creates a set of DTs from randomly selected subset of the training set. It then aggregates the votes from different DTs to decide the final class of the test data [30]. The term *random forest* comes from random decision forests initially proposed by Tin Kam Ho of Bell Labs in 1995. This method combines Breiman's 'bagging' idea and random feature selection. Bagging or *bootstrap aggregation* is a technique for reducing the variance of an estimated prediction function. A random forest classifier is in fact an extension to bagging which uses *de-correlated* trees.

The main advantages of random forest classifiers are that there is no need for pruning trees, accuracy and variable importance are generated automatically, overfitting is not a problem, they are not very sensitive to outliers in training data, and setting their parameters is easy.

However, there are some limitations. The regression process cannot predict beyond the available range in the training data and in regression the extreme values are often not predicted accurately. This causes underestimation of highs and overestimation of lows.

Unlike for traditional random forest which is a bagging technique, in *boosting* as the name suggests, one learns from another which in turn boosts the learning. The bagging scheme trains a bunch of individual models in a parallel way. Each model is trained by a random subset of the data. Conversely, boosting trains a bunch of individual models in a sequential way. Each individual model learns from mistakes made by the previous model. *AdaBoost* is a boosting ensemble model and performs well with the decision tree. AdaBoost learns from the mistakes by increasing the weight of misclassified data points. The optimization of Adaboost is sometimes by means of adaptive methods, such as gradient decent, and therefore, Adaboost and so-called *gradient boosting* are very similar.

In an application [31] Edla et al. used a random forest classifier for classification and analysis of simple first and second-order statistics representing various human mental states from a single-channel EEG. Some other applications of the random forest classifier to BCI have been detailed in [32].

7.3.3 Linear Discriminant Analysis

As a well established technique introduced by Vapnik [33], LDA is a method used to find a linear combination of features which characterizes or separates two or more classes of objects or events. In LDA, the data are projected onto a single dimension, where class assignment is made for a given test point. The resulting combination may be used as a linear classifier. In an LDA it is assumed that the classes have normal distributions. As for PCA, an LDA is used for both dimensionality reduction and data classification.

In a two-class dataset, given the a priori probabilities for class 1 and class 2 respectively as p_1 and p_2, class means and overall mean respectively as μ_1, μ_2, and μ, and the class variances as cov_1 and cov_2.

$$\mu = p_1 \times \mu_1 + p_2 \times \mu_2 \tag{7.10}$$

Then, within-class and between-class scatters are used to formulate the necessary criteria for class separability. Within-class scatter is the expected covariance of each of the classes. The scatter measures for multiclass case are computed as:

$$S_w = \sum_{j=1}^{C} p_j \times cov_j \tag{7.11}$$

where C refers to the number of classes and

$$cov_j = \left(\mathbf{x}_j - \mu_j\right)\left(\mathbf{x}_j - \mu_j\right)^T \tag{7.12}$$

Slightly differently, the inter-class scatter is estimated as:

$$S_b = \frac{1}{C}\sum_{j=1}^{C}\left(\mathbf{x}_j - \mu_j\right)\left(\mathbf{x}_j - \mu_j\right)^T \tag{7.13}$$

Then, the objective would be to find a discriminant plane, **w**, to maximize the ratio between inter-class and intra-class scatters (variances):

$$J_{LDA} = \frac{\mathbf{w}S_b\mathbf{w}^T}{\mathbf{w}S_w\mathbf{w}^T} \tag{7.14}$$

In practise, class means and covariances are not known but they can be estimated from the training set. In the above equations, either the maximum likelihood estimate or the maximum a posteriori estimate may be used instead of the exact values.

LDA has been widely used for EEG classification. In a study in [34] LDA has been used to classify the EEG spectral parameters of healthy elderly controls, Alzheimer's disease and vascular dementia. In this work the extracted spectral features were the absolute EEG delta power, decay from lower to higher frequencies, amplitude, centre and dispersion of the alpha power and baseline power of the entire EEG frequency spectrum. The regularization approach replaces the within-group sample covariance with a weighted average of the whole sample covariance using a shrinking intensity parameter. This parameter emphasizes larger eigenvalues of the covariance matrix while de-emphasizing smaller ones, therefore creating a pooled-covariance matrix that is corrected for the bias when estimating sample-based eigenvalues. The optimal shrinkage parameter is determined by cross-validation.

LDA is one of the most popular classification algorithms for BCI. One example can be seen in [35] where a two-class BCI has been designed where both classes have Gaussian distribution but the class variances are different.

In another BCI application [36] EEG mu/beta rhythm has been used as the input to an LDA classifier for deciding which rhythm can give the better classification performance. During this, the common spatial pattern (CSP) has been used to project the data in a way to maximize the ratio of projected energy of one class to that of the other class.

7.3.4 Support Vector Machines

Amongst all supervised classifiers, SVM is probably the most popular one for many applications including BCI while outperforming other classifiers in many linear (separable) and nonlinear (non-separable) cases. The SVM concept was introduced by Vapnik in 1979 [33, 37–41]. Unlike for LDA where the class distributions are Gaussian, in SVM there is no assumption on the data distribution. To understand the SVM concept, consider a binary classification for the simple case of a two-dimensional feature space of linearly separable training samples $S = \{(\mathbf{x}_1, y_1), (\mathbf{x}_2, y_2), \dots, (\mathbf{x}_m, y_m)\}$ (Figure 7.4) where $\mathbf{x} \in R^d$ is the input vector (where the feature space dimension $d = 2$ in this Figure) and $y_i \in \{1, -1\}$, $i = 1, \dots, m$, is the class label. A discriminating function is defined as:

$$f(\mathbf{x}) = sgn\left(\langle \mathbf{w}, \mathbf{x}\rangle + b\right) = \begin{cases} +1 & \mathbf{x} \in \textit{first class} \\ -1 & \mathbf{x} \in \textit{second class} \end{cases} \tag{7.15}$$

In this formulation $\mathbf{w}$ determines the orientation of a discriminant plane (or hyperplane). Clearly, there are an infinite number of possible planes that could correctly classify the training data. An optimal classifier finds the hyperplane for which the best generalizing hyperplane is equidistant or farthest from each set of points. Optimal separation is achieved when there is no separation error and the distance between the closest vector and the hyperplane is maximal.

One way to find the separating hyperplane in a separable case is by constructing the so-called *convex hulls* of each data set. The encompassed regions are the convex hulls for the data sets. By examining the hulls one can then determine the closest two points lying on the hulls of each class (note that these do not necessarily coincide with actual data points). By constructing a plane perpendicular and equivalent to these two points, an optimal hyperplane with robust classifier should result.

For an optimal separating hyperplane design, often few points, referred to as *support vectors* (SVs), are utilized (e.g. the three circled data feature points in Figure 7.4).

SVM formulation starts with the simplest case: linear machines are trained on separable data (for general case analysis; nonlinear machines trained on non-separable data result in a very similar quadratic programming (QP) problem). Then, the training data $\{\mathbf{x}_i, y_i\}$, $i = 1, \ldots, m, y_i \in \{1, -1\}$, $\mathbf{x}_i \in R^d$ are labelled. Consider having a hyperplane which separates the positive from the negative examples. Points $\mathbf{x}$ lying on the hyperplane satisfy $\langle \mathbf{w}, \mathbf{x} \rangle + b = \mathbf{0}$, where $\mathbf{w}$ is normal to the hyperplane, $b/\|\mathbf{w}\|_2$ is the perpendicular distance from the hyperplane to the origin, and $\|\mathbf{w}\|_2$ is the Euclidean norm of $\mathbf{w}$. Define the 'margin' of

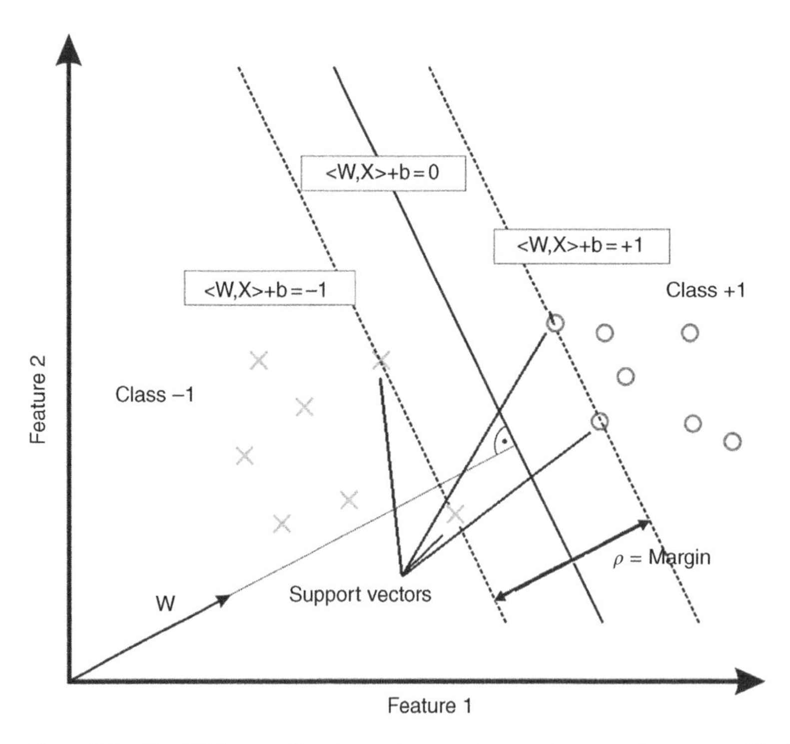

Figure 7.4 The SVM separating hyperplane and support vectors for a separable data case.

a separating hyperplane illustrated in Figure 7.4, and for the linearly separable case, the algorithm simply looks for the separating hyperplane with largest margin. Here, the approach is to reduce the problem to a convex optimization problem by minimizing a quadratic function under linear inequality constraints. To find a plane farthest from both classes of data, the margin between the supporting canonical hyperplanes for each class is maximized. The support planes are pushed apart until they meet the closest data points, which are then deemed to be the SVs (circled in Figure 7.4). Therefore, the SVM problem to find **w** is stated as:

$$\langle \mathbf{x}_i, \mathbf{w} \rangle + b \geq +1 \text{ for } y_i = +1 \tag{7.16a}$$

$$\langle \mathbf{x}_i, \mathbf{w} \rangle + b \geq -1 \text{ for } y_i = -1 \tag{7.16b}$$

which can be combined into one set of inequalities as $y_i(\langle \mathbf{x}_i, \mathbf{w} \rangle + b) - 1 \geq 0 \, \forall \, i$. The margin between these supporting planes (H_1 and H_2) can be shown to be $\gamma = 2/\|\mathbf{w}\|_2$. Therefore, to maximize this margin we therefore need to:

$$\begin{aligned} &\text{minimize } \langle \mathbf{w}, \mathbf{w} \rangle \\ &\text{subject to } y_i(\langle \mathbf{x}_i, \mathbf{w} \rangle + b) - 1 \geq 0 \; i = 1, \ldots, m \end{aligned} \tag{7.17}$$

This constrained optimization problem can be changed into an unconstrained problem by using Lagrange multipliers. This leads to minimization of an unconstrained empirical risk function (Lagrangian) which consequently results in a set of conditions called Kuhn–Tucker (KT) conditions. The new optimization problem brings about the so-called *primal form* as:

$$L(\mathbf{w}, b, \alpha) = \frac{1}{2}\langle \mathbf{w}, \mathbf{w} \rangle - \sum_{i=1}^{m} \alpha_i [y_i(\langle \mathbf{x}_i, \mathbf{w} \rangle + b) - 1] \tag{7.18}$$

where the α_i, $i = 1, \ldots, m$ are the Lagrangian multipliers. The Lagrangian primal has to be minimized with respect to **w**, and b and maximized with respect to $\alpha_i \geq 0$. Constructing the classical Lagrangian dual form facilitates this solution. This is achieved by setting the derivatives of the primal to zero and re-substituting them back into the primal. Hence, the dual form is derived as:

$$L(\mathbf{w}, b, \alpha) = \sum_{i=1}^{m} \alpha_i - \frac{1}{2}\sum_{i=1}^{m}\sum_{j=1}^{m} y_i y_j \alpha_i \alpha_j \langle \mathbf{x}_i \cdot \mathbf{x}_j \rangle \tag{7.19}$$

and

$$\mathbf{w} = \sum_{i=1}^{m} y_i \alpha_i \mathbf{x}_i \tag{7.20}$$

considering that $\sum_{i=1}^{m} y_i \alpha_i = 0$ and $\alpha_i \geq 0$. These equations can be solved using many different publicly available QP algorithms such as those proposed in [42, 43].

In non-separable cases where the classes have overlaps in the feature space, the maximum margin classifier described above is no longer applicable. With the help of some further mathematical derivations, we are often able to define a nonlinear hyperplane to accurately separate the datasets [6]. As we see later, this causes an overfitting problem which reduces the robustness of classifier. The ideal solution where no points are misclassified and no points lie within the margin is no longer feasible. This implies that we need to relax the constraints to allow for minimum misclassifications. In this case, the points that subsequently fall on the wrong side of the margin are considered to be errors. However, they have less influence on the location of the hyperplane (according to a pre-set *slack* variable) and as such are considered to be SVs. The classifier obtained in this way is called a *soft margin classifier*.

To optimize the soft margin classifier, we must allow violation of the margin constraints according to a pre-set *slack* variable ξ_i in the original constraints, which then become:

$$\langle \mathbf{x}_i, \mathbf{w} \rangle + b \geq +1 - \xi_i \text{ for } y_i = +1 \tag{7.21a}$$

$$\langle \mathbf{x}_i, \mathbf{w} \rangle + b \geq -1 + \xi_i \text{ for } y_i = -1 \tag{7.21b}$$

and $\xi_i \geq 0 \, \forall \, i$

For an error to occur, the corresponding ξ_i must exceed unity. Therefore, $\sum_{i=1}^{m} \xi_i$ is an upper bound on the number of training errors. Hence, a natural way to assign an extra cost for errors is to change the objective function to:

$$\min \langle \mathbf{w}, \mathbf{w} \rangle + C \sum_{i=1}^{m} \xi_i$$

$$\text{subject to } y_i(\langle \mathbf{x}_i, \mathbf{w} \rangle + b) \geq 1 - \xi_i \text{ and } \xi_i \geq 0 \textit{ for } i = 1, \ldots, m \;\; (7.22) \tag{7.22}$$

The primal form will then be:

$$L(\mathbf{w}, b, \xi, \alpha, \mathbf{r}) = \frac{1}{2} \langle \mathbf{w}, \mathbf{w} \rangle + C \sum_{i=1}^{m} \xi_i - \sum_{i=1}^{m} \alpha_i [y_i(\langle \mathbf{x}_i, \mathbf{w} \rangle + b) - 1 + \xi_i] - \sum_{i=1}^{m} r_i \xi_i \tag{7.23}$$

Hence, by differentiating this cost function with respect to $\mathbf{w}$, ξ, and b we can achieve:

$$\mathbf{w} = \sum_{i=1}^{m} y_i \alpha_i \mathbf{x}_i \tag{7.24}$$

and

$$\alpha_i + r_i = C \tag{7.25}$$

By substituting these into the primal form and again considering that $\sum_{i=1}^{m} y_i \alpha_i = 0$ and $\alpha_i \geq 0$, the dual form is derived as:

$$L(\mathbf{w}, b, \xi, \alpha, \mathbf{r}) = \sum_{i=1}^{m} a_i - \frac{1}{2} \sum_{i=1}^{m} \sum_{j=1}^{m} y_i y_j \alpha_i \alpha_j \langle \mathbf{x}_i \cdot \mathbf{x}_j \rangle \tag{7.26}$$

This is similar to the maximal marginal classifier with the only difference in that, here we have a new constraint of $a_i + r_i = C$, where $r_i \geq 0$, hence $0 \leq a_i \leq C$. This implies that the value C, sets an upper limit on the Lagrangian optimization variables a_i. This is sometimes referred to as the box constraint. The value of C offers a trade-off between accuracy of data fit and regularization. A small value of C (i.e. $C < 1$) significantly limits the influence of error points

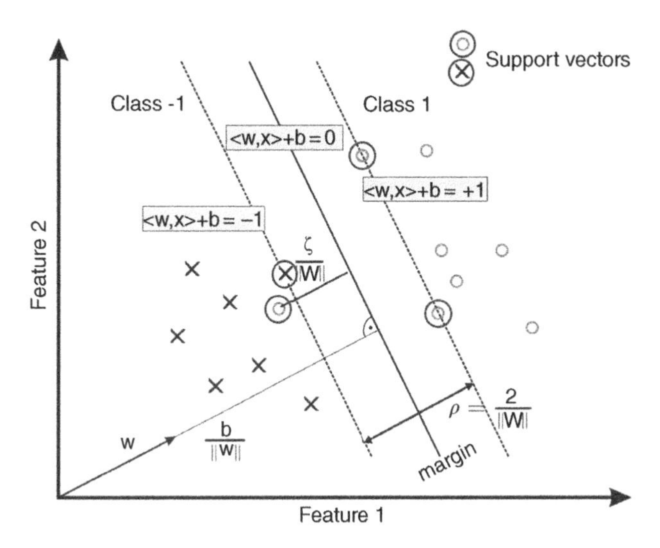

Figure 7.5 Soft margin and the concept of a slack parameter.

(or outliers), whereas if C is chosen to be very large (or infinite) then the soft margin (as in Figure 7.5) approach becomes identical to the maximal margin classifier. Therefore, in the use of the soft margin classifier, the choice of C strongly depends on the data. Appropriate selection of C is important and remains an area of research. One way to set C is by gradually increasing C from $\max(\alpha_i)$ for $\forall i$ and find the value for which the error (outliers, cross-validation, or number of misclassified points) is minimum. Eventually, C can be found empirically [44].

There will be no change in formulation of the SVM for the multidimensional cases. Only the dimension of the hyperplane changes depending on the number of feature types.

In many non-separable cases use of a nonlinear function may help to make the datasets separable. As can be seen in Figure 7.6, the datasets are separable if a nonlinear hyperplane is used. *Kernel mapping* offers an alternative solution by non-linearly projecting the data into a (usually) higher dimensional feature space to allow the separation of such cases.

The key success to kernel mapping is that special types of mapping that obey Mercer's theorem, sometimes called reproducing kernel Hilbert spaces (RKHSs) [33], offer an implicit mapping into feature space:

$$K(\mathbf{x}, \mathbf{z}) = \langle \varphi(\mathbf{x}), \varphi(\mathbf{z}) \rangle \tag{7.27}$$

This implies that there is no need to know the explicit mapping in advance, rather the inner product itself is sufficient to provide the mapping. This simplifies the computational burden significantly and in combination with the inherent generality of SVMs largely alleviates the dimensionality problem. Moreover, this means that the input feature inner product can simply be substituted with the appropriate kernel function to obtain the mapping while having no effect on the Lagrangian optimization theory. Hence:

$$L(\mathbf{w}, b, \xi, \alpha, \mathbf{r}) = \sum_{i=1}^{m} a_i - \frac{1}{2}\sum_{i=1}^{m}\sum_{j=1}^{m} y_i y_j \alpha_i \alpha_j K(\mathbf{x}_i \cdot \mathbf{x}_j) \tag{7.28}$$

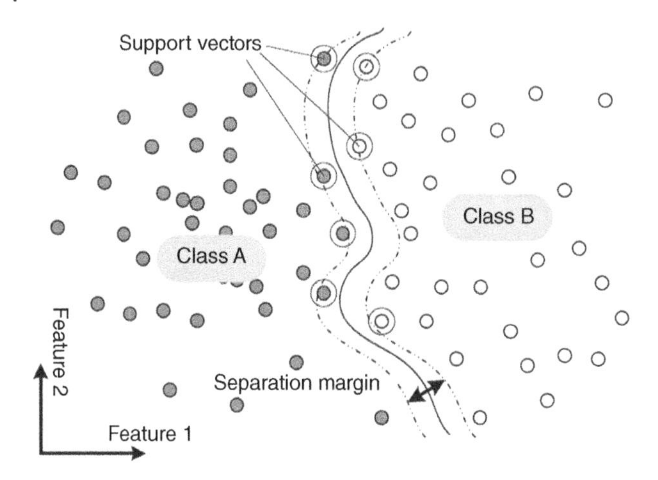

Figure 7.6 Nonlinear discriminant hyperplane (separation margin) for SVM.

The relevant classifier function then becomes:

$$f(\mathbf{x}) = sgn\left(\sum_{i=1}^{nSVs} y_i a_i K(\mathbf{x}_i \cdot \mathbf{x}_j) + b\right) \tag{7.29}$$

where $nSVs$ refers to the number of SVs. In this way all the benefits of the original linear SVM method are maintained. We can train a highly nonlinear classification function such as a polynomial or a radial basis function (RBF), or even a sigmoidal NN, using a robust and efficient algorithm that does not suffer from local minima. The use of kernel functions transforms a simple linear classifier into a powerful and general nonlinear classifier [44]. There are many nonlinearity functions used as kernels for SVM. However, the RBF is the most popular nonlinear kernel used in SVM for non-separable classes and is defined as:

$$K(u, v) = exp\left(\frac{\|u - v\|_2^2}{2\sigma^2}\right) \tag{7.30}$$

where σ^2 is the variance. As mentioned previously, a very accurate nonlinear hyperplane, estimated for a particular training dataset is unlikely to generalize well. This is mainly because the system may no longer be robust since a testing or new input can easily be misclassified.

Another issue related to the application of SVMs (as well as other classifiers) is the cross-validation problem. The classifier output distribution (without the hard limiter 'sign' in Eq. (7.24)) for a number of inputs for each class may be measured. The probability distributions of the results (which are centred at −1 for class '−1' and at +1 for class '+1') are plotted in the same figure. Less overlap between the distributions represents a better performance of the classifier. The choice of kernel influences the classifier performance in terms of cross-validation error. Figure 7.7 shows the output class distributions for zero and non-zero validation errors (here, the class distributions are considered Gaussian).

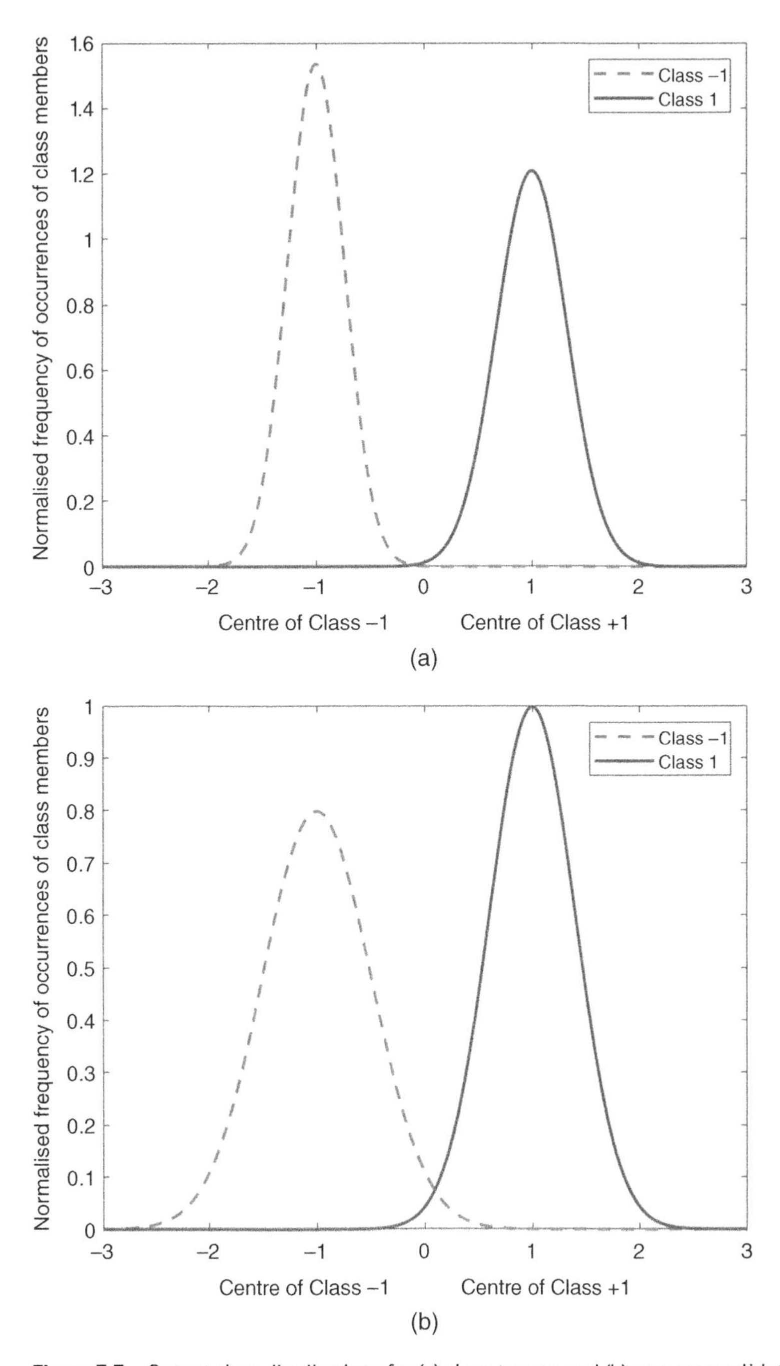

Figure 7.7 Output class distributions for (a) close to zero and (b) non-zero validation errors (here, the class distributions are considered Gaussian). The error is equivalent to the area under the overlapping part of the two distributions.

SVMs may be slightly modified to enable classification of multiclass data [45]. Currently there are two types of approaches for multiclass SVM. One is by constructing and combining several binary classifiers while the other is by considering all data in one optimization formulation directly. Moreover, some investigations have been undertaken to speed up the training step of the SVMs [46].

There are many applications of SVM for sensor networks [47]. An SVM has been utilized for target localization in sensor networks [48]. In this research, the algorithm partitions the sensor field using a fixed number of classes. The authors consider the problem of estimating the geographic locations of nodes in a wireless sensor network where most sensors are without an effective self-positioning functionality. Their SVM-based algorithm called Lagrangian support vector machine (LSVM), localizes the network merely based on connectivity information, addresses the border and coverage-hole problems, and performs in a distributed manner.

Yoo and Kim [49] introduced a semi-supervised online SVM, also called support vector regression (SVR), to alleviate the sensitivity of the classifier to noise and the variation in target localization using multiple wireless sensors. This is achieved by combining the core concepts of manifold regularization and the supervised online SVR.

Despite its excellent performance and very wide range of applications, SVM does not perform very well for large-scale samples and imbalanced data classes. Natural data are often imbalanced and consist of multiple categories or classes. Learning discriminative models from such datasets is challenging due to lack of representative data and the bias of traditional classifiers towards the majority class. Many attempts have been made to alleviate this problem. Sampling methods such as the synthetic minority oversampling technique (SMOTE) has been one of the popular methods in tackling this problem. Mathew et al. [50] tried to solve this problem by proposing a weighted kernel-based SMOTE that overcomes the limitation of SMOTE for nonlinear problems by oversampling in the SVM feature space. Compared to other baseline methods on multiple benchmark imbalanced datasets, their algorithm along with a cost-sensitive SVM formulation is shown to improve the performance. In addition, a hierarchical framework with progressive class order has been developed for multiclass imbalanced problems.

Kang et al. [51] proposed a weighted undersampling (WU) methodology for SVM based on space geometry distance. In their algorithm, the majority of samples are grouped into some sub-regions and different weights were assigned to them according to their Euclidean distance to the hyperplane. The samples in a sub-region with higher weight have more chance to be sampled and used in each learning iteration. This retains the data distribution information of the original data sets as much as possible.

SVM has been widely used in EEG classification particularly for two-class BCI systems. Here we look at only a very small number of examples.

Shoker et al. applied SVM to detect the independent components related to eye blink after decomposing the multichannel EEG signals using ICA-based blind source separation [9, 10]. Simple features indicating the power and morphology of eye-blink spikes have been used as the inputs. In [52] the SVM has been used to classify left-hand and right-hand movements from the EEGs at the C3 and C4 electrodes, for BCI applications.

In [53] a multiclass SVM has been proposed for EEG classification. Wavelet coefficients and Lyapunov exponents have been the input features. In this work an optimum classification scheme was proposed and the clues about the extracted features were inferred.

An interesting work on emotion recognition from 62-channel EEG has been presented in [54]. In this paper, stable EEG patterns over time for positive, neutral, and negative emotion recognition have been identified using SVM and the outcome compared with those of a number of other classifiers. Their results indicate that stable patterns exhibit consistency across sessions; the lateral temporal areas activate more for positive emotion than negative ones in the beta and gamma bands; the neural patterns of neutral emotion have higher alpha responses at parietal and occipital sites; and for negative emotion, the neural patterns have significantly higher delta responses at parietal and occipital sites and higher gamma responses at prefrontal sites.

In [55] kernel SVM has been used for EEG classification. However, in this method instead of using one kernel a sum of weighted kernels has been used. For mental task classification baseline, visual counting, mental letter composing, mathematical multiplication, and geometric figure rotation tasks have been classified using the above multikernel SVM. In their experiment wavelet coefficients and entropy have been used as the input features. For the cognitive task the EEG-ERP signals are decomposed into their constituent independent components and each independent component is further decomposed in the frequency domain using orthogonal empirical mode decomposition. Using these components the four-lobe brain connectivity is estimated using Granger causality in the theta band and the AR model coefficients (as features) are fed to the multikernel SVM.

7.3.5 k-Nearest Neighbour

As another supervised learning algorithm, k-nearest neighbour (KNN) classifies a data sample (called a query point) based on the labels (i.e. the output values) of the nearby data samples. For example, missing readings of an EEG electrode signal can be predicted using the average measurements of neighbouring electrodes. There are several functions to determine the nearest set of nodes. One simple method is to use the Euclidean distance between the signals from different electrodes. KNN has low computational cost since the function is computed relative to the local points (i.e. k-nearest points, where k is a small positive integer). This factor coupled with the correlated readings of neighbouring nodes makes KNN a suitable distributed learning algorithm for sensor networks. It has been shown that the KNN algorithm may provide inaccurate results when analyzing problems with high-dimensional spaces (more than 10–15 dimensions) as the distance to different data samples becomes invariant (i.e. the distances to the nearest and farthest neighbours are slightly similar) [56].

One potential problem inherent to the KNN approach is that it assumes that the k-nearest neighbours of a test example are located at roughly the same distance from it. In other words, the KNN method does not take into account the fact that the k-nearest neighbours of a test example might have largely differing distances from the test example. An intuitively appealing solution to this problem is to assign different degrees of importance to different nearest neighbours [57].

There have been many applications of KNN to EEG classification recently. In addition, since KNN is available in different platforms, it is often implemented by the research workers for comparison with other classification methods. In [58] KNN has been used in the design of an EEG-based concealed information test, which classifies the EEG data into two classes representing innocent and guilty.

In [59] KNN is used for BCI where the classification accuracy achievable has been investigated based on Dempster–Shafer theory (which is considered as a generalization of probability theory [60]). The subjects were asked to perform five different mental tasks during different trials: baseline or total relaxation, multiplication, i.e. silently multiplying two numbers (non-trivial), rotation, i.e. imagination of rotation of an imagined 3-D block, counting, i.e. visualization of numbers on an imaginary blackboard and incrementing, and finally letter composition, i.e. mental composition without vocalizing. To extract features from the EEG a sixth-order AR model and wavelet decomposition have been used after the eye-blink artefacts were removed using ICA. To test the classification method an EEG dataset containing signals recorded during the performance of five different mental tasks was used. It has been demonstrated that the Dempster–Shafer KNN classifier achieves more accurate classification results than the classical voting KNN classifier and the distance-weighted KNN classifier.

7.3.6 Gaussian Mixture Model

GMM is mainly used for estimating the observation probability given the features and follows the Bayes formula as:

$$p(\textit{object property} \mid \textit{context}) = p(O \mid v) = \frac{p(v \mid O)}{p(v)} p(O) \tag{7.31}$$

where $p(v)$ can easily be calculated as:

$$p(v) = p(v \mid O)p(O) + p(v \mid \overline{O})p(\overline{O}) \tag{7.32}$$

where $p(v \mid O)$ is the likelihood of the features given the observation (presence of an object) and $p(v \mid \overline{O})$ is the likelihood of the features when the object is absent. A GMM is then used to estimate these likelihoods as:

$$p(v \mid O) = \sum_{i=1}^{M} w_i \cdot G(v; \mu_i, \Sigma_i) \tag{7.33}$$

which is the sum of weighted Gaussians with unknown means and covariance matrices. The parameters w_i, μ_i, and Σ_i have to be estimated using a kind of optimization technique. Often expectation maximization [61] is used for this purpose.

GMM is a very popular modelling approach for representation of various waveforms including abnormal EEGs. In a study of neonatal seizure detection GMM has been shown to perform well [62].

7.3.7 Logistic Regression

Logistic regression is generally used to estimate the probability of an observation subject to an attribute. It estimates the observation probability from the given features. A logistic function $F(v)$ is computed from the features as:

$$\textit{Logit} = \log \frac{p(O \mid v)}{p(\overline{O} \mid v)} = F(v) \tag{7.34}$$

where $F(v)$ depends on the sequence of attributes (i.e. previous experiences):

$$F(v) = a_0 + \sum_{i=1}^{D} a_i \cdot v(i) \tag{7.35}$$

and for testing:

$$p(O \mid v) = \frac{1}{1 + e^{-F(v)}} \tag{7.36}$$

To give an example, consider:

Observation O: suffering from back pain and $\overline{O}$: not suffering from back pain

Attribute ν: age:

$$log \frac{p(O \mid v)}{p(\overline{O} \mid v)} = a_0 + a_1 \cdot (age - 20)$$

During the training stage a_0 (the log odds for a 20-year-old person) and a_1 (the log odds ratio when comparing two persons who differ by 1 year in age) are estimated. In the testing stage the probability of having back pain given the age is estimated as:

$$p(O \mid age) = \frac{1}{1 + e^{-a_0 + a_1(age - 20)}}$$

7.3.8 Reinforcement Learning

In reinforcement learning an agent learns a good behaviour. This means that it modifies or acquires new behaviours and skills incrementally [63]. It exploits the local information as well as the restrictions imposed on the application to maximize the influences of numerous tasks over a time period. Therefore, during the process of reinforcement learning, the machine learns by itself based on the penalties and rewards it receives in order to decide on its next step action. Thus, the learning process is sequential. With this algorithm, each node discovers the minimal needed resources to carry out its routine tasks with benefits allocated by the Q-learning technique. This technique is mainly used to assess the value or quality of action taken by the learning system (embedded in each sensor for the distributed cases) for estimation of the award to be given [64] and consequently undertaking the next step.

7.3.9 Artificial Neural Networks

An artificial neural network (ANN) is another decision-making system that attempts to mimics the human brain. The brain includes billions of neurons such as the one in Figure 7.8, which are all connected and provide a tremendous processing power. The larger the number of neurons and layers in such a network is, the more processing power and computational complexity will generally result.

ANNs are supervised nonlinear machine learning algorithms constructed by cascading chains of decision units (neurons followed by nonlinear functions and decision makers) used to recognize nonlinear and complex functions [8]. In sensor networks, using NNs

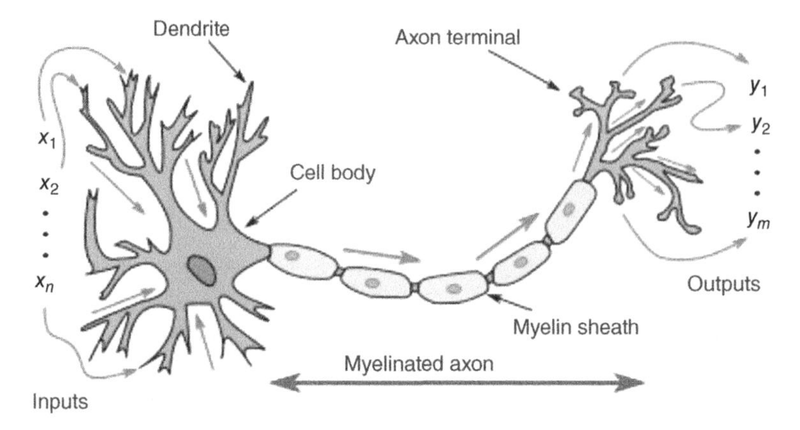

Figure 7.8 A biological neuron that expresses the fundamental elements of a neuron in an ANN.

in distributed manners is still not so pervasive due to the high computational requirements for learning the network weights and the high management overhead. However, in centralized solutions, NNs can learn multiple outputs and decision boundaries at once [65] which make them suitable for solving several network challenges using the same model.

As seen in Figure 7.9, each ANN consists of input and output layers plus a number of hidden layers. Each layer contains a number of neurons and each neuron performs a nonlinear operation on its input. Hard limiter (sign function), sigmoid, exponential, and tangent hyperbolic (tanh) functions are the most popular ones.

The major challenges in using ANNs are their optimization (i.e. estimation of the link weights) and selection of the number of layers and neurons. The backpropagation algorithm is often used in multilayer ANNs [66, 67]. This is very similar to optimization of adaptive filter coefficients, where the error between the achieved label (output) and the desired label

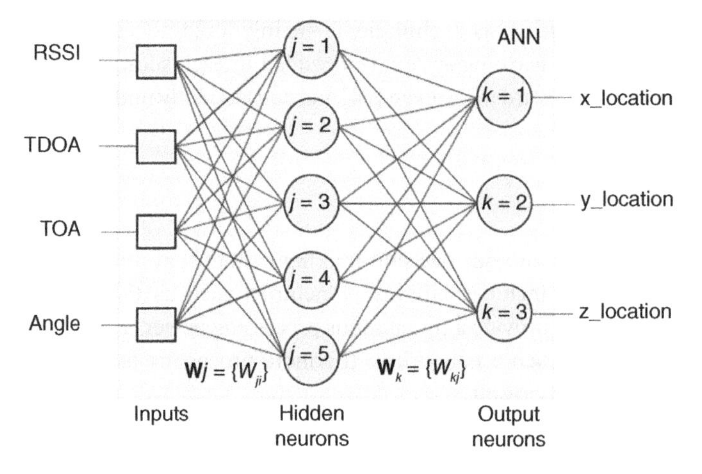

Figure 7.9 A simple three-layer NN for node localization in sensor networks in 3-D space.

is minimized in order to obtain the best set of link weights. The computational cost exponentially increases with the numbers of neurons and layers.

In an example of ANN application for sensor network node localization, the input variables are the propagating angle and distance measurements of the received signals from anchor nodes [68]. Such measurements may include received signal strength indicator (RSSI), time of arrival (TOA), and time difference of arrival (TDOA) as illustrated in Figure 7.9. After supervised training, the ANN generates an estimated node location as vector-valued coordinates in 3D space. The associated ANN algorithms include self-organizing map (or Kohonen's maps) and learning vector quantization (LVQ) (see [69] and references therein for an introduction to these methods). In addition to function estimation, one important application of NNs is in big data (high-dimensional and complex data set) feature detection, classification, and dimensionality reduction [70].

Weights and biases are usually initialized randomly, and then updated iteratively during training via backpropagation. Using backpropagation algorithm to train an ANN, the error between the output of each output neuron is compared with a target value (or the centre of a cluster) in order to find the link weights between various layers. For the two-layer network of Figure 7.9, assuming z_k is the output of the neuron k and t_k the target, the cost is defined as:

$$J(\mathbf{w}) = \frac{1}{2}\sum_{k=1}^{C}(t_k - z_k)^2 \tag{7.37}$$

where C represents the number of outputs (classes). Estimation of the output, z_k, is straightforward; assuming the input vector to the ANN is $\mathbf{x}$, the output of neuron k is:

$$z_k = f\left(\mathbf{w}_k^T f\left(\mathbf{w}_j^T \mathbf{x}\right)\right) = f\left(\sum_{j=1}^{n_1} w_{kj} f\left(\sum_{i=1}^{n_2} w_{ji}x_i + w_{j0}\right) + w_{k0}\right) \tag{7.38}$$

In the above equation (Eq. 7.38), n_1 and n_2 are the number of neurons in the first and second layers respectively, w_{j0} and w_{k0} are bias values (not shown in Figure 7.8), and $f(y)$, the neuron activation function is a continuous function which best approximates a hard limiter often called as rectified linear unit (ReLU). A popular example of activation function is the following exponential (sigmoid) function:

$$f(y) = \frac{1}{1 + e^{-y}} \tag{7.39}$$

which looks like the curve in Figure 7.10a. In Figure 7.10b an ReLU activation function can be seen too. Many other functions can also be used as stated previously. Different optimization techniques such as MSE minimization [71] can be used to minimize $J(\mathbf{w})$ and find the optimum w_{ji} and w_{kj} values.

The advantage of ReLU is that it does not saturate. So, it is better for multilayer networks avoiding propagation of saturation error over the network.

7.3.9.1 Deep Neural Networks

In parallel with development of powerful computers and accessing to local and remote memory clusters as well as cloud, DNNs have become widely popular. These classifiers

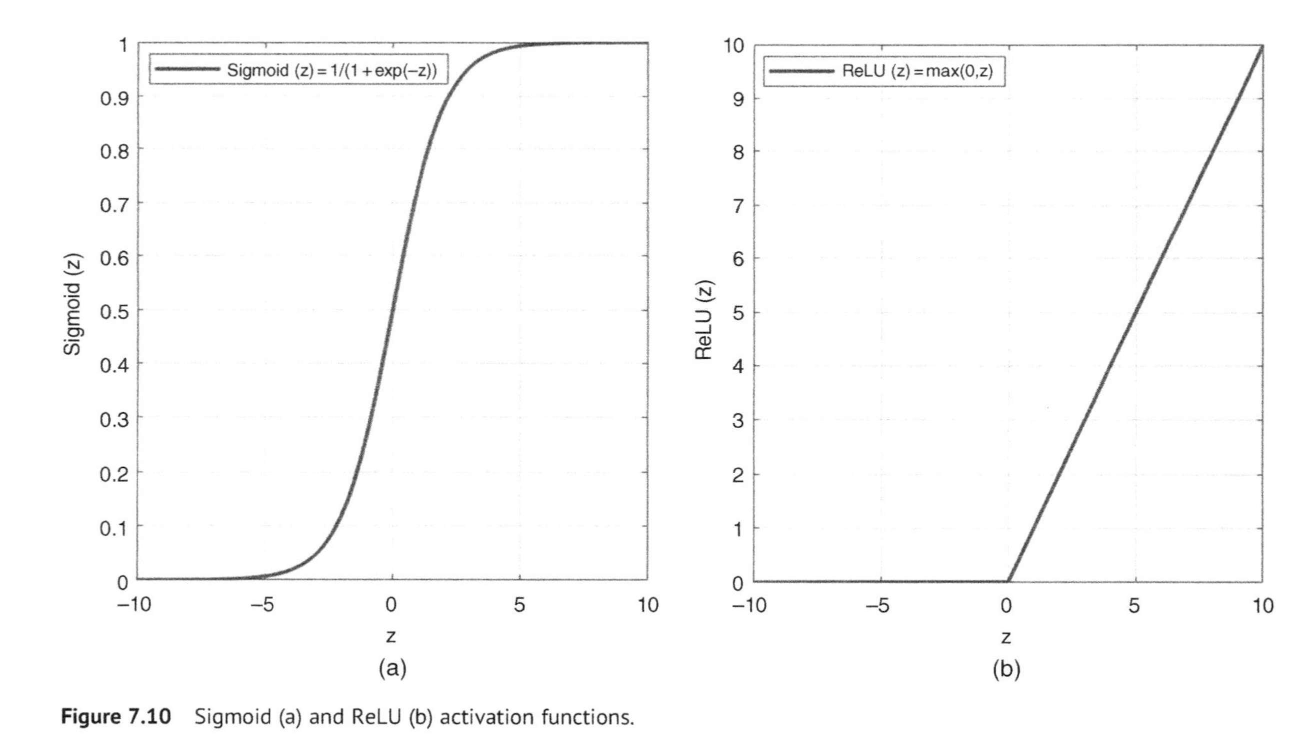

Figure 7.10 Sigmoid (a) and ReLU (b) activation functions.

often have larger number of neurons and layers, more ability to learn, more scalable, and further processing capability on their layers. Unlike in the past, currently, large data can be processed using DNNs in real-time. Deep learning using DNNs allows computational models that are composed of multiple processing layers to learn data representations with multiple abstraction levels.

With the advancement of deep learning algorithms, developers, analysts, and decision makers can explore and learn more about the data and their exposed relationships or hidden features. The new practices in developing data-driven application systems and decision-making algorithms require adaptation of deep learning algorithms and techniques in many application domains. In data-driven learning applications, the data governs the system behaviour.

Availability of deep structure as well as data-driven models paves the way for developing a new generation of deep networks called generative models. Unlike discriminative networks, which estimate a label for a test data, the generative networks assume they have the class label, and they wish to find the likelihood of particular features. They are often called generative adversarial networks (GANs) and include two nets, pitting one against the other (thus the term 'adversarial'). GANs were introduced by Goodfellow et al. [72] and were called the most interesting idea in machine learning in the last 10 years by LeCun, Facebook's AI research director.

In GANs, the generative models are designed via an adversarial process, in which two models are trained simultaneously: a generative model G that captures the data distribution, and a discriminative model D that estimates the probability that a sample is taken from the training data rather than G. To train the G model the probability of D making a mistake is maximized. This framework resembles a minimax two-player game. For arbitrary functions G and D, a unique solution exists, with G recovering the training data distribution and D equal to 0.5 everywhere. In the case where G and D are defined by multilayer perceptrons, the entire system can be trained with a backpropagation algorithm [72].

One particular type of deep feed-forward network is a CNN [73, 74]. This network is easier to train and can generalize much better than networks with full connectivity between adjacent layers. It often performs better in terms of both speed and accuracy for many computer vision and video processing applications. Recently, CNNs have found many new applications such as for detection of interictal epileptiform discharges (IEDs) from intracranial electroencephalogram (EEG) recordings [75, 76].

7.3.9.2 Convolutional Neural Networks

CNNs are designed to process multiple array data such as colour images composed of three 2-D arrays containing pixel intensities in the three colour channels. These networks have four main features that benefit from the properties of natural signals namely local connections, shared weights, pooling, and the use of numerous layers. A typical CNN architecture is structured as a series of stages (Figure 7.11). The first few stages are composed of two types of layers: convolutional and pooling layers. Units in a convolutional layer are organized in feature maps, within which each unit is connected to local patches in the feature maps of the previous layer through a set of weights called a filter bank. The result of this local weighted sum is then passed through a nonlinearity such as an ReLU.

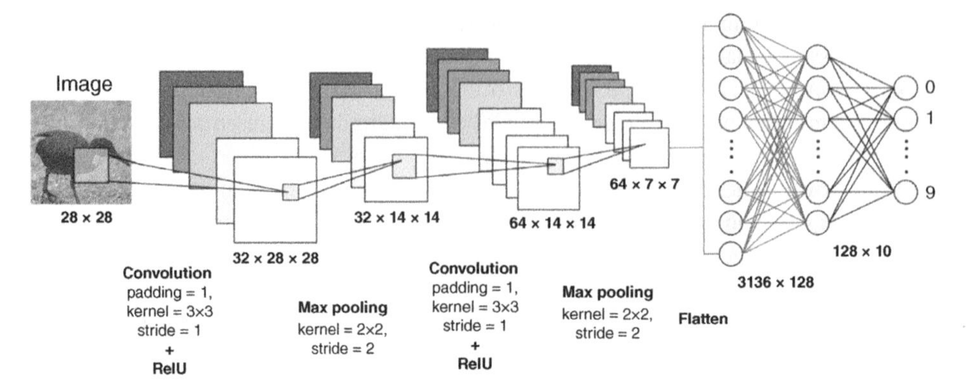

Figure 7.11 An example of a CNN and the operations in different layers.

All units in a feature map share the same filter bank. In a conventional CNN different feature maps in a layer use different filter banks. The reason for this architecture is twofold. First, in array data such as images, local groups of values are often highly correlated, forming distinctive local motifs that are easily detected. Second, the local statistics are invariant to location. In other words, if a motif can appear in one part of the image, it may appear anywhere in the image. Hence, we can have units at different locations sharing the same weights and detecting the same pattern in different parts of the array. From a mathematical point of view, the filtering operation performed by a feature map is a discrete convolution, hence the name.

Although the role of a convolutional layer is to detect local conjunctions of features from the previous layer, the role of the pooling layer is to merge semantically similar features into one. Since the relative positions of the features forming a motif can vary somewhat, its reliable detection can be performed by coarse-graining the position of each feature. A typical pooling unit computes the maximum of a local patch of units in one feature map (or in a few feature maps).

The neighbouring pooling units take input from patches that are shifted by more than one row or column, thereby reducing the representation dimension and creating an invariance to small shifts and distortions.

Stages of convolution, nonlinearity, and pooling are followed by more convolutional and fully connected layers. The backpropagation operation is then used for weight optimization in a CNN as for the normal multilayer ANN.

DNNs exploit the property that in many natural signals the higher-level features are obtained by composing lower-level ones. In images, local combinations of edges form motifs, motifs assemble into parts, and parts form objects. Similar hierarchies exist in speech and text from sounds to phones, phonemes, syllables, words, and sentences. The pooling allows representations to vary very little when elements in the previous layer vary in position and appearance.

The convolutional and pooling layers in CNNs are directly inspired by the classic notions of simple cells and complex cells in visual neuroscience [77], and the overall architecture is reminiscent of the LGN–V1–V2–V4–IT hierarchy (where LGN stands for lateral geniculate nucleus) in the visual cortex ventral pathway [78].

CNNs have their roots in the neocognitron [79], the architecture of which is somewhat similar, but does not have an end-to-end supervised learning algorithm such as backpropagation. Although CNNs were initially developed for image classification, an effective 1-D CNN was developed by Antoniades et al. [75, 76] for detection of IEDs from intracranial EEG recordings.

7.3.9.3 Autoencoders

Using two CNNs, one can build another type of DNN called an AE. The structure of an AE is seen in Figure 7.12. The *autoencoder* NN learns to copy its input to its output. It has a number of hidden layers, potentially with other encoder or decoder units between the layers, that describes a *code* used to represent the input. The first part of the AE, namely encoder, maps the input into the code, and the second part, namely decoder, maps the code to a reconstruction of the original input.

The fundamental concept behind the structure of AEs has been exploited in the area of ANNs for many years, with the first applications arounds the 1980s [80]. Before they become popular in learning generative models of data [81], the AEs had applications in dimensionality reduction or feature learning.

In its simplest form consider only one hidden layer, the encoder stage of an AE maps the input $\mathbf{x} \in R^d$ to $\mathbf{h} \in R^p$.

$$\mathbf{h} = \sigma(\mathbf{Wx} + \mathbf{b}) \tag{7.40}$$

where $\mathbf{h}$ is usually referred to as *code*, *latent variable*, or *latent representation*. Here, σ is an element-wise activation function. $\mathbf{W}$ is a weight matrix and $\mathbf{b}$ is a bias vector. Similar to other ANNs, these parameters are learned or updated using the backpropagation algorithm. In the decoder stage a similar process is followed to map the code $\mathbf{h}$ to $\mathbf{x}'$, i.e.:

$$\mathbf{x}' = \sigma'(\mathbf{W}'\mathbf{h} + \mathbf{b}') \tag{7.41}$$

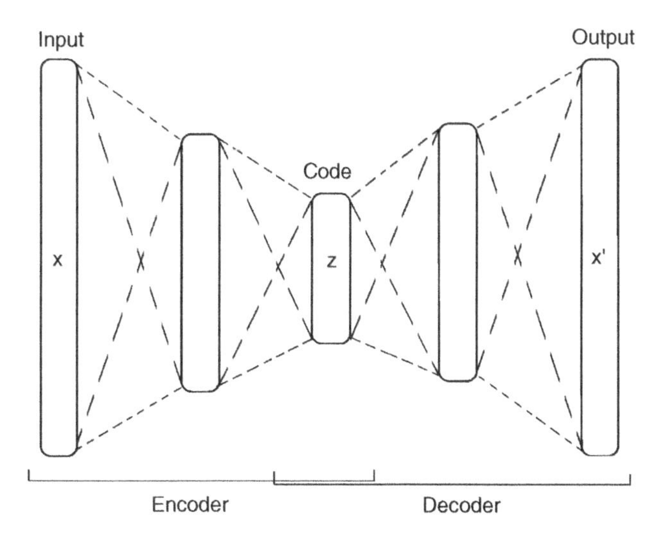

Figure 7.12 Structure of an autoencoder NN.

The objective is therefore to minimize the difference between the original and the reconstructed (decoded) data:

$$J(\mathbf{x},\mathbf{x}') = \|\mathbf{x}-\mathbf{x}'\|^2 = \|\mathbf{x}-\sigma'(\mathbf{W}'\sigma(\mathbf{W}\mathbf{x}+\mathbf{b})+\mathbf{b}')\|^2 \tag{7.42}$$

where, the input **x** is often denoised by averaging over a number of training sets. AEs are less sensitive to geometrical translation in the data, e.g. change in the position or size of the object in the scene or position change of an event-related potential (ERP) within the EEG segment.

To make the AEs more effective for particular applications they may be regularized. For example, in sparse autoencoders (SAEs) initially a larger number of hidden neurons are considered but only a small number of the hidden units are allowed to be active at once [82]. For a SAE, the cost function is penalized by a sparsity constraint on **W** used by the coder **h**.

Another regularized AE called denoising autoencoder (DAE) learns how to reconstruct a clean output from noisy versions of input [21]. A similar regularized method called contractive autoencoder (CAE) aims at making the ANN robust to slight variations in the input [83, 84].

7.3.9.4 Variational Autoencoder

Differently from a standard AE, a variational autoencoder (VAE) is a generative model similar to a GAN. VAEs have emerged as one of the most popular approaches to unsupervised learning of complicated distributions [85]. The mathematical concept behind a conventional VAE is very clear from the VAE schematic in Figure 7.13. Assuming the input having distribution, q, the mean $\boldsymbol{\mu}$ and covariance matrix $\sum$ are estimated. In the decoder the

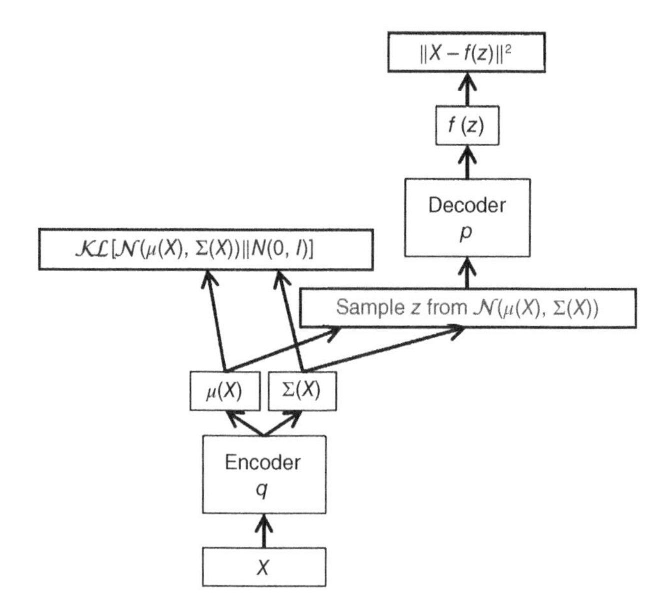

Figure 7.13 The schematic of the VAE structure. $\mathcal{KL}$ refers to Kullback–Leibler divergence.

output samples are drawn from a Gaussian distribution, p, with the same mean and variance. In the training optimization stage, the Euclidean distance between the reconstructed data, $f(z)$, sampled from distribution p, and the input is minimized. In order to characterize $f(z)$, its distance with a normal distribution of zero mean and identity variance is measured by means of Kullback–Leibler divergence denoted by $\mathcal{KL}$ [86].

The VAEs can be used in generation of surrogate data, in a similar fashion to GANs, mainly because the shape and distribution of the source data are both used in generation of surrogates.

7.3.9.5 Recent DNN Approaches

One of the early introduced DNNs was LeNet proposed by LeCun in 1988 for hand digit recognition [87]. LeNet is a feed-forward NN that consists of five consecutive layers of convolutional operations and pooling, followed by two fully connected layers. A later DNN approach was AlexNet [88]. This approach proposed by Krizhevesky et al., is considered as the first deep CNN architecture which showed ground-breaking results for image classification and recognition tasks.

Learning mechanism of CNN was largely based on trial and error, without deep understanding of the exact reason behind the improvement before 2013. This lack of understanding limited the performance of deep CNNs on complex images. In 2013, Zeiler and Fergus [89] proposed a multilayer deconvolutional ANN (DeconvNet), which became known as ZefNet to quantitatively visualize the network performance.

A deeper network namely visual geometry group (VGG), was later proposed [90]. VGG suggested that parallel placement of small size filters makes the receptive field as effective as that of large size filters. GoogleNet (also known as Inception-V1) is another CNN which won the 2014-ILSVRC competition [91]. The main objective of the GoogleNet architecture was to achieve high accuracy with a reduced computational cost. It introduced the new concept of inception module (block) in CNN, whereby multiscale convolutional transformations are incorporated using split, transform, and merge operations for feature extraction. Other deep networks such as ResNet [92], DenseNet [93], and many other DNN structures have been later introduced for increasing the speed, enhancing the accuracy of feature learning and classification.

In large ANNs, model pruning seeks to induce sparsity in a DNN's various connection matrices, thereby reducing the number of non-zero-valued parameters in the model. Weights in an ANN that are considered unimportant or rarely fire can be removed from the network with little to no consequence. Often, many neurons have a relatively small impact on the model performance, meaning we can achieve acceptable accuracy even when eliminating a large number of parameters. Reducing the number of parameters in a network becomes increasingly important as the neural architectures and datasets become larger in order to obtain reasonable execution times of models.

Among ANNs, recurrent neural networks (RNNs) and long short-term memory network (LSTM) rely on their previous states and therefore can learn state transitions for detection or recognition of particular/desired trends within the data. RNNs can be also used for prediction purposes. A shallow RNN however is not capable of using long-term dependencies between the data samples. LSTMs are a special kind of RNN, which are capable of learning long-term dependencies. They were introduced by Hochreiter and Schmidhuber [94] and were refined and popularized by many people such as [95]. Unlike RNN, which has only a single layer in each repeating module, the repeating module in an LSTM contains four interacting layers. A similar idea has been followed and explored for speech data generation. WaveNet DNN, as a probabilistic and AR DNN approach, has been designed and applied for speech waveform generation [96]. This architecture exploits the long-range temporal dependencies needed for raw audio generation.

Despite heavy DNN computational cost, some DNN structures such as ENet (efficient neural network), have been proposed to enable real-time applications. ENet is meant for spatial classification and segmentation of images, and uses significantly smaller number of parameters.

YOLOv3 [97] followed by YOLOv4 [98], aim at speeding up the object detection within images by choosing more effective features. In YOLOv4, new features including cross-stage-partial-connections, cross mini-batch normalization, self-adversarial training, Mish activation, mosaic data augmentation, DropBlock regularization [99], and complete intersection over union loss [100] are used and combined to achieve higher speed and accuracy [98].

7.3.9.6 Spike Neural Networks

Spike neural networks (SNNs) have appeared as the new generation of NNs, a more biological realistic approach by utilizing spikes, incorporating the concepts of space and time through *neural connectivity* and *plasticity*. They account for timing spikes being competitive with the traditional ANNs in terms of accuracy and computational power, and in some cases potentially better suited for hardware implementation due to their simple *integrate-and-fire* nature. They also better model the synaptic and neural currents and potentials. Figure 7.14a demonstrates the association of biological neuron activity and an artificial spiking neuron [101] and Figure 7.14b represents a simple SNN [102].

In generation of spike trains it is important to preserve the task-relevant information content of the input stimuli. Traditionally, it has been shown that most of the relevant information is included in the mean firing rate of neurons. There are two main encoding schemes: temporal encoding and rate-based encoding as depicted in Figure 7.15. The former is used over the latter one when patterns within the encoding window provide information about the stimulus that cannot be obtained from spike count. The rate-based encoding scheme is based on a spiking characteristic within a time interval (e.g. frequency), in the temporal encoding scheme the information is encoded in the time of the spikes. Rate-based encoding schemes ('rate as a spike count', 'rate as a spike density', and 'rate as a population activity') correspond to three different notions of mean firing rate (either an average over time, or an average over several repetitions of the experiment, or an average over a population of neurons respectively). Temporal encoding schemes are based on spike timing: 'time-to-first-spike' (when a code for the timing of the first spike contains all information about the new stimulus), 'phase' (when a 'time-to-first-spike' encoding scheme can be applied also where the reference signal is not a single event, but a periodic signal), and 'correlations and synchrony' (where spikes from other neurons are used as the reference signal for a spike code). The models for generation of spikes follow the methods described in Chapter 3 of this book. Therefore, application of SNN to EEG classification often requires the use of spike generation models [103, 104].

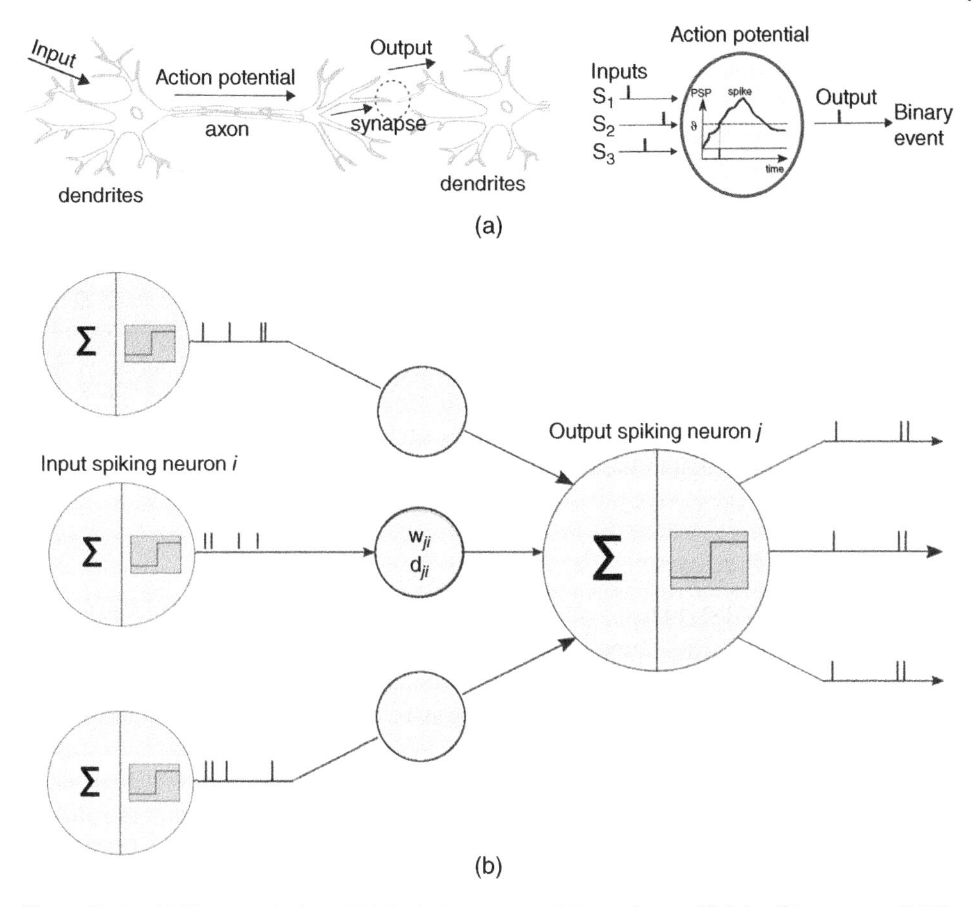

Figure 7.14 (a) The association of biological neuron activity and an artificial spiking neuron [101]. (b) A simple SNN [102].

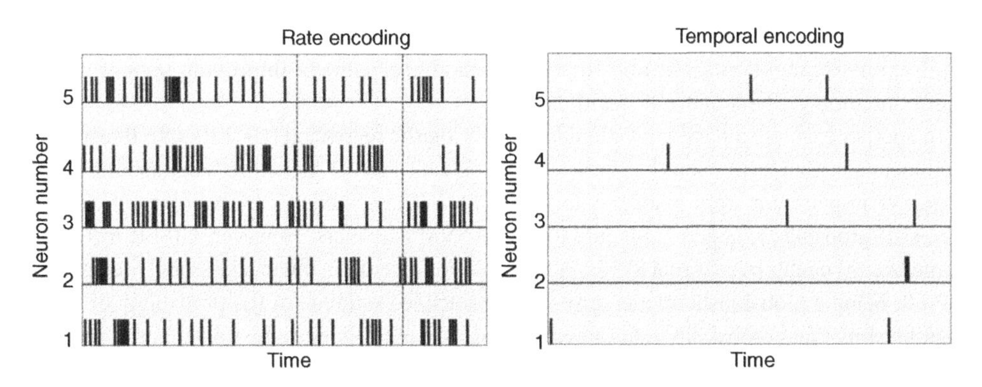

Figure 7.15 Rate-based encoding (on the left) and temporal encoding (on the right) [102].

7.3.9.7 Applications of DNNs to EEG

In some recently published papers [105] numerous applications of DNNs to EEG classification and identification have been addressed. The applications are many and include, but not limited to, classification of sleep stages and patterns from EEG signals [106], detection and classification of EEG patterns showing the onset of epileptic seizure [107], characterizing EEGs for attention deficit hyperactivity disorder (ADHD) [108], disorders of consciousness and brain injury [109, 110], depth of anaesthesia [111], mental fatigue [112], mental workload [113, 114], mood, or emotions [115], Alzheimer's [116], dementia [117], ischaemic stroke [118–121], Schizophrenia [122], emotions and affective states detection and recognition [123–139] and more widely in BCI to allow direct translation of brain activity into directives that affect the user's environment [114, 140, 141]. More recently, new DNNs have been designed to model the brain pathways which map scalp EEG into intracranial recordings for detection of IEDs from the scalp EEG [75, 76]. Most of these applications are discussed in detail in the related chapters of this book.

7.3.10 Gaussian Processes

Gaussian processes (GPs) [142] are other algorithms for solving *regression* and *probabilistic classification* problems. These supervised classifiers, though not effective for large dimension data, use prediction to smooth the observation and are versatile enough to use different kernels. Therefore, they are suitable for recovering and classifying biomedical signals which often have regular structure.

The GP approach performs inference using the noisy, potentially artefactual, data obtained from wearable sensors. Of fundamental importance is the GP notion as a distribution over functions, which is well suited to the analysis of patient physiological data time series, in which the inference over functions may be performed. This approach contrasts with conventional probabilistic approaches which define distributions over individual data points.

As an example, GPs have been used in an e-health platform for personalized healthcare to develop a patient-personalized system for analysis and inference in the presence of data uncertainty [143]. This uncertainty is typically caused by sensor artefact and data incompleteness. The method was used for clinical study and monitoring of 200 patients [143]. This provides the evidence that personalized e-health monitoring is feasible within an actual clinical environment, at scale, and that the method is capable of improving patient treatment outcome via personalized healthcare.

In another example a Bayesian Gaussian process logistic regression (GP-LR) with linear and nonlinear covariance functions has been used for classification of Alzheimer's disease and mild cognitive impairment from resting-state fMRI [144]. These models can be interpreted as a Bayesian probabilistic system analogue to kernel SVM classifiers. However, GP-LR methods confer some benefits over kernel SVMs. While SVMs only return a binary class label prediction, GP-LR, being a probabilistic model, provides a principled estimate of the probability of class membership. Class probability estimates are a measure of confidence the model has in its predictions. Such a confidence score can be very useful in the clinical setting.

7.3.11 Neural Processes

Garnelo et al. [145] introduced a class of neural latent variable models namely, neural processes (NPs), by combining GPs and ANNs. Like GPs, NPs define distributions over functions, are capable of rapid adaptation to new observations, and can estimate the uncertainty in their predictions. Like NNs, NPs are computationally efficient during training and evaluation but also learn to adapt their priors to data.

NPs perform regression by learning to map a context set of observed input–output pairs to a distribution over regression functions. Each function models the distribution of the output given an input, conditioned on the context. NPs have the benefit of fitting observed data efficiently with linear complexity in the number of context input–output pairs. They are able to learn a wide family of conditional distributions; they learn predictive distributions conditioned on context sets of arbitrary size.

Nonetheless, it has been shown that NPs suffer underfitting, giving inaccurate predictions at the inputs of the observed data they condition on [146]. This problem has been addressed by incorporating attention into NPs, allowing each input location to attend to the relevant context points for the prediction. It has been shown that, this greatly improves the accuracy of predictions, speeds up the training process, and expands the range of functions that can be modelled [146].

7.3.12 Graph Convolutional Networks

Graph convolutional networks (GCNs) are powerful NNs which enable machine learning on graphs. Even small size GCNs can produce useful feature representations of nodes in the network. In a GCN structure, the relative proximity of the network nodes is preserved in the two-dimensional representation even without any training. There are two different inputs to a GCN: one is the input feature matrix and the other a matrix representation of the graph structure such as the adjacency matrix of the graph. As an example, consider classification of EEG signals for healthy and dementia subjects. Using a GCN, the brain connectivity estimates can be incorporated into the training process as the proximity matrix.

As an example, for a two-layer GCN the following relation between the input and output can be realized [147]:

$$z = f(\mathbf{X}, \mathbf{A}) = softmax\left(\hat{\mathbf{A}}ReLU\left(\hat{\mathbf{A}}\mathbf{X}\mathbf{W}^{(0)}\right)\mathbf{W}^{(1)}\right) \tag{7.43}$$

where $\mathbf{A}$ is the adjacency matrix, $\hat{\mathbf{A}} = \widetilde{\mathbf{D}}^{-1/2}\widetilde{\mathbf{A}}\widetilde{\mathbf{D}}^{-1/2}$, $\hat{\mathbf{A}} = \mathbf{A} + \mathbf{I}_N$, $\mathbf{I}_N$ is an $N \times N$ identity matrix, and $\widetilde{D}_{ii} = \sum_j \widetilde{A}_{ij}$.

7.3.13 Naïve Bayes Classifier

The naïve Bayes classification method is a supervised learning algorithm applying Bayes' theorem with the 'naïve' assumption of independence between every pair of features. Therefore, the class label of the data can be estimated using a naïve Bayes classifier through the following procedure.

Given a class variable y and a dependent feature vector x_1 through x_n, Bayes' theorem states the following relationship:

$$p(y \mid x_1, \dots, x_n) = \frac{p(y)p(x_1, \dots, x_n \mid y)}{p(x_1, \dots, x_n)} \tag{7.44}$$

Using the naïve independence assumption:

$$p(y \mid x_1, \dots, x_n) = \frac{p(y)\prod_{i=1}^{n} p(x_i \mid y)}{p(x_1, \dots, x_n)} \tag{7.45}$$

As $p(x_1, \dots, x_n)$ is constant for any given input, the following classification rule can be deducted:

$$p(y \mid x_1, \dots, x_n) \propto p(y)\prod_{i=1}^{n} p(x_i \mid y) \tag{7.46}$$

The likelihood probability $p(x_i \mid y)$ is considered known either empirically or by assumption. Gaussian, multinomial, and Bernoulli are among the most popular distributions considered for these probabilities. This results in estimation of a class label as:

$$\hat{y} = arg\max_{y} p(y)\prod_{i=1}^{n} p(x_i \mid y) \tag{7.47}$$

Naïve Bayes classifiers are simple and work quite well in many real-world situations. They require a small amount of training data to estimate the necessary parameters [148].

Naïve Bayes learners and classifiers can be extremely fast compared to more computationally intensive methods. The decoupling of the class conditional feature distributions means that each distribution can be independently estimated as a one-dimensional distribution. This in turn helps alleviate the problem of the curse of dimensionality.

7.3.14 Hidden Markov Model

HMMs are state-space machines meant to detect a correct data sequence which may happen within a longer data sequence. HMMs are presented as the state diagrams. Therefore, for each application, an HMM has a certain number of states and knows the probability of each possible event happening in each state. Before moving to an application of HMM to EEG, a good example for better understanding the HMM concept is the detection of QRS in an electrocardiogram (ECG) signal depicted in Figure 7.16.

From Figure 7.16, an ECG is recognized when the symbols turn up in a correct sequence in the signal. Assuming the waveform in Figure 7.16 is the ECG of a healthy subject, the state diagram in Figure 7.17 classifies the subject as healthy as long as the above ECG sequence is detected in his/her ECG record.

In this diagram, state Q_0 is a wait state to detect the beginning of sequence starting from a bar (b). State Q_1 is a state which shows a bar has been detected. State Q_2 simply demonstrates that a bar and a small hump (p) have been already recognized. Q_3 represents the state of the system when a bar, a small hump, and a QRS wave are correctly sequenced.

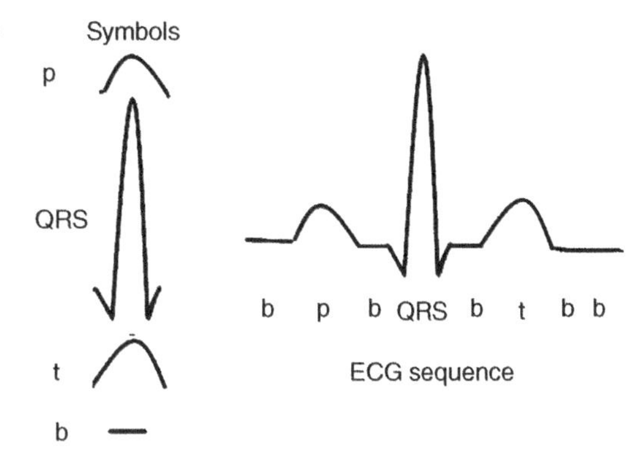

Figure 7.16 A synthetic ECG segment of a healthy individual and its corresponding symbols.

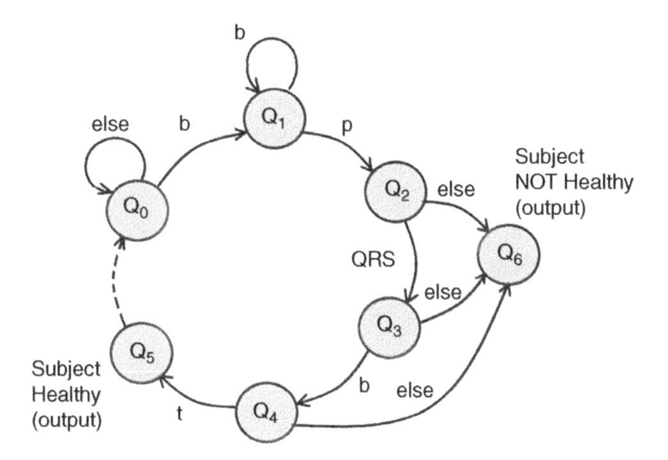

Figure 7.17 An HMM for detection of a healthy heart from an ECG sequence.

Q_4 and Q_5 similarly show the states after a bar and a large hump (t) have been respectively added to the data sequence. After detecting the correct sequence, the output is 'the subject is healthy'. After each state, if a wrong symbol or noise is detected, the system output is 'the subject is not healthy'.

In places where the objective is to detect a particular heart condition, a separate HMM should be designed to recognize that condition. Beside their applications in biomedical signal analysis and disease recognition, HMMs have a wide range of applications in speech, video, bioscience (such as genomics) and face recognition.

In order to use an HMM for classification, the initial state ($\boldsymbol{\pi} = \pi_i$, where $i = 1, \ldots, N$, where N is the number of states) the state probabilities ($\boldsymbol{A} = \{a_{ij}\}$, $i,j = 1, \ldots, N$) and the state transition probabilities ($\boldsymbol{B} = \{b_j(k)\}$, $j = 1, \ldots, N$, $k = 1, \ldots, T$, where T is the observation sequence length) should be known. For training the HMMs (i.e. to find the model parameters) however, a set of observations and the number of states should be known.

There are three major problems (objectives) in using or designing HMMs:

1) Given the observation sequence $O = o_1, o_2, ..., o_T$, and the model $\lambda = (\boldsymbol{A}, \boldsymbol{B}, \boldsymbol{\pi})$ what will be the probability of observation sequence, $p(O|\lambda)$?
2) Given the observation sequence and the model as above, how can we find the state sequence $Q = Q_1, ..., Q_N$ which generates such an observation sequence?
3) Given the observations and the number of states how the HMM model can be built up, i.e. how we find $\lambda = (\boldsymbol{A}, \boldsymbol{B}, \boldsymbol{\pi})$.

To solve the first problem, often, the iterative forward-backward algorithm is used while the Viterbi algorithm is employed to solve the second problem. Finally, a few algorithms, such as the most popular one, Baum-Welch (Leonard E Baum, 1960s) [149, 150], can be used to solve the third problem which is the most important one.

To go through forward and backward algorithms, set $\lambda = (\boldsymbol{A}, \boldsymbol{B}, \boldsymbol{\pi})$ with random initial conditions (unless some information about the parameters is available already). Then, follow the iterative procedures as shown below.

7.3.14.1 Forward Algorithm

Let $\alpha_i(t) = p(o_1 = y_1, ...o_t = y_t, Q_t = i \mid \lambda)$ the probability of observing $y_1, y_2, ... y_t$ and being in state i at time t:

$$\alpha_i(1) = \pi_i b_i(y_1) \tag{7.48}$$

and iterate over t:

$$\alpha_i(t+1) = b_i(y_{t+1}) \sum_{j=1}^{N} \alpha_j(t) a_{ji} \tag{7.49}$$

7.3.14.2 Backward Algorithm

Let $\beta_i(t) = p(o_{t+1} = y_{t+1}, ...o_T = y_T, Q_t = i \mid \lambda)$ the probability of observing $y_{t+1}, y_2, ... y_T$ partial sequence and being in state i at time t:

$$\beta_i(T) = 1, \tag{7.50}$$

and iterate over t:

$$\beta_i(t) = \sum_{j=1}^{N} \beta_j(t+1) a_{ij} b_j(y_{t+1}) \tag{7.51}$$

7.3.14.3 HMM Design

For the HMM design we need to calculate the following auxiliary variables following Bayes theorem:

$$\gamma_i(t) = p(Q_t = i \mid O, \lambda) = \frac{p(Q_t = i, O \mid \lambda)}{p(O \mid \lambda)} = \frac{\alpha_i(t)\beta_i(t)}{\sum_{j=1}^{N} \alpha_j(t)\beta_j(t)} \tag{7.52}$$

This is the probability of being in state i at time t for a given model λ and the observed sequence O. Then, the probability of being in states i and j at times t and $t + 1$ respectively for the given model λ and the observed sequence O is:

$$\xi_{ij}(t) = p(Q_t = i, Q_{t+1} = j \mid O, \lambda) = \frac{p(Q_t = i, Q_{t+1} = j, O \mid \lambda)}{p(O \mid \lambda)}$$
$$= \frac{\alpha_i(t)a_{ij}\beta_j(t+1)b_j(y_{t+1})}{\sum_{i=1}^{N}\sum_{j=1}^{N}\alpha_i(t)a_{ij}\beta_j(t+1)b_j(y_{t+1})} \tag{7.53}$$

The HMM parameters for the model λ are then updated using:

$$\hat{\pi}_i = \gamma_i(1) \tag{7.54}$$

$$\hat{a}_{ij} = \frac{\sum_{t=1}^{T-1}\xi_{ij}(t)}{\sum_{t=1}^{T-1}\gamma_i(t)} \tag{7.55}$$

$$\hat{b}_i(v_k) = \frac{\sum_{t=1,\; y_t = v_k}^{T}\gamma_i(t)}{\sum_{t=1}^{T}\gamma_i(t)} \tag{7.56}$$

$p(O|\lambda)$ is then calculated utilizing these parameters. After each iteration, this probability increases until the parameters $\boldsymbol{A}$, $\boldsymbol{B}$, and $\boldsymbol{\pi}$ reach their optimum values, where the algorithm terminates and there would not be any considerable change in the probability.

There have been reports of using HMMs for recognition or classification of EEGs. One may think of detecting an ERP waveform from the EEGs. Another example can be detection of epileptiform discharges from the intracranial EEG signals. These waveforms often have particular morphologies and a similar HMM used for QRS detection can be followed for their detections.

In a BCI application, the single-trial five-electrode EEG AR coefficients are estimated for four different movement tasks. PCA is then applied to reduce the number of parameters (as HMM is computationally very demanding for high-dimensional data), and then an HMM is used to classify the movement into one of the four tasks [151]. In this application mean and variance of the features are used to define the states. The AR order and the number of HMM states are set empirically. In a similar HMM approach sparse AR coefficients together with measures derived from them such as partial directed coherence (PDC) and directed transfer function (DTF) have been interpreted in terms of information flow between channels [152] and used. The AR model order was selected as the order corresponding to the lowest Bayesian information criterion (BIC), or Schwarz information criterion value [153], for a range of model orders. The number of states have also been selected as the number of clusters.

Obermaier et al. [154] used HMM for online classification of single-trial EEG during left and right imagery hand movement for a BCI application. In their design a three state HMM model is used where each state is characterized using the parameters of a GMM of three

Gaussians. In their experiments they achieved a lower error for HMM than a linear discriminant classifier.

In another application, EEG data have been used for estimation of mental fatigue using HMM [155]. In this work approximate entropy (ApEn) and Kolmogorov complexity (Kc) have been utilized to characterize the complexity and irregularity of EEG data in different mental fatigue states. Then the kernel principal component analysis (KPCA) and HMM have been combined to differentiate between the two states. The KPCA algorithm has been used to extract the nonlinear features from the EEG complexity parameters and improve the HMM generalization performance. Two HMMs, one representing the normal state and another one representing the fatigue state have been trained by using the EEG data segments recorded during the corresponding mental fatigue states.

There are also reports of using HMM for scoring sleep EEG. As an example, the spatiotemporal complexity of whole-brain networks and state transitions during sleep have been characterized in detail [156]. In order to obtain the most unbiased estimate of how the brain network states evolve through the human sleep cycle, an HMM has been used to analyze continuous simultaneous EEG and fMRI data. It has been claimed that the proposed HMM facilitated discovery of the dynamic choreography between different brain networks during the wake–non-rapid eye movement (REM) sleep cycle.

7.4 Common Spatial Patterns

CSP is a popular feature extraction and optimization method for classification of multichannel signals particularly EEG. Traditionally, CSP aims at estimation of spatial filters which discriminate between two classes based on their variances. CSP is one of the most successful feature extraction algorithms for BCI systems. A straightforward BCI application involves two-class classification of EEGs such as for classification of real or imagery left and right-hand movements. CSP ($\mathbf{w}$) minimizes the Rayleigh quotient of the spatial covariance matrices to achieve the variance imbalance between two classes of data $\mathbf{X}_1$ and $\mathbf{X}_2$. Before applying CSP, the signals are bandpass filtered and centred. The CSP goal is to find a spatial filter $\mathbf{w} \in \Re^c$ such that the variance of the projected samples of one class is maximized while the other's is minimized. The following maximization criterion is used for CSP estimation [157, 158]:

$$\mathbf{w}^{(CSP)} = arg \max_{\mathbf{w}} \frac{tr(\mathbf{w}^T \mathbf{C}_1 \mathbf{w})}{tr(\mathbf{w}^T \mathbf{C}_2 \mathbf{w})} \tag{7.57}$$

where $\mathbf{C}_1$ and $\mathbf{C}_2$ are covariance matrices of the two clusters $\mathbf{X}_1$ and $\mathbf{X}_2$. With k as any real constant, this optimization problem can be solved (though this is not the only way) by first observing that the function $J(\mathbf{w})$ remains unchanged even if the filter $\mathbf{w}$ is rescaled, i.e. $J(k\mathbf{w}) = J(\mathbf{w})$, Hence, extremizing $J(\mathbf{w})$ is equivalent to extremizing $\mathbf{w}^T\mathbf{C}_1\mathbf{w}$ subject to the constraint $\mathbf{w}^T\mathbf{C}_2\mathbf{w} = 1$ since it is always possible to find a rescaling of $\mathbf{w}$ such that $\mathbf{w}^T\mathbf{C}_2\mathbf{w} = 1$. Employing Lagrange multipliers, this constrained problem changes to an unconstrained problem as [159]:

$$L(\lambda, \mathbf{w}) = \mathbf{w}^T \mathbf{C}_1 \mathbf{w} - \lambda(\mathbf{w}^T \mathbf{C}_2 \mathbf{w} - 1) \tag{7.58}$$

A simple way to derive optimum $\mathbf{w}$ is to take derivatives of L and set to zero as follows:

$$\frac{\partial L}{\partial \mathbf{w}} = 0 \Rightarrow C_2^{-1}C_1\mathbf{w} = \lambda\mathbf{w} \tag{7.59}$$

Based on the standard eigenvalue problem, spatial filters extremizing Eq. (7.56) are then the eigenvectors of $\boldsymbol{M} = \boldsymbol{C}_2^{-1}\boldsymbol{C}_1$ corresponding to its largest and lowest eigenvalues. When using CSP, the extracted features are derived from the logarithm of the signal variance after projection to filters $\mathbf{w}$.

Eigenvalue λ measures the ratio of variances of the two classes. CSP is suitable for classification of both spectral [158] and spatial data since the power of the latent signal is larger for the first cluster than for the second cluster.

Applying this approach to separation of ERPs from EEG signals enhances one of the classes against the rest. This allows a better discrimination between the two classes, thus can be separated easier.

In early 2000 in a two-class BCI setup, Ramoser et al. [160] proposed application of CSP that learned to maximize the variance of bandpass filtered EEG signals from one class while minimizing their variance from the other class. Currently, CSP is widely used in BCI where evoked potentials or movement can cause alteration of signals.

The model uses the four most important CSP filters. The variance is then calculated from the CSP time series. Then, a log operation is applied and the weight vector obtained with the LDA is used to discriminate between left and right-hand movement imaginations. Finally, the signals are classified and based on that an output signal is produced to control the cursor on the computer screen. Patterns related to right-hand movement are shown in Figure 7.18a and those for left-hand movement in Figure 7.18b. These patterns are the strongest patterns.

Most of the existing CSP-based methods exploit covariance matrices on a subject-by-subject basis so that the inter-subject information is neglected. In that paper, the CSP has been modified for subject-to-subject transfer, where a linear combination of covariance matrices of subjects under consideration has been exploited. Two methods have been

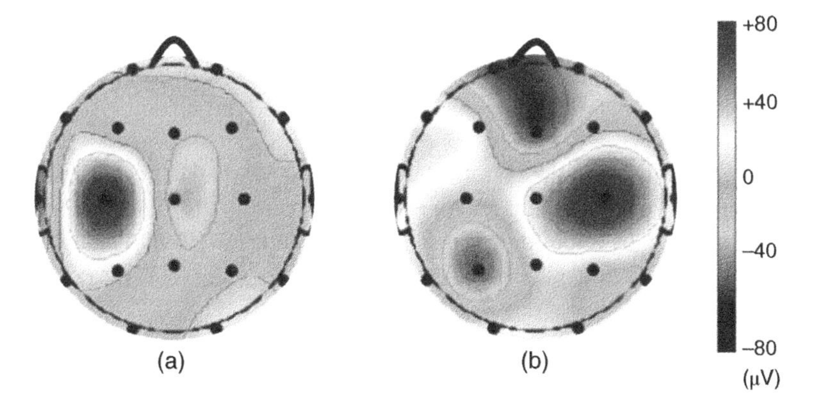

Figure 7.18 CSP patterns related to right-hand movement (a) and left-hand movement (b). The EEG channels are indicated by numbers that correspond to three rows of channels (electrodes) within the central and centro-parietal regions.

developed to determine a composite covariance matrix that is a weighted sum of covariance matrices involving subjects, leading to composite CSP [161].

Accurate estimation of CSPs under weak assumptions (mainly noisy cases) requires derivation of an asymptotically optimal solution. This in turn necessitates calculation of the loss in signal to noise-plus-interference ratio because of finite sample effect in a closed form. This is often important in detection/classification of ERPs since the EEG or magnetoencephalography (MEG) signals contain not only the spatiotemporal patterns bounded to the events but also ongoing brain activity as well as other artefacts such as eye blink and muscle movement artefacts. Therefore, to improve the estimation results the CSPs are applied to the recorded multichannel signals.

In Eq. (7.56), the trace values are equivalent to L_2-norm. Due to sensitivity of L_2-norm to outliers [162], it is replaced by L_1-norm. This changes Eq. (7.56) to

$$\mathbf{w}^{(CSP)} = arg \max_{\mathbf{w}} \frac{\|\mathbf{w}^T \mathbf{x}_1\|_1}{\|\mathbf{w}^T \mathbf{x}_2\|_1} = arg \max_{\mathbf{w}} \frac{\sum_{k=1}^{m} |\mathbf{w}^T \mathbf{x}_{1k}|}{\sum_{l=1}^{n} |\mathbf{w}^T \mathbf{x}_{2l}|} \tag{7.60}$$

where m and n are the total sampled points of the two classes and $\|\cdot\|_1$ refers to L_1-norm. It has been shown that using this approach, the effects of outliers are alleviated.

Most of the existing CSP-based methods exploit covariance matrices on a subject-by-subject basis so that inter-subject information is neglected. CSP and its variants have received much attention and have been one of the most efficient feature extraction methods for BCI. However, despite its straightforward mathematics, CSP overfits the data and is highly sensitive to noise.

To enhance CSP performance and address these shortcomings, recently it has been proposed to improve the CSP learning process using prior information in the form of regularization terms or constraints. Eleven different regularization methods for CSP have been categorized and compared by Lotte et al. [159]. These methods included regularization in the estimation of the EEG covariance matrix [163], composite CSP [161], regularized CSP with generic learning [164], regularized CSP with diagonal loading (DL) [158] and Invariant CSP [165]. All the algorithms were applied to a number of trials from 17 subjects and the CSP with Tikhonov regularization in which the optimization of $J(\mathbf{w})$ is penalized by minimizing $||\mathbf{w}||^2$, and consequently minimizing the influence of artefacts and outliers, was suggested as the best method.

To make CSP less sensitive to noise, the objective function in (7.59) may be regularized either during estimation of the covariance matrix for each class or during the minimization process of the CSP cost function by imposing priors to the spatial filters $\mathbf{w}$ [159]. A straight forward approach to regularize estimation of the covariance matrix for class i, i.e. $\hat{\mathbf{C}}_i$, is denoted as:

$$\hat{\mathbf{C}}_i = (1-\gamma)\mathbf{P} + \gamma\mathbf{I} \tag{7.61}$$

where $\mathbf{P}$ is estimated as:

$$\mathbf{P} = (1-\beta)\mathbf{C}_i + \beta\,\mathbf{G}_i \tag{7.62}$$

In these equations, $\mathbf{C}_i$ is the initial spatial covariance matrix for class i, $\mathbf{G}_i$, the so-called generic covariance matrix, computed by averaging the covariance matrices of a number of trials for class i, though it may be defined based on neurophysiological priors only, and γ and $\beta \in [0, 1]$ are the regularizing parameters. The generic matrix represents a given prior on how the covariance matrix for the mental state considered should be.

Conversely for regularization of the CSP objective function, a regularization term is added to the CSP objective function in order to penalize the resulting spatial filters that do not satisfy a given prior. This results in a slightly different objective function as [159]:

$$J(\mathbf{w}) = \frac{\mathbf{w}^T\mathbf{C}_1\mathbf{w}}{\mathbf{w}^T\mathbf{C}_2\mathbf{w} + \alpha\mathbf{Q}(\mathbf{w})} \tag{7.63}$$

The penalty function $\mathbf{Q}(\mathbf{w})$ weighted by the penalty (or Lagrange) parameter $\alpha \geq 0$ indicates how much the spatial filter $\mathbf{w}$ satisfies a given prior. Similar to the first term in the denominator of (7.62), this term needs to be minimized in order to maximize the cost function $J(\mathbf{w})$. In the literature, different quadratic and non-quadratic penalty functions have been defined. Some exemplar methods can be seen in [159, 165, 166]. A quadratic constraint, such as $\mathbf{Q}(\mathbf{w}) = \mathbf{w}^T\mathbf{K}\mathbf{w}$ results in solving a new set of eigenvalue problems as [159]:

$$(\mathbf{C}_2 + \alpha\mathbf{K})^{-1}\mathbf{C}_1\mathbf{w} = \lambda\mathbf{w} \text{ for the first pattern} \tag{7.64a}$$

$$(\mathbf{C}_1 + \alpha\mathbf{K})^{-1}\mathbf{C}_2\mathbf{w} = \lambda\mathbf{w} \text{ for the second pattern} \tag{7.64b}$$

The composite CSP algorithm proposed in [161] performs subject-to-subject transfer by regularizing the covariance matrices using other subjects' data. In this approach α and γ are zero and only β is non-zero. The generic covariance matrices $\mathbf{G}_i$ is defined according to covariance matrices of other subjects.

The approach based on regularized CSP with generic learning [164] uses both β and γ regularization terms, which means it intends to shrink the covariance matrix towards both the identity matrix and a generic covariance matrix $\mathbf{G}_i$. Here, similar to composite CSP, $\mathbf{G}_i$ is computed from the covariance matrices of other subjects.

DL is another form of covariance matrix regularization used in the BCI literature [158], which consists of shrinking the covariance matrix towards the identity matrix (sometimes called orthonormalization). Hence, in this approach, only γ is used $(\alpha = \beta = 0)$ and its value can automatically be identified using the Ledoit and Wolf's method [167] or by cross-validation [159].

In the approach based on Tikhonov regularization [159], the penalty term is defined as $P(\mathbf{w}) = \|\mathbf{w}\|^2 = \mathbf{w}^T\mathbf{w} = \mathbf{w}^T\mathbf{I}\mathbf{w}$ which is obtained by using $\mathbf{K} = \mathbf{I}$ in (7.64). Such regularization is expected to constrain the solution to filters with a small norm, hence mitigating the influence of artefacts and outliers.

Finally, in the weighted Tikhonov regularization method it is considered that some EEG channels are more important than others (as expected). Therefore, signals from these channels are weighted more than those of others. This consequently causes different penalizations for different channels. To perform this regularization method, define $P(\mathbf{w}) = \mathbf{w}^T\mathbf{D}_\mathbf{w}\mathbf{w}$, where $\mathbf{D}_\mathbf{w}$ is a diagonal matrix whose entries are defined based on the measures of CSPs from other subjects [159].

The obtained results in implementing 11 different regularization methods in [159] show that in places where the data are noisy, the regularization improves the results by approximately 3% on average. However, the best algorithm outperforms CSP by about 3–4% in mean classification accuracy and by almost 10% in median classification accuracy. The regularized methods are also more robust, i.e. they show lower variance across both classes and subjects. Among various approaches, the weighted Tikhonov regularized CSP proposed in [159] has been reported to have the best performance.

Regularization of **w** can also be used to reduce the number of EEG channels without compromising the classification score. Farquhar et al. [165] converted CSP into a quadratically constrained quadratic optimization problem with l_1–norm penalty and Arvaneh et al. [168] used an l_1/l_2-norm constraint. Recently, in [169] a computationally expensive quasi l_0-norm based principle has been applied to achieve a sparse solution for **w**. In [161] CSP has been modified for subject-to-subject transfer, to exploit a linear combination of covariance matrices of subjects under study. In this approach a composite covariance matrix, that is a weighted sum of covariance matrices involving subjects, has been used, leading to composite CSP. More recent applications of CSP to EEG signals, mostly for BCI, can be seen in [170]. The authors in this article reformulate the CSP as a constrained minimization problem and establish the equivalence of the reformulated and the original CSPs. The optimization problem is performed by alternately performing SVD and least squares. Under this new formulation, various regularization techniques for linear regression such as sparse CSP [159], transfer CSP [159, 171], and multisubject CSP [159] can be easily implemented to regularize the CSPs for different learning paradigms. A transfer CSP uses the transfer learning concept so the data for other subjects can be used for training. A multisubject CSP is used to reduce the calibration time to utilize the multitask learning framework to jointly learn CSP filters from the EEG signals from multiple subjects, each of which may only have very few EEG trials for training. While this learning paradigm has been actively studied in recent years, most previous work focuses on classification or regression and therefore cannot be directly utilized for CSP.

A multiclass CSP has been developed for vigilance detection from the EEG signals [172]. In this method a two-class CSP has been extended to multiclass by using joint approximate diagonalization (JAD). Vigilance or sustained attention is an important aspect for people who engaged in long time attention demanding tasks such as monotonous monitoring and driving. Vigilance detection has been an important topic in the field of BCI research. There is one drawback for the traditional CSP, that is, the CSP relies on the assumption that the data in each class follow a Gaussian distribution. However, this assumption is not always true for EEG data in practise, especially in the research of vigilance detection-based EEG (e.g. during sleep). Thus, traditional CSP suffers performance degradation in the case of non-Gaussian distributions. The traditional CSP has therefore been extended to the nonparametric CSP (NCSP) algorithm which does not explicitly rely on the assumption of the underlying class Gaussian distribution. This is then extended to nonparametric multiclass CSP (NMCSP) [172].

7.5 Summary

In this chapter, most popular machine learning methods and their applications for clustering or classification of biomedical signals with emphasis on EEG have been briefly introduced. Some machine learning algorithms have less sensitivity to noise and outliers.

These artefacts often make the classifier applications less reliable. Nonlinear techniques and regression-based algorithms are becoming more popular due to their advantages in classification of non-linearly separable data. New approaches in deep learning, which best mimic the human brain function, are capable of handling non-separable and nonlinear cases of very large datasets.

The trained large deep networks may be partially reused for classification of data of different nature through *transfer learning* [173]. Following this approach part of the network parameters may be kept fixed and the remaining, often fully connected, final layers trained using the new data.

In places where the training data size is limited, generation and application of surrogate data may boost the quality/accuracy of the results. In such cases, the classifiers should deal with the data distribution and statistics rather than the data samples. This is mainly because the methods for generation of surrogate data are able to generate new data with the same distribution as the data distribution. The surrogate data may not look like the data itself. In the field of deep networks, one option is to use a VAE which uses data distribution for learning [174].

Distributed and cooperative learning may become more popular in the near future due to the availability of on-sensor processing and decision making and the nature of multisensor scenario and the large data size. Often, signal processing methods boost machine learning techniques as they can extract and provide more meaningful data features prior to machine learning applications. Conversely, the need for on-board classification and state recognition requires many algorithms to be developed in Python, Java, and machine languages suitable for compact wearable devices, tablets, and mobile phones.

Moreover, real-time applications require faster systems which are able to process, communicate, and make decisions within one sample interval. Many regression-based classifiers can cope with real-time applications. In the area of DNN, a deep reinforcement learning approach, has been claimed to be able to perform real-time learning classification [175]. Nevertheless, these networks need considerable time for training.

Although current state-of-the-arts intelligent systems already exploit and mimic some human intelligences, still, emotional, social, attentional and moral–ethical intelligences are not implemented to their full potential. For example, current intelligent systems have the ability, to some extent, to detect and recognize human emotions, but so far, they do not possess self-awareness, self-management, self-assessment, social-awareness, and social skills to interact with other agents efficiently. Moreover, current intelligent systems still have limited cognitive skills in other domains and are not yet able to perform adequate decision making.

References

1 Prochazka, A., Kuchynka, J., Vysata, O. et al. (2018). Sleep scoring using polysomnography data features. *Springer Journal of Signal, Image and Video Processing (SIVP)* 12 (6): 1–9.

2 Akcay, S., Atapour-Abarghouei, A., and Breckon, T.P. (2018). GANomaly: semi-Supervised Anomaly Detection via Adversarial Training. arXiv:1805.06725 [cs.CV].

3 Sanei, S. (2013). *Adaptive Processing of Brain Signals*. Wiley.

4 Xindong, W., Vipin, K., Quinlan, R. et al. (2008). Top 10 algorithms in data mining. *Knowledge and Information Systems* 14 (1): 1–37.

5 Fu, L. and Medico, E. (2007). FLAME: a novel fuzzy clustering method for the analysis of DNA microarray data. *BMC Bioinformatics* 8 (3).

6 Vapnik, V. (1998). *Statistical Learning Theory*. Wiley.

7 Wan, J., Chen, M., Xia, F. et al. (2013). From machine-tomachine communications towards cyber-physical systems. *Computer Science and Information Systems* 10: 1105–1128.

8 Bengio, Y. (2009). Learning deep architectures for AI. *Foundations and Trends in Machine Learning* 2 (1): 1–127.

9 Shoker, L., Sanei, S., and Chambers, J. (2005). Artifact removal from electroencephalograms using a hybrid BSS-SVM algorithm. *IEEE Signal Processing Letters* 12 (10): 721–724.

10 Shoker, L., Sanei, S., Wang, W., and Chambers, J. (2004). Removal of eye blinking artifact from EEG incorporating a new constrained BSS algorithm. *IEE Journal of Medical and Biological Engineering and Computing* 43 (2): 290–295.

11 Kouchaki, S., Yang, Y., Walker, T.M. et al. (2019). Application of machine learning techniques to tuberculosis drug resistance analysis. *Bioinformatics* 35 (13): 2276–2282. https://doi.org/10.1093/bioinformatics/bty949.

12 Hartigan, J. and Wong, M. (1979). A k-mean clustering algorithm. *Applied Statistics* 28: 100–108.

13 Fellous, J.-M., Tiesinga, P.H.E., Thomas, P.J., and Sejnowski, T.J. (2004). Discovering spike patterns in neural responses. *The Journal of Neuroscience* 24 (12): 2989–3001.

14 Hastie, T., Tibshirani, R., and Walter, G. (2000). *Estimating the number of clusters in a dataset via the gap statistic*. Technical Report 208, Stanford University, Stanford, CA, USA.

15 Ester, M., Kriegel, H.-P., Sander, J., and Xu, X. (1996). A density-based algorithm for discovering clusters in large spatial databases with noise. KDD-96 Proceedings. AAAI 1996.

16 Ng, Y.A., Jordan, M.I., and Weiss, Y. (2002). On spectral clustering: analysis and an algorithm. *NIPS'01: Proceedings of the 14th International Conference on Neural Information Processing Systems: Natural and Synthetic* (January 2001), 849–856.

17 Ioffe, S. and Szegedy, C. (2015). Batch normalization: accelerating deep network training by reducing internal covariate shift. *International Conference on Machine Learning*, ICML 2015.

18 Lin, F., Cohen, W.W. (2010). Power iteration clustering. *Proceedings of the 27th International Conference on Machine Learning (ICML-10)*, Haifa, Israel (June 21–24, 2010).

19 Rasmus, A., Honkala, M., Berglund, M., and Raiko, T. (2015). Semi-supervised learning with ladder networks. arXiv:1507.02672 [cs.NE].

20 Bair, E. (2013). Semi-supervised clustering methods. *WIREs Computational Statistics* 5 (5): 349–361.

21 Vincent, P. and Larochelle, H. (2010). Stacked denoising autoencoders: learning useful representations in a deep network with a local denoising criterion. *Journal of Machine Learning Research* 11: 3371–3408.

22 Bezdek, J.C. (1981). *Pattern Recognition with Fuzzy Objective Function Algorithms*. Springer, 0-306-40671-3.

23 Jin, X. and Han, J. (2011). *K*-Medoids clustering. In: *Encyclopaedia of Machine Learning* (eds. C. Sammut and G.I. Webb). Boston, MA: Springer, 564–565. Online publication: 2017.

24 Li, W., Hu, X., Gravina, R., and Fortino, G. (2017). A neuro-fuzzy fatigue-tracking and classification system for wheelchair users. *IEEE Access* 5: 19420–19431.

25 Ayodele, T.O. (2010). Types of machine learning algorithms. In: *New Advances in Machine Learning* (ed. Y. Zhang), ISBN: 978-953-307-034-6, 20–48. InTech.

26 Safavian, S.R. and Landgrebe, D. (1991). A survey of decision tree classifier methodology. *IEEE Transactions on Systems, Man, and Cybernetics* 21 (3): 660–674.

27 Aydemir, O. and Kayikcioglu, T. (2014). Decision tree structure based classification of EEG signals recorded during two dimensional cursor movement imagery. *Journal of Neuroscience Methods* 229: 68–75.

28 Guan, S., Zhao, K., and Yang, S. (2019). Motor imagery EEG classification based on decision tree framework and Riemannian geometry. *Computational Intelligence and Neuroscience* 2019, Article ID 5627156: 13.

29 Barachant, A., Bonnet, S., Congedo, M., and Jutten, C. (2012). Multiclass brain-computer interface classification by Riemannian geometry. *IEEE Transactions on Biomedical Engineering* 59 (4): 920–928.

30 Breiman, L. (2001). Random forests. *Machine Learning* 45: 5–32.

31 Edla, D.R., Mangalorekar, K., Havalikar, D., and Dodia, S. (2018). Classification of EEG data for human mental state analysis using random forest classifier. *Procedia Computer Science* 132: 1523–1532.

32 Lotte, F., Bougrain, L., Cichocki, A. et al. (2018). A review of classification algorithms for EEG-based brain-computer interfaces: a 10-year update. *Journal of Neural Engineering* 15 (3): 55. https://doi.org/10.1088/1741-2552/aab2f2.hal-01846433.

33 Vapnik, V. (1995). *The Nature of Statistical Learning Theory*. New York: Springer.

34 Neto, E., Biessmann, F., Aurlien, H. et al. (2016). Regularized linear discriminant analysis of EEG features in dementia patients. *Frontiers in Aging Neuroscience* 8: 10. https://doi.org/10.3389/fnagi.2016.00273.

35 Zhang, R., Xu, P., Guo, L. et al. (2013). Z-score linear discriminant analysis for EEG based brain computer interfaces. *PLoS One* 8 (9): e74433.

36 Fu, R., Tian, Y., Bao, T. et al. (2019). Improvement motor imagery EEG classification based on regularized linear discriminant analysis. *Journal of Medical Systems* 43: 169. https://doi.org/10.1007/s10916-019-1270-0.

37 Bennet, K.P. and Campbell, C. (2000). Support vector machines: hype or hallelujah? *SIGKDD Explorations* 2 (2): 1–13.

38 Christianini, N. and Shawe-Taylor, J. (2000). *An Introduction to Support Vector Machines*. Cambridge University Press.

39 DeCoste, D. and Scholkopf, B. (2001). *Training Invariant Support Vector Machines. Machine Learning*. Kluwer Press.

40 Burges, C. (1998). A tutorial on support vector machines for pattern recognition. *Data Mining and Knowledge Discovery* 2: 121–167.

41 Gunn, S. (1998). *Support vector machines for classification and regression*. Technical Reports, Department of Electronics and Computer Science, Southampton University.

42 Chang, C.-C. and lin C.-J. LIBSVM – A Library for Support Vector Machines, last revised on April 14, 2021, https://www.csie.ntu.edu.tw/~cjlin/libsvm/ last accessed August 21 2021.

43 Cristianini, N., Jordan, M., and Schölkopf, B. et al. (2007). Kernel Machines, http://www.kernel-machines.org, last modified last modified 2007-02-01 15:16, (accessed 19 August 2021).

44 Chapelle, O. and Vapnik, V. (2002). Choosing multiple parameters for support vector machines. *Machine Learning* 46: 131–159.

45 Weston, J. and Watkins, C. (1999). Support vector machines for multi-class pattern recognition. *ESANN'1999 proceedings - European Symposium on Artificial Neural Networks Bruges (Belgium)* (21–23 April 1999), D-Facto public. 219–222, ISBN 2-600049-9-X.

46 Platt, J. (1998). Sequential minimal optimisation: a fast algorithm for training support vector machines. Technical Report, MSR-TR-98-14, Microsoft Research 1–21.

47 Gonzalez, B., Sanei, S., and Chambers, J. (2003). Support vector machines for seizure detection. *Proceedings of IEEE, ISSPIT2003*, 126–129. Germany.

48 Tran, D. and Nguyen, T. (2008). Localization in wireless sensor networks based on support vector machines. *IEEE Transactions on Parallel and Distributed Systems* 19 (7): 981–994.

49 Yoo, J. and Kim, H.J. (2015). Target localization in wireless sensor networks using online semi-supervised support vector regression. *Sensors* 15 (6): 12539–12559.

50 Mathew, J., Pang, C.K., Luo, M., and Leong, W.H. (2018). Classification of imbalanced data by oversampling in kernel space of support vector machines. *IEEE Transactions on Neural Networks and Learning Systems* 29 (9): 4065–4076.

51 Kang, Q., Shi, L., Zhou, M. et al. (2018). A distance-based weighted undersampling scheme for support vector machines and its application to imbalanced classification. *IEEE Transactions on Neural Networks and Learning Systems* 29 (9): 4152–4165.

52 Shoker, L., Sanei, S., and Sumich, A. (2005). *Distinguishing Between Left and Right Finger Movement from EEG using SVM, 2005 IEEE Engineering in Medicine and Biology 27th Annual Conference*, 5420–5423. Shanghai.

53 Guler, I. and Ubeyli, E.D. (2007). Multiclass support vector machines for EEG-signals classification. *IEEE Transactions on Information Technology in Biomedicine* 11 (2): 117–126.

54 Zheng, W.-L., Zhu, J.-Y., and Lu, B.-L. (2017). Identifying stable patterns over time for emotion recognition from EEG. *IEEE Transactions on Affective Computing* 10 (3): 417–429.

55 Li, X., Chen, X., Yan, Y. et al. (2014). Classification of EEG signals using a multiple kernel learning support vector machine. *Sensors* 2014 (14): 12784–12802.

56 Beyer, K., Goldstein, J., Ramakrishnan, R., and Shaft, U. (1999). *When is "nearest neighbor" meaningful? ICDT '99: Proceedings of the 7th International Conference on Database Theory* (January 1999), 217–235. Springer.

57 Dudani, S. (1976). The distance weighted k-nearest neighbor rule. *IEEE Transactions on Systems, Man, and Cybernetics* 6: 325–327.

58 Bablani, A., Edla, D.R., and Dodia, S. (2018). Classification of EEG data using k-nearest neighbor approach for concealed information test. *Procedia Computer Science* 143: 242–249.

59 Yazdani, A., Ebrahimi, T., and Hoffmann, U. (2009). Classification of EEG signals using Dempster Shafer theory and a k-nearest neighbor classifier, *2009 4th International IEEE/EMBS Conference on Neural Engineering*, 327–330. Antalya.

60 Dempster, A.P. (1967). Upper and lower probabilities induced by a multivalued mapping. *Annals of Mathematical Statistics* 38: 325–339.

61 Dempster, A.P., Laird, N.M., and Rubin, D.B. (1977). Maximum likelihood from incomplete data via the EM algorithm. *Journal of the Royal Statistical Society, Series B* 39 (1): 1–38.

62 Thomas, E.M., Temko, A., Lightbody, G. et al. (2010). Gaussian mixture models for classification of neonatal seizures using EEG. *Physiological Measurement* 31 (7): 1047–1064.

63 van Hasselt, H., Guez, A., and Silver, D. (2016). Deep reinforcement learning with double Q-learning. *Proceedings of the Thirtieth AAAI Conference on Artificial Intelligence*, AAAI-16, pp. 2094–2100.

64 Jin, C., Allen-Zhu, Z., Bubeck, S., and Jordan, M.I. (2018). Is Q-learning Provably Efficient? *Advances in Neural Information Processing Systems (NIPS)*, 31.

65 Lippmann, R. (1987). An introduction to computing with neural nets. *IEEE ASSP Magazine* 4 (2): 4–22.

66 Selfridge, O.G. (1958). *Pandemonium: a paradigm for learning in mechanisation of thought processes. Proceedings of the Symposium on Mechanisation of Thought Processes, 513–526.* London: Her Majesty's Stationery Office.

67 Rosenblatt, F. (1957). The Perceptron – A Perceiving and Recognizing Automaton. *Techical Reports 85-460-1*, Cornell Aeronautical Laboratory.

68 Dargie, W. and Poellabauer, C. (2010). *Localization*, 249–266. Wiley.

69 Kohonen, T. (2001). *Self-Organizing Maps*, Springer Series in Information Sciences, vol. 30. Berlin Heidelberg: Springer.

70 Hinton, G.E. and Salakhutdinov, R.R. (2006). Reducing the dimensionality of data with neural networks. *Science* 313 (5786): 504–507.

71 Lehmann, E.L. and Casella, G. (1998). *Theory of Point Estimation*, 2e. New York: Springer. ISBN 0-387-98502-6.

72 Goodfellow, I.J., Pouget-Abadie, J., Mirza, M. et al. (2014). Generative adversarial nets. *NIPS 2014 Proceedings of the 27th International Conference on Neural Information Processing Systems – Volume 2* (December 2014), 2672–2680.

73 LeCun, Y., Boser, B., Denker, J.S. et al. (1990). Handwritten digit recognition with a back-propagation network. *Advances in Neural Information Processing Systems* 2: 396–404.

74 Hubel, D.H. and Wiesel, T.N. (1962). Receptive fields, binocular interaction, and functional architecture in the cat's visual cortex. *The Journal of Physiology* 160: 106–154.

75 Antoniades, A., Spyrou, L., Martin-Lopez, D. et al. (2017). Detection of interictal discharges using convolutional neural networks from multichannel intracranial EEG. *IEEE Transactions on Neural Systems and Rehabilitation Engineering* 25 (12): 2285–2294.

76 Antoniades, A., Spyrou, L., Martin-Lopez, D. et al. (2018). Deep neural architectures for mapping scalp to intracranial EEG. *International Journal of Neural Systems* 28 (8): 1850009.

77 Hubel, D.H. and Wiesel, T.N. (1962). Receptive fields, binocular interaction, and functional architecture in the cat's visual cortex. *The Journal of Physiology* 160: 106–154.

78 Felleman, D.J. and Essen, D.C.V. (1991). Distributed hierarchical processing in the primate cerebral cortex. *Cerebral Cortex* 1: 1–47.

79 Fukushima, K. and Miyake, S. (1982). Neocognitron: a new algorithm for pattern recognition tolerant of deformations and shifts in position. *Pattern Recognition* 15: 455–469.

80 Hinton, G.E. and Zemel, R.S. (1994). Autoencoders, minimum description length and Helmholtz free energy. *Advances in Neural Information Processing Systems* 6 (NIPS 1993): 3–10.

81 Diederik, P.K., Welling, M. (2013). Auto-encoding variational Bayes. arXiv:1312.6114

82 Domingos, P. (2015). "4". The master algorithm: how the quest for the ultimate learning machine will remake our World. Basic Books. 352 pp. "Deeper into the Brain" subsection.

83 Goodfellow, I., Bengio, Y., and Courville, A. (2016). *Deep Learning*. MIT Press.

84 Bengio, Y., Yao, L., Alain, G., and Vincent, P. (2013). Generalized denoising auto-encoders as generative models. NIPS, pp. 899–907 https://papers.nips.cc/paper/2013/file/559cb990c9dffd8675f6bc2186971dc2-Paper.pdf (accessed 19 August 2021).

85 Kingma, D.P. and Welling, M. (2014). Auto-encoding variational Bayes. ICLR, 2014. arXiv:1312.6114 [stat.ML].

86 Walker, W., Doersch, C., Gupta, A., and Hebert, M. (2016). An uncertain future: forecasting from static images using variational autoencoders. arXiv:1606.07873.

87 LeCun, Y., Jackel, L.D., Bottou, L. et al. (1995). Learning algorithms for classification: a comparison on handwritten digit recognition. In: *Neural Networks: The Statistical Mechanics Perspective* (eds. J.H. Oh, C. Kwon and S. Cho), 261–276. World Scientific.

88 Krizhevsky, A., Sutskever, I., and Hinton, G.E. (2012). ImageNet classification with deep convolutional neural networks. *Advances in Neural Information Processing Systems NIPS'12* 1: 1–9.

89 Zeiler, M.D. and Fergus, R. (2014). Visualizing and understanding convolutional networks. *ECCV, Part I, LNCS* 8689, pp. 818–833.

90 Simonyan, K. and Zisserman, A. (2015). Very deep convolutional networks for large-scale image recognition. *ICLR* 75, 398–406.

91 Szegedy, C., Liu, W., Jia, Y. et al. (2015) Going deeper with convolutions. *IEEE Conference on Computer Vision and Pattern Recognition (CVPR)*, 1–9, Boston MA.

92 He, K., Zhang, X., Ren, S. and Sun, J. (2015). Deep residual learning for image recognition. *2016 IEEE Conference on Computer Vision and Pattern Recognition (CVPR)*, Las Vegas, NV (2016), 770–778.

93 Huang, G., Liu, Z., Maaten, L.V.D., and Weinberger, K.Q. (2017). Densely connected convolutional networks. *2017 IEEE Conference on Computer Vision and Pattern Recognition (CVPR)*, 2261–2269. Honolulu, HI.

94 Xu, K., Ba, J.L., Kiros, R. et al. (2015). Show, attend and tell: neural image caption generation with visual attention. *Proceedings of the 32nd International Conference on Machine Learning, (ICML 2015)*, Vol. 32, pp. 2048–2057.

95 Xu, K., Ba, J. L., Kiros, R., Cho, K., Courville, A., Salakhutdinov, R., Zemel, R. S., and Bengio, Y. (2015) Show, attend and tell: neural image caption generation with visual attention. Proceedings of the 32nd International Conference on Machine Learning, (ICML 2015), vol. 32, pp. 2048–2057.

96 Oord, A.V.D.N., Dieleman, S., Zen, H. et al. (2016). WaveNet: a generative model for raw audio. arXiv:1609.03499v2 [cs.SD].

97 Redmon, J. and Farhadi, A. (2018). YOLOv3: an incremental improvement. arXiv preprint arXiv:1804.02767.

98 Bochkovskiy, A., Wang, C.-Y., and Liao, H.-Y. (2020). YOLOv4: optimal speed and accuracy of object detection. arXiv:2004.10934.

99 Ghiasi, G., Lin, T.-Y., and Le, Q.V. (2018). DropBlock: A regularization method for convolutional networks. *Advances in Neural Information Processing Systems (NIPS)*: 10727–10737. arXiv:1810.12890.

100 Zheng, Z., Wang, P., Liu, W. et al. (2020). Distance-IoU Loss: faster and better learning for bounding box regression. *Proceedings of the AAAI Conference on Artificial Intelligence (AAAI)*, 34(07): 12993–13000.

101 Gerstner, W. and Kistler, W.M. (2002). *Spiking Neuron Models: Single Neurons, Populations, Plasticity*. Cambridge: Cambridge University Press https://doi.org/10.1017/CBO9780511815706.

102 Lobo, J.L., Del Ser, J., Bifet, A., and Kasabov, N. (2020). *Neural Networks* 121: 88–100.

103 Goel, P., Liu, H., Brown, D., and Datta, A. (2008). On the use of spiking neural network for EEG classification. *International Journal of Knowledge-based and Intelligent Engineering Systems* 12: 295–304.

104 Goel, P., Liu, H., Brown, D.J., and Datta, A. (2006). Spiking neural network based classification of task-evoked EEG signals. In: *Knowledge-Based Intelligent Information and Engineering Systems*, KES 2006. Lecture Notes in Computer Science, vol. 4251 (eds. B. Gabrys, R.J. Howlett and L.C. Jain). Berlin, Heidelberg: Springer.

105 Roy, Y., Banville, H., Albuquerque, I. et al. (2019) Deep learning-based electroencephalography analysis: a systematic review. arXiv:1901.05498v2.

106 Aboalayon, K.A.I., Faezipour, M., Almuhammadi, W.S., and Moslehpour, S. (2016). Sleep stage classification using EEG signal analysis: a comprehensive survey and new investigation. *Entropy* 18 (9): 272.

107 Acharya, U.R., Sree, S.V., Swapna, G. et al. (2013). Automated EEG analysis of epilepsy: a review. *Knowledge-Based Systems* 45: 147–165.

108 Arns, M., Conners, C.K., and Kraemer, H.C. (2013). A decade of EEG theta/beta ratio research in ADHD: a meta-analysis. *Journal of Attention Disorders* 17 (5): 374–383.

109 Giacino, J.T., Fins, J.J., Laureys, S., and Schiff, N.D. (2014). Disorders of consciousness after acquired brain injury: the state of the science. *Nature Reviews Neurology* 10 (2): 99.

110 Engemann, D.A., Raimondo, F., King, J.-R. et al. (2018). Robust EEG-based cross-site and cross-protocol classification of states of consciousness. *Brain* 141 (11): 3179–3192.

111 Hagihira, S. (2015). Changes in the electroencephalogram during anaesthesia and their physiological basis. *British Journal of Anaesthesia* 115 (suppl_1): i27–i31.

112 Hajinoroozi, M., Mao, Z., Jung, T.P. et al. (2016). EEG-based prediction of driver's cognitive performance by deep convolutional neural network. *Signal Processing: Image Communication* 47: 549–555.

113 Berka, C., Levendowski, D.J., Lumicao, M.N. et al. (2007). {EEG} correlates of task engagement and mental workload in vigilance, learning, and memory tasks. *Aviation, Space, and Environmental Medicine* 78 (5): B231–B244.

114 Thorsten, O.Z. and Christian, K. (2011). Towards passive brain–computer interfaces: applying brain–computer interface technology to human–machine systems in general. *Journal of Neural Engineering* 8 (2): 25005.

115 Al-Nafjan, A., Hosny, M., Al-Ohali, Y., and Al-Wabil, A. (2017). Review and classification of emotion recognition based on EEG brain-computer interface system research: a systematic review. *Applied Sciences* 7 (12): 1239.

116 Morabito, F.C., Campolo, M., Ieracitano, C. et al. (2016). Deep convolutional neural networks for classification of mild cognitive impaired and Alzheimer's disease patients from scalp EEG recordings. *2016 IEEE 2nd International Forum on Research and Technologies for Society and Industry Leveraging a better tomorrow (RTSI)*, 1–6, Bologna.

117 Morabito, F.C., Campolo, M., Mammone, N. et al. (2017). Deep learning representation from electroencephalography of early-stage Creutzfeldt-Jakob disease and features for differentiation from rapidly progressive dementia. *International Journal of Neural Systems* 27 (2): 1650039.

118 Page, A., Shea, C., and Mohsenin, T. (2016). Wearable seizure detection using convolutional neural networks with transfer learning. *2016 IEEE International Symposium on Circuits and Systems (ISCAS) (2016)*, 1086–1089.

119 Truong, N.D., Kuhlmann, L., Bonyadi, M.R., and Kavehei, O. (2018). Semi-supervised Seizure Prediction with Generative Adversarial Networks. 1–6. https://arxiv.org/pdf/1806.08235.pdf (accessed 19 August 2021).

120 Truong, N.D., Nguyen, A.D., Kuhlmann, L. et al. (2018). Convolutional neural networks for seizure prediction using intracranial and scalp electroencephalogram. *Neural Networks* 105: 104–111.

121 Tsiouris, K.M., Pezoulas, V.C., Zervakis, M. et al. (2018). A long short-term memory deep learning network for the prediction of epileptic seizures using EEG signals. *Computers in Biology and Medicine* 99: 24–37.

122 Chu, L., Qiu, R., Liu, H. et al. Individual recognition in schizophrenia using deep learning methods with random forest and voting classifiers: insights from resting state EEG streams. pp. 1–7. https://arxiv.org/abs/1707.03467 (accessed 19 August 2021).

123 Ben Said, A., Mohamed, A., Elfouly, T. et al. (2017). Multimodal deep learning approach for Joint EEG-EMG Data compression and classification. *IEEE Wireless Communications and Networking Conference, WCNC (2017)*. https://arxiv.org/pdf/1703.08970.pdf (accessed 19 August 2021).

124 Xu, H. and Plataniotis, K.N. (2016). Affective states classification using EEG and semi-supervised deep learning approaches. *2016 IEEE 18th International Workshop on Multimedia Signal Processing (MMSP)*, 1–6, Montreal, QC.

125 Liu, W., Zheng, W.L., and Lu, B.L. (2016). Emotion recognition using multimodal deep learning. In: *Neural Information Processing. ICONIP 2016*, Lecture Notes in Computer Science, vol. 9948 (eds. A. Hirose, S. Ozawa, K. Doya, et al.). Cham: Springer.

126 Jirayucharoensak, S., Pan-Ngum, S., and Israsena, P. (2014). EEG-based emotion recognition using deep learning network with principal component based covariate shift adaptation. *Hindawi Scientific World Journal* 2014. Vol. 2014, Article ID 627892, 10 pages.

127 Liao, C.-Y., Chen, R.-C., and Tai, S.-K. (2018). Emotion stress detection using EEG signal and deep learning technologies. *2018 IEEE International Conference on Applied System Invention (ICASI)*, 90–93, Chiba.

128 Lin, W., Li, C., and Sun, S. (2017). Deep convolutional neural network for emotion recognition using EEG and peripheral physiological signal. International Conference on Image and Graphics, 385–394.

129 Zheng, W.L. and Lu, B.L. (2015). Investigating critical frequency bands and channels for EEG-based emotion recognition with deep neural networks. *IEEE Transactions on Autonomous Mental Development* 7 (3): 162–175.

130 Zheng, W.L., Zhu, J.Y., Peng, Y., and Lu, B.L. (2014). EEG-based emotion classification using deep belief networks. *Proceedings of the IEEE International Conference on Multimedia and Expo* (1–6 September, 2014).

131 Li, K., Li, X., Zhang, Y., and Zhang, A. (2013). Affective state recognition from EEG with deep belief networks. *2013 IEEE International Conference on Bioinformatics and Biomedicine*, 305–310, Shanghai.

132 Teo, J., Hou, C.L., and Mountstephens, J. (2018). Preference classification using electroencephalography (EEG) and deep learning. *Journal of Telecommunication, Electronic and Computer Engineering (JTEC)* 10 (1): 87–91.

133 Frydenlund, A. and Rudzicz, F. (2015). Emotional affect estimation using video and EEG data in deep neural networks. *Lecture Notes in Computer Science (including subseries Lecture Notes in Artificial Intelligence and Lecture Notes in Bioinformatics)* 9091: 273–280.

134 Mehmood, R.M., Du, R., and Lee, H.J. (2017). Optimal feature selection and deep learning ensembles method for emotion recognition from human brain EEG sensors. *IEEE Access* 5: 14797–14806.

135 Kwon, Y., Nan, Y., and Kim, S.D. (2017). Transformation of EEG signal for emotion analysis and dataset construction for DNN learning. In: *Lecture Notes in Electrical Engineering*, vol. 474 (eds. J.J. Park, V. Loia, G. Yi and Y. Sung), 96–101. Springer.

136 Gao, Y., Lee, H.J., and Mehmood, R.M., Deep learning of EEG signals for emotion recognition. *2015 IEEE International Conference on Multimedia and Expo Workshops, ICMEW 2015* (June 2015). 1–5, IEEE.

137 Li, Z., Tian, X., Shu, L. et al. (2018). Emotion recognition from EEG using RASM and LSTM. *Internet Multimedia Computing and Service* 819: 310–318.

138 Zhang, T., Zheng, W., Cui, Z. et al. (2018). Spatial-temporal recurrent neural network for emotion recognition. *IEEE Transactions on Cybernetics* 1: 1–9.

139 Alhagry, S., Fahmy, A.A., and El-Khoribi, R.A. (2017). Emotion recognition based on EEG using LSTM recurrent neural network. *International Journal of Advanced Computer Science and Applications* 8 (10): 8–11.

140 Lotte, F., Bougrain, L., and Clerc, M. (2015). Electroencephalography (EEG)-based brain-computer inter-faces. *American Cancer Society*: 1–20.

141 Lotte, F., Bougrain, L., Cichocki, A. et al. (2018). A review of classification algorithms for EEG-based brain-computer interfaces: a 10-year update. *Journal of Neural Engineering* 15 (3): 031005.

142 Rasmussen, C. and Williams, C. (2006). *Gaussian Processes for Machine Learning*. Cambridge, MA: The MIT Press.

143 Clifton, L., Clifton, D.A., Pimentel, A.M.F. et al. (2013). Gaussian processes for personalized e-health monitoring with wearable sensors. *IEEE Transactions on Biomedical Engineering* 60 (1): 193–197.

144 Hurley, P., Serra, L., Bozzali, M. et al. (2015). Gaussian process classification of Alzheimer's disease and mild cognitive impairment from resting-state fMRI. *NeuroImage* 112: 232–243.

145 Garnelo, M., Schwarz, J., Rosenbaum, D. et al. (2018). Neural processes. arXiv:1807.01622v1 [cs.LG]

146 Kim, H., Mnih, A., Schwarz, J. et al. (2019). Attentive neural processes. *Proceedings of the International Conference on Learning Representations (ICLR)*, arXiv:1901.05761v2 [cs.LG].

147 Nipf, T.N. and Wellin, M. (2017). Semi-supervised classification with graph convolutional networks. *Proceedings of the International Conference on Learning Representations (ICLR), Conference paper* https://openreview.net/forum?id=SJU4ayYgl (accessed 19 August 2021).

148 Zhang, H. (2004). The optimality of naïve Bayes. *Proceedings of AAAI, FLAIRS*.

149 Rabiner, L. (2013). *First Hand: The Hidden Markov Model*. IEEE Global History Network.

150 Baum, L.E. and Petrie, T. (1966). Statistical inference for probabilistic functions of finite state Markov chains. *Annals of Mathematical Statistics* 37 (6): 1554–1563.

151 Argunsah, A.O. and Cetin, M. (2010). AR-PCA-HMM approach for sensorimotor task classification in EEG-based brain–computer interfaces, *2010 IEEE International Conference on Pattern Recognition. 20th International Conference on Pattern Recognition*, 113–116, Istanbul.

152 Williams, N.J., Daly, I., and Nasuto, S.J. (2018). Markov model-based method to analyse time-varying networks in EEG task-related data. *Frontiers in Computational Neuroscience* 12: 1–18.

153 Schwarz, G.E. (1978). Estimating the dimension of a model. *The Annals of Statistics* 6 (2): 461–464.

154 Obermaier, B., Gugera, C., Neuper, C., and Pfurtscheller, G. (2001). Hidden Markov models used for the offline classification of single trial EEG data. *Pattern Recognition Letters* 22 (12): 1299–1309.

155 Liu, J., Zhang, C., and Zheng, C. (2010). EEG-based estimation of mental fatigue by using KPCA–HMM and complexity parameters. *Biomedical Signal Processing and Control* 5: 124–130.

156 Stevner, A.B.A., Vidaurre, D., Cabral, J. et al. (2019). Discovery of key whole-brain transitions and dynamics during human wakefulness and non-REM sleep. *Nature Communications* 10 (1035): 1–14.

157 Koles, Z. (1991). The quantitative extraction and topographic mapping of the abnormal components in the clinical EEG. *Electroencephalography and Clinical Neurophysiology* 79 (6): 440–447.

158 Blankertz, B., Tomioka, R., Lemm, S. et al. (2008). Optimizing spatial filters for robust EEG single-trial analysis. *IEEE Signal Processing Magazine* 25 (1): 41–56.

159 Lotte, F. and Guan, C. (2011). Regularizing common spatial patterns to improve BCI designs: unified theory and new algorithms. *IEEE Transactions on Biomedical Engineering* 58 (2): 355–362.

160 Ramoser, H., Muller-Gerking, J., and Pfurtscheller, G. (2000). Optimal spatial filtering of single trial EEG during imagined hand movement. *IEEE Transactions on Rehabilitation Engineering* 8 (4): 441–446.

161 Kang, H., Nam, Y., and Choi, S. (2009). Composite common spatial pattern for subject-to-subject transfer. *IEEE Signal Processing Letters* 16 (8): 683–686.

162 Wang, H., Tang, Q., and Zheng, W. (2012). L1-norm-based common spatial patterns. *IEEE Transactions on Biomedical Engineering* 59 (3): 653–662.

163 Lu, H., Plataniotis, K., and Venetsanopoulos, A. (2009). Regularized common spatial patterns with generic learning for EEG signal classification. *Proceedings of IEEE Engineering in Medicine and Biology Conference*, 6599–6602, Minnesota, USA: EMBC.

164 Ledoit, O. and Wolf, M. (2004). A well-conditioned estimator for large dimensional covariance matrices. *Journal of Multivariate Analysis* 88 (2): 365–411.

165 Farquhar, J., Hill, N., Lal, T., and Schölkopf, B. (2006). Regularised CSP for sensor selection in BCI. *Proceedings of the 3rd International Brain-Computer Interface Workshop and Training Course* 2006, 14–15.

166 Yong, X., Ward, R., and Birch, G. (2008). Sparse spatial filter optimization for EEG channel reduction in brain-computer interface. *Proceedings of IEEE International Conference on Acoustic, Speech, and Signal Processing, 417–420*, Taiwan: ICASSP.

167 Ledoit, O. and Wolf, M. (2004). A well-conditioned estimator for large dimensional covariance matrices. *Journal of Multivariate Analysis* 88 (2): 365–411.

168 Arvaneh, M., Guan, C., Kai, A.K., and Chai, Q. (2011). Optimizing the channel selection and classification accuracy in EEG-based BCI. *IEEE Transactions on Biomedical Engineering* 58 (6): 1865–1873.

169 Goksu, F., Ince, N.F., Tewfik, A.H. (2011). Sparse common spatial patterns in brain computer interface applications. *Proceedings of the IEEE International Conference on Acoustics, Speech, and Signal Processing, 533–536*, Prague, Czech Republic: ICASSP.

170 Wang, B., Wong, C.M., Kang, Z. et al. (2020). Common spatial pattern reformulated for regularizations in brain–computer interfaces. *IEEE Transactions on Cybernetics*: 1–13. https://doi.org/10.1109/TCYB.2020.2982901.

171 Hatamikia, S. and Nasrabadi, A.M. (2015). Subject transfer BCI based on composite local temporal correlation common spatial pattern. *Computers in Biology and Medicine* 64: 1–11.

172 Yu, H., Lu, H., Wang, S. et al. (2019). A general common spatial patterns for EEG analysis with applications to vigilance detection. *IEEE Access* 7: 111102–111114.

173 Yosinski, J., Clune, J., Bengio, Y., and Lipson, H. (2014). How transferable are features in deep neural networks? arXiv:1411.1792.

174 Kingma, D. and Welling, M. (2014). Auto-encoding variational Bayes. ICLR. arXiv:1312.6114.

175 François-Lavet, V., Islam, R., Pineau, J. et al. (2018). An introduction to deep reinforcement learning. arXiv:1811.12560v2.

8

Brain Connectivity and Its Applications

8.1 Introduction

The history of brain connectivity research goes back to the time of modern neuroscience and Nobel prize winner, Santiago Ramon y Cajal (1852–1734), a Spanish neuroscientist, pathologist, and histologist specializing in neuroanatomy and the central nervous system (CNS). His detailed illustrations of cellular connections in the brain and staunch defence of neuron doctrine form the basis of the field. The work by his contemporary, Korbinian Brodmann (1868–1918), German neurologist who became famous for mapping the cerebral cortex and defining 52 distinct regions (each one of these regions has a number which follows his name, such as Brodmann area 4, which is the region of primary motor cortex), and some others, led to segmentation of the brain into distinct cytoarchitectonic regions. Many of these regions were further defined functionally by Wilder Penfield (1891–1976), a neurosurgeon, using electrophysiological mapping. The concept of brain network has been discussed in detail by Sporns [1] and the brain connectome projects provided an anatomical display of the pathways within the brain [2]. However, the study of connectivity is distinct from static brain mapping. General connectivity research is concerned with anatomical pathways, interactions, and communication between distinct units of the CNS. These units can be categorized into levels of microscale (individual neurons), mesoscale (columns), or macroscale (regions). Connectivity is also divided into structural and functional, each subdivided further into static and dynamic. Static components are defined by the regions and wiring in which communication and processing occurs. Dynamic components can be described by the functional relationship between static components. For example, functional connectivity, described as the temporal coherence between physically distant activity, and effective connectivity, described as networks of directional influences of one neural element over another. Static connectivity can be measured by anatomical properties using a number of imaging methods, including high-resolution magnetic resonance (MR) for showing the tissue anatomy, diffusion tensor imaging (DTI) for matter tractography, and histology (myelination). Conversely, dynamic connectivity can be measured by a wide variety of techniques, including methods based on causality estimation, often using electroencephalograms (EEG), or methods that can provide information about the spatial distribution and strength of dynamic connections, such as resting-state functional MRI (fMRI) connectivity.

Although an accurate and robust estimation of brain connectivity from EEG signals is still under research, the notion of brain connectivity has been well recognized by

EEG Signal Processing and Machine Learning, Second Edition. Saeid Sanei and Jonathon A. Chambers.

researchers in neuroimaging through noninvasive EEG, fMRI, and magnetoencephalogram (MEG) information. Among different neuroimaging methods, EEG and MEG directly reflect neuronal firing exhibiting a good temporal resolution (in milliseconds) despite a poor spatial resolution (of the order of few square centimetres). These data can therefore be used in estimation of the brain regions connectivity. Most brain functions rely on interactions between neuronal assemblies distributed within and across different cerebral regions.

According to an unpublished work called *EEG Connectivity: A Tutorial*, by Thatcher et al. the nature of brain sources has to be investigated first. In this work EEG sources are divided into two different types: one includes the electrical fields of brain operating at the speed of light where dipoles distributed in space turn on and off while oscillating at different amplitudes and the second type is the source of the electrical activity, which is an excitable medium of locally connected network. This has been nicely modelled by Hodgkin and Huxley who wrote the fundamental excitable medium equations of the brain in 1952 (as discussed in Chapter 3 of this book). The brain network comprises of axons, synapses, dendritic membranes, and ionic channels that behave like 'kindling' at the leading edge of a confluence of different fuels and excitations. Approximately, 80% of the cortex is excitatory, with recurrent loop connections. This obviously calls for stability of the whole network stemming from the fact that the refractory periods are relatively long and this allows for self-organizing and stability of the cortical activities.

Therefore, the brain can be considered as a connected network in many applications particularly those related to prolonged mental or physical activity. The connected network within the brain results in EEG connectivity. This connectivity is defined by the magnitude of coupling between neurons. Consequently, the brain connectivity is evaluated by measuring the magnitude strength, duration, and time delays from the electrical recording of electrical fields of the brain produced by the excitable medium.

Unlike dipole propagation, connectivity does not occur at the speed of light and is best measured when there are time delays; in fact, electrical volume conduction is not a property of the EEG excitable medium and occurs at zero time delay. This very important property of the excitable medium sources versus electrical properties means that the time delays determine whether or not and to what extent an excitable medium is responsible for the electrical potentials measured at the scalp surface.

So, volume conduction defined at zero phase lag is related to one type of EEG sources and the second type is lagged correlations related to the excitable medium. For the second type, the sources are often considered as coupled oscillators. This is mainly due to the fact that electrical potentials are ionic fluxes across polarized membranes of neurons with intrinsic rhythms and driven rhythms (self-sustained oscillations) [3–5].

Volume conduction involves near zero phase delays between any two points within the electrical field as collections of dipoles which oscillate in time [4]. Zero phase delay is one of the important properties of volume conduction. Therefore, measures such as the cross-spectrum, coherence, bicoherence, and coherence of phase delays become crucial and significant in evaluating brain connectivity independent of volume conduction.

For this purpose, correlation coefficient methods such as Pearson product correlation (e.g. 'comodulation' and 'Lexicor correlation') do not compute phase thus, are incapable of controlling volume conduction. The use of above approaches (bispectrum, etc.) for the study of brain connectivity is not only because of the ability to control volume conduction but also the need to measure the fine temporal details and temporal history of coupling or 'connectivity' within and between different regions of the brain. From physiology of the brain, both the thalamus and septo-hippocampal systems are approximately beneath and in the centre of the brain and contain 'pacemaker' neurons and neural circuits that regularly synchronize widely disparate groups of cortical neurons [6].

Cross-spectrum is the sum of both in-phase (i.e. cospectrum) and out-of-phase (i.e. quadspectrum) potentials. The in-phase component contains volume conduction and the synchronous activation of local neural generators. The out-of-phase component contains network or connectivity contributions from locations far from the location of a given source. In other words, cospectrum reflects volume conduction and quadspectrum represents non-volume conduction.

Cross-spectrum of coherence and phase difference can distinguish between volume conduction and network zero phase differences, for example, produced by the thalamus or the septal hippocampus-entorhinal cortex.

Pearson product correlation ('comodulation' and Lexicor 'spectral correlation coefficient') is one of the earliest way of measuring brain connectivity. This coefficient is often used to estimate the degree of association between EEG amplitudes or magnitudes over intervals of time and frequency [6]. The Pearson product correlation coefficient (PCC) is not used to calculate a cross-spectrum and therefore, it is neither used to calculate the phase nor involved in measurement of phase relationship consistency such as with coherence and bispectrum.

Coherence and Pearson PCCs, however, are statistical measures with similar accuracy and statistical significance. The Pearson coefficient is a valid and important normalized measure of coupling which has been used over 40 years. In recent works, application of Pearson PCC for magnitude has been called 'comodulation' [7].

'Spectral correlation' or 'spectral amplitude correlation' are more popular terms while comodulation, is a limited term since it fails to refer to the condition of a third source affecting the two other sources without these two latter sources being directly connected. In addition, comodulation cannot correct for volume conduction. 'Comodulation' has a different meaning than 'synchronization' [7] and to reduce confusion correlation or PCC term is used.

Further research by Blinowska et al. in the 1990s [8–10] demonstrated how directionality of cortical signal patterns changes within the brain using multivariate autoregressive (MVAR) modelling followed by directed transfer functions (DTFs). In some applications such as detection and classification of finger movement, it is very useful to discover how the associated movement signals propagate within the neural network of the brain. There is a consistent movement of the source signals from occipital to temporal regions. It is also apparent that during mental tasks different regions within the brain communicate with each other. The interaction and cross-talk among the EEG channels may be the only clue to understand this process. This requires recognition of transient periods of synchrony between various regions in the brain. These phenomena are not easy to observe by visual EEG inspection. Therefore, some signal processing techniques have to be used in order to infer such causal relationships. One time series is said to be causal to another if its inherent information enables the prediction of other time series. In some approaches to evaluate connectivity, coherency, or synchronization of brain regions the spatial statistics of scalp EEG are presented as coherence in individual frequency bands, these coherences result from both correlations among neocortical sources and volume conduction through tissues of the head, i.e. brain, cerebrospinal fluid, skull, and scalp.

One way for selection of the cortical areas is based on a list of regions of interest (ROIs) included in the so-called Parieto-Frontal Integration Theory (P-FIT [11]) which describes the regions of the human brain involved in intelligence and reasoning tasks. This P-FIT model includes the dorsolateral prefrontal cortex (i.e. Brodmann areas 10), the superior (Brodmann area 7) parietal lobe, and the anterior cingulate (Brodmann area 32). In addition, there is evidence that the most caudal region of the medial frontal cortex containing cingulate motor areas (CMA) is involved in movements of the hands and other body parts [12]. In particular, the activity in the regions including CMA has been also related directly to behavioural response rate [13].

Therefore, it is logical to have the most detailed brain connectivity map as the one achieved using cortical electrodes (i.e. electrocorticography [ECoG]) which are dense and often mounted around the motor area of the brain. However, this can be achieved invasively and has specific clinical use, such as for seizure or cortical stimulation.

8.2 Connectivity through Coherency

Spectral coherence [14] is a common method for determination of synchrony in EEG activity. Coherency is a normalized form of cross-spectrum and is given as:

$$Coh_{ij}^2(\omega) = \frac{E\left\{ |C_{ij}(\omega)|^2 \right\}}{E\{C_{ii}(\omega)\}E\{C_{jj}(\omega)\}} \tag{8.1}$$

where $C_{ij}(\omega) = X_i(\omega)X_j^*(\omega)$ is the Fourier transform of cross-correlation coefficients between channel i and channel j of the EEGs. Figure 8.1 shows an example of the cross-spectral coherence around one second prior to finger movement. A measure of this coherency, such as an average over a frequency band, is capable of detecting zero time-lag synchronization and fixed time non-zero time-lag synchronization, possibly occurring when a significant delay between the two neuronal population sites exists [15]. Nonetheless, it does not provide any information on directionality of the coupling between the two recording sites.

One important requirement of brain connectivity representation is the direction of information flow. In this respect, multivariate spectral techniques such as DTF or partial directed coherence (PDC) were proposed [16, 17] for determining the directional influences between any given pair of channels in a multivariate dataset. Both DTF and PDC [17, 18] rely on the key concept of Granger causality between time series [19], according to which an observed timeseries $x(t)$ causes another series $y(t)$ if the knowledge of $x(t)$'s past significantly improves the prediction of $y(t)$; this relation between time series is not reciprocal, i.e. $x(t)$ may cause $y(t)$ without $y(t)$ necessarily causing $x(t)$. This lack of reciprocity allows the evaluation of the direction of information flow between structures.

Granger causality (also called as Wiener–Granger causality) [19] attempts to extract and quantify directionality from EEGs. It is based on bivariate auto-regressive AR estimates of

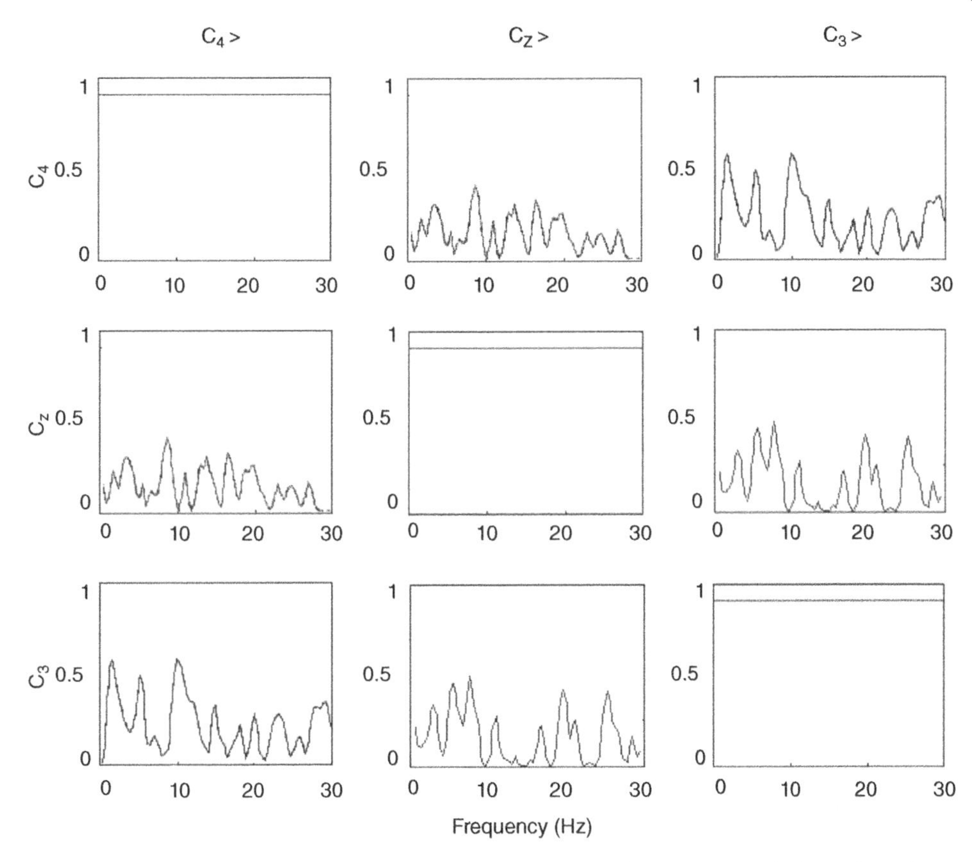

Figure 8.1 Cross-spectral coherence for a set of three electrode EEGs, one second prior to right-finger movement. Each block refers to one electrode. By careful inspection of this figure, one can observe the same waveform transferred from C_z to C_3.

the data. In a multichannel environment this causality is calculated from pair-wise combinations of electrodes. This method has been used to evaluate directionality of the source movement from the local field potential in a cat's visual system [20].

Based on Granger causality if past samples of a time series $y(t)$ can be used in prediction of another series $x(t)$ then, $y(t)$ is said to cause $x(t)$. In order to determine an estimate of this causality consider:

$$e(t) = x(t) - \sum_{i=1}^{p} \breve{a}(i)x(t-i) \tag{8.2}$$

and

$$e_1(t) = x(t) - \sum_{i=1}^{p} a_{11}(i)x(t-i) - \sum_{i=1}^{p} a_{12}(i)y(t-i) \tag{8.3}$$

which also implies:

$$e_2(t) = y(t) - \sum_{i=1}^{p} a_{22}(i)y(t-i) - \sum_{i=1}^{p} a_{21}(i)x(t-i) \tag{8.4}$$

Granger causality index is defined as:

$$GCI_{x(t)\rightarrow y(t)} = \ln \frac{\sigma_{e_2}^2}{\sigma_{e_1}^2} \tag{8.5}$$

where $\sigma_{e_1}^2$ and $\sigma_{e_2}^2$ are respectively the variances of first and second signals. Such a definition can easily be extended to multichannel case when the included channels change the residual variance ratios. This is a time domain measure. The Granger causality measure is, however, sensitive to arbitrary mixtures of independent noise.

8.3 Phase-Slope Index

To overcome the effect of noise and establish a more robust measure of connectivity phase-slope index (PSI) was introduced [21]. The main idea behind PSI is that the cause precedes the effect in time and hence the slope of phase of the cross-spectrum between two time series reflects the directionality [22]. The cross-spectrum of two signals $z_i(t)$ and $z_j(t)$ is defined as:

$$S_{ij}(f) = E\left[Z_i(f)Z_j^*(f)\right] \tag{8.6}$$

where $Z_i(f)$ and $Z_j(f)$ are respectively, the discrete (Fourier) frequency transform of $z_i(t)$ and $z_j(t)$ and $(.)^*$ denotes conjugate operation. Using this definition, the complex coherence is defined as:

$$C_{ij}(f) = \frac{S_{ij}(f)}{\sqrt{S_{ii}(f)S_{jj}(f)}} \tag{8.7}$$

The unnormalized PSI is then measured in terms of complex cross-spectrum as [21]:

$$\widetilde{P}_{ij} = Imag\left(\sum_{f \in B} C_{ij}^*(f)C_{ij}(f + \delta f)\right) \tag{8.8}$$

where B is the desired frequency band. Often PSI is normalized with respect to its variance as [22]:

$$P_{ij} = \widetilde{P}_{ij}/var\left(\widetilde{P}_{ij}\right) \tag{8.9}$$

It has been shown that absolute values of $P_{ij} > 2$ are significant [21]. However, this measure has been applied to ECoG signals rather than scalp EEG signals and it gives higher connectivity between the brain regions which are more synchronized.

8.4 Multivariate Directionality Estimation

Generally, application of the Granger causality for multivariate data in a multichannel recording is not computationally efficient [18, 20]. DTF [16], as an extension of Granger causality, is obtained from multichannel data and can be used to detect and quantify the coupling directions. Advantage of DTF over spectral coherence is that it can determine the directionality in coupling when frequency spectra of the two brain regions have overlapping spectra. The DTF has been adopted by some researchers to determine the coupling directionality [23, 24] since a directed flow of information or cross-talk between the sensors around the sensory motor area before finger movement has been demonstrated [25]. DTF is based on fitting the EEGs to an MVAR model. Assuming $\mathbf{x}(n)$ is an M-channel EEG signal, then in vector form, it can be modelled as:

$$\mathbf{x}(n) = -\sum_{k=1}^{p} \mathbf{L}_k \mathbf{x}(n-k) + \mathbf{v}(n) \quad (8.10)$$

where n is the discrete time index, p the prediction order, $\mathbf{v}(n)$ zero-mean noise vector and $\mathbf{L}_k$ is generally an $M \times p$ matrix of prediction coefficients. A similar method to the Durbin algorithm for single-channel signals, namely the Levinson–Wiggins–Robinson (LWR) algorithm is used to calculate MVAR coefficients [26]. The Akaike information criterion (AIC) [27] is also used for the estimation of prediction order p. By multiplying both sides of the above equation by $\mathbf{x}^T(n\text{-}k)$ and performing statistical expectation the following Yule–Walker equation is obtained [28]:

$$\sum_{k=0}^{p} \mathbf{L}_k \mathbf{R}(-k+p) = 0; \quad \mathbf{L}_0 = 1 \quad (8.11)$$

where $\mathbf{R}(q) = E[\mathbf{x}(n)\mathbf{x}^T(n+q)]$ is the covariance matrix of $\mathbf{x}(n)$. Cross-correlation of the signal and noise is zero since they are assumed uncorrelated. Similarly, the noise autocorrelation is zero for non-zero shift since the noise samples are uncorrelated. The data segment is considered short enough for the signal to remain statistically stationary within that interval and long enough to enable accurate measurement of the prediction coefficients.

8.4.1 Directed Transfer Function

Given the MVAR model coefficients, a multivariate spectrum can be obtained. Here it is assumed that the residual signal, $\mathbf{v}(n)$, is a white noise vector. Therefore

$$\boldsymbol{L}_f(\omega)\boldsymbol{X}(\omega) = \boldsymbol{V}(\omega) \quad (8.12)$$

where

$$\boldsymbol{L}_f(\omega) = \sum_{m=0}^{p} \boldsymbol{L}_m e^{-j\omega m} \quad (8.13)$$

and $\boldsymbol{L}_f(0) = \mathbf{I}$. Rearranging the above equation and replacing the noise by $\sigma_{\mathbf{v}}^2\mathbf{I}$ yields:

$$\boldsymbol{X}(\omega) = \boldsymbol{L}_f^{-1}(\omega) \times \sigma_{\mathbf{v}}^2\mathbf{I} = \boldsymbol{H}(\omega) \quad (8.14)$$

which represents the model spectrum of the signals or transfer matrix of the MVAR system. DTF or causal relationship between channel i and channel j can be defined directly from the transform coefficients [18] given by:

$$\Theta_{ij}^2(\omega) = \left|H_{ij}(\omega)\right|^2 \quad (8.15)$$

Electrode i is causal to j at frequency f if:

$$\Theta_{ij}^2(\omega) > 0 \tag{8.16}$$

The above DTF can be normalized as:

$$\gamma_{ij}^2(\omega) = \frac{|H_{ij}(\omega)|^2}{\sum_{m=1}^{N} |H_{im}(\omega)|^2} \tag{8.17}$$

Normalized DTF values are in the interval [0, 1], i.e.:

$$\sum_{k=0}^{N} \gamma_{ik}^2(\omega) = 1 \tag{8.18}$$

A time-varying DTF can also be generated (mainly to track the source signals) by calculating the DTF over short windows to achieve the short-time DTF (SDTF) [18].

As an important feature in classification of left and right-finger movements, or tracking the mental task-related sources, SDTF plays an important role. Some results of using SDTF for detection and classification of finger movement have been given in Chapter 17 of this book.

8.4.2 Direct DTF

In [29], and consequently in [30] another measure called direct DTF (dDTF) has been defined as a multiplication of a modified DTF by partial coherence:

$$\xi_{ij}^2(\omega) = \frac{C_{ij}^2(\omega)|H_{ij}(\omega)|^2}{\sum_f \sum_{m=1}^{k} |H_{im}(\omega)|^2} \tag{8.19}$$

where $C_{ij}(\omega)$ is the partial coherence defined as:

$$C_{ij}(\omega) = \frac{M_{ij}(\omega)}{\sqrt{M_{ii}(\omega)M_{jj}(\omega)}} \tag{8.20}$$

and $M_{ii}(\omega)$ and $M_{ij}(\omega)$ are respectively the spectra and cross-spectra. Moreover, distinction of direct from indirect transmission, in the case of signals from implanted electrodes, is essential. Both DTF and dDTF show propagation when there is a phase difference between the signals. Based on the phase coherence the delay between onsets of similar frequency components is estimated and it reveals the direction of propagation. The phase, however, should vary within 2π (or modulo 2π), otherwise the directionality maybe misjudged.

Partial coherence analysis has been used to determine graphical models for brain functional connectivity [31]. It has been investigated that the outcome of such analysis may be considerably influenced by factors such as the degree of spectral smoothing, line and interference removal, matrix inversion stabilization, and the suppression of effects caused by side-lobe leakage, combination of results from different epochs and people, and multiple hypothesis testing.

8.4.3 Partial Directed Coherence

Another similar measure of coherency, called PDC, has also been defined as [17, 30]:

$$P_{ij}(\omega) = \frac{A_{ij}(\omega)}{\sqrt{\mathbf{a}_j^*(\omega)\mathbf{a}_j(\omega)}} \tag{8.21}$$

where $A_{ij}(\omega)$ is an element of $\mathbf{A}(\omega)$, a Fourier transform of MVAR model coefficients $\mathbf{A}(\omega)$, where $\boldsymbol{a}_j(\omega)$ is jth column of $\mathbf{A}(\omega)$ and the asterisk denotes the vector Hermitian operator. PDC represents only direct flows between channels [30]. Finally, a generalized PDC approach has been defined as [30]:

$$GP_{ij}(\omega) = \frac{A_{ij}(\omega)}{\sum_{i=1}^{k} \left|A_{ij}(\omega)\right|^2} \tag{8.22}$$

which is used for connectivity estimation. These methods can be applied to consecutive segments of the signals for approximation of dynamics of the signal propagation. Alternatively, SDTF can be used instead of DFT.

Despite many applications in evaluation of degenerative neurological diseases such as Alzheimer's disease (AD), addressed later in this book, one application of PDC to EEG can be seen in analyzing the responses given in an interview and identify if the interviewee tells the truth [32]. In this paper, the EEG connectivity measures have been combined with some photoplethysmography (PPG) for better classification results.

8.5 Modelling the Connectivity by Structural Equation Modelling

As described in [33, 34], connectivity can be modelled and covariance between the node signals can be compared with real measurements.

Structural equation modelling (SEM) is meant to set up an a priori connectivity model and aims to answer: (i) the influence of a variable signal-to-noise ratio (SNR) level on accuracy of the pattern connectivity estimation obtained by SEM, (ii) the amount of data necessary to get good accuracy for connectivity estimation between cortical areas (iii) how the SEM performance is degraded by an imprecise anatomical model formulation and whether it is able to perform a good estimation of connectivity pattern when connections between the cortical areas are not correctly assumed, and (iv) which kind of errors should be avoided. SEM has also been used to model such activities from high-resolution (both spatial and temporal) EEG data. Anatomical and physiological constraints have been exploited to change an underdetermined set of equations to a determined one.

SEM consists of a set of linear structural equations containing observed variables and parameters defining causal relationships among the variables [35]. The variables can be endogenous (i.e. independent from the other variables in the model) or exogenous (independent from the model itself). Considering a set of variables (expressed as deviations from their means) with N observations, estimation of SEM for these variables may be expressed as:

$$y = \mathbf{B}\mathbf{y} + \boldsymbol{\Gamma}\mathbf{x} + \xi \tag{8.23}$$

where, $\mathbf{y}$ is an $m\times 1$ vector of dependent (endogenous) variables, $\mathbf{x}$ is an $n\times 1$ vector of independent (exogenous) variables, ξ is the $m\times 1$ vector of equation errors (random disturbances), and $\mathbf{B}$ $(m\times m)$ and $\boldsymbol{\Gamma}$ $(m\times \mathrm{n})$ are respectively the coefficient matrices of the endogenous and exogenous variables. ξ is assumed to be uncorrelated with the data and $\mathbf{B}$ to be a zero-diagonal matrix. If $\mathbf{z}$ is a vector containing all the $p = m + n$ exogenous and endogenous variables in the following order:

$$\mathbf{z}^T = [x_1 \cdots x_n y_1 \cdots y_m] \tag{8.24}$$

the observed covariances can be expressed as:

$$\boldsymbol{\Sigma}_{obs} = \frac{1}{N-1}\mathbf{Z}.\mathbf{Z}^T \tag{8.25}$$

where $\mathbf{Z}$ is a matrix of N observations of $\mathbf{z}$. The covariance matrix implied by the model can be obtained as follows:

$$\boldsymbol{\Sigma}_{mod} = E\left[\mathbf{z}\mathbf{z}^T\right] = \begin{bmatrix} E\left[\mathbf{x}\mathbf{x}^T\right] & E\left[\mathbf{x}\mathbf{y}^T\right] \\ E\left[\mathbf{y}\mathbf{x}^T\right] & E\left[\mathbf{y}\mathbf{y}^T\right] \end{bmatrix} \tag{8.26}$$

where if $E[\mathbf{x}\mathbf{x}^T] = \Phi$, then

$$E\left[\mathbf{x}\mathbf{y}^T\right] = \left((\mathbf{I}-\mathbf{B})^{-1}\Gamma\Phi\right)^T \tag{8.27}$$

$$E\left[\mathbf{y}\mathbf{x}^T\right] = (\mathbf{I}-\mathbf{B})^{-1}\Gamma\Phi \tag{8.28}$$

and

$$E\left[\mathbf{y}\mathbf{y}^T\right] = (\mathbf{I}-\mathbf{B})^{-1}\left(\Gamma\Phi\Gamma^T + \Psi\right)\left((\mathbf{I}-\mathbf{B})^{-1}\right)^T \tag{8.29}$$

where $\Psi = E[\xi\xi^T]$. With no constraints, the problem of minimization of the differences between the observed covariances and those implied by the model is underdetermined mainly because the number of variables ($\mathbf{B}$, $\boldsymbol{\Gamma}$, $\boldsymbol{\Phi}$, and $\boldsymbol{\Psi}$) is greater than the number of equations $(m+n)(m+n+1)/2$. The significance of SEM is that it eliminates some of the connections in the connectivity map based on some a priori anatomical and physiological (and possibly functional) information. For example, in Figure 8.2 if the connection a_{42} is not in the hypothesized model it may be set to zero.

So, if r is the number of parameters to be estimated, we will need to have $r \le (m+n)(m+n+1)/2$. The parameters are estimated by minimizing a function of the observed and implied covariances. The most widely used objective function for SEM is the maximum likelihood (ML) function [33]:

$$F_{ML} = \log|\boldsymbol{\Sigma}_{mod}| + \mathrm{tr}\left(\boldsymbol{\Sigma}_{obs}.\boldsymbol{\Sigma}_{mod}^{-1}\right) - \log|\boldsymbol{\Sigma}_{obs}| - p \tag{8.30}$$

where p is the number of observed variables (endogenous + exogenous). For multivariate normally distributed variables, the minimum of the ML function multiplied by $N-1$,

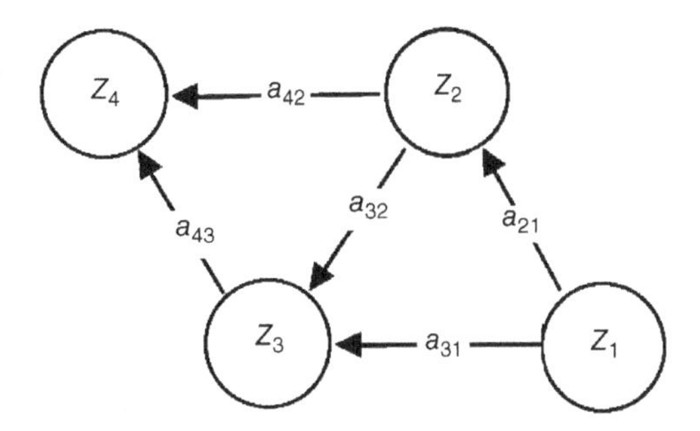

Figure 8.2 Connectivity pattern imposed in the generation of simulated signals. Values on the links represent the connection strength.

approximates a χ^2 distribution with $(p[p+1]/2)-t$ degrees of freedom, where t is the number of parameters to be estimated and p the total number of observed endogenous and exogenous variables. χ^2 statistics test can then be used to infer the statistical significance of the structural equation model obtained. LISREL [36], a publicly available software, may be used to implement the SEM for brain connectivity.

In [33] SEM has been applied to EEG, possibly for the first time, and the effect of noise and data length on accuracy of the results and modelling error have been examined.

In this experiment the authors investigated the influence of SNR on the accuracy of the pattern connectivity estimation obtained by SEM, the amount of EEG data necessary to achieve an acceptable accuracy of the estimation of connectivity between cortical areas, SEM performance degradation by an imprecise anatomical model formulation, and the error type which should be possibly avoided. In this experiment they used simulated models of connectivity between four brain cortical regions. It has been claimed that the proposed method retrieved the cortical connections between the areas under different experimental conditions. In the next trial they applied SEM to the data during a simple finger tapping experiment in humans, in order to underline the capability of the proposed methodology to draw patterns of cortical connectivity between brain areas during performing a simple motor task.

In most of the above approaches for brain connectivity measure involving distributed brain sources, application of a realistic head model is favourable. A true head model helps in accurate localization (and tracking) of the sources involved in the connectivity measure. Although solving the inverse problem such as in [5, 37] and using fMRI [38] have been purposed for this purpose, each one has its own drawbacks. The inverse problem is solved for localization of dipole sources and cannot track the distributed sources adequately. Conversely, fMRI reveals the information about oxygenation of blood vessels as a consequence of event-related activity and cortical activations. It also has a long time lag of two to five seconds with respect to the onset of the event or the activity (such as movement related) of the cortex.

In [39] both SEM and DTF have been implemented for estimation of connectivity from EEG movement-related potentials. The results of application of the SEM method for estimation of the connectivity shows the statistically significant cortical connectivity patterns obtained for the period preceding the movement onset in the alpha frequency band. The connectivity pattern during the period preceding the movement in the alpha band involves mainly the left parietal region, functionally connected with the left and right premotor cortical regions, the left sensorimotor area, and both the prefrontal brain regions.

The stronger functional connections correspond to the link between the left parietal and the premotor areas of both cerebral hemispheres. After the preparation and the beginning of finger movement, the changes in connectivity pattern can be noted. In particular, the origin of the functional connectivity links is positioned in the sensorimotor left cortical areas. From there, functional links are established with left prefrontal and both premotor areas. A functional link in this condition connects the right parietal area with the right sensorimotor area. The left parietal area, very active in the previous condition, was linked instead with the left sensorimotor and right premotor cortical areas.

As stated previously, the origin of functional connectivity links is positioned in the sensorimotor left cortical areas (SMl). From there, functional links are established between left prefrontal and both premotor areas (left and right) [39]. Model order, number, and locations of the ROIs are determined before applying the algorithm.

8.6 Stockwell Time–Frequency Transform for Connectivity Estimation

In very recent practises another effective method for the brain connectivity estimation has been proposed [40–42]. The Stockwell time–frequency transform (S-transform) [43] is an effective method for identification of link weights in a connected network. This method is more accurate and less sensitive to the changes in time–frequency parameters compared to the auto-regressive (AR)-based methods. It is defined as:

$$\mathbf{X}_k(\tau,f) = \int_{-\infty}^{\infty} \mathbf{x}_k(t)\frac{|f|}{\sqrt{2\pi}}e^{-\frac{(\tau-t)^2 f^2}{2}}e^{-i2\pi ft}dt \quad (8.31)$$

This is very similar to the short-term frequency transform where the signal is initially windowed by a Gaussian shape window. However, the variance of this Gaussian window is proportional to the inverse of frequency. Therefore, there is no need to define the variance. The cross-spectrum of the signal is defined as:

$$C_{kl}^{(ST)}(t,f) = \frac{S_{kl}^{(ST)}(t,f)}{\sqrt{S_{kk}^{(ST)}(t,f)S_{ll}^{(ST)}(t,f)}} \quad (8.32)$$

where

$$S_{kl}^{(ST)}(t,f) = \langle X_k(t,f)X_l^*(t,f)\rangle \quad (8.33)$$

where $\langle\rangle$ refers to sample expectation, is complex-valued. The imaginary part of S-coherency (ImSCoh) is related to the phase difference between the signals $\mathbf{x}_k(t)$ and $\mathbf{x}_l(t)$ at each frequency f, e.g. if ImSCoh is positive then, $\mathbf{x}_k(t)$ and $\mathbf{x}_l(t)$ are interacting and $\mathbf{x}_k(t)$ leads $\mathbf{x}_l(t)$. The cross-spectra values are then used in estimation of the combination weights as:

$$a_{kl}^{(ST)} = \frac{max\left(\mathrm{Im}\left(C_{kl}^{(ST)}(t,f),0\right)\right)}{\sum_{l\in N_k} max\left(\mathrm{Im}\left(C_{kl}^{(ST)}(t,f),0\right)\right)} \quad (8.34)$$

Figure 8.3 shows the algorithm performance for a simulated scenario where there are six noisy measurements. In this simulation zone 1 transmits its signal to zone 2 and another message is transmitted through zones 5, 4, and 3 respectively. The last zone includes an independent source without any leakage or transmission to other zones. The SNR is 10 dB and the results are averaged over three trials.

8.7 Inter-Subject EEG Connectivity

8.7.1 Objectives

The aim here is to estimate the connectivity between the brains of two or more subjects during performing cooperative or competitive tasks. Simultaneous EEG recording from more than one subject's brain is referred to as *EEG hyper-scanning*. Interpersonal body movement synchronization has widely been observed. For example, often it is experienced that one's footsteps unconsciously synchronized with those of his friend while walking together. Conversely, in a cooperative or competitive performance one's action is the response to the action of other(s). This can be examined in the case of team working or one competing against another player in a match. Such mechanisms of body movement coherency and their relation to implicit social interaction remain obscured.

8.7.2 Technological Relevance

Very recently, a study and measurement of fingertip movement between two participants while recording their EEG simultaneously have been introduced [44]. In particular, the aim has been to evaluate body movement synchrony and implicit social interactions between two or more participants and assess the underlying dynamics and the effective connectivity between and within their brain regions. The concurrent activity in multiple brains of the group may be estimated and the causal connections between regions of different brains (hyper-connectivity) indicated.

The study of concurrent and simultaneous brain activities of different subjects while they perform cooperative or competitive tasks is very important in investigating paradigms in many cases where the knowledge of simultaneous interactions between individuals has a value. A person's ability to concentrate, predict, cooperate, follow, compete, learn, and engage in long and tedious tasks for both normal subjects and patients with various brain abnormalities such as AD, mental fatigue, and during drug infusion may effectively be studied from simultaneous EEG records of subjects cooperating with or competing against each other.

Most recently, methods established for brain connectivity measures, from single subject EEG recordings, have been extended and applied to simultaneous multiple brain recordings [45–50]. Recorded data were processed in different ways. In a method based on the time–frequency domain approach, changes in amplitude as well as synchronization of the EEG rhythms were exploited in the conventional frequency bands of delta, theta, alpha, and beta. These approaches aim at estimation of coherency or synchronization between the recorded EEGs from multiple brains.

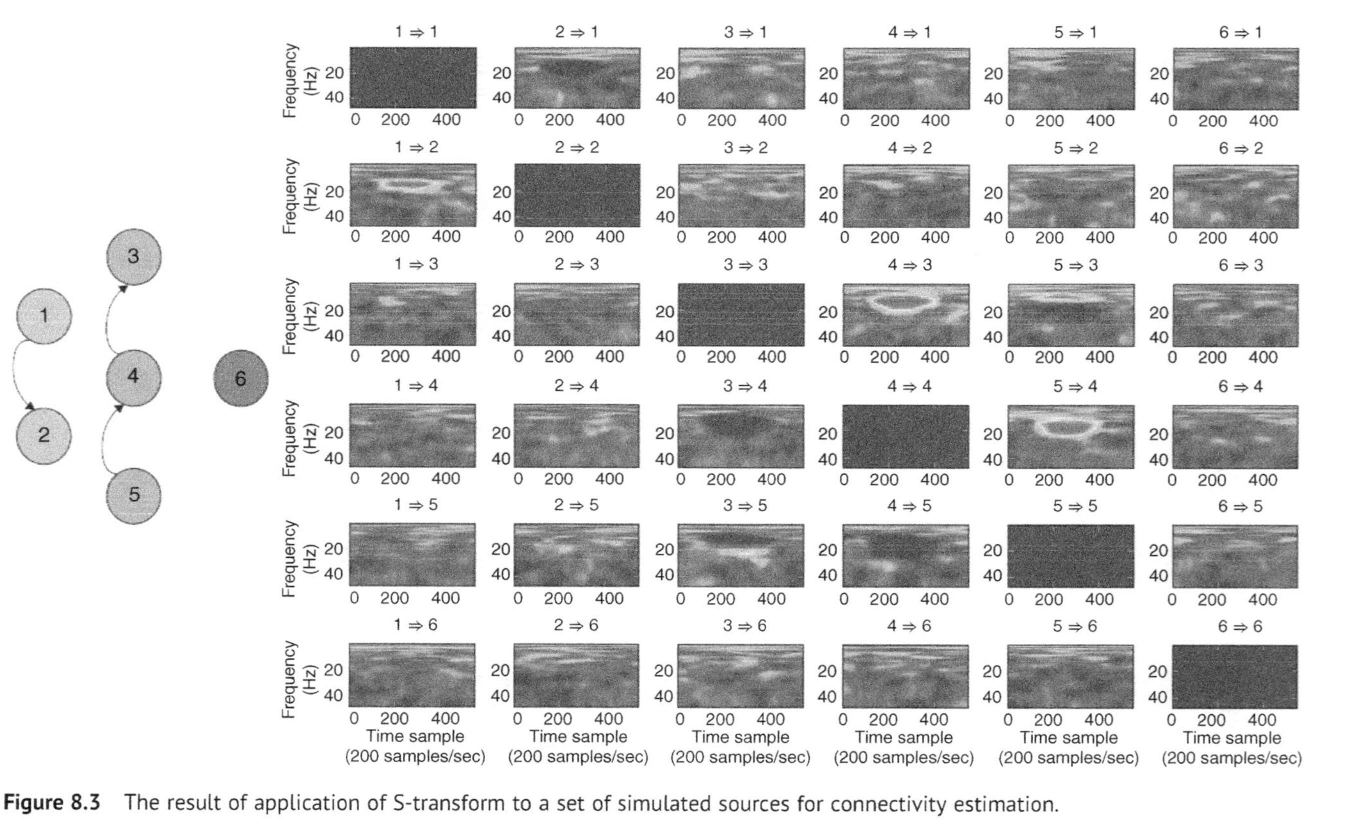

Figure 8.3 The result of application of S-transform to a set of simulated sources for connectivity estimation.

The second approach, which is based on connectivity or synchronization measures, allows for an estimation of concurrent activity in multiple brains and connections between regions of different brains (so-called hyper-connectivity). Granger causality was suggested to be used in this approach. Variety of techniques such as MVAR followed by DTF could also be used. Aforementioned approaches may also be combined to increase effectiveness of the system.

As a good example it has been shown that AD is closely related to alteration in the functional brain network, i.e. functional connectivity between different brain regions [51]. Both within-lobe connectivity and between-lobe connectivity as well as between hemisphere connectivity have been estimated for AD, normal control (NC), and mild cognitive impairment (MCI). The temporal lobe of AD has a significantly lesser amount of direct within-lobe connectivity than NC. This direct link within the temporal lobe may be attenuated by AD. It has also been shown that the hippocampus and parahippocampal zone are much more separated from other regions in AD than in NC. The temporal lobe of MCI, however, does not show a significant decrease in the amount of direct connections, compared with NC. The frontal lobe of AD shows considerably more connectivity than NC. This has been interpreted as compensatory reallocation or recruitment of cognitive resources as denoted in [51] and the references herein. There is no significant difference between AD, MCI, and NC in terms of connectivity within the parietal lobe and within the occipital lobe.

In terms of between-lobe connectivity, in general, human brains tend to have a less amount of between-lobe connections than within-lobe connections. In addition, AD has significantly more parietal-occipital direct connections than NC. Increase in the amount of connections between parietal and occipital lobes of AD has previously been reported in [52]. It may also be interpreted as a compensatory effect. Furthermore, MCI also shows increase in the amount of direct connections between parietal and occipital lobes, compared with NC, but the increase is not as significant as with AD. While the amount of direct connections between frontal and occipital lobes shows little difference between AD and NC, for MCI it shows a significant decrease. Also, AD causes less temporal-occipital and less frontal–parietal connections, but a higher parietal–temporal connectivity than NC.

Finally, between hemisphere connectivity estimation results indicate that AD disrupts the strong connection between the same regions in both left and right hemispheres, whereas this disruption is not significant in MCI [51].

For the same objectives another research was carried out by Escudero et al. [53] to differentiate between NC, AD, and MCI from MEG signals. This is discussed in Chapter 15 where manifestation of the neurodegenerative diseases in EEGs is explored.

8.8 State-Space Model for Estimation of Cortical Interactions

Most of the above approaches for estimation of brain cortical connectivity from EEG or MEG suffer from noise involved in the measurements. Therefore, it would be useful if a suitable technique could be developed to improve estimation of the prediction/connectivity parameters and be more robust against noise [54].

Nalatore et al. [55] developed a state-space approach to MVAR model estimation for invasive electrophysiological recordings and show that explicitly modelling noise results in improved connectivity estimates. Shumway and Stoffer [56] introduced an expectation–maximization approach to ML parameter estimation in state-space MVAR models.

In [57] a state-space model has been developed to estimate MVAR parameters. Such a model can represent the MVAR model of cortical dynamics, while an observation equation describes the physics relating the cortical signals to the measured EEG and the presence of spatially correlated noise. In this model it is assumed that the cortical signals originate from known regions of cortex, but the spatial distribution of activity within each region is unknown. Then, the MVAR parameters have been computed through an expectation–maximization approach. The algorithm also calculates the spatial activity distribution components, and the spatial covariance matrix of the noise from the measured EEG [57].

The state-space approach does not circumvent the limitations of EEG and MEG regarding noise, but it uses the ML criterion to solve for MVAR parameters from the signals. ML estimates are known to be asymptotically unbiased with variance approaching the Cramer–Rao lower bound as data length increases. Therefore, in a two-stage approach, the cortical sources are first estimated by solving a variant of the inverse problem and next, an MVAR model is fitted to the estimated cortical signals [57]. Mathematically, the process can be expressed as follows:

consider $\boldsymbol{x}_{n,j} = \left[x_{n,j}^{1}, \ldots, x_{n,j}^{M}\right]^{T}$, $n = 1, \ldots, N, j = 1, \ldots, J$ be the jth trial of an $M \times 1$ state vector representing samples of cortical signals from M regions at time n. An MVAR of order P for representing the cortical signals $\mathbf{x}_{n,\,j}$ can be described as:

$$\mathbf{x}_{n,j} = \sum_{p=1}^{P} \mathbf{A}_p \mathbf{x}_{n-p,j} + \mathbf{w}_{n,j} \tag{8.35}$$

where $\mathbf{A}_p$ is the matrix of prediction coefficients. This can be rewritten as:

$$\mathbf{x}_{n,j} = \mathbf{A}\mathbf{z}_{n-1,j} + \mathbf{w}_{n,j} \tag{8.36}$$

where $\mathbf{A} = [\mathbf{A}_1, \ldots, \mathbf{A}_P]$ is an $M \times MP$ matrix of prediction coefficients and $\mathbf{z}_{n-1,j} = \left[x_{n-1,j}^{T}, \ldots, x_{n-P,j}^{T}\right]^{T}$ as an $MP \times 1$ vector containing the past P state vectors. Here it is assumed that the initial vector $\mathbf{z}_{0,\,j}$ is Gaussian distributed with unknown mean $\boldsymbol{\mu}_0$ and unknown covariance matrix $\boldsymbol{\Sigma}_0$. The dynamical state-space model for an observation $\mathbf{y}_{n,\,j}$ is presented as:

$$\mathbf{y}_{n,j} = \mathbf{C}\boldsymbol{\Lambda}\,\mathbf{x}_{n,j} + \mathbf{v}_{n,j} \tag{8.37}$$

This can be combined with the MVAR model in (8.36) to have the state-space equations [57]:

$$\begin{cases} \mathbf{x}_{n,j} = \mathbf{A}\mathbf{z}_{n-1,j} + \mathbf{w}_{n,j} \\ \mathbf{y}_{n,j} = \mathbf{C}\boldsymbol{\Lambda}\,\mathbf{x}_{n,j} + \mathbf{v}_{n,j} \end{cases} \tag{8.38}$$

where $\mathbf{w}_{n,\,j}$ and $\mathbf{v}_{n,\,j}$ are respectively the $M \times 1$ and $L \times 1$ vectors of state and observation noises, $\mathbf{C}$ is the $L \times MF$ forward matrix of source to electrode mapping, and $\boldsymbol{\Lambda}$ is an $MF \times M$ block diagonal matrix with $F \times 1$ vectors $\boldsymbol{\lambda}^m$ on the diagonal representing any unknown parameters describing the spatial activity of the mth region [54]. Although such a presentation is not unique, the non-uniqueness problem does not affect the model parameter estimations [57]. An ML approach including expectation–maximization can then be employed to estimate the model (8.38) parameters including $\mathbf{A}$ which represent the connectivity parameters.

A similar state-space method has been applied to a radial basis function (RBF)-based connectivity model. An MVAR of order P for representing the cortical signals $\mathbf{x}_{n,\,j}$ can be described as in [58]. This model is important since it complies with the invariance property required to evaluate Granger causality. Also, the universal approximation theorem [59] states that an RBF network is capable of approximating any smooth function to an arbitrary degree of accuracy [54]. Therefore, RBF is able to model nonlinear dynamics of cortical signals. RBF has been used to estimate nonlinear Granger causality for intracranial EEG signal analysis [58]. To use RBF for estimation of the cortical connectivity between M cortical regions we have [54]:

$$\mathbf{x}_{n,j} = \boldsymbol{\Psi}\boldsymbol{\Phi}\left(\mathbf{Z}_{n-1,j}\right) + \mathbf{w}_{n,j} \tag{8.39}$$

where $\boldsymbol{\Psi}$ is an $M \times MI$ matrix of RBF weights, and $\boldsymbol{\Phi}\left(\mathbf{Z}_{n-1,j}\right) = \left[\varphi^1\left(\mathbf{z}^1_{n-1,j}\right)^T, \ldots, \varphi^M\left(\mathbf{z}^M_{n-1,j}\right)^T\right]^T$ is an $MI \times 1$ vector of nonlinear functions of P variables, where $\varphi^m = \left[\varphi^m_1, \ldots, \varphi^m_I\right]$ is an $I \times 1$ vector of RBF kernels [54]. The elements of $\boldsymbol{\Psi}$ reflect the coupling between brain cortical regions which can be estimated from the cortical signals by solving a system of linear equations.

In all the above methods there is no link between the brain connectivity and the task performed by the brain. Therefore, there is a need for modelling the pathways between the connectivity estimates and the performed task by the brain leading to a mental or physical action. In the following section, we see how the connectivity measures are incorporated into the design of an adaptive filter which models such pathways.

8.9 Application of Cooperative Adaptive Filters

None of the approaches mentioned above (including all previously used for brain connectivity estimation) takes into account any model of such processes, where the brain attempts to achieve certain objectives, in a concise manner, considering evolutions in both time and (electrode) space, particularly, when one brain sets the goals and the other brain follows.

Distributed sources assumption of movement-related cortical sources and the dynamics of synaptic currents motivates a new solution for investigation of brain connectivity. Recent evidences from a finger movement experiment show that the neurons in both contralateral and ipsilateral brain lobes are engaged in the movement [60]. However, there is no further investigation about dynamics of the movement in this literature. In this section a different and novel approach based on cooperative adaptive filtering theory, incorporating *diffusion adaptation* [61–63], is introduced. The fundamental aim of this research is to model the brain dynamics (i.e. varying connectivity in both time and space) for each particular body movement. There are enormous applications for this design such as brain behaviour during prolonged movement for healthy and Parkinson subjects and EEG hyper-scanning.

In the original work in diffusion adaptation [64–66] a simple strategy in sharing information was introduced. In this interesting work both time and space evolutions have been exploited in the design of an adaptive filter. Therefore, a response from a node in a network corresponds to not only the input (or stimulus) of the network but also the network model, including cooperation of nearby nodes, which evolves in time.

Adaptive networks in mobile communications consist of collection of nodes with learning and motion abilities that interact with each other locally in order to solve distributed processing and distributed inference problems in real-time. The objective of such a network is to estimate some filter coefficients in a fully distributed manner and in real-time, where each node is allowed to interact only with its neighbours.

In [61] a similar adaptation algorithm inspired by bacteria mobility, has been developed for adaptation over networks with mobile nodes. The nodes have limited functionalities and are allowed to cooperate with their neighbours to optimize a common objective function. In this elegant approach the nodes do not know the form of the cost function beforehand. They can only sense variations in values of the objective function as they diffuse through the space. A good nature inspired application example is in sensing the variation in the concentration of nutrients in the environment. In this model however, sharing the information has been limited to binary choices between run and tumble for the movement of mobile nodes.

Following the above work, a diffusion adaptation algorithm that exhibited self-organization properties was developed and applied to the model of cooperative hunting (of prey fish) among predators (sharks) while the fish herd and move towards a nutrition point [63]. In this new application the nodes of the network wish to track the location of food source and the location of the predator. The modelling equations are therefore applied to both targets.

To better estimate the parameters such as weights of the links between nodes, other previously described methods such as SEM can be used to initialize the nodes and their mutual connectivity (presented by the link weights) levels. Time evolution of the changes in these levels can be estimated while undertaking cooperative or competitive tasks (such as hand movement in a synchronized dance). Then, a constrained diffusion adaptation algorithm is developed to model the brain responses to these tasks performed by two or more human brains to achieve or avoid a particular objective (stimulus).

This approach requires definition of a number of bio-inspired and physiological constraints for each type of brain activity and also definition of regions of connectivity and the spatial connectivity likelihoods within each region. Eventually, depending on the brain responses, which regions of the brain are involved, the synchronization of the brains, and how they follow (or avoid) each other, the diffusion adaptation algorithm can be designed. Here, each measurement point (electrode) will be considered as a node and the expected time–space response of the brain to each stimulus becomes the target. Hence, from an algorithmic perspective a relatively complex constrained optimization problem is solved for which the optimum filter parameters are subject to decisions by the other cooperating or competing parties.

Subject to accurate definition of the constraints and pre-determination of time–space parameters, such a filter is able to more accurately model the time-varying interactions between two or more phenomena.

8.9.1 Use of Cooperative Kalman Filter

Most techniques in connectivity evaluation consider the signals to be stationary. They also consider the sources separately. Application of adaptive filtering in time and space in the so-called diffusion adaptation approach tracks a time–space varying process which results in a particular function. This involves sources from a number of neurons which communicate to each other through space and evolve in time. An important initial conclusion from this approach is that the signals are considered stationary, but often the system can cope with nonstationary data.

In [7, 67] a diffusion Kalman filtering approach to cooperative multi-agent sensor networking has been introduced. In most of these applications (excluding decentralized network) it is assumed that the correlation between estimates of two neighbouring agents is known to both agents. In addition, communication between the agents is required to be fast enough such that consensus can be reached between two consecutive Kalman filter updates. In [68] the filtering is based on covariance intersection (CI) method [69]. This approach fuses multiple consistent estimates with unknown correlations with a convex combination of the estimates and chooses the corresponding combination weights by minimizing the trace or determinant of an upper bound of the error covariance matrix [68].

In diffusion Kalman filtering approach it is assumed that $N_{k,i}$ nodes spatially distributed over a region, centred (so-called current node) at k at time i, communicate with each other (within limited range).

Being at state $\mathbf{x}_i$ the measurement at node k and time i is denoted by $\mathbf{y}_{k,\,i}$, the corresponding state-space model is of the form:

$$\begin{aligned} \mathbf{x}_{i+1} &= \mathbf{F}_i\mathbf{x}_i + \mathbf{w}_i \\ \mathbf{y}_{k,i} &= \mathbf{H}_{k,i}\mathbf{x}_i + \mathbf{v}_{k,i} \end{aligned} \tag{8.40}$$

where $\mathbf{w}_i$ and $\mathbf{v}_{k,\,i}$ are the state and measurement noises at time i and node k respectively. $\mathbf{F}_i$ and $\mathbf{H}_{k,\,i}$ are generally time-varying (but bounded) matrices and the noises are zero-mean, white and have [68]:

$$E\left[\begin{bmatrix} \mathbf{w}_i \\ \mathbf{v}_{k,i} \end{bmatrix}\begin{bmatrix} \mathbf{w}_i \\ \mathbf{v}_{k,i} \end{bmatrix}^T\right] = \begin{bmatrix} Q_i\delta_{ij} & 0 \\ 0 & R_{k,i}\delta_{kl}\delta ij \end{bmatrix} \tag{8.41}$$

where δ_{ij} is the Kronecker delta, Q_i and $R_{k,i}$ are assumed to be positive definite and bounded.

The diffusion Kalman filtering approach in [67], schematically presented as in Figure 8.4, aims at estimating stable state $\mathbf{x}_i$, while sharing data with its neighbours only. In this algorithm at every time instant i, node k transmits the quantities $H_{k,i}^T R_{k,i}^{-1} H_{k,i}$ and $H_{k,i}^T R_{k,i}^{-1} H_{k,i}$ to its neighbours in updating each intermediate estimate $\psi_{k,\,i}$. Information from the neighbouring nodes are combined using a $p \times p$ diffusion matrix $C_{k,\,l,\,i}$ subject to:

$$\sum_{l \in N_{k,i}} C_{k,l,i} = \boldsymbol{I}_p, C_{k,l,i} = 0 \text{ for } l \notin N_{k,i} \tag{8.42}$$

where $\mathbf{I}_p$ is a $p \times p$ identity matrix. The combination coefficients C_{kl} combine the information from the neighbouring nodes. Hence, their identification or estimation based on physical or physiological constraints, or an adaptive method such as [68], enhances the accuracy of overall diffusion filtering.

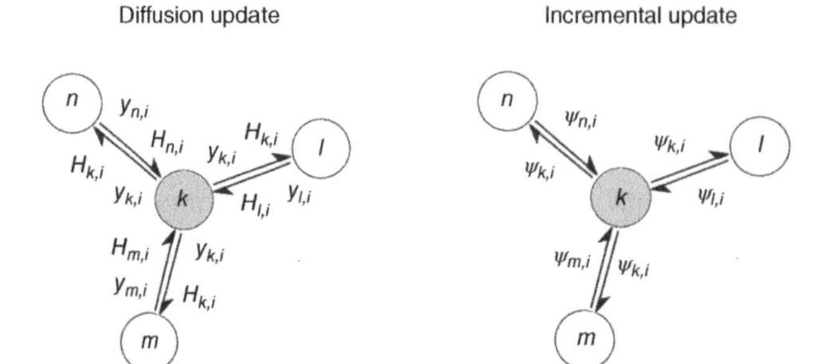

Figure 8.4 Representation of node k and its neighbouring nodes for diffusion Kalman filtering algorithm.

As mentioned earlier connectivity estimation methods such as SEM or MVAR may be used to define the link weights. This is currently under research. CI estimation, conversely, assumes that node k is to fuse the estimates $\hat{x}_{l,i|i-1}$ $(l \in N_{k,i})$ using its neighbours and through the estimates of their error covariance matrices $P_{l,\ i|i-1} > 0$. Based on the CI algorithm by Julier and Uhlman as reported in [68]:

$$\psi_{k,i} = C_{k,l,i}\hat{x}_{l,i|i-1} \tag{8.43}$$

where

$$\begin{aligned} C_{k,l,i} &= \beta_{k,l,i}\Lambda_{k,i}P_{l,i|i-1}^{-1} \\ \Lambda_{k,i} &= \left(\sum\nolimits_{l \in N_{k,i}} \beta_{k,l,i}P_{l,i|i-1}^{-1}\right)^{-1} \end{aligned} \tag{8.44}$$

$0 \le \beta_{k,\ l,\ i} \le 1$ where $\sum_{l \in N_{k,i}} \beta_{k,l,i} = 1$ is computed such that the trace or determinant of $\Lambda_{k,i}$ is minimized. The above optimization is often nonlinear. Therefore, some CI algorithms such as one presented in [68] have been proposed to estimate these parameters.

8.9.2 Task-Related Adaptive Connectivity

From the above connectivity estimation methods, it may be concluded that particular connectivity patterns are related to particular mental, cognitive, or movement activity. Such patterns are those which vary in both time and space (i.e. the engaged electrodes or actual brain source locations) and originate from distributed synaptic current sources. Modelling of such connectivity patterns can be of great importance in places where the neural activity corresponding to a movement or mental task such as in brain–computer interfacing (BCI), is to be learned.

Extension of adaptive filters, often used for single-channel applications, to cooperative adaptive filters where the input is a combination of information received from its neighbours can be a solution to application of adaptive filters for multichannel (usually correlated) data. This approach has become popular in communication networks. Incremental, where the information is transferred sequentially from node to node, or consensus and diffusion where the input is an aggregate of the information received from the neighbouring nodes, have become popular solutions to cooperative networks. Diffusion adaptation filters are similar to conventional adaptive filters where the inputs are sum of the contributions (or cooperation) between the input nodes (sources). The cooperation pattern between the nodes, itself, varies in time.

As a motivation for this study, recently, a study of brain connectivity and measurement of fingertip movement between two participants, while recording their EEG simultaneously, has been carried out [44]. In particular, the aim was evaluation of body movement synchrony during implicit social interactions between two or more participants, assessing the underlying dynamics and the effective connectivity between and within their brain, mainly cortical regions. Concurrent activity in multiple brains of a group of people may be estimated in order to indicate the causal connections between regions of different brains (hyper-connectivity).

Among the three approaches in cooperative adaptive filtering, i.e. incremental, consensus, and diffusion, the latter is more general and guarantees stability of the networks. Diffusion adaptation can be a very good candidate to model the evolution of neuronal network connectivity during a task performance. This method is more attractive when one subject performs a task as a model target and others follow. Unlike the network of agents, where usually the combination weights are unknown and therefore the weights considered equal (traditionally as an average of the link weights within the immediate neighbourhood, in the case of brain application, the brain connectivity values can be used for this purpose. Therefore, at each stage of adaptation, the connectivity coefficients, such as DTF, can be used to adjust the instantaneous cooperation weights among the nodes E electrodes or brain sources). To understand the concept the method is briefly reviewed here.

8.9.3 Diffusion Adaptation

In diffusion adaptation, the main goal is to estimate an $M \times 1$ unknown vector $\mathbf{w}$ from measurements collected at N nodes spread over a network. Each node k has access to time realizations $\{d_k(i), u_{k,i}\}$ of the data $\{\mathbf{d}_k, \mathbf{u}_k\}$ where $\mathbf{d}_k$ is a vector of scalar measurements and $\mathbf{u}_k$ is a $1 \times M$ row regression vector [62]. Considering these vectors from all the nodes, the data forms two global matrices $\mathbf{U}$ and $\mathbf{D}$ respectively. The objective is to estimate vector $\mathbf{w}$ that solves the minimization problem:

$$\min_{\mathbf{w}} E\|\mathbf{d} - \mathbf{U}\mathbf{w}\| \tag{8.45}$$

Assuming that by applying $\mathbf{w}$ the signal from node k, at time instance i, generates a signal $\psi_k^{(i)}$. Each node combines the information from its neighbouring nodes to produce a new input to the filter as:

$$\varphi_k^{(i-1)} = f_k\left(\psi_l^{(i-1)}; l \in N_k\right) \tag{8.46}$$

where f_k is a combining function and N_k is the number of nodes included in the neighbourhood of node k at time instance i. An option for selection of the above function is to consider it as a fixed linear averaging process. Another alternative is to consider it as a linear sum of weighted inputs from neighbouring nodes which can be adapted through time to minimize the global error [30]. In both cases adaptation can be presented in the following set of alternating equations:

$$\varphi_k^{(i-1)} = \sum_{l \in N_k} c_{kl}\psi_l^{(i-1)} \tag{8.47a}$$

$$\psi_k^{(i)} = \varphi_k^{(i-1)} + \mu_k u_{k,i}^* \left(d_k(i) - u_{k,i} \varphi_k^{(i-1)} \right) \quad (8.47b)$$

In Eq. (8.47a) the new input to the model is estimated using received signals from the neighbouring nodes. In (8.47b) the output of the filter is updated for iteration/time sample i. By implementing the above equations using alternating adaptation we can track the signal component in both space, using k and time, using i. Alternatively, by replacing $\psi_k^{(i)}$ by $w_{k,\,i}$ Eq. (8.47) can also be written as:

$$\psi_k^{(i)} = w_{k,i-1} + \mu_k u_{k,i}^* (d_k(i) - u_{k,i} w_{k,i-1}) \quad (8.48a)$$

$$w_{k,i} = \sum_{l \in N_k} C_{kl} \psi_l^{(i)} \text{ where } \sum_{l \in N_k} C_{kl} = 1 \quad (8.48b)$$

8.9.4 Brain Connectivity for Cooperative Adaptation

In application of cooperative adaptive algorithms to EEG signals the block diagram in Figure 8.5 can be followed.

Some examples of the above method have been reported for classification of EEG signals during clockwise and counter-clockwise hand drawings [40, 41]. In [41] the authors examine the flow of information among electrodes attached to the brain and use diffusion adaptation strategies to assess the brain cortical connectivity. The method uses the DTF technique to estimate the combination coefficients to drive the adaptation and learning process. The diffusion strategy is then applied to the problem of recognizing left-hand and right-hand movements and its superior performance is demonstrated relative to solutions that rely on stand-alone electrodes and do not exploit coordination among multiple electrodes. A connectivity pattern between the electrodes in the central region of the brain, as depicted in Figure 8.6 has been used in the above approach. In [40] S-coherency has been used instead of DTF and the diffusion adaptation algorithm has been designed to discriminate between clockwise and counter-clockwise hand movements during drawing a circle.

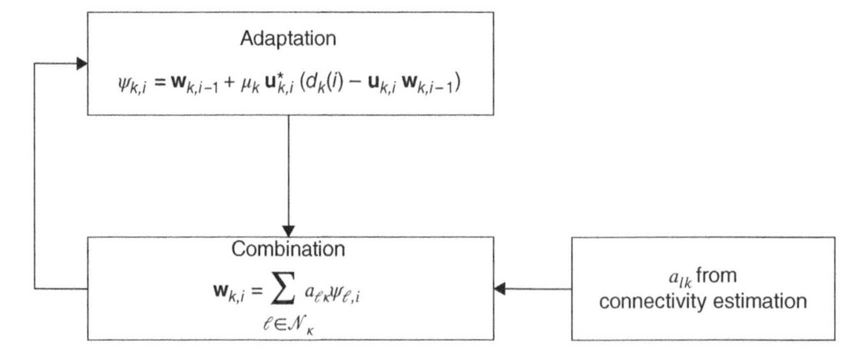

Figure 8.5 The use of brain connectivity for diffusion adaptation filtering.

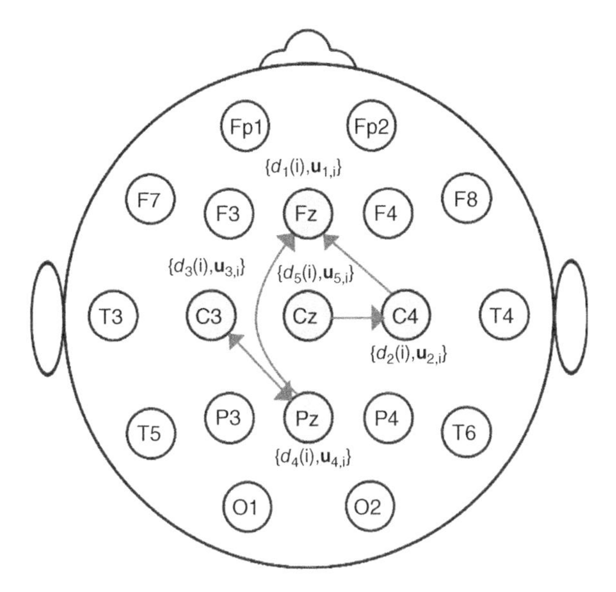

Figure 8.6 An illustration of brain connectivity pattern. EEG signals collected at the marked electrodes are used to train a cooperative network to estimate a model for left-hand and right-hand movements.

In an application to detection of body (hand) tremor from the EEG signals, the time intervals with the onset of hand tremor have been detected [42]. In this approach an adaptive multitask diffusion strategy has been introduced to estimate the underlying model between the gait information and the EEG signals. The method incorporates an S-transform-based connectivity measure that performs well even on a single-trial basis. The estimated connectivity values are then combined with the combination weights of the multitask diffusion strategy to model the relation between tremor and the brain signals. The outcome is an enhanced brain connectivity measure representing its time–space relation to the tremor.

Using the S-transform, a more robust estimation of brain connectivity can be achieved. Figure 8.7 represents the variation of combination weights (brain connectivity parameters) measured by the S-transform.

8.9.5 Other Applications of Cooperative Learning and Brain Connectivity Estimation

As explored in Chapter 9, related to brain event-related potentials (ERPs), tracking of P300 waveforms and their subcomponents P3a and P3b in time can be more accurate when the cooperation between the channels is also taken into account [70]. This exploits the spatial and temporal properties of the EEG waveforms in monitoring mental fatigue for healthy and dementia subjects more effectively. Hence, the symmetricity property of the brain left and right lobes in terms of generation of ERPs are used in the design of a collaborative adaptive particle filter. This system uses the link between C3, C4, P3, and P4 electrodes to better (compared to the non-collaborative method [71] developed for mental fatigue) estimate and track the amplitude and latency of P3a and P3b waveforms.

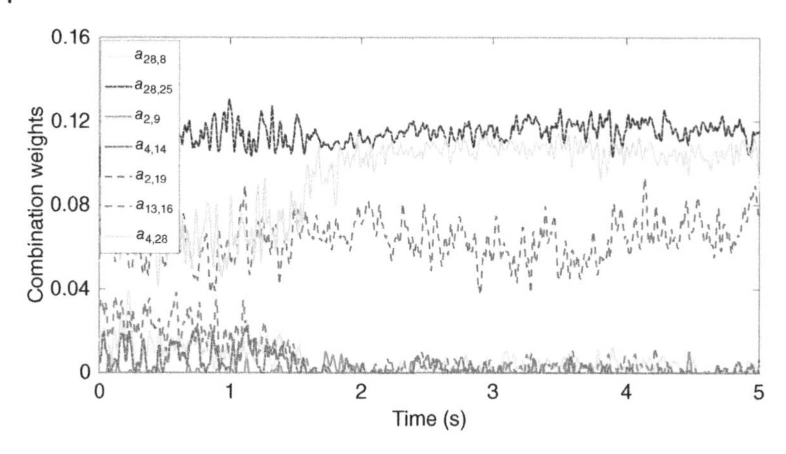

Figure 8.7 Variation of combination weights (brain connectivity parameters) measured by the S-transform. This is to show that the relative changes in the weights remain fairly consistent over a short time duration.

8.10 Graph Representation of Brain Connectivity

Graph theory studies the properties of mathematical structures, or graphs, designed to model the connections between a group of nodes which can be active sources. Graph theory was first applied by Leonhard Euler in 1736, who prefigured the idea of topology through the negative resolution for his famous Seven Bridges of Konigsberg problem. His assessment that there was no feasible way to walk through the city of Konigsberg by crossing each of the seven bridges only once, has helped him develop rigorous approaches to the field of graph theory (Euler 1741). Since then graph theory has been applied to many different areas including social sciences – used to understand the web of social relations, computer science, and currently, natural sciences with particular interest in the study of complex biological systems (Sporns 2011).

The theoretical network models have vastly contributed to the emerging field of brain connectomics, a field of analysis that combines some of the most advanced areas of study in medicine, neuroscience, and engineering to characterize and quantify the brain in terms of its structural and functional connections.

Topological changes in the brain's networks may be analyzed using graph theory, which represents the brain's connections as a set of nodes and edges.

A graph G consists of a set N_G of N nodes, and a set $\Im$ of edges such that if nodes m and n are linked, then $(m, n) \in \Im$. For undirected graphs, these node pairs are unordered. At the node level, a recorded signal $\mathbf{y} = [y_1, \ldots, y_N]^T$, assumed to be real valued, where y_n is the sample of signal $\mathbf{y}$ at node n. Also, the adjacency matrix $\mathbf{A}$ [72, 73], is defined as an $N \times N$ matrix whose entries $a_{n,m}$ are zero if $(m,n) \notin \Im$ and one otherwise. The adjacency matrix describes the interactions between entities and, by extension, can be considered as a tool for representing relationships between the data channels (vertices). In general, any one of the methods previously described in this chapter can be used to characterize or define a graph. For the particular case of brain connectivity estimation, the proposed method has the advantage of adaptability given the data are available [74]. In this application, the graph topology is considered unknown beforehand and has been estimated from the data.

Moreover, some graphs can be dynamic, such as brain activity supported by neurons or brain regions. A network structure is then proposed to capture online the nonlinear dependencies among streaming graph signals in the form of a possibly directed, adjacency matrix. By projecting the data into a higher- or infinite-dimension space, the nonlinear relationships between the electrode recordings (agents) are estimated. Kernel dictionaries are used to mitigate the increasing number of data points.

Consider that an N-node graph with adjacency matrix $\mathbf{A}$, as defined above, models a system such as the brain network. In this setting, the brain activity in each brain zone, $\boldsymbol{y}(i)$ can be measured at time instant i. For the case of brain connectivity, the existing nonlinearity in the connections motivates modelling the brain dynamics. Consider the following data model [74]:

$$\boldsymbol{y}(i) = \mathbf{A}\boldsymbol{f}(i) + \boldsymbol{v}(i) \tag{8.49}$$

Matrix $\mathbf{A}$ models how entries of $\boldsymbol{f}(i) \triangleq col\{ f_m(y_{L_m}(i))\}_{m=1}^{N}$, where $f_m : \Re^{L_m} \to \Re$, affect every node. $\boldsymbol{v}(i)$ accounts for additive noise. The least-mean-squares solution to this problem is:

$$\hat{\mathbf{A}} = \underset{\mathbf{A}}{\operatorname{argmin}} \frac{1}{2} E\|\boldsymbol{y}(i) - \mathbf{A}\boldsymbol{f}(i)\|^2 + \Psi(\mathbf{A}) \text{ subject to } a_{nm} \in \{0, 1\} \tag{8.50}$$

where $\Psi(\mathbf{A})$ is a regularization term to account for some prior knowledge of $\mathbf{A}$ such as symmetry or sparsity. The solution to this nonlinear problem is hard to achieve in its general form. However, by selecting f_m from the reproducing kernel Hilbert space (RKHS) the graph parameters can be estimated as [74]:

$$\hat{a}_n(i+1) = \hat{a}_n(i) + \mu_n\left[\boldsymbol{r}_{\widetilde{k}y} - \boldsymbol{R}_{\widetilde{k}\widetilde{k}}\hat{a}_n(i) - \eta_n\Gamma_n(i)\right] \tag{8.51}$$

where $\boldsymbol{R}_{\widetilde{k}\widetilde{k}} \approx \widetilde{k}(i)\widetilde{k}^T(i)$, $\widetilde{k}(i)$ is the selected RKHS nonlinearity function for $\boldsymbol{f}(i)$, and the entries of Γ_n, i.e. Γ_{nm} are:

$$\Gamma_{nm} = \begin{cases} \dfrac{\hat{a}_{nm}}{\|\hat{a}_{mn}\|_2} & \text{if } \|\hat{a}_{mn}\|_2 \neq 0 \\ 0 & \text{if } \|\hat{a}_{mn}\|_2 = 0 \end{cases} \tag{8.52}$$

In an example [74], seizure ECoG recordings from 76 cortical electrodes have been used. It has been found that the graph topology in terms both adjacency matrix and the connection weights differ for preictal and ictal periods.

8.11 Tensor Factorization Approach

Another interesting approach in analysis of connectivity is *link prediction*, which not only provides a measure of connectivity but also tracks the changes of the links between the nodes of a network. The problem of link prediction arises for groups of objects connected by multiple relation types. The solution should not only allow identification of the correlations (among the linked patterns) but also reveals the impact of various relations on prediction performance.

In [75] following the work in [76], it is assumed that the objects tend to form links with others having similar characteristics and related objects share the similar behaviour pattern. To formulate the link pattern prediction consider a set of N objects $X = \{x_1, x_2, ..., x_N\}$ and assume there are T different types of relations among object pairs defined by a set of T trajectory matrices [75]. An $N \times N \times T$ tensor $\underline{\mathbf{Y}}$ is then defined with entry $y_{i,j,t}$ representing the type t relation value between object pair x_i and x_j as [75]:

For $\forall i,j \in [1,... N]^2$ and $\forall t \in [1, ..., T]$

$$y_{i,j,t} = \begin{cases} 1 & \text{if } x_i \text{ links to } x_j \text{ by relation type } t \\ 0 & \text{otherwise} \end{cases} \tag{8.53}$$

Based on this $\underline{\mathbf{Y}}_{(i,j,:)}$ is defined as the *link pattern* involving T different types of relations between each pair of objects x_i and x_j. The problem to solve here is: given some observed link patterns information for a subset of related objects in the multi-relational network, how the unobserved link patterns for the remaining object pairs can be found? In Figure 8.8 an example of the problem and its tensor model [77].

To mathematically express the problem consider a nonnegative $N \times N$ matrix $\mathbf{W}$ such that its entries $w_{i,j}$ for $\forall i,j \in [1,..., N]^2$ are defined as [75]:

$$w_{i,j} = \begin{cases} 1 & \text{if link pattern between } x_i \text{ and } x_j \text{ is known} \\ 0 & \text{if link pattern between } x_i \text{ and } x_j \text{ is missing} \end{cases} \tag{8.54}$$

For the decomposition the following loss function is defined:

$$L(\underline{\mathbf{X}}) = \sum_{i=1}^{N} \sum_{j=1}^{N} w_{i,j} \|y_{i,j,:} - x_{i,j,:}\|^2 \tag{8.55}$$

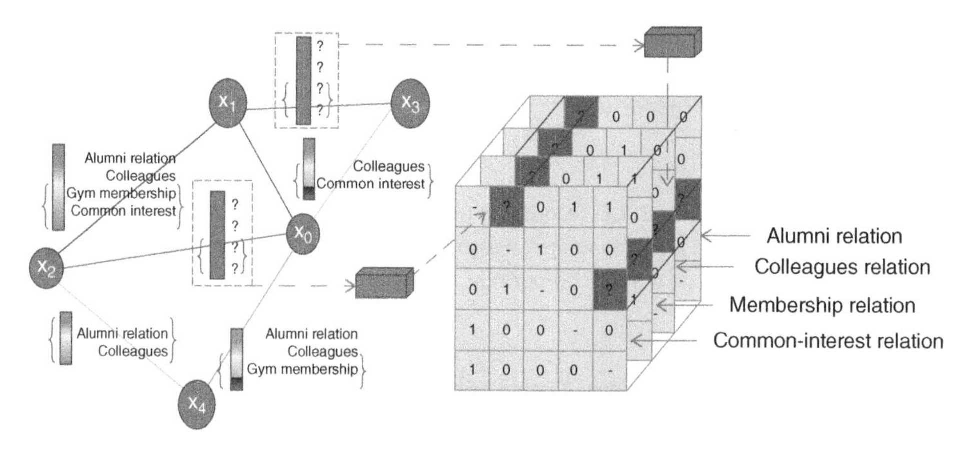

Figure 8.8 Example of modelling the multirelational social network as a tensor for predicting the unobserved link patterns in the network. To the right is the tensor representation of the network where each slice matrix represents one relation type and each tube fibre represents the link pattern between two nodes. Unknown link patterns are represented as '?' fibres.

where ||. || denotes Frobenius norm for a tensor. Following Canonical decomposition (CANDECOMP)/PARAFAC (CP) tensor decomposition and (8.55) we can solve the following optimization problem [75] using the conjugate gradient approach:

$$\arg\min_{\underline{\mathbf{X}}} L(\underline{\mathbf{X}}) \text{ subject to } x_{i,j,t} = \sum_{k=1}^{K} u_{i,k} v_{j,k} r_{t,k} \tag{8.56}$$

where K is rank of the tensor, and **U**, **V**, and **R** are the latent factor matrices for CP tensor decomposition. A popular approach to solve the problem is by alternating least-squares (ALS) as in [75]. This however requires estimation of a large number of parameters and consequently, needs high cost of computation and may result in over-fitting. To overcome these problems the above constrained problem can be changed to an unconstrained problem using regularization parameters (penalty functions), which leads to the following cost function:

$$\arg\min_{\mathbf{U},\mathbf{V},\mathbf{R},\gamma} L(\mathbf{U},\mathbf{V},\mathbf{R}) + \frac{\gamma_0}{2}\left(\|\mathbf{U}\|_F^2 + \|\mathbf{V}\|_F^2\right) + \frac{\gamma_T}{2}\left(\|\mathbf{R}\|_F^2\right) \tag{8.57}$$

where γ_0 and γ_T are the regularization parameters. The above approach has a high potential in brain connectivity estimation and estimating the unknown links between different brain regions.

In the linked multiway blind source separation (BSS), an approximate decomposition of a set of data tensors $\underline{\mathbf{X}}^{(s)} \in R^{I_1 \times I_2 \times \ldots \times I_N}$ $(s = 1, 2, \ldots, S)$ representing multiple subjects and/or multiple tasks (see Figure 8.9) is performed. The objective is to find a set of constrained factor matrices $\mathbf{U}^{(n,s)} = \left[\mathbf{U}_C^{(n)}, \mathbf{U}_I^{(n,s)}\right] \in \Re^{I_n \times J_n}$, $n = 1, 2, 3$) and core tensors $\underline{\mathbf{G}}(s) \in \Re^{I_1 \times J_2 \times J_3}$, which are partially linked or maximally correlated, i.e. they have the same common components or highly correlated components. In the case of EEG signals a 4-dimensional tensor $\underline{\mathbf{X}}$ can be defined in terms of its three-dimensional components as:

$$\underline{\mathbf{X}}^{(s)} = \underline{\mathbf{G}}^{(s)} \times_1 \mathbf{U}^{(1,s)} \ldots \times_N \mathbf{U}^{(N,s)} + \underline{\mathbf{E}}^{(s)} \tag{8.58}$$

where $\times_k$ denotes tensor multiplication along slab (dimension) k. Each factor $\mathbf{U}^{(n,s)}$ consists of two parts: one, $\mathbf{U}_C^{(n,s)}$, which are common bases for all subjects in the group and correspond to the same or maximally correlated components and another one, $\mathbf{U}_I^{(n,s)}$, which corresponds to stimuli/tasks independent individual characteristics. $\underline{\mathbf{X}}^{(s)}$ represents space (channel)–time–frequency data for the sth subject.

The factors are usually estimated using an alternating optimization method which minimizes the distance between the terms in the left and right of Eq. (8.54). In order to achieve unique solutions for all the factors some constraints are required. For the EEG signals such constraints can be the approximate source locations, sparsity in space or frequency domain, or a known structure of the core matrix, e.g. for independent factors the core matrix is diagonal.

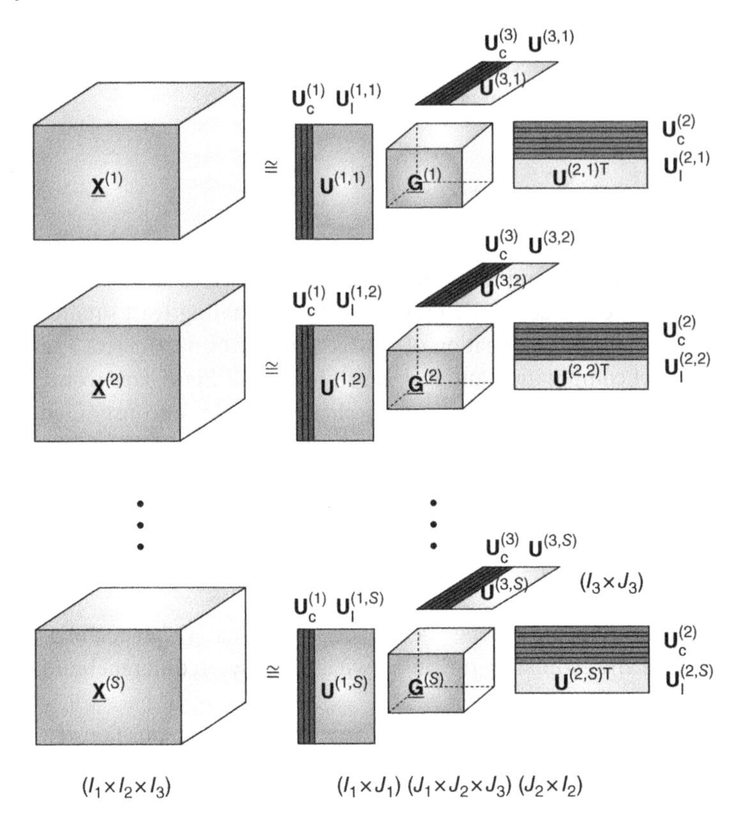

Figure 8.9 Conceptual model of tensors decomposition for linked multiway BSS. The objective is to find a set of constrained factor matrices $\mathbf{U}^{(n,s)} = \left[\mathbf{U}_C^{(n)}, \mathbf{U}_I^{(n,s)}\right] \in \Re^{I_n \times J_n}$, $n = 1, 2, 3)$ and core tensors $\underline{\mathbf{G}}(s) \in \Re^{J_1 \times J_2 \times J_3}$, which are partially linked or maximally correlated, i.e. they have the same common components or highly correlated components [77].

8.12 Summary

Estimation of the connectivity of the brain regions using EEG or MEG signals has its root in the works carried out approximately three decades ago. Although various methods such as MVAR and DTF, SEM, PDC, and dDTF have been popular for connectivity estimation, the new approaches such as by using the S-transform, diffusion adaptation, tensor factorization, and graph-based signal processing have opened new fronts in both modelling and estimation of the connectivity. Application of these recent methods to EEG signals is in its early stages and therefore, more research needs to be carried out to enable robust procedures for EEG or MEG connectivity measures. A major problem with most of these methods is that all the sources are considered cortical and thus, no attempt has been made to check the connectivity for deep sources. This requires localization of multiple deep sources which is another challenging problem. In a work by Thatcher et al. [78] low-resolution electromagnetic tomography algorithm (LORETA) source localization algorithm has been applied to determine the existing current sources within a large number of ROIs in the brain. Temporal and spatial correlations are later computed as the measure of connectivity. It also makes sense to restore the signals from noise and irrelevant information before estimation of connectivity and better deal with non-cortical sources. Certainly, the correlated sources cannot be separated using conventional BSS approaches. Moreover, application of connectivity estimation is not only useful for neurological diseases such as dementia, AD, and mild cognitive impairment (MCI), but also can assist in tracking the movement-related potentials within the brain. These concepts are discussed in future chapters of this book.

References

1. Sporns, O. (2011). *Networks of the Brain*. MIT Press.
2. Sporns, O. (2011). The human connectome: a complex network. *Annals of the New York Academy of Sciences* 1224: 109–125.
3. Steriade, M. (1995). Cellular substrates of brain rhythms. In: *Electroencephalography* (eds. Niedermeyer and L. da Silva), 27–62. Baltimore: Williams and Wilkins.
4. Nunez, P. (1981). *Electrical Fields of the Brain*. New York: Oxford University Press.
5. Nunez, P. (1994). *Neocortical Dynamics and Human EEG Rhythms*. New York: Oxford University Press.
6. Adey, W.R., Walter, D.O., and Hendrix, C.E. (1961). Computer techniques in correlation and spectral analyses of cerebral slow waves during discriminative behavior. *Experimental Neurology* 3: 501–524.
7. Sterman, M. and Kaiser, D. (2001). Comodulation: a new QEEG analysis metric for assessment of structural and functional disorders of the central nervous system. *Journal of Neurotherapy* 4 (3): 73–83.
8. Kaminski, M. and Blinowska, K.J. (1991). A new method of the description of the information flow in brain structure. *Biological Cybernetics* 65: 203–210.
9. Kaminski, M., Blinowska, K.J., and Szelenberger, W. (1997). Topographic analysis of coherence and propagation of EEG activity during sleep and wakefulness. *Electroencephalography and Clinical Neurophysiology* 102: 216–227.
10. Korzeniewska, A., Kasicki, S., Kaminski, M., and Blinowska, K.J. (1997). Information flow between hippocampus and related structures during various types of rat's behaviour. *Journal of Neuroscience Methods* 73: 49–60.
11. Jung, R.E. and Haier, R.J. (2007). The Parieto-frontal integration theory (P-FIT) of intelligence: converging neuroimaging evidence. *The Behavioral and Brain Sciences* 30 (2): 135–154.
12. Picard, N. and Strick, P.L. (1996). Motor areas of the medial wall: a review of their location and functional activation. *Cerebral Cortex* 6: 342–353.
13. Paus, T., Koski, L., Caramanos, Z., and Westbury, C. (1998). Regional differences in the effects of task difficulty and motor output on blood flow response in the human anterior cingulate cortex: a review of 107 PET activation studies. *Neuroreport* 9: R37–R47.
14. Gerloff, G., Richard, J., Hadley, J. et al. (1998). Functional coupling and regional activation of human cortical motor areas during simple, internally paced and externally paced finger movements. *Brain* 121 (8): 1513–1531.

15 Sharott, A., Magill, P.J., Bolam, J.P., and Brown, P. (2005). Directional analysis of coherent oscillatory field potentials in cerebral cortex and basal ganglia of the rat. *Journal of Physiology* 562 (3): 951–963.

16 Kaminski, M. and Blinowska, K. (1991). A new method of the description of information flow in the brain structures. *Biological Cybernetics* 65: 203–210.

17 Baccala, L.A. and Sameshima, K. (2001). Partial directed coherence: a new conception in neural structure determination. *Biological Cybernetics* 84: 463–474.

18 Kaminski, M., Ding, M., Truccolo, W., and Bressler, S. (2001). Evaluating causal relations in neural systems: Granger causality, directed transfer function, and statistical assessment of significance. *Biological Cybernetics* 85: 145–157.

19 Granger, C.W.J. (1969). Investigating causal relations in by econometric models and cross-spectral methods. *Econometrica* 37: 424–438.

20 Bernosconi, C. and König, P. (1999). On the directionality of cortical interactions studied by spectral analysis of electrophysiological recordings. *Biological Cybernetics* 81 (3): 199–210.

21 Nolte, G., Ziehe, A., Nikulin, V.V. et al. (2008). Robustly estimating the flow direction of information of information in complex physical systems. *Physical Review Letters* 100 (23): 1–4.

22 Rana, P., Lipor, J., Lee, H. et al. (2012). Seizure detection using the phase-slope index and multichannel ECoG. *IEEE Transactions on Biomedical Engineering* 4: 59.

23 Jing, H. and Takigawa, M. (2000). Observation of EEG coherence after repetitive transcranial magnetic stimulation. *Clinical Neurophysiology* 111: 1620–1631.

24 Kuś, R., Kaminski, M., and Blinowska, K. (2004). Determination of EEG activity propagation: pair-wise versus multichannel estimate. *IEEE Transactions on Biomedical Engineering* 51 (9): 1501–1510.

25 Ginter-Jr, J., Kaminski, M., Blinowska, K., and Durka, P. (2001). Phase and amplitude analysis in time-frequency-space; application to voluntary finger movement. *Journal of Neuroscience Methods* 110: 113–124.

26 Morf, M., Vieria, A., Lee, D., and Kailath, T. (1978). Recursive multichannel maximum entropy spectral estimation. *IEEE Transactions on Geoscience Electronics* 16: 85–94.

27 Akaike, H. (1974). A new look at statistical model order identification. *IEEE Transactions on Automatic Control* 19: 716–723.

28 Ding, M., Bressler, S.L., Yang, W., and Liang, H. (2000). Short-window spectral analysis of cortical event-related potentials by adaptive multivariate autoregressive modelling: data preprocessing, model validation, and variability assessment. *Biological Cybernetics* 83: 35–45.

29 Korzeniewska, A., Manszak, A., Kaminski, M. et al. Determination of information flow direction among brain structures by a modified directed transfer function method (dDTF). *Journal of Neuroscience Methods* 125: 195–207.

30 Blinowska, K.J. (2011). Review of the methods of determination of directed connectivity from multichannel data. *Medical & Biological Engineering & Computing* 49: 521–529.

31 Medkour, T., Walden, A.T., and Burgess, A. (2009). Graphical modelling for brain connectivity via partial coherence. *Journal of Neuroscience Methods* 180: 374–383.

32 Daneshi Kohan, M., Motie Nasrabadi, A., Shamsollahi, M.B., and Sharifi, A. (2020). EEG/PPG effective connectivity fusion for analyzing deception in interview. *Signal, Image and Video Processing* 14: 907–914.

33 Astolfi, L., Cincotti, F., Babiloni, C. et al. (May 2005). Estimation of the cortical connectivity by high resolution EEG and structural equation modeling: simulations and application to finger tapping data. *IEEE Transactions on Biomedical Engineering* 52 (5): 757–768.

34 David, O., Cosmelli, D., and Friston, K.J. (2004). Evaluation of different measures of functional connectivity using a neural mass model. *NeuroImage* 21: 659–673.
35 Bollen, K.A. (1989). *Structural Equations with Latent Variable*. New York: Wiley.
36 McIntosh, A.R. and Gonzalez-Lima, F. (1994). Structural equation modeling and its application to network analysis in functional brain imaging. *Human Brain Mapping* 2: 2–22.
37 Gevins, A. (1989). Dynamic functional topography of cognitive task. *Brain Topography* 2: 37–56.
38 Gevins, A., Brickett, P., Reutter, B., and Desmond, J. (1991). Seeing through the skull: advanced EEGs use MRIs to accurately measure cortical activity from the scalp. *Brain Topography* 4: 125–131.
39 Astolfi, L. and Babiloni, F. (2008). *Estimation of Cortical Connectivity in Humans: Advanced Signal Processing Techniques*. Morgan & Claypool.
40 Eftaxias, K. and Sanei, S. (2014). Discrimination of task-based EEG signals using diffusion adaptation and S-transform coherency. *Proceedings of the IEEE Workshop on Machine Learning for Signal Processing*. France: MLSP.
41 Eftaxias, K., Sanei, S., and Sayed, A. (2013). A new approach to evaluation and modeling of brain connectivity using diffusion adaptation, *Proceedings of the IEEE International Conference on Acoustics, Speech and Signal Processing*. Vancouver, Canada: ICASSP.
42 Monajemi, S., Eftaxias, K., Ong, S.-H., and Sanei, S. (2016). An informed multitask diffusion adaptation approach to study tremor in Parkinson's disease. *IEEE Journal of Selected Topics in Signal Processing*; Special Issue on Advanced Signal Processing in Brain Networks 10 (7): 1306–1314.
43 Stockwell, R., Mansinha, L., and Lowe, R. (1996). Localization of the complex spectrum: the s-transform. *IEEE Transactions on Signal Processing* 44 (4): 998–1001.
44 Astolfi, L., Toppi, J., De Vico Fallani, F. et al. (2010). Neuroelectrical hyperscanning measures simultaneous brain activity in humans. *Brain Topography* 23 (3): 243–256.
45 Dumas, G., Nadel, J., Soussignan, R. et al. (2010). Inter-brain synchronization during social interaction. *PLoS One* 5 (8): e12166.
46 Astolfi, L., Cincotti, F., Mattia, D. et al. (2009). Estimation of the cortical activity from simultaneous multi-subject recordings during the prisoner's dilemma. *Proceedings of the IEEE Engineering Medicine and Biology Society*: 1937–1939.
47 Lindenberger, U., Li, S.C., Gruber, W., and Müller, V. (2009). Brains swinging in concert: cortical phase synchronization while playing guitar. *BMC Neuroscience* 17: 10–22.
48 Babiloni, F., Cincotti, F., Mattia, D. et al. (2007). High resolution EEG hyperscanning during a card game. *Proceedings of the IEEE Engineering in Medicine and Biology Society*: 4957–4960.
49 Babiloni, F., Astolfi, L., Cincotti, F. et al. (2007). Cortical activity and connectivity of human brain during the prisoner's dilemma: an EEG hyperscanning study. *Proceedings of the IEEE Engineering in Medicine and Biology Society*: 4953–4956.
50 Babiloni, F., Cincotti, F., Mattia, D. et al. (2006). Hypermethods for EEG hyperscanning. *Proceedings of the IEEE Engineering in Medicine and Biology Society*: 3666–3669.
51 Huang, S., L, J., Liang Sun, J. et al. (2010). Learning brain connectivity of Alzheimer's disease by sparse inverse covariance estimation. *NeuroImage* 50: 935–949.
52 Supekar, K., Menon, V., Rubin, D. et al. (2008). Network analysis of intrinsic functional brain connectivity in Alzheimer's disease. *PLoS Computational Biology* 4 (6): e1000100. https://doi.org/10.1371/journal.pcbi.1000100.

53 Escodero, J., Sanei, S., Jarchi, D. et al. (Aug. 2011). Regional coherence evaluation in mild cognitive impairment and Alzheimer's disease based on adaptively extracted magnetoencephalogram rhythms. *Physiological Measurement* 32 (8): 1163–1180.

54 Cheong, B.L.P., Nowak, R., Lee, H.C. et al. (2012). Cross validation for selection of cortical interaction models from scalp EEG or MEG. *IEEE Transactions on Biomedical Engineering* 59 (2): 504–514.

55 Nalatore, H., Ding, M., and Rangarajan, G. (Apr. 2009). Denoising neural data with state-space smoothing: method and application. *Journal of Neuroscience Methods* 179: 131–141.

56 Shumway, R. and Stoffer, D. (1982). An approach to time series smoothing and forecasting using the EM algorithm. *Journal of Time Series Analysis* 3 (4): 253–264.

57 Cheong, B.L., Riedner, B., Tononi, G., and van Veen, B. (2010). Estimation of cortical connectivity from EEG using sate-space models. *IEEE Transactions on Biomedical Engineering* 57 (9): 2122–2134.

58 Ancona, N., Marrinazo, D., and Stramaglia, S. (2004). Radial basis function approach to nonlinear Granger causality of time series. *Physical Review E* 70: 56221–56227.

59 Park, J. and Sandberg, I.W. (1991). Universal approximation using radial-basis function networks. *Neural Computation* 3: 246–257.

60 Diedrichsen, J., Wiestler, T., and Krakauer, W. (May, 2012). Two distinct ipsilateral cortical representations for individuated finger movements. *Cerebral Cortex* https://doi.org/10.1093/cercor/bhs120.

61 Chen, J., Zhao, X., and Sayed, A.H. (Nov. 2010). *"Bacterial Motility Via Diffusion Adaptation," Proc. 44th Asilomar Conference on Signals*. Pacific Grove, CA: Systems and Computers.

62 Tu, S.-Y. and Sayed, A.H. (2010). Tracking behavior of mobile adaptive networks. *Proceedings of the 44th Asilomar Conference on Signals, Systems and Computers*. Pacific Grove, CA.

63 Tu, S.-Y. and Sayed, A.H. (2011). Cooperative prey herding based on diffusion adaptation. *Proceedings of the IEEE ICASSP 2011*. Prague, Czech Republic.

64 Cattivelli, F.S. and Sayed, A.H. (2010). Diffusion LMS strategies for distributed estimation. *IEEE Transactions on Signal Processing* 58 (3): 1035–1048.

65 Lopes, C. and Sayed, A.H. (2008). Diffusion least-mean squares over adaptive networks: formulation and performance analysis. *IEEE Transactions on Signal Processing* 56 (7): 3122–3136.

66 Takahashi, N., Yamada, I., and Sayed, A.H. (2010). Diffusion least-mean squares with adaptive combiners: formulation and performance analysis. *IEEE Transactions on Signal Processing* 58 (9): 4795–4810.

67 Cattivelli, F. and Sayed, A. (2010). Diffusion strategies for distributed Kalman filtering and smoothing. *IEEE Transactions on Automatic Control* 55 (9): 2069–2084.

68 Hu, J., Xie, L., and Zheng, C. (2012). Diffusion Kalman filtering based on covariance intersection. *IEEE Transactions on Signal Processing* 60 (2): 891–902.

69 Julier, S. and Uhlmann, J. (2001). General decentralised data fusion with covariance intersection (CI). In: *Handbook of Multiuser Data Fusion*, Chapter 12 (eds. D. Hall and J. Llinas), 25. CRC Press.

70 Monajemi, S., Jarchi, D., Ong, S.H., and Sanei, S. (2017). Cooperative particle filtering for detection and tracking of ERP subcomponents from multichannel EEG. *Journal of Entropy*, Special Issue on Entropy and Electroencephalography 19 (5): 199.

71 Jarchi, D., Sanei, S., and Lorist, M.M. (2011). Coupled particle filtering: a new approach for P300-based analysis of mental fatigue. *Biomedical Signal Processing and control* 6 (2): 175–185.

72 Biggs, N. (1993). *Algebraic Graph Theory*. Cambridge University Press.

73 Sandryhaila, A. and Moura, J.M.F. (2014). Big data analysis with signal processing on graphs: Representation and processing of massive data sets with irregular structure. *IEEE Signal Processing Magazine* 31 (5): 80–90.

74 Moscu, M., Borsoi, R., and Richard, C. (2020). Online graph topology inference with kernels for brain connectivity estimation. *Proceedings of IEEE International Conference on Acoustics, Speech and Signal Processing*, ICASSP, 2020. Barcelona, Spain.

75 Gao, S., Denoyer, L., and Gallinari, P. (2011). Link pattern prediction with tensor decomposition in multi-relational networks. *2011 IEEE Symposium on Computational Intelligence and Data Mining (CIDM)*. Paris, France.

76 Acar, E., Dunlavy, D.M., and Kolda, T.G. (2009). Link prediction on evolving data using matrix and tensor factorization. *ICDM Workshops*, 262–269.

77 Sanei, S., Ferdowsi, S., Nazarpour, K., and Cichocki, A. (2013). Advances in EEG Signal Processing. *IEEE Signal Processing Magazine*, Wiley Online Library.

78 Thatcher, R.W., Biver, C.J., and North, D. (2007). Spatial-temporal current source correlations and cortical connectivity. *Clinical EEG and Neuroscience* 38 (1): 35–48.

9

Event-Related Brain Responses

9.1 Introduction

The brain responses to sensory stimuli can be noninvasively recorded and clearly seen using averaging techniques first employed by Dawson in 1947 [1]. Peak amplitudes and latencies (in ms) are usually measured to quantify the evoked responses. These data provide quantitative extensions to the neurological examination. The clinical utility of evoked potentials (EPs) is based on their ability to: (i) demonstrate abnormal sensory system conduction, when the history and/or neurological examination is equivocal, (ii) reveal subclinical involvement of a sensory system ('silent' lesions), particularly when demyelination is suggested by symptoms and/or signs in another area of the central nervous system (CNS), (iii) help define the anatomic distribution and give some insight into the pathophysiology of a disease process, and (d) monitor the changes in a patient's neurological status [2]. Longer latency responses that are related to higher 'cognitive' functions such as event-related potentials (ERPs) which are a more general type of EPs. ERP is a measure of brain response that is the direct result of a specific sensory, cognitive, or motor event, or any stereotyped electrophysiological response to a stimulus. ERPs were first explained in 1964 [3, 4] and have remained as a useful diagnostic indicator in many applications in psychiatry and neurology as well as for brain–computer interfacing (BCI).

9.2 ERP Generation and Types

The brain generates different types of signals: (i) normal brain rhythms which constantly represent the state of the brain, (ii) ERPs, which are the brain responses to auditory, visual, or tactile stimulation, and (iii) movement-related potentials (MRPs) which appear due to intentional physical or imagery movements.

An ERP consists of a sequence of labelled positive and negative amplitude components. These components reflect various sensory, cognitive (e.g. stimulus evaluation) and motor processes that are classified on the basis of their scalp distribution and response to experimental stimulus. Therefore, ERPs (which can be detected from EEG or magnetoencephalogram (MEG) signals) directly measure the electrical response of the cortex to sensory, affective, or cognitive events. They are voltage fluctuations in the EEG induced within

EEG Signal Processing and Machine Learning, Second Edition. Saeid Sanei and Jonathon A. Chambers.

the brain, as a sum of a large number of action potentials (APs) that are time locked to sensory, motor, or cognitive events. They are typically generated in response to peripheral or external stimulations, and appear as somatosensory, visual, and auditory brain potentials, or as slowly evolving brain activity observed before voluntary movements or during anticipation of conditional stimulation.

Compared with the background EEG activity ERPs are quite small (1–30 μV). Therefore, traditionally they often need the use of a signal-averaging procedure for their elucidation. In addition, although evaluation of the ERP peaks may not be considered as a reliable diagnostic procedure, the application of ERP in psychiatry has been very common and widely followed.

An ERP waveform can be quantitatively characterized across three main dimensions, amplitude, latency, and scalp distribution [5]. In addition, an ERP signal may also be analyzed with respect to the relative latencies between its subcomponents. The amplitude provides an index of the extent of neural activity (and how it responds functionally to experimental variables), the latency (i.e. the time point at which peak amplitude occurs) reveals the timing of this activation, and the scalp distribution provides the pattern of the voltage gradient of a component over the scalp at any time instant.

ERP signals are either positive, represented by letter P, such as P300, or negative, represented by letter N, such as N100 and N400. The digits indicate the time in terms of milliseconds after the stimuli (audio, visual, or somatosensory). The amplitude and latency of the components occurring within 100 ms after stimulus onset are labelled oxogenous, and are influenced by physical attributes of stimuli such as intensity, modality, and presentation rate. Conversely, endogenous components such as P300 are nonobligatory responses to stimuli, and vary in amplitude, latency, and scalp distribution with strategies, expectancies, and other mental activities triggered by the event eliciting the ERP. These components are not influenced by the physical attributes of the stimuli.

Analysis of ERPs, such as visual evoked potentials (VEP), within the EEG signals is important in the clinical diagnosis of many psychiatric diseases such as dementia. Alzheimer's disease (AD), the most common cause of dementia, which is a degenerative disease of the cerebral cortex and subcortical structures. The relative severity of the pathological changes in the associated cortex accounts for the clinical finding of diminished visual interpretation skills with normal visual acuity. Impairment of visuo-cognitive skills often happens with this disease. This means that the patient may have difficulties with many complex visual tasks, such as tracing a target figure embedded in a more complex figure or identifying single letters that are presented briefly and followed by a pattern making stimulus. A specific indicator of dementia is the information obtained by using VEP. Several studies have confirmed that flash VEPs are abnormal while pattern VEPs are normal in patients with AD.

The most consistent abnormality in the flash VEP is the increase in the latency of the P100 component and an increase in its amplitude [6, 7]. Other ERP components such as N130, P165, N220, also have longer latencies in patients with AD [8, 9]. Figure 9.1 illustrates the shape of two normal and two abnormal P100 components. It is seen that the peak of the normal P100s (normally shown in reversed polarity) has a latency of approximately 106 ms whereas, the abnormal P100 peak has a latency of approximately 135 ms and has lower amplitude.

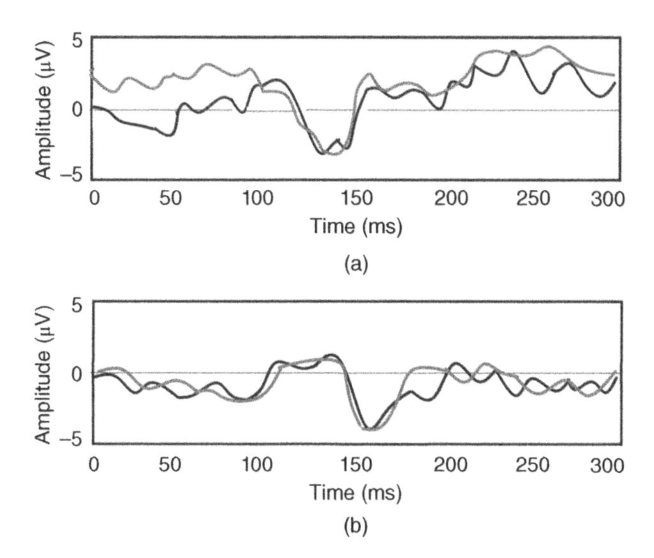

Figure 9.1 Four P100 components: (a) two normal P100 and (b) two abnormal P100 components. In (a) the P100 peak latency is at approximately 106 ms, whereas in (b) the P100 peak latency is at approximately 135 ms (waves are shown in reverse polarity).

Although determination of the locations of ERP sources within the brain is a difficult task the scalp distribution of an ERP component can often provide very useful and complementary information to that derived from amplitude and latency. Generally, two types of topographic maps can be generated: raw voltage (or surface potentials) and current source density (CSD), both derived from the electrode potentials.

The scalp-recorded ERP voltage activity reflects the summation of both cortical and subcortical neural activity within each time window. Conversely, CSD maps reflect primary cortical surface activity [10, 11]. CSD is obtained by spatially filtering the subcortical areas as well as cortical areas distal to the recording electrodes to remove the volume-conducted activity. CSD maps are useful for forming hypotheses about neural sources within the superficial cortex [12].

The functional MRI (fMRI) technique has become another alternative to investigate brain ERPs since it can detect the haemodynamics of the brain. However, there are at least three shortcomings with this brain imaging modality: firstly, the temporal resolution is low, secondly, the activated areas based on hemodynamic techniques do not necessarily correspond to the neural activity identified by ERP measures, and thirdly, the fMRI is not sensitive to the type of stimulus (e.g. target, standard, or novel). It is considered that the state of the subject changes due to differences in the density of different stimulus types across blocks of trials. By target P300, we refer to the P300 component elicited by events about which the subject has been instructed and to which the subject is required to generate some kinds of response. A novel stimulus indicates a sole or irregular stimulus.

The ERP parameters such as amplitude and latency are the indicators of the function of the brain neurochemical systems and can potentially be used as predictors of the response of an individual to psychopharmacotherapy [13]. ERPs are also related to the circumscribed

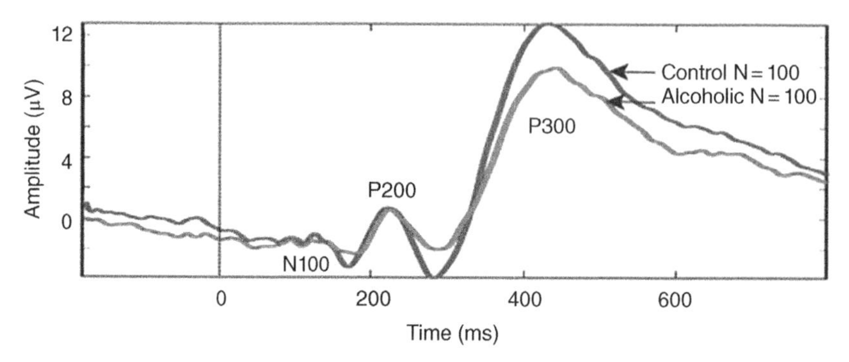

Figure 9.2 The average ERP signals for normal and alcoholic subjects. The curves show reduced P300 amplitude in alcoholics [14].

cognitive process. For example, there are interesting correlations between late evoked positivities and memory, N400 and semantic processes, or the latencies of ERPs and the timing of cognitive processes. Therefore, the ERP parameters can be used as indicators of cognitive processes and dysfunctions not accessible to behavioural testing. As an example, Figure 9.2 shows how the average (over N = 100 subjects) P300 differs for normal and alcoholic subjects.

The fine-grained temporal resolution of ERPs has been traditionally limited. In addition, overlapping components within ERPs, which represent specific stages of information processing, are difficult to distinguish [15, 16]. An example is the composite P300 wave, a positive ERP component, which occurs with a latency of about 300 ms after novel stimuli, or task relevant stimuli, which require an effortful response on the part of the individual under test [15–19].

The elicited ERPs are comprised of two main components: the mismatch negativity (MMN) and the novelty P300. By novelty P300, we refer to the P300 component elicited by events about which the subject has not been instructed prior to the experiment. The auditory MMN was discovered in 1978 by Näätänen, Gaillard, and Mäntysalo at the Institute for Perception, TNO in The Netherlands [20]. The MMN is the earliest ERP activity (which occurs within the first 10 ms after the stimulus) indicating that the brain has detected a change in a background of brain homogeneous events. The MMN is thought to be generated in and around the primary auditory cortex [21]. The amplitude of the MMN is directly proportional, and its latency inversely related, to the degree of difference between standard and deviant stimuli. It is most clearly seen by subtraction of the ERP elicited by the standard stimulus from that elicited by the deviant stimulus during a passive odd-ball paradigm (OP) when both of those stimuli are unattended or ignored. Therefore, it is relatively automatic.

The applications of ERP abnormality analysis in clinical research are not limited to those above and have been shown in many other neurological conditions such as attention deficit hyperactivity disorder (ADHD) [22, 23], dementia [24], Parkinson's disease (PD) [25], multiple sclerosis [26], head injuries [27], stroke [28], and obsessive–compulsive disorder [29].

9.2.1 P300 and its Subcomponents

The P300 wave represents cognitive functions involved in orientation of attention, contextual updating, response modulation, and response resolution [15, 17] and consists mainly of two overlapping subcomponents P3a and P3b [16, 19, 30]. P3a reflects an automatic orientation of attention to novel or salient stimuli independent of task relevance. Prefrontal, frontal, and anterior temporal brain regions play the main role in generating P3a giving it a frontocentral distribution [19]. In contrast, P3b has a greater centro-parietal distribution due to its reliance on posterior temporal, parietal and posterior cingulate cortex mechanisms [15, 16]. P3a is also characterized by a shorter latency and more rapid habituation than P3b [30].

A neural event is the frontal aspect of the novelty P300, i.e. P3a. For example, if the event is sufficiently deviant, the MMN is followed by the P3a. The eliciting events in this case are highly deviant environmental sounds such as a dog barking. Non-identifiable sounds elicit larger P3a than identifiable sounds. Accordingly, a bigger MMN results in a dominant P3a. Figure 9.3 shows typical P3a and P3b subcomponents.

P3b is also elicited by infrequent events but unlike P3a, it is task relevant, or involves a decision to evoke this component.

The ERPs are the responses to different stimuli, i.e. novel or salient. It is important to distinguish between the ERPs when they are the response to novel or salient (i.e. what has already been experienced) stimuli, or when the degree of novelty changes.

The orienting response engendered by deviant or unexpected events consists of a characteristic ERP pattern, which is comprised sequentially of the MMN and the novelty P300 or P3a. Each of P3a and P3b has different cognitive or possibly neurological base.

The orienting response [32, 33] is an involuntary shift of attention that appears to be a fundamental biological mechanism for survival. It is a rapid response to a new, unexpected, or unpredictable stimulus, which essentially functions as a what-is-it detector [34]. The plasticity of the orienting response has been demonstrated by showing that stimuli, which initially evoked the response no longer did so with repeated presentation [33]. Habituation

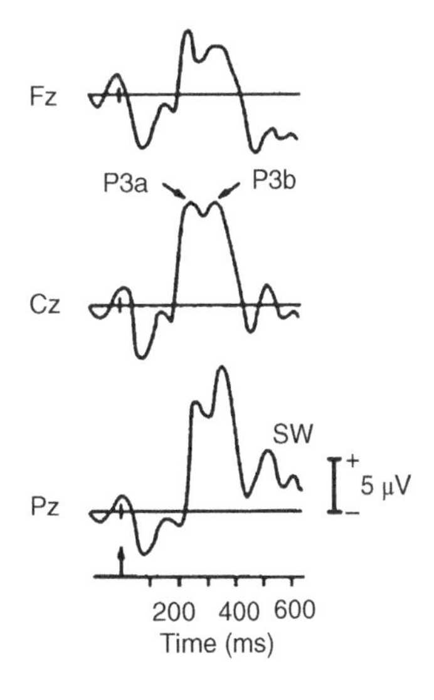

Figure 9.3 Typical P3a and P3b subcomponents of a P300 ERP signal viewed at Fz, Cz, and Pz electrodes [31].

of the response is proposed to indicate that some type of memory for these prior events has been formed, which modifies the response to the repeated incidences.

Abnormalities in P300 are found in several psychiatric and neurological conditions [18]. However, the impact of the diseases on P3a and P3b may be different. Both audio and visual P300 (i.e. a P300 signal produced earlier due to an audio or a visual stimulation) are used. Audio and visual P300 appear to be differently affected by illnesses and respond differently to their treatment. This suggests differences in the underlying structures and neurotransmitter systems [16]. P300 has significant diagnostic and prognostic potential especially when combined with other clinical symptoms and evidences.

In many applications such as human-computer interaction (HCI), muscular fatigue, visual fatigue, and mental fatigue are induced as a result of physical and mental activities. In order for the ERP signals and their subcomponents to be reliably used for clinical diagnosis, assessment of mental activities, fatigue during physical and mental activities, and for human-computer interfacing, very effective and reliable methods for their detection and parameter estimation have to be developed. In the following section a number of established methods for detection of ERP signals, especially P300 and its subcomponents, P3a and P3b, are described in sufficient detail.

9.3 Detection, Separation, and Classification of P300 Signals

Although in many applications, such as assessment of memory, a number of ERPs are involved, detection and tracking of P300 and its subcomponents' variability have been under more intensive research. Traditionally, EPs are synchronously averaged to enhance the evoked signal and suppress the background brain activity [35].

Step-wise discriminant analysis (SWDA) followed by peak picking and evaluation of the covariance, was first introduced by Farwell and Dounchin [36]. Later, the discrete wavelet transform (DWT) was also added to the SWDA to better localize the ERP components in both time and frequency [37].

Principal component analysis (PCA) has been employed to assess temporally overlapping EP components [38]. However, the resultant orthogonal representation does not necessarily coincide with the true component structure since the actual physiological components need not be orthogonal, i.e. the source signals may be correlated.

Independent component analysis (ICA) has been applied to ERP analysis by many researchers including Makeig et al. [39]. Infomax ICA [40] was used by Xu et al. [41] to detect the ERPs for the P300-based speller. In their approach those independent components (ICs) with relatively larger amplitudes in the latency range of P300 were kept, while the others were set to zero. Also, they exploited a priori knowledge about the spatial information of the ERPs and decided whether a component should be retained or wiped out. To manipulate the spatial information denote the ith row and jth column element in the inverse of the unmixing matrix, $\mathbf{W}^{-1}$, by w'_{ij}, therefore, $\mathbf{x}(n) = \mathbf{W}^{-1}\mathbf{u}(n)$. Then, denote the jth column of $\mathbf{W}^{-1}$ by $\mathbf{w}'_j$, which reflects the intensity distribution at each electrode for the jth IC u_j [39]. Then, for convenience the spatial pattern $\mathbf{W}^{-1}$ is transformed into an intensity order matrix $\mathbf{M} = \{m_{ij}\}$ with the same dimension. The value of the element m_{ij} in $\mathbf{M}$ is set to be the order number of the value w'_{ij} in the column vector $\mathbf{w}'_j$. For example $m_{ij} = 1$ if w'_{ij} has the largest value, $m_{ij} = 2$, if it has the second largest value, and so on. Based on the spatial distribution of the brain activities, an electrode set $\mathbf{Q} = \{q_k\}$ of interest is selected in which q_k is the electrode number and is equal to the row index of the multichannel EEG matrix $\mathbf{x}(n)$. For extraction of the P300 signals these electrodes are located around the vertex region (C_z, C_1, and C_2) since they are considered to have prominent P300. The spatial filtering of the ICs is simply performed as:

$$\widetilde{\mathbf{u}}_j(n) = \begin{cases} \mathbf{u}_j(n) & \text{if } \exists q_k \in \mathbf{Q} \text{ and } m_{q_k j} \leq T_r \\ 0 & \text{else} \end{cases} \tag{9.1}$$

where T_r is the threshold for the order numbers. T_r is introduced to retain the most prominent spatial information about the P300 signal. Therefore, $\widetilde{\mathbf{u}}_j(n)$ holds most of the source information about P300 while other irrelevant parts are set to zero. Finally, after the temporal and spatial manipulation of the ICs, the $\widetilde{\mathbf{u}}_j(n)$ s, $j = 1, 2, \ldots, M$, where M is the number of ICs, are back projected to the electrodes by using $\mathbf{W}^{-1}$ to obtain the scalp distribution of the P300 potential, i.e.:

$$\widetilde{\boldsymbol{x}}(n) = \mathbf{W}^{-1}\widetilde{\mathbf{u}}(n) \tag{9.2}$$

where $\widetilde{\mathbf{x}}(n)$ is the P300 enhanced EEG. The $\widetilde{\mathbf{x}}(t)$ features can then be measured for classification purposes [41].

9.3.1 Using ICA

Assuming ERP as an IC from other brain generated and artefact signals, ICA has been used by many researchers for detection and analysis of the ERP components from EEG, electromyogram (EMG) and fMRI (in the form of blood oxygen level dependence [BOLD]) data [42]. Four main assumptions underlie ICA decomposition of EEG or MEG time series: (i) signal conduction times are equal and summation of currents at the scalp sensors is linear, both are reasonable assumptions for currents carried to the scalp electrodes by volume conduction for EEG or for superposition of magnetic fields at SQUID sensors; (ii) spatial projections of components are fixed across time and conditions; (iii) source activations are temporally independent of one another across the input data; and (iv) statistical distributions of the component activation values are not Gaussian [42].

After application of ICA to the signals, the brain activities of interest accounted for by single or by multiple components can be obtained by projecting the selected ICA component(s) back onto the scalp, $\widetilde{\mathbf{X}} = \mathbf{W}^{-1}\mathbf{Y}^{(i)}$ where $\mathbf{W}$ is the estimated separating matrix using ICA and $\mathbf{Y}^{(i)}$ is the matrix of estimated/separated sources when all except source $\mathbf{y}_i$ have been set to zero. As the main conclusions here, ICA can separate stimulus-locked, response-locked, and non-event-related background EEG activities into separate components, allowing: (i) removal of pervasive artefacts of all types from single-trial EEG records, making possible analysis of highly contaminated EEG records from clinical populations; (ii) identification and segregation of stimulus- and response-locked event-

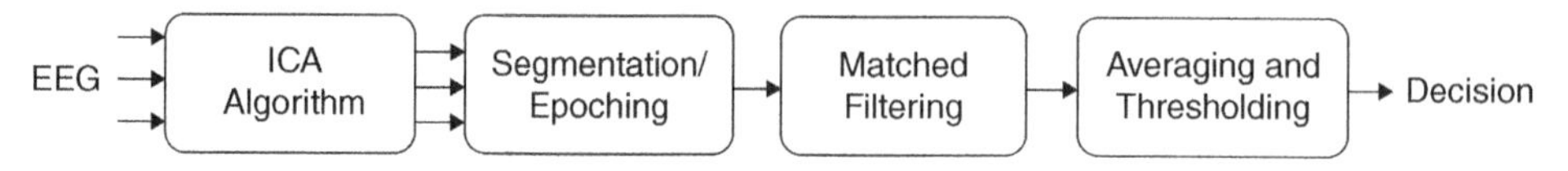

Figure 9.4 Block diagram of the ICA-based algorithm proposed in [43]. Three recorded data channels C_z, P_z, and F_z are the inputs to the ICA algorithm.

related activity in single-trail EEG epochs; (iii) separation of spatially overlapping EEG activities over the entire scalp and frequency band that may show a variety of distinct relationships to task events, rather than focusing on activity at single frequencies in single scalp channels or channel pairs; and (iv) investigation of the interaction between ERPs and ongoing EEG [42].

The work in [43] involved application of a matched filter together with averaging and applying a threshold for detecting the existence of the P300 signals. The block diagram in Figure 9.4 shows the method.

The IC corresponding to the P300 source is selected and segmented to form overlapping segments from 100 to 600 ms. Each segment is passed through a matched filter to give one feature that represents the maximum correlation between the segment and the average P300 template. However, a very obvious problem with this method is that the ICA system is very likely to be underdetermined (i.e. the number of sources is more than the number of sensors or observations) since only three mixtures are used. In this case the independent source signals are not separable.

Although ICA appears to be generally useful for EEG and fMRI analysis, it has some inherent limitations. First, ICA, in its general form, decomposes at most N sources from data collected at N scalp electrodes. Usually, the effective number of statistically independent source signals contributing to the scalp EEG is unknown and it is likely that the observed brain activity arises from more physically separable effective sources than the available number of EEG electrodes [42].

Second, ICA fails to separate correlated sources of ERP signals. The assumption of temporal independence used by ICA may not be satisfied when the training data set is small, or when separate topographically distinguishable phenomena nearly always co-occur in the data. In the latter case, simulations show that ICA may derive single components accounting for the co-occurring phenomena, along with additional components accounting for their brief periods of separate activation [42].

Third, ICA assumes that physical sources of artefacts and cerebral activity are spatially fixed over time. In general, this is not always true. Applying ICA to brain data shows the effectiveness of the independence assumption. ERP is generally the largest components (after eye-blink artefact,) within the scalp map emitted from one or two dipoles. This is unlikely to occur unless time courses of coherently synchronous neural activity in patches of cortical neurons generating the EEG are nearly independent of one another [44].

Fourth, application of ICA to fMRI does not necessarily provide the full information about the actual components of the fMRI sequences. However, by placing ICA in a regression framework, it is possible to combine some of the benefits of ICA with the hypothesis-testing approach of the general linear model (GLM) [45].

Current research on ICA algorithms has been focused on incorporating domain-specific constraints into the ICA framework. This would allow information maximization to be applied to the precise form and statistics of biomedical data [42].

9.3.2 Estimation of Single-Trial Brain Responses by Modelling the ERP Waveforms

Detection or estimation of EPs from only a single-trial EEG is favourable since (i) online processing of the signals can be performed and (ii) variability of ERP parameters across trials can be tracked. Unlike the averaging (multiple trial) [46] scheme, in this approach the shape of the ERPs is first approximated and then used to recover the actual signals. A few good examples, including mental fatigue, are investigated in Chapter 13 in which trial-to-trial variability of ERPs is of clinical importance.

A decomposition technique, which relies on the statistical nature of neural activity, is the one that efficiently separates the EEGs into their constituent components including the ERPs. A neural activity may be delayed when passing through a number of synaptic nodes each introducing a delay. Thus, the firing instants of many synchronized neurons may be assumed to be governed by Gaussian probability distributions. In a work by Lange et al. [35] the EP source waveform is assumed to consist of superposition of p components u_i delayed by τ_i:

$$s(n) = \sum_{i=1}^{p} k_i \cdot u_i(n-\tau_i) \tag{9.3}$$

The model equation for constructing the data from a number of delayed templates can be given in the z-domain as:

$$\mathbf{X}(z) = \sum_{i=1}^{p} \mathbf{B}_i(z) \cdot \mathbf{T}_i(z) + \mathbf{W}^{-1}(z)\mathbf{E}(z) \tag{9.4}$$

where $\mathbf{X}(z)$, $\mathbf{T}(z)$, and $\mathbf{E}(z)$ represent respectively, the z-domain observed (or measured) signals, the average EP, and a Gaussian white noise. Assuming that the background EEG is statistically stationary, $\mathbf{W}(z)$ is identified from the pre-stimulus data via auto-regressive (AR) modelling of the pre-stimulus interval and used for post-stimulus analysis [35]. The template and measured EEG are filtered through the identified $\mathbf{W}(z)$ to whiten the background EEG signal and thus, a closed-form least-squares (LS) solution of the model is formulated. Therefore, the only parameters to be identified are the matrices $\mathbf{B}_i(z)$. This avoids any iterative identification of the model parameters. This can be represented using a regression-type equation as:

$$\breve{x}(n) = \sum_{i=1}^{p} \sum_{j=-d}^{d} b_{i,j} \cdot \breve{T}_i(n-j) + e(n) \tag{9.5}$$

where d denotes the delay (latency), p is the number of templates, and $\breve{x}(n)$ $\breve{T}(n)$ are respectively, whitened versions of $x(n)$ and $T(n)$. Let $\mathbf{A}^T$ be the matrix of input templates and $\mathbf{b}^T$ the filter coefficients vector; then we construct:

$$\breve{\boldsymbol{x}} = [\breve{x}(d+1), \breve{x}(d+2), \ldots, \breve{x}(N-d)]^T, \tag{9.6}$$

$$A = \begin{bmatrix} \breve{T}_1(2d+1) & \breve{T}_1(2d+2) & \cdots & \breve{T}_1(N) \\ \breve{T}_1(2d) & \breve{T}_1(2d+1) & \cdots & \breve{T}_1(N-1) \\ \vdots & & & \\ \breve{T}_1(1) & \breve{T}_1(2) & \cdots & \breve{T}_1(N-2d) \\ \breve{T}_2(2d+1) & \breve{T}_2(2d+2) & & \breve{T}_2(N) \\ \vdots & & & \\ \breve{T}_2(1) & \breve{T}_2(2) & \cdots & \breve{T}_2(N-2d) \\ \vdots & & & \\ \breve{T}_p(2d+1) & \breve{T}_p(2d+2) & \cdots & \breve{T}_2(N) \\ \vdots & & & \\ \breve{T}_p(1) & \breve{T}_p(2) & \cdots & \breve{T}_p(N-2d) \end{bmatrix}^T, \tag{9.7}$$

and

$$b = \left[b_{1,-d}, b_{1,-d+1}, \ldots, b_{1,d}, b_{2,-d}, \ldots, b_{p,d}\right]^T. \tag{9.8}$$

The model can also be expressed in a vector form as:

$$\breve{\mathbf{x}}(n) = \mathbf{A}\mathbf{b}(n) + \boldsymbol{\varepsilon}(n) \tag{9.9}$$

where $\varepsilon(n)$ is the vector of prediction errors:

$$\boldsymbol{\varepsilon}(n) = [e(2d+1), e(2d+2), \ldots, e(N)]^T \tag{9.10}$$

To solve for the model parameters the sum squared error defined as:

$$\xi(\mathbf{b}) = \|\varepsilon(n)\|^2 \tag{9.11}$$

is minimized. Then, the optimal vector of parameters in the LS sense is obtained as:

$$\hat{\mathbf{b}} = \left(\mathbf{A}^T\mathbf{A}\right)^{-1}\mathbf{A}^T\breve{\mathbf{x}} \tag{9.12}$$

The model provides an easier way of analyzing the ERP components from single-trial EEG signals and facilitates tracking of amplitude and latency shifts of such components. The experimental results show that the main ERP components can be extracted with a high accuracy. The template signals may look like those in Figure 9.5.

A similar method has also been developed to detect and track the P300 subcomponents, P3a and P3b, of the ERP signals [47]. This is described in the following section.

9.3.3 ERP Source Tracking in Time

The major drawback of blind source separation (BSS) and dipole methods is that the number of sources needs to be known a priori to achieve good results. In a recent work [47] P300 signals are modelled as spike-shaped Gaussian signals. The latencies and variances of these signals are however subject to change. The spikes serve as reference signals onto which the

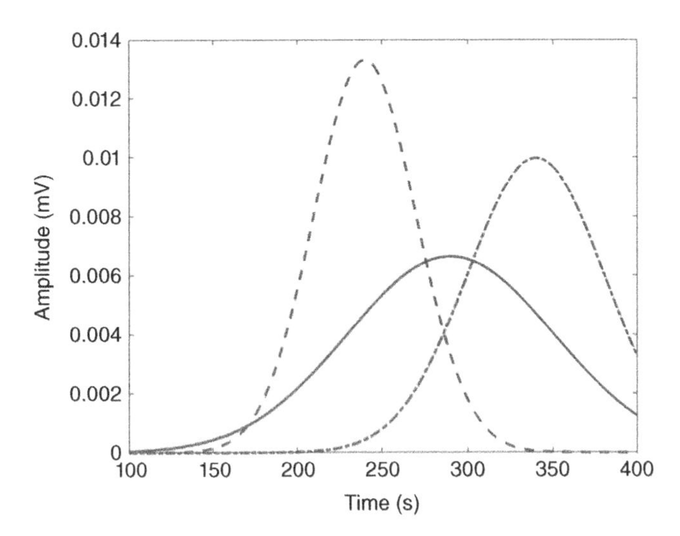

Figure 9.5 Synthetic ERP templates including a number of delayed Gaussian and exponential waveforms.

EEG data are projected. The existing spatiotemporal information is then used to find the closest representation of the reference in the data. The locations of all the extracted components within the brain are later computed using the LS method. Consider the $n_e \times T$ EEG signals $\mathbf{X}$, where n_e is the number of channels (electrodes) and T the number of time samples, then:

$$\mathbf{x}(n) = \mathbf{H}\mathbf{s}(n) = \sum_{i=1}^{m} \mathbf{h}_i s_i(n) \tag{9.13}$$

where $\mathbf{H}$ is the $n_e \times m$ forward mixing matrix and $s_i(n)$ are the source signals. Initially, the sources are all considered as the ERP components that are directly relevant and time locked to the stimulus and assumed to have transient spiky shape. m filters $\{\mathbf{w}_i\}$ are then needed (although the number of sources are not known beforehand) to satisfy:

$$s_i(n) = \mathbf{w}_i^T \mathbf{x}(n) \tag{9.14}$$

This can be achieved by the following minimization criterion:

$$\mathbf{w}_{opt} = arg \min_{\mathbf{w}_i} \left\| \mathbf{s}_i - \mathbf{w}_i^T \mathbf{X} \right\|_2^2 \tag{9.15}$$

where $\mathbf{X} = [\mathbf{x}_1, \mathbf{x}_2, \cdots, \mathbf{x}_{ne}]^T$, which however requires some knowledge about the shape of the sources. A Gaussian type spike defined as:

$$r(n) = \hat{s}_i(n) = e^{-\frac{(n-\tau_i)^2}{\sigma_i^2}} \tag{9.16}$$

where τ_i, $i = 1,2, \ldots T$, is the latency of the ith source and σ_i is its width, can be used as the model. The width is chosen as an approximation to the average width of the P3a and P3b subcomponents. Then, for each of the T filters ($T >> m$) we have:

$$\mathbf{w}_{lOpt}^T = arg \min_{\mathbf{w}_l} \left\| \mathbf{r}_l - \mathbf{w}_l^T \mathbf{X} \right\|_2^2, \quad \mathbf{y}_l = \mathbf{w}_l^T \mathbf{X} \tag{9.17}$$

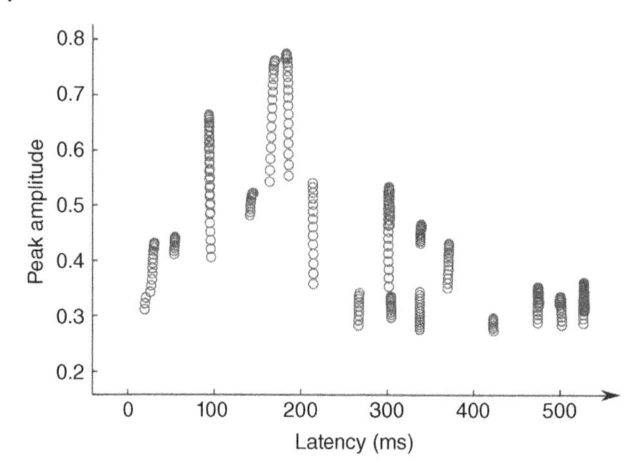

Figure 9.6 The results of the ERP detection algorithm [47]. The scatter plot shows the peaks at different latencies. The concentration of samples around the peaks make the ERP classification straightforward.

Therefore, each $\mathbf{y}_l$ has a similar latency to that of a source. Figure 9.6 shows the results of the ERP detection algorithm. It can be seen that all possible peaks are detected first. However, because the number of ERP signals is limited the outputs may be grouped into m clusters. To perform such classification we examine the following criteria for the latencies of the ith and $(i-1)$th estimated ERP components according to the following rule:

for $i = 1$ to T:

if $l(i) - l(i-1) < \beta$, where β is a small empirical threshold, then $\mathbf{y}_i$ and $\mathbf{y}_{i-1}$ are assumed to belong to the same cluster,

if $l(i) - l(i-1) > \beta$ then $\mathbf{y}_i$ and $\mathbf{y}_{i-1}$ belong to different clusters.

Here $l(i)$ denotes the latency of the ith component. The signals within each cluster are then averaged to obtain the related ERP signal, $\mathbf{y}_c$. $c = 1, 2,..., m$. Figure 9.7 presents the results after applying the above method to detect the subcomponents of a P300 signal from a single-trial EEG for a healthy individual (control) and a schizophrenic patient. From these results it can be seen that the mean latency of P3a for the patient is less than that of the control. The difference however is less for P3b. Conversely, the mean width of P3a is less for the control (healthy subject) than for the patient whereas the mean width of the P3b is more for the patient than for the control. This demonstrates the effectiveness of the technique in classification of the healthy and schizophrenic individuals.

9.3.4 Time–Frequency Domain Analysis

In places where some signal components are transient, i.e. a signal contains frequency components emerging and vanishing in time-limited intervals, localization of the active signal components within the time–frequency (TF) domain is very useful [48]. The traditional method proposed for such an analysis is application of the short time–frequency transform

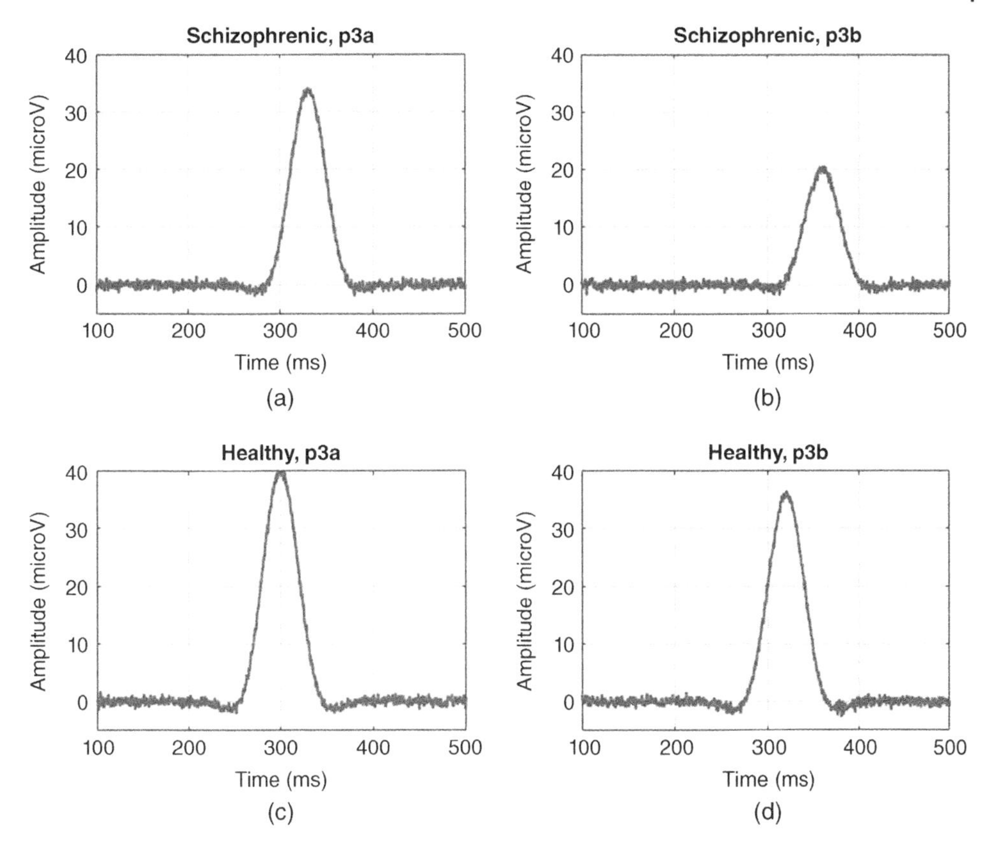

Figure 9.7 The average P3a and P3b for a schizophrenic patient (a) and (b) respectively, and for a healthy individual (c) and (d) respectively.

(STFT) [49]. The STFT enables the time localization of a certain sinusoidal frequency but with an inherent limitation due to Heisenberg's uncertainty principle which states that resolution in time and frequency cannot be both arbitrary small, because their product is lower bounded by $\Delta t \Delta \omega \geq 1/2$. Therefore, the wavelet transform (WT) has become very popular instead. An advantage of the WT over the STFT is that for the STFT the phase space is uniformly sampled, whereas in the WT the sampling in frequency is logarithmic which enables one to analyze higher frequencies in shorter windows and lower frequencies in longer windows in time [50].

As explained in Chapter 4 of this book a multi-resolution decomposition of signal $x(n)$ over I octaves is given as:

$$x(n) = \sum_{i=1}^{\infty} \sum_{k \in \overline{Z}} a_{i,k} g_i(n - 2^i k) + \sum_{k \in Z} b_{i,k} h_i(n - 2^i k) \tag{9.18}$$

where $\overline{Z}$ refers to integer values. The DWT computes the wavelet coefficients $a_{i,k}$ for $i = 1,\ldots, I$ and the scaling coefficients $b_{i,k}$ given by:

$$a_{i,k} = \text{DWT}\{x(n); 2^i, k2^i\} = \sum_{n=0}^{N} x(n) g_i^*(n - 2^i k) \tag{9.19}$$

and

$$b_{i,k} = \sum_{n=0}^{N} x(n) h_i^*\left(n - 2^i k\right) \tag{9.20}$$

where $(.)^*$ denotes conjugate. The functions $g(\bullet)$ and $h(\bullet)$ perform respectively as highpass and lowpass filters. Among the wavelets, the discrete B-Spline WTs have near-optimal TF localization [51]. Generally, they also have anti-symmetric properties, which make them more suitable for analysis of ERPs. The filters for an nth-order B-Spline wavelet multi-resolution decomposition, are computed as:

$$h(k) = \frac{1}{2}\left[b^{2n+1}\right]^{-1}(k)\uparrow_2 * b^{2n+1}(k) * p^n(k) \tag{9.21}$$

and

$$g(k+1) = \frac{1}{2}\left[b^{2n+1}\right]^{-1}(k)\uparrow_2 * (-1)^k p^n(k) \tag{9.22}$$

where $\uparrow_2$ indicates up-sampling by 2 and n is an integer. For the quadratic spline wavelet ($n = 2$) employed by Ademoglu et al. to analyze pattern reversal VEPs [50] substituting $n = 2$ in the above equations, the parameters can be derived mathematically as:

$$\left[\left(b^5\right)^{-1}\right](k) = Z^{-1}\left\{\frac{120}{z^2 + 26z + 66 + 26z^{-1} + z^{-2}}\right\} \tag{9.23}$$

$$p^2(k) = \frac{1}{2^2} Z^{-1}\left\{1 + 3z^{-1} + 3z^{-2} + z^{-3}\right\} \tag{9.24}$$

where Z^{-1} denotes inverse Z-transform. To find $[(b^5)^{-1}](k)$ the z-domain term may be factorized into:

$$\left[\left(b^5\right)^{-1}\right](k) = Z^{-1}\left[\frac{\alpha_1\alpha_2}{(1-\alpha_1 z^{-1})(1-\alpha_1 z)(1-\alpha_2 z^{-1})(1-\alpha_2 z)}\right] \tag{9.25}$$

where $\alpha_1 = -0.04309$ and $\alpha_2 = -0.43057$. Therefore:

$$\left[\left(b^5\right)^{-1}\right](k) = \frac{\alpha_1\alpha_2}{(1-\alpha_1^2)(1-\alpha_2^2)(1-\alpha_1\alpha_2)(\alpha_1-\alpha_2)} \cdot \left[\alpha_1\left(1-\alpha_2^2\right)\alpha_1^{|k|} - \alpha_2\left(1-\alpha_1^2\right)\alpha_2^{|k|}\right] \tag{9.26}$$

Conversely, $p(k)$ can also be computed easily by taking the inverse transform in Eq. (9.24). Therefore, the sample values of $h(n)$ and $g(n)$ can be obtained. Finally, to perform a WT only lowpass and highpass filtering followed by down-sampling are needed. In such a multi-resolution scheme the number of wavelet coefficients halves from one scale to the next. This requires and justifies longer time windows for lower frequencies and shorter time windows for higher frequencies.

In the above work the VEP waveform is recorded using a bipolar recording at electrode locations C_z and O_z of the conventional 10–20 EEG system. The sampling rate is 1 kHz. Therefore, each scale covers the following frequency bands: 250–500, 125–250, 62.5–125, 31.3–62.5, 15.5–31.5, 7.8–15.6 Hz (including N70, P100, and N130), and 0–7.8 Hz (residual scale). Although this method is not meant to discriminate between normal and pathological subjects, the ERPs can be highlighted and the differences between normal and abnormal cases are observed. By using the above spline wavelet analysis, it is observed that the main effect of a latency shift of the N70-P100-N130 complex is reflected in the sign and magnitude of the second, third, and fourth wavelet coefficients within the 7.8–15.6 Hz band. Many other versions of the WT approach have been introduced in the literature during the recent decade such as those in [52–54].

As an example, single-trial ERPs have been analyzed using wavelet networks (WNs) and the developed algorithm applied to the study of ADHD [52]. ADHD is a psychiatric and neurobehavioural disorder. It is characterized by significant occurrence of inattention or hyperactivity and impulsiveness. In this work it is assumed that the ERP and the background EEG are correlated, i.e. the parameters, which describe the signals before the stimulation, do not hold for the post-stimulation period [55]. In a WN topology they presented an ERP, $\hat{s}(t)$ as a linear combination of K modified wavelet functions $h(n)$:

$$\hat{s}(n) = \sum_{k=1}^{K} w_k h\left(\frac{n-b_k}{a_k}\right) \tag{9.27}$$

where b_k, a_k, and w_k, are respectively, the shift, scale, and weight parameters and $s(n)$ are called wavelet nodes. Figure 9.8 shows the topology of a WN for signal representation. In this figure $\psi_k = h((u - b_k)/a_k)$ in the case of a Morlet wavelet:

$$\hat{s}(n) = \sum_{k=1}^{K} w_{1,k} \cos\left(\omega_k\left(\frac{n-b_k}{a_k}\right)\right) + w_{2,k} \sin\left(\omega_k\left(\frac{n-b_k}{a_k}\right)\right) exp\left[-\frac{1}{2}\left(\frac{n-b_k}{a_k}\right)^2\right] \tag{9.28}$$

In this equation each of the frequency parameters ω_k and the corresponding scale parameter a_k define a node frequency $f_k = \omega_k/2\pi a_k$, and are optimized during the learning process.

In the above approach a WN is considered as a single hidden layer perceptron whose neuron activation function [56] is defined by wavelet functions. Therefore, the well-known back-propagation neural network algorithm [57] is used to simultaneously update the WN parameters (i.e. $w_{1,k}$, $w_{2,k}$, ω_k, a_k, and b_k of node k). The minimization is performed

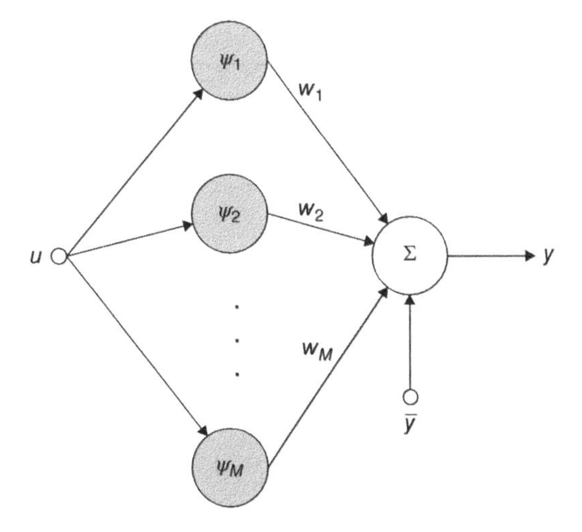

Figure 9.8 Construction of an ERP signal using a WN. The nodes in the hidden layer are represented by modified wavelet functions. The addition of the y value is to deal with functions whose means are non-zero.

on the error between the node outputs and the bandpass filtered version of the residual error obtained using auto-regressive moving average (ARMA) modelling. The bandpass filter and a tapering time window applied to ensure that the desired range is emphasized and the node sees its specific part of the TF plane [52]. It has been reported that for the study of ADHD this method enables detection of the N100 ERP components in places where the traditional time averaging approach fails due to the position variation of this component.

In a recent work there has been an attempt to create a wavelet-like filter from simulation of the neural activities which is then used in place of a specific wavelet to improve the performance of wavelet-based single-trial ERP detection. The main application of such a filter has been found to be in BCI [58].

9.3.5 Application of Kalman Filter

The main problem in linear time invariant filtering of ERP signals is that the signal and noise frequency components highly overlap. Another problem is that the ERP signals are transient and their frequency components may not fall into a certain range. Wiener filtering which provides optimum solution in the mean square (MS) sense for the majority of noise removal cases is not suitable for nonstationary signals such as ERPs.

With the assumption that the ERP signals are dynamically slowly varying processes the future realizations are predictable from the past realizations. These changes can be studied using a state-space model. Kalman filtering and generic observation models have been used to denoise the ERP signals [59]. Considering a single-channel EEG observation vector at stimulus time n, as $\mathbf{x}_n = [x_n(1), x_n(2), \ldots, x_n(N)]^T$, the objective is to estimate a vector of parameters such as $\boldsymbol{\theta}_n$ of length k for which the map from $\mathbf{x}_n$ to $\boldsymbol{\theta}_n$ is (preferably) a linear map. An optimum estimator, $\hat{\boldsymbol{\theta}}_{n,opt}$, which minimizes $E\left[\left\|\hat{\boldsymbol{\theta}}_n - \boldsymbol{\theta}_n\right\|^2\right]$, is $\hat{\boldsymbol{\theta}}_{n,opt} = E[\boldsymbol{\theta}_n \mid \mathbf{x}_n]$ and can be shown to be:

$$\hat{\boldsymbol{\theta}}_{n,opt} = \mu_{\theta_n} + \mathrm{C}_{\theta_n, x_n} \mathrm{C}_{x_n}^{-1} \left(\mathbf{x}_n - \mu_{x_n}\right) \tag{9.29}$$

where μ_{θ_n} and $\mu_{\mathbf{x}_n}$ are respectively, the means of $\boldsymbol{\theta}_n$ and $\mathbf{x}_t$, and C_{θ_n, x_n} is the cross-covariance matrix of the observations and the parameters to be estimated, and $C_{\mathbf{x}_n}$ is the covariance matrix of the column vector $\mathbf{x}_n$. Such an estimator is independent of the relationship between $\boldsymbol{\theta}_n$ and $\mathbf{x}_n$. The covariance matrix of the estimated error $\boldsymbol{\varepsilon}_\theta = \boldsymbol{\theta}_n - \hat{\boldsymbol{\theta}}_n$ can also be found as:

$$\mathrm{C}_\varepsilon = \mathrm{C}_{\theta n} - \mathrm{C}_{\theta_n, x_n} \mathrm{C}_{x_n}^{-1} \mathrm{C}_{x_n, \theta_n} \tag{9.30}$$

In order to evaluate C_{θ_n, x_n} some prior knowledge about the model is required. In this case the observations may be assumed to be of the form $\boldsymbol{x}_n = \boldsymbol{s}_n + \boldsymbol{v}_n$, where $\boldsymbol{s}_n$ and $\boldsymbol{v}_n$ are considered respectively as the response to the stimulus and the background EEG (irrelevant of the stimulus and ERP). Also, the ERP is modelled as:

$$\mathbf{x}_n = \mathbf{H}_n \boldsymbol{\theta}_n + \mathbf{v}_n \tag{9.31}$$

where $\mathbf{H}_n$ is a deterministic $N \times k$ observation model matrix. The estimated ERP $\hat{\mathbf{s}}_n$ can then be obtained as:

$$\hat{\mathbf{s}}_n = \mathbf{H}_n\hat{\boldsymbol{\theta}}_n \quad (9.32)$$

where the estimated parameters, $\hat{\theta}_n$, of the observation model $\mathbf{H}_n$, using a linear MS estimator and with the assumption of $\boldsymbol{\theta}_n$ and $\mathbf{v}_n$ being uncorrelated, is achieved as [60]:

$$\hat{\boldsymbol{\theta}}_n = \left(\mathbf{H}_n^T \mathrm{C}_{\boldsymbol{v}_n}^{-1}\mathbf{H}_n + \mathrm{C}_{\boldsymbol{\theta}_n}^{-1}\right)^{-1}\left(\mathbf{H}_n^T \mathrm{C}_{\boldsymbol{v}_n}^{-1}\mathbf{x}_n + \mathrm{C}_{\boldsymbol{\theta}_n}^{-1}\mu_{\boldsymbol{\theta}_n}\right) \quad (9.33)$$

This estimator is optimum if the joint distribution of $\boldsymbol{\theta}_n$ and $\mathbf{x}_n$ is Gaussian and the parameters and noise are uncorrelated.

However, in order to take into account the dynamics of the ERPs from trial to trial the evolution of $\boldsymbol{\theta}_n$ has to be taken into account. Such evolution may be denoted as:

$$\boldsymbol{\theta}_{n+1} = \mathbf{F}_n\boldsymbol{\theta}_n + \mathbf{G}_n\mathbf{w}_n \quad (9.34)$$

and some initial distribution for $\boldsymbol{\theta}_0$ assumed. Although the states are not observed directly, the parameters are related to the observations through Eq. (9.33). In this model it is assumed that $\mathbf{F}_n$, $\mathbf{H}_n$, and $\mathbf{G}_n$ are known matrices, $(\boldsymbol{\theta}_0, \boldsymbol{w}_n, \boldsymbol{v}_n)$ is a sequence of mutually uncorrelated random vectors with finite variance, $\mathrm{E}[\mathbf{w}_n] = \mathrm{E}[\mathbf{v}_n] = 0\ \forall n$, and the covariance matrices C_{w_n}, C_{v_n}, and C_{w_n,v_n} are known.

With the above assumptions the Kalman filtering algorithm can be employed to estimate $\boldsymbol{\theta}_n$ as [61]:

$$\hat{\theta}_n = \left(\boldsymbol{H}_n^T C_{\boldsymbol{v}_n}^{-1}\boldsymbol{H}_n + C_{\boldsymbol{\theta}_{n|n-1}}^{-1}\right)^{-1}\left(\boldsymbol{H}_n^T C_{\boldsymbol{v}_n}^{-1}\boldsymbol{x}_n + C_{\boldsymbol{\theta}_{n|n-1}}^{-1}\hat{\boldsymbol{\theta}}_{n|n-1}\right) \quad (9.35)$$

where $\hat{\boldsymbol{\theta}}_{n|n-1} = E[\boldsymbol{\theta}_n \mid \boldsymbol{x}_{n-1}, \dots, \boldsymbol{x}_1]$ is the prediction of $\boldsymbol{\theta}_n$ subject to $\hat{\boldsymbol{\theta}}_{n-1} = E[\boldsymbol{\theta}_{n-1} \mid \boldsymbol{x}_{n-1}, \dots, \boldsymbol{x}_1]$, which is the optimum MS estimate at $n-1$. Such an estimator is the best sequential estimator if the Gaussian assumption is valid and is the best linear estimator disregarding the distribution. The overall algorithm based on Kalman filtering for tracking of the ERP signal may be summarized as follows:

$$\mathrm{C}_{\hat{\theta}_0} = \mathrm{C}_{\theta_0} \quad (9.36)$$

$$\hat{\boldsymbol{\theta}}_0 = E[\boldsymbol{\theta}_0] \quad (9.37)$$

$$\hat{\boldsymbol{\theta}}_{n|n-1} = \mathbf{F}_{n-1}\hat{\boldsymbol{\theta}}_{n-1} \quad (9.38)$$

$$\mathrm{C}_{\hat{\theta}_{n|n-1}} = \mathbf{F}_{n-1}\mathrm{C}_{\hat{\theta}_{n-1}}\mathbf{F}_{n-1}^T + \mathbf{G}_{n-1}\mathrm{C}_{\mathbf{w}_{n-1}}\mathbf{G}_{n-1}^T \quad (9.39)$$

$$\mathbf{K}_t = \mathbf{C}_{\hat{\theta}_{n|n-1}}\mathbf{H}_n^T\left(\mathbf{H}_n\mathrm{C}_{\hat{\theta}_{n|n-1}}\mathbf{H}_n^T + \mathrm{C}_{\boldsymbol{v}_n}\right)^{-1} \quad (9.40)$$

$$\mathrm{C}_{\hat{\theta}_n} = (\mathbf{I} - \mathbf{K}_n\mathbf{H}_n)\mathrm{C}_{\hat{\theta}n|n-1} \quad (9.41)$$

$$\hat{\boldsymbol{\theta}}_n = \boldsymbol{\theta}_{n|n-1} + \mathbf{K}_n\left(\mathbf{x}_n - \mathbf{H}_n\hat{\boldsymbol{\theta}}_{n|n-1}\right) \quad (9.42)$$

In the work by Georgiadis et al. [59], however, a simpler observation model has been proposed. In this model the state-space equations have the forms:

$$\boldsymbol{\theta}_{n+1} = \boldsymbol{\theta}_n + \mathbf{w}_n \quad (9.43)$$

$$\mathbf{x}_n = \mathbf{H}_n\boldsymbol{\theta}_n + \mathbf{v}_n \quad (9.44)$$

Table 9.1 $\mathbf{K}_n$ and $\mathbf{P}_n$ for different recursive algorithms.

RLS	$\mathbf{K}_n = \mathbf{P}_{n-1}\mathbf{H}_n^T(\mathbf{H}_n\mathbf{P}_{n-1}\mathbf{H}_n^T + \lambda_n)^{-1}$ $\mathbf{P}_n = \lambda_n^{-1}(I - \mathbf{K}_n\mathbf{H}_n)\mathbf{P}_{n-1}$
LMS	$\mathbf{K}_n = \mu\mathbf{H}_n^T$ $\mathbf{P}_n = \mu(\mathbf{I} - \mu\mathbf{H}_n^T\mathbf{H}_n)^{-1}$
NLMS	$\mathbf{K}_n = \mu\mathbf{H}_n^T(\mu\mathbf{H}_n\mathbf{H}_n^T + 1)^{-1}$ $P_n = \mu I$
Kalman Filter	$\mathbf{K}_n = \mathbf{P}_{n-1}\mathbf{H}_n^T(\mathbf{H}_n\mathbf{P}_{n-1}\mathbf{H}_n^T + \mathrm{C}_{v_n})^{-1}$ $\mathbf{P}_n = (\mathbf{I} - \mathbf{K}_n\mathbf{H}_n)\mathbf{P}_{n-1} + \mathrm{C}_{\mathbf{w}_n}$

in which the state vector models a random walk process [62]. Now, assuming that the conditional covariance matrix of the parameter estimation error is $\mathbf{P}_t = \mathrm{C}_{\theta_t} + \mathrm{C}_{\mathbf{w}_t}$ the Kalman filter equations will be simplified to:

$$\mathbf{K}_n = \mathbf{P}_{n-1}\mathbf{H}_n^T(\mathbf{H}_n\mathbf{P}_{n-1}\mathbf{H}_n^T + C_{v_n})^{-1} \tag{9.45}$$

$$\mathbf{P}_n = (\mathbf{I} - \mathbf{K}_n\mathbf{H}_n)\mathbf{P}_{n-1} + C_{\mathbf{w}_n} \tag{9.46}$$

$$\hat{\theta}_n = \hat{\theta}_{n-1} + \mathbf{K}_n\left(x_n - \mathbf{H}_n\hat{\theta}_{n-1}\right) \tag{9.47}$$

where $\mathbf{P}_n$ and $\mathbf{K}_n$ are called respectively, the recursive covariance matrix estimate and Kalman-gain matrix. At this stage it is interesting to compare these parameters for the four well-known recursive estimation algorithms namely recursive least squares (RLS), least-mean-square (LMS), normalized least-mean-square (NLMS), and the above Kalman estimators for the above simple model. This is depicted in Table 9.1 [59].

In practise $\mathrm{C}_{\mathbf{w}_n}$ is considered as a diagonal matrix such as $\mathrm{C}_{\mathbf{w}_n} = 0.1\mathbf{I}$. Also it is logical to consider $\mathrm{C}_{v_n} = \mathbf{I}$, and, by assuming the background EEG to have Gaussian distribution it is possible to use the Kalman filter to dynamically estimate the ERP signals. As an example, it is possible to consider $\mathbf{H}_n = \mathbf{I}$, so that the ERPs are recursively estimated as:

$$\hat{s}_n = \hat{\theta}_n = \hat{\theta}_{n-1} + \mathbf{K}_n\left(x_n - \hat{\theta}_{n-1}\right) = \mathbf{K}_n x_n + (\mathbf{I} - \mathbf{K}_n)\hat{s}_{n-1} \tag{9.48}$$

Such assumptions make the application of the algorithms in Table 9.1 for ERP tracking and parameter estimation easier.

In a more advanced algorithm $\mathbf{H}_n$ may be adapted to the spatial topology of the electrodes. This means that the column vectors of $\mathbf{H}_n$ are weighted based on the expected locations of the ERP generators within the brain. Figure 9.9 compares the estimated ERPs for different stimuli for the LMS and recursive MS estimator (Kalman filter). From this figure it is clear that the estimates using the Kalman filtering appear to be more robust and show very similar patterns for different stimulation instants.

9.3.6 Particle Filtering and its Application to ERP Tracking

Particle filtering (PF) is an attractive but complicated state-space approach widely used for object tracking from video sequences, target tracking in airborne radar or generally multidimensional data. PF is more suitable in places where the linearity condition required for KF is not satisfied.

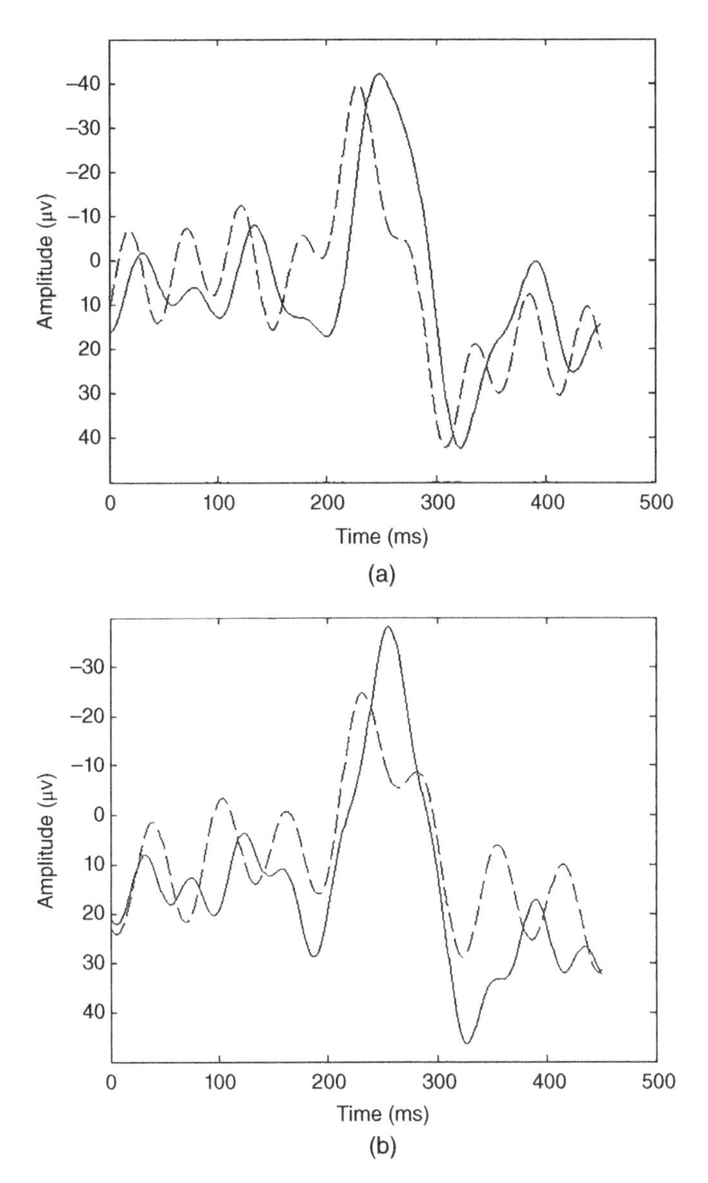

Figure 9.9 Dynamic variations of ERP signals. (a) First stimulus and (b) twentieth stimulus for the ERPs detected using LMS (dotted line) and Kalman filtering (thick line) approaches.

In a wavelet-based PF approach [63] the time-locked ERP in the wavelet domain in the kth trial is presented as $\mathbf{y}_k = [y_k(1), y_k(2), \ldots, y_k(M)]^T$

Following the fundamentals of PF, by modelling the wavelet domain signals (i.e. wavelet coefficients) in the state space, in general, the evolution of the state $\{\mathbf{x}_k\}$ and the relation between the state and the estimated wavelet coefficients (the measurement equation) are respectively given as follows.

$$\mathbf{x}_k = f_{k-1}(\mathbf{x}_{k-1}, \mathbf{w}_{k-1}) \tag{9.49}$$

$$\mathbf{y}_k = h_k(\mathbf{x}_k, \mathbf{v}_k) \tag{9.50}$$

where f_k and h_k are nonlinear functions of the state $\mathbf{x}_k$, and $\mathbf{w}_{k-1}$ and $\mathbf{v}_k$ are i.i.d. noise processes. In PF the filtered estimates of $\mathbf{x}_k$ are searched based on a set of all available wavelet coefficients. According to Bayes' rule, an available measurement at time k, i.e. $\mathbf{y}_k$ is iteratively used to update the posterior density:

$$p(\boldsymbol{x}_k \mid \boldsymbol{y}_{1:k}) = \frac{p(\boldsymbol{y}_k \mid \boldsymbol{x}_k)p(\boldsymbol{x}_k \mid \boldsymbol{y}_{1:k-1})}{p(\boldsymbol{y}_k \mid \boldsymbol{y}_{1:k-1})} \tag{9.51}$$

where $p(\boldsymbol{x}_k \mid \boldsymbol{y}_{1:k-1})$ is computed in the prediction stage using the Chapman-Kolmogorov equation and $p(\boldsymbol{y}_k \mid \boldsymbol{y}_{1:k-1})$ is a normalizing constant.

In PF, the posterior distributions are approximated by discrete random measures defined by particles $\{\mathbf{x}^{(n)}; n = 1,\ldots, N\}$ and their associated weights $\{\mathbf{w}^{(n)}; n = 1,\ldots, N\}$. The distribution based on these samples and weights at the kth trial is approximated as:

$$p(\mathbf{x}) \approx \sum\nolimits_{n=1}^{N} \mathbf{w}^{(n)} \delta\left(\mathbf{x} - \mathbf{x}^{(n)}\right) \tag{9.52}$$

where δ is the Dirac delta function and (n) refers to the nth weight. If the particles are generated according to distribution $p(\mathbf{x})$, the weights are identical and equal to $1/N$. When the distribution $p(\mathbf{x})$ is unknown, the particles are generated from a known distribution $\pi(\mathbf{x})$ called importance density. Following the concepts of importance sampling and following the Bayes rule, the particle weights can be iteratively estimated as:

$$\mathbf{w}_k^{(n)} \propto \mathbf{w}_{k-1}^{(n)} \frac{p\left(\boldsymbol{y}_k^{(n)} \mid \boldsymbol{x}_k^{(n)}\right) p\left(\boldsymbol{x}_k^{(n)} \mid \boldsymbol{x}_{k-1}^{(n)}\right)}{\pi\left(\boldsymbol{x}_k^{(n)} \mid \boldsymbol{x}_k^{(n)}, \boldsymbol{y}_{1:k}\right)} \tag{9.53}$$

The choice of importance density is crucial in the design of PF since it significantly affects the performance. This function must have the same support as the probability distribution to be approximated; the closer the importance function to the true distribution, the better the weight approximations. The most popular choice for the prior importance function, is given by

$$\pi\left(\mathbf{x}_k^{(n)} \mid \mathbf{x}_k^{(n)}, \mathbf{y}_{1:k}\right) \approx p\left(\mathbf{x}_k^{(n)} \mid \mathbf{x}_{k-1}^{(n)}\right) \tag{9.54}$$

This choice of importance density implies that we need to sample from $p\left(\mathbf{x}_k^{(n)} \mid \mathbf{x}_{k-1}^{(n)}\right)$. A sample can be obtained by generating a noise sample for $\mathbf{w}_{k-1}^{(n)}$, and setting

$$\mathbf{x}_k^{(n)} = f_{k-1}\left(\mathbf{x}_{k-1}^{(n)}, \mathbf{w}_{k-1}^{(n)}\right) \tag{9.55}$$

Therefore, the particle weights can be updated as [63]:

$$\mathbf{w}_k^{(n)} \propto \mathbf{w}_{k-1}^{(n)} p\left(\mathbf{y}_k^{(n)} \mid \mathbf{x}_k^{(n)}\right) \tag{9.56}$$

The importance sampling weights indicate the level of importance of the corresponding particle. Relatively small weights imply that the sample is drawn far from the main body of the posterior distribution and has a small contribution in the final estimation. Such a particle is said to be ineffective. If the number of ineffective particles is increased, the number of particles contributing to the estimation of states is decreased, so the performance of the filtering procedure deteriorates. The degeneracy can be avoided by resampling. During the resampling process the particles with small weights are eliminated and those with large weights are replicated.

An experiment using real EEG data over 60 trials, assuming Gaussian distributions for both state and observation noises and when $\mathbf{w}_{k-1}^{(n)}$ is sampled from a Gaussian distribution, using both KF and PF has been performed. The results are depicted in Figure 9.10 [64].

In another approach tracking variation of ERP parameters for both P3a and P3b subcomponents of P300 a Rao-Blackwellised PF (RBPF) approach has been developed [64]. In this approach the particles vary with respect to the changes in both P3a and P3b parameters simultaneously. RBPF combines KF and PF to track linear (such as amplitude in this work) and nonlinear (such as latency and width) parameters from a single EEG channel at the same time, i.e.:

$$\boldsymbol{x}_k = [\mathbf{a}_k(1)\ \mathbf{b}_k(1)\ \boldsymbol{s}_k(1)...\mathbf{a}_k(p)\ \boldsymbol{b}_k(p)\ \boldsymbol{s}_k(p)]^T \tag{9.57}$$

a, **b**, and **s** refer to the ERP parameters (amplitude, latency, and width) and p denotes the number of existing ERPs in a single-trial k. If all the ERPs are masked out and only P300 is retained, then P3a and P3b can be considered as the ERP components in (9.57), i.e. $p = 2$. Following RBPF, it is possible to take the amplitudes out and form a matrix and find a linear relation between the observation and amplitudes and a nonlinear relation with respect to other variables. Therefore, regarding the nonlinear portion of particles in Eq. (9.57) the state space and observation can be formulated as:

$$\boldsymbol{x}_k^{(\mathbf{1})} = [b_k(1)\ s_k(1)...b_k(p)\ s_k(p)]^T \tag{9.58}$$

$$\mathbf{x}_k^{(2)} = [a_k(1)...a_k(p)]^T \tag{9.59}$$

$$\mathbf{x}_k = \mathbf{x}_{k-1} + \mathbf{v}_{k-1} \tag{9.60}$$

$$\mathbf{z}_k = \left[\ f_1\left(\mathbf{x}_k^{(1)}\right)...f_p\left(\mathbf{x}_k^{(1)}\right)\ \right]\begin{bmatrix} a_k(1) \\ \vdots \\ a_k(p) \end{bmatrix} + \mathbf{n}_k \tag{9.61}$$

$$f_i\left(\mathbf{x}_k^{(1)}, t\right) = f_i(b_k(i), s_k(i), t)) = e^{-\frac{(t-b_k(i))^2}{2s_k^2(i)}} \tag{9.62}$$

where t denotes the time index and varies from beginning of the ERP component to the end of the ERP component. In each iteration the relation between the two RBPFs has to be evaluated. As long as the shapes of P3a or P3b are assumed to remain the same in every channel, we can relate the pairs of particles with the same width for P3a and P3b. Using the relation between two RBPFs, a constrained RBPF is built up which not only links the widths but also relies on having P3a before P3b. This is useful for removing the invalid particles. The weights of these invalid particles are then set to zero disabling them in predicting the new particles. These invalid particles are those whose latency of P3b is shorter than the latency of P3a.

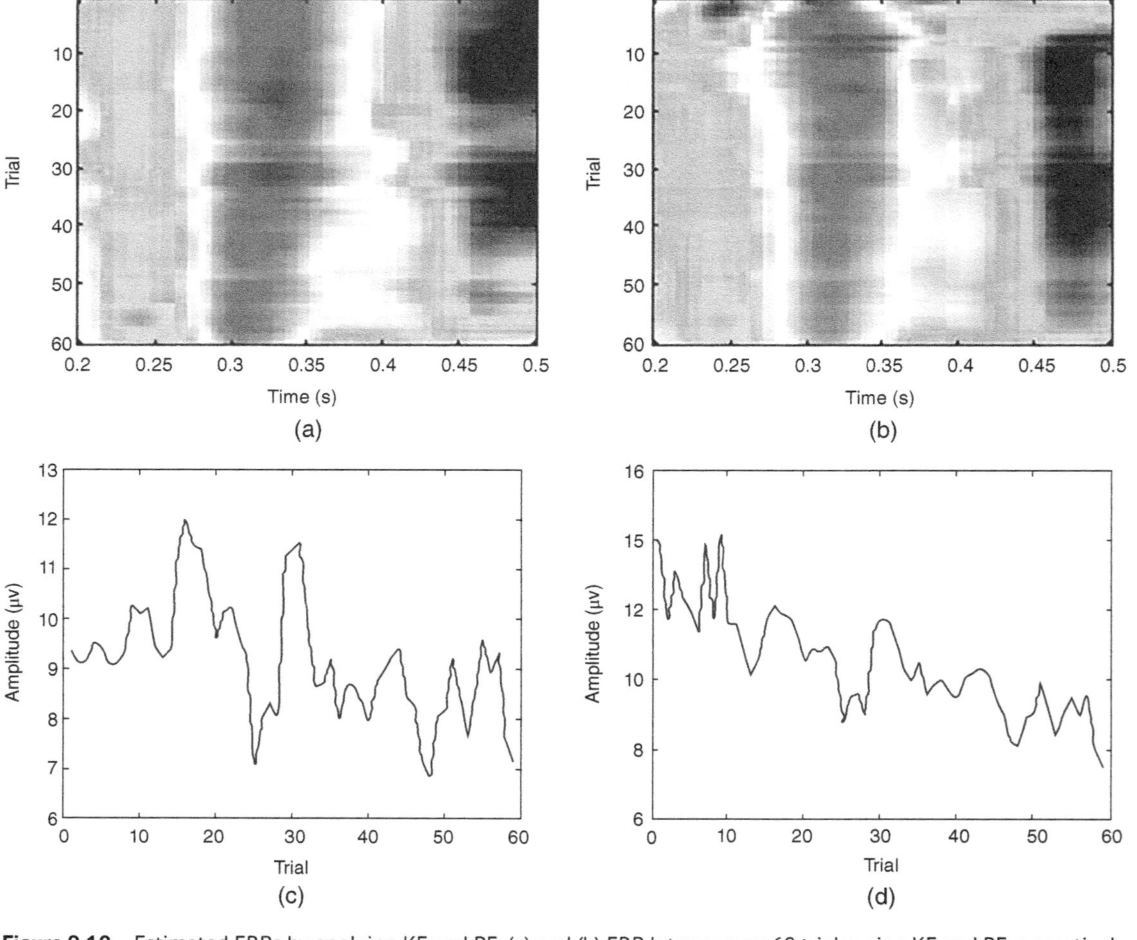

Figure 9.10 Estimated ERPs by applying KF and PF, (a) and (b) ERP latency over 60 trials using KF and PF respectively, and (c) and (d) amplitudes of the P300 waves for consecutive trials extracted by KF and PF respectively.

9.3.7 Variational Bayes Method

A very recent approach for ERP parameter estimation is by using variational Bayes. Variational Bayes methods, also called ensemble learning, can be seen as an extension of the expectation–maximization (EM) algorithm from maximum a posteriori (MAP) estimation of the single most probable value of each parameter to fully Bayesian estimation. Variational Bayes is employed to compute (an approximation to) the complete posterior distribution of the estimation parameters and latent variables [61]. It has been used effectively for spatiotemporal modelling of EEG and MEG systems for both distributed and dipole source models [63–65]. Variational Bayes has also been utilized for fast computation of source parameters [66]. The main shortcoming for this approach is that only specific priors (normally Gamma distributions) are used for the model parameters. This constraint often limits the performance of the method in practical applications. A spatiotemporal variational Bayes technique using a dipole source model has also been formulated in [67]. In these methods, the Markov Chain Monte Carlo (MCMC) technique was used to sample the unknown high-dimensional parameters from the marginalized posterior distribution. It is, however, likely for the solutions of this approach to trap in local maxima [67].

In [67] a new approach to detection and tracking of ERP subcomponents based on variational Bayes has been suggested. In this work the ERP subcomponent sources have been taken as equivalent current dipoles (ECDs), and their locations and parameters (amplitude, latency, and width) are estimated and tracked from trial to trial. Variational Bayes allows the parameters to be estimated separately using the likelihood function of each parameter.

The locations of ECDs can vary from trial to trial in a realistic head model. ERPs are assumed to be the superposition of a small number of ECDs and their temporal bases are modelled by Gaussian shape signals. The amplitudes, means, and variances of the Gaussian waves can be interpreted as the amplitudes, latencies, and widths of ERP subcomponents, respectively.

Variational Bayes shows that, when the prior distribution is unknown, maximizing the likelihood of each parameter (via separate estimation of each) is equivalent to minimizing the Kullback–Leibler distance between the estimated and the true posterior distributions.

The locations are estimated using the PF. Many studies have shown that the PF is one of the best methods when the relation between the desired parameter (states) and the measurement is nonlinear [36]. A closed-form solution for the amplitude is also given by the maximum likelihood (ML) approach.

Estimations of the amplitude and noise covariance matrix of the measurement are optimally estimated recursively by the ML approach, while estimations of the latency and width are obtained by a recursive Newton–Raphson technique [67].

The Newton–Raphson algorithm has rapid convergence. One challenge for this approach is that very low signal-to-noise ratio (SNR) in some trials can result in divergence of the filtering, which impacts negatively on the estimation of amplitudes, latencies, and widths. To compensate for this failure, recursive methods are suggested to improve the stability of the trial-to-trial filtering.

The main advantage of the method is the ability to track varying ECD locations.

To formulate the problem assume that there are q ECDs and their generated sources are combined over the L electrodes on the scalp, each consists of M samples, as [67]:

$$\mathbf{Y}_k = \sum_{i=1}^{q} \mathbf{H}(\rho_{k,i})\mathbf{a}_{k,i}\psi_{k,i} + \mathbf{N}_k \tag{9.63}$$

where $\mathbf{H}(\rho_{k,i})$ is the nonlinear forward matrix of size $L \times 3$ and is a nonlinear function of the ECD locations and $\mathbf{N}_k$ is an additive white zero-mean Gaussian noise of spatial variance $\mathbf{Q}_k$ which is unknown and temporal variance $\mathbf{I}$ (identity matrix). $\mathbf{a}_{k,i} \in \Re^{3\times 1}$ is the strength (amplitude) of the ECD moment in three dimensions and $\psi_{k,i}(t) = [\psi_{k,i}(1), \ldots, \psi_{k,i}(M)]$ refers to the temporal basis of the ith ECD moment.

$\mathbf{H}(\rho_{k,i})$ can be calculated in a spherical head model with three layers: skull, scalp, and skin (the medium within each layer is assumed to be homogenous) or can be obtained using a realistic head model. In the realistic head model, after considering each small region to be a cell (or vertex) of a fine mesh, a pre-estimated forward matrix $\mathbf{H}$ is given for each cell (vertex). Also, the covariance of noise can be denoted as $\mathbf{I} \otimes \mathbf{Q}_k$, where $\otimes$ denotes the Kronecker product.

Noise is assumed to be independent from the source activities and distributed identically across time, but not necessarily across sensors. These assumptions provide a fast and simple estimation of the noise covariance matrix from trial to trial.

Each $\psi_{k,i}(t)$ is given by a Gaussian wave as [67]:

$$\psi_{k,i}(t) = \frac{1}{\sigma_{k,i}\sqrt{2\pi}}\, exp\left(-\frac{(t-\mu_{k,i})^2}{2\sigma_{k,i}^2}\right) \tag{9.64}$$

The primary objective of this work is to recursively estimate the model parameters $\theta_{k,i} = \{\rho_{k,i}, \boldsymbol{a}_{k,i}, \boldsymbol{Q}_k, \mu_k, \sigma_k\}$ based on their previous estimates $\theta_{k-1,i}$ and the available measurements $\mathbf{Y}_k$. Furthermore, it is assumed that the changes in $\theta_{k-1,i}$ are smooth and it can be often considered to be a Markovian process [67]. In the recursive estimation the set of parameters is estimated first. Then, all the parameters are updated recursively.

The algorithm starts with estimating the posterior distribution $p(\boldsymbol{\theta}|\mathbf{Y})$ instead of $\boldsymbol{\theta}$ itself. The posterior distribution is often difficult to calculate particularly in nonlinear cases. Therefore, the MAP is approximated by a simpler distribution called *variational distribution*, $r(\boldsymbol{\theta})$. This can be found using the variational Bayes methods. To maximize the similarity between the MAP and $r(\boldsymbol{\theta})$ different criteria such as Kullback–Leibler or free energy can be utilized [67]. As the key restriction in variational Bayes, it is also assumed that the variational distributions are factorizable to the distributions of the constituent parameters in the set, i.e. $r(\boldsymbol{\theta}) = \prod_i r(\theta_i)$.

In the above approach, the locations are placed in one group and the latencies and the widths in another group. The parameter $\boldsymbol{\theta}$ is partitioned into $\boldsymbol{\rho}$ (location), $\mathbf{a}$ (amplitude), and $\mathbf{Q}$ (noise covariance matrix), and different methods can be employed for estimation of the posterior distribution of each sub-parameter $r(\theta_i)$ estimated as:

$$r(\theta_i) \propto p(\mathbf{Y} \mid \theta_i) \tag{9.65}$$

In the localization and tracking of multiple ECDs related to ERPs $\mathbf{R}$ can be considered as a matrix including the locations of all the dipoles, i.e. $\mathbf{R}_k = [\boldsymbol{\rho}_{k,1}, \ldots, \boldsymbol{\rho}_{k,q}] \in R^{3\times q}$. Then, following the process for estimation of particles in the previous section we have:

$$\mathbf{w}_k^{(n)} \propto \mathbf{w}_{k-1}^{(n)} p\left(\mathbf{y}_k^{(n)} \mid \mathbf{R}_k^{(n)}\right) \tag{9.66}$$

For tracking the ERP amplitude $\mathbf{a}_k$ and noise variance $\mathbf{Q}_k$ at trial k the log-likelihood of (9.63) is maximized. Accordingly, the negative log-likelihood to be minimized is defined as [67]:

$$f(\boldsymbol{\theta}, \mathbf{Y}_k) = tr\left\{ \left(\mathbf{Y}_k - \sum_{i=1}^{q} \mathbf{H}(\rho_{k,i}) \boldsymbol{a}_{k,i} \psi_{k,i} + \mathbf{N}_k\right)^T \mathbf{Q}_k^{-1} \times \left(\mathbf{Y}_k - \sum_{i=1}^{q} \mathbf{H}(\rho_{k,i}) \boldsymbol{a}_{k,i} \psi_{k,i} + \mathbf{N}_k\right)\right\} \tag{9.67}$$

An alternate minimization of the above equation results in stable values for $\mathbf{a}_k$ and $\mathbf{Q}_k$. The same function has also been minimized using the Newton–Raphson algorithm to find the latency and width of the ERP components [67].

The results of application of the method to a multichannel EEG dataset including 60 time-locked ERP trials for tracking P3a and P3b parameters can be seen in Figure 9.11. This figure shows the variations in latency and amplitude of both P3a and P3b. The decline in the P3a and P3b amplitude is mainly due to habituation. P3b is usually elicited as the response to target stimuli during an OP. Similarly, for a healthy subject, the latency of P3a increases due to habituation.

9.3.8 Prony's Approach for Detection of P300 Signals

Based on Prony's approach the source signals are modelled as a series expansion of damped sinusoidal basis functions. Prony's method is often used for estimation of pole positions when the noise is white. Also, in [68] the basis functions are estimated using the LS method, derived for coloured noise. This is mainly because after the set of basis functions are estimated using Prony's method, the optimal weights are calculated for each single-trial ERP by minimizing the squared error.

The amplitude and latency of the P300 signal give information about the level of attention. In many psychiatric diseases this level changes. Therefore, a reliable algorithm for more accurate estimation of these parameters is very useful.

In the above attempt, the ERP signal is initially divided into the background EEG and ERP, respectively, before and after the stimulus time instant. The ERP is also divided into two segments, the early brain response, which is a low-level high-frequency signal, and the late response, which is a high-level low-frequency signal, i.e.:

$$x(n) = \begin{cases} x_1(n) & -L \le n < 0 \\ s_e(n - n_0) + x_2(n) & 0 \le n \le L_1 \\ s_l(n - n_0 - L_1) + x_3(n) & L_1 \le n < N + L_1 \end{cases} \tag{9.68}$$

where $x_1(n)$, $x_2(n)$, and $x_3(n)$ are respectively the background EEGs before the stimulus, during the early brain response, and during the late brain response, and $s_e(n)$ and $s_l(n)$ are respectively, the early and late brain responses to a stimuli. Therefore, to analyze the signals

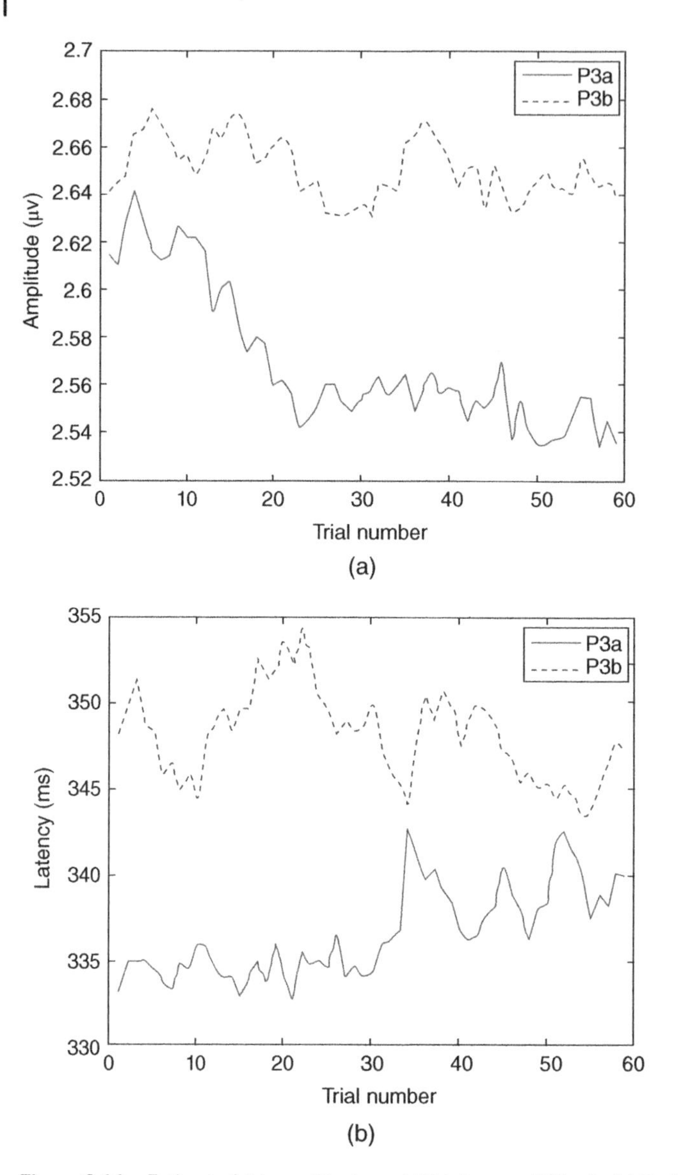

Figure 9.11 Estimated (a) amplitude and (b) latency of P3a (bold line) and P3b (dotted line) for real data.

they are windowed (using a window $g(n)$) within $L_1 \le n < N + L_1$within the duration the P300 signal exists. This gives:

$$x_g(n) = x(n)g(n) \approx \begin{cases} s_l(n - L_1) + x_3(n) & L_1 \le n < N + L_1 \\ 0 & \text{Otherwise} \end{cases} \tag{9.69}$$

We may consider $s(n) = s_l\,(n - L_l)$ for $L_1 \le n < N + L_1$, for simplicity. Based on Prony's algorithm the ERP signals are represented through the following overdetermined system:

$$s(n) = \sum\nolimits_{i=1}^{M} a_i \rho_i^n e^{\,j\omega_i n} = \sum\nolimits_{i=1}^{M} a_i z_i^n \tag{9.70}$$

where $z_i = \rho_i e^{j\omega_i}$ and $M < N$. Using vector notation we can write:

$$\mathbf{x}_g = \mathbf{Za} + \mathbf{x}_3 \tag{9.71}$$

where $\boldsymbol{x}_g = [x_g(L_1)...x_g(N + L_1 - 1)]^T$, $\boldsymbol{x}_3 = [x_3(L_1)...x_3(N + L_1 - 1)]^T$, $\boldsymbol{a} = [a_1...a_M]^T$ and

$$\mathbf{Z} = \begin{bmatrix} \rho_1 e^{j\omega_1} & \dots & \rho_M e^{j\omega_M} \\ \vdots & & \\ \rho_1 e^{j\omega_1 N} & \dots & \rho_M e^{j\omega_M N} \end{bmatrix} \tag{9.72}$$

Therefore, in order to estimate the waveform $\mathbf{s} = \mathbf{Za}$ both $\mathbf{Z}$ and $\mathbf{a}$ have to be estimated. The criterion to be minimized is the weighted LS criterion given by:

$$\min_{\mathrm{Z,a}} \left\{ (\mathbf{x}_g - \mathbf{Za})^H \mathbf{R}_{\boldsymbol{x}_3}^{-1} (\mathbf{x}_g - \mathbf{Za}) \right\} \tag{9.73}$$

where $\boldsymbol{R}_{\boldsymbol{x}_3}$ is the covariance matrix of the EEG (as a disturbing signal) after stimulation. For a fixed $\mathbf{Z}$ the optimum parameter values are obtained as:

$$\mathbf{a}_{opt} = \left(\mathbf{Z}^H \mathbf{R}_{x_3}^{-1} \mathbf{Z}\right)^{-1} \mathbf{Z}^H \mathbf{R}_{x_3}^{-1} \mathbf{x}_g \tag{9.74}$$

Then, the corresponding estimate of the single-trial ERP is achieved as:

$$\mathbf{s} = \mathbf{Za}_{opt} \tag{9.75}$$

Prony's method is employed to minimize (9.73) for coloured noise. In order to do that a matrix $\mathbf{F}$ orthogonal to $\mathbf{Z}$ is considered, i.e. $\mathbf{F}^H\mathbf{Z} = 0$ [69]. The ith column of $\mathbf{F}$ is $col\{\mathbf{F}\}_i = \left[\underbrace{0...0}_{i-1} f_0 \cdots f_M \underbrace{0...0}_{N-M-i} \right]^T$. The roots of the polynomial $f_0 + f_1 z^{-1} + ... + f_M z^{-M}$ are the elements $\mathbf{z}_i$ in $\mathbf{Z}$ [69]. The minimization in (9.73) is then converted to

$$\min_{\mathbf{F}} \left\{ \boldsymbol{x}_g^H \mathbf{F} (\mathbf{F}^H \boldsymbol{R}_{\boldsymbol{x}_3} \mathbf{F})^{-1} \mathbf{F}^H \mathbf{x}_g \right\} \tag{9.76}$$

In practise a linear model is constructed from the above nonlinear system using [69, 70]

$$\mathbf{F}(\mathbf{F}^H \mathbf{R}_{\boldsymbol{x}_3} \mathbf{F})^{-1} \mathbf{F}^H \approx \tilde{\mathbf{F}} \mathbf{R}_{\boldsymbol{x}_3}^{-1} \tilde{\mathbf{F}}^H \tag{9.77}$$

where the ith column of $\tilde{\mathbf{F}}$ is $col\{\tilde{\mathbf{F}}\}_i = \left[\underbrace{0...0}_{i-1} f_0 \cdots f_p \underbrace{0...0}_{N-p-i} \right]^T$ where $p >> M$ and therefore, it includes the roots of $\mathbf{F}$ plus p-M spurious poles. Then, the minimization in (9.74) changes to:

$$\min_{\tilde{\mathbf{F}}} \left\{ \mathbf{e}^H \mathbf{R}_{\boldsymbol{x}_3}^{-1} \mathbf{e} \right\} = \min_{\tilde{\mathbf{F}}} \left\{ \mathbf{x}_g^H \tilde{\mathbf{F}} \mathbf{R}_{\boldsymbol{x}_3}^{-1} \tilde{\mathbf{F}}^H \mathbf{x}_g \right\} \tag{9.78}$$

where $\mathbf{e} = \widetilde{\mathbf{F}}^H \mathbf{x}_g = \mathbf{X}\mathbf{f}$. In this equation

$$\mathbf{X} = \begin{bmatrix} x_g(p) & x_g(p-1) & \dots & x_g(0) \\ \vdots & \vdots & & \vdots \\ x_g(N-1) & x_g(N-2) & \dots & x_g(N-p-1) \end{bmatrix} = \begin{bmatrix} \tilde{\mathbf{x}} & \widetilde{\mathbf{X}} \end{bmatrix} \tag{9.79}$$

where $\tilde{x}$ is the first column of $\mathbf{X}$ and includes the spurious poles. The original Prony's method for solving (9.76) is found by replacing $\mathbf{R}_{\mathbf{x}_3} = \mathbf{I}$ with $f_0 = 1$ or $f_1 = 1$. However, the background EEG cannot be white noise, therefore, in this approach a coloured noise, $\boldsymbol{R}_{\boldsymbol{x}_3} \neq \boldsymbol{I}$, is considered. Hence the column vector $\mathbf{f}$ including the poles or the resonance frequencies of the basis functions in (9.68) is estimated as:

$$\mathbf{f} = \left[1; -\left[\widetilde{\mathbf{X}}^H \mathbf{R}_{\boldsymbol{x}_3}^{-1} \widetilde{\mathbf{X}}\right]_M^{-1} \left[\widetilde{\mathbf{X}}^H \mathbf{G}^{-H}\right]_M \mathbf{G}^{-1} \tilde{\mathbf{x}}\right] \tag{9.80}$$

where $\mathbf{G}\mathbf{G}^H = \boldsymbol{R}_{\mathbf{x}_3}$ and the notation $[.]_M$ denotes the rank M approximated matrix. The rank reduction is performed using a singular value decomposition (SVD). Although several realizations of the ERP signals may have similar frequency components, $\mathbf{a}$ may vary from trial to trial. Therefore, instead of averaging the data to find $\mathbf{a}$, the estimated covariance matrix is averaged over D realizations of the data. An analysis based on a forward-backward Prony solution is finally employed to separate the spurious poles from the actual poles. The parameter $\mathbf{a}$ can then be computed using (9.74).

Generally, the results achieved using Prony-based approaches for single-trial ERP extraction are rather reliable and consistent with the physiological and clinical expectations. The above algorithm has been examined for a number of cases including pre-stimulus, after stimulation, and during post-stimulus. Clear peaks corresponding to mainly P300 sources are evident in the extracted signals.

In another attempt, similar to the above approach, it is considered that for the ERP signals frequency, amplitude, and phase characteristics vary with time, therefore a piecewise Prony method (PPM) has been proposed [71]. The main advantages of using PPM are: firstly, this method enables modelling non-monotonically growing or decaying components with non-zero onset time. Therefore, by using PPM it is assumed that the sinusoidal components have growing or decaying envelopes, and abrupt changes in amplitude and phase. The PPM uses variable-length windows (previously suggested by [71, 72]) and variable sampling frequencies (performed also by Kulp [73] for adjusting sampling rate to the frequencies to be modelled) to overcome the limitations of the original Prony's method. Secondly, the window size is determined based on the signal characteristics. Also, the signals with large bandwidths are modelled in several steps, focusing on smaller frequency bands per step, as proposed in [74]. Thirdly, the varying-length windows try to include the conventional Prony components. Therefore, there is more consistency in detecting the ERP features (frequency, amplitude, and phase). Finally, it is reasonable to assume that the signal components obtained from adjacent windows with identical frequency are part of the same component and they can be combined into one component.

9.3.9 Adaptive Time–Frequency Methods

Combination of an adaptive signal estimation technique and TF signal representation can enhance the performance of the ERP and EP detections. This normally refers to modelling the signals with variable-parameter systems and application of adaptive estimators to estimate suitable parameters.

Using STFT the information about the duration of the activities are not exploited and therefore it cannot describe structures much shorter or much longer than the window length. WT can overcome this limitation by allowing for variable window length but there is still a reciprocal relation between the central frequency of the wavelet and its window length. Therefore, the WT does not precisely estimate the low-frequency components with short time duration or narrow-band high-frequency components.

In another work the adaptive chirplet transform (ACT) has been utilized to characterize the time-dependent behaviour of the VEPs from its transient to steady-state section [75]. Generally, this approach uses the matching pursuit (MP) algorithm to estimate the chirplets and a ML algorithm to refine the results and enable estimation of the signal with a low SNR. Using this method it is possible to visualize the early moments of a VEP response.

The ACT attempts to decompose the signals into Gaussian chirplet basis functions with four adjustable parameters of time-spread, chirp rate, time centre, and frequency centre. Moreover, it is shown that only three chirplets can be used to represent a VEP response.

In this approach the transient VEP which appears first following the onset of the visual stimulus [76] is analyzed together with the steady-state VEP [77, 78]. Identification of steady-state VEP has had many clinical applications and can help diagnosing sensory dysfunction [79, 80]. The model using only steady-state VEPs is however incomplete without considering the transient pattern. A true steady-state VEP is also difficult to achieve in the cases where the mental situation of the patient is not stable.

Chirplets (also called chirps) are windowed rapidly swept sinusoidal signals [81]. The bases for a Gaussian chirplet transform are derived from a simple Gaussian function $\pi^{-1/4} exp(-t^2/2)$ through four operations namely scaling, chirping, time- and frequency-shifting, which (as for wavelets) produces a family of wave packets with four adjustable parameters [81]. Such a continuous-time chirplet may be represented as:

$$g(t) = \pi^{-\frac{1}{4}} \Delta_t^{-\frac{1}{2}} e^{-\frac{1}{2}\left(\frac{t-t_c}{\Delta_t}\right)^2} e^{j[c(t-t_c)+\omega_c](t-t_c)} \tag{9.81}$$

where t_c is the time centre, ω_c is the frequency centre, $\Delta_t > 0$ is the effective time-spread, and c is the chirp rate which refers to the speed of changing the frequency. The chirplet transform of a signal $f(t)$ is then defined as:

$$a(t_c, \omega_c, c, \Delta_t) = \int_{-\infty}^{\infty} f(t) g^*(t) dt \tag{9.82}$$

Based on this approach, the signal $f(t)$ is reconstructed as a linear combination of Gaussian chirplets as:

$$f(t) = \sum_{n=1}^{p} a_n(t_c, \omega_c, c, \Delta_t) g_n(t) + e(t) \tag{9.83}$$

where p denotes the approximation order and $e(t)$ is the residual signal. Figure 9.12 illustrates the chirplets extracted from an EEG-type waveform. To estimate $g(t)$ at each round

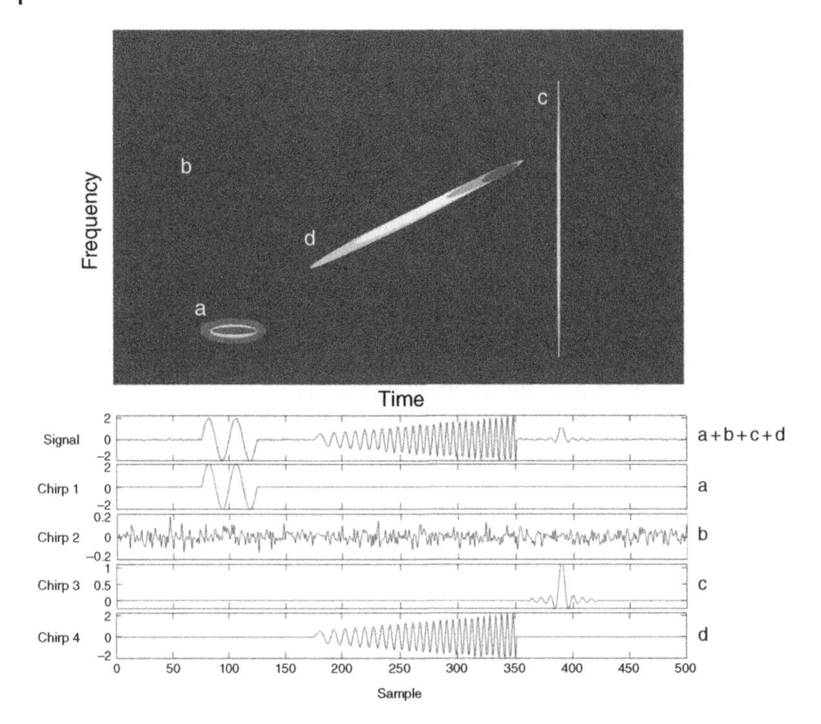

Figure 9.12 The chirplets extracted from a simulated EEG-type waveform [75]. (Bottom) The combined waveforms. (Top) T–F representation of the corresponding chirplets.

(iteration) six parameters have to be estimated (consider a complex). An optimal estimation of these parameters is generally not possible. However, there have been some suboptimal solutions such as in [82–85]. In the approach by Cui and Wong [75] a coarse estimation of the chirplets is obtained using the MP algorithm. Then, maximum likelihood estimation (MLE) is performed iteratively to refine the results.

To perform this the algorithm is initialized by setting $e(n) = f(n)$. A dictionary is constructed using a set of predefined chirplets to cover the entire T–F plane [82]. The MP algorithm projects the residual $e(n)$ to each chirplet in the dictionary and the optimal chirplet is decided based on the projection amplitude. In the next stage, the Newton–Raphson iterative algorithm is used to refine the results and achieve the optimum match.

Under a low SNR situation a post-processing step maybe needed to follow the MLE concept [86] followed by the EM algorithm proposed in [75].

9.4 Brain Activity Assessment Using ERP

Variations in P300 component parameters such as amplitude, latency, and duration, from trial to trial indicate the depth of cognitive information processing. For example, it has been reported that the P300 amplitude elicited by the mental task loading decreases with the increase in the perceptual/cognitive difficulty of the task [87]. Assessment of mental fatigue using P300 signals by measuring the amplitude and latency of the signals and also the alpha-band power have shown that [88] the P300 amplitude tends to decrease immediately after the experimental task whereas the latency decreases at this time. The amplitude decrease is an indicator of decreasing the attention level of the subject, which can be due to habituation. The increase in the latency might be indicative of the prolonged temporal processing due to the difficulty of cognitive information processing [87, 89, 90]. Therefore, the mental and physical fatigues can be due to this problem. In this work, it is also pointed out that one aspect of the mentally and physically fatigued could be because of decreased activity of the CNS that appears in both the temporal prolongation of cognitive information processing and the decrease in attention level. It has also been demonstrated that alpha-band power decreases due to attention to the task and beta band power rises immediately after the task. This indicates that the activity of the CNS is decelerated with accumulation of mental and physical fatigue. However, the appearance of fatigue is reflected more strongly in the amplitude and latency of the P300 signal and its subcomponents P3a and P3b rather than the alpha-band power. This is analyzed in detail in Chapter 13 of this book.

9.5 Application of P300 to BCI

Electrical cortical activities used in BCI may be divided into the following five categories: (i) beta (β) and mu (μ) rhythms: these activities range respectively within 8–12 and 12–30 Hz. These signals are associated with those cortical areas most directly connected to the motor output of the brain and can be willingly modulated with an imagery movement [91]; (ii) P300 EP: it is a late appearing component of an auditory, visual, or somatosensory ERP as explained before; (iii) visual N100 and P200: the ERPs with short latency that represent the exogenous response of the brain to a rapid visual stimulus. These potentials are used as clues indicating the direction of user gaze [92, 93]; (iv) steady-state visual EPs (SSVEP): these signals are natural responses to visual stimulations at specific frequencies. When the retina is excited by a visual stimulus ranging from 3.5 to 75 Hz, the brain generates an electrical activity at the same (or multiples of the) frequency of the visual stimulus. They are used for understanding which stimulus the subject is looking at in the case of stimuli with different flashing frequencies [94, 95]; and (v) slow cortical potentials (SCP): they are slow potential variations generated in the cortex after 0.5–10 seconds of presenting the stimulus. Negative SCPs are generated by movement whereas positive SCPs are associated with reduced cortical activation. Adequately trained users can control these potentials and use them to control the movement of a cursor on the computer screen [96].

To enable application of these waves especially P300 for a P300 based BCI the data have to be pre-processed to reduce noise and enforce P300 related information. A pattern recognition algorithm has to be later developed to check the presence of the P300 wave in the recorded ERP epochs and label it. Also, a feedback mechanism has to be established to send the user a visible signal on the monitor correlated to the recorded epoch. Finally, the parameters of the pattern recognition algorithm have to be made adaptable to the subject's characteristics.

The P300 speller, as the first P300-BCI, described by Farwell and Dounchin, adapted the OP as the operating principle of BCI [36]. In this paradigm the participant is presented with a Bernoulli sequence of events, each belonging to one of two categories. The participant is assigned a task that cannot be performed without a correct classification of the events, each belonging to one of two categories. Using this speller, for example, a 6×6 matrix is displayed to the subject. The system is operated by briefly intensifying each row and column of the matrix and the attended row and column elicits a P300 response. In a later work [97] it was found that some people who suffer from amyotrophic lateral sclerosis (ALS) can better respond to the system with a smaller size matrix (less than 6×6).

In addition, measurement and evaluation of the movement-related events such as event-related desynchronization (ERD) from the EEGs can improve the diagnosis of functional deficits in patients with cerebrovascular disorders and PD.

There is a high correlation between morphological, such as computed tomography (CT) and functional, such as EEG, findings in cerebrovascular disorders. For example, the ERD reduces over the affected hemisphere.

The pre-movement ERD in PD is less lateralized over the contralateral sensorimotor area and starts later than in control subjects. Also, post-movement beta event-related synchronization (ERS) is of smaller magnitude and delayed in PD as compared to controls. It has been shown [98] that based on only two parameters, namely the central ERD within 6–10 Hz, and post-movement ERS within 16–20 Hz, it is possible to discriminate PD patients with a Hoehn and Yahr scale of 1–3 from age-matched controls by a linear discriminant analysis. The Hoehn and Yahr scale is a commonly used system for describing how the symptoms of PD progress [66].

Also, following the above procedure in P300 detection and classification, two user adaptive BCI systems based on SSVEP and P300, have been proposed by Beverina et al. [99].

The P300 component has been detected (separated) using ICA for the BCI purpose in another attempt [41]. It has also been empirically confirmed that the visual spatial attention modulation of the SSVEP can be used as a control mechanism in a real-time independent BCI (i.e. when there is no dependency on peripheral muscles or nerves) [100].

P300 signals have also been used in the Wadsworth BCI development [101]. In this application a similar method to that proposed by Farwell and Donchin [36] is used for highlighting and detection of the P300 signals. It has been shown that this well established system can be used by severely disabled people in their homes with minimal ongoing technical oversight.

In another attempt, P300 signals have been used for the design of a speller (text-input application). Twenty-five channels around C3, C4, Cz, CPz, and FCz electrodes are manually selected to have the best result. In addition, P7 and P8 electrodes are also used. The number of channels is later reduced to 20 using PCA and selecting the largest eigenvalues. During this process the data are also whitened and the effect of the eye blink is removed. Support vector machines (SVM) are then used to classify the principal components for each character, using a Gaussian kernel [102].

9.6 Summary

ERP signals indicate the types and states of many brain abnormalities and mental disorders. In addition, they play an important role in some BCI approaches such as the SSVEP-based speller. These signals are characterized by their spatial, temporal, and spectrum locations.

Also, they are often considered as independent sources which can be separated from the background EEG using ICA. The ERP signals can be characterized by their amplitudes, latencies, source locations, and frequency contents. Automatic extraction and tracking of these signals from single EEG trials, however, requires sufficient knowledge and expertise in the development of mathematical and signal processing algorithms. Although there has not been any robust and well established method to detect and characterize these signals so far, some recently developed single-trial methods have high potentials, supported by analysis of real EEG signals for evaluation and tracking of more popular ERP components such as P300. More recent methods such as in [65] overcome the limiting conditions for instance, lack of correlation or independency of various ERP components in their separation procedure. More applications of ERP and EP are discussed in Chapter 13 for mental fatigue analysis, and in Chapter 16 for BCI.

References

1 Dawson, G.D. (1947). Investigations on a patient subject to myoclonic seizures after sensory stimulation. *Journal of Neurology, Neurosurgery, and Psychiatry* **10**: 141–162.

2 Walsh, P., Kane, N., and Butler, S. (2005). The clinical role of evoked potentials. *Journal of Neurology, Neurosurgery, and Psychiatry* **76** (Suppl. II): ii16–ii22.

3 Walter, W.G. (1964). Contingent negative variation: an electrical sign of sensorimotor association and expectancy in the human brain. *Nature* **203**: 380–384.

4 Sutton, S., Braren, M., Zoubin, J., and John, E.R. (1965). Evoked potential correlates of stimulus uncertainty. *Science* **150**: 1187–1188.

5 Johnson, R. Jr. (1992). Event-related brain potentials. In: *Progressive Supranuclear Palsy: Clinical and Research Approaches* (eds. I. Litvan and Y. Agid), 122–154. New York: Oxford University Press.

6 Visser, S.L., Stam, F.C., Van Tilburg, W. et al. (1976). Visual evoked response in senile and presenile dementia. *Electroencephalography and Clinical Neurophysiology* **40** (4): 385–392.

7 Cosi, V., Vitelli, E., Gozzoli, E. et al. (1982). Visual evoked potentials in aging of the brain. *Advances in Neurology* **32**: 109–115.

8 Coben, L.A., Danziger, W.L., and Hughes, C.P. (1981). Visual evoked potentials in mild senile dementia of Alzheimer type. *Electroencephalography and Clinical Neurophysiology* **52**: 100.

9 Visser, S.L., Van Tilburg, W., Hoojir, C. et al. (1985). Visual evoked potentials (VEP's) in senile dementia (Alzheimer type) and in nonorganic behavioural disorders in the elderly: comparison with EEG parameters. *Electroencephalography and Clinical Neurophysiology* **60** (2): 115–121.

10 Nunez, P.L. (1981). *Electric Fields of the Brain*. New York: Oxford University Press.

11 Picton, T.W., Lins, D.O., and Scherg, M. (1995). The recording and analysis of event-related potentials. In: *Hand Book of Neurophysiology*, vol. **10** (eds. F. Boller and J. Grafman), 3–73. Amsterdam: Elsevier.

12 Perrin, P., Pernier, J., Bertrand, O., and Echallier, J.F. (1989). Spherical splines for scalp potential and current density mapping. *Electroencephalography and Clinical Neurophysiology* **72**: 184–187.

13 Hegerl, U. (1999). Event-related potentials in psychiatry, Chap. 31. In: *Electroencephalography*, 4e (eds. E. Niedermayer and F.L. Da Silva), 621–636. Lippincott Williams & Wilkins.

14 Rangaswamy, M., Jones, K.A., Porjesz, B. et al. (2007). Delta and theta oscillations as risk markers in adolescent offspring of alcoholics. *International Journal of Psychophysiology* **63**: 3–15.

15 Diez, J., Spencer, K., and Donchin, E. (2003). Localization of the event-related potential novelty response as defined by principal component analysis. *Cognitive Brain Research* **17** (3): 637–650.

16 Frodl-Bauch, T., Bottlender, R., and Hegerl, U. (1999). Neurochemical substrates and neuroanatomical generators of the event-related P300. *Neuropsychobiology* **40**: 86–94.

17 Kok, A., Ramautar, J., De Ruiter, M. et al. (2004). ERP components associated with successful and unsuccessful stopping in a stop-signal task. *Psychophysiology* **41** (1): 9–20.

18 Polich, J. (2004). Clinical application of the P300 event-related brain potential. *Physical Medicine and Rehabilitation Clinics of North America* **15** (1): 133–161.

19 Friedman, D., Cycowics, Y., and Gaeta, H. (2001). The novelty P3: an event-related brain potential (ERP) sign of the brains evaluation of novelty. *Neuroscience and Biobehavioral Reviews* **25** (4): 355–373.

20 Näätänen, R., Paavilainen, P., Tiitinen, H. et al. (1993). Attention and mismatch negativity. *Psychophysiology* **30** (5): 436–450.

21 Kropotov, J.D., Alho, K., Näätänen, R. et al. (2000). Human auditory-cortex mechanisms of preattentive sound discrimination. *Neuroscience Letters* **280**: 87–90.

22 Johnstone, S.J., Barry, R.J., and Clarke, A.R. (2013). Ten years on: a follow-up review of ERP research in attention-deficit/hyperactivity disorder. *Clinical Neurophysiology* **124** (4): 644–657.

23 Barry, R.J., Johnstone, S.J., and Clarke, A.R. (2003). A review of electrophysiology in attention-deficit/hyperactivity disorder: II. Event-related potentials. *Clinical Neurophysiology* **114** (2): 184–198.

24 Ciuffini, R., Marrelli, A.M., Necozione, S. et al. (2014). Visual evoked potentials in Alzheimer's disease: electrophysiological study of the visual pathways and neuropsychological correlates. *Journal of Alzheimers Disease and Parkinsonism* **2014**: 1–4.

25 Prabhakar, S., Syal, P., and Srivastava, T. (2000). P300 in newly diagnosed non-dementing Parkinson's disease: effect of dopaminergic drugs. *Neurology India* **48** (3): 239–242.

26 Boose, M.A. and Cranford, J.L. (1996). Auditory event-related potentials in multiple sclerosis. *Otology and Neurotology* **17** (1): 165–170.

27 Duncan, C.C., Kosmidis, M.H., and Mirsky, A.F. (2008). Event–related potential assessment of information processing after closed head injury. *Psychophysiology* **40** (1): 45–59.

28 D'Arcy, R.C.N., Marchand, Y., Eskes, G.A. et al. (2003). Electrophysiological assessment of language function following stroke. *Clinical Neurophysiology* **114** (4): 662–672.

29 Hanna, G.L., Carrasco, M., Harbin, S.M. et al. (2012). Error-related negativity and tic history in pediatric obsessive-compulsive disorder. *Journal of the American Academy of Child and Adolescent Psychiatry* **51** (9): 902–910.

30 Comerchero, M. and Polich, J. (1999). P3a and P3b from typical auditory and visual stimuli. *Clinical Neurophysiology* **110** (1): 24–30.

31 Friedman, D. and Squires-Wheeler, E. (1994). Event-related potentials (ERPs) as indicators of risk for schizophrenia. *Schizophrenia Bulletin* **20** (1): 63–74.

32 Luria, A.R. (1973). *The Working Brain*. New York: Basic Books.

33 Sokolov, E.N. (1963). *Perception and the Conditioned Reflex*. Oxford, UK: Pergamon Press.

34 Friedman, D., Cycowicz, Y.M., and Gaeta, H. (2001). The novelty P3: an event-related brain potential (ERP) sign of the brain's evaluation of novelty. *Neuroscience and Biobehavioral Reviews* **25**: 355–373.

35 Lange, D.H., Pratt, H., and Inbar, G.F. (1997). Modeling and estimation of single evoked brain potential components. *IEEE Transactions on Biomedical Engineering* **44** (9): 791–799.

36 Farwell, L.A. and Dounchin, E. (1998). Talking off the top of your heard: toward a mental prosthesis utilizing event-related brain potentials. *Electroencephalography and Clinical Neurophysiology* **70**: 510–523.

37 Donchin, E., Spencer, K.M., and Wijesingle, R. (2000). The mental prosthesis: assessing the speed of a P300-based brain-computer interface. *IEEE Transactions on Rehabilitation Engineering* **8**: 174–179.

38 McGillem, C.D. and Aunon, J.I. (1977). Measurement of signal components in single visually evoked brain potentials. *IEEE Transactions on Biomedical Engineering* **24**: 232–241.

39 Makeig, S., Jung, T.P., Bell, A.J., and Sejnowsky, T.J. (1997). Blind separation of auditory event-related brain responses into independent components. *Proceedings of the National Academy of Sciences of the United States of America* **94**: 10979–10984.

40 Bell, A.J. and Sejnowsky, T.J. (1995). An information maximisation approach to blind separation and blind deconvolution. *Neural Computation* **7**: 1129–1159.

41 Xu, N., Gao, X., Hong, B. et al. (2004). BCI competition 2003-data set IIb: enhancing P300 wave detection using ICA-based subspace projections for BCI applications. *IEEE Transactions on Biomedical Engineering* **51** (6): 1067–1072.

42 Jung, T.-P., Makeig, S., Mckeown, M. et al. (2001). Imaging brain dynamics using independent component analysis. *Proceedings of the IEEE* **89** (7): 1107–1122.

43 Serby, H., Yom-Tov, E., and Inbar, G.F. (2005). An improved P300-based brain-computer interface. *IEEE Transactions on Neural Systems and Rehabilitation Engineering* **13** (1): 89–98.

44 Makeig, S., Enghoff, S., Jung, T.-P., and Sejnowski, T.J. (2000). Moving-window ICA decomposition of EEG data reveals event-related changes in oscillatory brain activity. In: *2nd International Workshop on Independent Component Analysis and Signal Separation*, 627–632.

45 McKeown, M.J. (2000). Detection of consistently task-related activations in fMRI data with hybrid independent component analysis. *NeuroImage* **11**: 24–35.

46 Aunon, J.I., McGillem, C.D., and Childers, D.G. (1981). Signal processing in evoked potential research: averaging and modelling. *CRC Critical Reviews in Bioengineering* **5**: 323–367.

47 Spyrou, L., Sanei, S., and Cheong Took, C. (2007). Estimation and location tracking of the P300 subcomponents from single-trial EEG. *Proceedings of the IEEE, International Conference on Acoustics, Speech, and Signal Processing*. USA: ICASSP.

48 Marple, S.L. Jr. (1987). *Digital Spectral Analysis with Applications*. Englewood Cliffs, NJ: Prentice Hall.

49 Bertrand, O., Bohorquez, J., and Pernier, J. (1994). Time-frequency digital filtering based on an invertible wavelet transform: an application to evoked potentials. *IEEE Transactions on Biomedical Engineering* **41** (1): 77–88.

50 Ademoglu, A., Micheli-Tzanakou, E., and Istefanopulos, Y. (1997). Analysis of pattern reversal visual evoked potentials (PRVEP's) by spline wavelets. *IEEE Transactions on Biomedical Engineering* **44** (9): 881–890.

51 Unser, M., Aldroubi, A., and Eden, M. (1992). On the asymptotic convergence of B-spline wavelets to Gabor functions. *IEEE Transactions on Information Theory* **38** (2): 864–872.

52 Heinrich, H., Dickhaus, H., Rothenberger, A. et al. (1999). Single sweep analysis of event-related potentials by wavelet networks – methodological basis and clinical application. *IEEE Transactions on Biomedical Engineering* **46** (7): 867–878.

53 Quiroga, R.Q. and Garcia, H. (2003). Single-trial event-related potentials with wavelet denoising. *Clinical Neurophysiology* **114**: 376–390.

54 Bartnik, E.A., Blinowska, K., and Durka, P.J. (1992). Single evoked potential reconstruction by means of wavelet transform. *Biological Cybernetics* **67**: 175–181.

55 Basar, E. (1988). EEG dynamics and evoked potentials in sensory and cognitive processing by brain. In: *Dynamics of Sensory and Cognitive Processing By Brain* (ed. E. Basar), 30–55. Springer.

56 Zhang, Q. and Benveniste, A. (1992). Wavelet networks. *IEEE Transactions on Neural Networks* **3**: 889–898.

57 Rumelhart, D., Hinton, G.E., and Williams, R.J. (1986). Learning internal representations by error propagation. In: *Parallel Distributed Processing*, vol. **1** (eds. D. Rumelhart and J.L. McClelland), 318–362. Cambridge, MA: MIT Press.

58 Glassman, E.L. (2005). A wavelet-like filter based on neuron action potentials for analysis of human scalp electroencephalographs. *IEEE Transactions on Biomedical Engineering* **52** (11): 1851–1862.

59 Georgiadis, S.D., Ranta-aho, P.O., Tarvainen, M.P., and Karjalainen, P.A. (2005). Single-trial dynamical estimation of event-related potentials: a Kalman filter-based approach. *IEEE Transactions on Biomedical Engineering* **52** (8): 1397–1406.

60 Sorenson, H.W. (1980). *Parameter Estimation: Principles and Problems*. New York: Marcel Dekker.

61 Melsa, J. and Cohn, D. (1978). *Decision and Estimation Theory*. New York: McGraw-Hill.

62 Yates, R.D. and Goodman, D.J. (2005). *Probability and Stochastic Processes*, 2e. Wiley.

63 Mohseni, H.R., Wilding, E.L., and Sanei, S. (2008). Single trial estimation of event-related potentials using particle filtering. *Proceedings of the IEEE International Conference on Acoustics, Speech, and Signal Processing*. Taiwan: ICASSP.

64 Jarchi, D., Makkiabadi, B., and Sanei, S. (2009). Estimation of trial to trial variability of p300 subcomponents by coupled rao-blackwellised particle filtering. *Proceedings of Statistical Signal Processing Workshop*. Cardiff, UK.

65 Jarchi, D., Sanei, S., Principe, J.C., and Makkiabadi, B. (2011). A new spatiotemporal filtering method for single-trial ERP subcomponent estimation. *IEEE Transactions on Biomedical Engineering* **58** (1): 132–143.

66 Hoehn, M. and Yahr, M. (1967). Parkinsonism: onset, progression and mortality. *Neurology* **17** (5): 427–442.

67 Mohseni, H.R., Ghaderi, F., Wilding, E.L., and Sanei, S. (2010). Variational Bayes for spatiotemporal identification of event-related potential subcomponents. *IEEE Transactions on Biomedical Engineering* **57** (10): 2413–2428.

68 Hansson, M., Gansler, T., and Salomonsson, C. (1996). Estimation of single event-related potentials utilizing the Prony method. *IEEE Transactions on Biomedical Engineering* **43** (10): 973–978.

69 Scharf, L.L. (1991). *Statistical Signal Processing*. Reading, MA: Addison-Wesley.

70 Garoosi, V. and Jansen, B.H. (2000). Development and evaluation of the piecewise Prony method for evoked potential analysis. *IEEE Transactions on Biomedical Engineering* **47** (12): 1549–1554.

71 Barone, P., Massaro, E., and Polichetti, A. (1989). The segmented Prony method for the analysis of nonstationary time series. *Astronomy and Astrophysics* **209**: 435–444.

72 Meyer, J.U., Burkhard, P.M., Secomb, T.W., and Intaglietta, M. (1989). The Prony spectral line estimation (PSLE) method for the analysis of vascular oscillations. *IEEE Transactions on Biomedical Engineering* **36**: 968–971.

73 Kulp, R.W. (1981). An optimum sampling procedure for use with the Prony method. *IEEE Transactions on Electromagnetic Compatibility* **23**: 67–71.

74 Steedly, W.M., Ying, C.J., and Moses, R.L. (1994). A modified TLS-Prony method using data decimation. *IEEE Transactions on Signal Processing* **42**: 2292–2303.

75 Cui, J. and Wong, W. (2006). The adaptive chirplet transform and visual evoked potentials. *IEEE Transactions on Biomedical Engineering* **53** (7): 1378–1384.

76 Regan, D. (1989). *Human Brain Electrophysiology: Evoked Potentials and Evoked Magnetic Fields in Science and Medicine*. New York: Elsevier.

77 Middendorf, M., McMillan, G., Galhoum, G., and Jones, K.S. (2000). Brain computer interfaces based on steady-state visual-evoked responses. *IEEE Transactions on Rehabilitation Engineering* **8** (2): 211–214.

78 Cheng, M., Gao, X.R., Gao, S.G., and Xu, D.F. (2002). Design and implementation of a brain-computer interface with high transfer rates. *IEEE Transactions on Biomedical Engineering* **49** (10): 1181–1186.

79 Holliday, A.M. (1992). *Evoked Potentials in Clinical Testing*, 2e. UK: Churchill Livingston.

80 Heckenlively, J.R. and Arden, J.B. (1991). *Principles and Practice of Clinical Electrophysiology of Vision*. St Louis, MO: Mosby Year Book.

81 Mann, S. and Haykin, S. (1995). The chirplet transform – physical considerations. *IEEE Transactions on Signal Processing* **43** (11): 2745–2761.

82 Bultan, A. (1999). A four-parameter atomic decomposition of chirplets. *IEEE Transactions on Signal Processing* **47** (3): 731–745.

83 Mallat, S.G. and Zhang, Z. (1993). Matching-pursuit with time-frequency dictionaries. *IEEE Transactions on Signal Processing* **41** (12): 3397–3415.

84 Gribonval, R. (2001). Fast matching pursuit with a multiscale dictionary of Gaussian chirps. *IEEE Transactions on Signal Processing* **49** (5): 994–1001.

85 Qian, S., Chen, D.P., and Yin, Q.Y. (1998). Adaptive chirplet based signal approximation. *Proceedings of the IEEE International Conference on Acoustics, Speech, and Signal Processing*. 1–6: 1781–1784. ICASSP.

86 O'Neil, J.C. and Flandrin, P. (1998). Chirp hunting. Proceedings of the IEEE-SP International Symposium on Time–Frequency and Time–Scale Analysis, 425–428.

87 Ullsperger, P., Metz, A.M., and Gille, H.G. (1998). The P300 component of the event-related brain potential and mental effort. *Ergonomics* **31**: 1127–1137.

88 Uetake, A. and Murata, A. (2000). Assessment of mental fatigue during VDT task using event-related potential (P300). *Proceedings of the IEEE International Workshop on Root and Human Interactive Communication*, 235–240. Osaka, Japan.

89 Ulsperger, P., Neumann, U., Gille, H.G., and Pictschan, M. (Elsevier, 1986). P300 component of the ERP as an index of processing difficulty. In: *Human Memory and Cognitive Capabilities* (eds. F. Flix and H. Hagendorf), 723–773. Amsterdam.

90 Neumann, U., Ulsperger, P., and Erdman, U. (1986). Effects of graduated processing difficulty on P300 component of the event-related potential. *Zeitschrift für Psychologie* **194**: 25–37.

91 Mc Farland, D.J., Miner, L.A., Vaugan, T.M., and Wolpaw, J.R. (2000). Mu and beta rhythms topographies during motor imagery and actual movement. *Brain Topography* **3**: 177–186.

92 Vidal, J.J. (1977). Real-time detection of brain events in EEG. *Proceedings of the IEEE* **65**: 633–664, Special Issue on Biological Signal Processing and Analysis.

93 Sutter, E.E. (1992). The brain response interface communication through visually induced electrical brain response. *Journal of Microcomputer Applications* **15**: 31–45.

94 Morgan, S.T., Hansen, J.C., and Hillyard, S.A. (1996). Selective attention to the stimulus location modulates the steady state visual evoked potential. *Neurobiology* **93**: 4770–4774.

95 Muller, M.M. and Hillyard, S.A. (1997). Effect of spatial selective attention on the steady-state visual evoked potential in the 20-28 Hz range. *Cognitive Brain Research* **6**: 249–261.

96 Birbaumer, N., Hinterberger, T., Kubler, A., and Neumann, N. (2003). The thought-translation device (ttd): neurobehavioral mechanisms and clinical outcome. *IEEE Transactions on Neural Systems and Rehabilitation Engineering* **11**: 120–123.

97 Sellers, E.W., Kubler, A., and Donchin, E. (2006). Brain-computer interface research at the University of South Florida Cognitive Psychology Laboratory: the P300 speller. *IEEE Transactions on Neural Systems and Rehabilitation Engineering* **14** (2): 221–224.

98 Diez, J., Pfurtscheller, G., Reisecker, F. et al. (1997). Event-related desynchronization and synchronization in idiopathic Parkinson's disease. *Electroencephalography and Clinical Neurophysiology* **103** (1): 155–155.

99 Beverina, F., Palmas, G., Silvoni, S. et al. (2003). User adaptive BCIs: SSVEP and P300 based interfaces. *Journal of PsychNology* **1** (4): 331–354.

100 Kelly, S.P., Lalor, E.C., Finucane, C. et al. (2005). Visual spatial attention control in an independent brain-computer interface. *IEEE Transactions on Biomedical Engineering* **52** (9): 1588–1596.

101 Vaughan, T.M., McFarland, D.J., Schalk, G. et al. (2006). The Wandsworth BCI Research and Development Program: at home with BCI. *IEEE Transactions on Neural Systems and Rehabilitation Engineering* **14** (2): 229–233.

102 Thulasides, M., Guan, C., and Wu, J. (2006). Robust classification of EEG signal for brain-computer interface. *IEEE Transactions on Neural Systems and Rehabilitation Engineering* **14** (1): 24–29.

10

Localization of Brain Sources

10.1 Introduction

The human brain consists of a large number of regions each when active, generates a local magnetic field or synaptic electric current. The brain activities can be considered to constitute signal sources which are either spontaneous, corresponding to the normal rhythms of the brain, a result of brain stimulation, or are related to physical movements. Localization of brain signal sources from solely EEGs has been an active area of research during the last two decades.

Source localization is necessary to study brain physiological, mental, pathological, and functional abnormalities, and even problems related to various body disabilities, and ultimately to specify the cortical and deep sources of abnormalities.

Radiological imaging modalities have also been widely used for this purpose. However, these techniques are often unable to locate the abnormalities when they stem from the abnormal brain functions. Moreover, they are costly and some may not be accessible for all patients at the time they are needed.

The functional magnetic resonance imaging (fMRI) modality is able to show the location of anatomical disorders such as tumours and functional changes in the brain as the results of changes in blood oxygen level. This can be due to event-related and movement-related stimulations or some abnormalities such as epileptic seizure. From the fMRI, one may detect the effect of blood oxygen level dependence (BOLD) during metabolic changes, such as those caused by interictal seizure, in the form of white patches. The major drawbacks of fMRI, however, are its poor temporal resolution and its limitations in detecting the details of functional and mental activities. As a result, despite the cost limitation for this imaging modality, it has been reported that in 40–60% of the cases with interictal activity in EEG, fMRI cannot locate any foci even when the recording is simultaneous with EEG.

Functional brain imaging and source localization based on scalp potentials require a solution to an ill-posed inverse problem with many possible solutions. Selection of a particular solution often requires a priori knowledge acquired from the overall physiology of the brain and the status of the subject.

EEG Signal Processing and Machine Learning, Second Edition. Saeid Sanei and Jonathon A. Chambers.

Although in general localization of the brain sources is a difficult task, there are some simple situations where the localization problem can be simplified and accomplished:

- In places where the objective is to find the proximity of the actual source locations over the scalp. A simple method is to attempt to somehow separate the sources using principal component analysis (PCA) or independent component analysis (ICA), and backproject the sources of interest onto the electrode space and look at the scalp topography.
- Situations, where the aim is to localize a certain source of interest within the brain, e.g. a source of evoked potential (EP) or a source of movement-related potential (as widely used in the context of brain–computer interfacing [BCI]). It is easy to fit a single dipole at various locations over a coarse spatial sampling of the source space, and then choose the location producing the best match to the electrode signals (mixtures of the sources) as the focus of a spatially constrained, but more finely sampled search. The major problem in this case is that the medium is not linear and this causes error especially for sources deep inside the brain or where there is no prior knowledge about the shape of the source signal.

10.2 General Approaches to Source Localization

Brain source localization is probably the most challenging and difficult operation in dealing with EEG signals due to three main reasons: firstly, the human head is nonhomogeneous; secondly, human heads are neither spherical nor have similarity to each other, and finally, the sources can be multiple, distributed, or correlated (temporally, spatially, or both). Moreover, understanding of the methods requires deep study of the mathematics involved in magnetic and electric field propagations.

In order to localize multiple sources within the brain using EEG or magnetoencephalography (MEG) two general approaches have been proposed by researchers namely:

- equivalent current dipole (ECD)
- linear distributed (LD) approaches.

In the ECD approach the signals are assumed to be generated by a relatively small number of focal sources [1–4]. In the LD approaches all possible source locations are considered simultaneously [5–12]. Current source density (CSD) models, i.e. standardized low-resolution electromagnetic tomography algorithm (sLORETA), minimum norm least squares (MNLS), and low-resolution electromagnetic tomography algorithm (LORETA), were developed to overcome the limitations of dipole models, which assume a single or a small number of dipoles can represent the source of EEG [13].

Compared with EEG, MEG signals are more useful and reliable for localization of brain dipole sources [14] mainly due to the transparency of head tissue against magnetic fields and consequently linearity of the localization parameters (such as attenuation). Unfortunately, MEG is a very expensive system and much less accessible for brain screening.

In the inverse methods using the dipole source model the sources are considered as a number of discrete magnetic dipoles located in certain places in a three-dimensional space within the brain. The dipoles have fixed orientations and variable amplitudes.

Conversely, in the current distributed-source reconstruction (CDR) methods, there is no need for any knowledge about the number of sources. Generally, this problem is considered as an underdetermined inverse problem. An L_p-norm solution is the most popular regulation operator to solve this problem. This regularized method is based upon minimizing the cost function:

$$\psi = \|\mathbf{Ls} - \mathbf{x}\|_p + \lambda \|\mathbf{Ws}\|_p, \tag{10.1}$$

where $\mathbf{s}$ is the vector of source currents, $\mathbf{L}$ is the lead field matrix; $\mathbf{x}$ is the EEG measurements, $\mathbf{W}$ is a diagonal location weighting matrix, λ is the regulation parameter, and $1 \leq p \leq 2$, the norm, is the measure in complete normed vector (Banach) space [15]. A minimum L_p-norm method refers to the above criterion when $\mathbf{W}$ is equal to the identity matrix.

10.2.1 Dipole Assumption

For a dipole at location $\mathbf{L}$, the magnetic field observed at electrode i at location $\mathbf{R}(i)$ is achieved as (Figure 10.1)

$$\mathbf{B}(i) = \frac{\mu}{4\pi} \frac{\mathbf{Q} \times (\mathbf{R}(i) - \mathbf{L})}{|\mathbf{R}(i) - \mathbf{L}|} \text{ for } i = 1, \ldots, n_e \tag{10.2}$$

where $\mathbf{Q}$ is the dipole moment, | . | denotes absolute value, and $\times$ represents outer vector product. This is frequently used as the model for magneto-encephalographic (MEG) data observed by magnetometers. This can be extended to the effect of a volume containing m dipoles at each one of the n_e electrodes (Figure 10.2).

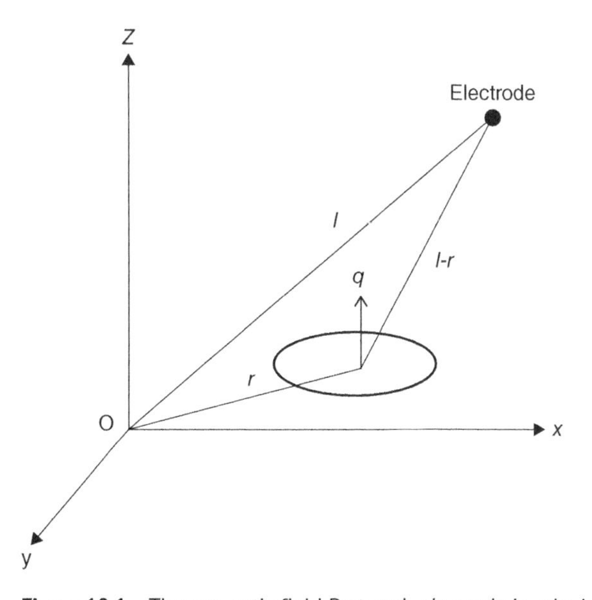

Figure 10.1 The magnetic field B at each electrode is calculated with respect to the moment of the dipole and the distance between the centre of the dipole and the electrode.

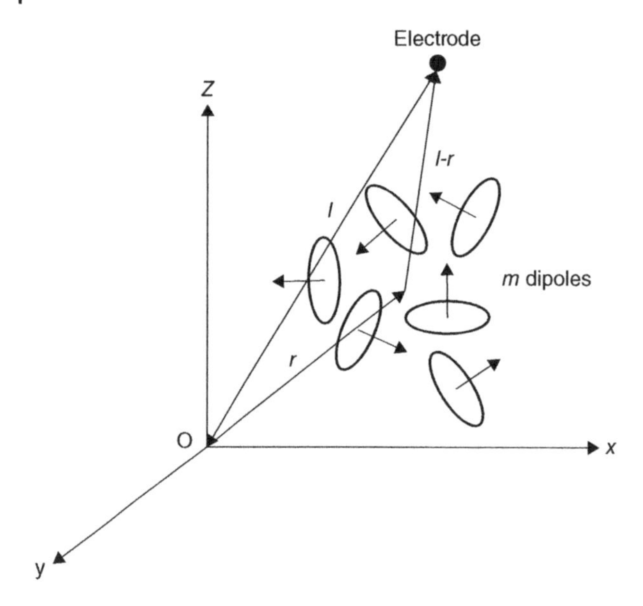

Figure 10.2 The magnetic field B at each electrode is calculated with respect to the accumulated moments of the *m* dipoles and the distance between the centre of the dipoles' volume and the electrode.

In the *m*-dipole case the magnetic field at point *j* is obtained as:

$$\mathbf{B}(i) = \frac{\mu_0}{4\pi}\sum_{j=1}^{m}\frac{\mathbf{Q}_j \times (\mathbf{R}(i) - \mathbf{L}_j)}{|\mathbf{R}(i) - \mathbf{L}_j|} \quad \text{for } i = 1, \ldots, n_e \tag{10.3}$$

where n_e is the number of electrodes and $\mathbf{L}_j$ represents the location of the *j*th dipole. The matrix **B** can be considered as $\mathbf{B} = [\mathbf{B}(1), \mathbf{B}(2)\ldots, \mathbf{B}(n_e)]$. Conversely, the dipole moments can be factorized into the product of their unit orientation moments and strengths, i.e. $\mathbf{B} = \mathbf{GQ}$ (normalized with respect to $\mu_0/4\pi$) where $\mathbf{G} = [\mathbf{g}(1), \mathbf{g}(2),\ldots \mathbf{g}(m)]$ is the propagating medium (mixing matrix) and $\mathbf{Q} = [\mathbf{Q}_1, \mathbf{Q}_2, \ldots \mathbf{Q}_m]^T$. The vector $\mathbf{g}(i)$ has dimension 1×3 (thus **G** is $m \times 3$). Therefore, $\mathbf{B} = \mathbf{GMS}$. **GM** can be written as a function of location and orientation such as **H(L,M)**, and therefore $\mathbf{B} = \mathbf{H(L,M)S}$. The initial solution to this problem was by using a least-squares (LS) search that minimizes the difference between the estimated and the measured data:

$$J_{ls} = \|\mathbf{X} - \mathbf{H}(\mathbf{L}, \mathbf{M})\mathbf{S}\|_F^2 \tag{10.4}$$

where **X** is the magnetic (potential) field over the electrodes. The parameters to be estimated are location, dipole orientation, and magnitude for each dipole. This is subject to knowing the number of the sources (dipoles). If too few dipoles are selected then the resulting parameters are influenced by the missing dipoles. Conversely, if too many dipoles are selected the accuracy will decrease since some of them are not valid brain sources. Also, the computation cost is high due to optimizing a number of parameters simultaneously. One way to overcome this is by converting this problem into a projection minimization problem as:

$$J_{ls} = \|\mathbf{X} - \mathbf{H}(\mathbf{L}, \mathbf{M})\mathbf{S}\|_F^2 = \left\|\mathbf{P}_\mathbf{H}^\perp \mathbf{X}\right\|_F^2 \tag{10.5}$$

The matrix $\boldsymbol{P}_H^\perp$ projects the data onto the orthogonal complement of the column space of $\mathbf{H}(\mathbf{L}, \mathbf{M})$. $\mathbf{X}$ can be reformed by singular value decomposition (SVD), i.e. $\mathbf{X} = \mathbf{U}\boldsymbol{\Sigma}\mathbf{V}^\mathrm{T}$. Therefore

$$J_{ls} = \left\|\mathbf{P}_\mathrm{H}^\perp \mathbf{U}\boldsymbol{\Sigma}\mathbf{V}^\mathrm{T}\right\|_F^2 \tag{10.6}$$

In this case orthogonal matrices preserve the Frobenius norm. The matrix $\mathbf{Z} = \mathbf{U}\boldsymbol{\Sigma}$ is $m \times m$ unlike $\mathbf{X}$ which is $n_e \times T$, where T is the number of samples. Generally, the rank of $\mathbf{X}$ satisfies $rank(\mathbf{X}) \leq \mathrm{m}$, and $T >> m$, and $\boldsymbol{\Sigma}$ can have only m non-zero singular values. This means that the overall computation cost has been reduced by a large amount. The SVD can also be used to reduce the computation cost for $\mathbf{P}_\mathbf{H}^\perp = \left(\mathbf{I} - \mathbf{H}\mathbf{H}^\dagger\right)$. The pseudo-inverse $\boldsymbol{H}^\dagger$ can be decomposed as $\mathbf{V}_H \boldsymbol{\Sigma}_H^\dagger \mathbf{U}_H^T$, where $\mathbf{H} = \mathbf{U}_H \boldsymbol{\Sigma}_H^\dagger \mathbf{V}_H^T$.

The dipole model, however, requires a priori knowledge about the number of sources, which is usually unknown.

Many experimental studies and clinical experiments have examined the developed source localization algorithms. Yao and Dewald [16] have evaluated different cortical source localization methods such as the moving dipole (MDP) method [17], minimum *Lp* norm [18], and low-resolution tomography (LRT) as for LORETA [19] using simulated and experimental EEG data. In their study, only the scalp potentials have been taken into consideration.

In this study some other source localization methods such as the cortical potential imaging method [20] and the three-dimensional resultant vector method [21] have not been included. These methods, however, follow similar steps as minimum norm methods and dipole methods respectively.

10.3 Head Model

One of the requirements for brain source localization, particularly for the forward model, is the information about the head model. The head model is the model for which the EEG forward solution is calculated. The forward solution determines how much a given electrical source in the brain will impact each electrode on the scalp. It provides the lead field matrix from which the inverse problem will be solved. For a more accurate solution the individual's MRI is often used to construct the head model, particularly in clinical studies where the source localization is used to guide surgery as for example in epilepsy or in functional mapping of the eloquent cortex. In a less precise application where the MRI is not available, a template MRI can be used (for example the MRI brain) [22]. The MRI needs several preprocessing steps in order to achieve a proper delineation of the grey matter in which the source activity is estimated, and to describe the different head tissue layers with different conductivity parameters. Since the electric field that spreads from the sources to the scalp surface is attenuated by these layers (particularly by the skull), a proper incorporation of the head shape and the conductivity parameters in the head model is essential during the localization/reconstruction process. Once the MRI is pre-processed, the electrodes have to be positioned on the head corresponding to how they were positioned during the recordings. It is evident that the localization accuracy depends on how accurately the electrode positions correspond to the real position from which the signals are recorded during the experiment.

For an exact model where simultaneous MRI and EEG are recorded, the imposed magnetic field artefacts have to be removed. Since this is only for one still MRI, the rest of the EEG remains unaffected.

As soon as the MRI is captured the lead field has to be calculated. The lead field determines how the electric activity at a certain electrode is related to the activity of the different sources in the brain. The more precise and anatomically correct this lead field is determined, the more precise the source localization will be. In order to calculate the lead field, a head model has to be created that incorporates as realistically as possible the shape of the head and the conductivity parameters of the different tissues between the current sources in the brain and the potential on the scalp. There have been many attempts to construct realistic head models. Yet, even the most complicated approaches are simplified descriptions of the complex structure of head tissues. The often-used realistic models are the boundary element model (BEM) and the finite element model (FEM). These models are much superior to the simple three-shell spherical head models [23–25] when applied to real data [26, 27]. The downside of these sophisticated head models is an increased computational load to achieve numerical solutions. They are also more sensitive to any mishap happening during the brain and grey matter extraction, as there would be more brain tissues and more parameters involved. Given the low temporal resolution of MRI, Figure 10.3 shows the steps in building up the head model [22].

Nevertheless, there are other parameters which affect the usable human head model. One more important one is age. The conductivity of skull decreases nonlinearly with age. In recent studies for a 10 year old subject the conductivity ratio between the brain and skull is approximately 10 to 1, whereas for a 55 year old subject this ratio is approximately 25 to 1. For elderlies this ratio can be 50–80 to 1 [28, 29].

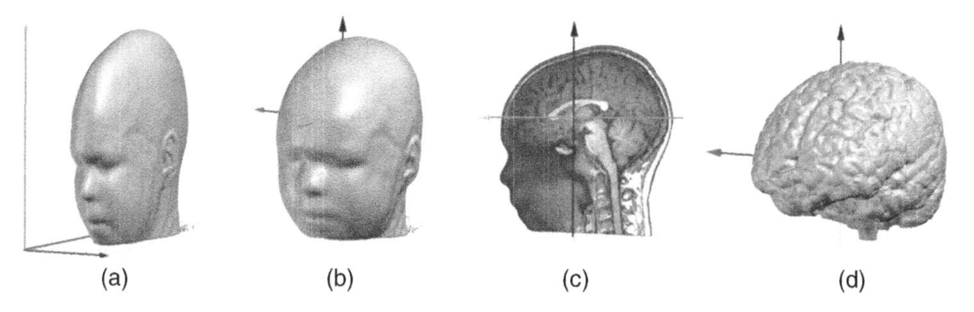

Figure 10.3 The steps in using MRI data to build up a head model. (a) Original anisotropic MRI. (b) Result of up-sampling and rotation with red, green, and blue axis pointing, respectively to x,y,z. (c) Adjustment of the cutting planes and setting the correct geometrical coordinate origin. (d) The result of skull-stripping to isolate the brain.

10.4 Most Popular Brain Source Localization Approaches

Although most of the methods described here have been used from both EEG and MEG signals, those specifically used for dipole source localization perform much more accurately for MEG signals since the brain is more linear and homogenous with respect to its magnetic field compared with its electric fields used for EEG registration.

The traditional approaches for brain source localization use linear or recursive signal processing techniques. These techniques exploit the multichannel nature of EEG or MEG data, the spatial or in some cases temporal or spectral sparsity of the sources, and if available an approximate shape of the desired source. However, in recent years where powerful computers and graphic cards with large memory allow the use of very computationally intensive machine learning methods, application of deep neural networks (DNNs) in source localization seem inevitable. The use of DNNs in solving similar problems such as sound source separation [30, 31] and communication channel estimation [32] have become popular in recent years.

In such methods the DNN learns the best arrangement of the cortical source locations for having particular patterns of EEG recordings over the scalp. The inherent nonlinearity property of the activation functions and the kernel-type nature of the multilayer system elegantly accommodates the nonlinearity imposed on the brain EEG sources by the head nonhomogeneous layers.

10.4.1 EEG Source Localization Using Independent Component Analysis

In a simple though not very accurate approach ICA is used to separate the EEG sources [33]. The correlation values of the estimated sources and the mixtures are then used to build up the model of the mixing medium. The LS approach is then used to find the sources using the inverse of these correlations, i.e. $d_{ij} \approx 1/(C_{ij})^{0.5}$, where d_{ij} shows the distance between source i and electrode j, and C_{ij} shows the correlation of their signals. The use of correlation values makes the scheme a simple approach to inverse problem based localization. The method has been applied to separate and localize the sources of P3a and P3b subcomponents for five healthy subjects and five schizophrenic patients. Study of ERP components such as P300 and its constituent subcomponents is very important in analysis, diagnosing, and monitoring of mental and psychiatric disorders. Source location, amplitude, and latency of these components have to be quantified and used in the classification process. Figure 10.4 illustrates the results overlayed on two MRI templates for the above two cases. It can be seen that for the healthy subjects the subcomponents are well apart whereas for the schizophrenic patients they are geometrically mixed.

10.4.2 MUSIC Algorithm

Multiple signal classification (MUSIC) [34] has been used for localization of the magnetic dipoles within the brain [35–38] using EEG signals. In an early development of this algorithm a single-dipole model within a three-dimensional head volume is scanned and projections onto an estimated signal subspace are computed [39]. To locate the sources, the user must search the head volume for multiple local peaks in the projection metric.

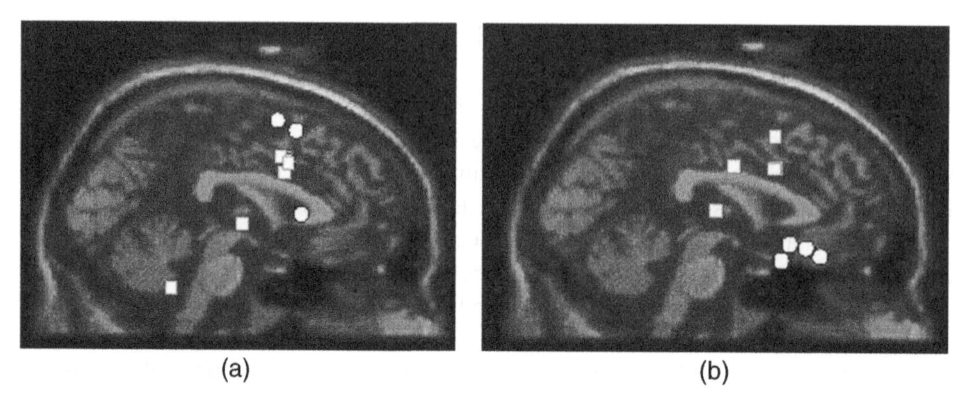

Figure 10.4 Localization results for (a) the schizophrenic patients and (b) the normal subjects. The circles represent P3a and the squares represent P3b.

In an attempt to overcome the exhaustive search by the MUSIC algorithm a recursive MUSIC algorithm was developed [35]. Following this approach, the locations of the fixed, rotating, or synchronous dipole sources are automatically extracted through a recursive use of subspace estimation. This approach tries to solve the problem of how to choose the locations, which give the best projection on to the signal (EEG) space. In the absence of noise and using a perfect head and sensors model, the forward model for the source at the correct location projects entirely into the signal subspace. In practise, however, there are estimation errors due to noise. In finding the solution to the above problem there are two assumptions which may not be always true: first, the data are corrupted by additive spatially white noise and second, the data are produced by a set of asynchronous dipolar sources. These assumptions are waved in the proposed recursive MUSIC algorithm [35].

The original MUSIC algorithm for the estimation of brain sources may be described as follows. Consider the head model for transferring the dipole field to the electrodes to be $\mathbf{A}(\boldsymbol{\rho},\boldsymbol{\theta})$, where $\boldsymbol{\rho}$ and $\boldsymbol{\theta}$ are respectively the dipole location and direction parameters. The relationship between the observations (EEG), $\mathbf{X}$, the model $\mathbf{A}$, and the sources $\mathbf{S}$ is given as:

$$\mathbf{X} = \mathbf{A}\mathbf{S}^T + \mathbf{E} \tag{10.7}$$

where $\mathbf{E}$ is the noise matrix. The goal is to estimate the parameters $\{\boldsymbol{\rho}, \boldsymbol{\theta}, \mathbf{S}\}$, given the data set $\mathbf{X}$. The correlation matrix of $\mathbf{X}$, i.e. $\mathbf{R}_x$ can be decomposed as:

$$\mathbf{X}_F = [\boldsymbol{\Phi}_s, \boldsymbol{\Phi}_e]\begin{bmatrix} \boldsymbol{\Lambda} + n_e\sigma_e^2\mathbf{I} & 0 \\ 0 & n_e\sigma_e^2\mathbf{I} \end{bmatrix}[\boldsymbol{\Phi}_s, \boldsymbol{\Phi}_e]^T \tag{10.8}$$

or

$$\mathbf{R}_x = \boldsymbol{\Phi}_s\boldsymbol{\Lambda}_s\boldsymbol{\Phi}_s^T + \boldsymbol{\Phi}_e\boldsymbol{\Lambda}_e\boldsymbol{\Phi}_e^T \tag{10.9}$$

where $\boldsymbol{\Lambda}_s = \boldsymbol{\Lambda} + n_e\sigma_e^2\mathbf{I}$ is the $m \times m$ diagonal matrix combining both the model and noise eigenvalues, and $\boldsymbol{\Lambda}_e = n_e\sigma_e^2\mathbf{I}$ is the $(n_e - m) \times (n_e - m)$ diagonal matrix of noise-only eigenvalues. Therefore, the signal subspace span ($\boldsymbol{\Phi}_s$) and noise-only subspace span ($\boldsymbol{\Phi}_e$) are orthogonal. In practise, T samples of the data are used to estimate the above parameters, i.e.

$$\hat{\mathbf{R}}_x = \mathbf{X}\mathbf{X}^T = \hat{\mathbf{\Phi}}_s\hat{\mathbf{\Lambda}}_s\hat{\mathbf{\Phi}}_s^{\ T} + \hat{\mathbf{\Phi}}_e\hat{\mathbf{\Lambda}}_e\hat{\mathbf{\Phi}}_e^{\ T} \tag{10.10}$$

where the first m left singular vectors of the decomposition are designated as $\hat{\mathbf{\Phi}}_s$ and the remaining eigenvectors as $\hat{\mathbf{\Phi}}_e$. Accordingly, the diagonal matrix $\hat{\Lambda}_s$ contains the first m eigenvalues and $\hat{\mathbf{\Lambda}}_e$ contains the remainder. To estimate the above parameters the general rule using least-squares (LS) fitting is:

$$\left\{\hat{\boldsymbol{\rho}}, \hat{\boldsymbol{\theta}}, \hat{\mathbf{S}}\right\} = arg \min_{\boldsymbol{\rho}, \boldsymbol{\theta}, \mathbf{S}} \|\mathbf{X} - \mathbf{A}(\boldsymbol{\rho}, \boldsymbol{\theta})\mathbf{S}^T\|_F^2 \tag{10.11}$$

where $\| \,.\, \|_F$ denotes Frobenius norm. By optimal substitution [40] we achieve:

$$\left\{\hat{\boldsymbol{\rho}}, \hat{\boldsymbol{\theta}}\right\} = arg \min_{\boldsymbol{\rho}, \boldsymbol{\theta}} \|\mathbf{X} - \mathbf{A}\mathbf{A}^{\dagger}\mathbf{X}\|_F^2 \tag{10.12}$$

where $\mathbf{A}^{\dagger}$ is the Moore-Penrose pseudo-inverse of $\mathbf{A}$ [41]. Given that the rank of $\mathbf{A}(\boldsymbol{\rho}, \boldsymbol{\theta})$ is m and the rank of $\hat{\mathbf{\Phi}}_S$ is at least m, the smallest subspace correlation value, $C_m = \text{subcorr}\{\mathbf{A}(\boldsymbol{\rho}, \boldsymbol{\theta}), \hat{\mathbf{\Phi}}_s\}_m$, represents the minimum subspace correlation (maximum principal angle) between principal vectors in the column space of $\mathbf{A}(\boldsymbol{\rho},\boldsymbol{\theta})$ and the signal subspace $\hat{\mathbf{\Phi}}_s$. In MUSIC, the subspace correlation of any individual column of $\mathbf{A}(\boldsymbol{\rho}, \boldsymbol{\theta})$, i.e. $\mathbf{a}(\boldsymbol{\rho}_i,\boldsymbol{\theta}_i)$ with the signal subspace must therefore equal or exceed this minimum subspace correlation:

$$C_i = \text{subcorr}\left\{\mathbf{a}(\boldsymbol{\rho}_i, \boldsymbol{\theta}_i), \hat{\boldsymbol{\Phi}}_s\right\} \geq C_m \tag{10.13}$$

$\hat{\mathbf{\Phi}}_s$ approaches $\mathbf{\Phi}_s$ when the number of samples increases or the signal-to-noise ratio (SNR) becomes higher. Then the minimum correlation approaches unity when the correct parameter set is identified, such that the m distinct sets of parameters $(\boldsymbol{\rho}_i,\boldsymbol{\theta}_i)$ have subspace correlations approaching unity. Therefore a search strategy is followed to find the m peaks of the metric [35]:

$$subcorr^2\left\{\mathbf{a}(\boldsymbol{\rho}, \boldsymbol{\theta}), \hat{\mathbf{\Phi}}_s\right\} = \frac{\boldsymbol{a}^T(\rho, \theta)\hat{\boldsymbol{\Phi}}_S\hat{\boldsymbol{\Phi}}_S^T\boldsymbol{a}(\rho, \theta)}{\|\boldsymbol{a}^T(\rho, \theta)\|_2^2} \tag{10.14}$$

where $\|\cdot\|_2^2$ denotes squared Euclidean norm. For a perfect estimation of the signal subspace m global maxima equal to unity are found. This requires searching for both sets of parameters $\boldsymbol{\rho}$ and $\boldsymbol{\theta}$. In a quasilinear approach [35] it is assumed that $\mathbf{a}(\boldsymbol{\rho}_i,\boldsymbol{\theta}_i) = \mathbf{G}(\boldsymbol{\rho}_i)\boldsymbol{\theta}_i$ where $\mathbf{G}(.)$ is called the gain matrix. In the EEG source localization application therefore, we first find the dipole parameters $\boldsymbol{\rho}_i$ which maximize the $subcorr\{\mathbf{G}(\boldsymbol{\rho}_i),\hat{\mathbf{\Phi}}_s\}$ and then extract the corresponding quasilinear $\boldsymbol{\theta}_i$ which maximize this subspace correlation. This avoids explicitly searching for these quasilinear parameters, reducing the overall complexity of the nonlinear search. Therefore, the overall localization of the EEG sources using classic MUSIC can be summarized to the following steps: (i) Decompose $\mathbf{X}$ or $\mathbf{XX}^T$ and select the rank of the signal subspace to obtain $\hat{\mathbf{\Phi}}_S$. Slightly overspecifying the rank has little effect on the performance whereas, underspecifying it can dramatically reduce the performance; (ii) form the gain matrix $\mathbf{G}$ at each point (node) of a dense grid of dipolar source locations and calculate the subspace correlations $subcorr\{\mathbf{G}(\boldsymbol{\rho}_i),\hat{\mathbf{\Phi}}_s\}$; and (iii) find the peaks of the plot $\sqrt{1 - C_1^2}$, where C_1 is the maximum subspace correlation. Locate m or fewer peaks in the grid. At each peak, refine the search grid to improve the location accuracy and check the second subspace correlation. A large second subspace correlation is an indication of a rotating dipole [35]. Unfortunately, often there are errors in estimating the signal subspace and the correlations are computed at only a finite set of grid points. A recursive MUSIC algorithm overcomes this problem by recursively building up the independent topography model. In this model the number of dipoles is initially considered as one and the search is carried out to locate a single dipole. A second dipole is then added and the dipole point that maximizes the second subspace correlation, C_2, is found. At this stage there is no need to recalculate C_1. The number of dipoles is increased and the new subspace correlations are computed. If m topographies comprise m_1 single-dipolar topographies and m_2 two-dipolar topographies, then, the recursive MUSIC will first extract the m_1 single-dipolar models. At the $(m_1 + 1)$th iteration, no single-dipole location that correlates well with the subspace will be found. By increasing the number of dipole elements to two, the searches for both have to be carried out simultaneously such that the subspace correlation is maximized for C_{m_1+1}. The procedure continues to build the remaining m_2 two-dipolar topographies. After finding each pair of dipoles that maximizes the appropriate subspace correlation, the corresponding correlations are also calculated.

Another extension of MUSIC-based source localization for diversely polarized sources, namely recursively applied and projected (RAP) MUSIC [37], uses each successively located source to form an intermediate array gain matrix (similar to the recursive MUSIC algorithm) and projects both the array manifold and the signal subspace estimate into its orthogonal complement. In this case the subspace is reduced. Then, the MUSIC projection to find the next source is performed. This method was initially applied to localization of MEG sources [37, 38].

In another study by Xu et al. [42], an approach to EEG three-dimensional (3D) dipole source localization by using a non-recursive subspace algorithm, called FINES, has been proposed. The approach employs projections onto a subspace spanned by a small set of particular vectors in the estimated noise-only subspace, instead of the entire estimated noise-only subspace in the case of classic MUSIC. The subspace spanned by this vector set is, in the sense of principal angle, closest to the subspace spanned by the array manifold associated with a particular brain region. By incorporating knowledge of the array manifold in identifying the FINES vector sets in the estimated noise-only subspace for different brain regions, this approach is claimed to be able to estimate sources with enhanced accuracy and spatial resolution, thus enhancing the capability of resolving closely spaced sources and reducing estimation errors. The simulation results show that, compared to classic MUSIC, FINES has a better resolvability of two closely spaced dipolar sources and also a better estimation accuracy of source locations. In comparison with RAP-MUSIC, the performance of the FINES is also better for the cases where the noise level is high and/or correlations among dipole sources exist [42].

A method for using a generic head model, in the form of an anatomical atlas, has also been proposed to produce EEG source localization [43]. The atlas is fitted to the subject by a nonrigid warp using a set of surface landmarks. The warped atlas is used to compute a FEM of the forward mapping or lead-fields between neural current generators and the EEG electrodes. These lead-fields are used to localize current sources from the EEG data of the subject and the sources are then mapped back to the anatomical atlas. This approach provides a mechanism for comparing source localizations across subjects in an atlas-based coordinate system.

10.4.3 LORETA Algorithm

The LORETA for localization of brain sources has already been commercialized. In this method, the electrode potentials and matrix **X**, are considered to be related as:

$$\mathbf{X} = \mathbf{LS} \tag{10.15}$$

where **S** is the actual (current) source amplitudes (densities) and **L** is an $n_e \times 3m$ matrix representing the forward transmission coefficients from each source to the array of sensors. **L** has also been referred to as the system response kernel or the lead field matrix [44]. Each column of **L** contains the potentials observed at the electrodes when the source vector has unit amplitude at one location and orientation and is zero at all others. This requires the potentials to be measured linearly with respect to the source amplitudes based on the superposition principle. Generally, this is not true and therefore such an assumption inherently creates some error. The fitted source amplitudes, **S**, can be approximately estimated using an exhaustive search through the inverse LS solution, i.e.:

$$\mathbf{S} = \left(\mathbf{L}^{\mathrm{T}}\mathbf{L}\right)^{-1}\mathbf{L}^{\mathrm{T}}\mathbf{X} \tag{10.16}$$

L may be approximated by 3D simulations of current flow in the head, which requires a solution to the well-known Poisson equation [45]:

$$\nabla.\sigma\nabla\mathbf{X} = -\boldsymbol{\rho} \tag{10.17}$$

where σ is the conductivity of the head volume $(\Omega\mathrm{m})^{-1}$ and $\boldsymbol{\rho}$ is the source volume current density $(\mathrm{A/m^3})$. An FEM or BEM may be used to solve this equation. In such models the geometrical information about the brain layers and their conductivities [46] have to be known. Unless some a priori knowledge can be used in the formulation, the analytic model is ill-posed and a unique solution is hard to achieve.

Conversely, the number of sources, m, is typically much larger than the number of sensors, n_e, and the system in (10.15) is underdetermined. Also, in the applications where more concentrated focal sources are to be estimated such methods fail. As we see later, using a minimum norm approach, some researchers choose the solution which satisfies some constraints such as the smoothness of the inverse solution.

One approach is the minimum norm solution, which minimizes the norm of S under the constraint of the forward problem:

$$\min \|\mathrm{S}\|_2^2, \text{subject to } \mathbf{X} = \mathbf{LS} \tag{10.18}$$

with a solution as:

$$\mathbf{S} = \mathbf{L}^{\mathrm{T}}\left(\mathbf{LL}^{\mathrm{T}}\right)^{\dagger}\mathbf{X} \tag{10.19}$$

The motivation of the minimum norm solution is to create a sparse solution with zero contribution from most of the sources. This method has the serious drawback of poor localization performance in three-dimensional (3D) space. An extension to this method is the weighted minimum norm (WMN) solution, which compensates for deep sources and hence performs better in 3D space. In this case the norms of the columns of **L** are normalized. Hence the constrained WMN is formulated as:

$$\min \|\mathbf{WS}\|_2^2, \text{subject to} \mathbf{X} = \mathbf{LS} \tag{10.20}$$

with a solution as:

$$\mathbf{S} = \mathbf{W}^{-1}\mathbf{L}^{\mathrm{T}}\left(\mathbf{L}\mathbf{W}^{-1}\mathbf{L}^{\mathrm{T}}\right)^{\dagger}\mathbf{X} \tag{10.21}$$

where **W** is a diagonal $3m \times 3m$ weighting matrix, which compensates for deep sources in the following way:

$$\mathbf{W} = diag\left[\frac{1}{\|\mathbf{L}_1\|_2}, \frac{1}{\|\mathbf{L}_2\|_2}, \ldots, \frac{1}{\|\mathbf{L}_{3m}\|_2}\right] \tag{10.22}$$

where $\|\mathbf{L}_i\|_2$ denotes the Euclidean norm of the ith column of **L**, i.e. **W** corresponds to the inverse of the distances between sources and electrodes.

In another similar approach a smoothing Laplacian operator is employed. This operator produces a spatially smooth solution agreeing with the physiological assumption mentioned earlier. The function of interest is then:

$$\min \|\mathrm{BWS}\|_2^2, \text{subject to} \mathbf{X} = \mathbf{LS} \tag{10.23}$$

where **B** is the Laplacian operator. This minimum norm approach produces a smooth topography in which the peaks representing the source locations are accurately located.

An FEM has been used to achieve a more anatomically realistic volume conductor model of the head, in an approach called adaptive standardized LORETA/FOCUSS underdetermined system solver (FOCUSS) [47]. It is claimed that when using this application-specific method a number of different resolution solutions using different mesh intensities can be combined to achieve the localization of sources with less computational complexities. Initially, FEM is used to approximate the solutions to (10.23) with a realistic representation of the conductor volume based on magnetic resonance (MR) images of the human head. The dipolar sources are presented using the approximate Laplace method [48].

10.4.4 FOCUSS Algorithm

The FOCUSS [47] algorithm is based on a high-resolution iterative WMN method that uses the information from the previous iterations as:

$$\min \|\mathbf{CS}\|_2^2, \text{subject to} \mathbf{X} = \mathbf{LS} \tag{10.24}$$

where $\boldsymbol{C} = (\boldsymbol{Q}^{-1})^T\boldsymbol{Q}^{-1}$ and $\boldsymbol{Q}_i = \boldsymbol{W}\boldsymbol{Q}_{i-1}[diag(\boldsymbol{S}_{i-1}(1)\cdots\boldsymbol{S}_{i-1}(3m)]$ the solution at iteration i becomes:

$$\mathbf{S}_i = \mathbf{Q}_i\mathbf{Q}_i^T\mathbf{L}^T\left(\mathbf{L}\mathbf{Q}_i\mathbf{Q}_i^T\mathbf{L}^T\right)^+\mathbf{X} \tag{10.25}$$

The iterations stop when there is no significant change in the estimation. The result of FOCUSS is highly dependent on the initialization of the algorithm. In practise, the algorithm converges close to the initialization point and may easily become stuck in some local minimum. A clever initialization of FOCUSS has been suggested to be the solution to LORETA [47].

10.4.5 Standardized LORETA

As another option referred to as sLORETA, a unique solution to the inverse problem is found. It uses a different cost function, which is:

$$\min_{\mathbf{S},\mathbf{L}} \left(\|\mathbf{X} - \mathbf{LS}\|_2^2 + \lambda \|\mathbf{S}\|_2^2 \right) \tag{10.26}$$

Hence, sLORETA uses a zero-order Tikhonov-Phillips regularization [49, 50], which provides a possible solution to the ill-posed inverse problems:

$$\mathbf{s}_i = \mathbf{L}_i^T \left[\mathbf{L}_i \mathbf{L}_i^T + \lambda_i \mathbf{I} \right]^{-1} \mathbf{X} = \mathbf{R}_i \mathbf{S} \tag{10.27}$$

where $\mathbf{s}_i$ indicates the candidate sources and $\mathbf{S}$ are the actual sources. $\mathbf{R}_i$ is the resolution matrix defined as:

$$\mathbf{R}_i = \mathbf{L}_i^T \left[\mathbf{L}_i \mathbf{L}_i^T + \lambda_i \mathbf{I} \right]^{-1} \mathbf{L}_i \tag{10.28}$$

The reconstruction of multiple sources performed by the final iteration of sLORETA is used as an initialization for the combined ALF and WMN (or FOCUSS) algorithms [48]. The number of sources is reduced each time and (10.19) is modified to:

$$\mathbf{s}_i = \mathbf{W}_i \mathbf{W}_i^T \mathbf{L}_f^T \left[\mathbf{L}_f \mathbf{W}_i \mathbf{W}_i^T \mathbf{L}_f^T + \lambda \mathbf{I} \right]^{-1} \mathbf{X} \tag{10.29}$$

$\mathbf{L}_f$ indicates the final $n \times m$ lead field returned by the sLORETA. $\mathbf{W}_i$ is a diagonal $3m_f \times 3m_f$ matrix, which is recursively refined based on the current density estimated by the previous step:

$$\mathbf{W}_i = diag\left(s_{i-1}(1), s_{i-1}(2), \ldots, s_{i-1}(3n_f)\right) \tag{10.30}$$

and the resolution matrix in (10.28) after each iteration changes to:

$$\mathbf{R}_i = \mathbf{W}_i \mathbf{W}_i^T \mathbf{L}_f^T \left[\mathbf{L}_f \mathbf{W}_i \mathbf{W}_i^T \mathbf{L}_f^T + \lambda \mathbf{I} \right]^{-1} \boldsymbol{L}_f \tag{10.31}$$

Iterations are continued until the solution does not significantly change. In another approach by Liu et al. [51] called shrinking standard LORETA–FOCUSS (SSLOFO), sLORETA is used for initialization. Then, it uses the re-WMN of FOCUSS. During the process the localization results are further improved by involving the above standardization technique. However, FOCUSS normally creates an increasingly sparse solution during iteration. Therefore, it is better to eliminate the nodes with no source activities or recover those active nodes which might be discarded by mistake. The algorithm proposed in [52] shrinks the source space after each iteration of FOCUSS, hence reducing the computational load [51]. For the algorithm not to be trapped in a local minimum a smoothing operation is performed. The overall SSLOFO is therefore summarized by the following steps [51, 52]:

1) Estimate the current density $\hat{\mathbf{S}}_0$ using sLORETA.
2) Initialize the weighting matrix as $\mathbf{Q}_0 = diag[\hat{\mathbf{S}}_0(1), \hat{\mathbf{S}}_0(2) \ldots \hat{\mathbf{S}}_0(3m)]$.
3) Estimate the source power using standardized FOCUSS.
4) Retain the prominent nodes and their neighbouring nodes. Adjust the values on these nodes through smoothing.

5) Redefine the solution space to contain only the retained nodes, i.e. only the corresponding elements in **S** and the corresponding column in **L**.
6) Update the weighting matrix.
7) Repeat steps 3–6 until a stopping condition is satisfied.
8) The final solution is the result of the last step before smoothing.

The stopping condition maybe by defining a threshold, or when there is no considerable change in the weights in further iterations.

10.4.6 Other Weighted Minimum Norm Solutions

In an LD approach by Phillips et al. [5] a WMN solution (or Tikhonov regularization) method [53] has been proposed. The solution has been regularized by imposing some anatomical and physiological information upon the overall cost function in the form of constraints. The squared error costs are weighted based on spatial and temporal properties. The information such as hemodynamic measures of brain activity from other imaging modalities such as fMRI, are used as constraints (or priors) together with the proposed cost function. In this approach it is assumed that the sources are sufficiently densely distributed and the sources are oriented orthogonal to the cortical sheet.

The instantaneous EEG source localization problem using a multivariate linear model and the observations, **X**, as electrode potentials, is generally formulated on the basis of the observation model:

$$\mathbf{X} = \Im(\mathbf{r}, \mathbf{J}) + \mathbf{V} \tag{10.32}$$

where, $\mathbf{X} = [\mathbf{x}_1, \mathbf{x}_2, \ldots, \mathbf{x}_{ne}]^T$ has dimension $n_e \times T$, where T represents the length of the data in samples and n_e is the number of electrodes, $\mathbf{r}$ and $\mathbf{J} = [\mathbf{j}_1, \mathbf{j}_2, \ldots, \mathbf{j}_m]$ are respectively source locations and moments of the sources, and $\mathbf{V}$ is the additive noise matrix. $\Im$ is the function linking the sources to the electrode potentials. In calculation of $\Im$ a suitable three-layer head model is normally considered [54, 55]. A structural MR image of the head may be segmented into three isotropic regions namely, brain, skull, and scalp, of the same conductivity [56] and used as the model. Most of these models consider the head as a sphere for simplicity.

However, in [5] the EEG sources are modelled by a fixed and uniform three-dimensional grid of current dipoles spread within the entire brain volume. Also, the problem is an underdetermined linear problem as:

$$\mathbf{X} = \mathbf{LJ} + \mathbf{V} \tag{10.33}$$

where, **L** is the head field matrix, which inter-relates the dipoles to the electrode potentials. To achieve a unique solution for the above underdetermined equation some constraints have to be imposed. The proposed regularization method constrains the reconstructed source distribution by jointly minimizing a linear mixture of some weighted norm $\|\mathbf{Hj}\|_2$ of the current source $\mathbf{j}$ and the main cost function of the inverse solution. Assuming the noise is Gaussian with a covariance matrix $\mathbf{C}_v$ then:

$$\hat{\mathbf{j}} = arg\min_{j}\left\{\left\|\mathbf{C}_v^{-1/2}(\mathbf{Lj} - \mathbf{x})\right\|_2^2 + \lambda^2\|\mathbf{Hj}\|_2^2\right\} \tag{10.34}$$

where the Lagrange multiplier λ has to be adjusted to make a balance between the main cost function and the constraint $\|\mathbf{Hj}\|_2$. The covariance matrix is scaled in such a way to have trace($\mathbf{C}_v$) = rank($\mathbf{C}_v$) (recall that trace of a matrix is sum of its diagonal elements, and rank of a matrix refers to its number of independent columns [rows]). This can be stated as an overdetermined LS problem [20]. The solution to the minimization of (10.34) for a given λ is in the form of:

$$\hat{\mathbf{J}} = \mathbf{BX} \tag{10.35}$$

where

$$\mathbf{B} = \left[\mathbf{L}^T\mathbf{C}_v^{-1}\mathbf{L} + \lambda^2\left(\mathbf{H}^T\mathbf{H}\right)\right]^{-1}\mathbf{L}^T\mathbf{C}_v^{-1} = \left(\mathbf{H}^T\mathbf{H}\right)]^{-1}\mathbf{L}^T\left[\mathbf{L}\left(\mathbf{H}^T\mathbf{H}\right)^{-1}\mathbf{L}^T + \lambda^2\mathbf{C}_v\right]^{-1} \tag{10.36}$$

These equations describe the WMN solution to the localization problem. However, this is not complete unless a suitable spatial or temporal constraint is imposed. Theoretically, any number of constraints can be added to the main cost function in the same way and the hyperparameters such as Lagrange multipliers, λ, can be calculated by expectation–maximization [57]. However, more assumptions such as those about the covariance of the sources have to be implied in order to find effectively $\mathbf{L}$, which includes the information about the locations and the moments. One assumption, in the form of constraint, can be based on the fact that $(diag(\mathbf{L}^T\mathbf{L}))^{-1}$, which is proportional to the covariance components, should be normalized. Another constraint is based on the spatial fMRI information which appears as BOLD when the sources are active. Evoked responses can also be used as temporal constraints.

Various inverse problem methods for localization of EEG sources can be seen in Figure 10.5.

In reality, EEG and MEG recordings are the spatial integration of the activity from large and multiple remotely located populations of neurons. Functionally connected brain networks are those temporally correlated and spatially remote neurological sources. Such functional networks have been described in brains engaged in complex cognitive tasks that necessarily require the transient integration of numerous functional areas widely distributed over the brain and in continuous interaction.

One way of doing this integration can be by neural phase synchrony. This refers to neuronal groups, oscillating in specific frequency bands, that are engaged into precise phase-locking over a limited period of time. In turn, this has motivated the search for robust methods to directly measure such phase synchrony from experimentally recorded biological signals. Some applications of phase synchrony can be found for monitoring mental fatigue as described in Chapter 13 of this book.

Generally, massive diffusive effects and poor SNR preclude accurate estimation of indices related to the cortical dynamics from non-averaged scalp recordings.

In [58] a similar WMN approach used for dipole source localization has been utilized for estimation of neural dynamics of cortical sources. In their method surrogate data have been used to reduce the source space in order to reconstruct such data.

The activity is expected to be produced by a large number of brain regions interacting together in different dynamic networks. The distributed-source model considers a great

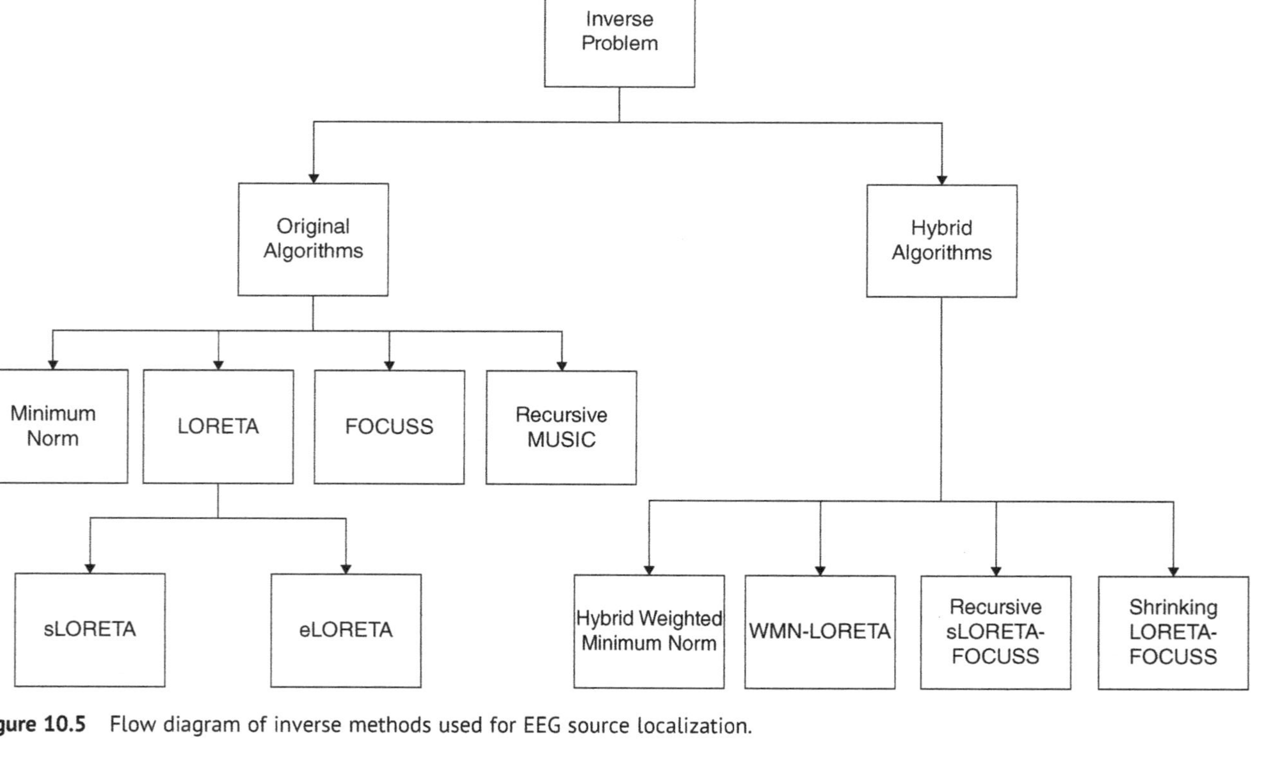

Figure 10.5 Flow diagram of inverse methods used for EEG source localization.

number of current dipoles placed on the whole cortical surface. Similar to the single-dipole source model, a WMN solution associated with the cortical distributed-source model is proposed here [58].

10.4.7 Evaluation Indices

The accuracy of inverse solution using the simulated EEG data has been evaluated by three indices: (i) the error distance (ED), i.e. the distance between the actual and estimated locations of the sources, (ii) the undetected source number percentage (USP), and (iii) the falsely detected source number percentage (FSP). Obviously, these quantifications are based on the simulated models and data. For real EEG data it is hard to quantify and evaluate the results obtained by different inverse methods.

ED between the estimated source locations, $\tilde{\mathbf{s}}$, and the actual source locations, $\mathbf{s}$, is defined as:

$$\mathrm{ED} = \frac{1}{N_d}\sum_{i=1}^{N_d} \min_j \{\|\tilde{\mathbf{s}}_i - \mathbf{s}_j\|\} + \frac{1}{N_{ud}}\sum_{j=1}^{N_{ud}} \min_i \{\|\tilde{\mathbf{s}}_i - \mathbf{s}_j\|\} \tag{10.37}$$

where i and j are the indices of locations of the estimated and actual sources, and N_d and N_{ud} are respectively the total numbers of estimated and undetected sources.

USP and FSP are respectively defined as USP $= N_{un}/Nreal \times 100\%$ and FSP $= N_{false}/N_{estimated} \times 100\%$ where N_{un},N_{real}, N_{false}, and $N_{estimated}$ are respectively the numbers of undetected, real, falsely detected, and estimated sources.

In practise three types of head volume conductor models can be used: a homogeneous sphere head volume conductor model, a BEM or a FEM model. Since the FEM is computationally very intensive, the subject specific BEM model, albeit for an oversimplifying sphere head model, is currently used [59].

In terms of ED, USP, and FSP, LRT1 (i.e. p = 1) has been verified to give the best localization results.

Use of the temporal properties of brain signals to improve the localization performance has also been attempted. An additional temporal constraint can be added assuming that for each location the change in the source amplitude with time is minimal. The constraint to be added is $\min \|\mathbf{s}(n) - \mathbf{s}(n-1)\|_F^2$, where n denotes the time index.

10.4.8 Joint ICA-LORETA Approach

In another study [60] Infomax ICA-based blind source separation (BSS) has been implemented as a preprocessing scheme before the application of LORETA to localize the sources underlying the mismatch negativity (MMN). MMN is an involuntary auditory ERP, which peaks at 100–200 ms when there is a violation of a regular pattern. This ERP appears to correspond to a primitive intelligence. MMN signals are mainly generated in the supramental cortex [61–65].

The LORETA analysis was performed with the scalp maps associated with selected ICA components to find the generators of these maps. Only values greater than 2.5 times the standard deviation of the standardized data (in the LORETA spatial resolution) were accepted as activations.

The inverse problem has also been tackled within a Bayesian framework [66]. In such methods some information about the prior probabilities are normally essential. Again we may consider the EEG generation model as:

$$\mathbf{x}(n) = \mathbf{H}\mathbf{s}(n) + \mathrm{v}(n) \tag{10.38}$$

where $\mathbf{x}(n)$ is an $n_e \times 1$ vector containing the EEG sample values at time n, $\mathbf{H}$ is an $n_e \times m$ matrix representing the head medium model, $\mathbf{s}(n)$ includes the $m \times 1$ vector sample values of the sources at time n and $\mathbf{v}(n)$ is the $n_e \times 1$ vector of noise samples at time n. The a priori information about the sources imposes some constraints on their locations and their temporal properties. The estimation may be performed using a maximum a posteriori (MAP) criterion in which the estimator tries to find $\mathbf{s}(n)$ that maximizes the probability distribution of $\mathbf{s}(n)$ given the measurements $\mathbf{x}(n)$. The estimator is denoted as:

$$\hat{\mathbf{s}}(n) = max\,[p(\mathbf{s}(n) \mid \boldsymbol{x}(n))] \tag{10.39}$$

and following Bayes' rule where the posterior probability is:

$$p(\mathbf{s}(n) \mid \boldsymbol{x}(n)) = p(\boldsymbol{x}(n) \mid \mathbf{s}(n))p(\mathbf{s}(n))/p(\boldsymbol{x}(n)) \tag{10.40}$$

where $p(\boldsymbol{x}(n) \mid \mathbf{s}(n))$ is the likelihood,$p(\boldsymbol{x}(n))$ is the marginal distribution of the measurements, and $p(\mathbf{s}(n))$ is the prior probability. The posterior can be written in terms of energy functions, i.e.:

$$p(\mathbf{s}(n) \mid z(n)) = \frac{1}{z(n)} exp\,[-U(\mathbf{s}(n))] \tag{10.41}$$

and $U(\mathbf{s}(n)) = (1-\lambda)U_1(\mathbf{s}(n)) + \lambda U_2(\mathbf{s}(n))$ where U_1 and U_2 correspond to the likelihood and the prior respectively, and $0 \le \lambda \le 1$. The prior may be separated into two functions, spatial priors, U_s, and temporal priors, U_t. The spatial prior function can take into account the smoothness of the spatial variation of the sources. A cost function that determines the spatial smoothness is:

$$\Phi(u) = \frac{u^2}{1 + (u/K)^2} \tag{10.42}$$

where K is the scaling factor which determines the required smoothness. Therefore, the prior function for the spatial constraints can be written as:

$$U_s(\mathbf{s}(n)) = \sum_{k=1}^{n_e} \left[\Phi_k^x(\nabla_x \mathbf{s}(n) \mid k) + \Phi_k^y(\nabla_y \mathbf{s}(n) \mid k)\right] \tag{10.43}$$

where the indices x and y correspond to horizontal and vertical gradients respectively.

The temporal constraints are imposed by assuming that the projection of $\mathbf{s}(n)$ to the space perpendicular to $\mathbf{s}(n-1)$ is small. Thus, the temporal prior function, as the second constraint, can be written as:

$$U_t(\mathbf{s}(n)) = \left\|\mathbf{P}_{n-1}^{\perp}\mathbf{s}(n)\right\|^2 \tag{10.44}$$

where $\boldsymbol{P}_{n-1}^{\perp}$ is the projection onto the perpendicular space to $\mathbf{s}(n-1)$. Therefore, the overall minimization criterion for estimation of $\mathbf{s}(n)$ becomes:

$$\hat{\mathbf{s}}(n) = arg\min_{\mathbf{S}} \left\{ \|\mathbf{x}(n) - \mathbf{H}\mathbf{s}(n)\|^2 + \alpha \sum_{k=1}^{n_e} \left[\Phi_k^x(\nabla_x \mathbf{s}(n) \mid k) + \Phi_k^y(\nabla_y \mathbf{s}(n) \mid k) \right] + \beta \|\mathbf{P}_{n-1}^{\perp} \mathbf{s}(n)\|^2 \right\} \quad (10.45)$$

where α and β are the penalty terms (regularization parameters).

According to the results of this study the independent components can be generated by one or more spatially separated source. This confirms that each dipole is somehow associated with one dipole generator [67]. In addition, it is claimed that a specific brain structure can participate in different components, working simultaneously in different observations. The combination of ICA and LORETA exploits spatiotemporal dynamics of the brain as well as localization of the sources.

In [51] four different inverse methods namely, WMN, sLORETA, FOCUSS, and SSLOFO have been compared (based on a spherical head assumption and in the absence of noise). It has been shown that in their original forms SSLOFO gives more accurate localization when WMN and sLORETA fail in achieving unique and accurate solutions.

Conversely, while sLORETA has the best performance when only one source is present, for two or more sources LORETA with p equal to 1.5 outperforms other methods. When the relative strength of one of the sources is decreased, all algorithms have more difficulty reconstructing that source. However, LORETA with $p = 1.5$ continues to outperform other algorithms. If only the strongest source is of interest sLORETA is recommended, while LORETA with $p = 1.5$ is recommended if two or more of the cortical sources are of interest. In [68] these results are used as a guidance for choosing a CSD algorithm to locate multiple cortical sources of EEG and for interpreting the results of these algorithms.

10.5 Forward Solutions to the Localization Problem

In these approaches often an approximation of the shape (morphology) of the desired source is available. Some examples are as follows.

10.5.1 Partially Constrained BSS Method

In a recent work [69] the locations of the known sources such as some normal brain rhythms have been used as a prior in order to find the locations of the abnormal or the other brain source signals using constrained BSS. The cost function of the BSS algorithm is constrained by this information and the known sources are iteratively calculated. Consider $\tilde{\mathbf{A}} = \left[\mathbf{A}_k \vdots \mathbf{A}_{uk}\right]$ is the mixing matrix including the geometrical information about the known, $\mathbf{A}_k$, and unknown, $\mathbf{A}_{uk}$, sources. $\mathbf{A}_k$ is an $n_e \times k$ matrix and $\mathbf{A}_{uk}$ is an $n_e \times (m-k)$ matrix. Given $\mathbf{A}_k$, $\mathbf{A}_{uk}$ may be estimated as follows:

$$\mathbf{A}_{uk_{n+1}} = \mathbf{A}_{uk_n} - \xi \nabla_{\mathbf{A}_{uk}} J_c \quad (10.46)$$

where

$$\nabla_{\mathbf{A}_{uk}} J_c = 2\left(\left[\mathbf{A}_k \vdots \mathbf{A}_{uk}\right] - \mathbf{R}_{n+1}\mathbf{W}_{n+1}^{-1}\right) \tag{10.47}$$

$$\mathbf{R}_{n+1} = \mathbf{R}_n - \gamma \nabla_{\mathbf{R}}(J_c) \tag{10.48}$$

and

$$J_c = \left\|\widetilde{\mathbf{A}}_n - \mathbf{R}_{n+1}\mathbf{W}_{n+1}^{-1}\right\|_F^2 \tag{10.49}$$

$$\nabla_{\mathbf{R}}(J_c) = 2\left(\mathbf{W}_{n+1}^{-1}\mathbf{A}_{k_n} + \mathbf{R}_n\mathbf{W}_{n+1}^{-1}\left(\mathbf{W}_{n+1}^{-1}\right)^T - \mathbf{W}_{n+1}^{-1}\mathbf{A}_{uk_n}\right) \tag{10.50}$$

with

$$\mathbf{W}_{n+1} = \mathbf{W}_n - \mu \nabla_{\mathbf{W}} J \tag{10.51}$$

where $J(\mathbf{W}) = J_m(\mathbf{W}) + \lambda J_c(\mathbf{W})$; J_m is the main BSS cost function. The parameters μ, γ, and ζ are either set empirically or changed iteratively; they decrease when the convergence error decreases and recall that $\mathbf{A}_k$ is known and remains constant.

10.5.2 Constrained Least-Squares Method for Localization of P3a and P3b

Often some information about the shape of the source helps in its localization. ERPs look like positive and negative humps in the EEG signals. Such humps may be approximated by Gaussian or Laplacian functions. A method based on LS can be followed [70]. Using this method, the scalp maps (the column vectors of the forward matrix $\mathbf{H}$) are estimated. Consider:

$$\mathbf{R} = \mathbf{X}\mathbf{Y}^T = \mathbf{H}\mathbf{S}\mathbf{Y}^T \tag{10.52}$$

where $\mathbf{Y}$ is a matrix with rows equal to $\mathbf{y_c}$ and $\mathbf{Y} = \mathbf{D}\mathbf{S}$, where $\mathbf{D}$ is a diagonal scaling matrix:

$$D = \begin{bmatrix} d_1 & 0 \ldots 0 \\ 0 & d_2 \ldots 0 \\ \vdots & \ddots \vdots \\ 0 \quad 0 & d_m \end{bmatrix} \tag{10.53}$$

Post-multiplying $\mathbf{R}$ by $\mathbf{R}_y^{-1} = \mathbf{Y}\mathbf{Y}^{-1}$:

$$\mathbf{R}\mathbf{R}_y^{-1} = \mathbf{H}\mathbf{S}\mathbf{Y}^T\left(\mathbf{Y}\mathbf{Y}^T\right)^{-1} = \mathbf{H}\mathbf{D}^{-1}\mathbf{Y}\mathbf{Y}^T\left(\mathbf{Y}\mathbf{Y}^T\right)^{-1} = \mathbf{H}\mathbf{D}^{-1} \tag{10.54}$$

The order of the sources is arbitrary. Therefore, the permutation does not affect the overall process. Hence, the ith scaled scalp map corresponds to the scaled ith source. An LS method may exploit the information about the scalp maps to localize the ERP sources within the brain.

Assuming an isotropic propagation model of the head, the sources are attenuated with the third power of the distance [71], i.e. $d_j = 1/h_j^{1/3}$, where h_j is the jth element of a specific column of the **H** matrix. The source locations **q** are found as the solution to the following LS problem:

$$\min_{\mathbf{q},M} E(\mathbf{q},M) = \min_{\mathbf{q},M} \sum_{j=1}^{n_e} \left[M\|\mathbf{q}-\mathbf{a}_j\|_2 - \mathbf{d}_j \right]^2 \tag{10.55}$$

where $\mathbf{a}_j$ are the positions of the electrodes, $\mathbf{d}_j$s are the scaled distances, and M (scalar) is the scale to be estimated together with **q**. M and **q** are iteratively estimated according to:

$$\mathbf{q}_{\rho+1} = \mathbf{q}_\rho - \mu_1 \nabla_{\mathbf{q}} E\big|_{q=q_\rho} \tag{10.56}$$

and

$$M_{\rho+1} = M_\rho - \mu_2 \nabla_M E\big|_{M=M_\rho} \tag{10.57}$$

where μ_1 and μ_2 are the learning rates and $\nabla_{\mathbf{q}}$ and ∇_M are respectively the gradients with respect to **q** and M which are computed as [71]

$$\nabla_{\mathbf{q}} E(\mathbf{q},M) = 2\sum_{j=1}^{n_e} (\mathbf{q}-\mathbf{a}_j)\left(M^2 - M\frac{\mathbf{d}_j}{\|\mathbf{q}-\mathbf{a}_j\|_2} \right) \tag{10.58}$$

$$\nabla_M E(\mathbf{q},M) = 2\sum_{j=1}^{ne} M\|\mathbf{q}-\mathbf{a}_j\|_2^2 - \|\mathbf{q}-\mathbf{a}_j\|_2 \mathbf{d}_j \tag{10.59}$$

The solutions to **q** and M are unique given $n_e \geq 3$ and $n_e > m$. Using the above localization algorithm 10 sets of EEGs from five patients and five controls, each divided into 20 segments, are examined. Figure 10.6 illustrates the locations of P3a and P3b sources for the patients.

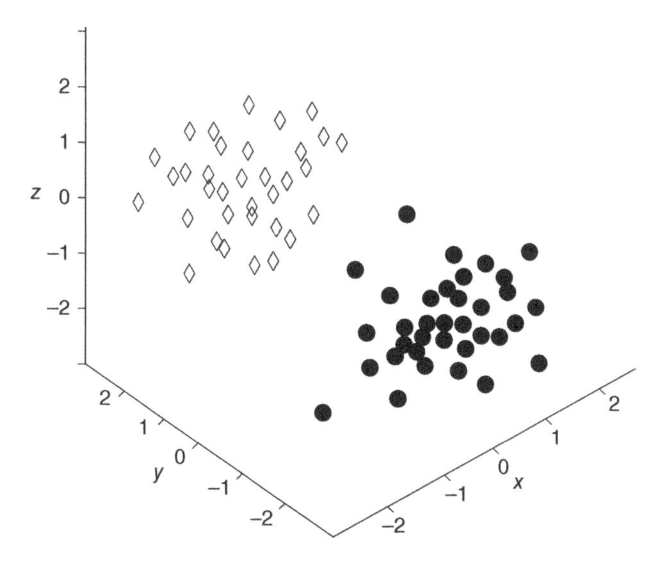

Figure 10.6 The locations of the P3a and P3b sources for five patients in a number of trials. The diamonds ⋄ are the locations of the P3a sources and the circles • show the locations of P3b sources. The *x*-axis denotes right to left, the *y*-axis shows front to back, and *z*-axis denotes up and down. It is clear that the classes are distinct.

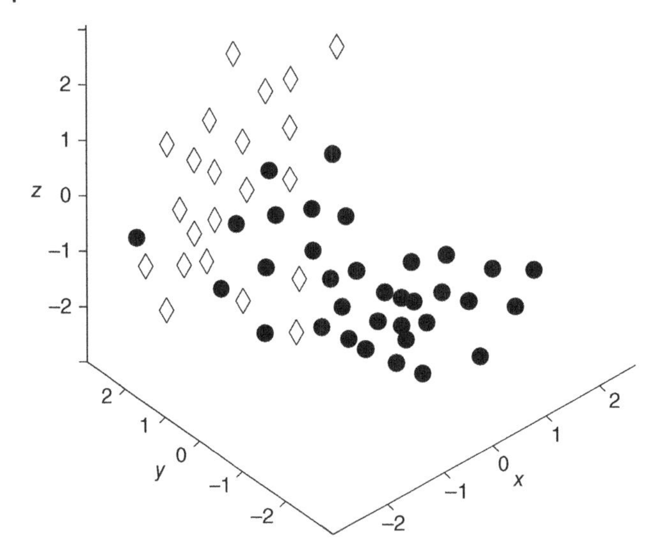

Figure 10.7 The locations of the P3a and P3b sources for five healthy individuals in a number of trials. The diamonds ◊ are the locations of the P3a sources and the circles • show the locations of P3b sources. The *x*-axis denotes right to left, the *y*-axis shows front to back, and *z*-axis denotes up and down. The overlap between the cluster makes the space linearly non-separable.

From this figure it is clear that the clusters representing the P3a and P3b locations are distinct. Figure 10.7 conversely, presents the sources for the control subjects. Unlike in Figure 10.6, the P3a and P3b sources are randomly located within the brain.

10.5.3 Spatial Notch Filtering Approach

For the brain the propagation media is not homogenous. The current flow is strongly attenuated by the skull due to its high resistivity. This attenuation has to be properly modelled when solving the so-called forward problem, i.e. determining the potential at each scalp electrode generated by a known source in the brain. In many cases, such as in localization of ERPs or epileptiform discharges, the morphology of the source of interest is vaguely known. The following approach is a solution to such a forward problem.

A spatial notch filter (SNF), also called a null beamformer (BF), has been designed to maximize the similarity between the extracted ERP source and a generic template which approximates an ERP shape, and at the same time enforce a null at the location of the source [72]. This method is based on a head and source model, which describes the propagation of the brain sources. The sources are modelled as magnetic dipoles, potentially located at some vertices within the brain, and their propagation to the sensors is mathematically described by an appropriate forward model. The considered 3D grid can be fine or coarse depending on the required accuracy of the results. In this model the EEG signal is considered as an $n_e \times T$ matrix, where n_e is the number of electrodes and T is the number of time samples for an EEG channel signal block:

$$\mathbf{X} = \mathbf{HMS} + \mathbf{N} = \sum_{i=1}^{m} \mathbf{H}_i \mathbf{m}_i \mathbf{s}_i + N \tag{10.60}$$

where $\mathbf{H}$ is an $n_e \times 3m$ matrix describing the forward model of the m sources to the n_e electrodes. $\mathbf{H}$ is further decomposed into m matrices $\mathbf{H}_i$, $i = 1, \ldots, m$ as:

$$\mathbf{H} = [\mathbf{H}_1 \cdots \mathbf{H}_i \cdots \mathbf{H}_m] \tag{10.61}$$

where $\mathbf{H}_i$ are $n_e \times 3$ matrices whose each column describes the potential at the electrodes due to the ith dipole for each of the three orthogonal orientations. For example, the first column of $\mathbf{H}_i$ describes the forward model of the x component of the ith dipole when y and z components are zero. Similarly, $\mathbf{M}$ is a $3m \times m$ matrix describing the orientation of the m dipoles and is decomposed as:

$$\mathbf{M} = \begin{bmatrix} \mathbf{m}_1 & 0 & 0 & 0 & 0 \\ 0 & \cdots & 0 & 0 & 0 \\ 0 & 0 & \mathbf{m}_i & 0 & 0 \\ 0 & 0 & 0 & \cdots & 0 \\ 0 & 0 & 0 & 0 & \mathbf{m}_m \end{bmatrix} \tag{10.62}$$

where $\mathbf{m}_i$ is a 3×1 vector describing the orientation of the ith dipole; $\mathbf{s}_i$, a $1 \times T$ vector, is the time signal originated from the ith dipole, and $\mathbf{N}$ is the measurement noise and the modelling error. In addition, the ERP signal $\mathbf{r}_i$ for the ith source has been modelled as a Gaussian shape template with variable width and latency.

A constrained optimization procedure is then followed in which the primary cost function is the Euclidean distance between the reference signal and the filtered EEG [72], i.e.:

$$f_d(\mathbf{w}) = \left\| \mathbf{r}_i - \mathbf{w}^T \mathbf{X} \right\|_2^2 \tag{10.63}$$

The minimum point for this can be obtained by the classic LS minimization and is given by:

$$\mathbf{w}_{opt} = \left(\mathbf{X}\mathbf{X}^T \right)^{-1} \mathbf{X}\mathbf{r}^T \tag{10.64}$$

Following this method an $n_e \times 1$ filter $\mathbf{w}_{opt}$ is designed which gives an estimate of the reference signal when applied to EEGs. However, this procedure does not include any spatial information unless $\mathbf{w}_i$s for all the sources are taken into account. In this way, we can construct a matrix $\mathbf{W}$, similar to the separating matrix in an ICA framework, which could be converted to the forward matrix $\mathbf{H}$. To proceed with this idea a constraint function is defined as:

$$f_c(\mathbf{w}) = \mathbf{w}^T \mathbf{H}(p) = 0 \tag{10.65}$$

where $\mathbf{H}(p)$ is the forward matrix of a dipole at location p and perform a grid search over a number of locations. The constrained optimization problem is then stated as [72]:

$$\min f_d(\mathbf{w}) \text{ subject to } f_c(\mathbf{w}) = 0 \tag{10.66}$$

without going through the details the SNF $\mathbf{w}$ for extraction of a source at location j, as the desired source (ERP component), when there is neither noise or correlation among the components, is described as:

$$\mathbf{w}^T = \mathbf{r}\sum_{i \neq j} \mathbf{s}_i^T \mathbf{m}_i^T \mathbf{H}_i^T \mathbf{C}_x^{-1} \mathbf{H}_j \left(\mathbf{H}_j^T \mathbf{C}_x^{-1} \mathbf{H}_j\right)^{-1} \mathbf{H}_j^T \mathbf{C}_x^{-1} \quad (10.67)$$

where $\mathbf{C}_x = \mathbf{X}\mathbf{X}^T$. In the case of correlation among the sources, the BF, $\mathbf{w}_c$, for extraction of a source at location j, includes another term as [72]:

$$\mathbf{w}_c^T = \mathbf{w}^T + r\widetilde{\mathbf{X}}^T \mathbf{C}_x^{-1} \mathbf{H}(p) \left(\mathbf{H}^T(p) \mathbf{C}_x^{-1} \mathbf{H}(p)\right)^{-1} \mathbf{H}^T(p) \mathbf{C}_x^{-1} \quad (10.68)$$

where $\widetilde{\mathbf{X}}$ includes all the sources correlated with the desired source and $\mathbf{w}$ is given in Eq. (10.67). A similar expression can also be given for when the noise is involved [72].

The SNF algorithm finds the location of the desired brain ERP components (by manifesting a sharp null at the position of the desired source) with a high accuracy even in the presence of interference and noise and where the correlation between the components is considerable.

The ability of the algorithm for correct localization of the sources in various scenarios has been investigated. A forward model has also been built up using the Brainstorm software [73]. Brainstorm is a collaborative, open-source application for MEG and EEG data analysis (visualization, processing, and advanced source modelling).

A three-layer spherical head model with conductivities of 0.33, 0.0042, 0.33 μS/cm, for the scalp, skull, and brain, respectively has been used. 32 Gaussian pulses in 32 different locations with random orientations were created, peaking at different latencies and using 30 electrodes. This represents an underdetermined system. Several different cases have been examined in order to evaluate the effect of noise and correlation between the sources. 2700 voxels have been considered. The source of interest is originally placed at location numbered 1000. For the simple case of no noise and uncorrelated sources, an accurate localization of the source has been achieved as depicted in Figure 10.8. In this figure, the numbers on the horizontal axis show the location of the vertices within the grid inside the brain. Also, to apply the proposed spatial notch filtering method to real data a Gaussian shape (which could be Laplacian shape) template has been selected and used as $\mathbf{r}$.

The method has been compared to the standard linearly constrained minimum variance (LCMV) BF for various levels of noise and correlation between the ERP and the background EEG. The results of this comparison can be viewed in Figures 10.9 and 10.10.

To examine the algorithm for real EEG data the reference electrodes were linked to the earlobes, the sampling frequency set to $Fs = 2$ kHz, and the data were subsequently band-pass filtered (0.1–70 Hz). The EEG data were recorded for control and schizophrenic patients. The stimuli were presented through ear plugs inserted in the ear. Forty rare tones (1 kHz) were randomly distributed amongst 160 frequent tones (2 kHz). Their intensity was 65 dB with 10 and 50 ms duration for rare and frequent tones, respectively. The subject was asked to press a button as soon as he heard a low tone (1 kHz). The ability to distinguish between low and high tones was confirmed before the start of the experiment. The task is designed to assess basic memory processes for both healthy subjects and schizophrenic patients. ERP components measured in this task included N100, P200, N200, and P3a and P3b. However, the results for the P3a and P3b have been demonstrated [72]. Figures 10.11 and 10.12 show the results of P3a and P3b source localization for the control

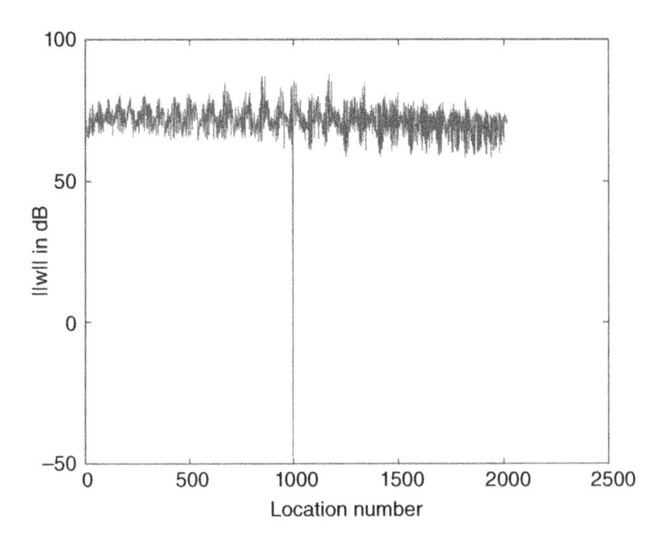

Figure 10.8 Localization plot for one source uncorrelated with other sources in a noise free environment. The location number refers to a geometrical location in a 3-D grid within the brain [72].

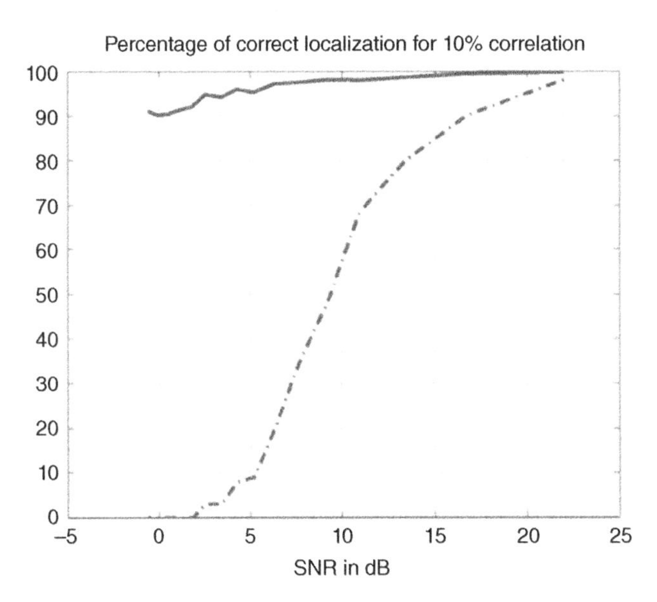

Figure 10.9 Percentage of successful localizations for various SNRs for the SNF algorithm (solid bold line) and the LCMV (dashed line). The purpose is to evaluate the performance of the algorithm for different orientations of the sources. The same noise sequence has been used for 1000 different orientations and various SNR values. Here, the correlation is 10% [72].

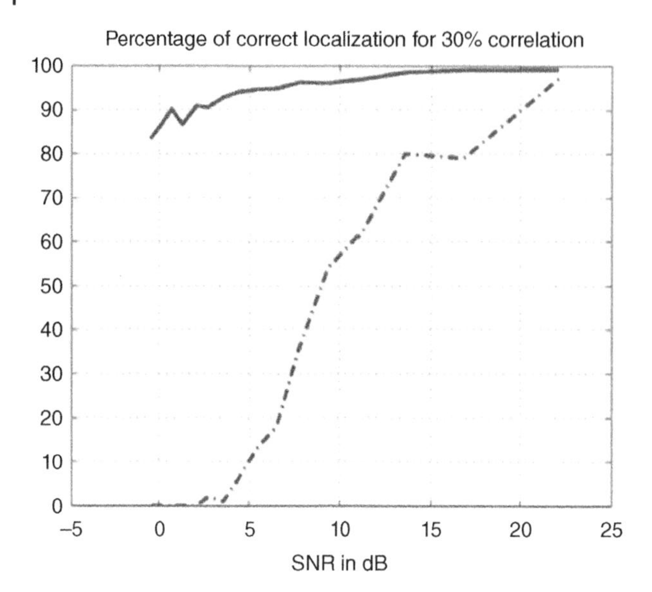

Figure 10.10 Percentage of successful localizations for various SNRs for the SNF algorithm (solid bold line) and the LCMV (dashed line). The purpose is to evaluate the performance of the algorithm for different orientations of the sources. The same noise sequence has been used for 1000 different orientations and various SNR values. Here, the correlation is 30% [72].

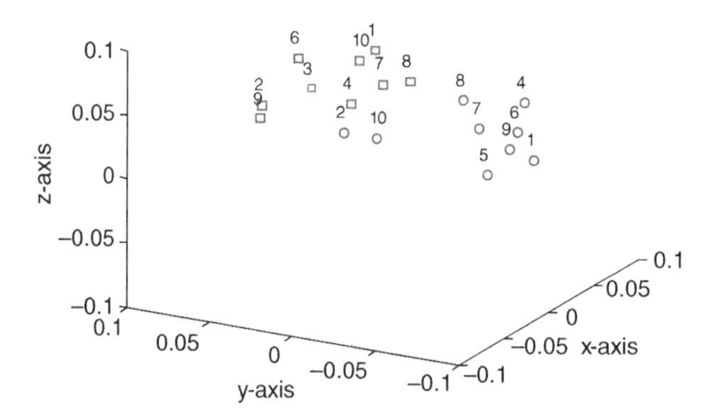

Figure 10.11 Localization plot for P3a, circles, o, and P3b, squares, □, for 10 normal people. The numbers correspond to the control subject's number (e.g. o1 shows the location of the P3a for control subject number 1). The three axes refer to the geometrical coordinates in metres. The *y*-axis determines front–back of the head, *x*-axis is left–right, and *z*-axis is the vertical position. Units are in metres [72].

and schizophrenic patients. In these figures the numbers next to the circles and squares show the patient number.

It can be seen that the locations of the P3a and P3b for the schizophrenic patients are less distinct than the locations for the healthy individuals [72].

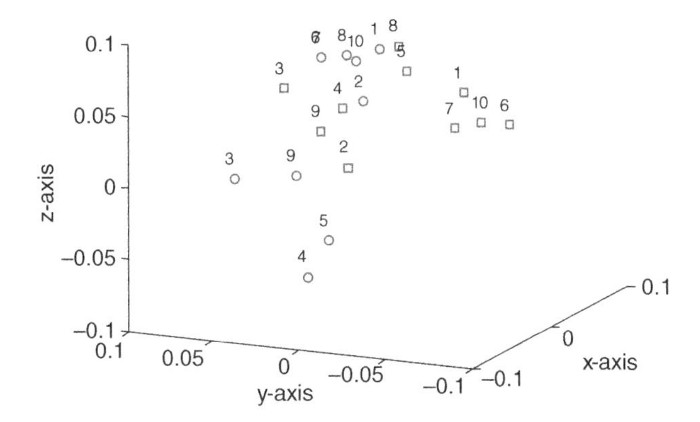

Figure 10.12 Localization plot for the P3a, circles, o, and P3b, squares, □, for 10 schizophrenic patients. The numbers correspond to the control subject's number (e.g. o1 shows the location of the P3a for control subject number 1). The three axes refer to the geometrical coordinates in metres. The *y*-axis determines front–back of the head, *x*-axis is left–right, and *z*-axis is the vertical position. Units are in metres [72].

10.6 The Methods Based on Source Tracking

10.6.1 Deflation Beamforming Approach for EEG/MEG Multiple Source Localization

Poor spatial resolution of EEG/MEG motivates research into methods that can more accurately localize the sources from the recordings using these modalities. A popular strategy in brain source localization is by using the dipole source assumption [74]. Physiologically, this assumption may not be always true. Under this assumption, electric current dipoles are specified by their three-dimensional locations and their three-dimensional moments. The LCMV BF is a well established method in MEG dipole source localization [75]. BF provides an adaptive method which places nulls, using some linear constraints, at positions corresponding to the noise sources. The transient and often correlated nature of the neural activation in different parts of the brain often, however, limits the BF performance. As an example, since MEG has variable sensitivity to source locations, the noise gain of the filter varies with the changes in location. One strategy to account for this effect is to normalize the output power of BF with respect to the estimated power in the presence of only noise [76]. Beamspace transformations for dimension reduction which preserve source activity located within a given region of interest have also been presented in [77]. For more details of different methods for MEG source localization, consider [78, 79].

In [74] the BF approach has been generalized to include more constraints. The first constraint additionally stops the noise power at the output of the BF. This method, the so-called *deflation beamformer,* assumes the covariance matrix of the noise is known a priori. The second constraint places nulls at known locations where other dipoles have been detected previously to improve the detection of the unidentified dipoles. In this approach, a dipole is located by finding the grid node which has the maximum power, while deflating the power of any dipoles that have already been identified. Using a regularized optimization approach by means of penalty functions such as Lagrange multipliers, the multiple-constraint problem is changed to an unconstrained problem and solved. During this process the power at each location is estimated and normalized with respect to the power in the presence of only noise. The deflation BF method helps to overcome two main problems of multiple dipole source localization using the BF, i.e. its shortcoming of dipole localization to the dominant sources, and its inaccurate performance in the presence of highly correlated sources. In addition, to improve the performance of the method further, an iterative approach for deflation and localization of dipoles has been developed [74]. To formulate the problem let **y** be one of the measurements from N scalp electrodes. Each dipole is specified by its three-dimensional location ρ and its three-dimensional moment **m**. The medium between the sources and the electrodes is assumed to be homogeneous, and the MEG field **y** to be the superposition of the fields from q dipoles. Therefore, we may write:

$$\mathbf{y} = \sum_{i=1}^{q} \mathbf{H}(\rho_i)\mathbf{m}(\rho_i) + \mathbf{n} \tag{10.69}$$

where **n** is the noise mutually uncorrelated with the signals. For a two-source problem, assuming source ρ_1 is already localized, the deflation beamforming problem for locating source ρ_2 is stated as the following constrained optimization problem for **W** [74]:

$$\min_{\mathbf{W}^T} \left\|\mathbf{W}^T\mathbf{y}\right\| \quad \text{subject to } \min_{\mathbf{W}^T} \left\|\mathbf{W}^T\mathbf{n}\right\|, \quad \mathbf{W}^T\mathbf{H}(\rho_1) = 0, \quad \text{and } \mathbf{W}^T\boldsymbol{H}(\rho_2) = \mathbf{I} \tag{10.70}$$

The term $\mathbf{W}^T\mathbf{H}(\rho_1)$ deflates the first source. For higher number of sources more sources should be deflated before the new source can be detected. The optimum solution for **W** can then be found as [74]:

$$\mathbf{W}^T = \tilde{\mathbf{I}}_d^T \left(\tilde{\mathbf{H}}^T (\mathbf{C}_y + \lambda\mathbf{C}_n)^{-1}\tilde{\mathbf{H}}\right)^{-1} \tilde{\mathbf{H}}^T \left(\mathbf{C}_y + \lambda\mathbf{C}_n\right)^{-1} \tag{10.71}$$

where $\tilde{\mathbf{I}}_d^T = [\mathbf{0}\ \mathbf{I}]$ and $\tilde{\mathbf{H}}^T = [\mathbf{H}(\rho_1)\ \mathbf{H}(\rho_2)]$. $\mathbf{C}_n$ and $\mathbf{C}_y$ are the covariance matrices of noise and background EEG including the first source.

In many practical cases, we have no prior knowledge about the true source locations. Under these circumstances, if we locate the first source incorrectly, localization of other sources may also fail. To circumvent this problem, the sources are found via a number of iterations such that in each round the other sources are deflated while searching for the current source location; the sources then swap and source 1 is estimated while previously estimated source 2 is deflated.

To examine the method on real MEG, event-related fields (ERFs) were recorded in an auditory paradigm. 1500 auditory stimuli were delivered every 0.5 second bilaterally to the subject. The stimulus was a broadband noise lasting 0.1 second. This data set was selected because the stimulus train generates activity in the left and right primary auditory cortices. The data were acquired from a 28 year old male subject. Whole head MEG recordings were made using a 275-channel radial gradiometer system. The sampling rate was 1000 Hz and recordings were filtered off-line with a bandpass of 0.03–40 Hz. After visual rejection of trials containing eye-blink and movement artefacts, the remaining trials were averaged. The estimated noise covariance matrix **C***n* used in both the BF and deflation BF approaches, was calculated from the 0.1 second pre-stimulus segments. In this experiment, two dipoles were considered: one in the right and one in the left hemisphere of the brain).

Figure 10.13a and b show the power profiles of the BF and estimated locations in axial and coronal views. The estimated locations are shown with cross markers. The estimated power profile and estimated locations using the deflation BF method are shown in Figure 10.13c–f. Figure 10.13c and d show the location of the first dipole while deflating the second dipole, and

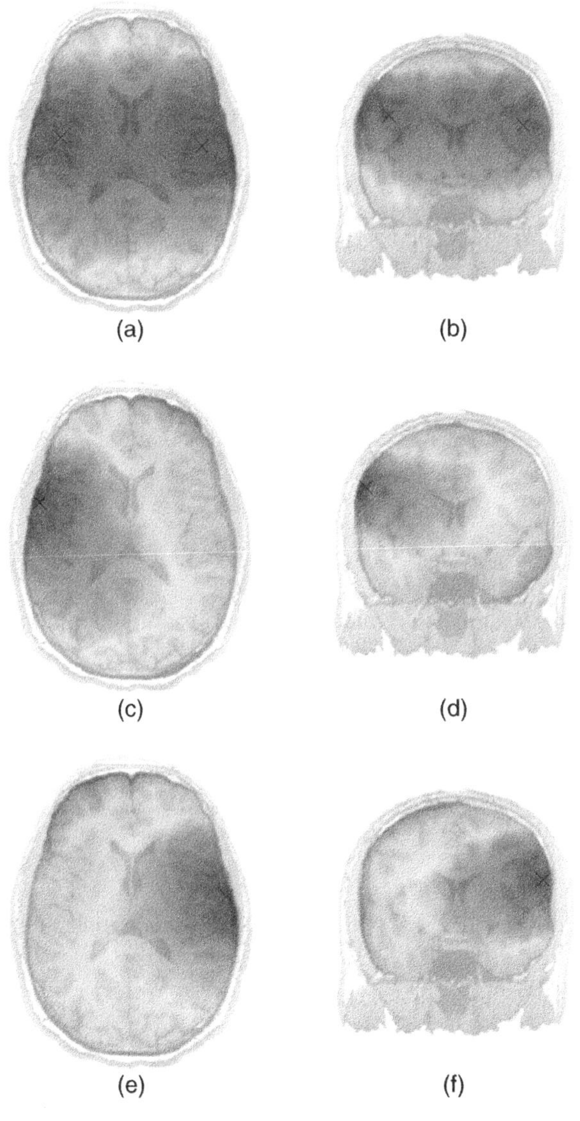

Figure 10.13 Topographies or power profiles of real MEG data obtained using (a) BF method in axial view, (b) BF method in coronal view, (c) deflation BF while deflating the second source in axial view, (d) deflation BF while deflating the second source in coronal view, (e) deflation BF while deflating the first source in axial view, (f) deflation BF while deflating the first source in coronal view [74].

Figure 10.13e and f show the location of the second dipole while deflating the first dipole. The deflation BF method converged after three iterations. It is seen from Figure 10.13a and b that for the BF the sources are bleeding towards the centre of the head due to the partial correlation between the sources. Conversely, the deflation BF places the sources at biologically plausible locations in the primary auditory cortices in the left and right hemisphere. The power obtained using the deflation BF is also more focal than the power obtained using the BF approach. Furthermore, unlike for the BF, no spurious activations near the centre of the sphere model were observed when deflation BF was implemented [74].

10.6.2 Hybrid Beamforming – Particle Filtering

In EEG or MEG source localization often the number of unknown parameters is higher than the number of known measurements and therefore the system is underdetermined. *Dipole source localization* assumes that one or multiple current dipoles represent the electric sources [80]. To express and formulate the problem, suppose the measured multichannel signals $\mathbf{y}_k$ from M sensors at time k are produced by q dipoles, so we can model $\mathbf{y}_k$ as:

$$\mathbf{y}_k = \sum_{i=1}^{q} \mathbf{F}(\rho_k(i))\mathbf{m}_k(i) + \mathbf{v}_k \tag{10.72}$$

where $\rho_k(i)$ is a three-dimensional location vector, $\mathbf{m}_k(i)$ is a three-dimensional moment vector of the ith dipole and $\mathbf{v}_k$ is the observation noise. It is assumed here that the number of dipoles is known a priori and there locations are to be found. Hence, the states are defined as [80]:

$$\mathbf{L}_k = [\rho_k(1), \ldots, \rho_k(q)] \tag{10.73}$$

By introducing an $M \times 3q$ matrix of location $\mathbf{F}(\mathbf{L}_k)$ and a $3q \times 1$ vector of moments $\underline{\mathbf{m}}_k$ as [80]:

$$\mathbf{F}(\mathbf{L}_k) = \left[\mathbf{F}(\rho_1), \ldots, \mathbf{F}\left(\rho_q\right)\right] \tag{10.74}$$

$$\underline{\mathbf{m}}_k = \left[\mathbf{m}_k^T(1), \ldots, \mathbf{m}_k^T(q)\right]^T \tag{10.75}$$

In that case a matrix form of Eq. (10.72) is given as:

$$\mathbf{y}_k = \mathbf{F}(\mathbf{L}_k)\underline{\mathbf{m}}_k + \mathbf{v}_k$$

where $\mathbf{F}$ is a nonlinear function of q dipoles. The state-space equations of the particle filter (PF) take the following form:

$$\mathbf{L}_k = \mathbf{L}_{k-1} + \mathbf{w}_{k-1} \tag{10.76a}$$

$$\mathbf{y}_k = \mathbf{F}(\mathbf{L}_k)\underline{\mathbf{m}}_k + \mathbf{v}_k \tag{10.76b}$$

In which $\mathbf{w}_k$ is the state noise. In these equations to estimate $\mathbf{L}_k$ however, $\underline{\mathbf{m}}_k$ has to be known. The vector of moments linearly corresponds to the EEG (or MEG) measured data and under no noise assumption, $\underline{\mathbf{m}}_k$, can be recursively calculated as [80]:

$$\underline{\mathbf{m}}_k = \mathbf{F}^{\dagger}(\mathbf{L}_{k-1})\mathbf{y}_k \tag{10.77}$$

where $\mathbf{F}^{\dagger} = (\mathbf{F}^T\mathbf{F})^{-1}\mathbf{F}^T$ which is the pseudo-inverse of $\mathbf{F}$. Note that in Eq. (10.77), $\underline{\mathbf{m}}_k$ is estimated from the location matrix $\mathbf{L}_{k-1}$ at the previous step and measurements $\mathbf{y}_k$ at the current step. This method is a grid-based method meaning that the sources are considered to be in one of the predefined vertices of a 3D grid inside the head. Therefore, after the prediction stage, the locations might need to be adjusted to the nearest cells' locations.

To use a BF, assume that there are G grid points out of which q of them are the source locations and the EEG/MEG signals can be decomposed as [80]:

$$\mathbf{y}_k = \sum_{i=1}^{G} \mathbf{F}(g_i)\mathbf{m}_k(i) \tag{10.78}$$

To select the source of interest coming from location $\breve{\rho}$, a linear spatial filter $\mathbf{W}_{\breve{\rho}}$ is used to have the following ideal response:

$$\mathbf{W}_{\breve{\rho}}^T\mathbf{F}(g_i) = \begin{cases} \mathbf{F}(g_i) \; g_i \in \breve{\rho} \\ \mathbf{O} \text{ else} \end{cases} \tag{10.79}$$

where $\mathbf{O}$ is a $3 \times M$ null matrix. Using two new matrices $\underline{\mathbf{F}} = [\mathbf{F}(g_1)\ldots\mathbf{F}(g_G)]$ and $\mathbf{F_O} = [\mathbf{O}\ldots\mathbf{F}(\rho(1))\ldots\mathbf{F}(\rho(q))\ldots\mathbf{O}]$ which has zero entries except at locations of the dipoles we can have:

$$\mathbf{W}_{\breve{\rho}}^T\underline{\mathbf{F}} = \mathbf{F_O} \tag{10.80}$$

From this an optimum solution can be found as [80]:

$$\mathbf{W}_{\breve{\rho}}^T = \underline{\mathbf{F}}^{\dagger^T}\mathbf{F}_{\mathbf{O}}^T \tag{10.81}$$

To construct the beamforming PF the spatially filtered data are used instead of the original measurements to compute the moment vector $\underline{\mathbf{m}}_k$. The filter in (10.81) now can be applied to the measurements to have

$$\underline{\mathbf{m}}_k = \mathbf{F}^{\dagger}(\mathbf{L}_{k-1})\mathbf{W}_{\breve{\rho}}^T\mathbf{y}_k \tag{10.82}$$

Matrix $\mathbf{F_O}$ in Eq. (10.81) is constructed for each particle using locations indicated by particles from the previous step. Moreover, since matrix $\mathbf{F}$ is the matrix of all gains and is independent of the desired locations, for computational efficiency, its pseudo-inverse can be calculated only one time. This algorithm has been applied to synthetic and real sources and its good performance demonstrated [80].

10.7 Determination of the Number of Sources from the EEG/MEG Signals

For the majority of applications in source localization a priori information about the number of sources is required. The problem of detection of the number of independent (or uncorrelated) sources can be defined as analyzing the structure of the covariance matrix of the observation matrix. This matrix can be expressed as $\mathbf{C} = \mathbf{C}_{sig} + \mathbf{C}_{noise}$, where $\mathbf{C}_{sig}$ and $\mathbf{C}_{noise}$ are respectively, the covariance of source signals and the covariance of noise. PCA and SVD may perform well if the noise level is low. In this case the number of dominant eigenvalues represents the number of sources. In the case of white noise, the covariance matrix $\mathbf{C}_{noise}$ can be expressed as $\mathbf{C}_{noise} = \sigma_n^2\mathbf{I}$, where σ_n^2 is the noise variance and $\mathbf{I}$ is the identity matrix. In the case of coloured noise some similar methods can be implemented if the noise covariance is known apart from σ_n^2. The noise covariance matrix is a symmetric positive definite matrix $\mathbf{C}_{cnoise} = \sigma_n^2\boldsymbol{\Psi}$. Then, a non-singular square matrix $\boldsymbol{\psi}$ $(m \times m)$ exists such that $\boldsymbol{\Psi} = \boldsymbol{\psi}\boldsymbol{\psi}^T$ [81].

For both white and coloured noises the eigenvalues can be calculated from the observation covariance matrices, and then analyzed by implementing the information theoretic criterion to estimate the number of independent sources [82].

However, it has been shown that this approach is suboptimal when the sources are temporally correlated [83]. Selection of an appropriate model for EEG analysis and source localization has been investigated by many researchers, and several criteria have been established to solve this problem. In most of these methods the amplitudes of the sources are tested to establish whether they are significantly larger than zero, in which case the sources are included in the model. Alternatively, the locations of the sources can be tested to determine whether they differ from each other significantly, in which case these sources should also be included in the model.

PCA and ICA may separate the signals into their uncorrelated and independent components respectively. By backprojecting the individual components to the scalp electrodes both the above criteria may be tested. Practically, the number of distinct active regions within the backprojected information may denote the number of sources. The accuracy of the estimation also increases when the regions are clustered based on their frequency contents. However, due to the existence of noise with unknown distribution the accuracy of the results is still under question.

In [84] the methods based on residual variance (RV), Akaike information criterion (AIC), Bayesian information criterion (BIC), and Wald tests on amplitudes (WA) and locations (WL) have been discussed. These methods have been later examined on MEG data [85] for both pure white error and coloured error cases. The same methods can be implemented for the EEG data too. In this test the MEG data from m sensors and T samples are collected for each independent trial $j = 1, \ldots, n_e$ in the $m \times T$ matrix $\mathbf{Y}_j = (\mathbf{y}_{1j}, \ldots, \mathbf{y}_{Tj})$, with $\mathbf{y}_{ij} = (y_{1ij}, \ldots, y_{mij})^T$. Considering the average over number of trials n as $\bar{\mathbf{Y}} = \frac{1}{n}\sum_{j=1}^{n}\mathbf{Y}_j$, the model for the averaged data can be given as:

$$\bar{\mathbf{Y}} = \mathbf{GA} + \mathbf{E} \tag{10.83}$$

where $\mathbf{G}$ includes the sensor gains of the sources of unit amplitudes. Matrix $\mathbf{G}$ depends on the location and orientation parameters of the dipolar sources. Based on this model the tests for model selection are as follows:

The RV test defined as [84]:

$$\mathrm{RV} = 100\frac{tr\left[(\bar{\mathbf{Y}} - \mathbf{GA})(\bar{\mathbf{Y}} - \mathbf{GA})^T\right]}{tr[\bar{\mathbf{Y}}\bar{\mathbf{Y}}]} \tag{10.84}$$

compares the squared residuals to the squared data for all sensors and samples simultaneously. The RV decreases as a function of the number of parameters, and therefore over-fits easily. The model is said to fit if the RV is below a certain threshold [86].

The AIC method penalizes the log-likelihood function for additional parameters required to describe the data. These parameters may somehow describe the sources [86]. The number of sources has been kept limited for this test since at some point any additional source hardly decreases the log-likelihood function but increases the penalty. The AIC is defined as:

$$\mathrm{AIC} = nmT\ln\left(\frac{\pi s^2}{n}\right) + \frac{1}{ns^2}\mathrm{tr}\left[(\bar{\mathbf{Y}} - \mathbf{GA})(\bar{\mathbf{Y}} - \mathbf{GA})^T\right] + 2p \tag{10.85}$$

In this equation s^2 is the average of diagonal elements of the spatial covariance matrix [84].

The BIC test resembles the AIC method with more emphasis on the additional parameters. Therefore, less over-fitting is expected when using BIC. This criterion is defined as [87]:

$$\mathrm{BIC} = n_e mT\ln\left(\frac{\pi s^2}{n_e}\right) + \frac{1}{n_e s^2}\mathrm{tr}\left[(\bar{\mathbf{Y}} - \mathbf{GA})(\bar{\mathbf{Y}} - \mathbf{GA})^T\right] + pln(mT) \tag{10.86}$$

Similarly, the model with the minimum BIC is selected.

The Wald test is another important criterion, which gives the opportunity to test a hypothesis on a specific subset of the parameters [88]. Both amplitudes and locations of the sources can be tested using this criterion. If $\mathbf{r}$ is a q vector function of the source parameters (i.e. the amplitude and location), $\mathbf{r}_h$ the q vector of fixed hypothesized value of $\mathbf{r}$, $\mathbf{R}$ the $q \times k$ Jacobian matrix of $\mathbf{r}$ with respect to k parameters, and $\mathbf{C}$ the $k \times k$ covariance matrix of source parameters, then the Wald test is defined as [89]:

$$\mathbf{W} = \frac{1}{q}(\mathbf{r} - \mathbf{r}_h)^T\left(\mathbf{RC}^{-1}\mathbf{R}'\right)^{-1}(\mathbf{r} - \mathbf{r}_h) \tag{10.87}$$

An advantage of using the WA technique in spatiotemporal analysis is the possibility of checking the univariate significance levels to determine at which samples the sources are active.

The tests carried out for two synthetic sources and different noise components [85] showed that thc WL test has superior overall performance, and the AIC and WA perform well when the sources are close together.

These tests have been based on simulations of the sources and noise. It is also assumed that the locations and the orientations of the source dipoles are fixed and only the amplitudes change. For real EEG (or MEG) data, however, such information may be subject to change and generally unknown.

Probably the most robust approach for detection of the number of sources is that developed by Bai and He [81]. In this approach an information criterion method in which the penalty functions are selected based on the spatiotemporal source model, has been developed to estimate the number of independent dipole sources from EEG or MEG In this approach the following steps are followed:

1) Calculating the covariance matrix $\mathbf{C_X}$ from the measured data matrix $\mathbf{X}$.
2) Applying the SVD to decompose $\mathbf{C_X}$ to obtain the eigenvalues λ_i, where $\lambda_1 > \ldots > \lambda_m$.

3) Calculating the information criterion value using the eigenvalues. The information criterion (IC_k) can be calculated using either:

$$IC = w(n_e - k) \log \frac{1}{n_e - k} \sum_{i=k+1}^{n_e} \lambda_1 - w \sum_{i=k+1}^{n_e} \log \lambda_1 + 2d(k, n_e)\beta(w) \tag{10.88}$$

when the noise information is accurate. In Eq. (10.88) $d(k, m) = k(2m - k + 1)/2$. In places where the noise statistics is unknown or:

$$IC = -\left(w - 1 - k - \frac{2(n_e - k)^2 + n_e - k + 2}{6(n_e - k)} + \sum_{i=1}^{k} \frac{\bar{\lambda}_{n_e - k}^2}{(\lambda_i - \bar{\lambda}_{n_e - k})^2}\right) \cdot \log\left(\bar{\lambda}_{n_e - k}^{-(n_e - k)} \prod_{i=k+1}^{n_e} \lambda_i\right) + 2d(k, n_e)\beta(w - 1) \tag{10.89}$$

where in this equation $d(k, n_e) = k(n_e - k + 2)(n_e - k - 1)/2$, $\bar{\lambda}_{n_e - k}$ is the average of the $n_e - k$ smallest eigenvalues. In both equations n_e is the number of electrode signals, k is the number of assumed dipole sources to be estimated, w is the number of time points to form a spatiotemporal data matrix, and $\beta(w)$ is the penalty function which can have constant or logarithmic values [81].

4) According to the rule of information criterion [81] the number of sources with minimum IC is selected as the estimated number of sources.

In various experiments it has been shown that in moderate noise environment, which is the case of EEG signals, the accuracy of the method is above 80%.

10.8 Other Hybrid Methods

In some other works the researchers have tried to better focus on the source of interest by applying some preprocessing of the data. As an example in [90], empirical mode decomposition followed by the wavelet transform have been used to filter the signals and focus on the sources within particular frequency bands active in certain time intervals. Then, multiple sparse priors (MSP) and iterative regularization algorithms have been used to solve the inverse localization problem.

In a recent review by Michel and Brunet [91] the different steps in brain source localization have been explained and a stand-alone freely available academic software called Cartool has been used to illustrate the results of their analysis. In this article a brief table has been provided to illustrate various academic and commercial tools which can be used for source localization (see Table 10.1). Most of these tools are used in many other EEG or MEG analyzes too.

10.9 Application of Machine Learning for EEG/MEG Source Localization

As stated in the previous sections of this chapter, the brain functional activity is associated with the generation of currents and resultant voltages which may be observed on the scalp as the EEG. The properties of the dipole sources may be studied by solving either the

Table 10.1 Most popular academic and commercial software packages that can be used for inverse solution to EEG source localization problems [91].

Name	Website (last visited in September 2020)	Inverse models
	Academic Software Packages	
Brainstorm	https://neuroimage.usc.edu/brainstorm/	Dipole modelling, Beamformer, sLORETA, dSPM
Cartool	https://sites.google.com/site/cartoolcommunity	Minimum Norm, LORETA, LAURA, Epifocus
EEGLab	https://sccn.ucsd.edu/eeglab/index.php	Dipole modelling
FieldTrip	http://www.fieldtriptoolbox.org	Dipole modelling, Beamformer, Minimum Norm
LORETA	http://www.uzh.ch/keyinst/loreta.htm	LORETA, sLORETA, eLORETA
MNE	https://mne.tools/stable/index.html	MNE, dSPM, sLORETA, eLORETA
MUTMEG	https://www.nitrc.org/projects/nutmeg	Beamformer
SPM (mainly for fMRI)	https://www.fil.ion.ucl.ac.uk/spm	dSMP
	Commercial Software Packages	
BESA	http://www.besa.de/products/besa-research/besa-research-overview	Dipole modelling, RAP-MUSIC, LORETA, sLORETA, LAURA, SSLOFO
Brainvision analyzer	https://www.brainproducts.com	LORETA
BrainVoyager	https://www.brainvoyager.com	Beamformer, Minimum Norm, LORETA, LAURA
GeoSource	https://www.egi.com/research-division/electrical-source-imaging/geosource	Minimum Norm, LORETA, sLORETA, LAURA
CURRY	https://compumedicsneuroscan.com/curry-source-reconstruction	Dipole modelling, MUSIC, Beamformer, Minimum norm, sLORETA, eLORETA, SWARM

forward or inverse problems. The forward problem utilizes a volume conductor model for the head, in which the potentials on the conductor surface are computed based on an assumed current dipole at an arbitrary location, orientation, and strength. In the inverse problem, conversely, a current dipole, or a group of dipoles, is identified based on the observed EEG. Both the forward and inverse problems are typically solved by numerical procedures, such as a BEM and an optimization algorithm. The forward solutions require accurate head model and the inverse solutions are inaccurate, cannot handle multiple sources easily, and are computationally intensive.

In machine learning-based methods, particularly by using DNNs, the learning system is trained to find the best arrangement of the cortical source locations for having a particular pattern of multichannel scalp EEG recordings [92–98].

An early work on using artificial neural networks (ANNs) for brain source localization has been carried out by Abeyratne et al. [99]. In their design, they use the head, data, and sensor models as a priori information while a simple scenario of spherical head model and single source has been considered. The network is trained on the data files assuming the position of the dipole is known. During the training phase, a dipole-parameter-vector (X_p, Y_p, Z_p, M_{xp}, M_{yp}, M_{zp}) from the training-output file, and its corresponding scalp-voltage-vector ($V1_p$, $V2_p$, ..., Vm_p) from the training-input file, are applied to the ANN simultaneously. Scalp-voltage-vector is set as the input and the dipole-parameter-vector is set as the target output.

In [100] feed-forward neural networks trained to solve the inverse single-dipole localization problem have an input layer of 27 neurons, corresponding to the $M = 27$ potentials at the electrodes. The input layer is followed by one or two hidden layers having N neurons. The output layer finally contains the neurons that produce the desired dipole parameters. The neurons of the input layers have linear activation functions and the neurons of the hidden and output layers use hyperbolic tangent activation function. Each neuron of the hidden and output layers is connected at its input to a bias neuron with constant output equal to one, to permit a non-zero threshold of the activation function. Every potential distribution presented to the network is first average referenced by subtracting the average of all potential values. Subsequently, the average referenced potentials are normalized by dividing them by the magnitude of the largest. The dipole location parameters are normalized to 1 with respect to the radius of the outer head boundary in the simple spherical head model (9.2 cm radius). In the case of a realistically shaped head model, the location parameters are normalized with respect to the radius of the best-fitting sphere for the scalp–air interface. The optimal dipole orientation (in the LS sense) for a given location can be calculated in a straightforward manner. Therefore, here, the neural network is employed to estimate only the dipole location parameters.

To test the system, it is trained using a large number of inputs with the dipoles spread randomly over the cortex. The authors have achieved an average localization error of approximately 3.5 mm for a network of two hidden layers.

In [101] both forward and inverse problems have been attempted using ANNs which are trained off-line using back-propagation techniques to learn the complex source-potential relationships of head volume conduction. Once trained, these networks are able to generalize their knowledge to localize functional activity within the brain with reasonable computational cost.

10.10 Summary

Source localization, from only EEG and MEG signals, is an ill-posed optimization problem. This is mainly due to the fact that the number of sources is unknown. This number may change from time-to-time, especially when the objective is to investigate the EP or movement-related sources. Most of the proposed algorithms fall under one of the two methods of ECD and LD approaches. Some of the above methods such as sLORETA have been commercialized and reported to have reasonable outcome for many applications. A hybrid system of different approaches seems to give better results. Localization may also be more accurate if the proposed cost functions can be constrained by some additional information stem from clinical findings or from certain geometrical boundaries. Non-homogeneity of the head medium is another major problem for EEG source localization. This is less troublesome for MEG; comprehensive medical and physical experimental studies have to be carried out to find an accurate model of the head. Using or fusion of other modalities such as fMRI will indeed enhance the accuracy of the localization results. There are many potential applications for brain source localization such as for localization of the ERP signals [33] and MRPs useful for BCI [102], and seizure source localization [103]. Recent advances in EEG and MEG source localization including application of computationally intensive machine learning systems such as DNNs, allow for more accurate localization of multiple sources. Head model parameters, electrode positions, and the number of sources still remain as the main requirements for machine learning-based source localization techniques.

References

1. Miltner, W., Braun, C., Johnson, R.E., and Rutchkin, A.D.S. (1994). A test of brain electrical source analysis (BESA): a simulation study. *Electroencephalography and Clinical Neurophysiology* 91: 295–310.
2. Scherg, M. and Ebersole, J.S. (1994). Brain source imaging of focal and multifocal epileptiform EEG activity. *Clinical Neurophysiology* 24: 51–60.
3. Scherg, M., Best, T., and Berg, P. (1999). Multiple source analysis of interictal spikes: goals, requirements, and clinical values. *Journal of Clinical Neurophysiology* 16: 214–224.
4. Aine, C., Huang, M., Stephen, J., and Christopher, R. (2000). Multistart algorithms for MEG empirical data analysis reliably characterize locations and time courses of multiple sources. *NeuroImage* 12: 159–179.
5. Phillips, C., Rugg, M.D., and Friston, K.J. (2002). Systematic regularization of linear inverse solutions of the EEG source localization problem. *NeuroImage* 17: 287–301.
6. Backus, G.E. and Gilbert, J.F. (1970). Uniqueness in the inversion of inaccurate gross earth data. *Philosophical Transactions of the Royal Society of London, Series A* 266: 123–192.
7. Sarvas, J. (1987). Basic mathematical and electromagnetic concepts of the biomagnetic inverse problem. *Physics in Medicine and Biology* 32: 11–22.
8. Hamalainen, M.S. and Llmoniemi, R.J. (1994). Interpreting magnetic fields of the brain: minimum norm estimates. *Medical and Biological Engineering and Computing* 32: 35–42.
9. Menendez, R.G. and Andino, S.G. (1999). Backus and Gilbert method for vector fields. *Human Brain Mapping* 7: 161–165.
10. Pascual-Marqui, R.D. (1999). Review of methods for solving the EEG inverse problem. *International Journal of Bioelectromagnetism* 1 (1): 75–86.
11. Uutela, K., Hamalainen, M.S., and Somersalo, E. (1999). Visualization of magnetoencephalographic data using minimum current estimates. *NeuroImage* 10: 173–180.
12. Phillips, C., Rugg, M.D., and Friston, K.J. (2002). Anatomically informed basis functions for EEG source localisation: combining functional and anatomical constraints. *NeuroImage* 16: 678–695.

13 Dattola, S., Morabito, F.C., Mammone, N., and La Foresta, F. (2020). Findings about LORETA Applied to High-Density EEG—A Review. *Electronics* 9: 660, 18 pages.

14 Hämäläinen, M., Hari, R., Ilmoniemi, R.J. et al. (1993). Magnetoencephalography – theory, instrumentation, and applications to noninvasive studies of the working human brain. *Reviews of Modern Physics* 65: 413.

15 Banach, S. (1932). *Théorie des opérations linéaires*, vol. 1. Warsaw: Virtual Library of Science Math. – Phys. Collection.

16 Yao, J. and Dewald, J.P.A. (2005). Evaluation of different cortical source localization methods using simulated and experimental EEG data. *NeuroImage* 25: 369–382.

17 Yetik, I.S., Nehorai, A., Lewine, J.D., and Muravchik, C.H. (2005). Distinguishing between moving and stationary sources using EEG/MEG measurements with an application to epilepsy. *IEEE Transactions on Biomedical Engineering* 52 (3): 471–479.

18 Xu, P., Tian, Y., Chen, H., and Yao, D. (2007). Lp norm iterative sparse solution for EEG source localization. *IEEE Transactions on Biomedical Engineering* 54 (3): 400–409.

19 Pascual-Marqui, R.D., Michel, C.M., and Lehmann, D. (1994). Low resolution electromagnetic tomography; a new method for localizing electrical activity in the brain. *International Journal of Psychophysiology* 18 (65): 49.

20 He, B., Zhang, X., Lian, J. et al. (2002). Boundary element method-based cortical potential imaging of somatosensory evoked potentials using subjects' magnetic resonance images. *NeuroImage* 16: 564–576.

21 Ricamato, A., Dhaher, Y., and Dewald, J. (2003). Estimation of active cortical current source regions using a vector representation scanning approach. *Journal of Clinical Neurophysiology* 20: 326–344.

22 Brodbeck, V., Spinelli, L., Lascano, A.M. et al. (2011). Electroencephalographic source imaging: a prospective study of 152 operated epileptic patients. *Brain* 134: 2887–2897. https://doi.org/10.1093/brain/awr243.

23 Akalin Acar, Z. and Makeig, S. (2013). Effects of forward model errors on EEG source localization. *Brain Topography* 26: 378–396. https://doi.org/10.1007/s10548-012-0274-6.

24 Baillet, S., Riera, J.J., Marin, G. et al. (2001). Evaluation of inverse methods and head models for EEG source localization using a human skull phantom. *Physics in Medicine and Biology* 46: 77–96. https://doi.org/10.1088/0031-9155/46/1/306.

25 Fuchs, M., Wagner, M., and Kastner, J. (2007). Development of volume conductor and source models to localize epileptic foci. *Journal of Clinical Neurophysiology* 24: 101–119. https://doi.org/10.1097/WNP.0b013e318038fb3e.

26 Guggisberg, A.G., Dalal, S.S., Zumer, J.M. et al. (2011). Localization of cortico-peripheral coherence with electroencephalography. *NeuroImage* 57: 1348–1357. https://doi.org/10.1016/j.neuroimage.2011.05.076.

27 Wang, G., Worrell, G., Yang, L. et al. (2011). Interictal spike analysis of high-density EEG in patients with partial epilepsy. *Clinical Neurophysiology* 122: 1098–1105. https://doi.org/10.1016/j.clinph.2010.10.043.

28 Hoekema, R., Wieneke, G.H., Leijten, F.S. et al. (2003). Measurement of the conductivity of skull, temporarily removed during epilepsy surgery. *Brain Topography* 16: 29–38. https://doi.org/10.1023/A:1025606415858.

29 Latikka, J., Kuurne, T., and Eskola, H. (2001). Conductivity of living intracranial tissues. *Physics in Medicine and Biology* 46: 1611–1616. https://doi.org/10.1088/0031-9155/46/6/302.

30 Wang, D. and Chen, J. (2018). Supervised speech separation based on deep learning: an overview. *IEEE/ACM Transactions on Audio, Speech, and Language Processing* 26 (10): 1702–1706.
31 Abeßer, J. (2020). A review of deep learning based methods for acoustic scene classification. *Applied Sciences* 10 (6): 16.
32 Aldossari, S.M. and Chen, K.-C. (2019). Machine learning for wireless communication channel modeling: an overview. *Wireless Personal Communications* 106: 41–70.
33 Spyrou, L., Jing, M., Sanei, S., and Sumich, A. (2007). Separation and localization of P300 sources and their subcomponents using constrained blind source separation. *EURASIP Journal of Applied Signal Processing*: 82912 (2006).
34 Schmit, R.O. (1986). Multiple emitter location and signal parameter estimation. *IEEE Transactions on Antennas and Propagation* 34: 276–280; reprint of the original paper presented at RADC Spectrum Estimation Workshop, 1979.
35 Mosher, J.C. and Leahy, R.M. (1998). Recursive Music: a framework for EEG and MEG source localization. *IEEE Transactions on Biomedical Engineering* 45 (11): 1342–1354.
36 Mosher, J.C., Leahy, R.M., and Lewis, P.S. (1999). EEG and MEG: forward solutions for inverse methods. *IEEE Transactions on Biomedical Engineering* 46 (3): 245–259.
37 Mosher, J.C. and Leahy, R.M. (1999). Source localization using recursively applied and projected (RAP) MUSIC. *IEEE Transactions on Biomedical Engineering* 47 (2): 332–340.
38 Ermer, J.J., Mosher, J.C., Baillet, S., and Leahy, R.M. (2001). Rapidly recomputable EEG forward models for realisable head shapes. *Physics in Medicine and Biology* 46: 1265–1281.
39 Mosher, J.C., Lewis, P.S., and Leahy, R.M. (1992). Multiple dipole modelling and localization from spatio-temporal MEG data. *IEEE Transactions on Biomedical Engineering* 39: 541–557.
40 Golub, G.H. and Pereyra, V. (1973). The differentiation of pseudo-inverses and nonlinear least squares problems whose variables separate. *SIAM Journal on Numerical Analysis* 10: 413–432.
41 Golub, G.H. and Van Loan, C.F. (1984). *Matrix Computations*, 2e. Baltimore, MD: John Hopkins University Press.
42 Xu, X.-L., Xu, B., and He, B. (2004). An alternative subspace approach to EEG dipole source localization. *Physics in Medicine and Biology* 49: 327–343.
43 Darvas, F., Ermer, J.J., Mosher, J.C., and Leahy, R.M. (2006). Generic head models for atlas-based EEG source analysis. *Human Brain Mapping* 27 (2): 129–143.
44 Buchner, H., Knoll, G., Fuchs, M. et al. (1997). Inverse localization of electric dipole current sources in finite element models of the human head. *Electroencephalography and Clinical Neurophysiology* 102 (4): 267–278.
45 Steele, C.W. (1996). *Numerical Computation of Electric and Magnetic Fields*. Kluwer Academic Publishers.
46 Geddes, A. and Baker, L.E. (1967). The specific resistance of biological material-A compendium of data for the biomedical engineer and physiologist. *Medical and Biological Engineering* 5: 271–293.
47 Schimpf, P.H., Liu, H., Ramon, C., and Haueisen, J. (2005). Efficient electromagnetic source imaging with adaptive standardized LORETA/FOCUSS. *IEEE Transactions on Biomedical Engineering* 52 (5): 901–908.
48 Gorodnitsky, I.F., George, J.S., and Rao, B.D. (1999). Neuromagnetic source imaging with FOCUSS: a recursive weighted minimum norm algorithm. *Journal of Clinical Neurophysiology* 16 (3): 265–295.

49 Pascual-Marqui, R.D. (2002). Standardized low resolution brain electromagnetic tomography (sLORETA): technical details. *Methods and Findings in Experimental and Clinical Pharmacology* 24D: 5–12.

50 Hanson, P.C. (1998). *Rank-Efficient and Discrete Ill-Posed Problems*. Philadelphia, PA: SIAM.

51 Liu, H., Schimpf, P.H., Dong, G. et al. (2005). Standardized shrinking LORETA-FOCUSS (SSLOFO): a new algorithm for spatio-temporal EEG source reconstruction. *IEEE Transactions on Biomedical Engineering* 52 (10): 1681–1691.

52 Liu, H., Gao, X., Schimpf, P.H. et al. (2004). A recursive algorithm for the three-dimensional imaging of brain electric activity: shrinking LORETA-FOCUSS. *IEEE Transactions on Biomedical Engineering* 51 (10): 1794–1802.

53 Tikhonov, A.N. and Arsenin, V.Y. (1997). *Solution of Ill Posed Problems*. New York: Wiley.

54 Ferguson, A.S. and Stronik, G. (1997). Factors affecting the accuracy of the boundary element method in the forward problem: I. Calculating surface potentials. *IEEE Transactions on Biomedical Engineering* 44: 440–448.

55 Buchner, H., Knoll, G., Fuchs, M. et al. (1997). Inverse localisation of electric dipole current sources in finite element models of the human head. *Electroencephalography and Clinical Neurophysiology* 102 (4): 267–278.

56 Ashburner, J. and Friston, K.J. (1997). Multimodal image coregistration and partitioning – a unified framework. *NeuroImage* 6: 209–217.

57 Vapnic, V. (1998). *Statistical Learning Theory*. Wiley.

58 David, O., Garnero, L., Cosmelli, D., and Varela, F.J. (2002). Estimation of neural dynamics from MEG/EEG cortical current density maps: application to the reconstruction of large-scale cortical synchrony. *IEEE Transactions on Biomedical Engineering* 49 (9): 975–987.

59 Fuchs, M., Drenckhahn, R., Wichmann, H.A., and Wager, M. (1998). An improved boundary element method for realistic volume-conductor modelling. *IEEE Transactions on Biomedical Engineering* 45: 980–977.

60 Macro-Pallares, J., Grau, C., and Ruffini, G. (2005). Combined ICA-LORETA analysis of mismatch negativity. *NeuroImage* 25: 471–477.

61 Alain, C., Woods, D.L., and Night, R.T. (1998). A distributed cortical network for auditory sensory memory in humans. *Brain Research* 812: 23–27.

62 Rosburg, T., Haueisen, J., and Kreitschmann-Andermahr, I. (2004). The dipole location shift within the auditory evoked neuromagnetic field components N100m and mismatch negativity (MMNm). *Clinical Neurophysiology* 308: 107–110.

63 Jaaskelainen, I.P., Ahveninen, J., Bonmassar, G., and Dale, A.M. (2004). Human posterior auditory cortex gates novel sounds to consciousness. *Proceedings of the National Academy of Sciences of the United States of America* 101: 6809–6814.

64 Kircher, T.T.J., Rapp, A., Grodd, W. et al. (2004). Mismach negativity responses in schizophrenia: a combined fMRI and whole-head MEG study. *The American Journal of Psychiatry* 161: 294–304.

65 Muller, B.W., Juptner, M., Jentzen, W., and Muller, S.P. (2002). Cortical activation to auditory mismatch elicited by frequency deviant and complex novel sounds: a PET study. *NeuroImage* 17: 231–239.

66 Serinagaoglu, Y., Brooks, D.H., and Macleod, R.S. (2005). Bayesian solutions and performance analysis in bioelectric inverse problems. *IEEE Transactions on Biomedical Engineering* 52 (6): 1009–1020.

67 Makeig, S., Debener, S., Onton, J., and Delorme, A. (2004). Mining event related brain dynamic trends. *Cognitive Science* 134: 9–21.

68 Bradley, A., Yao, J., Dewald, J., and Richter, C.-P. (2016). Evaluation of electroencephalography source localization algorithms with multiple cortical sources. *PLoS One* 11 (1): e0147266.

69 Latif, M.A., Sanei, S., Chambers, J.A., and Shoker, L. (2006). Localization of abnormal EEG sources using blind source separation partially constrained by the locations of known sources. *IEEE Signal Processing Letters* 13 (3): 117–120.

70 Spyrou, L., Sanei, S., and Cheong Took, C. (2007). Estimation and location tracking of the P300 subcomponents from single-trial EEG. *IEEE Int. Conf on Acoustics, Speech, and Signal Processing*. ICASSP, USA.

71 Sarvas, J. (1987). Basic mathematical and electromagnetic concepts of the biomagnetic inverse problem. *Physics in Medicine and Biology* 32 (1): 11–22.

72 Spyrou, L. and Sanei, S. (2008). Source localisation of event related potentials incorporating spatial notch filters. *IEEE Transactions on Biomedical Engineering* 55 (9): 2232–2239.

73 Tadel, F., Baillet, S., Mosher, J.C. et al. (2011). *Brainstorm: a user-friendly application for MEG/EEG analysis. Computational Intelligence and Neuroscience* 2011: 879716. https://doi.org/10.1155/2011/879716.

74 Mohseni, H.R. and Sanei, S. (2010). A new beamforming-based MEG dipole source localization method. *Proceedings of the IEEE International Conference on Acoustics, Speech and Signal Processing*. ICASSP, USA.

75 Van Veen, B.D., Van Dronglen, W., Yuchtman, M., and Suzuki, A. (1997). Localization of brain electrical activity via linearly constrained minimum variance spatial filtering. *IEEE Transactions on Biomedical Engineering* 44: 867–880.

76 Robinson, S.E. and Vrba, J. (1999). Functional neuroimaging by synthetic aperture magnetometry (SAM). In: *Recent Advances in Biomagnetism* (eds. T. Yoshimoto, M. Kotani, S. Kuriki, et al.), 302–305. Japan: Tohoku University Press.

77 Gutierrez, D., Nehorai, A., and Dogandzic, A. (2006). Performance analysis of reduced-rank beamformers for estimating dipole source signals using EEG/MEG. *IEEE Transactions on Biomedical Engineering* 53 (5): 840–844.

78 Baillet, S., Mosher, J.C., and Leahy, R.M. (2001). Electromagnetic brain mapping. *IEEE Signal Processing Magazine* 18: 14–30.

79 Michela, C.M., Murraya, M.M., Lantza, G. et al. (2004). EEG source imaging. *Clinical Neurophysiology* 115: 2195–2222.

80 Mohseni, H.R., Ghaderi, F., Wilding, E., and Sanei, S. (2009). A beamforming particle filter for EEG dipole source localization. *Proceedings of the IEEE International Conference on Acoustics, Speech and Signal Processing*. ICASSP, Taiwan.

81 Bai, X. and He, B. (2006). Estimation of number of independent brain electric sources from the scalp EEGs. *IEEE Transactions on Biomedical Engineering* 53 (10): 1883–1892.

82 Knösche, T., Brends, E., Jagers, H., and Peters, M. (1998). Determining the number of independent sources of the EEG: a simulation study on information criteria. *Brain Topography* 11: 111–124.

83 Stoica, P. and Nehorai, A. (1989). MUSIC, maximum likelihood, and Cramer-Rao bound. *IEEE Transactions on Acoustics, Speech, and Signal Processing* 37 (5): 720–741.

84 Waldorp, L.J., Huizenga, H.M., Nehorai, A., and Grasman, R.P. (2002). Model selection in electromagnetic source analysis with an application to VEFs. *IEEE Transactions on Biomedical Engineering* 49 (10): 1121–1129.
85 Waldorp, L.J., Huizenga, H.M., Nehorai, A., and Grasman, R.P. (2005). Model selection in spatio-temporal electromagnetic source analysis. *IEEE Transactions on Biomedical Engineering* 52 (3): 414–420.
86 Akaike, H. (1973). Information and an extension of the maximum likelihood principle. *Proceedings of the 2nd International Symposium on Information Theory, Supplement Problem Controls in Information Theory*, 267–281.
87 Chow, G.C. (1981). A comparison of information and posterior probability criteria for model selection. *Journal of Econometrics* 16: 21–33.
88 Huzenga, H.M., Heslenfeld, D.J., and Molennar, P.C.M. (2002). Optimal measurement conditions for spatiotemporal EEG/MEG source analysis. *Psychometrika* 67: 299–313.
89 Seber, G.A.F. and Wild, C.J. (1989). *Nonlinear Regression*. Toronto, Canada: Wiley.
90 Andrés Muñoz-Gutiérrez, P., Giraldo, E., Bueno-López, M., and Molinas, M. (2018). Localization of active brain sources from EEG signals using empirical mode decomposition: a comparative study. *Frontiers in Integrative Neuroscience* 12: 55. https://doi.org/10.3389/fnint.2018.00055.
91 Michel, C.M. and Brunet, D. (2019). EEG source imaging: a practical review of the analysis steps. *Frontiers in Neurology* 10: 325.
92 Sun, M. and Sclabassi, R.J. (2000). The forward EEG solutions can be computed using artificial neural networks. *IEEE Transactions on Biomedical Engineering* 47 (8).
93 Abeyratne, U.R., Zhang, G., and Saratchand, P. (2001). EEG source localization: comparative study of classical and neural network methods. *International Journal of Neural Systems* 11 (4): 349–359.
94 Zhang, Q., Bai, X., Akutagawa, M. et al. (2002). A method for two EEG sources localization by combining BP neural networks with nonlinear least square method. *7th International Conference on Control, Automation, Robotics And Vision (lCARCV'O2)*, (December 2002). Singapore.
95 Li, Z., Akutagawa, M., and Kinouchi, Y. (2005). Brain source localization for two dipoles using a combined method from 32-channel EEGs. *Proceedings of the 2005 IEEE Engineering in Medicine and Biology 27th Annual Conference Shanghai*, China (1–4 September 2005).
96 Li, J-W., Wang, Y-H., Wu, Q. et al. (2008). EEG source localization of ERP based on multidimensional support vector regression approach. *Proceedings of the Seventh International Conference on Machine Learning and Cybernetics*, Kunming (12–15 July 2008).
97 Li, J-W., Wang, Y-H., Wu, Q. et al. (2009). Support vector machine method using in EEG signals study of epileptic spike. *Proceedings of the Eighth International Conference on Machine Learning and Cybernetics*, Baoding (12–15 July 2009).
98 McNay, D., Michielssen, E., Rogers, R.L. et al. (1996). Multiple source localization using genetic algorithms. *Journal of Neuroscience Methods* 64 (2): 163–172.
99 Abeyratne, U.R., Kinouchi, Y., Oki, H. et al. (1991). Artificial neural networks for source localization in the human brain. *Brain Topography* 4 (1): 3–21.
100 Van Hoey, G., De Clercq, J., Vanrumste, B. et al. (2000). EEG dipole source localization using artificial neural networks. *Physics in Medicine and Biology* 45: 997–1011.

101 Robert, J.S., Sonmez, M., and Sun, M. (2013). EEG source localization: a neural network approach. *Neurological Research* 23: 457–464.

102 Wentrup, M.G., Gramann, K., Wascher, E., and Buss, M. (2005). EEG source localization for brain-computer-interfaces. *Proceedings of the IEEE EMBS Conference*, 128–131.

103 Ding, L., Worrell, G.A., Lagerlund, T.D., and He, B. (2006). 3D source localization of Interictal spikes in epilepsy patients with MRI lesions. *Physics in Medicine and Biology* 51: 4047–4062.

11

Epileptic Seizure Prediction, Detection, and Localization

11.1 Introduction

Epileptic seizure was possibly the main motivation for measuring the electrical potentials of the brain which led to the invention of electroencephalogram (EEG) systems and analysis of EEG patterns. The fundamental concepts of epilepsy were refined and developed in ancient Indian medicine during the Vedic period of 4500–1500 BCE. In the Ayurvedic literature of Charaka Samhita (around 400 BCE and the oldest existing description of the complete Ayurvedic medical system), epilepsy is described as *'apasmara'* which means 'loss of consciousness'. During 460–379 BCE, Hippocrates discussed epilepsy as a 'disturbance of the brain'. The word 'epilepsy' being derived from the Greek word *'epilambanein'*, which means 'to seize or attack'. We now know, however, that seizures are the result of sudden, usually brief, excessive electrical discharges in a group of brain cells (neurones) and that different parts of the brain can be the site of such discharges. The clinical manifestations of seizures therefore vary and depend on where in the brain the disturbance first starts and how far it spreads. Transient symptoms can occur, such as loss of awareness or consciousness and disturbances of movement, sensation (including vision, hearing and taste), mood, or mental function.

The literature of Charaka Samhita contains abundant references to all aspects of epilepsy including symptomatology, aetiology, diagnosis, and treatment. Another ancient and detailed account of epilepsy is on a Babylonian tablet in the British Museum in London. This is a chapter from a Babylonian textbook of medicine comprising 40 tablets dating as far back as 2000 BCE. The tablet accurately records many of the different seizure types we recognize today. In contrast to the Ayurvedic medicine of Charaka Samhita, however, it emphasizes the supernatural nature of epilepsy, with each seizure type associated with the name of a spirit or god – usually evil. Treatment was, therefore, largely a spiritual matter. The Babylonian view was the forerunner of the Greek concept of *'the sacred disease'*, as described in the famous treatise by Hippocrates (dated to the fifth century BCE). The term *'seleniazetai'* was also often used to describe people with epilepsy because they were thought to be affected by the phases of the moon or by the moon god (Selene), and hence the notion of *'moonstruck'* or *'lunatic'* (the Latinized version) arose. Hippocrates, however, believed that epilepsy was not sacred, but a disorder of the brain. He recommended physical treatments and stated that if the disease became chronic, it was incurable.

EEG Signal Processing and Machine Learning, Second Edition. Saeid Sanei and Jonathon A. Chambers.

Avicenna (980–1037 CE), a famous Persian clinician and philosopher, considered different types of epilepsy and their specific treatments. He classifies the causes of epilepsy into two, those caused by brain diseases and those associated with the diseases of other organs. He believed that organs such as the stomach can influence the brain and can cause epilepsy [1].

However, the perception that epilepsy was a brain disorder did not begin to take root until the eighteenth and nineteenth centuries CE. The intervening 2000 years were dominated by more supernatural views. In Europe, for example, St Valentine has been the patron saint of people with epilepsy since mediaeval times. During this time people with epilepsy were viewed with fear, suspicion, and misunderstanding and were subjected to enormous social stigma. People with epilepsy were treated as outcasts and punished. Some, however, succeeded and became famous the world over. Among these people were Julius Caesar, Czar Peter the Great of Russia, Pope Pius IX, the writer Fyodor Dostoevsky and the poet Lord Byron.

During the nineteenth century, as neurology emerged as a new discipline distinct from psychiatry, the concept of epilepsy as a brain disorder became more widely accepted, especially in Europe and the United States of America (USA). This helped to reduce the stigma associated with the disorder. Bromide, introduced in 1857 as the world's first effective antiepileptic drug, became widely used in Europe and the USA during the second half of the last century.

The first hospital centre for the *'paralyzed and epileptic'* was established in London in 1857. At the same time a more humanitarian approach to the social problems of epilepsy resulted in the establishment of epilepsy 'colonies' for care and employment.

The new understanding of epilepsy (pathophysiology) was also established in the nineteenth century with the work of neurologist Hughlings Jackson in 1873 who proposed that seizures were the result of sudden brief electrochemical discharges in the brain. He also suggested that the character of the seizures depended on the location and function of the site of the discharges. Soon afterwards the electrical excitability of the brain in animals and man was discovered by David Ferrier in London, together with Gustav Theodor Fritsch and Eduard Hitzig in Germany.

The important application of EEG from the 1930s onwards was in the field of epilepsy. The EEG revealed the presence of electrical discharges in the brain. It also showed different patterns of brainwave discharges associated with different seizure types. The EEG also helped to locate the site of seizure discharges and expanded the possibilities of neurosurgical treatments, which became much more widely available from the 1950s onwards in London, Montreal, and Paris.

During the first half of the twentieth century the main drugs for treatment were phenobarbitone (first used in 1912) and phenytoin (first used in 1938). Since the 1960s, there has been an accelerating process of drug discovery, based in part on a much greater understanding of the electrochemical activities of the brain, especially the excitatory and inhibitory neurotransmitters. In developed countries in recent years, several new drugs have come into the market and seizures can now be controlled in 70–80% of newly diagnosed children and adults.

Neuroimaging techniques such as fMRI and positron emission tomography (PET) boost the success in diagnosis of epilepsy. Such technology has revealed many of the more subtle brain lesions responsible for epilepsy. Several brain lesions such as trauma, congenital, developmental, infection, vascular, and tumour might lead to epilepsy in some people. Of the 50 million people in the world with epilepsy, some 35 million have no access to appropriate treatment. This is either because services are nonexistent or because epilepsy is not viewed as a medical problem or a treatable brain disorder.

As a major campaign for the treatment of epilepsy, the International League Against Epilepsy (ILAE) and the International Bureau for Epilepsy (IBE) joined forces with the World Health Organization in 1997 to establish the Global Campaign Against Epilepsy to address these issues.

Epilepsy is a sudden and recurrent brain malfunction and is a disease which reflects an excessive and hypersynchronous activity of the neurons within the brain and is probably the most prevalent brain disorder among adults and children, second only to stroke. Over 50 million people worldwide are diagnosed with epilepsy whose hallmark is recurrent seizures [2]. The prevalence of epileptic seizures changes from one geographic area to another [3].

The seizures occur at random to impair the normal function of the brain. Epilepsy can be treated in many cases and the most important treatment today is pharmacological. The patient takes anticonvulsant drugs on a daily basis trying to achieve a steady-state concentration in the blood, chosen to provide the most effective seizure control. Surgical intervention is an alternative for carefully selected cases that are refractory to medical therapy. However, in almost 25% of the total number of patients diagnosed with epilepsy, seizures cannot be controlled by any available therapy. Furthermore, side effects from both pharmacological and surgical treatments have been reported.

Looking at scalp EEG, an epileptic seizure can be characterized by paroxysmal occurrence of synchronous oscillations. Strong synchronization of neural activities (characterized by the algebraic connectivity metric) is caused by abnormal neuronal firing during a seizure onset [4]. Such seizures can be classified into two main categories depending on the extent of involvement of various brain regions: focal (or partial) and generalized. Generalized seizures involve most areas of the brain whereas focal seizures originate from a circumscribed region of the brain, often called epileptic foci [5]. Figure 11.1 shows two segments of EEG signals involving generalized and focal seizures respectively.

Successful surgical treatment of focal epilepsies requires exact localization of the epileptic focus and its delineation from functionally relevant areas [6]. The physiological aspects of seizure generation and the treatment and monitoring of a seizure including presurgical examinations have been well established [5] and the medical literature provided.

EEG, magnetoencephalograms (MEG), and recently functional magnetic resonance imaging (fMRI) are the major neuroimaging modalities used for seizure detection. The blood oxygenation level dependent (BOLD) regions in fMRI of the head clearly show the epileptic foci. However, the number of fMRI machines is limited in each area, they are costly, and a full-body scan is time consuming. Therefore, using fMRI for all patients at all times is not feasible. MEG, conversely, is noisy and since the patient under care has to be steady during the recording, it is hard to achieve clear data for moderate and severe cases using current MEG machines.

Therefore, EEG remains the most useful and cost effective modality for the study of epilepsy. Although for generalized seizure the duration of seizure can be easily detected using a naked eye, for most focal epilepsies however, such intervals are difficult to recognize.

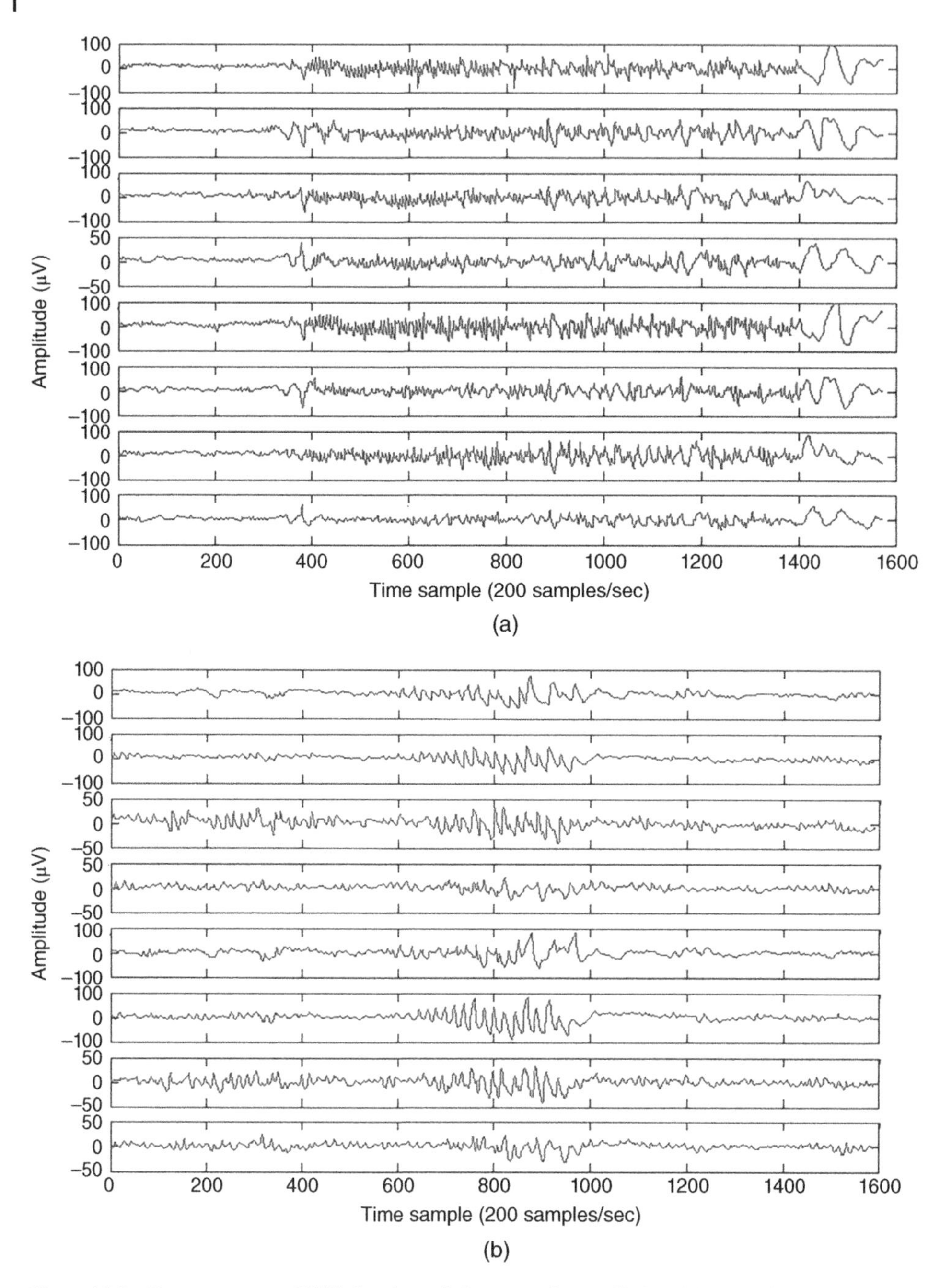

Figure 11.1 Two segments of EEG signals each from a patient suffering: (a) generalized seizure and (b) focal seizure onset.

From the pathological point of view, there are clear classification schemes for seizures. 'Partial' is used to describe isolated phenomena that reflect focal cortical activity, either evident clinically or by EEG. The term 'simple' indicates that consciousness is not impaired. For example, a seizure visible as a momentarily twitching upper extremity, which subsides, would be termed a simple partial seizure with motor activity. Partial seizures may have motor, somatosensory, psychic, or autonomic symptoms [7].

The term 'complex' defines an alteration of consciousness associated with the seizure. 'Generalization' is a term used to denote spread from a focal area of the cortex, which could be evident clinically by EEG, and involves all areas of the cortex with resulting generalized motor convulsion. It is known that in adults the most common seizure type is that of initial activation of one area of the cortex with subsequent spread to all areas of the cortex; frequently this occurs too quickly to be appreciated by bedside observation.

The other major grouping of seizure types is for generalized seizures, which may be termed convulsive or nonconvulsive. For this type, all areas of the cortex are activated at once. This, for example, is seen with absence seizures and myoclonic seizures [8].

Tonic–clonic (grand mal) seizures are more common and because of that other types of seizures may escape detection. They were diagnosed for the first time in 1827 by Bravais [9]. The recurrent behaviour of the spike waveforms in the EEG signals triggers investigation for a possible seizure disorder. Seizures of frontal or temporal cortical origin with nonclassical motor movements are fairly commonly encountered. The patient may show some seemingly organized motor activity without the usually in-phase jerking movements more typical of generalized seizures. Also complicating the problem is that clouding or alteration of consciousness may occur without complete loss of consciousness.

The term aura is used to represent ongoing seizure activity limited to a focal area of the cortex; in this case the abnormal electrical activity associated with the seizure does not spread or generalize to the entire cerebral cortex but remains localized and persists in one abnormal focus.

One seizure type may evolve into another seizure type. For example, a simple motor seizure may evolve into a complex partial seizure with altered consciousness; the terminology for this would be 'partial complex status epilepticus' [8].

Absence seizures (also known as petit mal) are a primarily generalized seizure type involving all cortical areas at once; this is typically a seizure disorder of childhood with a characteristic EEG pattern [8]. At times, absence seizures may persist with minimal motor movements and altered consciousness for hours or days. Absence epileptic seizure and complex partial epileptic seizure are often grouped under the term 'nonconvulsive status epilepticus' and are referred to at times as twilight or fugue states.

The term 'subtle status epilepticus' is more correctly used to indicate patients that have evolved from generalized convulsive epileptic seizure or are in a comatose state with epileptiform activity.

The character of an epileptic seizure is determined based on the region of the brain involved, and the underlying basic epileptic condition, which are mostly age-determined. As a conclusion, clinical classification of epileptic seizures is summarized as:

- Partial (focal) seizures: these seizures arise from an electric discharge of one or more localized areas of the brain regardless of whether the seizure is secondarily generalized. Depending on their type, they may or may not impair consciousness. Whether seizures are partial or focal, they begin in a localized area of the brain, but then may spread to the whole brain causing a generalized seizure. They are divided to:
  - simple partial
  - complex partial

  - simple partial onset followed by alteration of consciousness
  - partial evolving to secondarily generalized.
- Generalized seizures (convulsive and nonconvulsive): the electrical discharge which leads to these seizures involves the whole brain and may cause loss of consciousness and/or muscle contractions or stiffness. They include what used to be known as 'grand mal' convulsion and also the brief 'petit mal' absence of consciousness. These seizures are further divided into:
  - absence (typical and atypical)
  - clonic, tonic, or tonic–clonic (grand mal)
  - myoclonic
  - atonic (astatic).
- Epileptic seizures with unknown waveform patterns.
- Seizures precipitated by external triggering events [8].

Moreover, epileptic seizures may be chronic or acute. However, so far no automatic classification of these seizures based on the EEG waveforms has been reported.

Most traditional epilepsy analysis methods, based on the EEG, are focused on the detection and classification of epileptic seizures among which the best method of analysis is still the visual inspection of the EEG by a highly skilled electroencephalographer or neurophysiologist. However, with the advent of new signal processing methodologies several computerized techniques have been proposed to detect and localize epileptic seizures. [10–13]. In addition, based on the mathematical theory of nonlinear dynamics [14], there has been an increased interest in the analysis of the EEG for prediction of epileptic seizures. In recent years in parallel with availability of enhanced computational power using graphic cards, high capacity core memory, and over-cloud computation, the strategy in EEG analysis shifted back towards machine learning. These techniques highly benefit from feature learning, modelling, and classification capabilities of deep neural networks (DNNs).

Researchers have tried to highlight different signal characteristics within various domains and classify the signal segments based on the measured features. Adult seizure is more definable than neonate (newborn) seizure although the morphology of epileptiform spikes originating from deep brain layers is more pronounced and clear. Therefore, its detection, labelling, and classification is not very difficult. Conversely, neonate seizure is not always epileptic and can have many other physiological or pathological reasons such as lack of oxygen before or during birth due to placental abruption, a difficult or prolonged labour, or compression of the umbilical cord. Therefore, various automated spike detection approaches have been developed [15–20].

Recently, predictability of seizure from long EEG recordings has attracted many researchers. In one hand, it has been shown that epileptic sources gradually tend to be less chaotic from a few minutes before the seizure onset. This finding is clinically very important since the patients do not need to be under anticonvulsants administration permanently, but from just a few minutes before seizure. Conversely, with the help of new signal processing and machine learning techniques, the interictal epileptiform discharges (IEDs) can be detected and characterized more accurately. The information can be effectively used for prediction of seizure onset. In the following sections the major research topics in the areas of epileptic seizure detection and prediction are discussed using both scalp and intracranial (cortical and subdural) EEG recordings.

Nevertheless, not all seizures are epileptic. Some people experience symptoms similar to those of an epileptic seizure but without any unusual electrical activity in the brain. This is known as a nonepileptic seizure (NES). NES is most often caused by mental stress or a physical condition and can be divided into two types: organic NESs and psychogenic seizures.

An organic NES has a physical cause (relating to the body), such as fainting (syncope) and metabolic causes such as diabetes. Therefore, they may be relatively easy to diagnose and the underlying cause can be found. For example, a faint may be diagnosed as being caused by a physical problem in the heart and can be treated to stop the seizure.

A psychogenic NES is caused by mental or emotional processes, rather than by a physical cause. Psychogenic seizures may happen when someone's reaction to painful or difficult thoughts and feelings affects them physically. These seizures include dissociative seizures which happen unconsciously, panic attacks which can happen in frightening situations, and can cause sweating, palpitations (being able to feel your heartbeat), trembling, and difficulty breathing. The person may also lose consciousness and may shake (convulse). Factitious seizures are ones for which the person has some level of conscious control over them. An example of this is when seizures form part of Münchausen's syndrome, a rare psychiatric condition where a person is driven by a need to have medical investigations and treatments.

11.2 Seizure Detection

The human body is highly affected by the seizure onset. Therefore, various sensor and measurement modalities may be used to detect it [21, 22]. Here, however, the focus is on the use of brain waves mainly EEG for seizure detection.

11.2.1 Adult Seizure Detection from EEGs

In clinics, for patients with medically intractable partial epilepsies, time-consuming video-EEG monitoring of spontaneous seizures is often necessary [12]. Visual analysis of interictal EEG is however time intensive. Application of invasive methods for monitoring the seizure signals and identification of an epileptic zone is hazardous and involves risk for the patient.

The majority of seizure detection algorithms involve two parts: one is to process the data and extract the necessary and useful features and the second one is to employ machine learning for classification of those features. Before designing any automated seizure detection system the characteristics of the EEG signals before, in between two seizure onsets, during, and after the seizure onset have to be determined and evaluated. Several features have been identified to better describe the features. These may represent the static behaviour of the signals within a short time interval, such as signal energy, or the dynamic behaviour of the signals such as chaoticity and the change in frequency during the seizure onset.

In some early works in seizure spike detection [23–26] a number of parameters such as relative amplitude, sharpness, and duration of EEG waves were measured from the EEG signals and evaluated. The method is sensitive to various artefacts. In these attempts different states such as active wakefulness or desynchronized EEG were defined, in which typical nonepileptic transients were supposed to occur [25, 26]. A multistage system to detect the epileptiform activities from the EEGs was developed by Dingle et al. [27]. They combined a mimetic approach with a rule-based expert system, and thereby considered and exploited both the spatial and temporal systems. In another approach [28] multichannel EEGs were used and a correlation-based algorithm was attempted to reduce the muscle artefacts. Following this method approximately 67% of the spikes can be detected. By incorporating both multichannel temporal and spatial information, and including the electrocardiogram (ECG), electromyogram (EMG), and electro-oculogram (ECoG) information into a rule-based system [21] a higher detection rate was achieved. A two-stage automatic system was developed by Davey et al. [29]. In the first stage a feature extractor and in the second stage a classifier were introduced. A 70% sensitivity was claimed for this system.

Artificial neural networks (ANNs) have been used for seizure detection by many researchers [30, 31]. The Kohonen self-organizing feature map ANN [32, 33] was used for spike detection by Kurth et al. [31]. In this work for each patient three different-sized neural networks (NN) have been examined. The training vector included a number of signals with typical spikes, a number of eye-blink artefact signals, some signals of muscle artefacts, and also background EEG signals. The major problem with these methods is that the epileptic seizure signals do not follow similar patterns. Presenting all types of seizure patterns to the ANN, conversely, reduces the sensitivity of the overall detection system. Therefore, a clever feature detection followed by a robust classifier often provides an improved result.

In [34] the authors proposed a deep learning-based method to automatically detect the epileptic seizure onsets and offsets in multichannel EEG signals. A convolutional neural network (CNN) was designed to identify occurrences of seizures in EEG epochs and a method to determine the seizure onsets and offsets. They applied their method to the Children's Hospital Boston-Massachusetts Institute of Technology (CHB-MIT) Scalp EEG database containing 24 cases [35]. Their proposed filter captures the temporal and spatial patterns in the EEG epochs. Their method follows the clinical decision criteria and is claimed to have over 90% accuracy.

Among recent works, time–frequency (TF) approaches effectively use the fact that the seizure sources are localized in the TF domain. Most of these methods are mainly for detection of neural spikes [36] of different types. Different TF methods following different classification strategies have been proposed by many researchers [37, 38] in this area. The methods are especially useful since the EEG signals are statistically nonstationary. The discrete wavelet transform (DWT) obtains a better TF representation than the TF based on the short-term Fourier transform due to its multiscale (multilevel) characteristics, i.e. it can model the signal according to its coarseness. The DWT analyzes the signal over different frequency bands, with different resolutions, by decomposing the signal into a coarse approximation and detail information. In a recent approach by Subasi [38], a DWT-based TF method followed by an ANN has been suggested. The ANN classifies the energy of various resolution (detail) levels. Using this technique, it is possible to detect more than 80% of adult epileptic seizures. Other TF distributions such as pseudo-Wigner–Ville can also be used for the same purpose [39].

In an established work by Osorio et al. [40, 41] a digital seizure detection algorithm has been proposed and implemented. The system is capable of accurate real-time detection, quantitative analysis, and very short-term prediction of clinical onset of seizure. This system computes a measure, namely 'foreground', of the median signal energy in the frequencies between 8 and 42 Hz in a short window of specific length (e.g. two seconds). The foreground is calculated through the following steps: (i) decomposing the signals into epileptiform (containing epileptic seizures) and nonepileptiform (without any seizure) components using a 22-coefficient wavelet filter (DAUB4, level 3) which separates the frequency sub-bands from 8 to 40 Hz; (ii) the epileptiform component is squared; and (iii) the squared components are median filtered. Conversely, a 'background' reference signal is obtained as an estimate of the median energy of a longer time (approximately 30 minutes) of the signal. A large ratio between the foreground and background will then show the event of seizure [42]. An analogue system was later developed to improve the technical drawbacks of the above system such as speed and noise [42].

There are many features that can be detected/measured from the EEGs for detection of epileptic seizures. Often seizures increase the average energy of the signals during the onset. For a windowed segment of the signal this can be measured as:

$$E(n) = \frac{1}{L}\sum_{p=t-L/2}^{n-1+L/2} x^2(p) \tag{11.1}$$

where L is the window length and the time index n is the window centre. The seizure signals have a major cyclic component and therefore generally exhibit a dominant peak in the frequency domain. The frequency of this peak however decays with time during the onset of seizure. Therefore, the slope of decay is a significant factor in the detection of seizure onset. Considering $X(f,n)$ the estimated spectrum of the windowed signal $x(n)$ centred at n, the peak at time n will be:

$$f_d(n) = arg\max_f (|X(f,n)|) \tag{11.2}$$

The spectrum is commonly estimated using autoregressive modelling [43]. From the spectrum the peak frequency is measured. The slope of the decay in the peak frequency can then be measured and used as a feature. The cyclic nature of the EEG signals can also be measured and used as an indication of seizure. This can be best identified by incorporating certain higher-order statistics of the data. One such indicator is related to the second and fourth order statistics of the measurements as follows [44]:

$$I = \left|C_2^0(0)\right|^{-4}\sum_{\alpha\neq 0}|P^\alpha|^2 \tag{11.3}$$

where $P^a = C_4^\alpha(0,0,0)$ represents the Fourier coefficients of the fourth order cyclic cumulant at zero lag and can be estimated as follows:

$$\hat{C}_4^\alpha(0,0,0) = \frac{1}{N}\sum_{n=0}^{N-1} x_c^4(n)e^{-\frac{j2\pi n\alpha}{N}} - 3\sum_{\beta=0}^{\alpha} C_2^{\alpha-\beta}(0)C_2^\beta(0) \tag{11.4}$$

where $x_c(n)$ is the zeroed mean version of x(n), and an estimation of $C_2^\alpha(0)$ is calculated as:

$$\hat{C}_2^\alpha(0) = \frac{1}{N}\sum_{n=0}^{N-1} x_c(n)e^{-\frac{j2\pi n\alpha}{N}} \tag{11.5}$$

This indictor is also measured with respect to time index n since it is calculated for each signal window centred at n. I measures, for a frame centred at n, the spread of the energy. Over the range of frequencies before seizure onset, the EEG is chaotic and no frequency appears to control its trace. During seizure, the EEG becomes more ordered (rhythmic) and therefore the spectrum has a large peak.

The above features have been measured and classified using a support vector machine (SVM) classifier [43]. It has been illustrated that for both tonic–clonic and complex partial seizures the classification rate can be as high as 100%.

In a robust detection of seizure, however, all statistical measures from the EEGs together with all other symptoms such as blood morphology, body movement, pain, changes in metabolism, heart rate variability, respiration, before, during, and after seizure, have to be quantified and effectively taken into account. This is a challenge for future signal processing/data fusion-based approaches. Nevertheless, the advances in sensor technology and body sensor networking [45] facilitate multimodal measurements and data fusion.

In another attempt for seizure detection a cascade of classifiers based on ANNs has been used [46]. The classification is performed in three stages. In the first stage six features related to geometrical shape of a seizure spike are fed into two perceptrons to classify peaks into *definite epileptiform*, *definite nonepileptiform*, and *unknown* waveforms. The features are selected based on the expected shape of an epileptic spike.

Since three outputs are needed after the first stage (also called pre-classifier) two single layer perceptrons are used in parallel. One perceptron is trained to give +1 for *definite nonepileptiform* and −1 otherwise. The second network produces +1 for *definite epileptiform* and −1 for otherwise. A segment, which produces −1 at the output of both networks is assigned to the *unknown* group.

In the second stage the *unknown* waveforms (spikes) are classified using a radial basis function (RBF) neural network. An RBF has been empirically shown to have a better performance than other ANNs for this stage of classification. The inputs are the segments of actual waveforms. The output is selected by hard-limiting the output of the last layer (after normalization).

Finally, in the third stage a multi-dimensional SVM with a nonlinear (RBF) kernel is used to process the data from its multichannel input. An accuracy of up to 100% has been reported for detection of generalized and most focal seizures [46].

A hidden Markov model (HMM) has also been used for seizure detection in [47]. In this approach an automated method called adaptive slope of wavelet coefficient counts over various thresholds (ASCOT) has been developed to classify patient episodes as seizure waveforms. ASCOT involves extracting the feature matrix by calculating the mean slope of wavelet coefficient counts over various thresholds in each frequency sub-band. They also evaluated the performance of the method considering various window sizes and concluded that better results come with shorter window sizes.

Sakai et al. [48] proposed a DNN to detect spatiotemporal abnormal intervals from EEGs of epilepsy patients caused by epilepsy. Their CNN detects the abnormal spatiotemporal intervals. These intervals are estimated not only from the electrode close to the focal region but those in the surrounding regions. Before applying the CNN, the signals obtained from 16 electrodes at one time slot can be treated as a 5×4 matrix (16 electrodes +4 missing corner values). So, the spatiotemporal EEG recordings are configured as a third-order tensor of

$5 \times 4 \times n$ (n is the data length in time). Then, each tensor segment is an input of the proposed system. In their CNN the numbers of inputs and outputs are the same and the abnormality is recognized for each channel. The network with $n = 500$ is depicted in Figure 11.2 [48].

Nevertheless, identification of the exact shape of an epileptic discharge, which is often far more complicated than that considered in [46] needs a thorough study of intracranial neuron activities and is studied in the later sections of this chapter. It has been shown that the morphology of these waveforms can not only change across subjects but it changes with age too.

In [49] it has been demonstrated that the IEDs have age-dependent characteristics. With increasing age, focal IEDs are less sharp, have lower amplitudes, have less prominent slow waves and they become more lateralized. Their findings can help EEG readers in detecting and correctly describing IEDs in patients of various age. Some examples of such waveforms, namely, IED, recorded using foramen ovale (FO) subdural electrodes and scalp EEG electrodes can be seen in Figure 11.3a and b respectively. Therefore, EEG readers should always consider patient age when interpreting IEDs.

These IEDs, however, are smeared and highly attenuated by the head tissues before reaching the EEG electrodes. Normally, less than 10% of these IEDs are visible from the scalp EEGs. They can be better detected by using either the temporo-lateral electrodes or pulled-down T7 and T8 electrodes as in Maudsley electrode setup.

Epilepsy is often characterized by the sudden occurrence of synchronous activity within relatively large brain neuronal networks that disturb the normal working of the brain. Therefore, some measures of dynamical change have also been used for seizure detection. These measures significantly change in the transition between preictal to ictal state or even in the transition between interictal and ictal state. In the latter case the transition can occur either as a continuous change in phase such as in some cases of mesial temporal lobe epilepsy (MTLE) or as a sudden leap, for example in most cases of absence seizures. In the approaches based on chaos measurement by estimation of attractor dimension (as discussed later in this chapter), for the first case, the attractor of the system gradually deforms from an interictal to an ictal attractor. In the second case, where a sharp critical transition takes place, we can assume that the system has at least two simultaneous interictal and ictal attractors all the time. In a study by Lopes da Silva et al. [50] three states (routes) have been characterized as illustrative models of epileptic seizures: (i) an abrupt transition of the bifurcation type, caused by a random perturbation. An interesting example of which is the epileptic syndrome characterized by paroxysmal spike-and-wave discharges in the EEG and nonconvulsive absence type of seizures; (ii) a route where a deformation of the attractor is caused by an external perturbation (photosensitive epilepsy); and (iii) a deformation of the attractor leading to a gradual evolution into the ictal state, e.g. temporal lobe epilepsy (TLE). The authors concluded that under these routes it is possible to understand those circumstances where the transition from the ongoing (interictal) activity mode to the ictal (seizure) mode may, or may not, be anticipated. Also, any of the three routes is possible, depending on the type of the underlying epilepsy. Seizures may be generally unpredictable, as most often in absence-type seizures of idiopathic (primary) generalized epilepsy, or predictable, preceded by a gradual change in dynamics, detectable some time before the manifestation of seizure, as in TLE.

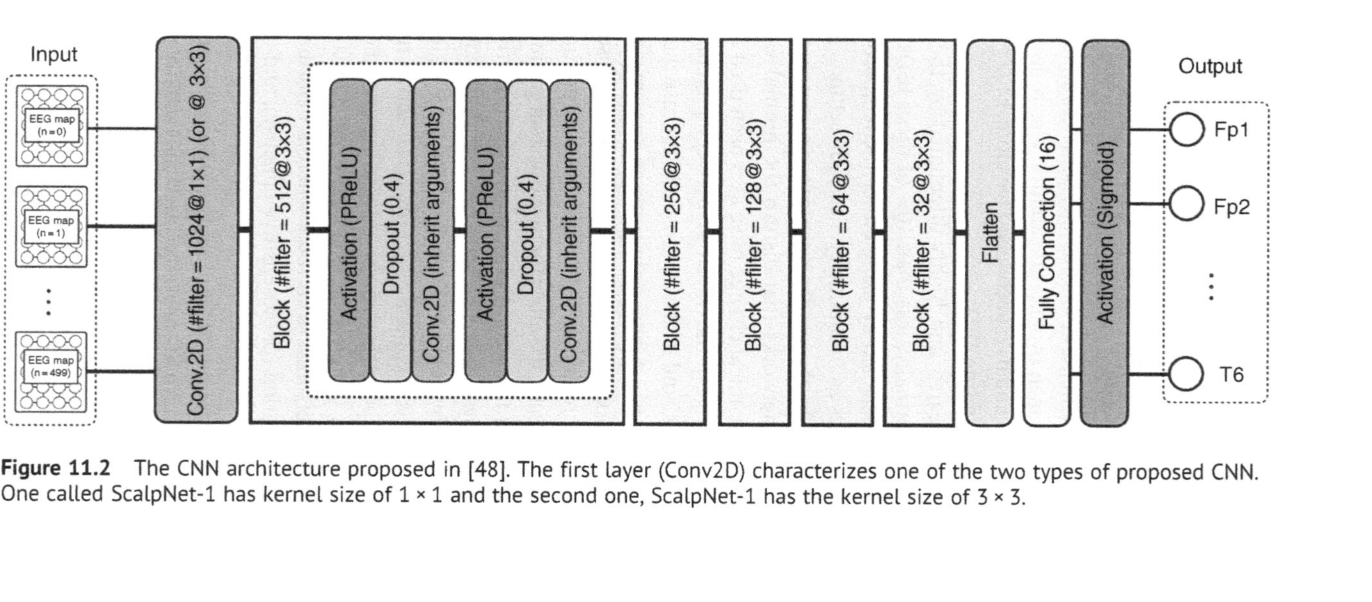

Figure 11.2 The CNN architecture proposed in [48]. The first layer (Conv2D) characterizes one of the two types of proposed CNN. One called ScalpNet-1 has kernel size of 1 × 1 and the second one, ScalpNet-1 has the kernel size of 3 × 3.

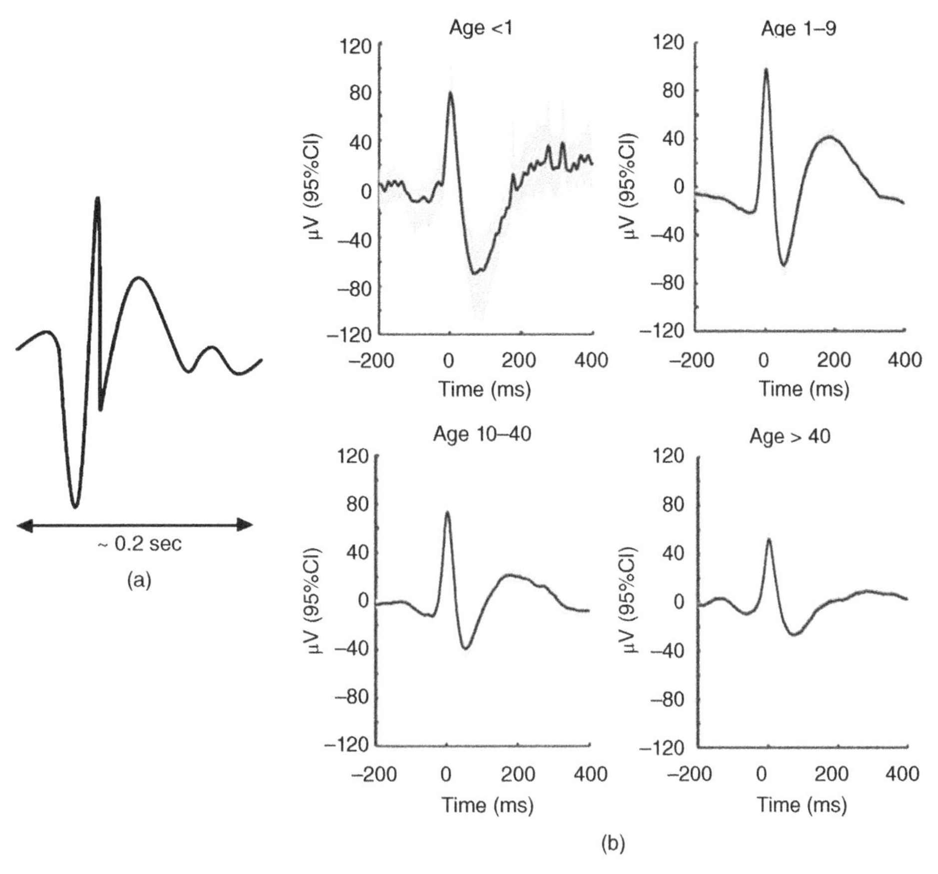

Figure 11.3 (a) A sample of IED recorded using intracranial foramen ovale electrodes. (b) The change in IED with respect to age change seen over the scalp EEG electrodes. The shaded region represents the variability of the IED.

Chaotic behaviour of the EEGs is discussed in Section 11.3 and is the most important approach for seizure onset prediction.

11.2.2 Detection of Neonatal Seizure

Seizure occurs in approximately 0.5% of newborn (the first four weeks of life) babies. It represents a distinctive indicator of abnormality in the central nervous system (CNS). There are many causes for this abnormality with the majority due to biochemical imbalances within the CNS, intracranial haemorrhage and infection, developmental (structural) defects, and passive drug addiction and withdrawal [51]. Analysis of neonatal EEG is a very difficult issue in biomedical signal processing context. Unlike in adults, the presence of spikes may not be the indication of seizure. The clinical signs in the newborn are not always as obvious as those for an adult, where seizure is often accompanied by uncontrollable, repetitive, or jerky movement of the body, or the tonic flexion of muscles. The less obvious symptoms in the newborn, i.e. subtle seizures, may include sustained eye opening with ocular fixation, repetitive blinking or fluttering of eyelids, drooling, sucking, or other slight facial expressions or body movements. Therefore, detection of epileptic seizures for the newborn is far more complex than for adults and so far only a few approaches have been attempted. These approaches are briefly discussed in this section.

Although there is significant spectral range for the newborn seizure signal [52] in most of the cases the seizure frequency band lies within the delta and theta bands (1–7 Hz). However, TF approaches are more popular due to the statistical nonstationarity of the data but they can also be inadequate since the spikes may have less amplitude than the average amplitude of a normal EEG signal.

In most cases the signals are preprocessed before application of any seizure detection algorithm. Eye blinking and body movements are the major sources of artefacts [53]. Conventional adaptive filtering methods (with or without a reference signal) may be implemented to remove interference [54]. This may be followed by the calculation of a TF representation of seizure [55] or used in a variable-time model of the epileptiform signal.

In an early work on neonatal seizure detection [56] Gotman et al. developed a number of real-time approaches for detecting a wide range of patterns, including rhythmic paroxysmal discharges at a wide range of frequencies, as well as repetitive spike patterns, even when they are not very rhythmic. The EEGs obtained from 55 newborns, recorded at three hospitals were used. A total of 281 hours of recordings containing 679 seizures were analyzed. They indicated that 71% of the seizures and 78% of seizure clusters (group of seizures separated by less than 90 seconds) were detected, with a false detection rate (FDR) of 1.7/h.

In their first classification algorithm they used dominant frequency, width of the dominant spectral peak, power ratio between full width half-maximum band of the dominant peak to the power of the background signal within the same band, and the stability of current epoch related to the maximum divided by the minimum peak within 10 seconds epochs [56]. They then classified the segments by comparing the above features with predefined threshold levels. In their second method they detected multiple spikes by applying a highpass filter and mark the signal segment as a seizure segment if six or more spikes are detected within 10 seconds epochs. In their third method, they lowpass filtered the neonate EEG to enable detection of very slow rhythmic discharges [56].

A model-based approach was proposed to model a seizure segment and the model parameters were estimated [57]. The steps are as follows:

1) The periodogram of the observed vector of an EEG signal, $\mathbf{x} = [x(1), x(2),..., x(N)]^T$, is given as:

$$\mathrm{I}_{xx}(k) = \frac{1}{2\pi N}\left|\sum_{n=1}^{N} e^{-j\lambda_k n}x(n)\right|^2 \tag{11.6}$$

where $\lambda_k = \frac{2\pi k}{N}$.

2) A discrete approximation to the log-likelihood function for estimation of a parameter vector of the model, $\boldsymbol{\theta}$, is computed as:

$$L_N(\mathbf{x},\boldsymbol{\theta}) = -\sum_{k=0}^{\left\lfloor\frac{N-1}{2}\right\rfloor}\left[\log(2\pi)^2 S_{xx}(\lambda_k,\boldsymbol{\theta}) + \frac{I_{xx}(\lambda_k)}{S_{xx}(\lambda_k,\boldsymbol{\theta})}\right] \tag{11.7}$$

where, $S_{xx}(\lambda_k, \theta)$ is considered as the spectral density of a Gaussian vector process **x** with parameters $\boldsymbol{\theta}$, and $\lambda_k = 2\pi k/N$. L_N needs to be maximized with respect to $\boldsymbol{\theta}$. As $N \to \infty$ this approximation approaches that of Whittle's [58].
The parameter estimate of Whittle is given by:

$$\hat{\boldsymbol{\theta}} = arg \max_{\boldsymbol{\theta}} (L_N(\mathbf{x}, \boldsymbol{\theta}); \boldsymbol{\theta} \in \boldsymbol{\Theta}) \tag{11.8}$$

The parameter space $\boldsymbol{\Theta}$ may include any property of either the background EEG or the seizure, such as nonnegative values for post-synaptic pulse shaping parameters. This is the major drawback of the method since these parameters are highly dependent on the model of both seizure spikes and the background EEG. A self-generating model [59] followed by a quadratic discriminant function was used to distinguish between the spectrum density of background EEG segments ($S_k^{Background}$) and seizure segments ($S_k^{Seizure}$); k represents the segment number. This model has already been explained in Chapter 2.

3) Based on the model above the power of the background EEG and the seizure segment is calculated [57] and their ratio is tested against an empirical threshold level, say γ, i.e.:

$$\Gamma = \frac{P_{Seizure}}{P_{Background}}$$

where

$$P_{Background} = \sum_{k=0}^{N-1} S_k^{Background} \text{ and } P_{Seizure} = \sum_{k=0}^{N-1} S_k^{Seizure} \tag{11.9}$$

It is claimed that for some well-adjusted model parameters a false alarm percentage as low as 20% can be achieved [57].

A TF approach for neonate seizure detection has been suggested [60] in which a template in the TF domain is defined as:

$$Z_{ref}(\tau, \nu) = \sum_{i=1}^{L} exp\left(\frac{-(\nu - \alpha_i \tau)^2}{2\sigma^2}\right) \tag{11.10}$$

where L is the possible number of seizure spikes in a signal segment and the time scales α_is and variance σ^2 can change respectively, with the position and width of the template. $Z_{ref}(n, f)$, with variable position and variance in the TF domain, resembles a seizure waveform and is used as a template. This template is convolved with the TF domain EEG, $X(n, f)$, in both the time and frequency domains, i.e.:

$$\eta_{\alpha,\sigma^2}(n,f) = X(n,f) ** Z_{ref}(n,f) \tag{11.11}$$

where, $**$ represents the 2D convolution with respect to discrete time and frequency, and $X(n, k)$ for a real discrete signal x is defined as:

$$X(n,k) = \sum_{q=\frac{n}{2}}^{N-\frac{n}{2}} x\left(q + \frac{n}{2}\right) x\left(q - \frac{n}{2}\right) e^{-j2\pi kq} \tag{11.12}$$

where N is the signal length in terms of samples. Assuming the variance is fixed, the α_is may be estimated in order to maximize a test criterion to minimize the difference between Z_{ref} and the desired waveform. This criterion is tested against a threshold level to decide whether there is a seizure. The method has been applied to real neonate EEGs and a FDR as low as 15% in some cases has been reported [60].

The neonate EEG signals do not unfortunately manifest any distinct TF pattern at the onset of seizure. Moreover, they include many seizure type spikes due to immature mental activities of newborns and their rapid changes in brain metabolism. Therefore, although the above methods work well for some synthetic data, due to the nature of neonate seizures, as described before, they oftentimes fail to detect effectively all types of neonate seizures.

In a work by Karayiannis et al. a cascaded rule-based neural network algorithm for detection of epileptic seizure segments in neonatal EEG has been developed [61]. In this method it is assumed that the neonate seizures are manifested as subtle but somehow stereotype repetitive waveforms that evolve in amplitude and frequency before eventually decaying. Three different morphologies of seizure patterns for, respectively, pseudosinusoidal, complex morphology, and rhythmic runs of spike-like waves have been considered. These patterns are illustrated in Figure 11.4 [61].

The automated detection of neonate seizure is then carried out in three stages. Each EEG channel is treated separately, and spatial and inter-channel information are not exploited. In the first stage the spectrum amplitude is used to separate bursts of rhythmic activities. In the second stage the artefacts, such as the results of patting, sucking, respiratory function, and electrocardiograms (ECG) are mitigated, and in the last stage a clustering operation is performed to distinguish between epileptic seizures, nonepileptic seizures, and the normal EEG affected by different artefacts. As a result of this stage, isolated and inconsistent candidate seizure segments are eliminated, and the final seizure segments are recognized.

In another machine learning approach, the performances of conventional feed-forward neural networks (FFNN) [62] and quantum neural networks (QNN) [63] have been compared for classification of some frequency domain features in all the above stages. These features are almost identical to those used by Gotman et al. [56] and denoted as first dominant frequency, second dominant frequency, width of dominant frequency, percentage of power contributed to the first dominant frequency, percentage of power contributed to the second dominant frequency, peak ratio, and stability ratio (a time domain parameter that measures the amplitude stability of the EEG segment). It has also been shown that there is no significant difference in using an FFNN or QNN and the results will be approximately the same [61]. The overall algorithm is very straightforward to implement and its both sensitivity and specificity have been shown to be above 80%.

In [64] the neonate EEGs are preprocessed before wavelet domain features are extracted, ranked, and classified using different classifiers.

Temko et al. [65] used only nine EEG channels namely F_3, F_4, C_3, C_4, Cz, T_3, T_4, O_1, and O_2 for their neonatal seizure detection system. They used a large number of statistical and dynamical features and used SVMs and Gaussian mixture models (GMMs) to classify them.

11.3 Chaotic Behaviour of Seizure EEG

Nonlinear analysis techniques provide insights into many processes, which cannot be directly formulated or exactly modelled using state machines. This requires time-series analysis of long sequences. A state-space reconstruction of the chaotic data may be

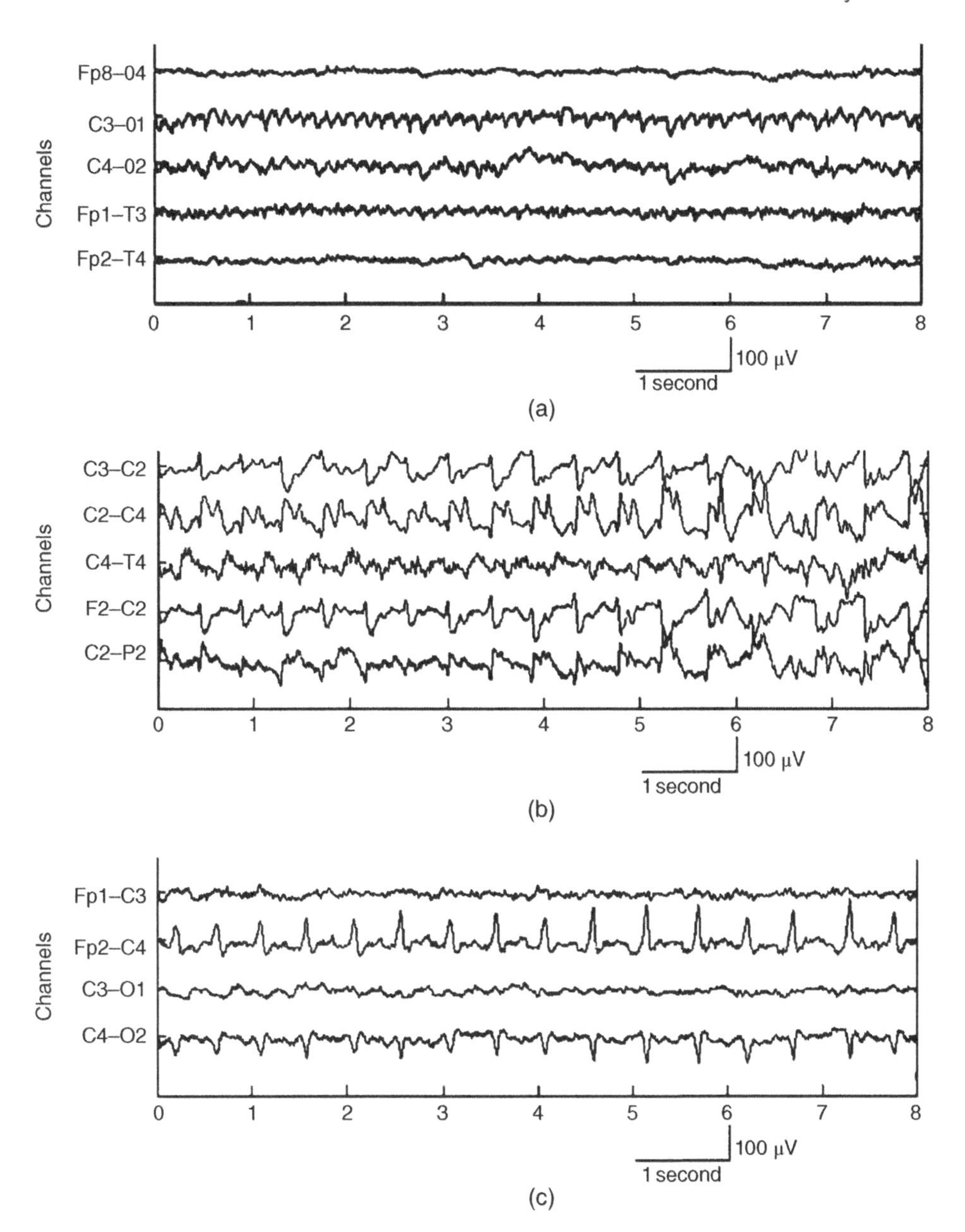

Figure 11.4 The main three different neonate seizure patterns. (a) Low amplitude depressed brain type discharge around the channels C3-O1 and FP1-T3. (b) Repetitive complex slow waves with superimposed higher frequencies in all the channels. (c) Repetitive or periodic runs of sharp transients in channels FP2-C4 and C4-O2.

performed based on embedding methods [66, 67]. Using the original time series and its time-delayed copies, i.e. $\mathbf{x}(n) = [x(n), x(n+T), \ldots, x(n+(d_E-1)T)]$, an appropriate state space can be reconstructed. Here $\mathbf{x}(n)$ is the original one-dimensional data, T is the time delay, and d_E is the embedding dimension. The time delay T is calculated from the first minimum of the average mutual information (AMI) function [68]. The minimum point of the AMI function provides adjacent delay coordinates with a minimum of redundancy. Embedding dimension (d_E) can be computed from a global false nearest neighbours (GFNN) analysis [69], which compares the distances between neighbouring trajectories at successively higher dimensions. The false neighbours occur when trajectories that overlap in dimension d_i are distinguished in dimension d_{i+1}. As i increases, the total percentage of false neighbours declines and d_E is chosen where this percentage approaches zero.

The stationarity of the EEG patterns may be investigated and established by evaluating recurrence plots generated by calculation of the Euclidean distances between all pairs of points $\mathbf{x}(i)$ and $\mathbf{x}(j)$ in the embedded state-space and then such points are plotted in the (i,j) plane where δ_{ij} is less than a specific radius ρ:

$$\delta_{ij} = \|\mathbf{x}(i) - \mathbf{x}(j)\|_2 < \rho \tag{11.13}$$

where $\| . \|_2$ denotes the Euclidean distance. Since i and j are time instants, the recurrence plots convey natural and subtle information about temporal correlations in the original time series [70]. Nonstationarities in the time series are manifested as gross homogenities in the recurrent plot. The value of ρ for each time series is normally taken as a small percentage of the total data set size.

The scale-invariant self-similarity, as one of the hallmarks of low-dimensional deterministic chaos [71] results in a linear variation of the logarithm of the correlation sum, $\log[C(r,N)]$, with respect to $\log(r)$ as $r \rightarrow 0$. In places where such similar dynamics exist, the *correlation dimension*, *D*, is defined as:

$$D = \lim_{N \rightarrow \infty} \lim_{r \rightarrow 0} \frac{\log\left[C(r,N)\right]}{\log(r)} \tag{11.14}$$

where $C(r,N)$ for each signal segment starting from time sample n is calculated as:

$$C(r,N) = \frac{1}{(N-n)(N-n-1)} \times \sum_{i=1}^{N} \sum_{j=1}^{N} H(r - \|\mathbf{x}(i) - \mathbf{x}(j)\|_2) \tag{11.15}$$

where r represents the volume of points being considered and $H(.)$ is a heaviside step function. Computation of $C(r,N)$ is susceptible to noise and nonstationarities. It is also dominated by the finite length of the data set. The linearity of the relationship between $\log[C(r,N)]$ and $\log(r)$ can be examined from the local slopes of $\log[C(r,N)]$ versus $\log(r)$.

Generally, traditional methods such as the Kolmogorov entropy, the correlation dimension, or Lyapunov exponents [7] can be used to quantify the dynamical changes of the brain.

Finite-time Lyapunov exponents quantify the average exponential rate of divergence of neighbouring trajectories in state space, and thus provide a direct measure of the sensitivity of the system to infinitesimal perturbations.

A method based on the calculation of the largest Lyapunov exponent (LLE) has been often used for evaluation of chaos in the intracranial EEG (iEEG) signals. In a p-dimensional system there are p different Lyapunov exponents, λ_i. They measure the exponential rate of convergence or divergence of the different directions in the phase space. If one of the exponents is positive, the system is chaotic. Thus, two close initial conditions will diverge exponentially in the direction defined by that positive exponent. Since these exponents are ordered, $\lambda_1 \geq \lambda_2 \ldots \geq \lambda_d$, to study the chaotic behaviour of a system it is sufficient to study the changes in the LLE, λ_1. Therefore, we focus on the changes in the value of λ_1 as the epileptic brain moves from one state to another.

The maximum Lyapunov exponent (MLE) (λ_1) for a dynamical system can be defined from [72]:

$$d(n) = d_0 e^{\lambda_1 n}, \tag{11.16}$$

where $d(n)$ is the mean divergence between neighbouring trajectories in state space at time n and d_0 is the initial separation between neighbouring points. *Finite-time* exponents (λ^*) are distinguished from true Lyapunov exponents λ_1, which are strictly defined only in the dual limit as $n \rightarrow \infty$ and $d_0 \rightarrow 0$ in Eq. (11.17). For the finite length observation λ^* is the average of the Lyapunov exponents.

A practical procedure for the estimation of λ_1 from a time series was proposed by Wolf et al. [73]. This procedure gives a global estimate of λ_1 for stationary data. Since the EEG data are nonstationary [74], the algorithm to estimate λ_1 from the EEG should be capable of automatically identifying and appropriately weighting the transients of the EEG signals. Therefore, a modification of Wolf's algorithm, proposed in [75], which modifies mainly the searching procedure to account for the nonstationarity of the EEG data may be used. This estimate is called the short-term LLE, STL_{max} and the changes of the brain dynamics can be studied by the time evolution of the STL_{max} values at different electrode sites. Estimation of the STL_{max} for time sequences using Wolf's algorithm has already been explained in Chapter 6.

11.4 Seizure Detection from Brain Connectivity

ECoG is an effective brain screening system for seizure IED detection and localization. In [76] a seizure detection method based on the phase-slope index (PSI) of directed influence applied to multichannel ECoG has been proposed.

The PSI metric, which is simply the normalized imaginary part of the spectral coherency defined in Chapter 8, used for the brain connectivity measure, identifies increases in the spatiotemporal interactions between channels that clearly distinguish seizure from interictal activity. This connectivity measure is compared with a threshold to detect the presence of seizures. The threshold is chosen based on a moving average of recent activity to accommodate differences between patients and slow changes within each patient over time.

11.5 Prediction of Seizure Onset from EEG

Although most seizures are not life threatening, they are an unpredictable source of annoyance and embarrassment. They occur when a massive group of neurons in the cerebral cortex begins to discharge in a very organized way, leading to a temporary synchronized electrical activity that disrupts the normal activity of the brain. Sometimes, such disruption manifests itself in a brief impairment of consciousness, but it can also produce a more or less complex series of abnormal sensory and motor manifestations. Most of the epileptic patients are continuously on anticonvulsant drugs. Therefore, there is great interest in the development of devices that incorporate algorithms capable of detecting early onset of seizures or even predicting them hours before they occur. A seizure prediction system can detect seizures prior to their occurrence and allow clinicians to provide timely treatment for patients with epilepsy.

The brain is assumed to be a dynamical system, since epileptic neuronal networks are essentially complex nonlinear structures and their interactions are thus expected to exhibit nonlinear behaviour. These methods have substantiated the hypothesis that quantification of the brain's dynamical changes from the EEG might enable prediction of epileptic seizures, while traditional methods of analysis have failed to recognize specific changes prior to seizure.

Iasemidis et al. [77] were the first group to apply nonlinear dynamics to clinical epilepsy. The main concept in their studies is that a seizure represents a transition of the epileptic brain from chaotic to a more ordered state, and therefore the spatiotemporal dynamical properties of the epileptic brain are different for different clinical states. Further studies of the same group, based on the temporal evolution of the short-term LLE (a modification of the LLE to account for the nonstationarity of the EEG) for patients with TLE [75], suggested that the EEG activity becomes progressively less chaotic as the seizure approaches. Therefore, the idea that seizures were abrupt transitions in and out of an abnormal state was substituted by the idea that the brain follows a dynamical transition to seizure for at least some kinds of epilepsy. Since these pioneering studies, nonlinear methods derived from the theory of dynamical systems have been employed to quantify the changes in the brain dynamics before the onset of seizures, providing evidence to the hypothesis of a *route* to seizure. Lehnertz and Elger [78] focused their studies on the decrease of complexity in neuronal networks prior to seizure. They used the information provided by changes in the *neuronal complexity loss* that summarizes the complex information content of the correlation dimension profiles in just a single number. Lerner [79] observed that changes in the correlation integral could be used to track accurately the onset of seizure for a patient with TLE. However, Osorio et al. [80] demonstrated that these changes in the correlation integral could be perfectly explained by changes in the amplitude and frequency of the EEG signals. Van Quyen et al. [81] found a decrease in the dynamical similarity during the period prior to seizure and that this behaviour became more and more pronounced as the onset of seizure approached. Moser et al. [82] employed four different nonlinear quantities within the framework of the Lyapunov theory and found strongly significant preictal changes. Litt et al. [83] demonstrated that the energy of the EEG signals increases as seizure approaches. In their later works, they provided evidence of seizure predictability based on the selection of different linear and nonlinear features of the EEG [84]. Iasemidis et al. [85, 86], by using the spatiotemporal evolution of the short-term LLE, demonstrated that minutes or even hours before seizure, multiple regions of the cerebral cortex progressively approach a similar degree of chaoticity of their dynamical states. They called it *dynamical entrainment* and hypothesized that several critical sites have to be locked with the epileptogenic focus over a common period of time in order for a seizure to take place. Based on this hypothesis they presented an adaptive seizure prediction algorithm that analyzes continuous EEG recordings for prediction of TLE when only the occurrence of the first seizure is known [87].

Most of these studies for prediction of epilepsy are based on iEEG recordings. Two main challenges face the previous methods in their application to scalp EEG data: (i) the scalp signals are more subject to environmental noise and artefacts than the iEEG, and (ii) the meaningful signals are attenuated and mixed in their propagation through soft tissue and bone. Traditional nonlinear methods (TNMs), such as the Kolmogorov entropy or the Lyapunov exponents may be affected by the above two difficulties and therefore they may not distinguish between slightly different chaotic regimes of the scalp EEG [88]. One approach to circumvent these difficulties is based on the definition of different nonlinear measures that yield better performance over the TNM for the scalp EEG. This is the approach followed by Hively and Protopopescu [89]. They proposed a method based on the phase-space dissimilarity measures (PSDM) for forewarning of epileptic events from scalp EEG. The approach of Iasemidis et al. of dynamical entrainment has also been shown to work well on scalp unfiltered EEG data for seizure predictability [90–92].

In principle, a nonlinear system can lie in a high-dimensional or infinite-dimensional phase space. Nonetheless, when the system comes into a steady state, portions of the phase space are revisited over time and the system lies in a subset of the phase space with a finite and generally small dimension, called an *attractor*. When this attractor has sensitive dependence to initial conditions (it is chaotic), it is termed as a *strange* attractor and its geometrical complexity is reflected by its dimension, D_a. In practise, the system's equations are not available and we only have discrete measurements of a single observable, $u(n)$, representing the system. If the system comes into such a steady state, a p-dimensional phase space can be reconstructed by generating p different scalar signals, $x_i(n)$, from the original observable, $u(n)$, and embedding them into a p-dimensional vector:

$$\mathbf{x}(n) = \left[x_1(n), x_2(n), \ldots, x_p(n)\right]^T \tag{11.17}$$

According to Takens [93], if p is chosen large enough, we shall generally obtain a good phase portrait of the attractor and therefore good estimates of the nonlinear quantities. In particular, Takens' theorem states that the embedding dimension, p, should be at least equal to $2 \times D_a + 1$. The easiest and probably the best way to obtain the embedding vector $x(n)$ from $u(n)$ is by *the method of delays*. According to this method, p different time delays, $n_0 = 0$, $n_1 = \tau$, $n_2 = 2\tau, \ldots, n_p = (p-1)\tau$, are selected and the p different scalar signals are obtained as $x_i(n) = x(n + n_i)$ for $i = 0, \ldots, p-1$. If τ is chosen carefully, we obtain a good phase portrait of the attractor and therefore good estimates of the parameters of nonlinear behaviour.

Since the brain is a nonstationary system, it is never in a steady state in the strictly dynamical sense. However, it can be considered as a dynamical system that constantly moves from one stable steady state to another. Therefore, local estimates of nonlinear measures should be possible and the changes of these quantities should be representative of the dynamical changes of the brain.

Previous studies have demonstrated a more ordered state of the epileptic brain during seizure than before or after it. The correlation dimension has been used to estimate the dimension, d, of the ictal state [77]. The values obtained ranged between 2 and 3, demonstrating the existence of a low-dimensional attractor. Therefore, an embedding dimension of seven should be enough to obtain a good image of this attractor and a good space portrait of the ictal state. As concluded in Chapter 2, increasing the value of p more than what is strictly necessary increases the effect of noise and thus higher values of p are not recommended.

In a new approach [94] it has been shown that the TNM can be applied to the scalp EEGs indirectly. This requires the underlying sources of the brain to be correctly separated from the observed electrode signals without any a priori information about the source signals or the way the signals are combined. The effects of noise and other internal and external artefacts are also highly mitigated by following the same strategy. It is expected to obtain signals similar to the intracranial recordings to which TNM can be applied. To do so, the signal segments are initially separated into their constituent sources using blind source separation (BSS) (assuming the sources are independent). Since for a practical prediction algorithm, the nonlinear dynamics have to be quantified over long-term EEG recordings. After using a block-based BSS algorithm the continuity has to be maintained for the entire recording. This problem turns out not to be easy due to the two inherent ambiguities of BSS: (i) The variances (energies) of the independent components are unknown, and (ii) Due to the inherent permutation problem of the BSS algorithms the order of the independent components cannot be determined.

The first ambiguity states that the sources can be estimated up to a scalar factor. Therefore, when moving from one block to another, the source amplitudes are generally different and the signals may be inverted. This ambiguity can be solved as explained below, so its effect can be avoided. The nonlinear dynamics are quantified by the LLE λ_1. The λ_1 estimate for each block is based on ratios of distances between points within the block. Consequently, as long as λ_1 is estimated for the sources obtained by applying BSS to each block of EEG data individually, there is no need to adjust the energy of the sources.

The second ambiguity, however, severely affects the algorithm. The order in which the estimated sources appear, as a result of the BSS algorithm, changes from block-to-block. Therefore, we need a procedure to reorder the signals to align the same signal from one block to another and maintain the continuity for the entire recording. The next section explains the approach followed in our algorithm for this purpose.

We have followed an overlap window approach to maintain the continuity of the estimated sources, solving both indeterminacies simultaneously. Instead of dividing the EEG recordings into sequential and discontinuous blocks, we employ a sliding window of fixed length, L, with an overlap of L-N samples ($N < L$) and apply the BSS algorithm to the block of data within that window. Therefore, we assume that $\mathbf{x}(n) = [x_1(n), x_2(n), \ldots, x_m(n)]^T$ represents the entire scalp EEG recording, where m is the number of sensors. Two consecutive windows of data are selected as $\mathbf{x}_1(n) = \mathbf{x}(n_0 + n)$ and $\mathbf{x}_2(n) = \mathbf{x}(n_0 + N + n)$ for $t = 1, \ldots, L$, where $n_0 \geq 0$. Therefore

$$\mathbf{x}_2(n) = \mathbf{x}_1(N + n) \text{ for } n = 1, \ldots, L - N. \tag{11.18}$$

Once the BSS algorithm has been applied to $\mathbf{X}_1(n)$ and $\mathbf{X}_2(n)$, two windows of estimated sources $\hat{\mathbf{s}}_1(n) = [\hat{s}_1(n), \hat{s}_2(n), \ldots, \hat{s}_m(n)]^T$ and $\hat{\mathbf{s}}_2(n) = \left[\hat{s}'_1(n), \hat{s}'_2(n), \ldots, \hat{s}'_m(n)\right]^T$ will be obtained respectively, where m is the number of sources. These two windows overlap within a time interval, but due to the inherent ambiguities of BSS $\hat{\mathbf{s}}_1(n)$ and $\hat{\mathbf{s}}_2(n)$ are not equal in this interval. Instead:

$$\hat{\mathbf{s}}_2(n) = \mathbf{P} \cdot \mathbf{D} \cdot \hat{\mathbf{s}}_1(n + N) \text{ for } n = 1, \ldots, L - N, \tag{11.19}$$

where $\mathbf{P}$ is an $n \times n$ permutation matrix and $\mathbf{D} = diag\{d_1, d_2, \ldots, d_n\}$ is the scaling matrix. Therefore, $\hat{\mathbf{s}}_2(t)$ is just a copy of $\hat{\mathbf{s}}_1(t)$ in the overlap block, with the rows (sources) permuted, and each of them is scaled by a real number d_i that accounts for the scaling ambiguity of BSS. A measure of similarity has also been used between the rows of $\hat{\mathbf{s}}_1(t)$ and $\hat{\mathbf{s}}_2(t)$ within the overlap region for this purpose. Cross-correlation. The cross-correlation between two zero-mean wide sense stationary random signals $x(t)$ and $y(t)$ is defined as:

$$r_{xy}(\tau) = E[x(n)y(n+\tau)] \tag{11.20}$$

where $E[.]$ denotes the expectation operation. This measure gives an idea of the similarity between $x(n)$ and $y(n)$, but its values are not bounded and depend on the amplitude of the signal. Therefore, it is preferable to use a normalization of r_{xy} given by the cross-correlation coefficient:

$$\rho_{xy} = \frac{r_{xy}}{\sigma_x \sigma_y} \tag{11.21}$$

where σ_x and σ_y are the standard deviations of $x(n)$ and $y(n)$ respectively, and the cross-correlation coefficient satisfies:

$$-1 \leq \rho_{xy} \leq 1 \tag{11.22}$$

Furthermore, if $\rho_{xy} = 1$, then $y = ax$, with $a > 0$, and $x(t)$ and $y(t)$ are perfectly correlated; and if $\rho_{xy} = -1$, then $y = -ax$ and $x(n)$ and $y(n)$ are perfectly anti-correlated. When the two signals have no information in common, $\rho_{xy} = 0$, they are said to be uncorrelated.

Theoretically, the BSS algorithm gives independent sources at the output. Since two independent signals are uncorrelated, if the cross-correlation coefficient ρ is calculated between one row $\hat{s}_i(n)$ of the overlap block of $\hat{\mathbf{S}}_1(n)$ and all the rows $\hat{s}'_j(t)$ of the overlap block of $\hat{\mathbf{S}}_2(n)$, we should obtain all the values equal to zero except one of them for which $|\rho_{ij}| = 1$. In other words, if we define the matrix $\mathbf{\Gamma} = \{\gamma_{ij}\}$, with elements equal to the absolute value of ρ between the ith row of the overlap segment of $\hat{\mathbf{S}}_1(n)$ and the jth row of the overlap segment of $\hat{\mathbf{S}}_2(n)$, then $\mathbf{\Gamma} = \mathbf{P}^{\mathrm{T}}$ and the permutation problem can be solved.

Once the permutation problem has been solved, each of the signals $\hat{s}_i(n)$ corresponds to only one of the signals $\hat{s}'_j(n)$, but the latter signals are scaled and possibly inverted versions of the former signals, due to the first inherent ambiguity of BSS. The BSS algorithm sets the variances of the output sources to one and therefore $\hat{s}_i(n)$ and $\hat{s}'_j(n)$ both have equal variance. Since the signals only share an overlap of L-N samples, the energy of the overlap segment of these signals will generally be different and therefore can be used to solve the amplitude ambiguity. In particular:

$$\hat{s}_i(n+N) = sign\left(\rho_{ij}\right) \frac{\sigma_i}{\sigma'_j} \hat{s}'_j(n), \qquad \text{for } n = 1, \ldots, L-N \tag{11.23}$$

where ρ_{ij} is calculated for $\hat{s}_i(n)$ and $\hat{s}'_j(n)$ within the overlap segment, and σ_i and σ'_j are the standard deviations of $\hat{s}_i(n)$ and $\hat{s}'_j(n)$ respectively, within the overlap segment. This should solve the scaling ambiguity of the BSS algorithm.

In practise, the estimated sources are not completely uncorrelated and therefore $\boldsymbol{\Gamma} \neq \mathbf{P}^{\mathrm{T}}$. However, for each row it is expected to obtain only one of the elements γ_{ij} close to unity corresponding to $j = j_0$, and the rest close to zero. Therefore, the algorithm can still be applied to maintain the continuity of the signals.

After the sources are estimated TNM can be applied to track the dynamics of the estimated source signals.

Using simultaneous scalp and iEEG recordings the performance of the above system has been observed. The intracranial recordings were obtained from multicontact FO electrodes. Electrode bundles are introduced bilaterally through the FO under fluoroscopic guidance. The deepest electrodes within each bundle lie next to medial temporal structures, whereas the most superficial electrodes lie at or just below the FO [95]. As FO electrodes are introduced via anatomical holes, they provide a unique opportunity to record simultaneously from scalp and medial temporal structures without disrupting the conducting properties of the brain coverings by burr holes and wounds, which can otherwise make simultaneous scalp and intracranial recordings unrepresentative of the habitual EEG [96]. Simultaneously, scalp EEG recordings were obtained from standard silver cup electrodes applied according to the 'Maudsley' electrode placement system [97], which is a modification of the extended 10–20 system. The advantage of the Maudsley system with respect to the standard 10–20 system is that it provides a more extensive coverage of the lower part of the cerebral convexity, increasing the sensitivity for the recording from basal subtemporal structures.

For the scalp EEG the overlap window approach was used to maintain the continuity of the underlying sources and once the continuity was maintained, the resulting sources were divided into nonoverlapping segments of 2048 samples. The LLE, λ_1 was estimated for each of these segments. The intracranial signals were also divided into segments of the same size and λ_1 was estimated for each of these segments. In both cases, the parameters used for the estimation of λ_1 were those used by Iasemidis et al. for the estimation of STL_{max}, as explained in [85]. After the λ_1s are calculated for different segments, they are included in a time sequence and smoothed by time averaging.

Within the simultaneous intracranial and scalp EEG recording of 5 minutes and 38 seconds containing a focal seizure, the seizure is discernible in the intracranial electrodes (Figure 11.5), from around 308 seconds, and the ictal state lasts throughout the observation. Figure 11.5 shows a segment of the signals recorded by the scalp electrodes during the seizure. The signals are contaminated by noise and artefact signals such as eye blinking and ECG, and the seizure is not clearly discernible. Figure 11.6 shows the signals obtained after applying the BSS-based prediction algorithm to the same segment of scalp EEGs. The first and second estimated sources seem to record the seizure components while the noise and artefacts are separated into the other two sources. Figure 11.7a and b illustrate the smoothed λ_1 variations for two intracranial electrodes located in the focal area. The smoothed λ_1 is calculated by averaging the current value of λ_1 and the previous two values. These two electrodes show a clear drop in the value of λ_1 at the occurrence of seizure, starting prior to the onset. However, the iEEG was contaminated by a high-frequency activity that causes fluctuations of λ_1 for the entire recording. Figure 11.7c–f illustrate the smoothed λ_1 evolution for four scalp electrodes once the baseline was removed. The value of λ_1 presents large fluctuations that can be due to the presence of noise and artefacts. Although the values seem to be lower as the seizure approaches, there is not a clear trend before seizure in any of the electrodes.

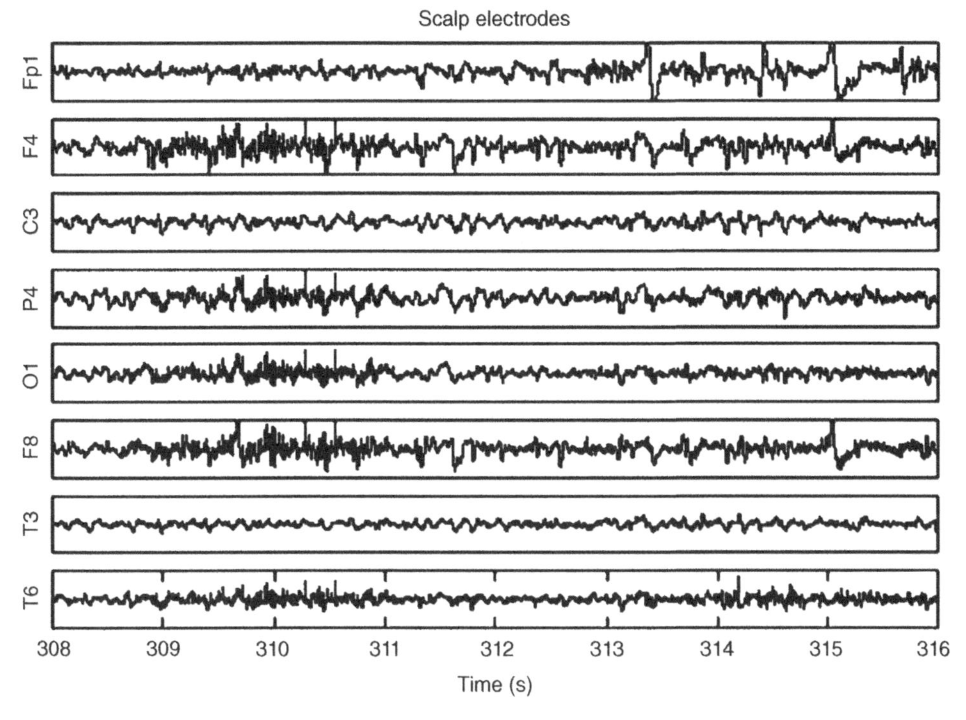

Figure 11.5 Eight seconds of EEG signals from eight out of 16 scalp electrodes during a seizure (direct current (DC) removed).

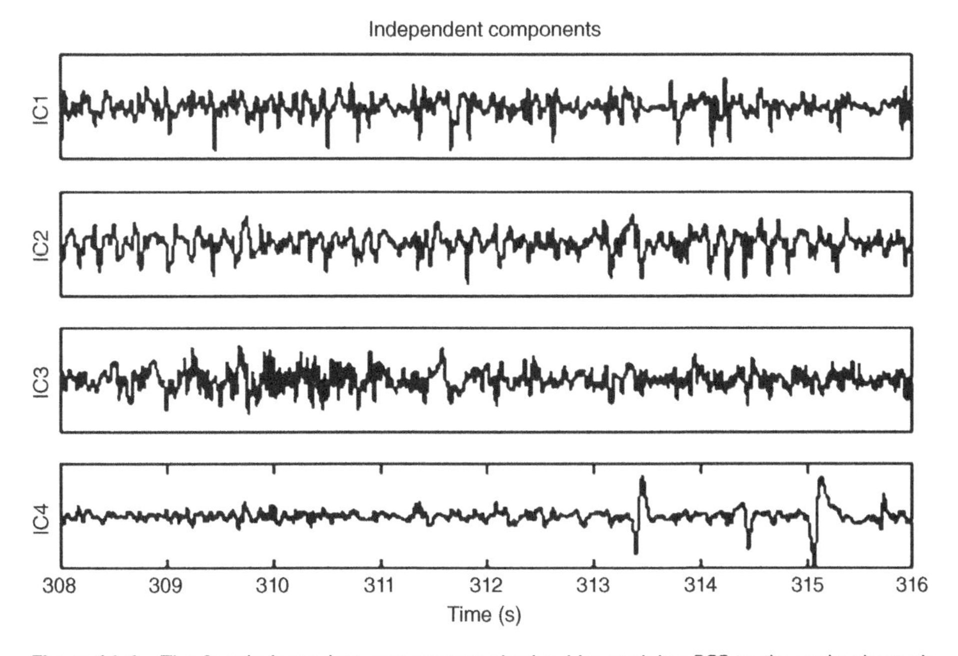

Figure 11.6 The four independent components obtained by applying BSS to the scalp electrode signals shown in Figure 11.5.

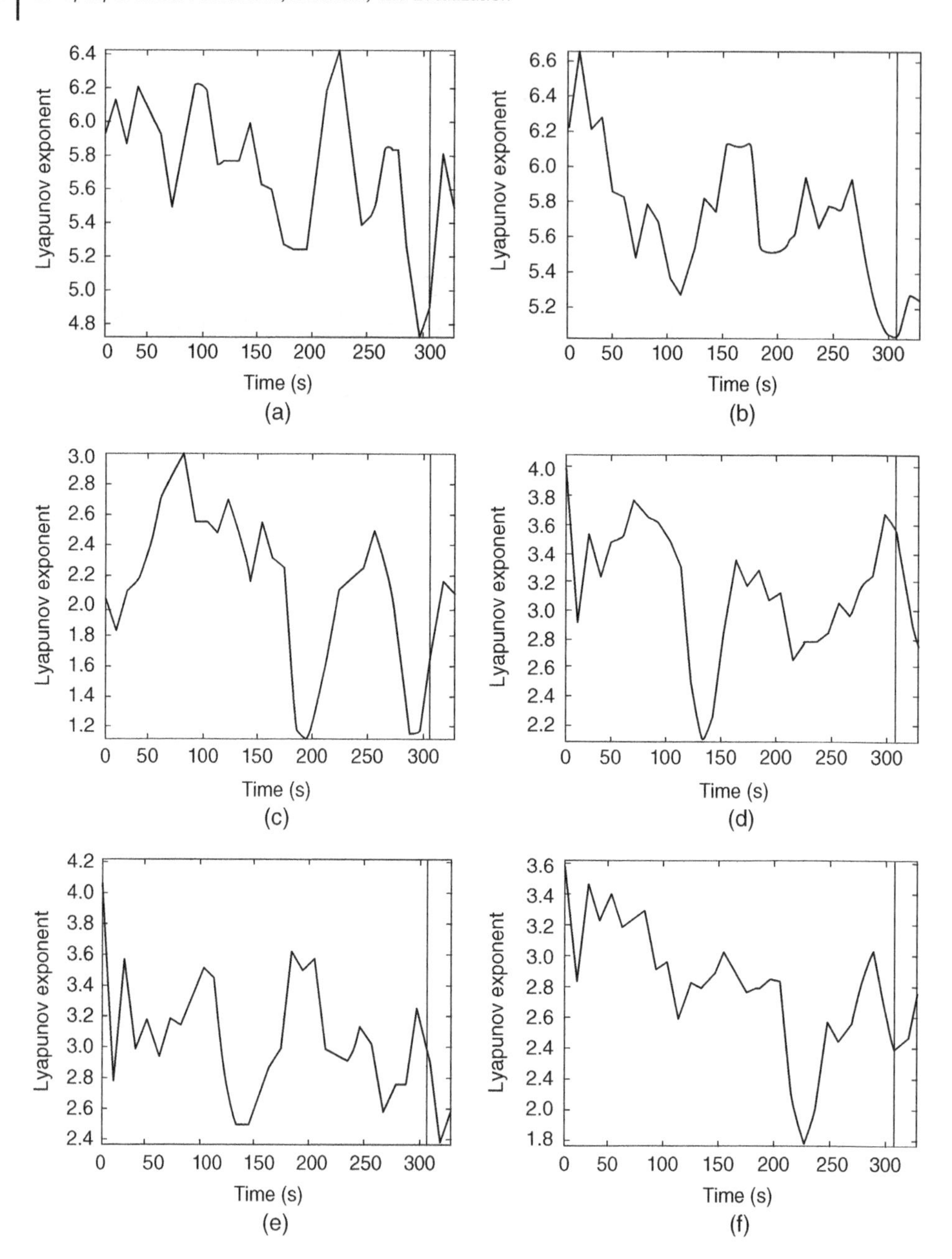

Figure 11.7 The smoothed λ_1 evolution over time for two intracranial electrodes located in the focal area. (a) For the LF4 electrode. (b) For the LF6 electrode. (c–f) The smoothed λ_1 evolutions for four scalp electrodes. The length of the recording is 338 seconds and the seizure occurs at 306 seconds (marked by the vertical line).

Figure 11.8 shows the results obtained for two of the estimated sources after the application of the proposed BSS algorithm. The algorithm efficiently separates the underlying sources from the eye blinking artefacts and noise. Both figures show how the BSS algorithm efficiently separates the epileptic components. The value of λ_1 reaches the minimum value

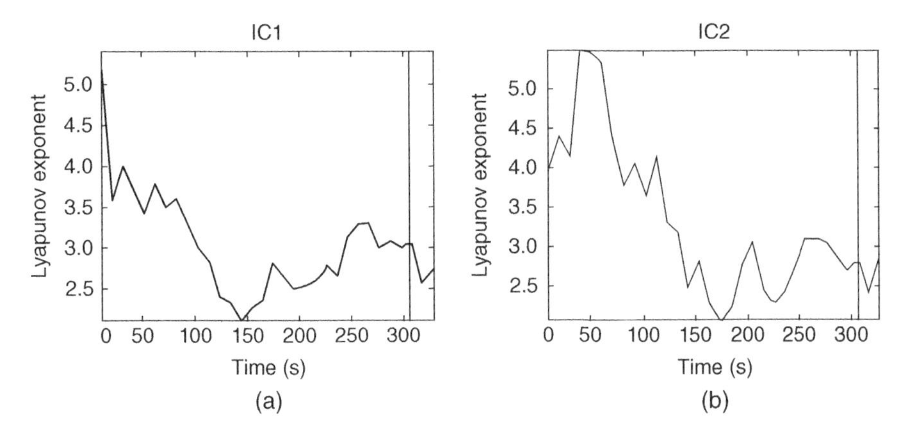

Figure 11.8 Smoothed λ_1 evolution over time for two independent components IC1 and IC2, for which ictal activity is prominent (see Figure 11.4).

more than one minute prior to seizure, remaining low until the end of the recording. This corresponds with the seizure lasting until the end of the recording.

Figures 11.9–11.11 illustrate the results obtained from a recording of duration 5 minutes and 34 seconds. In this particular case the epileptic component was not clearly visible by visual inspection of the intracranial electrode signals. The intracranial electrodes may not have recorded the electrical activity of the epileptic focus because of their location. Figure 11.9a shows a segment of the signals recorded by eight scalp electrodes during the seizure. Although the signals are contaminated by noise and artefacts, the seizure components are discernible in several electrodes. Figure 11.9b illustrates the signals obtained for the same segment of data after the BSS algorithm. In this case the seizure component seems to be separated from noise and artefacts in the third estimated source. Figure 11.10a displays the evolution of the smoothed λ_1 for four different intracranial electrodes. The values fluctuate during the recording but there is a gradual drop in λ_1 starting at the beginning of the recording. A large drop in the value of λ_1 is observed for the four electrodes around 250 seconds and reaches a minimum value around 275 seconds. However, the onset of seizure occurs around 225 seconds and therefore none of the intracranial electrodes is able to predict the seizure. Figure 11.10b shows the variation of smoothed λ_1 for four scalp electrodes. Likewise, for the intracranial electrodes, λ_1 values have large fluctuations but present a gradual drop towards seizure. Similarly, the drop to the lowest value of λ_1 starts after 250 seconds and therefore the signals from these electrodes are not used for seizure prediction.

Figure 11.11 illustrates the changes in the smoothed λ_1 for the third estimated source obtained after the application of BSS. λ_1 starts decreasing approximately two minutes before the onset of seizure. The minimum of λ_1 is obtained around the same time as for the intracranial and scalp electrodes. However, a local minimum is clear at the onset of seizure and the values are clearly lower during the seizure than at the beginning of the recording. The BSS algorithm seems to separate the epileptic component in one of the estimated sources allowing the prediction of seizure even when this is not possible from the intracranial electrodes.

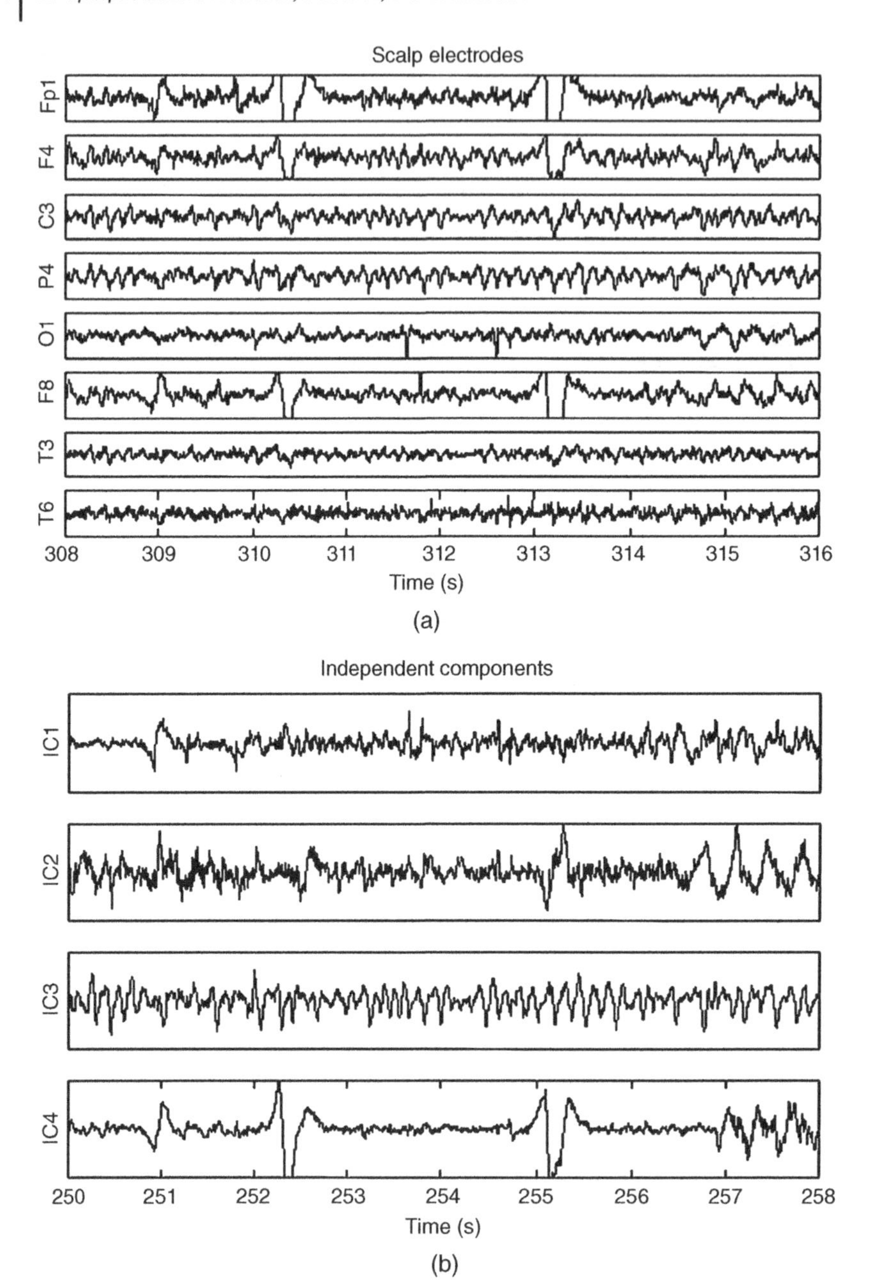

Figure 11.9 (a) A segment of eight seconds of EEG signals (with zero mean) for eight out of 16 scalp electrodes during the seizure. (b) The four independent components obtained by applying BSS to the scalp electrode signals shown in (a). The epileptic activity seems to be discernible in IC3 and its λ_1 evolution is shown in Figure 11.11.

Figure 11.12a and b show the results obtained for a third EEG recording lasting 5 minutes and 37 seconds. The electrodes recorded a generalized seizure. Figure 11.12a illustrates the results for four intracranial electrodes. The value of λ_1 does not show any clear decrease until 250 seconds when there is a sudden drop in λ_1 for all the electrodes. The minimum

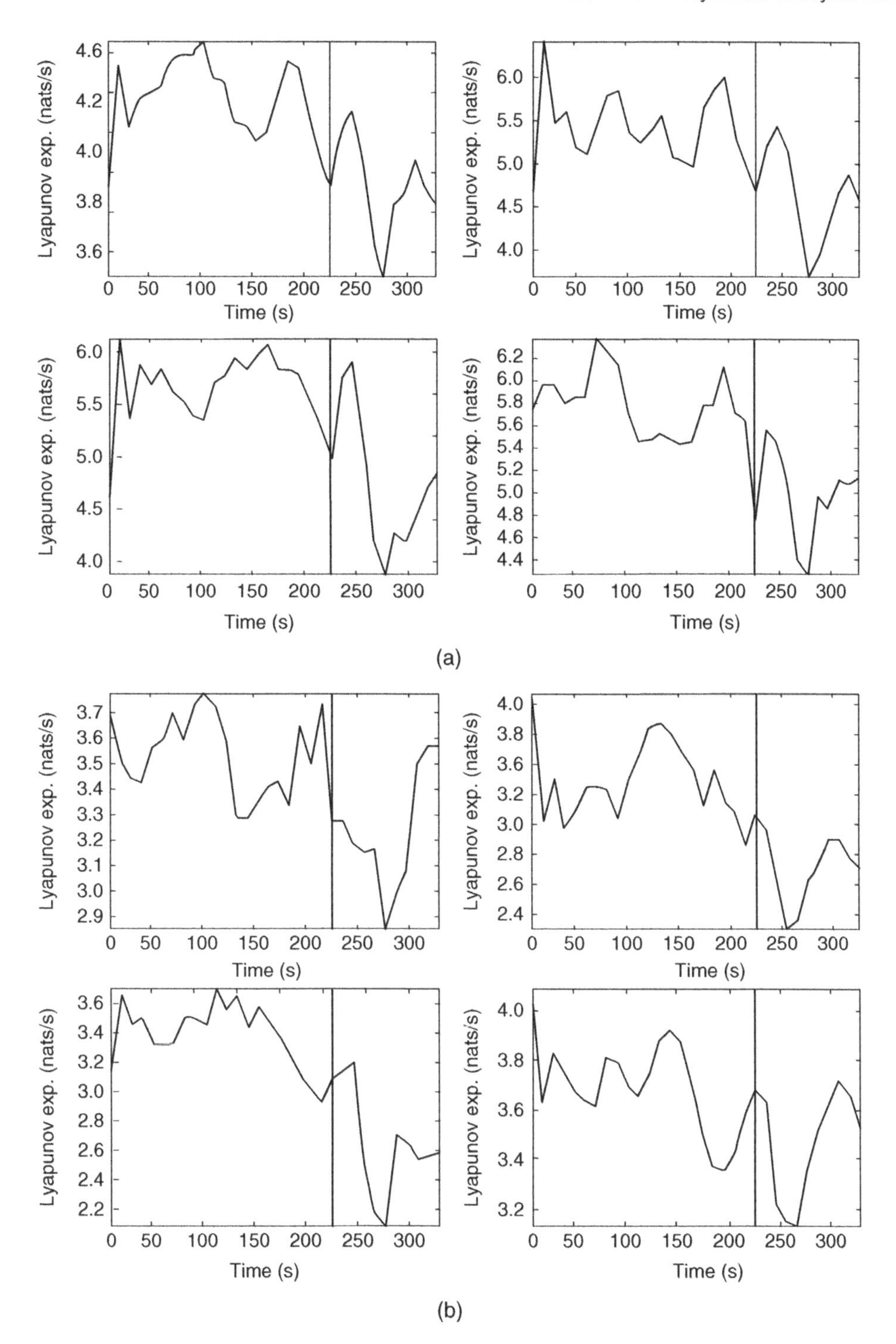

Figure 11.10 (a) Intracranial EEG analysis: three-point smoothed λ_1 evolution from focal seizure. In this case the electrical activity of the epileptic focus seemed not to be directly recorded by the intracranial electrodes. (b) Scalp EEG analysis: the smoothed λ_1 evolution for four scalp electrodes. The length of the recording is 334 seconds. The seizure occurs at 225 seconds (marked by the vertical line).

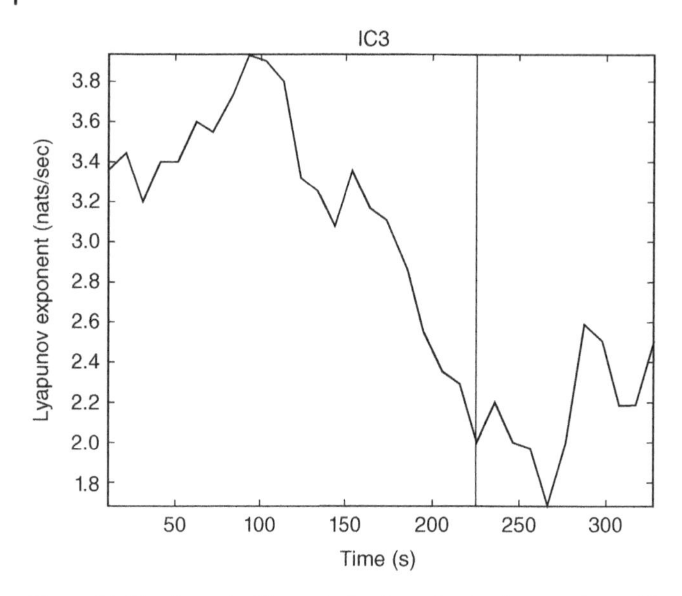

Figure 11.11 Smoothed λ_1 evolution for a focal seizure estimated from the independent component IC3 (i.e. the one where ictal activity is discernible in Figure 11.9b). The electrical activity of the epileptic focus was not directly recorded by the intracranial electrodes. The length of the recording is 334 seconds, the seizure occurs at 225 seconds (marked by the vertical line).

value is obtained several seconds later; however, the onset of seizure was clearly discernible from the intracranial electrodes around 236 seconds. Therefore, as an important conclusion, the iEEG is not able to predict the onset of seizure in such cases and they are only able to detect the seizure after its onset. There is a clear drop in the value of λ_1 but it does not occur soon enough to predict the seizure.

Figure 11.12b shows the results obtained after the application of BSS. The evolution of λ_1 is similar to the evolution for intracranial recordings. However, the drop in the value of λ_1 for the estimated source seems to start decreasing before it does for the intracranial electrodes. The minimum λ_1 for the estimated source occurs before such a minimum for λ_1 is achieved for the intracranial electrodes. This means that by preprocessing the estimated sources using the BSS-based method, the occurrence time of seizure can be estimated more accurately.

There is no doubt that local epileptic seizures are predictable from the EEGs. Scalp EEG recordings seem to contain enough information about the seizure; however, this information is mixed with the signals from the other sources within the brain and it is buried in noise and artefacts. Incorporating a suitable preprocessing technique such as BSS algorithm separates the seizure signal (long before the seizure) from the rest of the sources, noise, and artefacts within the brain. Therefore, the well-known TNMs for evaluation of the chaotic behaviour of the EEG signals can be applied. Using BSS, prediction of seizure might be possible from scalp EEG data, even when this is not possible with iEEG, especially when the epileptic activity is spread over a large portion of the brain in most of the cases. However, the results obtained from three sets with generalized seizure support the idea of unpredictability of this type of seizure, since they are preceded and followed by normal EEG activity.

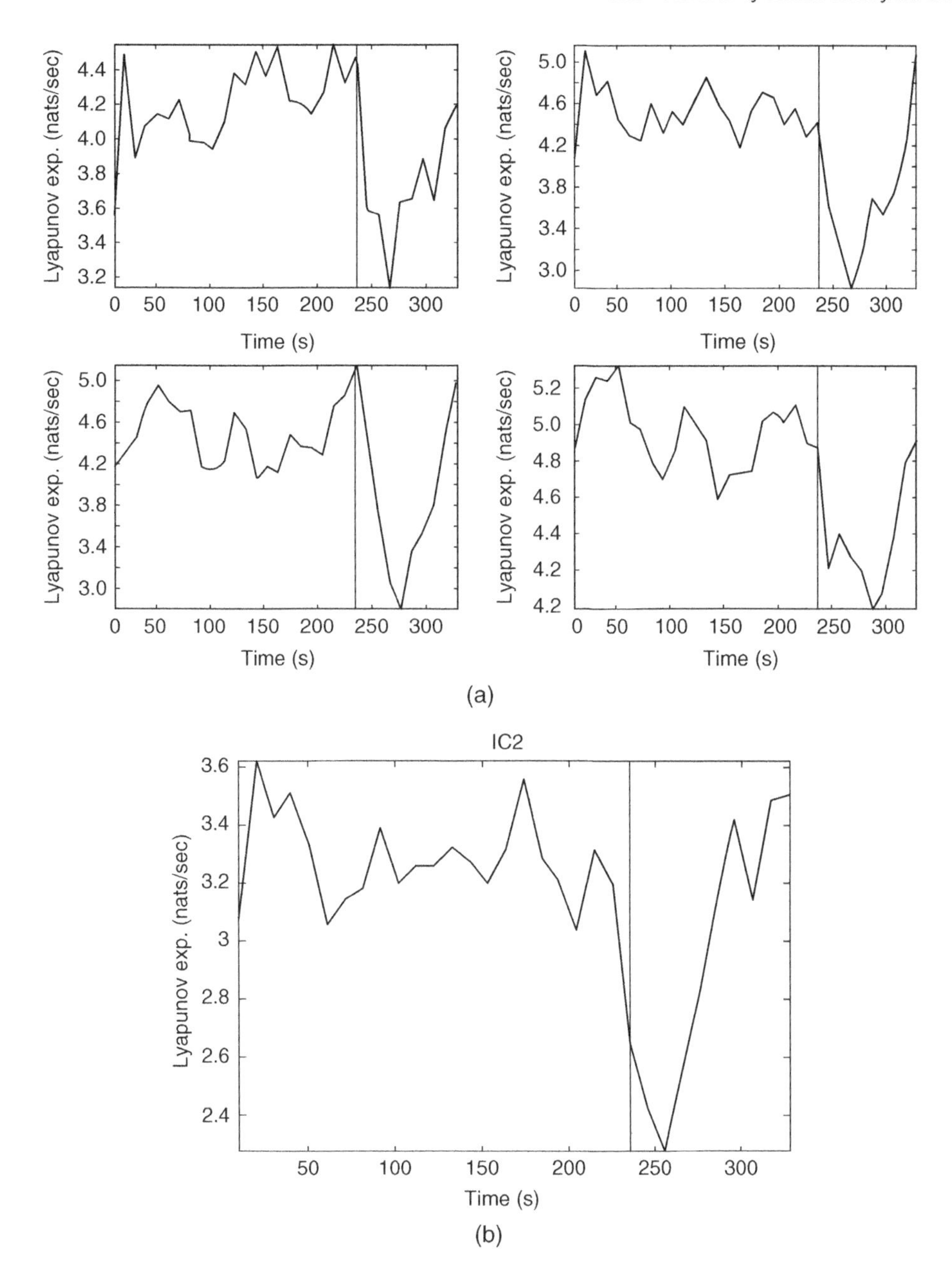

Figure 11.12 (a) Smoothed λ_1 evolution of four intracranial electrodes for a generalized seizure. The length of the recording is 337 seconds and it records a generalized seizure starting at 236 seconds (marked by the vertical line). (b) Smoothed λ_1 evolution of IC2 component estimated from the corresponding scalp EEG. The length of the recording is 337 seconds and it records a generalized seizure at 236 seconds (marked by the vertical line).

The results obtained by the analysis of the scalp EEGs, although they are very promising, are subject to several limitations. The principal limitation is due to the length of the recordings. These recordings allowed the comparison between scalp and iEEG; however, they were of relatively short duration. Therefore, it is basically assumed that the epileptic component is active during the entire recording. Longer recordings are needed to better examine the value of BSS in the study of seizure predictability. For such recordings it can be hardly assumed that the underlying sources are active during the entire recording and therefore the algorithm needs to detect the beginning and end of these activities. Furthermore, the algorithm employed to maintain the continuity fails in some cases where a segment of scalp EEG is corrupted or the electrical activity is not correctly recorded. The number of corrupted segments increases for longer recordings and therefore a new methodology to maintain the continuity of the estimated sources for these particular segments should be combined with the overlap window approach. Another limitation arises from the fixed number of output signals selected for the BSS algorithm.

Generally, a seizure can be predicted if the signals are observed during a previous seizure and by analyzing the interictal period. Otherwise, there remains a long way to go to be able to predict accurately the seizure from only EEG. Conversely, seizure is predictable if the EEG information is combined with other information such as that gained from video sequence of the patients, heart rate variability, and respiration.

In a recent review by Carney et al. [98], different approaches for seizure prediction have been listed. The first group of methods is called univariate methods which include: (i) time–frequency domain analysis using the short-term Fourier transform or wavelet transform to capture any significant peak in frequency over time representing brain synchronization before the seizure onset. These methods have been mostly applied to ECoG signals; (ii) change in signal distribution and higher-order moments of the EEG time series; (iii) estimation of the correlation dimension and intensity to identify the randomness level in EEG time series; (iv) autocorrelation and autoregressive modelling to evaluate the brain connectivity changes during preictal to ictal to interictal periods; (v) Kolmogorov entropy as a more classic approach to the variation in randomness during preictal or interictal periods where the signals shifts from chaotic to synchronized; (vi) the use of dynamical similarity index which in some particular cases has been shown that it can track spatiotemporal changes in brain dynamics minutes in advance of an impending seizure. The measure is computed by phase-space reconstruction of the EEG time series by using time intervals between two positive zero crossings and the measure of the dynamical similarity between the reference and test windows, respectively, using the cross-correlation integral; (vii) loss of recurrence and local flow representing the degree of nonstationarity; and (viii) Lyapunov exponent to evaluate the chaos, as discussed before.

The second group includes multivariate methods. Multivariate time-series analyses consist of more than one observation recorded sequentially over time. These methods, with a wide range of applications, are used to assess the interaction between the different components of the system under consideration. EEGs, as multichannel data, can be conceptualized as a series of numerical values (voltages) over time and space (gathered from multiple electrodes), called a multivariate time series. The methods for multivariate EEG analysis include: [98] (i) Simple synchronization measure showing that during seizures, highly synchronous activity is seen, either focally or in a more generalized pattern. It has been suggested that this activity may begin hours before the initiation of a seizure; (ii) the correlation among EEG channels; (iii) phase synchrony based on spectral coherence; (iv) multivariate autoregressive (MVAR) which is often used for brain connectivity estimation; and (v) short-term LLE as explained before.

In a more recent review of seizure prediction schemes [99] a comprehensive study of various EEG signal processing methods as well as data conditioning techniques has been presented. In addition, the role and application of various machine learning techniques such as SVM, ANN including DNN, and logistic regression for seizure prediction have been discussed.

The authors in [100] used iEEG data of patients and trained a DNN classifier to distinguish between preictal and interictal signals. For this, the classifier learns the iEEG patterns before the seizure onset. Therefore, the DNN can perform as a seizure predictor and the user can tune its sensitivity or warning time.

Also, the authors in [101] proposed a long-term recurrent neural network (RNN). They converted the preprocessed (artefact removed) EEG time series into two-dimensional images for multichannel fusion. A long-term recurrent convolutional network (LRCN) was proposed to create a spatiotemporal deep learning model for predicting epileptic seizures. The convolutional network block was used to automatically extract deep features from the data. Then, a long short-term memory (LSTM) block was incorporated to enable learning a time sequence for identification of the preictal segments. Figure 11.13 shows the structure of their seizure prediction system.

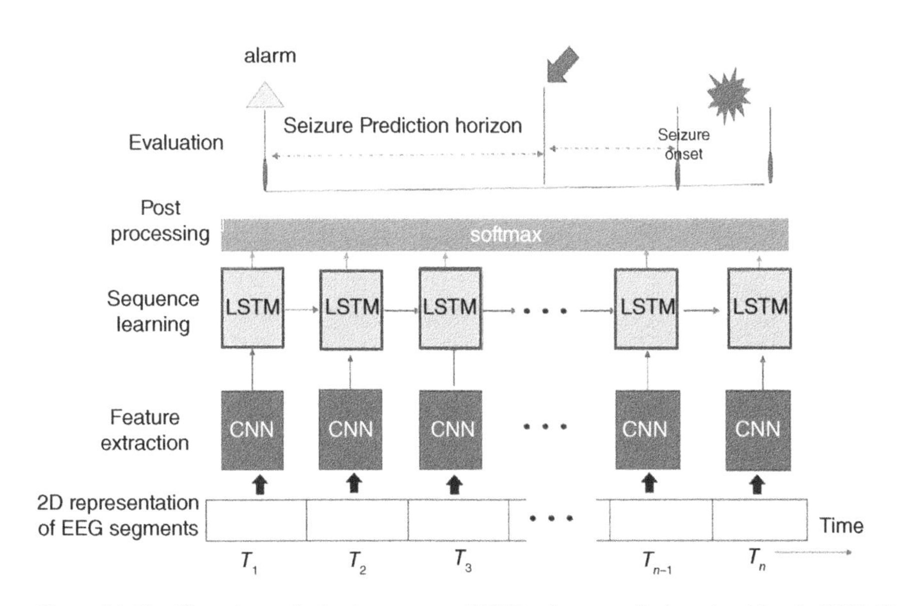

Figure 11.13 The schematic for the proposed LRCN seizure prediction algorithm in [101]. The LRCN processes the (possibly) variable-length feature input with a CNN, whose outputs are fed into a stack of LSTMs that decide if the seizure onset is approaching.

11.6 Intracranial and Joint Scalp–Intracranial Recordings for IED Detection

11.6.1 Introduction to IED

IEDs originate from deep within the brain or hippocampus and in many cases are descriptive of the epileptic condition. They are the most reliable biomarkers and are widely used in clinical evaluations. Detection of IEDs particularly from scalp EEG has attracted the interest of researchers in both machine learning and neurosciences and a variety of algorithms, mostly by using machine learning, have been developed [102]. These algorithms are based on methods such as template matching [103, 104], utilizing various classification tools [105–107], using wavelet TF analysis [108] dictionary learning [109], differential operator applied to the data directly recorded from the cortex using ECoG [110], spike rate estimation from intracranial signals for seizure prediction purposes [111] and other methods common in the well established field of spike detection [112, 113] and clustering [114]. The common characteristic of all these methods is that a description of an IED/spike signal is obtained either through modelling or a kind of similarity measurement with features of interest. This is often facilitated by obtaining useful representations of the signal that can better exploit its structure. Finding the optimal features for IED detection is a challenging yet important problem.

In the joint scalp–intracranial recordings while scalp EEG is mounted, subdural electrodes are inserted into the brain hippocampus through FO holes [115]. These holes can be seen in Figure 11.14. Conversely, the joint scalp–intracranial electrode setup can be seen in Figure 11.15, which shows an x-ray image of the head with both electrode sets installed.

In Figure 11.16 a segment of concurrent data can be observed. The first 20 signals belong to scalp EEG and the last 12 signals recorded using FO electrodes, belong to iEEG.

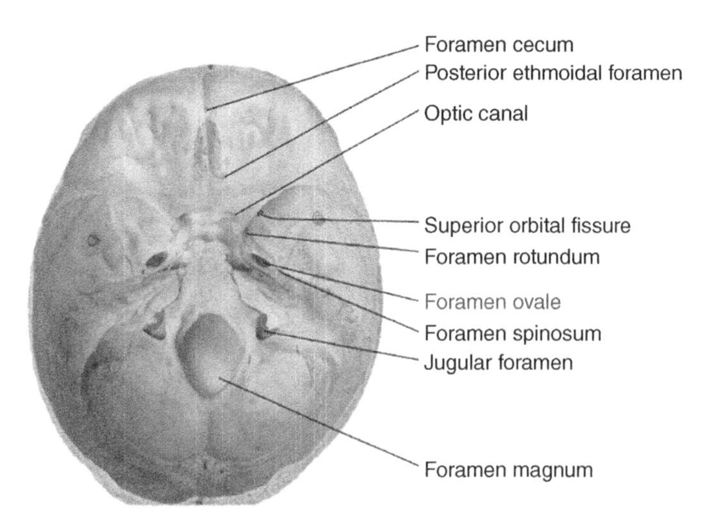

Figure 11.14 Foramen ovale holes where the subdural electrodes are inserted into the brain hippocampus.

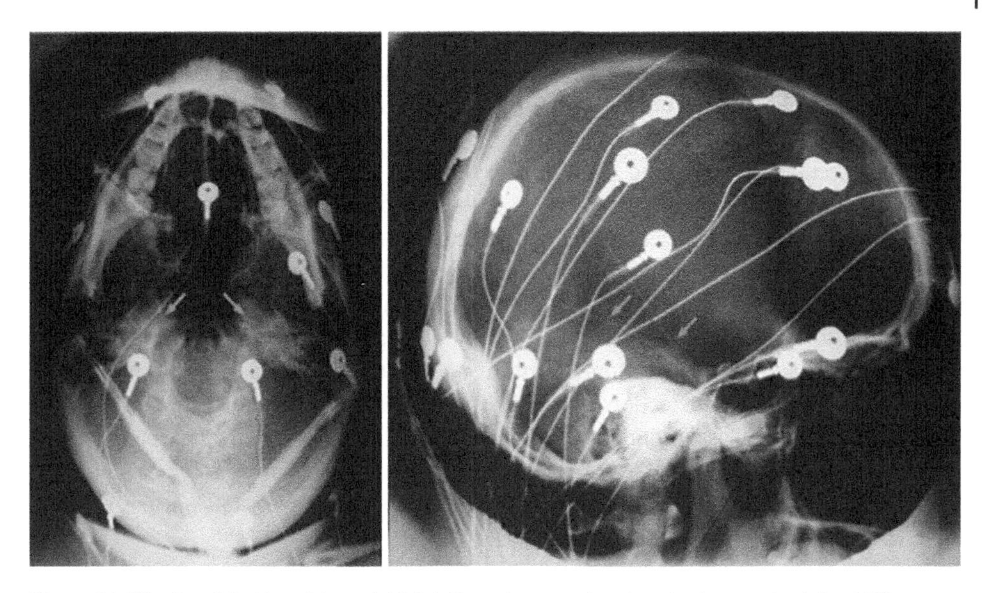

Figure 11.15 Basal (left) and lateral (right) X-ray images showing the inserted subdural FO electrodes (shown by arrows) and scalp EEG electrodes.

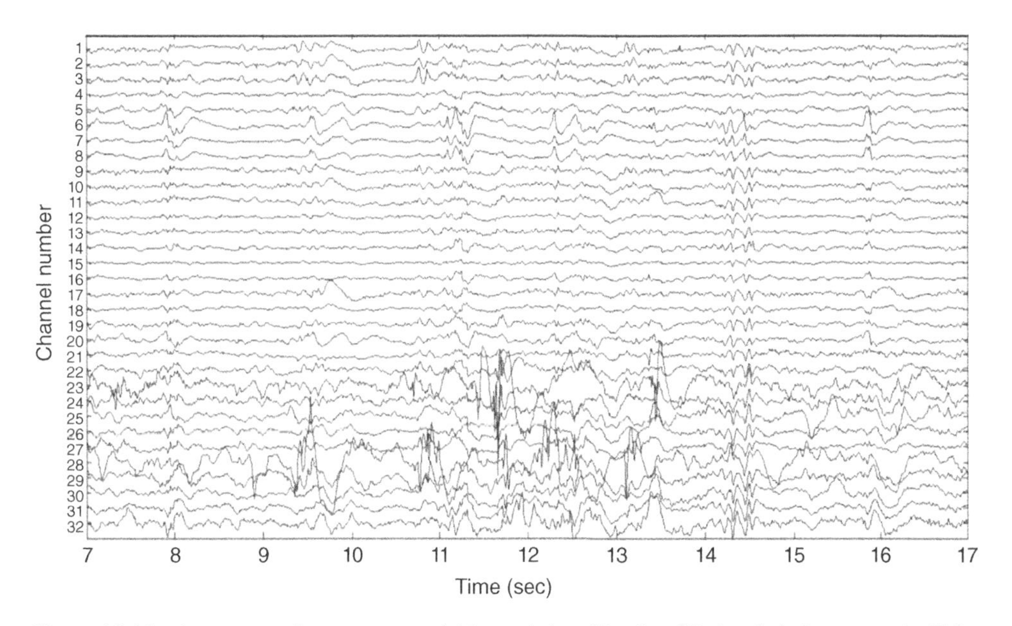

Figure 11.16 A segment of concurrent multichannel data. The first 22 signals belong to scalp EEG and the last 12 signals, recorded using FO electrodes, belong to intracranial EEG (iEEG).

To enable detection of IEDs from scalp EEG using the joint recordings, primarily the intracranial data should be analyzed and the IEDs identified either manually or automatically. Obviously, detection of IEDs from very long (recorded in hours or days) intracranial data is very cumbersome. Therefore, establishing a reliable method in automatic IED detection is always desirable. Unfortunately, there is uncertainty in detecting IEDs even if it is done by expert clinicians mainly because of their variabilities in morphology.

Often inspection of IEDs from the scalp EEG is that for a subset of epileptic patients, there are no visually discernible IEDs on the scalp, rendering the above procedures ineffective, both for detection purposes and algorithm evaluation. Conversely, intracranially placed electrodes yield a much higher incidence of visible IEDs as compared to concurrent scalp electrodes. Spyrou et al. utilized concurrent scalp and iEEG from a group of TLE patients with low number of scalp-visible IEDs [116]. They tried to find out if by using the timing information of the IEDs from iEEG, the resulting concurrent scalp EEG contains enough information for the IED segments to be reliably distinguished from non-IED segments. Their proposed automatic detection algorithm was tested in a leave-subject-out fashion, where each test subject's detection algorithm is based on the other patients' data.

Here, a fundamental assumption is that although a spike shape may not be clearly visible in the single-trial scalp EEG, IED information is indeed present in those cases since the IEDs captured by averaging iEEG-timed scalp EEGs over multiple trials are more likely to be visible.

The relation between the amplitude, location and propagation characteristics of IEDs in both recording modalities have been analyzed extensively [117–120].

11.6.2 iEED-Times IED Detection from Scalp EEG

Spyrou et al. [116] investigated 20 minute simultaneous scalp and iEEG recordings from 18 TLE patients (11 males, seven females, mean age 25.2 years, range 13–37). For the iEEG multicontact FO electrode bundles and for the scalp EEG the Maudsley EEG electrode setting were used. Patients suffering from seizures arising from mesial temporal structures was submitted for telemetry recording with FO when history, interictal scalp EEG, neuroimaging and neuropsychological studies were not able to confidently determine the side of seizure onset, or there were doubts about a lateral temporal or extra-temporal seizure onset. In 10 patients the seizure onset was identified within mesial temporal structures preceding in at least 2 ms the scalp changes while in eight patients it was located outside the mesial temporal region (lateral temporal). Most EEG recordings only contained wakefulness (13 patients) but some (5 patients) also included slow-wave sleep.

The raw EEG data were sliced in a $\pm$162.5 ms ($\pm$32 samples) window centred at the points scored as spike from intracranial data and baselined on the preceding 162.5 ms with the resulting signal finally being linearly detrended to remove undesired drifts. Non-IED segments were also obtained from time segments where there are no scored IEDs.

In their work, TF features were used. A fundamental assumption is that although a spike shape may not be visible or clear on the single-trial scalp EEG, the IED information is indeed present in such cases since averaging iEEG-timed scalp EEG IEDs provides visible spike structures. Wavelet features were obtained by performing a discrete wavelet decomposition using the 'db4' mother wavelet since it provides the highest correlation with IEDs.

Through five level wavelet decomposition they used the maximum, the minimum, the mean, and the standard deviation of the wavelet coefficients in each of the five sub-bands as features. Also, they used the time domain representations of the wavelet coefficients for the five sub-bands. In addition, chirplet features were obtained using matching pursuit with a variable number of chirplets [121]. Chirplets can represent signals with variable frequency over time [121]. A chirplet is defined as:

$$ch(t) = g_{s,a}(t-u)e^{jvt} \tag{11.24}$$

with:

$$g_{s,a}(t) = \frac{1}{\pi^{1/4}\sqrt{\sigma(s,a)}} exp\left(-\left(\frac{1}{\sigma^2(s,a)} - jl(s,a)\right)\frac{t^2}{2}\right) \tag{11.25}$$

Each chirplet consists of four parameters controlling the scaling (s), chirp rate (a), through the functions $\sigma()$ and $l()$, the time shift (u) and frequency shift (v) [121]. For each data segment only three chirplets were fitted to represent the waveform.

The between-subject differences were decreased [116] and the scalp and intracranial segments were normalized to have unit norm per electrode channel. These reduce the individual differences between subjects enhancing the performance of the classifier.

Next, both scalp-visible and non-scalp-visible IEDs are used in the classifier training and for analyzing the performance of the algorithm. A regularized linear logistic regression approach [122, 123] is used to train classifiers that can distinguish between IED and non-IED events. Logistic regression has been widely used for EEG classification. Linear classifiers work by estimating a weight row vector $\mathbf{w}$ and a threshold (or bias) b that operate on the data $\mathbf{x}$ in the following way:

$$f(\mathbf{x}) = \sum_{j=1}^{L} \mathbf{w}(j)\mathbf{x}(j) + b \tag{11.26}$$

where L is the number of features and j the feature index. For the two classes of IED and non-IED segments here, the sign of $f(\mathbf{x})$ denotes the prediction of data $\mathbf{x}$ class membership. One benefit of logistic regression is that it provides a natural way to express the class membership probability from the observed data. The magnitude of $f(\mathbf{x})$ corresponds to class membership probability, $p(\text{IED}|\mathbf{x})$, denoting the probability that a segment of data $\mathbf{x}$ contains an IED as expressed by the logistic function:

$$p(\text{IED} \mid \mathbf{x}) = \frac{1}{1 + \exp(-f(\mathbf{x}))} \tag{11.27}$$

and

$$p(\text{IED} \mid \mathbf{x}) = 1 - p(\text{nonIED} \mid \mathbf{x}).$$

Weight vector $\mathbf{w}$ is estimated by minimizing a quadratically regularized logistic regression loss function where there is a trade-off between classification accuracy and over-fitting and generalization performance. The regularization parameter, which controls the trade-off, is estimated by cross-validation [124].

The classifier is trained for each subject separately using a leave-subject-out detection strategy which removes a subject's data (test subject) and combines the rest of the subjects' classifiers in a linear voting scheme. The individual classifiers are assigned equal weights. Such a weighting procedure corresponds to a naïve Bayes combination of the individual classifiers [125]. The accuracy of the classifier resulting from the linear voting scheme is obtained for the sliced IED and non-IED data segments of the test subject. Individual classifiers whose within-subject or between-subject classification performance was less than

55% were excluded from the ensemble. The performance of each subject's classifier is estimated through the accuracy of the best regularization parameter. Detection of IEDs is performed by sweeping through the data and applying the classifier to a ±162.5 ms signal window of the scalp EEG.

To obtain the best training set, each identified IED pattern from the iEEG data was given a score by the expert epileptologist related to the certainty that the pattern is an IED. This results in a histogram of Figure 11.17. From this histogram the numbers higher than five denote sufficient likelihood for being an epileptic spike and the segment can be used in the IED training set.

Figure 11.18 shows the classification accuracy of the ensemble classifier with respect to the number of training subjects. For each number of training subjects 100 different subsets of the full set were used in the ensemble. It can be observed that for approximately 10 subjects a ceiling level is reached, however the standard deviation (grey area) of the accuracy between different subsets decreases as more subjects are used for training.

The algorithm obtained a 65% accuracy in recognizing scalp IED from non-IED segments with 68% accuracy when trained and tested on the same subject. Also, it was able to identify the scalp-invisible IED events for most patients with a low number of false-positive detections. The results represent a proof of concept that the IED information is contained in the scalp EEG even if they are not visually identifiable and also that between-subject differences in the IED topology and shape are small enough such that a generic algorithm can be used.

In Figure 11.19 the ratio of detected IEDs (from the scalp EEG), to the total number of IEDs that were identified by the expert epileptologist per score value for high (red) to low (black) detection thresholds. The higher the detection threshold the smaller the number of detections that are considered IEDs. At the same time the false positives are reduced too.

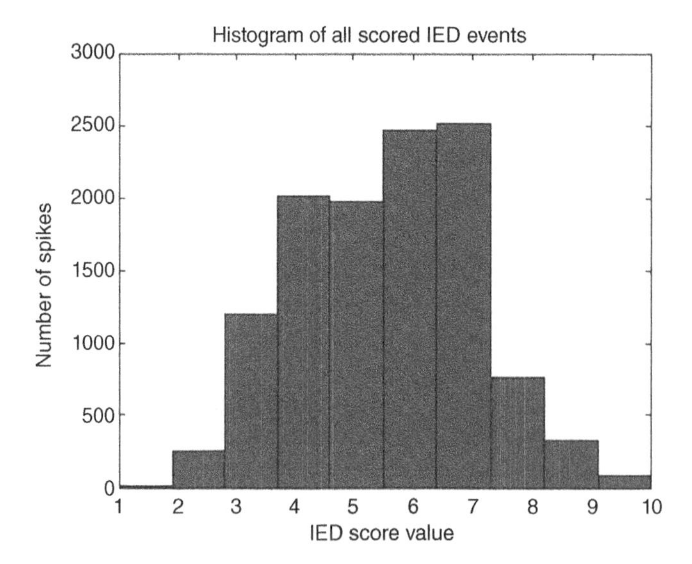

Figure 11.17 The scoring histogram provided by an expert in epilepsy.

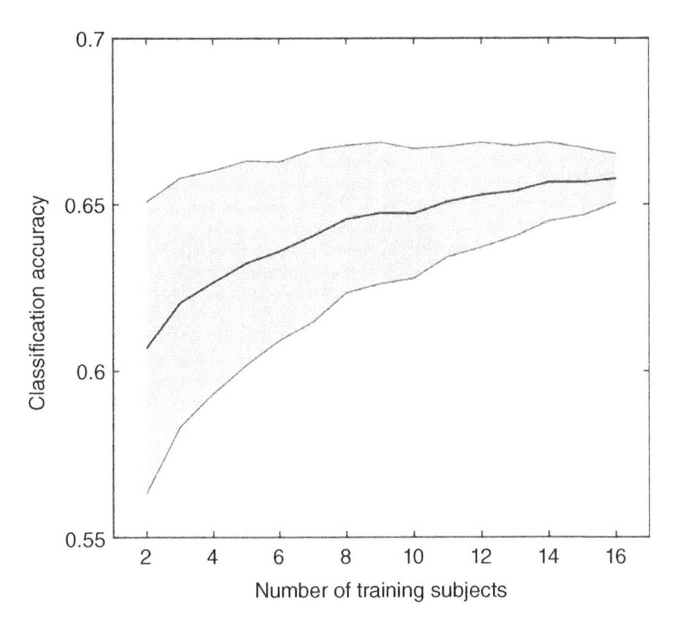

Figure 11.18 Classification accuracy of the ensemble classifier with respect to the number of training subjects.

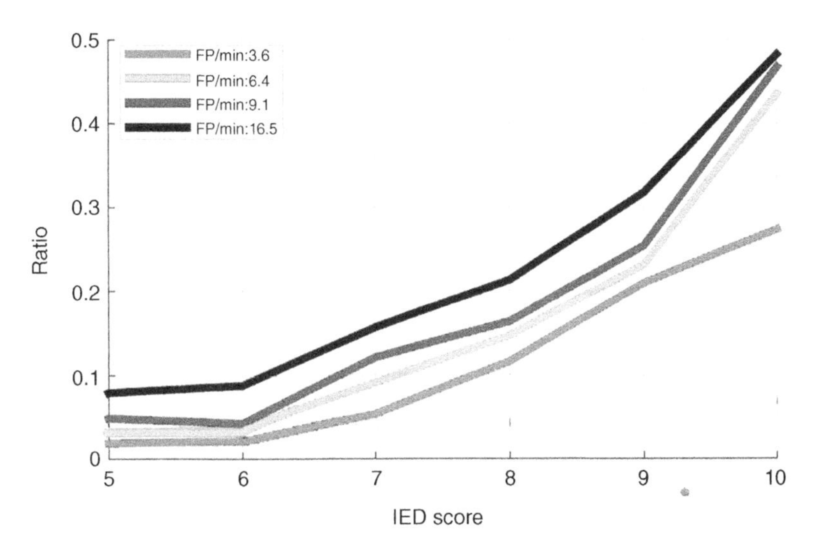

Figure 11.19 The ratio of detected IEDs (from the scalp EEG), to the total number of IEDs that were identified from the iEEGs by the expert epileptologist for each score value for high (red) to low (black) detection thresholds. FP refers to percentage of false positives.

For example, 21% of IEDs with an IED score of 9 were detected by the algorithm for a detection threshold that produced 3.6 FP/min. In is noteworthy that less than 9% of the spikes were visible from the scalp.

The method used to confirm true and false positives consisted of projecting the detected components onto the scalp topography as shown in Figure 11.20.

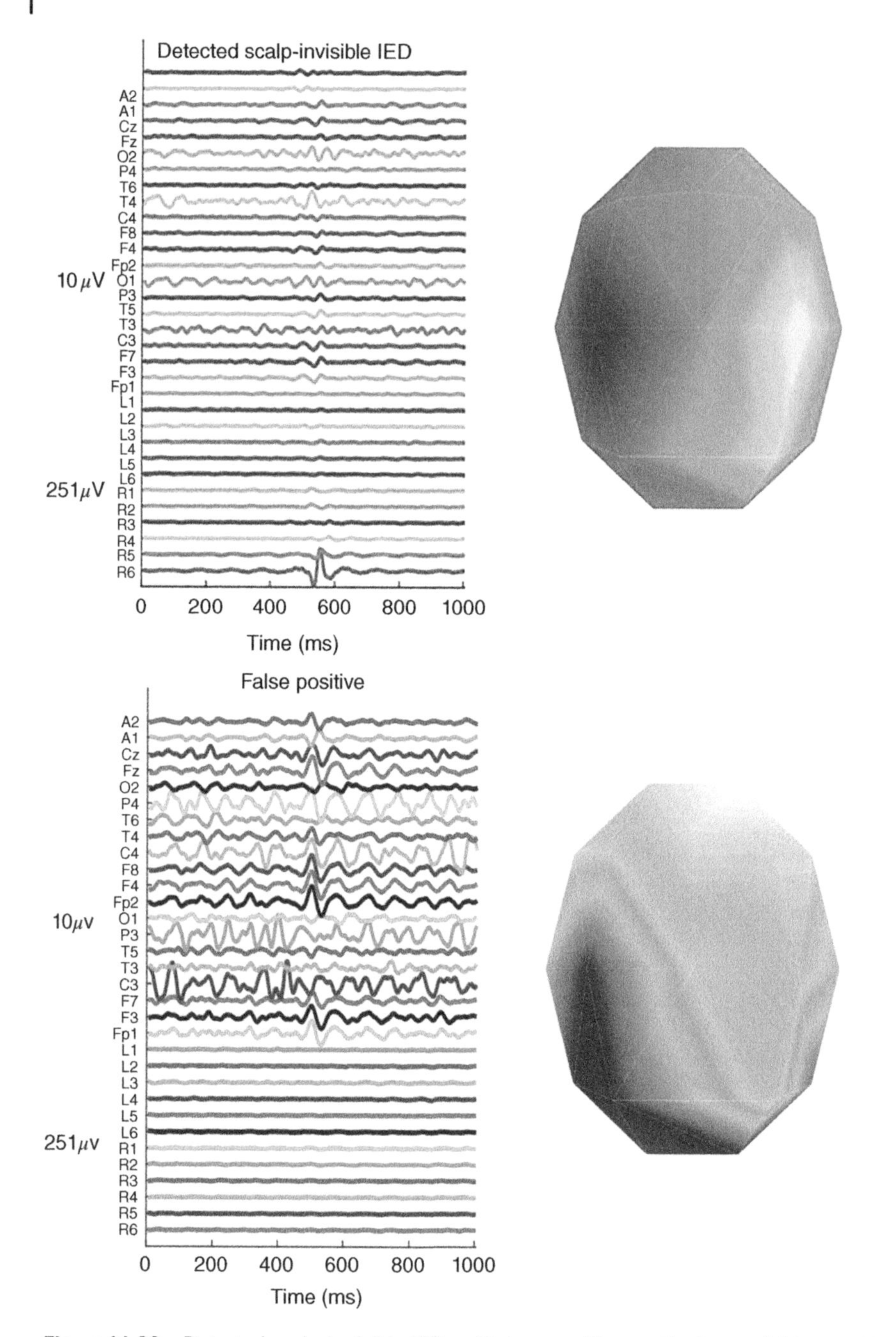

Figure 11.20 Detected scalp-invisible IEDs with true positive on the top and false positive at the bottom. The distributions of the sources in the right-hand side topographs verify the claim.

11.6.3 A Multiview Approach to IED Detection

Using the above data, a multiview approach using Tucker tensor decomposition for detection of IEDs has been attempted in [126]. In their work, they applied tensor factorization to a set of EEG data from a group of epileptic patients and factorizes the data into three modes: space, time, and frequency with each mode containing a number of components or signatures. Separate classifiers were then trained on various feature sets corresponding to complementary combinations of those modes and components and test the classification accuracy of each set. The relative influence on the classification accuracy of the respective spatial, temporal, or frequency signatures were then analyzed and useful interpretations made. Additionally, it was shown that through tensor factorization the dimensionality is reduced by evaluating the classification performance with respect to the number of mode components and by rejecting components with insignificant contribution to the classification accuracy.

11.6.4 Coupled Dictionary Learning for IED Detection

In a new coupled dictionary learning with sparse approximation (CDLSA) algorithm for detection of IEDs from scalp EEG the concurrent intracranial and scalp EEG dataset are used to learn a dictionary and a mapping function between the two modalities [127]. The aim is to infer the IEDs from only the scalp EEG by using that dictionary and mapping function. To briefly explain the method, let $\mathbf{x}_i \in \Re^N$ be the high quality (iEEG) signal and $\mathbf{s}_i \in \Re^M$ be the low quality (scalp EEG) signal. Then

$$\mathbf{x}_s = \mathbf{C}\mathbf{x}_i + \mathbf{v} \tag{11.28}$$

where $\mathbf{C} \in \Re^{N \times M}$is the mapping between the two modalities and $\mathbf{v}$ is the measurement noise and interference. We wish to estimate a dictionary $\mathbf{D} \in \Re^{N \times m}$ that is common between the two modalities in a similar fashion. Hence, assuming the IEDs, $\mathbf{a}$, to be temporally sparse, the aim is to minimize the following objective function:

$$J(\mathbf{D}, \mathbf{a}) = \|\mathbf{x}_s - \mathbf{C}\mathbf{D}\mathbf{a}\|_2^2 + \|\mathbf{x}_i - \mathbf{D}\mathbf{a}\|_2^2 + \lambda\|\mathbf{a}\|_1 \tag{11.29}$$

Since both modalities use the same sparse coefficients, the above cost function can be written in a closed form as:

$$J(\hat{\mathbf{D}}, \mathbf{a}) = \|\hat{\mathbf{x}} - \hat{\mathbf{D}}\mathbf{a}\|_2^2 + \lambda\|\mathbf{a}\|_1 \tag{11.30}$$

where $\hat{\mathrm{x}} = \begin{bmatrix} \mathbf{x}_s \\ \mathbf{x}_i \end{bmatrix}$ and $\hat{\mathbf{D}} = \begin{bmatrix} \mathbf{C} \\ \mathbf{I} \end{bmatrix} \mathbf{D} = \hat{\mathbf{C}}\mathbf{D}$. Considering $\mathbf{A} = \{\mathbf{a}\}$ for all the segments, the solutions are derived from the following quadratic and linear equations [127]:

$$\mathbf{D} = \left(\hat{\mathbf{C}}^T\hat{\mathbf{C}}\right)^{-1}\hat{\mathbf{X}}\mathbf{A}^T\left(\mathbf{A}\mathbf{A}^T\right)^{-1} \tag{11.31}$$

Then, for each $\mathbf{a} \in \mathbf{A}$ the following standard nonnegative quadratic programme (NNQP) is solved:

$$\mathbf{a}^T\left(\hat{\mathbf{D}}^T\hat{\mathbf{D}}\right)\mathbf{a} - \mathbf{a}^T\left(\hat{\mathbf{D}}^T\hat{\mathbf{x}} - \lambda\right) = 0 \quad \text{subject to } \mathbf{a} \geq 0 \tag{11.32}$$

where λ controls the sparsity of the solution. The algorithm alternates between estimating **D** and **A** until convergence. In each alternation, **C** can also be estimated as:

$$\mathbf{C} = \mathbf{X}_s(\mathbf{DA})^{\dagger} \tag{11.33}$$

where † denotes the pseudo-inverse. This method is called DLSA+C. The above algorithm was further improved by sparse coding of the measured low-resolution test data and changing it to high-resolution data using $\mathbf{D}_s = \mathbf{CD}$ only [127]. This leads to an improved algorithm called CDLSA which enhances the result considerably.

Figure 11.21 represent the IEDs from four subjects detected using the above CDLSA method and other benchmark approaches [127].

11.6.5 A Deep Learning Approach to IED Detection

An ensemble deep learning architecture for nonlinearly mapping scalp to iEEG data is proposed in [129]. The proposed architecture exploits the information from a limited number of joint scalp–intracranial recordings to establish a new methodology for detecting the epileptic discharges from the scalp EEG for the general population for which the intracranial data are not available.

To achieve this, the neural network in Figure 11.22 is used. The network comprises of a fully connected network with different numbers of inputs and outputs in the left and an autoencoder (AE) in the right. During the training the scalp EEG segments are applied to the left and the corresponding segments of intracranial signals (iEEG) to the right. The middle layer then learns how to map the scalp data (in the left) to the patterns similar to the iEEGs in the right. The fundamental idea behind this design is that a new scalp EEG test pattern is first projected (mapped) to the iEEG domain and then classified. Figure 11.23 represents the concept.

Statistical tests and qualitative analysis have revealed that the generated pseudo-intracranial data are highly correlated with the true intracranial data. This facilitated the detection of IEDs in scalp recordings where such waveforms are not easily visible. The IED detection accuracy of above 70% for over 80% of the subjects has been achieved. Figure 11.24 shows how accurately the intracranial waveforms can be learned from the scalp EEGs.

As a real-world clinical application, these pseudo-iEEG are then used by a CNN for the automated classification of IED and non-IED waveforms. This circumvents the unavailability of iEEG and the limitations of scalp EEG in IED visualization.

In another attempt by the same research group a CNN has been designed to perform feature learning and detect the IEDs [130]. The CNNs are trained in a subject independent fashion to demonstrate how meaningful features are automatically learned in a hierarchical process. They illustrate how the convolved filters in the deepest layers provide insight towards the different types of IEDs within the group, as confirmed by expert clinicians.

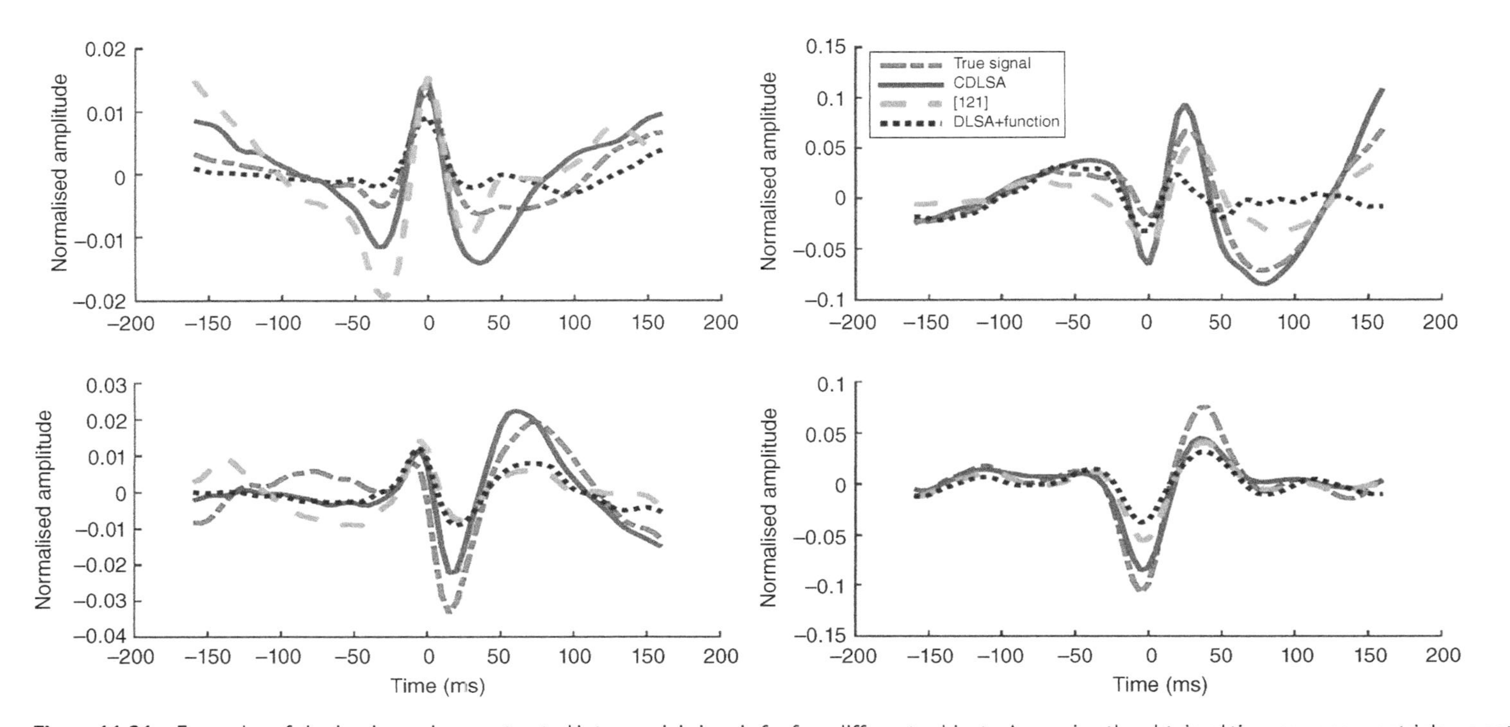

Figure 11.21 Examples of single-channel reconstructed intracranial signals for four different subjects. Averaging the obtained time courses over trials we get waveforms for the true test signal, the CDLSA method, the method in [128] and the DLSA+C method.

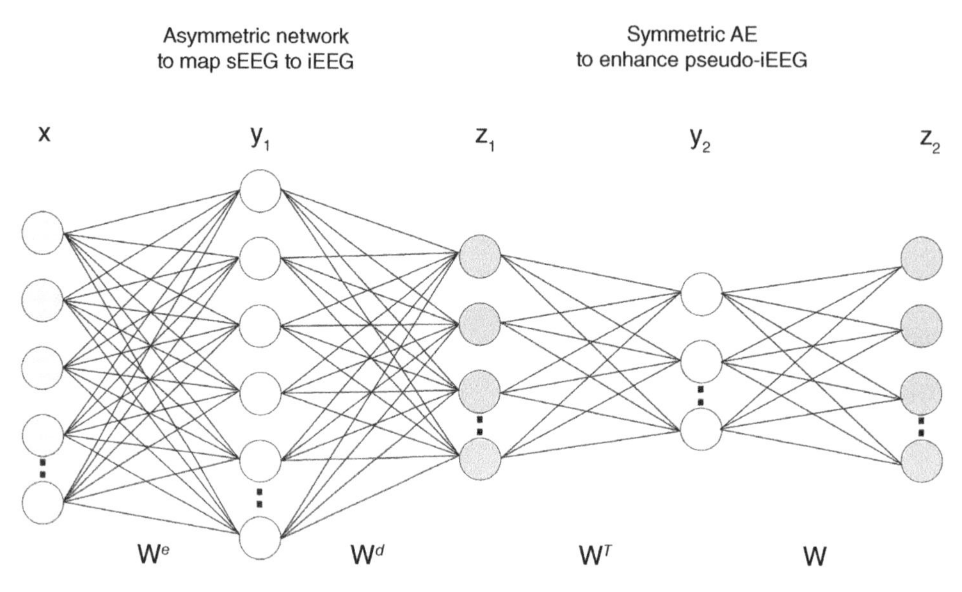

Figure 11.22 Topology of the DNN for mapping scalp to iEEGs. **X** is the scalp EEG, $\mathbf{y}_1$ is the hidden layer of the asymmetric fully connected network in the left, $\mathbf{z}_1$; $\mathbf{z}_2$ are the estimated sources of iEEG, $\mathbf{y}_2$ is the hidden layer of the AE, $\mathbf{W}^e$, and $\mathbf{W}^d$ are the weights of the AE and **W** are the tied weights of the AE.

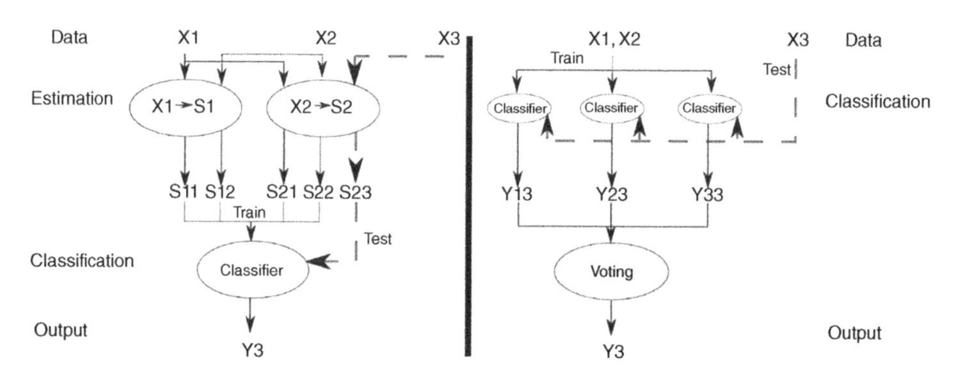

Figure 11.23 A schematic comparison between the proposed method (left) and an ensemble model (right). In this scenario, consider X2 to be more similar to X3 than X1. On the right side, both X1 and X2 are used to train the classifiers in order to predict X3. Whereas, the left model only uses the estimator trained with X2 to generate the unknown signal S3. This estimation (S23) is used to predict the classes of Y3.

The IED morphologies found in the filters can help evaluate the treatment of a patient. To improve the learning of the deep model, moderately different score classes are utilized as opposed to binary IED and non-IED labels. The resulting model achieves state of the art classification performance and is also invariant to time differences between the IEDs. This study suggests that deep learning is suitable for automatic feature generation from iEEG data, while also providing insight into the data.

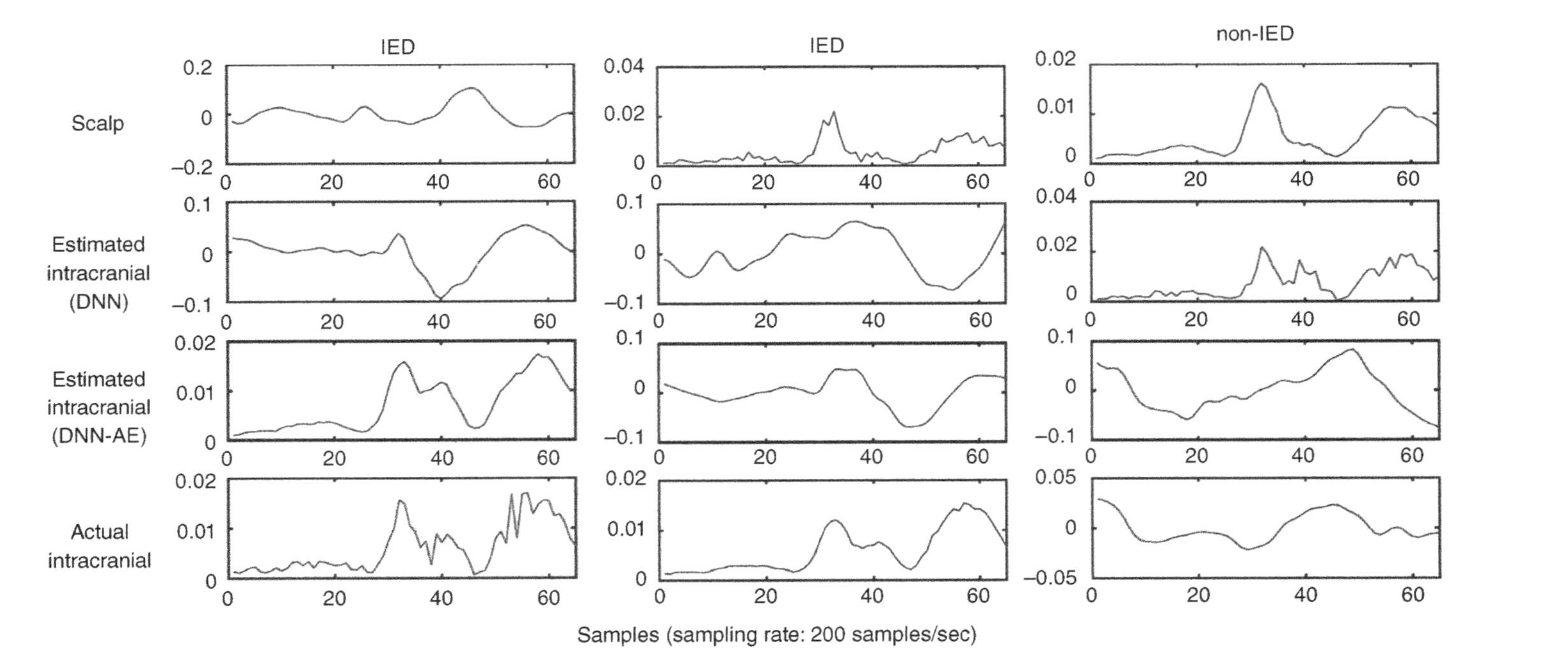

Figure 11.24 Estimation of iEEG for two IED segments and a non-IED segment (averaged over all channels) using the fully connected DNN only and the DNN-AE. In the latter case the additional symmetric layer led to a smoother estimation of the intracranial data.

11.7 Fusion of EEG–fMRI Data for Seizure Prediction

Clinical MRI has been of primary importance for visualization/detection of brain tumours, stroke, and multiple sclerosis. In 1990, Ogawa [131] showed that MRI can be sensitized to cerebral oxygenation, using deoxyhaemoglobin as an endogenous susceptibility contrast agent. Using gradient-echo imaging, a form of MRI image encoding sensitive to local inhomogeneity of the static magnetic field, he demonstrated (for an animal) that the appearance of the blood vessels of the brain changed with blood oxygenation. In some later papers published by his group they presented the detection of human brain activations using this BOLD [132, 133] by performing so-called fMRI. Now, it is established that by an increase in neuronal activity, local blood flow increases. The increase in perfusion, in excess of that needed to support the increased oxygen consumption due to neuronal activation, results in a local decrease in the concentration of deoxyhaemoglobin. Since deoxyhaemoglobin is paramagnetic, a reduction in its concentration results in an increase in the homogeneity of the static magnetic field, which yields an increase in the gradient-echo MRI signal. Although the BOLD fMRI does not measure brain activity directly, it relies on neurovascular coupling to encode the information about the brain function into detectible hemodynamic signals. It will clearly be useful to exploit this information in localization of seizure during the ictal period.

As will be stated in Chapter 10, brain sources can be localized within the brain. The seizure signals however, are primarily not known either in time or in space. Therefore, if the onset of seizure is detected within one ictal period, the location of the source may be estimated and the source which originates from that location tracked. Using fMRI it is easy to detect the BOLD regions corresponding to the seizure sources. Therefore, in separation of the sources using BSS, it can be established that the source signals originated from those regions which represent seizure signals. This process can be repeated for all the estimated source segments and thereby the permutation problem can be solved and the continuity of the signals maintained.

Unfortunately, in a simultaneous EEG–fMRI recording the EEG signals are highly distorted by the fMRI effects including gradient artefact due to the magnetic field, which is periodic, and the ballistocardiogram artefact, which is quasi-periodic. Figure 11.25 shows a multichannel EEG affected by fMRI scanner ballistocardiogram artefact. The gradient artefact has been already removed. An effective preprocessing technique is then required to remove this artefact. Also, in a long-term recording it is difficult to keep the patient steady under the fMRI system. This also causes severe distortion of the fMRI data and new methods are required for its removal.

In [134] a joint EEG–fMRI seizure dataset has been recorded from a number of patients and analyzed. In patients with medically refractory focal epilepsy who are candidates for epilepsy surgery, concordant noninvasive neuroimaging data are useful to guide invasive EEG recordings or surgical resection. Simultaneous EEG–fMRI recordings can reveal regions of haemodynamic fluctuations related to epileptic activity and help localize its generators. However, many of these studies (40–70%) remain inconclusive, principally due to the absence of IEDs during simultaneous recordings, or lack of haemodynamic changes correlated to IEDs. The researchers investigated whether the presence of epilepsy-specific voltage maps on scalp EEG correlated with haemodynamic changes and could help localize the

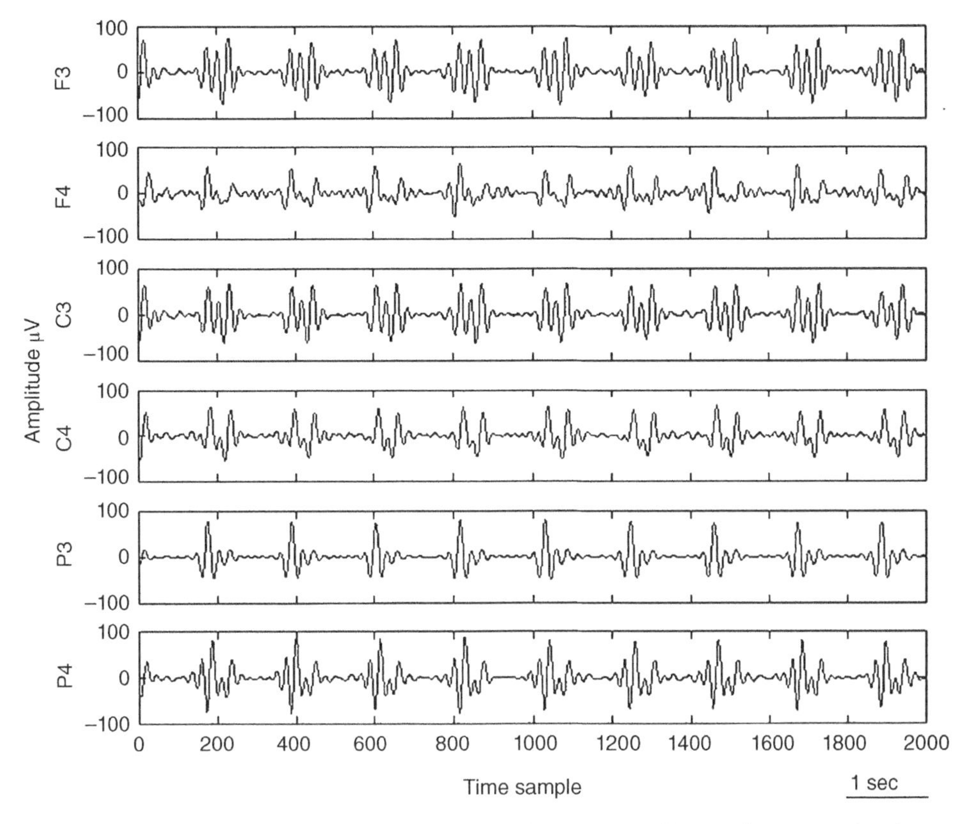

Figure 11.25 A segment of EEG signals affected by the scanner ballistocardiogram artefact in a simultaneous EEG–fMRI recording.

epileptic focus. In 23 patients with focal epilepsy, they built epilepsy-specific EEG voltage maps using averaged IEDs recorded during long-term clinical monitoring outside the scanner and computed the correlation of this map with the EEG recordings in the scanner for each time frame. The time course of this correlation coefficient was used as a regressor for fMRI analysis to map the haemodynamic changes related to these epilepsy-specific maps (topography-related haemodynamic changes). The overall procedure has been depicted in Figure 11.26 [134].

The method was first validated in five patients with significant haemodynamic changes correlated to IEDs on conventional analysis. They then applied the method to 18 patients who had inconclusive simultaneous EEG–fMRI studies due to the absence of IEDs or absence of significant correlated haemodynamic changes.

Bai et al. [135] acquired simultaneous EEG–fMRI for typical childhood absence seizures from nine paediatric patients. Absence seizure is a brief 5–10 seconds episode of impaired consciousness and 3–4 Hz generalized spike–wave discharges, which begin and end abruptly. Their EEG time–frequency analysis revealed abrupt onset and end of 3–4 Hz spike–wave discharges with a mean duration of 6.6 seconds. Behavioural analysis also showed rapid onset and end of deficits associated with electrographic seizure start and

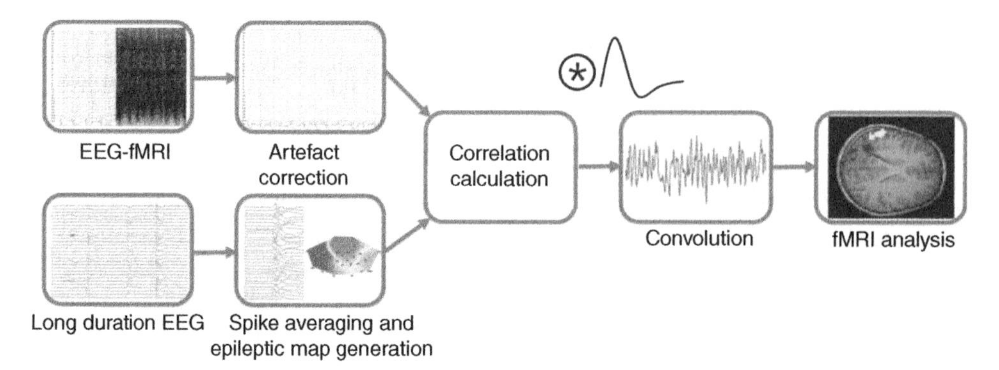

Figure 11.26 Schematic diagram of the topographic map correlation procedure. The EEG artefacts are removed from the EEGs (top left). Typical spikes derived from long-term EEG are averaged to build the epilepsy-specific EEG map (epileptic map bottom left). The time course of the correlation of this map with the intra-scanner EEG is calculated and this result is then convolved with the haemodynamic response function to build a regressor for fMRI analysis.

end. Nevertheless, they observed small early fMRI BOLD increases in the orbital/medial frontal and medial/lateral parietal cortex more than 5 seconds before the seizure onset followed by profound fMRI decrease continuing up to more than 20 seconds after the seizure stops. This time course differed markedly from the hemodynamic response function (HRF) model used in conventional fMRI analysis, consisting of large increases beginning after the electrical event onset, followed by small fMRI BOLD decrease. Other regions, such as the lateral frontal cortex, showed more balanced increase in BOLD followed by approximately equal decreases. The thalamus showed delayed increase after the seizure onset followed by a small decrease, most closely following the HRF model. These findings reveal a complex and long-lasting sequence of fMRI changes during absence seizures, which are not detectable by conventional HRF modelling in many regions. Their results demonstrate a complex sequence of fMRI changes in absence seizures, which are not detectable by predicted HRF modelling. In addition, BOLD fMRI increases have been independently found in the orbital, medial frontal, and parietal regions before the seizure onset, and large, long-lasting fMRI decreases in widespread brain regions after seizure termination.

11.8 Summary

Detection of epileptic seizures using different techniques has been successful particularly for detection of adult seizures. The false alarm rate for detection of seizure in the newborn is however, still too high and therefore, the design of a reliable newborn seizure detection system is still an open problem. Conversely, although predictability of seizure from the EEGs (both scalp and intracranial) has been approved, more research is necessary to increase accuracy. Fusion of different neuroimaging modalities may indeed pave the path for more reliable seizure detection and prediction systems. Detection of an increasing number of IEDs by the use of new DNNs enables: (i) better prediction of seizure onset, mainly by investigating the temporal distribution and frequency of occurrence of IEDs, (ii) application of simultaneous EEG–fMRI analysis for both seizure detection and localization, and (iii) enhances the usability and effectiveness of deep brain stimulation (DBS) for seizure suppression. With no doubt, these applications will remain important future research topics.

References

1 Zali, F., Bahrami, M., and Akhtari, E. (2020). Uterine epilepsy: a historical report from Avicenna's point of view. *Neurological Sciences* 41: 229–232.

2 Iasemidis, L.D. (2003). Epileptic seizure prediction and control. *IEEE Transactions on Biomedical Engineering* 50: 549–558.

3 Annegers, J.F. (1993). The epidemiology of epilepsy. In: *The Treatment of Epilepsy* (ed. E. Wyllie), 157–164. Philadelphia: Lea and Febiger.

4 Bomela, W., Wang, S., Chou, C.-A., and Li, S. Jr. (2020). Real-time inference and detection of disruptive EEG networks for epileptic seizures. *Scientific Reports, Nature Research* 10: 8653. https://doi.org/10.1038/s41598-020-65401-6.

5 Engel, J. Jr. and Pedley, T.A. (1997). *Epilepsy: A Comprehensive Text-Book*. Philadelphia: Lippinott-Ravon.

6 Lehnertz, K., Mormann, F., Kreuz, T. et al. (2003). Seizure prediction by nonlinear EEG analysis. *IEEE Engineering in Medicine and Biology Magazine* 22 (1): 57–63.

7 Peitgen, H. (2004). *Chaos and Fractals: New Frontiers of Science*. New York: Springer-Verlag.

8 Niedermeyer, E. and Da Silva, F.L. (1999). *Electroencephalography, Basic Principles, Clinical Applications, and Retailed Fields*, 4e. Lippincott, Williams &Wilkins.

9 Bravais, L.F. (1827). *Researchers sur les Symptoms et le Traitement de l'Épilesie Hémiplégique*, Paris: Thése de Paris, no. 118.

10 Chee, M.W.L., Morris, H.H., Antar, M.A. et al. (1993). Presurgical evaluation of temporal lobe epilepsy using interictal temporal spikes and positron emission tomography. *Archives of Neurology* 50: 45–48.

11 Godoy, J., Luders, H.O., Dinner, D.S. et al. (1992). Significance of sharp waves in routine EEGs after epilepsy surgery. *Epilepsia* 33: 513–288.

12 Kanner, A.M., Morris, H.H., Luders, H.O. et al. (1993). Usefulness of unilateral interictal sharp waves of temporal lobe origin in prolonged video EEG monitoring studies. *Epilepsia* 34: 884–889.

13 Steinhoff, B.J., So, N.K., Lim, S., and Luders, H.O. (1995). Ictal scalp EEG in temporal lobe epilepsy with unitemporal versus bitemporal interictal epileptiform discharges. *Neurology* 45: 889–896.

14 Akay, M. (2001). Nonlinear biomedical signal processing. In: *Dynamic Analysis and Modelling*, vol. II (ed. M. Akay). IEEE Press.

15 Walter, D.O., Muller, H.F., and Jell, R.M. (1973). Semiautomatic quantification of sharpness of EEG phenomenon. *IEEE Transactions on Biomedical Engineering* BME-20: 53–54.

16 Ktonas, P.Y. and Smith, J.R. (1974). Quantification of abnormal EEG spike characteristics. *Computers in Biology and Medicine* 4: 157–163.

17 Ktonas, P.Y. (1987). Automated spike and sharp wave (SSW) detection. In: *Methods of Analysis of Brain Electrical and Magnetic Signals, EEG Handbook*, vol. 1 (eds. A.S. Gevins and A. Remond), 211–241. Amsterdam, The Netherlands: Elsevier.

18 Ozdamar, O., Yaylali, I., Jayakar, P., and Lopez, C.N. (1991). Multilevel neural network system for EEG spike detection. *Computer-Based Medical Systems, Proceedings of the 4th Annual IEEE Symposium* (eds. I.N. Bankmann and J.E. Tsitlik). Washington DC: IEEE Computer Society.

19 Webber, W.R.S., Litt, B., Lesser, R.P. et al. (1993). Automatic EEG spike detection: what should the computer imitate. *Electroencephalography and Clinical Neurophysiology* 87: 364–373.

20 Wilson, S.B., Harner, R.N., Duffy, B.R. et al. (1996). Spike detection I. Correlation and reliability of human experts. *Electroencephalography and Clinical Neurophysiology* 98: 186–198.

21 Glover, J.R. Jr., Raghavan, N., Ktonas, P.Y., and Frost, J.D. (1989). Context-based automated detection of epileptogenic sharp transients in the EEG: elimination of false positives. *IEEE Transactions on Biomedical Engineering* 36: 519–527.

22 Conradsen, I., Beniczky, S., Hoppe, K. et al. (2012). Automated algorithm for generalized tonic–clonic epileptic seizure onset detection based on sEMG zero-crossing rate. *IEEE Transactions on Biomedical Engineering* 59 (2): 579–585.

23 Gotman, J. and Gloor, P. (1976). Automatic recognition and quantification of interictal epileptic activity in the human scalp EEG. *Electroencephalography and Clinical Neurophysiology* 41: 513–529.

24 Gotman, J., Ives, J.R., and Gloor, R. (1979). Automatic recognition of interictal epileptic activity in prolonged EEG recordings. *Electroencephalography and Clinical Neurophysiology* 46: 510–520.

25 Gotman, J. and Wang, L.Y. (1991). State-dependent spike detection: concepts and preliminary results. *Electroencephalography and Clinical Neurophysiology* 79: 11–19.

26 Gotman, J. and Wang, L.Y. (1992). State-dependent spike detection: validation. *Electroencephalography and Clinical Neurophysiology* 83: 12–18.

27 Dingle, A.A., Jones, R.D., Caroll, G.J., and Fright, W.R. (1993). A multistage system to detect epileptiform activity in the EEG. *IEEE Transactions on Biomedical Engineering* 40: 1260–1268.

28 Glover, J.R. Jr., Ktonas, P.Y., Raghavan, N. et al. (1986). A multichannel signal processor for the detection of epileptogenic sharp transients in the EEG. *IEEE Transactions on Biomedical Engineering* 33: 1121–1128.

29 Davey, B.L.K., Fright, W.R., Caroll, G.J., and Jones, R.D. (1989). Expert system approach to detection of epileptiform activity in the EEG. *Medical & Biological Engineering & Computing* 27: 365–370.

30 Webber, W.R.S., Litt, B., Wilson, K., and Lesser, R.P. (1994). Practical detection of epileptiform discharges (EDs) in the EEG using an artificial neural network: a comparison of raw and parameterised EEG data. *Electroencephalography and Clinical Neurophysiology* 91: 194–204.

31 Kurth, C., Gilliam, F., and Steinhoff, B.J. (2000). EEG spike detection with a Kohonen feature map. *Annals of Biomedical Engineering* 28: 1362–1369.

32 Kohonen, T. The self-organizing map. *Proceedings of the IEEE* 78: 1464–1480.

33 Kohonen, T. (1997). *The Self-Organizing Maps*, 2e. New York: Springer.
34 Boonyakitanont, P., Lek-uthai, A., and Songsiri, J. (2020). *Automatic epileptic seizure onset-offset detection based on CNN in scalp EEG. ICASSP 2020 – 2020 IEEE International Conference on Acoustics, Speech and Signal Processing (ICASSP)*, 1225–1229. Barcelona, Spain.
35 Goldberger, A.L., Amaral, L.A.N., Glass, L. et al. (2000). PhysioBank, PhysioToolkit, and PhysioNet. *Circulation* 101 (23): e215–e220.
36 Nenadic, Z. and Burdick, J.W. (2005). Spike detection using the continuous wavelet transform. *IEEE Transactions on Biomedical Engineering* 52 (1): 74–87.
37 Boashash, B., Mesbah, M., and Colditz, P. (2003). *Time-Frequency Detection of EEG Abnormalities*," Chap. 15, Article 15.5, 663–669. Elsevier.
38 Subasi, A. (2005). Epileptic seizure detection using dynamic wavelet network. *Expert Systems with Applications* 29: 343–355.
39 Lutz, A., Lachaux, J.-P., Martinerie, J., and Varela, F.J. (2002). Guiding the study of brain dynamics by using first-person data: synchrony patterns correlate with ongoing conscious states during a simple visual task. *Proceedings of the National Academy of Sciences* 99 (3): 1586–1591.
40 Osorio, I. (1999). System for the prediction, rapid detection, warning, prevention, or control of changes in activity states in the brain of subject. *US Patent* 5, 995, 868.
41 Osorio, I., Frei, M.G., and Wilkinson, S.B. (1998). Real-time automated detection and quantitative analysis of seizures and short-term prediction of clinical onset. *Epilepsia* 39 (6): 615–627.
42 Bhavaraju, N.C., Frei, M.G., and Osorio, I. (2006). Analogue seizure detection and performance evaluation. *IEEE Transactions on Biomedical Engineering* 53 (2): 238–245.
43 Gonzalez-Vellon, B., Sanei, S., and Chambers, J. (1993). Support vector machines for seizure detection. *Proceedings of ISSPIT*, 126–129. Darmstadt, Germany.
44 Raad, A., Antoni, J., Sidahmed, M. (2003). Indicators of cyclostationarity: proposal, statistical evaluation and application to diagnosis. *Proceedings of IEEE ICASSP*, VI: 757–760.
45 Sanei, S., Jarchi, D., and Constantinides, A.G. (2020). *Body Sensor Networking Design and Algorithms*. Wiley.
46 Acir, N., Oztura, I., Kuntalp, M. et al. (2005). Automatic detection of epileptiform events in EEG by a three-stage procedure based on artificial neural network. *IEEE Transactions on Biomedical Engineering* 52 (1): 30–40.
47 Lee, M., Ryu, J., and Kim, D.H. (2020). Automated epileptic seizure waveform detection method based on the feature of the mean slope of wavelet coefficient counts using a hidden Markov model and EEG signals. *Wiley ETRI Journal* 42 (2): 217–229.
48 Sakai, T., Shoji, T., Yoshida, N. et al. (2020). ScalpNet: detection of spatiotemporal abnormal intervals in epileptic EEG using convolutional neural networks. *Proceedings of IEEE Conference on Acoustics, Speech, and Signal Processing, ICASSP 2020*. Barcelona, Spain.
49 Aanestad, E., Gilhus, N.E., and Brogger, J. (2020). Interictal epileptiform discharges vary across age groups. *Clinical Neurophysiology* 131: 25–33.
50 da Silva, F.H.L. et al. (2003). Dynamical diseases of brain systems: different routes to epileptic seizures. *IEEE Transactions on Biomedical Engineering* 59 (5): 540–548.
51 Volpe, J.J. (1987). *Neurology of the Newborn*. Philadelphia, PA: Saunders.

52 Liu, A., Hahn, J.S., Heldt, G.P., and Coen, R.W. (1992). Detection of neonatal seizures through computerized EEG analysis. *Electroencephalography and Clinical Neurophysiology* 82: 30–37.

53 Shoker, L., Sanei, S., Wang, W., and Chambers, J. (2004). Removal of eye blinking artifact from EEG incorporating a new constrained BSS algorithm. *IEE Journal of Medical and Biological Engineering and Computing* 43: 290–295.

54 Celka, P., Boashash, B., and Colditz, P. (2001). Preprocessing and time-frequency analysis of newborn EEG seizures. *IEEE Engineering in Medicine and Biology Magazine* 20: 30–39.

55 Pfurtscheller, G. and Fischer, G. (1977). A new approach to spike detection using a combination of inverse and matched filter techniques. *Electroencephalography and Clinical Neurophysiology* 44: 243–247.

56 Gotman, J., Flanagan, D., Zhang, J., and Rosenblatt, B. (1997). Automatic seizure detection in the newborn: methods and initial evaluation. *Electroencephalography and Clinical Neurophysiology* 103: 356–362.

57 Roessgen, M., Zoubir, A.M., and Boashash, B. (1998). Seizure detection of newborn EEG using a model-based approach. *IEEE Transactions on Biomedical Engineering* 45 (6): 673–685.

58 Choudhuri, N., Ghosal, S., and Roy, A. (2004). Contiguity of the whittle measure for a Gaussian time series. *Biometrika* 91 (1): 211–218.

59 da Silva, F.H.L., Hoeks, A., Smits, H., and Zetterberg, L.H. (1974). Model of brain rhythmic activity; the alpha rhythm of the thalamus. *Kybernetik* 15: 27–37.

60 O'Toole, J., Mesbah, M., and Boashash, B. (2005). Neonatal EEG seizure detection using a time-frequency matched filter with a reduced template set. *Proceedings of the 8th International Symposium on Signal Processing and its Applications*, 215–218.

61 Karayiannis, N.B. et al. (2006). Detection of pseudosinusoidal epileptic seizure segments in the neonatal EEG by cascading a rule-based algorithm with a neural network. *IEEE Transactions on Biomedical Engineering* 53 (4): 633–641.

62 Bishop, C.M. (1995). *Neural Networks for Pattern Recognition*. New York: Oxford University Press.

63 Purushothaman, G. and Karayiannis, N.B. (1997). Quantum neural networks (QNNs): inherently fuzzy feedforward neural networks. *IEEE Transactions on Neural Networks* 8 (3): 679–693.

64 Faust, O., Acharya, U.R., Adeli, H., and Adeli, A. (2015). Wavelet-based EEG processing for computer-aided seizure detection and epilepsy diagnosis. *Journal of Seizure* 26: 56–64.

65 Temko, A., Lightbody, G., Thomas, E.M. et al. (2012). Instantaneous measure of EEG Channel importance for improved patient-adaptive neonatal seizure detection. *IEEE Transactions on Biomedical Engineering* 59 (3): 717–727.

66 Takens, F. (1981). Detecting strange attractors in turbulence. In: *Dynamical Systems and Turbulence* (eds. D. Rand and L.-S. Young), 366–381. Berlin: Springer-Verlag.

67 Sauer, T., Yurke, J.A., and Casdagli, M. (1991). Embedology. *Journal of Statistical Physics* 65 (3/4): 579–616.

68 Fraser, A.M. and Swinney, H.L. (1986). Independent coordinates for strange attractors from mutual information. *Physical Review A* 33: 1134–1140.

69 Kennel, M.B., Brown, R., and Abarbanel, H.D.I. (1992). Determining minimum embedding dimension using a geometrical construction. *Physical Review A* 45: 3403–3411.

70 Casdagli, M.C. (1997). Recurrence plots revisited. *Physica D* 108 (1): 12–44.

71 Kantz, H. and Schreiber, T. (1995). Dimension estimates and physiological data. *Chaos* 5 (1): 143–154.

72 Rosenstein, M.T., Collins, J.J., and Deluca, C.J. (1993). A practical method for calculating largest Lyapunov exponents from small data sets. *Physica D* 65: 117–134.

73 Wolf, A., Swift, J.B., Swinney, H.L., and Vastano, J.A. (1985). Determining Lyapunov exponents from a time series. *Physica D* 16: 285–317.

74 Kawabata, N. (1973). A nonstationary analysis of the electroencephalogram. *IEEE Transactions on Biomedical Engineering* 20: 444–452.

75 Iasemidis, L.D., Sackellares, J.C., Zaveri, H.P., and Willians, W.J. (1990). Phase space topography and the Lyapunov exponent of electrocorticograms in partial seizures. *Brain Topography* 2: 187–201.

76 Rana, P., Lipor, J., Lee, H. et al. (2012). Seizure detection using the phase-slope index and multichannel ECoG. *IEEE Transactions on Biomedical Engineering* 59 (4): 1125–1134.

77 Iasemidis, L.D., Shiau, D.-S., Sackellares, J.C. et al. (2004). A dynamical resetting of the human brain at epileptic seizures: application of nonlinear dynamics and global optimization techniques. *IEEE Transactions on Biomedical Engineering* 51 (3): 493–506.

78 Lehnertz, K. and Elger, C.E. (1995). Spatio-temporal dynamics of the primary epileptogenic area in temporal lobe epilepsy characterized by neuronal complexity loss. *Electroencephalography and Clinical Neurophysiology* 95: 108–117.

79 Lerner, D.E. (1996). Monitoring changing dynamics with correlation integrals: case study of an epileptic seizure. *Physica D* 97: 563–576.

80 Osorio, I., Harrison, M.A.F., Lai, Y.C., and Frei, M.G. (2001). Observations on the application of the correlation dimension and correlation integral to the prediction of seizures. *Journal of Clinical Neurophysiology* 18: 269–274.

81 Quyen, M.L.V., Martinerie, J., Baulac, M., and Varela, F.J. (1999). Anticipating epileptic seizures in real time by a non-linear analysis of similarity between {EEG} recordings. *NeuroReport* 10: 2149–2155.

82 Moser, H.R., Weber, B., Wieser, H.G., and Meier, P.F. (1999). Electroencephalogram in epilepsy: analysis and seizure prediction within the framework of Lyapunov theory. *Physica D* 130: 291–305.

83 Litt, B., Estellera, R., Echauz, J. et al. (2001). Epileptic seizures may begin hours in advance of clinical onset: a report of five patients. *Neuron* 30: 51–64.

84 D'Alessandro, M., Esteller, R., Vachtsevanos, G. et al. (2003). Epileptic seizure prediction using hybrid feature selection over multiple intracranial EEG electrode contacts: a report of four patients. *IEEE Transactions on Biomedical Engineering* 50: 603–615.

85 Iasemidis, L.D., Principe, J.C., and Sackellares, J.C. (2000). Measurement and quantification of spatio-temporal dynamics of human epileptic seizures. In: *Nonlinear Biomedical Signal Processing* (ed. M. Akay), 296–318. IEEE Press.

86 Sackellares, J.C., Iasemidis, L.D., Shiau, D.S. et al. (2000). Epilepsy-when chaos fails. In: *Chaos in the Brain?* (eds. K. Lehnertz and C.E. Elger), 112–133. Singapore: World Scientific.

87 Iasemidis, L.D., Shiau, D., Chaovalitwongse, W. et al. (2003). Adaptive epileptic seizure prediction system. *IEEE Transactions on Biomedical Engineering* 50: 616–627.

88 Hively, L.M., Protopopescu, V.A., and Gailey, P.C. (2000). Timely detection of dynamical change in scalp {EEG} signals. *Chaos* 10: 864–875.

89 Hively, L.M. and Protopopescu, V.A. (2003). Channel-consistent forewarning of epileptic events from scalp EEG. *IEEE Transactions on Biomedical Engineering* 50: 584–593.

90 Iasemidis, L., Principe, J., Czaplewski, J. et al. (1997). Spatiotemporal transition to epileptic seizures: a nonlinear dynamical analysis of scalp and intracranial EEG recordings. In: *Spatiotemporal Models in Biological and Artificial Systems* (eds. F. Silva, J. Principe and L. Almeida), 81–88. Amsterdam: IOS Press.

91 Sackellares, J., Iasemidis, L., Shiau, D. et al. (1999). Detection of the preictal transition from scalp EEG recordings. *Epilepsia* 40 (S7): 176.

92 Shiau, D., Iasemidis, L., Suharitdamrong, W. et al. (2003). Detection of the preictal period by dynamical analysis of scalp EEG. *Epilepsia* 44 (S9): 233–234.

93 Takens, F. (1981). Detecting strange attractors in turbulence. In: *Lectures Notes in Mathematics, Dynamical Systems and Turbulence, Warwick 1980* (eds. D.A. Rand and L.S. Young), 366–381. Berlin: Springer-Verlag.

94 Corsini, J., Shoker, L., Sanei, S., and Alarcon, G. (2006). Epileptic seizure predictability from scalp EEG incorporating constrained blind source separation. *IEEE Transactions on Biomedical Engineering* 53 (5): 790–799.

95 Fernandez, J., Alarcon, G., Binnie, C.D., and Polkey, C.E. (1999). Comparison of sphenoidal, foramen ovale and anterior temporal placements for detecting interictal epileptiform discharges in presurgical assessment for temporal lobe epilepsy. *Clinical Neurophysiology* 110: 895–904.

96 Nayak, D., Valentín, A., Alarcon, G. et al. (2004). Characteristics of scalp electrical fields associated with deep medial temporal epileptiform discharges. *Clinical Neurophysiology* 115: 1423–1435.

97 Margerison, J.H., Binnie, C.D., and McCaul, I.R. (1970). Electroencephalographic signs employed in the location of ruptured intracranial arterial aneurysms. *Electroencephalography and Clinical Neurophysiology* 28: 296–306.

98 Carney, P.R., Myers, S., and Geyer, J.D. (2011). Seizure prediction: methods. *Journal of Epilepsy and Behavior* 22: S94–S101.

99 Assi, E.B., Nguyenb, D.K., Rihanac, S., and Sawan, M. (2017). Towards accurate prediction of epileptic seizures: a review. *Journal of Biomedical Signal Processing and Control* 34: 144–157.

100 Kiral-Kornek, I., Roy, S., Nurse, E. et al. (2018). Epileptic seizure prediction using big data and deep learning: toward a mobile system. *Journal of EBioMedicine* 27: 103–111.

101 Weia, X., Zhoub, L., Zhangc, Z. et al. (2019). Early prediction of epileptic seizures using a long-term recurrent convolutional network. *Journal of Neuroscience Methods* 327: 108395.

102 Lodder, S.S. and van Putten, M.J.A.M. (2014). A self-adapting system for the automated detection of inter-ictal epileptiform discharges. *PLoS One* 9: e85180.

103 Zhanfeng, J., Sugi, T., Goto, S. et al. (2011). Multichannel template extraction for automatic EEG spike detection. *International Conference on Complex Medical Engineering, Harbin Heilongjiang*, 179–184. China.

104 Vijayalakshmi, K. and Abhishek, A.M. (2010). Spike detection in epileptic Patients' EEG data using template matching technique. *International Journal of Computers and Applications* 2: 5–8.

105 Zhou, J., Schalkoff, R.J., Dean, B.C., and Halford, J.J. (2012). Morphology based wavelet features and multiple mother wavelet strategy for spike classification in EEG signals.

International *Conference of the IEEE Engineering in Medicine and Biology Society*, 3959–3962.

106 Yinxia, L., Weidong, Z., Qi, Y., and Shuangshuang, C. (2012). Automatic seizure detection using wavelet transform and SVM in long-term intracranial EEG. *IEEE Transactions on Neural Systems and Rehabilitation Engineering* 20 (6): 749–755.

107 Makeyev, O., Liu, X., Luna-Munguia, H. et al. (2012). Toward a noninvasive automatic seizure control system in rats with transcranial focal stimulations via tripolar concentric ring electrodes. *IEEE Transactions on Neural Systems and Rehabilitation Engineering* 20: 422–431.

108 Ghosh-Dastidar, S., Adeli, H., and Dadmehr, N. (2007). Mixed-band wavelet-chaos-neural network methodology for epilepsy and epileptic seizure detection. *IEEE Transactions on Biomedical Engineering* 54 (9): 1545–1551.

109 Spyrou, L. and Sanei, S. (2016). Coupled dictionary learning for multimodal data: an application to concurrent intracranial and scalp EEG. *IEEE International Conference on Acoustics, Speech and Signal Processing*. 2349–2353.

110 Majumdar, K. and Vardhan, P. (2011). Automatic seizure detection in ECoG by differential operator and windowed variance. *IEEE Transactions on Neural Systems and Rehabilitation Engineering* 19 (4): 356–365.

111 Li, S., Zhou, W., Yuan, Q., and Liu, Y. (2013). Seizure prediction using spike rate of intracranial EEG. *IEEE Transactions on Neural Systems and Rehabilitation Engineering* 21 (6): 880–886.

112 Tzallas, A.T., Tsipouras, M.G., Tsalikakis, D.G. et al. (2009). Automated epileptic seizure detection methods: a review study. *Electroencephalographic and Psychological Aspects* (ed. D. Stevanovic). InTech Open http://www.intechopen.com/books/epilepsyhistologicalelectroencephalographic-andpsychological-aspects/automatedepileptic-seizuredetection-methods-a-review-study (accessed 19 August 2021).

113 Wilson, S.B. and Emerson, R. (2002). Spike detection: a review and comparison of algorithms. *IEEE Transactions on Neural Networks* 113: 1873–1881.

114 Wilson, S.B., Turner, C.A., Emerson, R.G., and Scheuer, M.L. (1999). Spike detection II: automatic, perception-based detection and clustering. *Clinical Neurophysiology* 110: 404–411.

115 Fernandez Torre, J., Alarcon, G., Binnie, C., and Polkey, C. (1999). Comparison of sphenoidal, foramen ovale and anterior temporal placements for detecting interictal epileptiform discharges in presurgical assessment for temporal lobe epilepsy. *Clinical Neurophysiology* 110: 895–904.

116 Spyrou, L., Martín-Lopez, D., Valentín, A. et al. (2016). Detection of intracranial signatures of interictal epileptiform discharges from concurrent scalp EEG. *International Journal of Neural Systems* 26 (4): 1650016. https://doi.org/10.1142/S0129065716500167.

117 Alarcon, G., Garcia Seoane, J.J., Binnie, C.D. et al. (1997). Origin and propagation of interictal discharges in the acute electrocorticogram. Implications for pathophysiology and surgical treatment of temporal lobe epilepsy. *Brain* 120 (1): 2259–2282.

118 Fernandez Torre, J.L., Alarcon, G., Binnie, C.D. et al. (1999). Generation of scalp discharges in temporal lobe epilepsy as suggested by intraoperative electrocorticographic recordings. *Journal of Neurology, Neurosurgery, and Psychiatry* 67 (1): 51–58.

119 von Ellenrieder, N., Beltrachini, L., Perucca, P., and Gotman, J. (2014). Size of cortical generators of epileptic interictal events and visibility on scalp EEG. *NeuroImage* 94: 47–54.

120 Ramantani, G., Dumpelmann, M., Koessler, L. et al. (2014). Simultaneous subdural and scalp EEG correlates of frontal lobe epileptic sources. *Epilepsia* 55 (2): 278–288.

121 Bultan, A. (1999). A four-parameter atomic decomposition of chirplets. *IEEE Transactions on Signal Processing* 47: 731–745.

122 Hastie, T., Tibshirani, R., and Friedman, J. (2009). *The Elements of Statistical Learning*. Springer.

123 Farquhar, J. and Hill, J. (2013). Interactions between preprocessing and classification methods for event related-potential classification: best-practice guidelines for brain-computer interfacing. *Neuroinformatics* 11: 175–192.

124 Schaffer, C. (1993). Selecting a classification method by cross-validation. *Machine Learning* 13 (1): 135–143.

125 Blokland, Y., Spyrou, L., Thijssen, D. et al. (2014). Combined EEG-fNIRS decoding of motor attempt and imagery for brain switch control: an offline study in patients with tetraplegia. *IEEE Transactions on Neural Systems and Rehabilitation Engineering* 22 (2): 1–8.

126 Spyrou, L., Kouchaki, S., and Sanei, S. (2018). Multiview classification and dimensionality reduction of EEG data through tensor factorisation. *Journal of Signal Processing Systems* 90: 273–284. https://doi.org/10.1007/s11265-016-1164-z.

127 Spyrou, L. and Sanei, S. (2016). Coupled dictionary learning for multimodal data: an application to concurrent intracranial and scalp EEG. *Proceedings of the IEEE International Conference on Acoustics, Speech and Signal Processing*, 2349–2353. Shanghai, China.

128 Yang, J., Wright, J., Huang, T.S., and Ma, Y. (2010). Image super-resolution via sparse representation. *IEEE Transactions on Image Processing* 19: 2861–2873.

129 Antoniades, A., Spyrou, L., Martín-Lopez, D. et al. (2018). Deep neural architectures for mapping scalp to intracranial EEG. *International Journal of Neural Systems* 28 (8): 1850009. https://doi.org/10.1142/S0129065718500090.

130 Antoniades, A., Spyrou, L., Martín-Lopez, D. et al. (2017). Detection of interictal discharges using convolutional neural networks from multichannel intracranial EEG. *IEEE Transactions on Neural Systems and Rehabilitation Engineering* 25 (12): 2285–2294.

131 Menon, R.S. (2002). Postacquisition suppression of large-vessel BOLD signals in high-resolution fMRI. *Magnetic Resonance in Medicine* 47 (1): 1–9.

132 Huettel, S.A., Song, A.W., and McCarthy, G. (2004). *Functional Magnetic Resonance Imaging*. Sunderland, MA: Sinauer Associates.

133 Toga, A.W. and Mazziotta, J.C. (2002). *Brain Mapping: The Methods*, 2e. San diego, CA: Academic.

134 Grouiller, F., Thornton, R., Groening, K. et al. (2011). With or without spikes: localization of focal epileptic activity by simultaneous electroencephalography and functional magnetic resonance imaging. *Brain* 134: 2867–2886.

135 Bai, X., Vestal, M., Berman, R. et al. (2010). Dynamic time course of typical childhood absence seizures: EEG, behavior, and functional magnetic resonance imaging. *The Journal of Neuroscience* 30 (17): 5884–5893.

12

Sleep Recognition, Scoring, and Abnormalities

12.1 Introduction

12.1.1 Definition of Sleep

Sleep has been described by early scientists as a passive condition where the brain is isolated from the rest of the body. Alcmaeon claimed that sleep is caused by the blood receding from the blood vessels in the skin to the interior parts of the body. Philosopher and scientist Aristotle suggested that while food is being digested, vapours rise from the stomach and penetrate into the head. As the brain cools, the vapours condense, flow downward, and then cool the heart which causes sleep. Some others still claim that toxins that poison the brain cause sleep [1]. With the discovery of brain waves and later the discovery of electroencephalogram (EEG) system, the way sleep was studied changed. EEG enables a researcher to record the detailed electrical activity of the brain during sleep.

Sleep is the state of natural rest observed in humans and animals, and even in vertebrates such as the fruit fly Drosophila. Lack of sleep seriously influences our brain's ability to function. With continued lack of sufficient sleep, the parts of the brain that control language, memory, planning, and sense of time are severely affected whereby their judgement deteriorates. Sleep is an interesting and not perfectly known physiological phenomenon. The sleep state has provided important evidence for diagnosing mental disease and psychological abnormality. Sleep is characterized by a reduction in voluntary body movement, decreased reaction to external stimuli, an increased rate of anabolism (the synthesis of cell structures), and a decreased rate of catabolism (the breakdown of cell structures). Hence, sleep is necessary for the life of most creatures and is a basic human need and is critical to both physical and mental health. The capability for arousal from sleep is a protective mechanism and also necessary for health and survival. In terms of physiological changes in the body and particularly changes in the state of the brain sleep is different from unconsciousness [2]. However, in manifestation, sleep is defined as a state of unconsciousness from which a person can be aroused. In this state, the brain is relatively more responsive to internal stimuli than external stimuli. Sleep should be distinguished from coma. Coma is an unconscious state from which a person cannot be aroused.

Although historically, sleep was thought to be a passive state, it is now known to be a dynamic process, and our brains are active during sleep. The quality of sleep affects our physical and mental health and the immune system.

EEG Signal Processing and Machine Learning, Second Edition. Saeid Sanei and Jonathon A. Chambers.

States of brain activity during sleep and wakefulness result from different activating and inhibiting forces that are generated within the brain. Neurotransmitters (chemicals involved in nerve signalling) control whether one is asleep or awake by acting on nerve cells (neurons) in different parts of the brain. Neurons located in the brainstem actively cause sleep by inhibiting other parts of the brain that keep a person awake.

For a human being, it has been demonstrated that the metabolic activity of the brain decreases significantly after 24 hours of sustained wakefulness. Sleep deprivation results in a decrease in body temperature, a decrease in immune system function as measured by white blood cell count (the soldiers of the body), and a decrease in the release of growth hormone. Sleep deprivation can also cause increased heart rate variability [3].

Sleep is necessary for the brain to remain healthy. Sleep deprivation makes a person drowsy and unable to concentrate. It also leads to impairment of memory and physical performance and reduced ability to carry out mathematical calculations and other mental tasks. If sleep deprivation continues, hallucinations and mood swings may develop. Sleep deprivation not only has a major impact on cognitive functioning but also on emotional and physical health. Disorders such as sleep apnoea which result in excessive daytime sleepiness have been linked to stress and high blood pressure. Research has also suggested that sleep loss may increase the risk of obesity because chemicals and hormones that play a key role in controlling appetite and weight gain are released during sleep.

Release of growth hormone in children and young adults takes place during deep sleep. Most cells of the body show increased production and reduced breakdown of proteins during deep sleep. Sleep helps humans maintain optimal emotional and social functioning while we are awake by giving rest during sleep to the parts of the brain that control emotions and social interactions.

12.1.2 Sleep Disorder

Lack of sleep and too much sleep are linked to many chronic health problems, such as heart disease and diabetes. Sleep disturbances can also be a warning sign for medical and neurological problems, such as congestive heart failure, osteoarthritis and Parkinson's disease.

Sleep disorders involve problems with the quality, timing, and amount of sleep, which cause problems with functioning and distress during the daytime. There are several different types of sleep disorders, of which insomnia is the most common. Insomnia is also considered as a psychiatric disorder and therefore it is discussed in Chapter 16 of this book too. Other sleep disorders are narcolepsy, obstructive sleep apnoea (OSA) and restless leg syndrome.

Sleep difficulties are linked to both physical and emotional problems. Sleep problems can both contribute to or exacerbate mental health conditions and be a symptom of other mental health conditions.

Not all sleep abnormalities have been investigated through the study of EEG but with other screening modalities. At the end of this chapter we address some sleep disorders monitored using a multimodal approach called polysomnography (PSG) which has been a well established method for sleep analysis. The updates of the American Academic of Sleep Medicine (AASM) initiated in 2007 [4] are mainly based on the PSG parameters. The latest updates can be seen in the AASM website (https://aasm.org/clinical-resources/scoring-manual).

12.2 Stages of Sleep

Sleep is a dynamic process. Loomis provided the earliest detailed description of various stages of sleep in the mid-1930s, and in the early 1950s Aserinsky and Kleitman identified rapid eye movement (REM) sleep [2]. There are two distinct states that alternate in cycles and reflect differing levels of neuronal activity. Each state is characterized by a different type of EEG activity. Generally, sleep consists of non-rapid eye movement (NREM) and REM sleep states. NREM is further subdivided into four stages of I (Drowsiness), II (Light Sleep), III (Deep Sleep) and IV (Very Deep Sleep).

During the night the NREM and REM stages of sleep alternate. Stages I, II, III, and IV are followed by REM sleep. A complete sleep cycle, from the beginning of stage I to the end of REM sleep, usually takes about one and a half hours. However, generally, the ensuing sleep is relatively short and, in practise, often 10–30 minutes duration suffices.

12.2.1 NREM Sleep

The first stage, *Stage* I, is the stage of drowsiness and very light sleep, which is considered as a transition between wakefulness and sleep. During this stage, the muscles begin to relax. It occurs upon falling asleep and during brief arousal periods within sleep, and usually accounts for 5–10% of total sleep time. An individual can be easily awakened during this stage. Drowsiness shows marked age-determined changes. Hypnagogic rhythmical (4–6 waves s^{-1}) theta activity of late infancy and early childhood is a significant characteristic of such ages. Later in childhood, and in several cases, in the declining years of life, the drowsiness onset involves larger amounts of slow activity mixed with the posterior alpha rhythm [5]. In adults however, the onset of drowsiness is characterized by gradual or brisk alpha dropout [5]. The slow activity increases as the drowsiness becomes deeper. Other findings show that in light drowsiness the P300 response increases in latency and decreases in amplitude [6], and the inter- and intra-hemispheric EEG coherence alter [7]. Figure 12.1 shows a set of EEG signals recorded during the state of drowsiness. The seizure type activity within the signal is very clear.

Deep drowsiness involves appearance of vertex waves. Before the appearance of the first spindle trains, vertex waves occur (transition from stages I to II). These sharp waves are also known as parietal humps [8]. The vertex wave is a compound potential; a small spike discharge of positive polarity followed by a large negative wave, which is a typical discharge wave. It may occur as an isolated event with larger amplitude than that of normal EEG. In aged individuals they may become small, inconspicuous, and hardly visible. Another signal feature for deep drowsiness is the positive occipital sharp transients of sleep (POST).

Spindles (also called sigma activity), the trains of barbiturate-induced beta activity, occur independently in approximately 18–25 Hz predominantly in the frontal lobe of the brain. They may be identified as a 'group of rhythmic waves characterized by progressively increasing, then gradually decreasing amplitude' [5]. However, the use of middle electrodes shows a very definite maximum of the spindles over the vertex during the early stages of sleep.

Stage II of sleep occurs throughout the sleep period and represents 40–50% of the total sleep time. During stage II, brain waves slow down with occasional bursts of rapid waves.

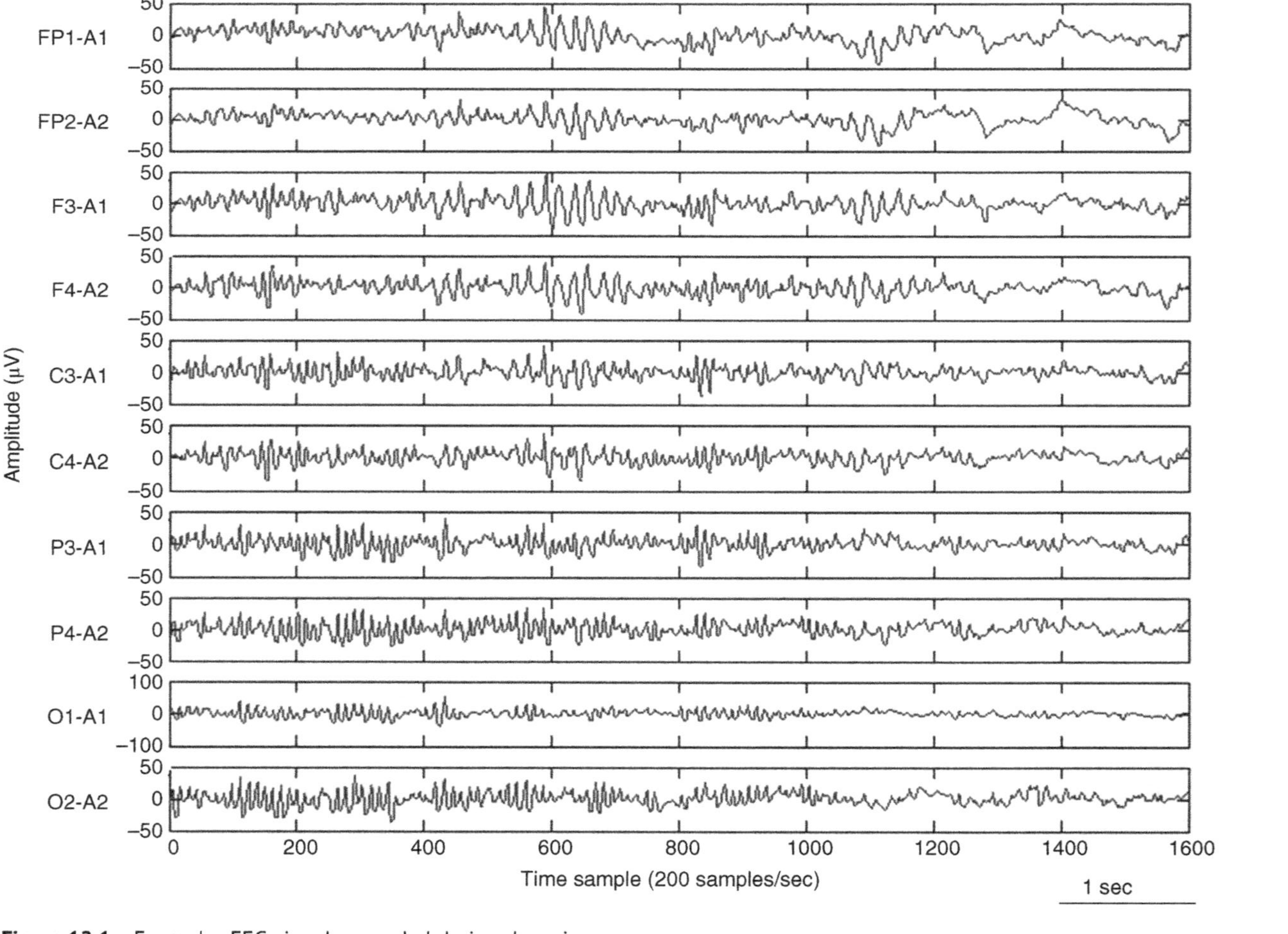

Figure 12.1 Exemplar EEG signals recorded during drowsiness.

Eye movement stops during this stage. Slow frequencies ranging from 0.7 to 4 Hz are usually predominant; their voltage is high with a very prominent occipital peak in small children and gradually falls when they become older.

K-complexes appear in stage II and constitute a significant response to arousing stimuli. As to the topographical distribution over the brain, the K-complex shows a maximum over the vertex and has presence around the frontal midline [5]. As to the wave morphology, the K-complex consists of an initial sharp component, followed by a slow component that fuses with a superimposed fast component.

In Stage III, delta waves begin to appear. They are interspersed with smaller, faster waves. Sleep spindles are still present in approximately 12–14 Hz but gradually disappear as the sleep becomes deeper. Figure 12.2 illustrates the brain waves during this stage of sleep.

In Stage IV, delta waves are the primary waves recorded from the brain. Delta or slow-wave sleep (SWS) usually is not seen during routine EEG [9]. However, it is seen during prolonged (>24 hours) EEG monitoring.

Stages III and IV are often distinguished from each other only by the percentage of delta activity. Together, they represent up to 20% of total sleep time. During Stages III and IV all eye and muscle movement ceases. It is difficult to wake up someone during these two stages. If someone is awakened during deep sleep; he does not adjust immediately and often feels groggy and disoriented for several minutes after waking up. Generally, analysis of EEG morphology during Stage IV has been of less interest since the brain functionality cannot be examined easily.

12.2.2 REM Sleep

REM sleep including 20–25% of the total sleep follows NREM sleep and occurs four to five times during a normal eight- to nine-hour sleep period. The first REM period of the night may be less than 10 minutes in duration, while the last period may exceed 60 minutes.

In an extremely sleepy individual, the duration of each bout of REM sleep is very short or it may even be absent. REM sleep is usually associated with dreaming. During REM sleep, the eyeballs move rapidly, the heart rate and breathing become rapid and irregular, the blood pressure rises, and there is loss of muscle tone (paralysis), i.e. the muscles of the body are virtually paralyzed. The brain is highly active during REM sleep, and the overall brain metabolism may be increased by as much as 20%. The EEG activity recorded in the brain during REM sleep is similar to that recorded during wakefulness.

In a patient with REM sleep behaviour disorder (RBD), the paralysis is incomplete or absent, allowing the person to act out his dreams which can be vivid, intense, and violent. These dream-acting behaviours include talking, yelling, punching, kicking, sitting, jumping from the bed, arm flailing and grabbing. Although the RBD may occur in association with different degenerative neurological conditions the main cause is still unknown.

Evaluation of REM sleep involves a long waiting period since the first phase of REM does not appear before 60–90 minutes after the start of sleep. The EEG in the REM stage shows low voltage activity with slower rate of alpha. Figure 12.3 shows the brain waves during REM sleep.

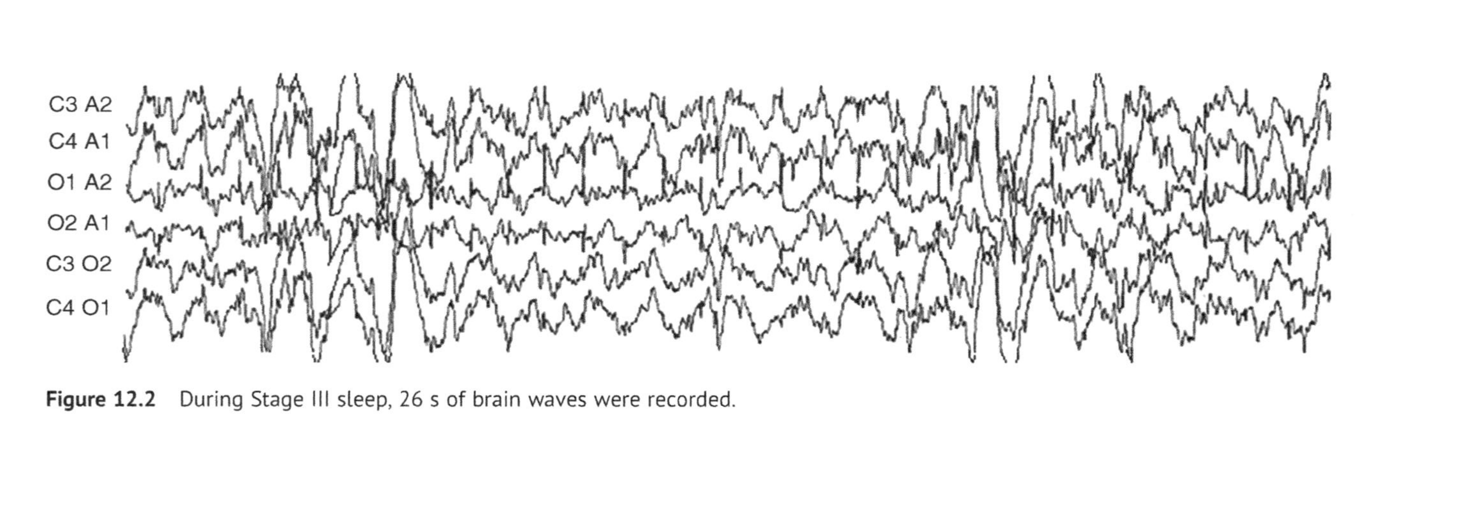

Figure 12.2 During Stage III sleep, 26 s of brain waves were recorded.

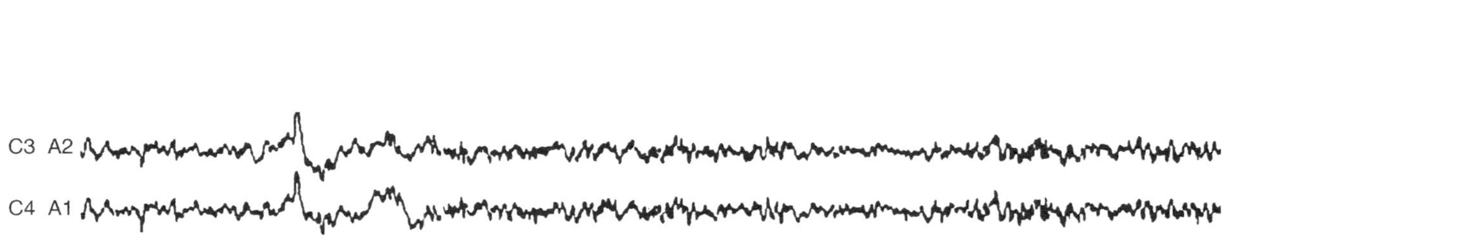

Figure 12.3 Twenty-six seconds of brain waves recorded during the REM state.

12.3 The Influence of Circadian Rhythms

Biological variations that occur in the course of 24 hours are called circadian rhythms. Circadian rhythms are controlled by the biological clock of the body. Many bodily functions follow the biological clock, but sleep and wakefulness comprise the most important circadian rhythm. Circadian sleep rhythm is one of the several body rhythms modulated by hypothalamus.

Light directly affects the circadian sleep rhythm. Light is called *zeitgeber*, a German word meaning time-giver, because it sets the biological clock.

Body temperature cycles are also under control of the hypothalamus. An increase in body temperature is seen during the course of the day and a decrease is observed during the night. The temperature peaks and troughs are thought to mirror the sleep rhythm. People who are alert late in the evening (i.e. evening types) have body temperature peaks late in the evening, while those who find themselves most alert early in the morning (i.e. morning types) have body temperature peaks early in the morning.

Melatonin (a chemical produced by the pineal gland in the brain and a hormone associated with sleep) has been implicated as a modulator of light entrainment. It is secreted maximally during the night. Prolactin, testosterone, and growth hormone also demonstrate circadian rhythms, with maximal secretion during the night. Figure 12.4 shows a typical concentration of melatonin in a healthy adult man.

Sleep and wakefulness are influenced by different neurotransmitters in the brain. Some substances can change the balance of these neurotransmitters and affect our sleep and wakefulness. Caffeinated drinks (for example, coffee) and medicines (for example, diet pills) stimulate some parts of the brain and can cause difficulty in falling asleep. Many drugs prescribed for the treatment of depression suppress REM sleep.

Heavy smokers who smoke heavily often sleep very lightly and have reduced duration of REM sleep. They tend to wake up after three or four hours of sleep due to nicotine

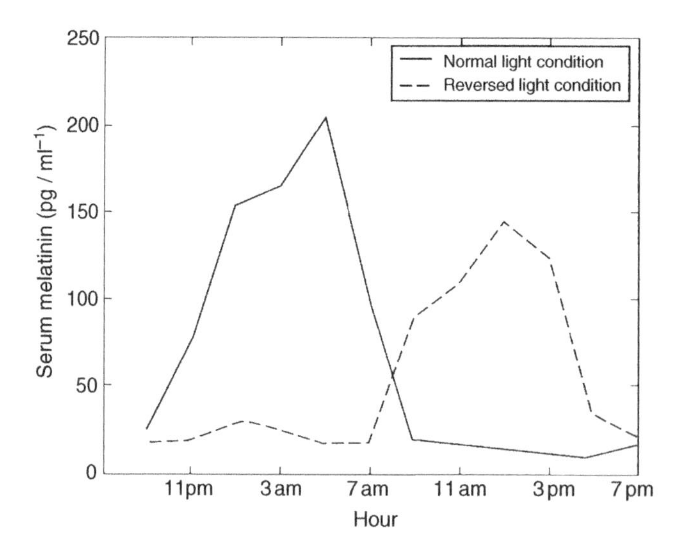

Figure 12.4 A typical concentration of melatonin in a healthy adult man (extracted from [10]).

withdrawal. Some people who have insomnia may use alcohol. Even though alcohol may help people to fall into light sleep, it deprives them of REM sleep and the deeper and more restorative stages of sleep. Alcohol keeps them in the lighter stages of sleep from which they can be awakened easily. During REM sleep, we lose some of our ability to regulate our body temperature. Therefore, abnormally hot or cold temperatures can disrupt our REM sleep. If our REM sleep is disturbed, the normal sleep cycle progression is affected during the next sleeping time, and there is a possibility of slipping directly into REM sleep and going through long periods of REM sleep until the duration of REM sleep that is lost is caught up.

Generally, sleep disruption by any cause can be a reason for an increase in seizure frequency or severity. It can also have negative effect on short-term memory, concentration, and mood. Seizure, itself, during the night can disrupt sleep. Also, using any anticonvulsant drug may affect sleep in different ways. Approximately 90 different sleep disorders, including snoring, obstructive sleep apnoea hypopnea syndrome (OSAHS), insomnia, narcolepsy, bruxism, restless leg syndrome have been reported by the International Classification of Sleep Disorder [11].

Both the frequency of seizure and the locality of seizure sources within the brain may change in different sleep stages and wakefulness.

12.4 Sleep Deprivation

Sleep deprivation is evaluated in terms of the tasks impaired and the average duration. In tasks requiring judgement, increasingly risky behaviours emerge as the total sleep duration is limited to five hours per night. The high cost of an action is seemingly ignored as the sleep-deprived person focuses on limited benefits. These findings can be explained by the fact that metabolism in the prefrontal and parietal associational areas of the brain decrease in individuals deprived of sleep for 24 hours. These brain areas are important for judgement, impulse control, attention, and visual association.

Sleep deprivation is a relative concept. Small amounts of sleep loss (for example, one hour per night over many nights) produce subtle cognitive impairment, which may go unrecognized. More severe restriction of sleep for a week leads to profound cognitive deficits, which may also go unrecognized by the individual. If one feels drowsy during the day, falls asleep for very short periods of time (five minutes or so), or regularly falls asleep immediately after lying down, then that individual is probably sleep-deprived.

Many studies have made it clear that sleep deprivation is dangerous. With decreased sleep, higher-order cognitive tasks are impaired early and disproportionately. Sleep-deprived people perform as poorly as or worse than people who are intoxicated on tasks used for testing coordination,. Total sleep duration of seven hours per night over one week has resulted in decreased speed in tasks of both simple reaction time and more demanding computer-generated mathematical problem solving. Total sleep duration of five hours per night over one week shows both a decrease in speed and the beginning of accuracy failure.

Using sleep deprivation for detection and diagnosis of some brain abnormalities has been reported by some researchers [12–14]. It consists of sleep loss for 24–26 hours. This was used by Klingler et al. [15] to detect the epileptic discharges that could otherwise be missed.

Based on these studies it has also been concluded that sleep depravation is a genuine activation method [16]. Its efficacy in provoking abnormal EEG discharges is not due to drowsiness. Using the information in Stage III of sleep, the focal and generalized seizure may be classified [17].

12.5 Psychological Effects

In the majority of sleep measurements and studies EEG has been used in combination with a variety of other physiological parameters. Insomnia is probably the most popular psychological effect which directly influence the quality of sleep and the changes in NREM/REM distribution overnight. Moreover, EEG studies have documented abnormalities in sleep patterns in psychiatric patients with suicidal behaviour, including longer sleep latency, increased REM time, and increased phasic REM activity. Sabo et al. [18] compared sleep EEG characteristics of adult depressives with and without a history of suicide attempts and noted that those who attempted suicide had consistently more REM time and phasic activity in the second REM period but less delta wave counts in the fourth non-REM period. Another study [19] conducted at the same laboratory replicated the findings with psychotic patients. On the basis of two studies, the authors [19] suggest that the association between REM sleep and suicidality may cut across diagnostic boundaries and that sleep EEG changes may have a predictive value for future suicidal behaviour. REM sleep changes were later replicated by other studies in suicidal schizophrenia [20] and depression [21].

Three cross-sectional studies examined the relationship between sleep EEG and suicidality in depressed adolescents. Dahl et al. [22] compared sleep EEG between a depressed suicidal group, a depressed nonsuicidal group, and normal controls. Their results indicated that suicidal depressed patients had significantly prolonged sleep latency and increased REM phasic activity with a trend for reduced REM latency compared to both nonsuicidal depressed and control groups. Goetz et al. [23] and McCracken et al. [24] replicated the finding of greater REM density among depressive suicidal adolescents.

Study of normal ageing and transient cognitive disorders in the elderly has also shown that the most frequent abnormality in the EEG of elderly subjects is slowing of alpha frequency whereas most healthy individuals maintain alpha activity within 9–11 Hz [25, 26].

12.6 EEG Sleep Analysis and Scoring

12.6.1 Detection of the Rhythmic Waveforms and Spindles Employing Blind Source Separation

After the EEGs are recorded, to analyze and monitor sleep disorders the main stage is detection of the waveforms corresponding to different stages of sleep. In order to facilitate recording of the EEGs during sleep with a small number of electrodes a method to best select the electrode positions and remove the effects of electrocardiogram (ECG), electro-oculogram (EOG), and electromyogram (EMG) may be useful [21].

Recording over a long period is often necessary to investigate adequately the sleep signals and establish a night sleep profile, the electrophysiological activities manifested within the above signals have to be recognized and studied. A blind source separation (BSS) algorithm is sought to separate the four desired signals of EEG, two EOGs, and one ECG. In order to maintain the continuity of the estimated sources in the consecutive blocks of data the estimated independent components have been cross-correlated with the electrode signals and those of consecutive signal segments most correlated with each particular electrode signal are considered the segments of the same source. As a result of this work the alpha activity may not be seen consistently since the BSS system is generally underdetermined and therefore it cannot separate alpha activity from the other brain activities. However, the EMG complexes and REM are noticeable in the separated sources.

12.6.2 Time–Frequency Analysis of Sleep EEG Using Matching Pursuit

Some extensions to the Rechtschaffen and Kales (R&K) system using the conventional EEG signal processing methods have been proposed by Malinowska et al. [27]. These extensions include a finer timescale than the division into 20–30 seconds epochs, a measure of spindle intensity, and the differentiation of single and randomly evoked K-complexes in response to stimuli from spontaneous periodic ones. Figure 12.5 illustrates some typical spindles and K-complex waveforms.

The adaptive time–frequency (T–F) approximation of signals using matching pursuit (MP) introduced initially by Mallat and Zhang [29] has been used in developing a method to investigate the above extensions [27]. MP has been reviewed in Chapter 2 of this book. The required dictionary of waveforms consists of Gabor functions mainly because these functions provide optimal joint T-F localization [30]. Real valued continuous-time Gabor functions can be represented as:

$$g_\gamma(t) = K(\gamma)e^{-\pi\left(\frac{t-u}{s}\right)^2} \cos\left(\omega(t-u) + \phi\right) \tag{12.1}$$

where $K(\gamma)$ is a normalizing factor of g_γ and the parameters of the Gabor function $\gamma = \{u, \omega, s\}$ provide a three-dimensional continuous space from which a finite dictionary must be chosen. In this application these parameters are drawn from uniform distributions over the signal range and correspond to the dictionary size. These parameters are fitted to the signal by the MP algorithm and often are directly used for analysis.

Due to spectral and temporal variation of EEG during the changes in the state of sleep, Gabor functions are used to exploit the changes in the above parameters, especially the T-F amplitude and phase indicators.

In this work the deep-sleep stages (III and IV) are detected from the EEG signals based on the classical R&K criteria; derivation of the continuous description of SWS fully compatible with the R&K criteria has been attempted, a measure of spindling activity has been followed, and finally a procedure for detection of arousal has been presented. MP has been shown to well separate various waveforms within the sleep EEGs. Assuming the slow-wave activity (SWA) and sleep spindle to have the characteristics in Table 12.1, they can be automatically detected by MP decomposition.

As the result of decomposition and examination of the components, if between 20 and 50% of the duration of an epoch is occupied by SWA it corresponds to Stage III, and if above

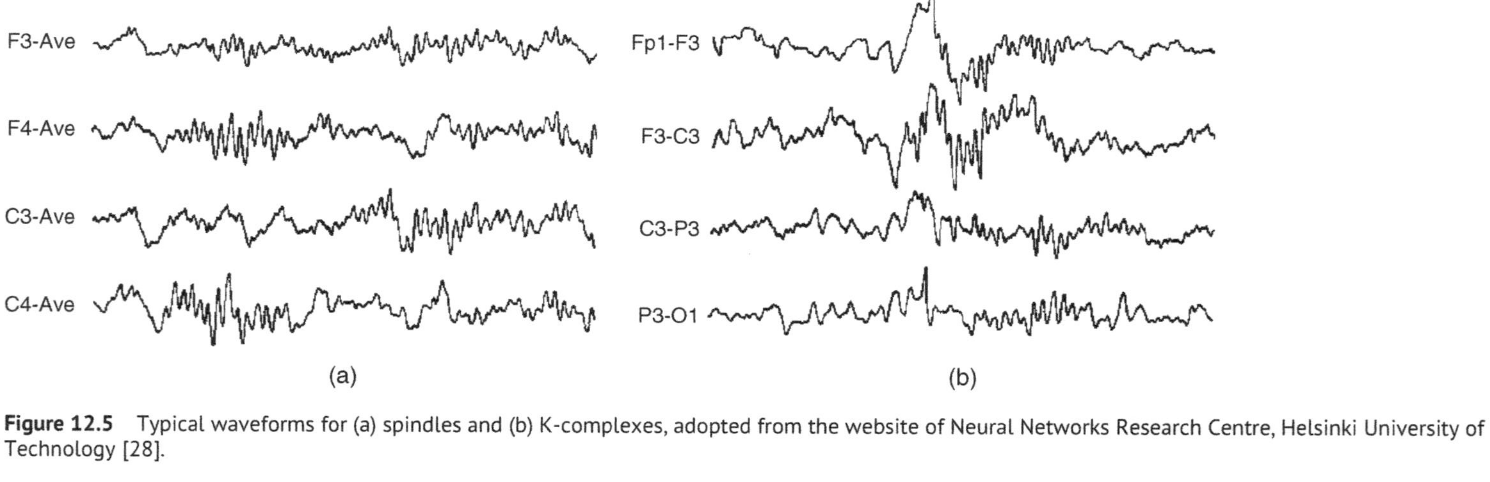

Figure 12.5 Typical waveforms for (a) spindles and (b) K-complexes, adopted from the website of Neural Networks Research Centre, Helsinki University of Technology [28].

Table 12.1 The time, frequency, and amplitudes of both SWA and sleep spindles [27].

	Time Duration	Frequency	Min. Amplitude
SWA	0.5 - ∞ s	0.5–4 Hz	$0.99 \times V_{EEG} + 28.18$ μV
Sleep Spindles	0.5–2.5 s	11–15 Hz	15 μV

50% of the duration of an epoch is occupied by SWA it corresponds to Stage IV. Therefore, Stages III and IV of sleep can be recognized by applying the R&K criteria.

The first approach to the automatic detection of arousals was based upon the MP decomposition of only the C3 –A2 single EEG channel and the standard deviation of EMG by implementing the rules established by the American Sleep Disorders Association (ASDA) [31]. The MP structure shows the frequency shifts within different frequency bands. Such shifts lasting three seconds or longer are related to arousal. To score a second arousal a minimum of ten seconds of intervening sleep is necessary [27]. In Figure 12.6 the results of applying the MP algorithm using Gabor functions for the detection of both rhythms and transients can be viewed. Each blob in the TF energy map corresponds to one Gabor function. The 8–12 Hz alpha wave, sleep spindles (and one K-complex), and SWA of stages II and IV are respectively presented in Figure 12.6 (a), (b), (c), and (d).

The sleep spindles exhibited inversely relate to the SWA [32]. The detected arousals decrease in relation to the amount of light NREM sleep, with particular concentration before the REM episodes [33].

The MP algorithm has also been used in the differentiation of single randomly evoked K-complexes in response to stimuli from spontaneous periodic ones [27].

The tools and algorithms developed for recognition and detection of sleep stages can be applied to diagnosis of many sleep disorders such as apnoea and the disturbances leading to arousal.

The MP-based time–frequency domain analysis was later improved by filtering the signals using singular spectrum analysis (SSA) before feeding the signal into the time–frequency analysis block [34]. During the SSA operation on a single EEG channel of C3-A2: (i) the eigentriples with the eigenvalue λ_j is rejected if $j > \eta$ where:

$$\eta = \min\left\{ h : \frac{\sum_{i=1}^{h} \lambda_i}{\sum_{i=1}^{M} \lambda_i} \right\} \tag{12.2}$$

where M is the total number of eigenvalues or embedding dimension; and (ii) among the remaining eigenvalues only those with a pair of similar amplitudes representing sinusoidal components are selected. After applying the MP-based method a multiclass support vector machine (SVM) is used to classify the sleep stage. A block diagram of the overall system is depicted in Figure 12.7.

The scoring result compared with manual scoring by a sleep expert can be seen in Figure 12.8, which is reasonably accurate and acceptable for being used in clinic.

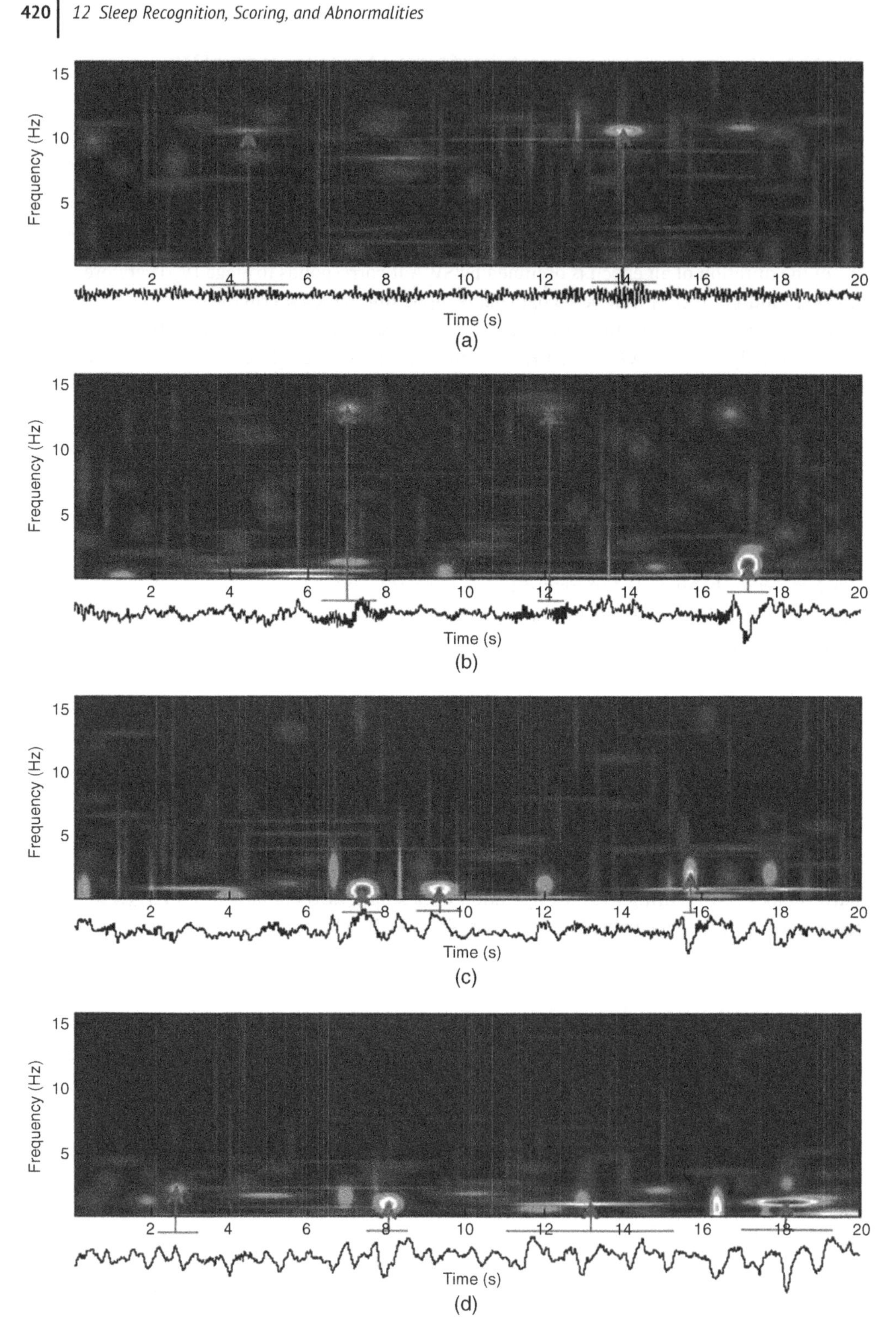

Figure 12.6 Time–frequency energy map of 20 seconds epochs of sleep EEG in different stages. The arrows point to the corresponding blobs for (a) awake alpha, (b) spindles and K-complex related to stage II, and (c) and (d) SWAs related, respectively, to Stages III and IV [27].

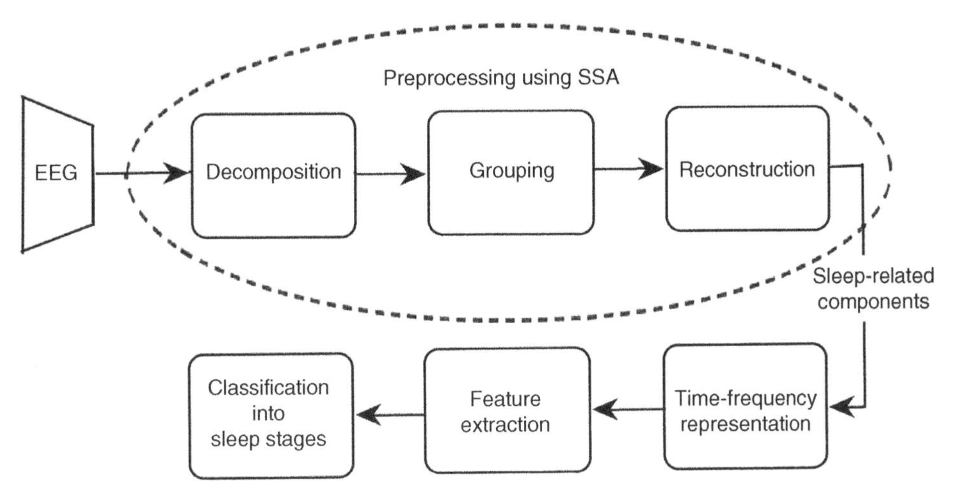

Figure 12.7 Block diagram of the sleep scoring system proposed in [34].

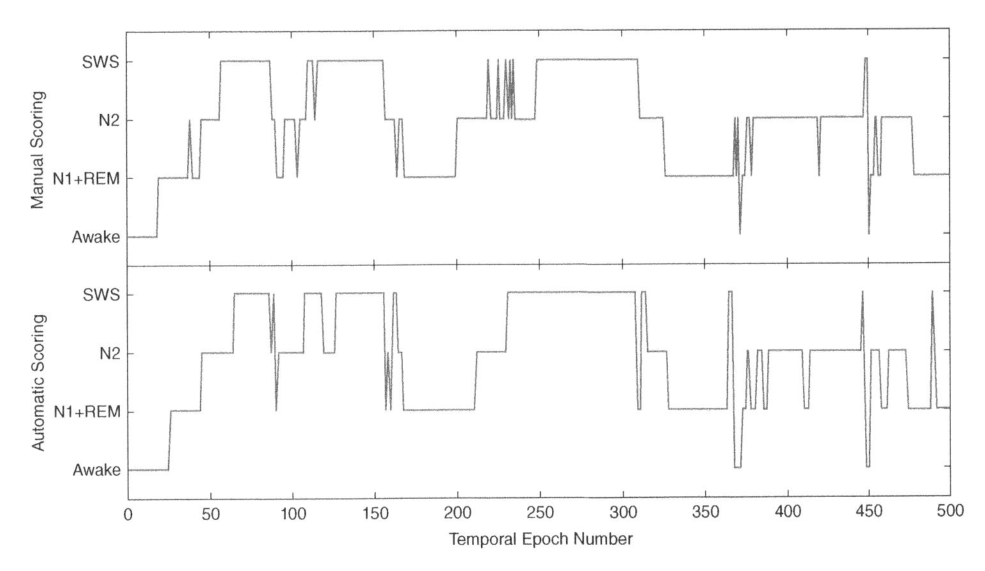

Figure 12.8 The scoring result of the proposed system in [34] (bottom) compared with manual scoring by a sleep expert (top).

12.6.3 Detection of Normal Rhythms and Spindles Using Higher-Order Statistics

Use of long-term spectrum analysis for detection and characterization of sleep EEG waveforms has been very popular [35, 36]. However, these methods are unable to detect transient and isolated characteristic waves such as humps and K-complexes accurately.

In one approach, higher-order statistics (HOS) of the time domain signals together with the spectra of the EEG signals during the sleep have been utilized to characterize the dynamics of sleep spindles [37]. The spindles are considered as periodic oscillations with steady-state behaviour which can be modelled as a linear system with sinusoidal input, or nonlinear system with a limit cycle.

In this work, second and third-order correlations of the time domain signals are combined to determine the stationarity of periodic spindle rhythms to detect transitions between multiple activities. The spectra (normalized spectrum and bispectrum) of the signals, conversely, describe frequency interactions associated with nonlinearities occurring in the EEGs.

The power spectrum of the wide sense stationary discrete signal, $x(n)$, is the power spectrum of its autocorrelation function given by:

$$P(\omega) = \sum_{n=-\infty}^{\infty} R(n)e^{-jn\omega} \cong \frac{1}{N}\sum_{i=1}^{N} X_i(\omega)X_i^*(\omega) \tag{12.3}$$

where $X_i(\omega)$ is the Fourier transform of the ith segment of one EEG channel. Also, the bispectrum of data is defined as:

$$B(\omega_1, \omega_2) = \sum_{n_1=-\infty}^{\infty}\sum_{n_2=-\infty}^{\infty} x(n)x(n+n_1)x(n+n_2)e^{-j(\omega_1 n_1 + \omega_2 n_2)}$$

$$\cong \frac{1}{N}\sum_{i=1}^{N} X_i(\omega_1)X_i(\omega_2)X_i(\omega_1+\omega_2) \tag{12.4}$$

where N is the number of segments of each EEG channel. Using Eqs. (12.2) and (12.3), a normalized bispectrum (also referred to as bicoherence, second-order coherency, or bicoherency index) is defined as [38]:

$$b^2(\omega_1, \omega_2) = \frac{|B(\omega_1, \omega_2)|^2}{P(\omega_1)P(\omega_2)P(\omega_1+\omega_2)} \tag{12.5}$$

which is an important tool for evaluating signal nonlinearities [39]. This measure (and the measure in (12.4)) has been widely used for detection of coupled periodicities. Eq. (12.5) acts as the discriminant of a linear process from a nonlinear one. For example, b^2 is constant for either linear systems [39] or fully coupled frequencies [40] and $b^2 = 0$ for either Gaussian signals or random phase relations where no quadratic coupling occurs. Coupling of the frequencies occurs when the values of normalized bispectrum vary between zero and one ($0 < b^2 < 1$). The coherency value of one refers to quadratic interaction and an approximate zero value refers to either low or absent interactions [38].

To find the spindle periods, another method similar to the average magnitude difference function (AMDF) algorithm, often used for pitch detection from speech signals, has been applied to the short intervals of the EEG segments. The procedure has been applied to both second-order and third-order statistical measures as [37]:

$$D_n(k) = 1 - \frac{\gamma_n(k)}{\sigma_{\gamma_n}} \tag{12.6}$$

where

$$\gamma_n(k) = \sum_{m=-\infty}^{\infty} |x(n+m)w(m) - x(n+m-k)w(m-k)| \tag{12.7}$$

and $\sigma_{\gamma_n} = \sqrt{\sum_i \gamma_n^2(i)}$ is the normalization factor and $w(m)$ is the window function, and

$$Q_n(k) = 1 - \frac{\phi_n(k)}{\sigma_{\phi_n}} \tag{12.8}$$

where

$$\phi_n(k) = \sum_{m=-\infty}^{\infty} |q(n+m)w(m) - q(n+m-k)w(m-k)| \tag{12.9}$$

and $q(n)$ is the inverse two-dimensional Fourier transform of the bispectrum and $\sigma_{\varphi_n} = \sqrt{\sum_i \varphi_n^2(i)}$. The measures are used together to estimate the periodicity of the spindles. For purely periodic activities we expect these estimates to give similar results. In this case Eq. (12.8) manifests peaks (as in AMDF) where the first peak denotes the spindle frequency.

Based on this investigation in summary it has been shown that (*i*) spindle activity may not uniformly dominate all regions of the brain, (*ii*) during the spindle activity frontal recordings still exhibit rich mixtures in frequency content and coupling. Conversely, a poor coupling may be observed at the posterior regions while showing dominant activity of the spindles, and (*iii*) it is concluded that the spindle activity may be modelled using at least second-order nonlinearity.

12.6.4 Sleep Scoring Using Tensor Factorization

A new supervised approach for decomposition of single-channel sleep EEG is introduced in [41]. The performance of the traditional SSA algorithm is significantly improved by applying tensor decomposition instead of a traditional singular value decomposition (SVD). In this approach, the inherent frequency diversity of the data has also been effectively exploited to highlight the subspace of interest.

Traditional SSA explained in Chapter 5 does not exploit the inherent nonstationarity and therefore may fail in actual data decomposition. Tensor-based SSA (TSSA) has been proposed as a solution to this problem [41]. Like SSA, the first stage of TSSA includes an embedding operation followed by a tensor decomposition method instead of SVD. In the embedding stage a 1-D time series $\mathbf{x}$ with length n is mapped into tensor $\underline{\mathbf{X}}$ To do that, first $\mathbf{x}$ is segmented using a nonoverlapping window of size l and a $\lfloor n/l \rfloor \times l$ matrix $\mathbf{X}$ is obtained from $\mathbf{x}$:

$$\mathbf{X} = \begin{bmatrix} x_1 & x_2 & \cdots & x_l \\ x_{l+1} & x_{l+2} & \cdots & x_{2l} \\ \vdots & & \ddots & \vdots \\ x_{(I-1)l} & x_{(I-1)l+1} & \cdots & x_{Il} \end{bmatrix} \tag{12.10}$$

where $I = \lfloor n/l \rfloor$. Then, this matrix is converted to tensor $\underline{\mathbf{X}}$ as demonstrated in Figure 12.9 by considering each slab of the tensor as a windowed version of $\mathbf{X}$. Converting a matrix to tensor can be explained by the following equation:

$$\underline{\mathbf{X}}_{i::} = \mathbf{X}(i, (j-1)o + 1 : (j-1)o + l_1) \tag{12.11}$$

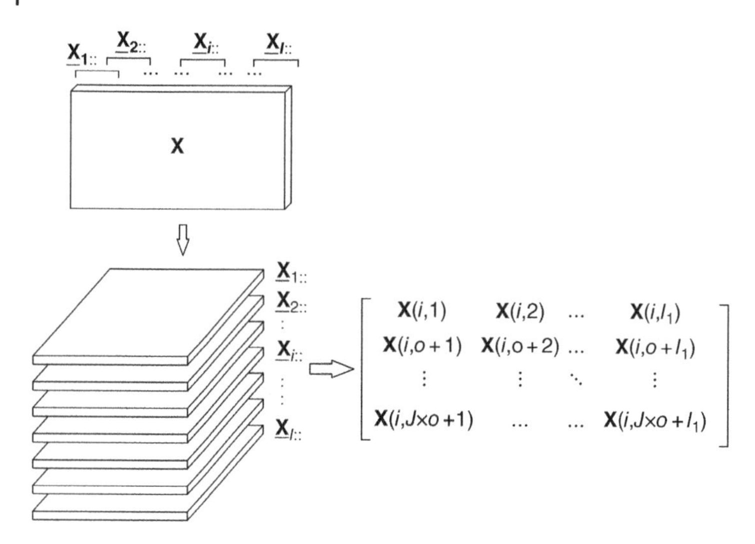

Figure 12.9 $I \times K$ matrix **X** is converted to tensor $\underline{\mathbf{X}}$ where J is the number of segments obtained by window size l_1 and overlapping interval $l_1 - o$.

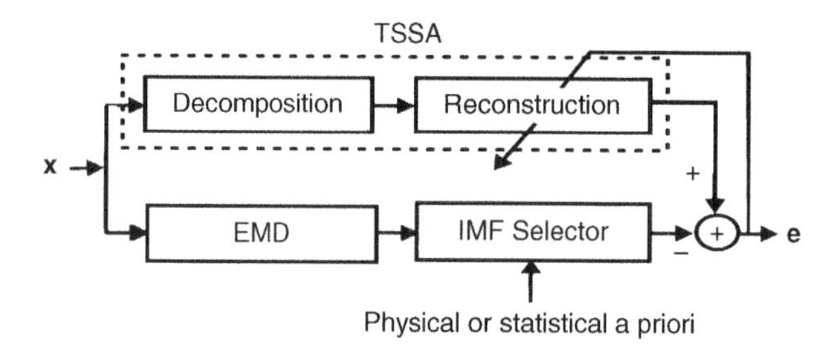

Figure 12.10 Block diagram of the single-channel source separation system using an adaptive procedure for selecting the desired subspace. This is carried out by tuning a set of weights governed by the EMD process.

where $j = 1, 2, ..., J$; $J = \left\lfloor \frac{l - l_1}{o} + 1 \right\rfloor$, and $i = 1, 2, ..., I$

where $l_1 - o$ is the overlapping interval for the successive windows and l_1 is the window size. Then, the factors of the constructed tensor are estimated using alternating least squares to best fit the PARAFAC model.

The same as for SSA, in TSSA subgroups are defined as $I = I_1 \cup I_2 \cup ... I_m$ and $\underline{\mathbf{X}} = \sum_{i=1}^{m} \underline{\mathbf{X}}_{I_i}$. In this work empirical mode decomposition (EMD) has been used to select the subgroups with intrinsic mode functions (IMFs) within the frequency range of interest [41]. Figure 12.10 represents a block diagram of the overall system. Finally, a reconstruction of the original signal can be obtained by Hankelization which is performed by converting the signal segments to a Hankel matrix. Then, the data are reconstructed by one-to-one correspondence.

The stages of sleep for the subjects in normal condition, with sleep restriction, and with sleep extension/depravation have been estimated and compared with the scores given by clinical experts. The performed objective measures have also demonstrated the superior performance of the method compared to a number of benchmarks.

12.6.5 Application of Neural Networks

As a nonlinear multiple-input multiple-output classifier, neural networks (NNs) can be used to classify different waveforms for recognition of various stages of sleep and also the type of mental illnesses. NNs have been widely used to analyze complicated systems without accurately modelling them in advance [42]. Often no a priori information about the data statistics, such as distribution, is necessary in developing an NN. The input to NN classifiers can be the raw data, compressed data, or the most descriptive features measured or estimated from the data in the original or transformed domains. In the case of sleep data, a number of typical waveforms from the sleep EEG can be used for training and classification. They include spindle, hump, alpha wave, hump train (although not present generally in the EEGs), and background wave. Each of these features manifests itself differently in the T-F domain.

Time delay NNs (TDNN) may be used to detect the wavelets with roughly known positions on the time axis [43]. In such networks the deviation in location of the wavelet in time has to be small. For EEGs, however, a shift larger than the wavelet duration must be compensated since the occurrence times of the waveforms are not known.

In order to recognize the time-shifted pattern, another approach named as sleep EEG recognition NN (SRNN) has been proposed by Shimada et al. [44]. This NN has one input layer, two hidden layers, and one output layer. From the algorithmic point of view and the input–output connections, SRNN, and TDNN are very similar. As the main difference, in a TDNN each row of the second hidden layer is connected to a single cell of the output layer, while in a SRNN similar cells from both layers are connected.

In order to use an SRNN the data are transformed into the TF domain. Instead of moving a sliding window over time, however, overlapped blocks of data are considered in this approach. Two-dimensional blocks with horizontal axis of time and vertical axis of frequency are considered as the inputs to the NN. Considering $y_{j,c}$ and $d_{j,c}$ to be, respectively, the jth output neuron and the desired pattern for the input pattern c then:

$$E = \frac{1}{2} \sum_{j=1}^{\text{Output neurons}} \sum_{c=1}^{\text{Input patterns}} \left(y_{j,c} - d_{j,c} \right)^{1/2} \tag{12.12}$$

Therefore, the learning rule minimizes the cost function by using the following gradient:

$$\Delta w_{p,q} = \mu \frac{\partial E}{\partial w_{p,q}} \tag{12.13}$$

where $w_{p,q}$ are the link weights between neurons p and q and μ is the learning rate. In the learning phase the procedure [44] performs two passes: forward and backward, through the network. In the forward pass the inputs are applied and the outputs are computed. In the backward pass, the outputs are compared with the desired patterns and an error is calculated. The error is then back projected through the network and the connection weight is changed by gradient descent of the mean squared error (MSE) as a function of weights. The optimum weights $w_{p,\ q}$ are obtained when the learning algorithm converges. The weights are then used to classify a new waveform, i.e. to perform generalization. Further details of the performance of this scheme can be found in [44].

12.6.6 Model-Based Analysis

Model-based approaches rely on an a priori knowledge about the mechanism of generation of the data or the data itself. Characterizing a physiological signal generation model for NREM has also been under study by several researchers [45–47]. These models describe how the depth of NREM sleep is related to the neuronal mechanism that generates slow waves. This mechanism is essentially feedback through closed loops in neuronal networks or through the interplay between ion currents in single cells. It is established that the depth of NREM sleep modulates the gain of the feedback loops [48]. According to this model, the sleep-related variations in the slow-wave power (SWP) result from variations in the feedback gain. Therefore, increasing depth of sleep is related to an increasing gain in the neuronal feedback loops that generates the low-frequency EEG.

In [49] a model-based estimator of the slow-wave-feedback gain has been proposed. The initial model depicted in Figure 12.11 is analogue and a discrete time approximation of that has been built up. In the analogue model, $G(f)$ is the complex frequency transfer function of a bandpass (resonance) filter as:

$$G(f) = \frac{1}{1 + j \cdot Y(f)} \tag{12.14}$$

where $j = \sqrt{-1}$
and

$$Y(f) = \frac{f_0}{B} \cdot \left(\frac{f}{f_0} - \frac{f_0}{f}\right) \tag{12.15}$$

where the resonance frequency f_0 and the bandwidth B are approximately 1 and 1.5 Hz. The closed loop equation can be written as:

$$\dot{u}(t) = p(t) \cdot s(t) + \dot{w}(t) \tag{12.16}$$

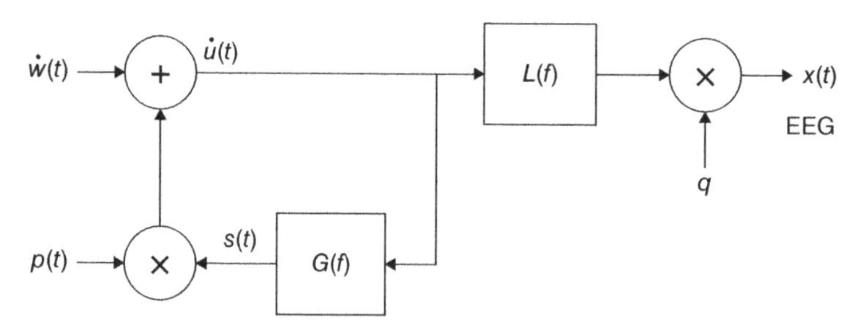

Figure 12.11 A model for neuronal slow-wave generation; $\dot{u}(t)$ is the derivative of the EEG source, $G(f)$ denotes the frequency-selective feedback of the slow-wave, $s(t)$. The signal $p(t)$ is the feedback gain to be identified, $w(t)$ is the white noise, $L(f)$ is the lowpass filter representing the path from source to electrodes and q is the attenuation effect by the skull.

where $\dot{u}(t) = du(t)/dt$ and $\dot{w}(t) = dw(t)/dt$. It is seen that the transfer function depends on $p(t)$. For $p(t)$ constant:

$$U(f) = \frac{1}{1 - p \cdot G(f)} \tag{12.17}$$

and therefore:

$$|U(f)|^2 = \frac{1 + Y^2(f)}{(1-p)^2 + Y^2(f)} \tag{12.18}$$

In which case:

$$S(f) = G(f) \cdot U(f) = \frac{G(f)}{1 - p \cdot G(f)} = \frac{1}{1 - p + j \cdot Y(f)} \tag{12.19}$$

The output $x(t)$ is the lowpass copy of $\dot{u}(t)$ attenuated by the factor of q. For $p = 0$ there is no feedback and for $p = 1$ there is an infinite peak at $f = f_0$. The lowpass filter $L(f)$ is considered known with a cut-off frequency of approximately 1.8 Hz.

The feedback gain of the model $p(t)$ represents the sleep depth. Therefore, the objective would be to estimate the feedback gain. To do that, the observation $du(t)$ is sampled by a 50 Hz sampler (a sampling interval of $\Delta = 0.02$ seconds). Then, define $Du(k\Delta) = u(k\Delta + \Delta) - u(k\Delta)$ over an interval $[0 \le k\Delta < N\Delta - \Delta]$, with $N\Delta = T$. Also, it is considered that $p(t) = p$, constant over the interval T. Eq. (12.13) then becomes

$$Du(k\Delta) = p.s(k\Delta).\Delta + Dw(k\Delta) \tag{12.20}$$

where $Dw(k\Delta) = w(k\Delta + \Delta) - w(k\Delta)$ is the increment of the standard continuous-time Wiener process $w(t)$. Assuming the initial state for the feedback filter $G(f)$ to be G_0 and $w(k\Delta)$ to have Gaussian distribution, the likelihood of $Du(k\Delta)$ can be factorized according to Bayes' rule as:

$$\begin{aligned} P[Du(k\Delta) : 0 \le k < N-1 \mid G_0, p] &= \prod_{k=0}^{N-1} [Du(k\Delta) \mid [Du(m\Delta) : 0 \le m < k], G_0, p] \\ &= \prod_{k=0}^{N-1} [Du(k\Delta) \mid s(k\Delta), p] \\ &= \prod_{k=0}^{N-1} \left\{ \frac{1}{\sqrt{2\pi\Delta}} exp\left(-[Du(k\Delta) - p.s(k\Delta).\Delta]^2/2\Delta \right) \right\} \\ &= \frac{1}{\sqrt{2\pi\Delta}} exp\left\{ \sum_{k=0}^{N-1} \{ -[Du(k\Delta) - p.s[k\Delta].\Delta]^2/2\Delta \} \right\} \end{aligned} \tag{12.21}$$

To maximize this likelihood it is easy to conclude that the last term in the square brackets has to be maximized. This gives [49]:

$$\hat{p} = \frac{\sum_{k=0}^{N-1} [s(k\Delta).Du(k\Delta)]}{\sum_{k=0}^{N-1} s^2(k\Delta).\Delta} \tag{12.22}$$

Hence, $\hat{p}$ approximates the amount of slow wave and is often represented as its percentage. This completes the model and therefore the sleep EEG may now be constructed.

12.7 Detection and Monitoring of Brain Abnormalities during Sleep by EEG and Multimodal PSG Analysis

EEG provides important and unique information about the sleeping brain. PSG has been a well established method of sleep analysis and the main diagnostic tool in sleep medicine, which interprets the sleep signal macrostructure based on the criteria explained by R&K [50]. Polysomnography or sleep study is a multi-parametric test and a comprehensive recording of the biophysiological changes occurring during sleep. It is usually performed at night. The PSG monitors many body functions including brain (EEG), eye movements (EOG), muscle activity or skeletal muscle activation (EMG) and heart rhythm (ECG) during sleep.

12.7.1 Analysis of Sleep Apnoea

Sleep apnoea syndrome (SAS), chronic snoring, and daytime excessive sleepiness cause problems in daily life and have short-term and long-term physiological and psychological effects. After the identification of sleep apnoea in the 1970s, PSG has been used as a gold standard to combine and evaluate breathing, peripheral pulse oximetry, heartrate, and two channels of EEG for the assessment of these disorders. Later the recorded sound of snoring has been the main focus of research. Generally, it has been shown that not all the signals recorded by the PSG are needed for the assessment of apnoea or hypopnea [51–53]. In [51, 52] it has been shown that the snoring sound and the SpO2 value from the output of an oximeter can be utilized for the analysis of apnoea. There are also reports of using only one of these two modalities, i.e. the sound signals [53] or evaluation of hypoxemia using oximeters [54] is enough for the assessment of apnoea. However, the spindles and slow-wave activities, arousals, and associated activities can be detected from the EEG signals and monitored during the sleep. For analysis of apnoea description of these activities often requires temporal segmentation of the signals into fixed segments of 20–30 seconds.

SAS with a high prevalence of approximately 2% in women and 4% in men between the ages of 30 to 60 years is the cause of many physiological and psychological problems which consequently change the life style of the patient [19, 55]. This syndrome is often treated by means of continuous positive airway pressure therapy or by surgery. SAS refers to sleep disorders which cause breathing pauses during sleep at night. SAS can be the result of different anatomical or physiological factors. Two common SASs namely central sleep apnoea (CSA) and OSA have been classified in a recent research [56]. In case of the CSA, sleep apnoea originates from the central nervous system. If during sleep the duration of pauses in breathing is at least 10s at upper respiratory tract and the pulmonary system at the same time then the CSA is diagnosed. In the case of OSA, the reason for the breathing pauses is the respiratory tract obstruction.

Using a standard PSG has been a popular approach for diagnosis and monitoring of this disease. The measurement is often overnight to record the sleep stage, respiratory efforts, oronasal airflow, electrocardiographic findings, and oxyhaemoglobin saturation parameters in an attended laboratory setting [20]. However, PSG is not easily accessible, it is expensive and intrusive. Therefore, researchers try to extract the necessary diagnostic information from a smaller number of recording modalities as stated above. Beyond snoring sound and the oxygen level, EEG is another modality used for the assessment of sleep apnoea.

In recent assessment of sleep and sleep apnoea EEG has been used for the detection of SASs. Coherence function (CF) and mutual information (MI) measures have been used as the EEG signal features in discriminating the CSA and OSA from controls. The sleep EEG series recorded from patients and healthy volunteers are classified by using a feed-forward neural network (FFNN) with two hidden layers and utilizing the synchronic activities between C3 and C4 channels of the EEG recordings. Among the sleep stages, Stage II is considered in tests. The results show that the degree of central EEG synchronization during night sleep is closely related to sleep disorders like CSA and OSA. The MI and CF have been shown to give cooperatively meaningful information to support clinical findings [56].

Diagnosis of sleep disorders and other related abnormalities may not be complete unless other physiological symptoms are studied. These symptoms manifest themselves within other physiological and pneumological signals such as respiratory airflow, position of the patients, EMG signal, hypnogram, level of SaO_2, abdominal effort, and thoracic effort which may also be considered in the classification system.

A simple system to use the features extracted from the sleep signals in classification of the apnoea and detection of the SAS has been suggested and used [12]. This system processes the signals in three phases. In the first phase the relevant characteristics of the signals are extracted and a segmentation based on significant time intervals of variable length is performed. The intermediate phase consists of assigning suitable labels to these intervals and combining these symbolic information sources with contextual information in order to build the necessary structures that will identify clinically significant events. Finally, all the relevant information is collected and a ruled-based system is established to exploit the above data, provide the induction, and produce a set of conclusions.

In a rule-based system for detection of the SAS, two kinds of cerebral activities are detected and characterized from the EEG signals: rhythmic (alpha, beta, theta, and delta rhythms) and transitory (K-complexes and spindles) [57]. The magnitude and the change in the magnitude (evolution) are measured and the corresponding numerical values are classified together with the other features based on clinical and heuristic [13] criteria.

To complete this classifier the slow eye movement, very frequent during sleep stage I, and REM are measured using electro-oculogram (EOG) signals. The distinction between the above two eye movements is based on the synchrony, amplitude, and slope of the EOG signals [12].

In another approach for sleep staging of the patients with OSA the EEG features are used, classified, and compared with the results from cardiorespiratory features [14].

Prochazka et al. used PSG in a machine learning approach to study the sleep abnormalities. They analyzed polysomnographic records of 33 healthy individuals, 25 individuals with sleep apnoea, and 18 individuals with sleep apnoea together with restless leg syndrome [58] recorded overnight. A screenshot of the PSG they used can be seen in Figure 12.12.

A sixth-order Savitzky–Golay filter [59] has been employed using 80 overlapping frames to reduce the fluctuations of the estimated spectral components. Then, they computed time and wavelet domain features and classified them using SVMs, k-nearest neighbours, decision tree, and NNs. In almost all the cases decision tree and NN classifiers achieved highest accuracy (within 85.6–97.5%) for classification of wake, REM, and all NREM cases, i.e. sleep

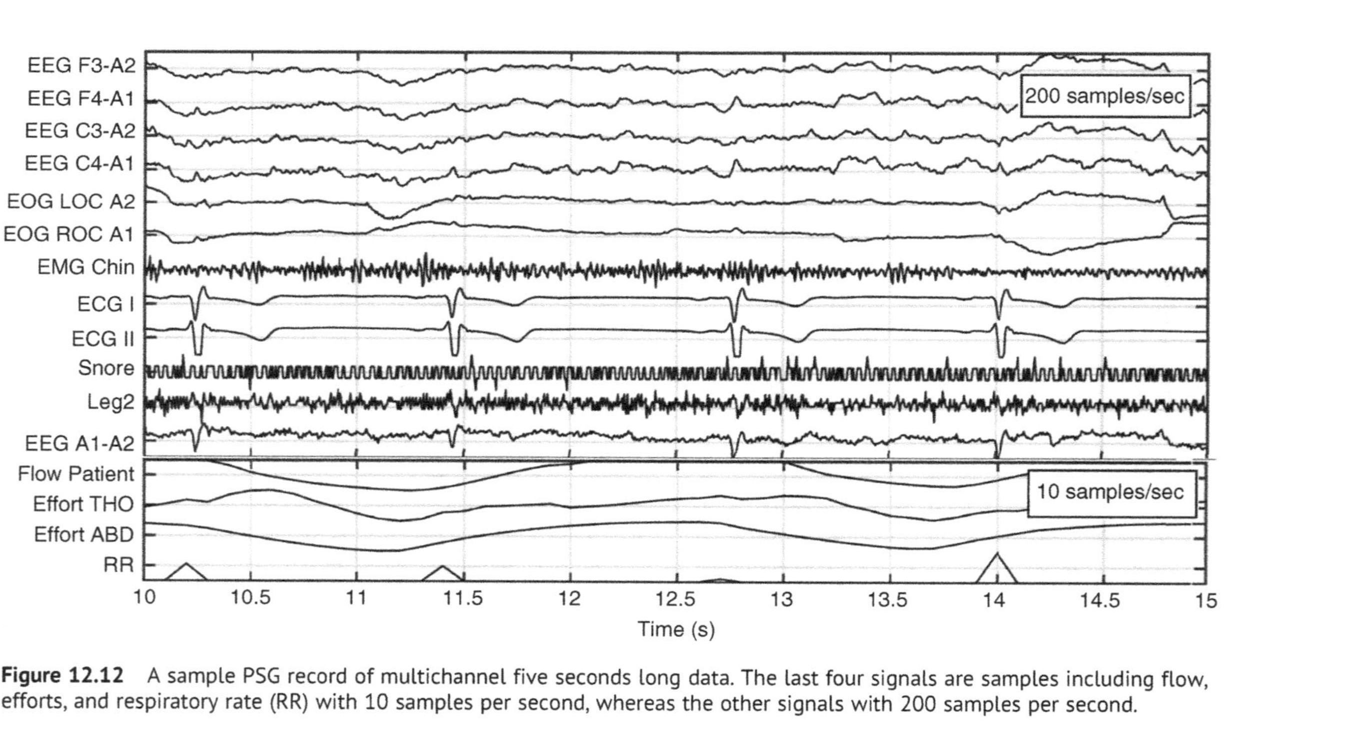

Figure 12.12 A sample PSG record of multichannel five seconds long data. The last four signals are samples including flow, efforts, and respiratory rate (RR) with 10 samples per second, whereas the other signals with 200 samples per second.

stages. They also concluded that as age increases, the occurrences of wake stages increase and those of REM stages decrease.

In another machine learning approach, Jarchi et al. used ECG and EMG components of PSG for recognition of breathing and movement-related sleep disorders [60]. They extracted EMG features exploiting entropy and statistical moments and developed an iterative pulse peak detection algorithm using synchro-squeezed wavelet transform (SSWT) for reliable extraction of heart rate and breathing-related features from ECG. They used a very comprehensive deep learning approach using a combination of RNNs and deep neural networks (DNNs) to make accurate classification of healthy subjects, patients with OSA, patients with restless leg syndrome and those with both OSA and leg syndrome. However, in this approach EEG has not been utilized.

In another research, sleep apnoea and hypopnea have been analyzed using a single-channel ECG signal and the respiratory-related modulation [61]. The signals have been analyzed using SSWT within the respiratory frequency range. The time–frequency ridge was estimated to provide a robust estimation of instantaneous respiratory frequency and detect those regions with/without sleep apnoea/hypopnea events. Signal reconstruction using the Inverse synchro-squeezing wavelet transform (ISSWT) has been performed for which the highlighted peaks can identify and measure the duration of sleep apnoea/hypopnea events.

12.7.2 EEG and Fibromyalgia Syndrome

Fibromyalgia syndrome (FMS) is defined by the existence of chronic, often full-body pain accompanied by a variety of additional symptoms such as widespread musculoskeletal pain, fatigue, and pain in the tendons ligaments, and sleep disturbances. With the diversity and complexity of these symptoms, fibromyalgia is often misdiagnosed, leaving sufferers frustrated and confused. However, sleep apnoea commonly coexists with fibromyalgia. The sleep disorder known as alpha-EEG anomaly is prevalent in FMS patients. This condition is defined by interruption of deep sleep by awake-like brain activity.

Alpha waves indicate you are awake but relaxed. In the corresponding EEG records, a sleeping brain typically displays delta waves during the third stage of sleep, which is considered 'deep' sleep. In the alpha-EEG anomaly, alpha waves intrude into deep sleep, indicating that the brain is not resting like it should. Alpha wave intrusion occurs when alpha waves appear with non-REM sleep when delta activity is expected. Some researchers theorize that this anomaly may explain the unrefreshing sleep that is characteristic of fibromyalgia [62, 63]. Therefore, an evaluation of EEG can reveal some valuable information for diagnosis of this disorder.

12.7.3 Sleep Disorders of Neonates

Recently, the changes in EEG associated with apnoeic episodes in neonates have been studied, e.g. in [64–66]. In [64] it has been found that the prolonged apnoeic episodes accompanied by hypoxia and bradycardia can be associated with altered cerebral function in the neonate. This study can help clinicians to know the lowest limit of oxygen saturation required to suppress EEG activity. It has also been shown that only apnoeic events with oxygen desaturations below 20% are associated with complete EEG suppression.

Despite the frequency of occurrence of apnoea in the neonate and the concern about adverse long-term effects [66], few studies have used EEG to examine the effects of apnoeic episodes [64]. From the sleep EEG it can be discovered if the apnoea is related or correlated with neonate seizure, hypoxia, or any other neurological or even physiological disorders.

12.8 Dreams and Nightmares

It is difficult to deny that we all dream. People awakened from REM periods in sleep experiments report they have been dreaming 80–100% of the time. REM dreams are considered to be more perceptual and emotional as opposed to NREM dreams. Content of NREM dreams is often a recreation of some psychologically important event. According to Freud [67] REM dreams are like primary-process thinking which is often unrealistic and emotional, and NREM dreams are like secondary-process thinking which is more realistic. There are mainly three theories on the meaning of dreams.

Based on Freud theory, we dream to satisfy unconscious desires or wishes, especially those involving sex and aggression. If we were to fulfil these wishes during day time it would create too much anxiety. Freud stated that the wishes are represented with symbols since they would otherwise be anxiety producing. Based on this theory a therapist must interpret these symbols to help clients discover unconscious desires.

The theory states that the hindbrain (the lower part of the brainstem, comprising the cerebellum, pons, and medulla oblongata) transmits chaotic patterns of signals to the cerebral cortex, and then higher-level cognitive processes in the cerebral cortex try to integrate these signals into a dream plot [68].

Dreams can also be viewed as extensions of waking life, which include thoughts and concerns especially emotional ones. Then, in a sense dreams provide clues to the person's problems, concerns, and emotions [69]. Different from nightmare, dreaming may be a sign of good sleeping. Although there are many theories and hypothesis about dreams, there are still many unanswered questions about the reasons for dreams, their exact functions, and how physiological processes are involved. In The Interpretation of Dreams [70] Freud predicted that 'Deeper research will one day trace the path further and discover an organic basis for the mental event'. Recent works begin to fulfil this prediction through any possible approach including EEG analysis.

Sleep EEG records are likely to present alpha attenuation as an indicator of visual activity during dreaming [71]. Conversely, unlike dreams in general, nightmares are known to exclusively occur during REM sleep and during the longer and later REM phases in the latter part of the sleep cycle.

From the neuroscience viewpoint, dream research could be considered as the study of cognitive processes during sleep. However, neurobiological evidence is intrinsically limited by the impossibility to directly access dreaming, and by the necessity to focus on recall of dream experience collected during post-awakening wakefulness (i.e. dream recall). Numerous studies underlined the continuity between mechanisms that regulate cognitive functions in wakefulness and sleep [72]. In particular, it has been stressed that EEG studies have provided empirical data to sustain the predictive role played by theta oscillations in the retrieval of episodic information both in wakefulness and sleep, considering dream recall as an episodic mnestic trace. To enable monitoring all possible physiological symptoms during sleep and dreaming, PSG is suggested as a gold standard tool for this purpose [72].

12.9 EEG and Consciousness

EEG has increasingly become a tool of choice for monitoring the course of anaesthesia and state of consciousness. Frontal brain activity more significantly changes by varying the level of anaesthesia and can be a good biomarker for monitoring the depth of consciousness during clinical operation. This monitors the brain state during propofol infusion to adjust the correct dose.

Hence, consciousness assessment methods such as behaviour scale or index derived from the frontal EEG have been widely adopted in clinical settings [73, 74] although no consensus has been reached as to which marker is superior for evaluating the patient's level of consciousness.

During unconsciousness slow waves elicit from the frontal cortex and spread over the whole brain. Unlike sleep, there are no spindles or K-complexes. Therefore, in monitoring the unconsciousness level, assessing spectral features such as power spectrum of low-frequency range (mainly within delta and theta bands) around the frontal zone (facilitated by spatial filtering) seems to be adequate.

Bispectral index (BIS) monitor is one of the systems commonly utilized to assess depth of sedation when administering sedative, hypnotic, or anaesthetic agents during surgical and medical procedures. BIS is a statistically based, empirically derived complex parameter that is composed of a combination of time domain, frequency domain, and high order spectral subparameters.

BIS is a number (or percentage) from 0 (flat EEG) to 100 (for fully awake). It is unique in the sense that it integrates several disparate descriptors of the EEG into a single variable, based on a large volume of clinical data to synthesize a combination that correlates behavioural assessments of sedation and hypnosis, yet is insensitive to the specific anaesthetic or sedative agent chosen [75]. The BIS values are known to be calculated from four EEG subparameters: burst suppression ratio (BSR), QUAZI suppression index, relative beta ratio (RBR), and SyncFastSlow (SFS), using multiple regression equations with different weights according to the depth of anaesthesia [76, 77].

In the analysis reported in [78] the use of BIS measure from EEGs in operation theatres and ICUs is recommended but states that this should not be the only measure for taking the final decision about the depth of anaesthesia.

12.10 Functional Brain Connectivity for Sleep Analysis

As broadly considered in this chapter, sleep study is popularly based on the classification of sleep into stages defined by their EEG signatures, and various physiological and mental brain disorders manifest themselves differently in the sleep EEG signals but the underlying brain dynamics remain unclear. In [79] the authors argue that the use of fixed scoring windows of 10–30 seconds and only a few EEG electrodes means that PSG involves considerable averaging of brain activity in both time and space which leads to an incomplete representation of brain activity.

They build upon a recent study showing that individual PSG stages can be extracted from functional magnetic resonance imaging (fMRI) recordings in a data-driven way [80], they leveraged the full spatiotemporal resolution of blood oxygenation level dependent (BOLD) signals to find large-scale networks in sleep, applying a hidden Markov model (HMM) on fMRI simultaneously recorded with EEG.

They characterize the spatiotemporal complexity of whole-brain networks and state transitions during sleep using a Markovian data-driven analysis of continuous simultaneous fMRI and EEG during sleep. The proposed HMM facilitates discovery of the dynamic choreography between different whole-brain networks across the wake–non-REM sleep cycle.

An HMM with a limited number of states is used to explicitly model the transition probabilities between its inferred states. They show that this information can be used to discover new whole-brain aspects of sleep, complementing the traditional segmentation of brain activity offered by PSG.

The results reveal key trajectories to switch within and between EEG-based sleep stages, while highlighting the heterogeneities of stage N1 sleep and wakefulness before and after sleep.

In [81] the authors showed that the EEG coherence has high values within the same hemisphere at wake and sleep stage N1. During stages N2, N3, and N4 and REM sleep, more regions became correlated, particularly for slow oscillations. In the delta band, the coherence values differed between right and left cerebral hemispheres during all sleep stages. In other frequency bands, no difference was observed between the two sides. In the right hemisphere, coherence values changed across sleep stages in delta frequency but remained unchanged for other sleep EEG frequencies. In the left hemisphere, there was no change in coherence across sleep stages and each frequency except for sigma and beta.

The change in the brain functional connectivity within five areas of frontal, motor, temporal, somatosensory, and visual areas during transition between wakefulness and sleep has also been observed using near infrared spectroscopy (NIRS) simultaneously recorded with the EEG signals [82]. Further research using fMRI revealed that poor sleep quality associates with decreased functional and structural brain connectivity in normative ageing study (which studies the effects of ageing on various health issues) [83]. In another fMRI study for characterizing functional connectivity and network topology in a large cohort of patients it was found that the functional brain connectivity alterations in restless legs syndrome are modulated by dopaminergic medication [84].

12.11 Summary

Study of sleep EEG has opened a new direction to investigate the psychology of a human being and various sleep disorders. Different stages of sleep may be identified using simple established tools in signal processing and machine learning. An accurate scoring process requires detection of changes in background cortical activity as well as spindles, K-complexes, and vertex waves. Generally, it has been shown that sleep signals can be scored rather accurately. Sleep disorders such as apnoea and fibromyalgia syndrome can be studied and monitored through analysis of EEG. Detection and classification of mental diseases from the sleep EEG signals, however, require deeper analysis of the data by developing and utilizing advanced processing and learning techniques. The analysis becomes more challenging when other parameters such as age are involved. For example, in neonates many different types of complex waveforms may be observed for which the origin and causes are still unknown. Conversely, there are some similarities between the normal rhythms within the sleep EEG signals and the EEGs of abnormal rhythms such as epileptic seizure and hyperventilation. An efficient algorithm (based on sole EEG or combined with other physiological signals) should be able to differentiate between these cases too.

References

1 Lavie, P. (1996). *Enchanted World of Sleep* (trans. Berris),. London: Yale University.

2 Steriade, M. (1992). Basic mechanisms of sleep generation. *Neurology* 42 (Suppl. 6): 9–17.

3 Kubicki, S., Scheuler, W., and Wittenbecher, H. (1991). Short-term sleep EEG recordings after partial sleep deprivation as a routine procedure in order to uncover epileptic phenomena: an evaluation of 719 EEG recordings. *Epilepsy Research. Supplement* 2: 217–230.

4 Iber, C., Ancoli-Israel, S., Chesson, A., and Quan, S.F. (2007). *The AASM Manual for the Scoring of Sleep and Associated Events: Rules, Terminology and Technical Specifications*. Westchester, IL: American Academy of Sleep Medicine.

5 Niedermeyer, E. (1999). Sleep and EEG. In: *Electroencephalography Basic Principles, Clinical Applications, and Related Fields* (eds. E. Niedermeyer and F.L. Da Silva), 174–188. Lippincott, Williams & Wilkins.

6 Koshino, Y., Nishio, M., Murata, T. et al. (1993). The influence of light drowsiness on the latency and amplitude of P300. *Clinical Electroencephalography* 24: 110–113.

7 Wada, Y., Nanbu, Y., Koshino, Y. et al. (1996). Inter- and intrahemispheric EEG coherence during light drowsiness. *Clinical Electroencephalography* 27: 24–88.

8 Niedermeyer, E. (1999). Maturation of EEG: development of walking and sleep patterns. In: *Electroencephalography, Basic Principles, Clinical Applications, and Retailed Field* (eds. E. Niedermeyer and F.L. de Silva). Lippincott, Williams and Wilkins.

9 Bonanni, E., Di Coscio, E., Maestri, M. et al. (2012). Differences in EEG delta frequency characteristics and patterns in slow-wave sleep between dementia patients and controls: a pilot study. *Journal of Clinical Neurophysiology* 29 (1): 50–54.

10 Brzezinski, A. (1997). Melatonin in humans. *New England Journal of Medicine* 336: 186–195.

11 Shiomi, F.K., Pisa, I.T., and de Campos, C.J.R. (2011). Computerized analysis of snoring in sleep apnea syndrome. *Brazilian Journal of Otorhinolaryngology* 77 (4): 488–498.

12 Cabrero-Canosa, M., Hernandez-Pereira, E., and Moret-Bonillo, V. (2004). Intelligent dignosis of sleep apnea syndrome. *IEEE Engineering in Medicine and Biology Magazine* 23 (2): 72–81.

13 Karskadon, M.A. and Rechtschaffen, A. (1989). *Priniciples and Practice of Sleep Medicine*, 665–683. Philadelphia: Saunders.

14 Redmond, S.J. and Heneghan, C. (2006). Cardiorespiratory-based sleep staging in subjects with obstructive sleep apnea. *IEEE Transactions on Biomedical Engineering* 53 (3): 485–496.
15 Klingler, D., Tragner, H., and Deisenhammer, E. (1991). The nature of the influence of sleep deprivation on the EEG. *Epilepsy Research. Supplement* 2: 231–234.
16 Jovanovic, U.J. (1991). General considerations of sleep and sleep deprivation. *Epilepsy Research. Supplement* 2: 205–215.
17 Naitoh, P., Kelly, T.L., and Englund, C. (1990 Apr-Jun). Health effects of sleep deprivation. *Occupational Medicine* 5 (2): 209–237.
18 Sabo, E., Reynolds, C.F., Kupfer, D.J., and Berman, S.R. (1991). Sleep, depression, and suicide. *Psychiatry Research* 36 (3): 265–277.
19 Weitzenblum, E. and Racineux, J.-L. (2004). *Syndrome d'Apnées Obstructives du Sommeil*, 2e. Paris, France: Masson.
20 Man, G.C. and Kang, B.V. (1995). Validation of portable sleep apnea monitoring device. *Chest* 108 (2): 388–393.
21 Porée, F., Kachenoura, A., Gavrit, H. et al. (2006). Blind source separation for ambulatory sleep recording. *IEEE Transactions on Information Technology in Biomedicine* 10 (2): 293–301.
22 Dahl, R.E., Puig-Antich, J., Ryan, N.D. et al. (1990). EEG sleep in adolescents with major depression: the role of suicidality and inpatient status. *Journal of Affective Disorders* 19 (1): 63–75.
23 Goetz, R.R., Puig-Antich, J., Dahl, R.E. et al. (1991). EEG sleep of young adults with major depression: a controlled study. *Journal of Affective Disorders* 22 (1–2): 91–100.
24 McCracken, J.T., Poland, R.E., Lutchmansingh, P., and Edwards, C. (1997). Sleep electroencephalographic abnormalities in adolescent depressives: effects of scopolamine. *Biological Psychiatry* 42: 577–584.
25 Van Swededn, B., Wauquier, A., and Niedermeyer, E. (1999). Normal aging and transient cognitive disorders in elderly," Chap. 18. In: *Electroencephalography* (eds. E. Niedermeyer and F.L. de Silva). Williams and Wilkins.
26 Klass, D.W. and Brenner, R.P. (1995). Electroencephalography in the elderly. *Journal of Clinical Neurophysiology* 12: 116–131.
27 Malinowska, U., Durka, P.J., Blinowska, K.J. et al. (2006). Micro- and macrostructure of sleep EEG; a universal, adaptive time-frequency parametrization. *IEEE Engineering in Medicine and Biology Magazine* 25 (4): 26–31.
28 Laboratory of Computer and Information Science (CIS), Department of Computer Science and Engineering, Helsinki University of Technology, http://www.cis.hut.fi (access 19 August 2021).
29 Mallat, S. and Zhang, Z. (1993). Matching pursuit with time-frequency dictionaries. *IEEE Transactions on Signal Processing* 41: 3397–3415.
30 Mallat, S. (1999). *A Wavelet Tour of Signal Processing*, 2e. NY: Academic.
31 American Sleep Disorder Association (1992). EEG arousals: scoring rules and examples. A preliminary report from the sleep disorder task force of the American sleep disorder association. *Sleep* 15 (2): 174–184.
32 Aeschbach, D. and Borb'ely, A.A. (1993). All-night dynamics of the human sleep EEG. *Journal of Sleep Research* 2 (2): 70–81.
33 Terzano, M.G., Parrino, L., Rosa, A. et al. (2002). CAP and arousals in the structural development of sleep: an integrative perspective. *Sleep Medicine* 3 (3): 221–229.
34 Mahvash Mohammadi, S., Kouchaki, S., Ghavami, M., and Sanei, S. (2016). Improving time–frequency domain sleep EEG classification via singular spectrum analysis. *Elsevier Journal of Neuroscience Methods* 273: 96–106.

35 Principe, J.C. and Smith, J.R. (Oct. 1986). SAMICOS-A sleep analysing microcomputer system. *IEEE Transactions on Biomedical Engineering* BME-33: 935–941.
36 Principe, J.C., Gala, S.K., and Chang, T.G. (May 1989). Sleep staging automation based on the theory of evidence. *IEEE Transactions on Biomedical Engineering* 36: 503–509.
37 Akgül, T., Sun, M., Sclabassi, R.J., and Çetin, A.E. (2000). Characterization of sleep spindles using higher order statistics and spectra. *IEEE Transactions on Biomedical Engineering* 47 (8): 997–1009.
38 Nikias, C.L. and Petropulu, A. (1993). *Higher Order Spectra Analysis, A Nonlinear Signal Processing Framework*. Englewood Cliffs, NJ, USA: Prentice Hall.
39 Rao, T.S. (1993). Bispectral analysis of nonlinear stationary time series. In: *Handbook of Statistics*, vol. 3 (eds. D.R. Brillinger and P.R. Krishnaiah). Amsterdam, Netherland: Elsevier.
40 Michel, O. and Flandrin, P. (1996). Higher order statistics in chaotic signal analysis. In: *Computer Techniques and Algortihms in Digital Signal Processing*, vol. 75, 105–154. Academic Press.
41 Kouchaki, S., Sanei, S., Arbon, E.L., and Dijk, D.-J. (JANUARY 2015). Tensor based singular spectrum analysis for automatic scoring of sleep EEG. *IEEE Transactions on Neural Systems and Rehabilitation Engineering* 23 (1): 1–9.
42 Lippmann, R.P. (Apr. 1987). An introduction to computing with neural nets. *IEEE, Acoustics, Speech, and Signal Processing (ASSP) Magazine* 4 (2): 4–22.
43 Weibel, A., Hanazawa, T., Hinton, G., and Lang, K. (Mar. 1989). Phoneme recognition using time-delay neural networks. *IEEE Transactions on Acoustics, Speech, and Signal Processing* 37: 328–339.
44 Shimada, T., Shiina, T., and Saito, Y. (March 2000). Detection of characteristic waves of sleep EEG by neural network analysis. *IEEE Transactions on Biomedical Engineering* 47 (3): 369–379.
45 Kemp, B., Zwinderman, A.H., Tuk, B. et al. (2000). Analysis of a sleep-dependent neuronal feedback loop: the slow-wave microcontinuity of the EEG. *IEEE Transactions on Biomedical Engineering* 47 (9): 1185–1194.
46 Kemp, B. (1996). NREM sleep depth = neuronal feedback = slow-wave shape. *Journal of Sleep Research* 5: S106.
47 Merica, H. and Fortune, R.D. (2003). A unique pattern of sleep structure is found to be identical at all cortical sites: a neurobiological interpretation. *Cerebral Cortex* 13 (10): 044–1050.
48 Steriade, M., McCormick, D.A., and Sejnowski, T.J. (1993). Thalamocortical oscillations in the sleeping and aroused brain. *Science* 262: 679–685.
49 Kemp, B., Zwinderman, A.H., Tuk, B. et al. (2000). Analysis of a sleep-dependent neuronal feedback loop: the slow-wave microcontinuity of the EEG. *IEEE Transactions on Biomedical Engineering* 47 (9): 1185–1194.
50 Rechtschaffen, A. and Kales, A. (eds.) (1986). *A Manual of Standardized Terminology and Scoring System for Sleep Stages in Human Subjects, National Institutes of Health Publications*. Washington DC: US Government Printing Office, no. 204.
51 Nobuyuki, A., Yasuhiro, N., Taiki, T. et al. (2009). Trial of measurement of sleep apnea syndrome with sound monitoring and SpO2 at home. *11th International Conference on e-Health Networking, Applications and Services (Healthcom), Sydney, NSW, 2009*: 66–79.
52 Ydollahi, A. and Giannouli, E. (2010). Sleep apnea monitoring and diagnosis based on pulse oximetery and tracheal signals. *Medical & Biological Engineering & Computing* 48: 1987–1097.

53 Sola-Soler, J., Fiz, J.A., Morea, J., and Jane, R. (2012). Multiclass classification of subjects with sleep apnoea-hypopnoea syndrome through snoring analysis. *Medical Engineering & Physics* JJBE-2043 https://doi.org/10.1016/j.medenegphy.2011.12.008.

54 Hang, L.W., Wang, J.-F., Yen, C.-W., and Lin, C.-L. (2009). EEG arousal prediction via hypoxemia indicator in patients with obstructive sleep apnea syndrome. *Internet Journal of Medical Update* 4 (2): 24–28.

55 Young, T., Palta, M., Dempsey, J. et al. (1993). The occurrence of sleep-disordered breathing among middle-aged adults. *The New England Journal of Medicine* 328: 1230–1235.

56 Aksahin, M., Aydın, S., Fırat, H., and Erogul, O. (2012). Artificial apnea classification with quantitative sleep EEG synchronization. *Journal of Medical Systems* 36 (1): 139–144.

57 Steriade, M., Gloor, P., Llinas, R.R. et al. (1990). Basic mechanisms of cerebral rhythmic activities. *Electroencephalography and Clinical Neurophysiology* 76: 481–508.

58 Procházka, A., Kuchynka, J., Vyšata, O. et al. (2018). Sleep scoring using polysomnography data features. *Signal, Image and Video Processing* 12: 1043–1051.

59 Schafer, R. (2011). What is a Savitzky–Golay filter? *IEEE Signal Processing Magazine* 28 (4): 111–117.

60 Jarchi, D., Andreu-Perez, J., Kiani, M. et al. (2020). Recognition of patient groups with sleep related disorders using bio-signal processing and deep learning. *Sensors* 20: 2594. https://doi.org/10.3390/s20092594.

61 Jarchi, D., Sanei, S., and Prochazka, A. (2019). Detection of aleep apnea/hypopnea events using synchrosqueezed wavelet transform. *Proceedings of the IEEE International Conference on Acoustics, Speech, and Signal Processing*. Brighton, UK: ICASSP.

62 Mueller, H.H., Donaldson, C.C.S., Nelson, D.V., and Layman, M. (2001). Treatment of fibromyalgia incorporating EEG-driven stimulation: a clinical outcomes study. *Journal of Clinical Psychology* 57 (7): 933–952.

63 Hammond, D.C. (2001). Treatment of chronic fatigue with neurofeedback and self-hypnosis. *Journal of Neuroengineering and Rehabilitation* 16: 295–300.

64 Low, E., Dempsey, E.M., Ryan, C.A. et al. (2012). EEG suppression associated with apneic episodes in a neonate. *Case Reports in Neurological Medicine* 2012: 250801, 7 pages, doi: https://doi.org/10.1155/2012/250801.

65 Murray, D.M., Boylan, G.B., Ryan, C.A., and Connolly, S. (2009). Early EEG findings in hypoxic-ischemic encephalopathy predict outcomes at 2 years. *Pediatrics* 124 (3): e459–e467.

66 Janvier, A., Khairy, M., Kokkotis, A. et al. (2004). Apnea is associated with neurodevelopmental impairment in very low birth weight infants. *Journal of Perinatology* 24 (12): 763–768.

67 Franken, R.E. (1988). *Human Motivation*. California: Brooks/Cole Publishing Co.

68 Hobson, J.A. and Stickgold, R. (1995). The conscious state paradigm: a neurocognitive approach to waking, sleeping, and dreaming. In: *The Cognitive Neurosciences* (ed. M.S. Gazzaniga), 1373–1389. Cambridge, MA: MIT Press.

69 Plotnik, R. (1995). *Introduction to Psychology*, 3e. California: Brooks/Cole.

70 Freud, S. (1900). *The Interpretation of Dreams*. New York: Modern Library.

71 Bértolo, H., Paiva, T., Pessoa, L. et al. (2003). Visual dream content, graphical representation and EEG alpha activity in congenitally blind subjects. *Brain Research. Cognitive Brain Research* 15 (3): 277–284.

72 Scarpelli, S., D'Atri, A., Gorgoni, M. et al. (2015). EEG oscillations during sleep and dream recall: state- or trait-like individual differences? *Frontiers in Psychology* 6: 605, 10 pages.

73 Gaskell, A.L., Hight, D.F., Winders, J. et al. (2017). Frontal alpha-delta EEG does not preclude volitional response during anaesthesia: prospective cohort study of the isolated forearm technique. *British Journal of Anaesthesia* 119: 664–673.

74 Boly, M., Massimini, M., Tsuchiya, N. et al. (2017). Are the neural correlates of consciousness in the front or in the back of the cerebral cortex? Clinical and neuroimaging evidence. *The Journal of Neuroscience* 37: 9603–9613.

75 Kaul, H.L. and Bharti, N. (2002). Monitoring depth of anaesthesia. *Indian Journal of Anaesthesia* 46 (4): 323–332.

76 Rampil, I.J. (1998). A primer for EEG signal processing in anesthesia. *Anesthesiology* 89: 980–1002.

77 Glass, P.S., Bloom, M., Kearse, L. et al. (1997). Bispectral analysis measures sedation and memory effects of propofol, midazolam, isoflurane, and alfentanil in healthy volunteers. *Anesthesiology* 86: 836–847.

78 Hajat, Z., Ahmad, N., and Andrzejowski, J. (2017). The role and limitations of EEG -based depth of anaesthesia monitoring in theatres and intensive care. *Anaesthesia* 72 (Suppl. 1): 38–47.

79 Stevner, A.B.A., Vidaurre, D., Cabral, J. et al. (2019). Discovery of key whole-brain transitions and dynamics during human wakefulness and non-REM sleep. *Nature Communications* 10: 1035, 14 pages.

80 Haimovici, A., Tagliazucchi, E., Balenzuela, P., and Laufs, H. (2017). On wakefulness fluctuations as a source of BOLD functional connectivity dynamics. *Scientific Reports* 7: 5908.

81 Jurysta, F., Van Wettere, L., Lanquart, J.-P. et al. (2013). Description of the brain functional connectivity across sleep stages using a coherence analysis. *European Psychiatry* 28 (Supplement 1): 1.

82 Nguyen, T., Babawale, O., Kim, T. et al. (2018). Exploring brain functional connectivity in rest and sleep states: a fNIRS study. *Scientific Reports* 8:16144, 10 pages.

83 Amorim, L., Magalhães, R., Coelho, A. et al. (20 November 2018). Poor sleep quality associates with decreased functional and structural brain connectivity in normative aging: a MRI multimodal approach. *Frontiers in Aging Neuroscience* 10, Article 375: 1–16.

84 Tuovinen, N., Stefani, A., Mitterling, T. et al. (April 2020). Functional brain connectivity alterations in restless legs syndrome are modulated by dopaminergic medication. *Sleep* 43 (Supplement_1): A4.

13

EEG-Based Mental Fatigue Monitoring

13.1 Introduction

Fatigue is a state of body stability and awareness also called exhaustion, lethargy, languidness, languor, lassitude, and listlessness, associated with physical or mental weakness. *Physical fatigue* is the inability to continue normal body functioning. Conversely, *mental fatigue* can be defined as the state of reduced brain activity, ability to respond to various stimuli, and associated cognitive functions arising during continuous mental activity. Mental fatigue can manifest itself both as somnolence (decreased wakefulness), or only as diminished attention not necessarily including sleepiness [1]. Fatigued people often experience difficulties in concentration and appear more easily distracted. Objective cognitive testing may differentiate various states of mental fatigues. Unfortunately, neurocognitive mechanisms underlying the effects of mental fatigue are not yet well understood.

Evaluation of mental fatigue has turned out to be a necessary and attractive research subject due to its effects on personnel who require extended attention in execution of their work such as drivers, pilots, security guards, and because of its effects on many other human biometrics. These effects can be recognized by examining and analyzing the brain activity either directly using electroencephalograms (EEG) and magnetoencephalograms (MEG) or indirectly using functional magnetic resonance images (fMRIs).

Mental fatigue can be the result of boredom, depression, disease, jet lag, lack of sleep [2, 3], mental stress, overestimation and understimulation, thinking, or prolonged working. Most of these causes lead to temporary mental fatigue. However, some diseases such as uraemia, blood disorder, and chronic fatigue syndrome (CFS) may cause chronic fatigue which can last even for a few months. Therefore, rarely, there is need for full clinical check up to find out the causes of chronic fatigue. The majority of people with chronic fatigue do not have any underlying cause discovered after a year with a condition. Conversely, there is not any need for any clinical examination when the subject is under temporary mental fatigue.

Temporary mental fatigue may become an issue if the subject involves in sensitive and critical jobs such as driving or guarding. Consequently, and unlike physical fatigue, mental fatigue can seriously affect normal sleep and often causes various sleep disorders [2, 3]. Therefore, an inclusive study of mental fatigue from the EEG signals requires analysis of both background EEGs and event-related brain responses (ERPs).

EEG Signal Processing and Machine Learning, Second Edition. Saeid Sanei and Jonathon A. Chambers.

In a study by Lorist et al. [4] it was examined whether: (i) error-related brain activity, (ii) indexing performance monitoring by the anterior cingulate cortex, and (iii) strategic behavioural adjustments were modulated by mental fatigue. In this experiment the hand response time to follow a visual stimulus and its variance were measured and evaluated. As a result, it was found that firstly, mental fatigue is associated with compromised performance monitoring and inadequate performance adjustments after errors. Secondly, monitoring functions of anterior cingulate cortex and striatum rely on dopaminergic inputs from the midbrain. Thirdly, patients with striatal dopamine deficiencies show symptomatic mental fatigue. Based on these collective results it was concluded that mental fatigue results from a failure to maintain adequate levels of dopaminergic transmission to the striatum and the anterior cingulate cortex, resulting in impaired cognitive control.

Traditionally, there are three main approaches for estimation of fatigue level. These are based on psychometrics, video, and physiological measurements. Psychometrics-based approaches involve questionnaires, filled up by subjects at respective intervals based on which the fatigue level is estimated [5, 6]. This type of approach is less reliable, as it can be biased by the subjective nature of questionnaires. Video-based measurements mostly to capture the facial expressions have also been used as a marker for mental fatigue [7]. Video-based methods are not accurate either as they are subjective and changes with mode, age, and environment. More reliable approaches are by the use of physiological and neurophysiological measures, such as EEG [8–10], electrocardiography (ECG), and electro-oculography (EOG) [11, 12].

From the results of study by Craig et al. [8] it can be concluded that the normal brain rhythms do not significantly change due to fatigue and therefore, the change in power spectrum of various conventional EEG frequency bands may not be a good indicator of mental fatigue. Nevertheless, in many fatigue classification algorithms they use the spectral features. Also, in severe mode change such as high concentration or anxiety due to approaching or facing an accident in driving [11] a sudden change in beta activity (or EOG) is inevitable. This means advanced processing and learning techniques are required to extract or learn the necessary features from the brain responses which are manifested in the ERP amplitudes and latencies as well as developing and using suitable methods for estimating the brain synchronization and connectivity.

It is common to measure various time and frequency domain features from the EEG data recorded over several hours for the assessment of mental fatigue. Then, different classifiers such as random forest [13], support vector machines (SVM) [14–16], and hidden Markov model (HMM) [17, 18] can be used to classify them. The changes in some statistical measures such as Kolmogorov entropy have also been evaluated for the same purpose [19].

Synchronization, coherency, and connectivity of the brain regions may be examined during the normal-to-fatigue transition. In the case of mental fatigue, these parameters for different conventional frequency bands can gradually change when the brain moves from alert/normal-to-fatigue states.

Conversely, based on ERP studies, the ability of subjects to focus their attention on a task decreases with time [1, 4]. Attention is an important feature of dynamic human behaviour; it allows one to (*i*) directly process the incoming information in order to focus on the relevant information for achieving certain goals and (*ii*) actively ignore irrelevant information that may interfere with the goals [1]. Therefore, in the analysis of mental fatigue one key direction is to examine how mental fatigue affects these attentional processes. This can be achieved by analysis of different ERP components and subcomponents before and during the fatigue state. Some variations in attention-related ERP components induced by mental fatigue have provided strong evidence that attentional processes are indeed influenced by mental fatigue [1].

In this chapter some attempts in detection, quantification, and monitoring of mental fatigue using either brain rhythms or ERPs and their subcomponents are described in detail. Some mathematical and signal processing techniques with potential application to fatigue analysis are also explored. Finally, the above approaches based on EEG and ERP are linked to pave the path for a unified approach to analysis of mental fatigue.

13.2 Feature-Based Machine Learning Approaches

There are many proposals for classification of selected EEG features for monitoring or prediction of mental fatigue. Unfortunately, many of them tried to extract EEG spectral features and classify them without looking at the underlying physiological states of the brain. Nevertheless, some paid more attention to the changes in complexity of the signals and tried to classify such measures. In the following subsections only some samples of different feature-based methods are addressed.

13.2.1 Hidden Markov Model Application

In [20] a dynamic fatigue detection model based on a HMM is proposed in this paper. Driver fatigue can be estimated by this model in a probabilistic way using various forms of physiological and contextual information. As the physiological information, EEG, electromyogram (EMG), and respiration signals were simultaneously recorded by wireless wearable sensors during the real driving.

In their work, the raw EEG signals were divided into the conventional EEG frequency bands by means of a Fast Fourier transformation. The power of theta (4–8 Hz), alpha (8–13 Hz), and beta (13–30 Hz) bands were used as the EEG information also used to calculate the power ratio $P_{theta+alpha}/P_{alpha+beta}$. They also used root mean square of the filtered EMG within the frequency band of 25–500 Hz as another feature. A respiration feature was considered as the mean frequency power of the respiration signal.

As contextual features, the following have been considered: (i) driver-related, such as personality, sleep quality, circadian rhythm, and physical condition, (ii) vehicle-related, i.e. noise, seating comfort degree, and temperature, and (iii) road-related, including monotony of road, density of vehicles, and the number of lanes.

A kernel distribution estimate over different time segments is then used to estimate the fatigue likelihood. Contextual information offered by specific environmental factors was used as a prior for fatigue. The posterior probability is captured by an HMM-based fatigue recognition method in the final stage. The HMM states are conditioned by the contextual information.

13.2.2 Kernel Principal Component Analysis and Hidden Markov Model

Approximate entropy (ApEn) and Kolmogorov complexity (Kc) are estimated from the EEG signals and utilized to characterize the complexity and irregularity of EEG data under different mental fatigue states [18]. The kernel principal component analysis (KPCA) algorithm is employed to extract the nonlinear features from the complexity parameters and improve the generalization performance of the HMM.

In KPCA the covariance matrix is computed over a nonlinear function of the signal (such as radial basis function [RBF]) rather that the signal itself, i.e. $C_x = \frac{1}{L}\sum_{i=1}^{L}\varphi(x_i)\varphi(x_i)^T$. The remaining feature selection and data reduction stages follows the traditional principal component analysis (PCA).

Two HMMs, one representing the normal/alert state (HMM_N) and another one the fatigue state (HMM_F) are trained by using the EEG data segments recorded during the corresponding mental fatigue states. The parameters of the models are estimated by the given training data and are then used to classify the same training data. Finally, HMM_N and HMM_F are estimated by using the correct classified trials. Classification of an unknown EEG data segment is based on a selection of the maximum single best path probability $p(\overline{V} \mid HMM)$ calculated via the Viterbi algorithm [21]. Calculating $p(\overline{V} \mid HMM_N)$ and $p(\overline{V} \mid HMM_F)$ for all EEG segments will result in a propagation of these probabilities, which allows sample by sample classification.

It has been concluded that both complexity parameters significantly decrease as the mental fatigue level increases suggesting that these parameters can be good indicators of mental fatigue level. Using this approach, a classification accuracy of 84% is achieved.

Moreover, the joint KPCA–HMM method can significantly reduce the dimensionality of the feature vectors.

13.2.3 Regression-Based Fatigue Estimation

EEG spectral features have been previously used to score the fatigue state. As an example, in [22] the authors combine EEG log sub-band power spectrum, correlation analysis, PCA, and linear regression models to indirectly estimate driver's drowsiness level in a virtual-reality-based driving simulator. Regression techniques such as support vector regression are suitable for online classification, monitoring and estimation of fatigue or drowsiness which can be applied to driver or pilot fatigue and prevent the associated hazards. In [5] the EEG power spectral density was used to extract spectral features using Welch's method (Hamming window and 50% overlap). The power in each of the five EEG frequency bands, i.e. delta, theta, alpha, beta, and gamma, is calculated and the ratios $P_{theta+alpha}/P_{beta}$ and P_{theta}/P_{beta} were extracted from each channel, resulting in a total of 168 features for 24 channels. These features have been used as the inputs to their regression models and the results compared.

They also proposed dynamic time warping as a similarity measure between the estimated fatigue level obtained from the regression output and the actual fatigue level considered as the reaction time. The lowest similarity has been considered as the warping distance value and the highest as the similarity between the signals.

They concluded that by using a RBF kernel and sliding five-second window the best performance can be achieved.

13.2.4 Regularized Regression

The authors in [23] tackle the problem of online driver drowsiness estimation from EEG signals by integrating fuzzy sets with domain adaptation. They attempt online calibration to mitigate the inter-subject variability before applying their fuzzy sets for classification.

To formulate the problem, for a feature space a domain D in transfer learning consists of a d-dimensional feature space X and a marginal probability distribution $P(\mathbf{x})$, i.e. D = {X, $P(\mathbf{x})$}, where $\mathbf{x} \in$ X.

Two domains D^z and D^t are different if $\chi^z \neq \chi^t$ or $P^z(\mathbf{x}) \neq P^t(\mathbf{x})$. In transfer learning a task T consists of an output space Y and a conditional probability distribution Q(y|$\mathbf{x}$). Two tasks T^z and T^t are different if $Y^z \neq Y^t$, or $Q^z(y \mid \mathbf{x}) \neq Q^t(y \mid \mathbf{x})$.

Given the zth source domain D^z with n_z samples $(\mathbf{x}_i^z, y_i^z), i = 1, \dots, n_z$, and a target domain D^t with m calibration samples $\left(\mathbf{x}_j^t, y_j^t\right), j = 1, \dots, m$, domain adaptation aims to learn a target prediction function $f(\mathbf{x}): \mathbf{x} \rightarrow y$ with low expected error on D^t, under the assumptions that $\chi^z = \chi^t$, $Y^z = Y^t$, $P^z(\mathbf{x}) \neq P^t(\mathbf{x})$, and $Q^z(y \mid \mathbf{x}) \neq Q^t(y \mid \mathbf{x})$.

In driver drowsiness estimation, EEG signals from a new subject are in the target domain, while EEG signals from the zth existing subject are in the zth source domain [23]. An EEG epoch consists of the feature vector in either domain. Though the source and target domain features are extracted in the same way, their marginal and conditional probability distributions are generally different, i.e. $P^z(\mathbf{x}) \neq P^t(\mathbf{x})$ and $Q^z(y \mid \mathbf{x}) \neq Q^t(y \mid \mathbf{x})$, because different subjects often behave similarly despite having distinct drowsy neural responses. As a result, data from a source domain cannot represent data in the target domain accurately and must be integrated with some target domain data for inducing the target domain regression function.

Based on the above problem description, with the help of fuzzy sets, the regression problem is transformed into a 'classification' problem and hence conditional probability distribution adaptation is performed as explained in [23]. To do that, the above four assumptions are linked and formulated using regularized parameters and solved using regularized optimization.

13.2.5 Other Feature-Based Approaches

In [24] the authors applied independent component analysis (ICA) by entropy rate bound minimization (ICA-ERBM) to the EEG signals for source separation. This method exploits both the non-Gaussian property and sample correlation during the minimization of mutual information [25]. Then, autoregressive based short-term power spectra of the ICAs are computed. The signal powers over short sliding windows of 2 seconds duration with 1.75 seconds overlap are used as the features and then applied to a multilayer perceptron classifier with two outputs denoting alert and fatigue.

Wavelet packet energy (WPE) values have been used as the input features to a random forest classifier in [26]. In another attempt, the researchers used EEG to validate the Karolinska sleepiness scale (KSS) based on the Karolinska drowsiness test (KDT) [27]. In their method, the occipital alpha-theta power difference and the frontal and occipital scores on the second principal component of the EEG spectrum were calculated for each one-minute interval of five minutes eyes closed section of the record. These values were used to validate the match between KDT and KSS [28].

13.3 Measurement of Brain Synchronization and Coherency

Mental fatigue affects the neural communication pathways between various brain zones. Accordingly, it changes the coherency and synchronization between the brain regions. In the following subsections, several different metal fatigue indicators based on the above changes are introduced.

13.3.1 Linear Measure of Synchronization

Incorporation of brain synchronization and coherency measures between the cerebral regions [29] as estimators of fatigue is another major approach which has been followed by a number of researchers such as in [30, 31]. A functional relationship between different brain regions is generally associated with synchronous electrical activities in these regions. Recorded EEGs can be used for measuring synchronization of different brain regions. In the approach by our group [32] linear and nonlinear synchronization of different brain regions are evaluated by exploiting the empirical mode decomposition (EMD) algorithm [33]. The synchronization measure can reveal useful information about the changes in the functional connectivity of brain regions during the fatigue state.

Therefore, first a brain region is selected and then the EMD algorithm, as a signal-dependent decomposition method, is applied to one channel of the EEG time series in the selected region to decompose it to waveforms modulated in amplitude and frequency. The iterative extraction of these components called intrinsic mode functions (IMFs), is based on local representation of the signal as the sum of a local oscillating component and a local trend. The IMFs can be considered as the reference signals for the brain rhythmic activities. Then, the linear and nonlinear synchronization measures can be estimated using the selected IMFs in different brain regions for the subject before and during fatigue state. These measures are important for detecting and evaluating the mental fatigue in real-world applications. One important issue is that the extracted IMFs might be noisy. This happens for the first few IMFs. In several research works conventional filtering was suggested to remove the noise from the IMFs. This, however, may result in loss of phase information.

As an effective option, an adaptive line enhancer (ALE) [34] can be applied to the resulting (narrowband) IMF which may contain wideband noise. Since the ALE is traditionally meant for detection of periodic signals with known period, it exploits the cyclic nature of the IMFs to restore the noise. The filtered IMFs are then used in the measurement of synchronization.

The ALE is used in many applications such as sonar, biomedical, and speech signal processing where the aim is to detect periodic signal components. The ALE can be considered as a degenerate form of adaptive noise canceller. The reference signal in the ALE, instead of being derived separately, as in the adaptive noise canceller, is a delayed version of the input signal [35]. Therefore, the delay Δ is the prediction depth of the ALE, measured in units of the sampling period. The reference input $\mathbf{u}(n-\Delta)$ is processed by a transversal filter resulting in an error signal $e(n)$, defined as the difference between the actual input $\mathbf{u}(n)$ and the ALE's output. The error signal is used to activate the adaptive algorithm by adjusting the weights of the transversal filter, mainly according to Widrow's least-mean-square (LMS) algorithm. The tap-weight adaptation can then be obtained through:

$$\mathbf{w}(n+1) = \mathbf{w}(n) + \mu\mathbf{u}(n)e(n) \tag{13.1}$$

where n is the iteration time. In this equation μ is the step size, $e(n)$ is the adaptation error and $\mathbf{u}(n)$ is the input vector, respectively, at time n. In the case of a fixed step size LMS algorithm, μ is chosen as a constant. After applying the ALE to the extracted IMF by the EMD algorithm, it is possible to measure the synchronization measures of the different enhanced IMFs obtained from different parts of the brain. Given a clean IMF around the frequency of interest (such as alpha or beta brain rhythms), linear, and nonlinear synchronization parameters can be evaluated as follows.

Suppose we have two IMFs simultaneously obtained from different channels. Every IMF is a real valued signal. The discrete Hilbert transform (HT) [33] is used to compute the analytic signal for an IMF. The discrete HT denoted by $H_d[\bullet 3]$ of signal $x(t)$ is given by:

$$H_d[x(t)] = \sum_{\substack{\delta = -\infty \\ \delta \neq 0}}^{+\infty} \frac{x(\delta)}{t-\delta} \tag{13.2}$$

The analytic function for the ith IMF $d_i(t)$ is defined as:

$$C_i(t) = d_i(t) + jH_d[d_i(t)] = a_i(t)e^{j\theta_i(t)} \tag{13.3}$$

where $a_i(t)$ and $\theta_i(t)$ are the instantaneous amplitude and phase of the ith IMF respectively. The analytic signal is utilized in determining the instantaneous quantities such as energy, phase, and frequency. The discrete time instantaneous frequency (IF) of the ith IMF is then given as the derivative of the phase $\theta_i(t)$ calculated at t:

$$\omega_i(t) = \frac{d\theta_i(t)}{dt} \tag{13.4}$$

Conversely, the Hilbert spectrum represents the distribution of the signal energy as a function of time and frequency. If we partition the frequency range into k frequency bins, the instantaneous amplitude of the ith IMF at kth frequency bin can be defined as:

$$H_i(k,t) = a_i(t)v_i^k(t) \tag{13.5}$$

where $v_i^k(t)$ takes the value 1 if $\omega_i(t)$ falls within kth band. The marginal spectrum corresponding to the Hilbert spectrum $H_i(k,t)$ of the ith IMF is defined as:

$$\psi_i(k) = \sum_{t=1}^{T} H_i(k,t) \tag{13.6}$$

where T is the data length. Using this information, the coherence function, as a measure of synchronization between the two IMFs, is defined as:

$$\zeta_{ij}(k) = \frac{|\chi_{ij}(k)|^2}{\psi_i(k)\psi_j(k)} \qquad k = 1, 2, \ldots, B \tag{13.7}$$

where B is the number of frequency bins, $\chi_{ij}(k)$ is the cross-spectrum between the ith and jth IMFs, $\psi_i(\mathrm{k})$ and $\psi_j(\mathrm{k})$ are the marginal power spectra of the ith and jth IMFs respectively, and $\zeta_{ij}(k)$ is a quantity which shows how much the ith and jth IMFs are correlated.

13.3.2 Nonlinear Measure of Synchronization

It is likely to have two dynamic systems for modelling the activities of two brain lobes with synchronized phases while their amplitudes are uncorrelated [30]. Therefore, it is useful to consider phase synchronization as a nonlinear measure of synchronization. Two signals are phase synchronous if the difference of their phases remains constant across time [31]. After finding the instantaneous phase of the ith and jth IMFs, the phase synchronization measure can be defined as:

$$\gamma_{ij} = \left|\left\langle e^{\,j\theta_{ij}(t)}\right\rangle_t\right| = \sqrt{\left\langle \cos\theta_{ij}(t)\right\rangle_t^2 + \left\langle \sin\theta_{ij}(t)\right\rangle_t^2} \tag{13.8}$$

where $\langle\,.\,\rangle$ denotes average over time, $\theta_{ij}(t) = \theta_i(t) - \theta_j(t)$, and $\theta_i(t)$ is the instantaneous phase of the ith IMF obtained by the HT. In this case, γ_{ij} will be zero if the phases are not synchronized at all and will be one when the phase difference is constant (perfect synchronization). The key feature of γ_{ij} is that it is only sensitive to phases, irrespective of the amplitude of each signal.

The above technique has been applied to a set of real data. The real EEG data used here belong to one of the subjects that participated in a continuous visual experiment. The stimuli were presented in the centre of a computer screen positioned at a viewing distance of 80 cm. In each trial, the participants were presented with a horizontal array of three uppercase letters, the central one of which was the target letter and the remaining letters were the flankers. The participants were instructed to make a quick left-hand response if the central letter was an H and a right-hand response if the central letter was an S. The experimental sessions lasted three hours. EEG was recorded from 22 scalp sites, using Sn electrodes attached to an electrode cap (ElectroCap International). Standard 10–20 sites were F7, F3, Fz, F4, F8, T7, C3, Cz, C4, T8, P7, P3, Pz, P4, P8, O1, Oz, and O2. Additional intermediate sites were FC5, FC1, FC2, and FC6. The two separate segments of data from the first half an hour and the last half an hour were extracted. These segments are the representation of alert and fatigue states. Each segment is analyzed separately to measure the linear synchronization (coherency) and nonlinear synchronization (phase synchronization) between the left and right hemispheres. The alpha rhythm is extracted from F7 and F4, the beta rhythm is extracted from FC5 and FC6 and the theta rhythm is extracted from P7 and P8. Overlapped windows with a length of four seconds were considered for applying the HT and estimating the coherency and phase synchronization. The choice of channels in each hemisphere for extracting each rhythm is based on the HT of the resulting IMFs in several electrodes in the right and left part of the brain. The IMF is selected from the channel with IMF (in a certain frequency band) having more continuous frequency traces in the time–frequency distribution. Because the resulting beta rhythm was rather noisy, the ALE was used to enhance it. The results of the estimated coherency and phase synchronization values for the two segments of the data at three more effective frequency bands are shown in Figures 13.1 and 13.2 respectively. The algorithm is repeated over more trials for alert and fatigue states. In almost

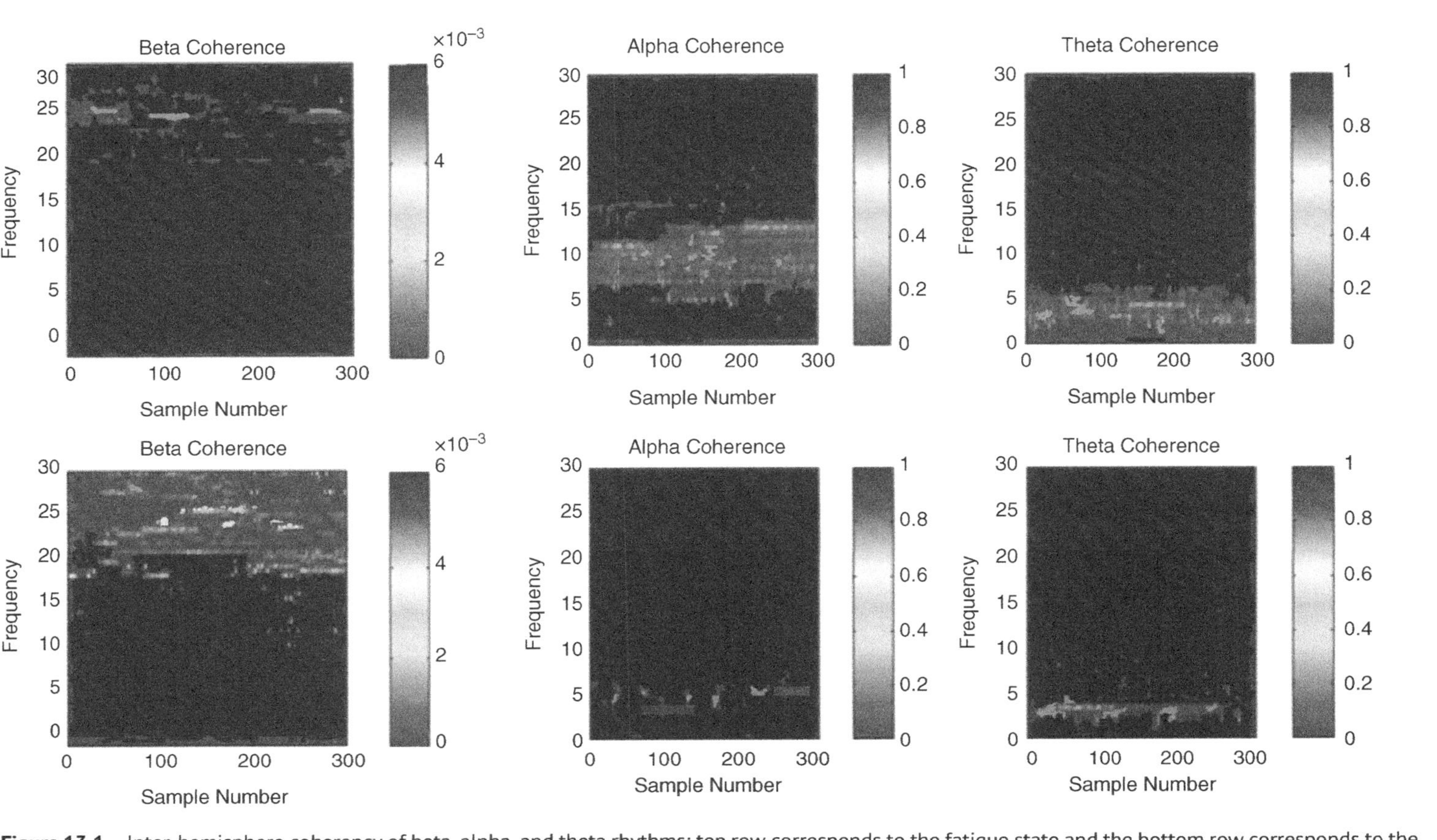

Figure 13.1 Inter-hemisphere coherency of beta, alpha, and theta rhythms; top row corresponds to the fatigue state and the bottom row corresponds to the alert state.

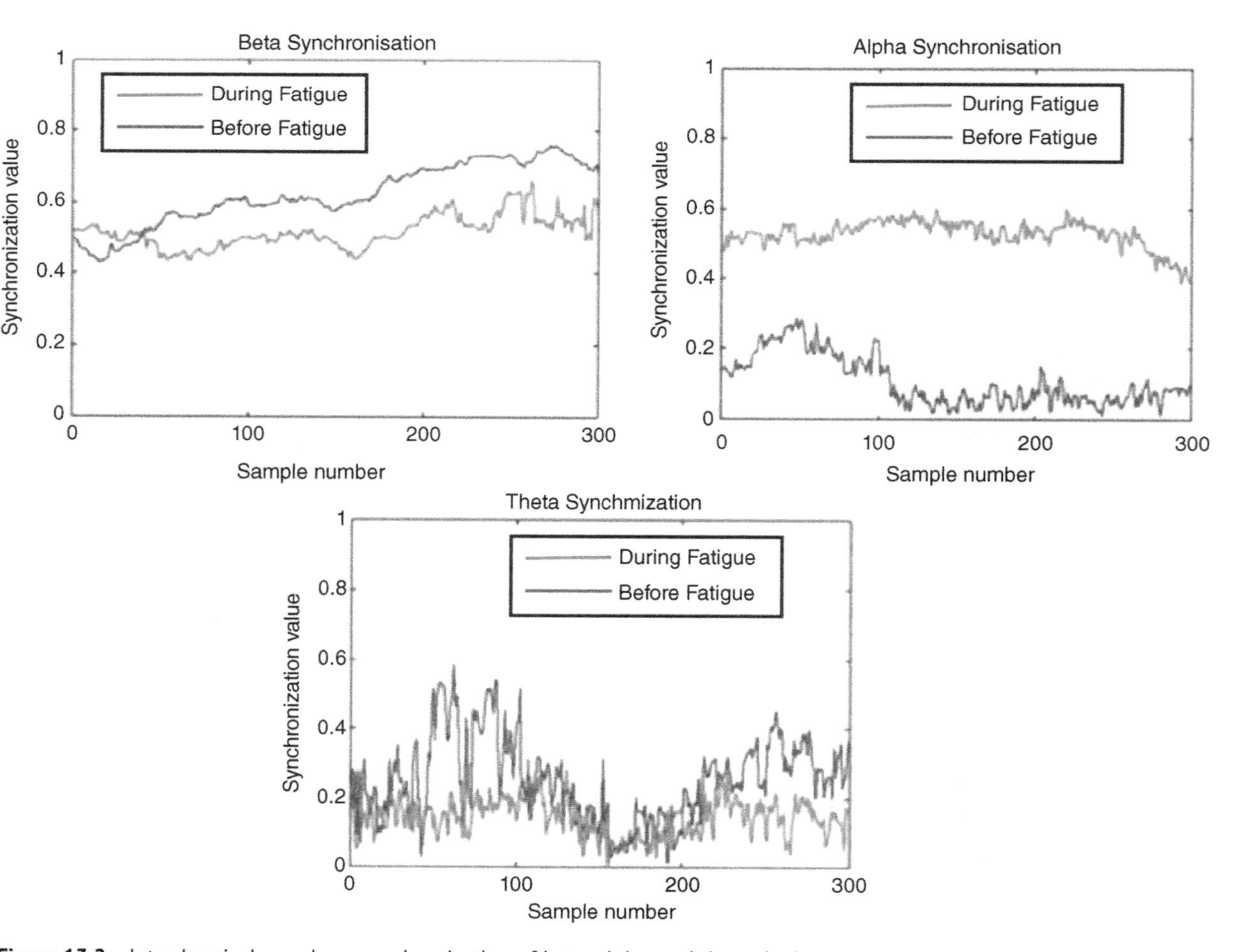

Figure 13.2 Inter-hemisphere phase synchronization of beta, alpha, and theta rhythms.

all the trials the coherency of beta rhythms between the left and right hemispheres decreases and the coherencies for alpha and theta rhythms increase from the alert-to-fatigue state. The estimated phase synchronization values between the rhythms have been shown to be different across different trials especially for the theta rhythm. However, in most trials, the phase synchronization of beta rhythms decreases while for alpha rhythms it increases from the alert-to-fatigue state.

Part of this work has been repeated by some other researchers such as those in [36] in the context of the assessment of driving fatigue. They used phase synchronization between the regions and also the synchronization within each region (named functional units) to set the features for the fatigue assessment. They concluded that the degree of global synchronization increases when moving from alert to fatigue and the synchronization values of the delta and alpha bands of the EEG signals from frontal and posterior occipital lobes significantly increase as the result of driving fatigue.

13.4 Evaluation of ERP for Mental Fatigue

Assessment of the changes in ERP parameters due to mental fatigue is another general approach to fatigue analysis. As mentioned in previous chapters of this book, the most popular method which also happens to be the simplest approach to extract the ERP components from single-trial measurements is averaging over time-locked single trials. In this approach the characteristics of the ERP wave (amplitude, latency, and width) are considered constant across trials and averaging over single trials leads to the attenuation of background EEG which is considered as a random process. In many real applications, this assumption is not true. For example, when there is a change in the degree of mental fatigue, habituation, or the level of attention, the ERP waveform changes from trial to trial. This variability of ERP components is ignored by averaging the ERP over a number of trials. Inter-trial variability of ERP components is an important parameter in order to investigate brain abnormalities. Several methods have been developed for single-trial estimation of ERP components based on statistical signal processing. These methods include Wiener [37], maximum a posteriori (MAP) [34] and Kalman filtering [30, 35] approaches. The objective of all these methods is to estimate ERP components from single-trial measurements and track the variability of their parameters across trials. However, these methods may fail in situations where the ERP signal to background noise ratio is very low or where the measured ERPs vary from trial to trial. An effective analysis of ERPs should thus be based on a robust single-trial estimation in which inter-trial variability of ERPs is also taken into account. A recent work in [38] formulates wavelet coefficients of the time-locked measured ERPs in a state space and then estimates the ERP components using particle filtering. It is shown that the formulated particle filtering outperforms Kalman filtering. The method, however fails to estimate ERP subcomponents. In this case particle filtering is a better solution to this problem. The proposed method here considers the temporal correlations between the subcomponents recorded from different sites of the brain and enables separation and identification of the ERP subcomponents.

Variation of different ERPs with time has been shown in some experiments. In these experiments the subjects were unable to prevent automatic shifting of attention [1] to irrelevant stimuli, reflected by a larger negativity in the N1 latency range (160–220 ms) for irrelevant, compared to relevant stimuli. This difference in negativity was unaffected by the time spent on a task. However, N1 and N2b (350–410 ms) amplitude did change with time: the N1 amplitude decreased and the difference in N2b amplitudes between relevant and irrelevant stimuli (larger N2b amplitude evoked by relevant stimuli) decreased with time on task. The results of these experiments indicated that the effects of mental fatigue on goal-directed (top-down) and stimulus driven (bottom-up) attention are disassociated: mental fatigue results in a reduction in goal-directed attention, leaving subjects performing in a more stimulus driven fashion.

The P300 component has been found useful in identifying the depth of cognitive information processing. It has been reported that the P300 amplitude elicited by mental task loading decreases with an increase in the perceptual or cognitive difficulty of the task and its latency increases when the stimulus is cognitively difficult to process [17, 32, 33, 37, 39, 40]. P300 contains two subcomponents: P3a and P3b. These subcomponents usually overlap over the scalp and have temporal correlation [37]. The P3a is an unintentional response of the brain to a salient stimulus independent of the task [14, 34]. Prefrontal, frontal, and anterior temporal brain regions play an important role in generating P3a, giving it a frontocentral distribution. P3b is more distributed over the centroparietal region as it is mainly generated by posterior temporal, parietal, and posterior cingulate mechanisms [35]. Moreover, P3a is different from P3b in its shorter latency and more rapid habituation [14, 35]. P3a has a more frontal cortical distribution while P3b has a more parietal cortical distribution [30]. It has been suggested in [38, 39] to use the P300 subcomponents for mental fatigue analysis since the averaged P300 does not always correspond to manifestation or appearance of mental fatigue. Evaluation of mental fatigue based on ERP is therefore recommended to be conducted from multiple perspectives: using not only P300 amplitude and latency but also feature parameters related to its subcomponents such as P3a and P3b.

This has been investigated by a number of researchers. In the work by Jarchi et al. [41] and the references herein the ERP components and their variations in amplitude, latency, and duration have been evaluated. A new coupled particle filtering for tracking variability of P300 subcomponents, i.e. P3a and P3b, across trials has been developed in this work. The latency, amplitude, and width of each subcomponent, as the main varying parameters, have been modelled using the state-space system. In this model the observation is modelled as a linear function of amplitude and a nonlinear function of latency and width. Two Rao-Blackwellised particle filters (RBPF) are then coupled and employed for recursive estimation of the state of the system across trials. By including some simple physiologically-based constraints, the proposed technique prevents generation of invalid particles during estimation of the state of the system. The constraints exploit the relative geometrical topographies of P3a and P3b. The main advantage of the algorithm compared with other single-trial-based methods is its robustness in the low signal-to-noise ratio situations.

In this approach the variables describing the states of the state-space approach have been defined as the amplitudes, latencies, and widths of both the P3a and P3b subcomponents. Empirically, the amplitude varies linearly and the other two variables change nonlinearly with the changes in brain response. Based on the RBPF concept a Kalman filter tracks the linear component whereas the changes in the nonlinear parameters are tracked by the particle filter.

To formulate the problem using the RBPF consider that the nonlinear state is denoted by $\boldsymbol{x}_k^1$ and the linear (under some possible conditions) state by $\boldsymbol{x}_k^2$. RBPF may be used for tracking the ERP in a single-channel signal. In the initialization stage, instead of generating random particles, it is feasible to use the averaged ERP of several consecutive trials and to generate some particles which contribute more to the posterior density. In the subsequent trials, due to the existence of noise and artefact in the data, the correct track of the ERP subcomponent parameters may be lost. Following this work [41], some physiologically meaningful constraints on the state variables have been imposed in order to remove invalid particles while they contribute to estimation of the posterior density. The constraints can be set using the prior knowledge about the ERP subcomponents' specifications. This can be expanded to more EEG channels. This allows incorporation of more statistical and physiological constraints within the estimation process.

The constrained BRPF (CBRPF) consists of two inter-connected co-operating BRPFs. This co-operation allows for discarding irrelevant particles and refining the state transitions [41].

P300 subcomponents (P3a and P3b) are spatially disjoint. They usually overlap temporally over the scalp [35]. We recall that P3a is stronger over the frontal electrodes whereas P3b has larger amplitude over the parietal electrodes. The Fz electrode in the frontal and Pz electrode in the parietal site are therefore good choices for this analysis. Two inter-connected RBPFs are developed. The first one (RBPF1) is applied to the Fz channel while the second one (RBPF2) to the Pz channel. Such assumptions and using the above channels is in line with our objectives in characterizing or tracking the P300 subcomponents.

A simple approach to mathematically describe the state variables of RBPF1 and RBPF2 is as follows:

$$\mathbf{x}_k^1 = \mathbf{R}_k^i = \left[\mathbf{b}_k^i \;\; \mathbf{s}_k^i \;\; \widetilde{\mathbf{b}}_k^i \;\; \widetilde{\mathbf{s}}_k^i\right]^T \tag{13.9}$$

and

$$\mathbf{x}_k^2 = \mathbf{A}_k^i = \left[\mathbf{a}_k^i \;\; \widetilde{\mathbf{a}}_k^i\right]^T \tag{13.10}$$

where $\mathbf{a}_k^i$, $\mathbf{b}_k^i$, and $\mathbf{s}_k^i$; $i = 1, 2$, are respectively the amplitude, latency, and width of P3a and $\widetilde{\boldsymbol{a}}_k^i$, $\widetilde{\boldsymbol{b}}_k^i$, and $\widetilde{\mathbf{s}}_k^i$; $i = 1, 2$, are respectively the amplitude, latency, and width of P3b for the kth trial of the ith RBPF. For such a system the state-space equations are therefore:

$$\begin{aligned} \mathbf{R}_k^i &= \mathbf{R}_{k-1}^i + \mathbf{w}_{k-1}^{\upsilon i} \\ \mathbf{A}_k^i &= \mathbf{A}_{k-1}^i + \mathbf{w}_{k-1}^{Ai} \end{aligned} \tag{13.11}$$

where $\mathbf{w}_{k-1}^{\upsilon i}$, and $\mathbf{w}_{k-1}^{Ai}$ are zero-mean Gaussian white noise (GWN) with known covariance matrices. It is seen that tracking two temporally correlated P300 subcomponents from a single channel does not guarantee the correct estimation of the components. This is mainly because the formulated RBPF considers the sum of two Gaussians plus noise and tracks the changes of Gaussian parameters across trials.

Therefore, the variations in the parameters may not be well tracked if any sudden change in any one of the parameters of the Gaussians occurs. There is also possibility of modelling a segment of noise as a Gaussian. Therefore, we should use two RBPFs. Exploiting the logical links between these RBPFs, it is possible to prevent deviation of the estimation from the actual values. In order to make a connection between the two RBPFs, it is assumed that the latencies and widths of P3a and P3b in the brain within both Fz and Pz channels are the same and only the amplitudes can be different. This assumption may not be exactly true in the real case, but it helps to couple two RBPFs in order to simultaneously track the changes in a system where there is an uncertainty in occurrence of the signals and their changes across trials because of the very low amplitude of the signal and inter-trial variability. Therefore, the benefit of simultaneous tracking is to reduce the uncertainty in the system by considering some meaningful physiological-based constraints.

In addition, we can assume a small delay in the latencies of P3a and P3b within the Fz and Pz channels. In the initialization stage, considering the averaged ERP we generate a rather large number of particles with relatively large variances for the latency, amplitude, and width. These initial particles are the same for RBPF1 and RBPF2. In the subsequent iterations, the particles with small weights are replaced by the particles with large weights.

Furthermore, it should be noted that in each iteration the required relation between the two RBPFs should be taken into account for drawing samples from the prior density. Since the shapes of P3a or P3b are the same for the Fz and Pz channels, the pairs of particles with the same width for P3a and P3b can be coupled. The basis of the designed CRBPF is in coupling particle pairs.

Initializations of the particles for RBPF1 and RBPF2 are the same [41]. In particle pairs, the width is the same for P3a in both RBPF1 and RBPF2, and also the same for P3b. Therefore, it may be sufficient to consider the same width for every subcomponent in both RBPFs in each iteration. However, the amplitudes of the subcomponents are different and the latencies differ with small delay. The widths and latencies of the subcomponents can be generated from the prior density. Then, Kalman filtering can be applied for estimation of amplitudes of the subcomponents. In this stage, some invalid particles may be generated. It is effective to detect these invalid particles and remove them by setting their weights to zero so that they do not have any contribution in the estimation of state of the system. One class of invalid particles are those where the estimated amplitude of P3a at Pz channel is larger than the amplitude of P3a at Fz channel. Also, the particles in which the estimated amplitude of P3b at Fz channel is larger than the amplitude of P3b at the Pz channel are considered invalid. Usually, P3a and P3b overlap over the scalp but nearly always P3a has a shorter latency than P3b. Therefore, it is possible to have the P3a and P3b latencies close to each other and the new estimated latencies iteratively substituted for P3a and P3b. The particles in which the latency of P3b is shorter than the latency of P3a are marked invalid. Mutual influence between the P3a and P3b is also used to adjust the weights accordingly [41].

The performance of this system has been examined using EEG recorded data from subjects who were required to be nonsmokers, to have normal sleep patterns, not to work night shifts, and not to be under medication. As described in Section 13.3.2, in each trial a horizontal array of three uppercase letters, in which the central one was the target letter and the remaining letters were the flankers, were presented to the subjects. The subjects were instructed to make a left-hand response as quickly as possible if the central letter was an H and a right-hand response if the central letter was an S. The letter array remained on the screen until a response was given by the subject or disappeared after 1200 ms. The subjects were also allowed to correct their incorrect responses within 500 ms following the initial response. In 50% of the trials, the target letter was presented in red or green background while in the other half of the trials it was presented in black background. In half of the trials, the flankers had the same type and colour as the target letter (e.g. HHH or SSS: compatible) while in the other half of the trials they had a different type and colour than the target letter (e.g. SHS or HSH: incompatible). The stimuli of different types were presented randomly with equal probability. About 1000 ms before appearance of the three letters, a pre-cue was presented for 150 ms, specifying either the colour of the target letter (the Dutch word for 'red' or 'green') or the response hand (the Dutch word for 'left' or 'right'). The hand and colour cues were presented randomly with equal probability. In 80% of the trials, the cues were valid. During each trial, a fixation mark remained visible on the screen (an asterisk of 0.5×0.5 cm).

The interval between the initial response to one trial and the beginning of pre-cue presentation on the following trial varied randomly between 900 and 1100 ms. The subjects were instructed to respond quickly and accurately and to minimize eye movements and blinking during task performance. The task comprised of a two-hour experimental block without breaks followed by a practise block of 80 trials. EEG data was recorded from 22 scalp positions and the eye-blink artefacts removed. In some previous research using this dataset, the P300 was detected from the averaged ERP [38]. The CRBPF was applied to the Fz and Pz channels considering 50 trials from the first and fourth half an hour. In the first half an hour the subject was not under fatigue, while in the fourth half an hour the subject was under fatigue. In both cases, before and during fatigue, the averaged ERP of 20 trials was used to initialize the particles of the CRBPF. The results of tracking P3a and P3b variability are shown in Figure 13.3. This figure illustrates the variability of P3a and P3a amplitudes, latencies, and widths before and during the fatigue state. Based on the results, it is possible to draw some conclusions about the effect of mental fatigue on P300.

Generally, in some other researches on mental fatigue [17, 32, 33, 39, 40], the decrease in the overall P300 amplitude and increase in its latency have been reported. But using the above method, we could separate P300 into its constituent subcomponents and evaluate the effect of fatigue on each subcomponent separately. Based on the results, the latencies of both P3a and P3b increase with time over the task interval. During the fatigue state, the increase of the latency of P3a is slightly more than that for the P3b. Indeed, the amplitude of P3a decreases more than that of the P3b. The width of P3b remains approximately constant in the fatigue state while the width of P3a decreases. The P3a amplitude seems to become smaller with time on the task, while that of the P3b seems to increase somewhat.

During the first half an hour, a clear difference between P3a and P3b scalp positions is observed; P3a is more prominent at Fz, as expected. During the last half an hour the P3a became very small at Pz, while the P3b is prominent at this position. P3a is related to novelty. However, it is possible that the decrease in the P3a amplitude with time on the task is related to practise or habituation to the task. Therefore, mental fatigue can be related to the increase in P3a and P3b latencies and decrease in P3a amplitude and width. In order to investigate the effect of fatigue in more details, it is useful to apply the method to more signals and also to more electrodes in the frontal and posterior scalp regions.

Although monitoring ERPs influenced by mental fatigue is a valuable diagnostic tool, but it may not be robust enough to mark the details of gradual changes. This can apply for

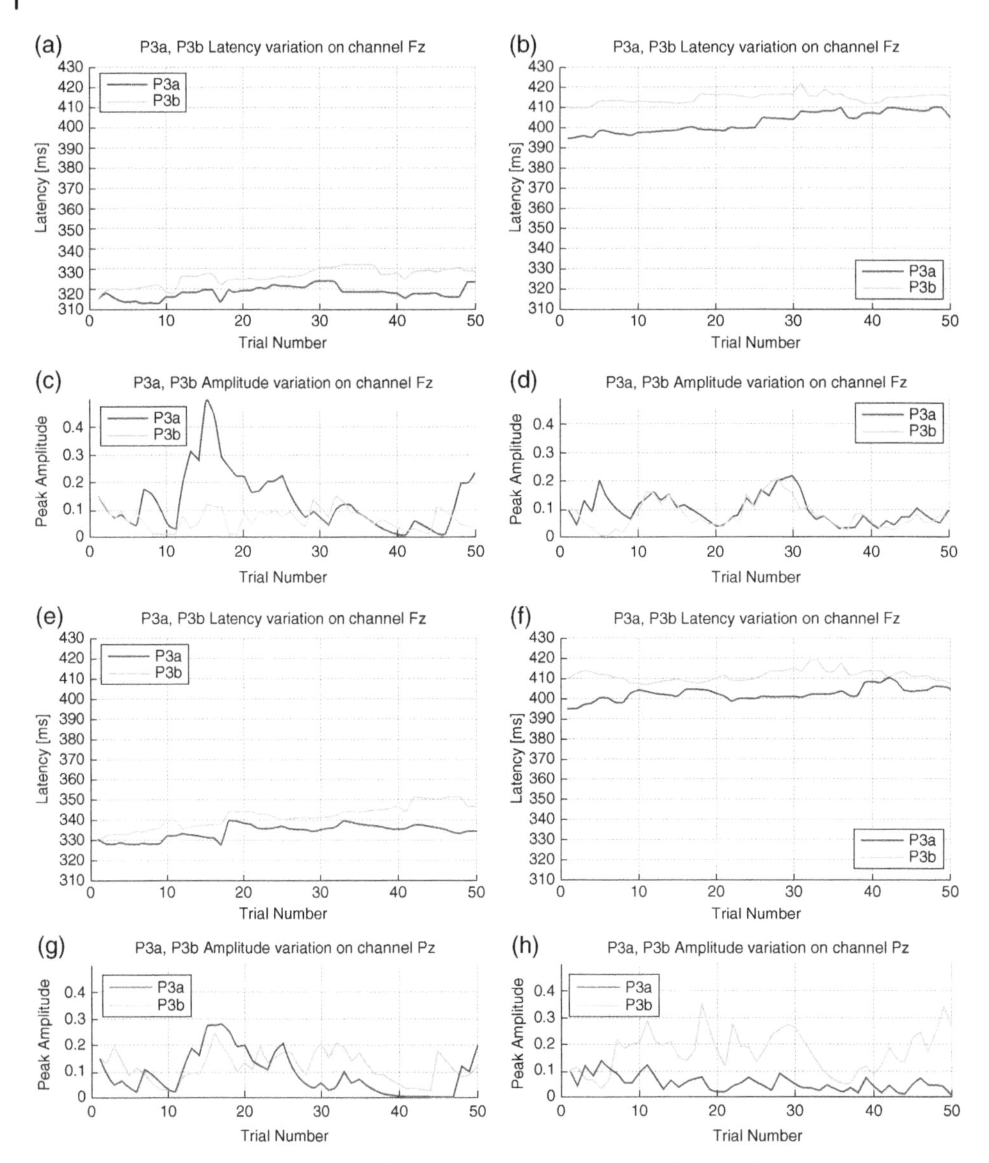

Figure 13.3 Tracking variability of P3a and P3b before and during fatigue; the left column corresponds to the parameters measured before fatigue (a), (c), (e), (g), (i), (k) while the right column to the parameters measured during the fatigue (b), (d), (f), (h), (j), (l).

evaluation of coherency and synchronization too. However, for the sake of robustness, it would be extremely useful if both approaches could be applied simultaneously to the same data sets recorded from the same subjects. This requires a suitable recording protocol which elicits both ERP and the changes in the brain rhythms. For this purpose, one way is to stimulate the brain after a longer time interval to allow for complete settling of the brain after each stimulus and enable evaluation of synchronization and coherency within these intervals.

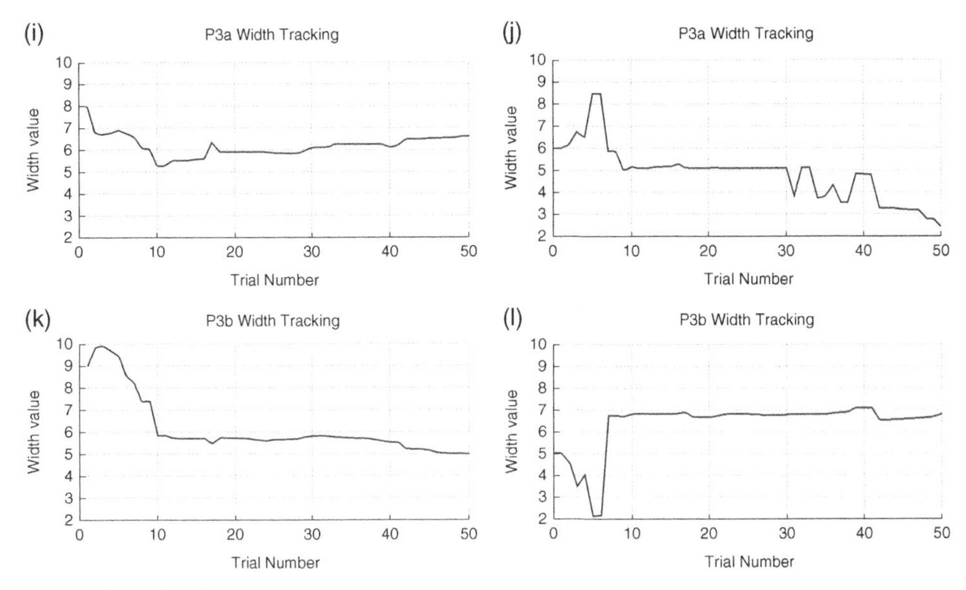

Figure 13.3 (Continued)

13.5 Separation of P3a and P3b

To enable evaluation of the parameters of P3a and P3b one may attempt to separate them as two different source components. The most important issue in dealing with this problem is the fact that the two components originate from the same activation or one is the consequence of the other. P3a reflects an unintentional response of the brain to a novel or salient stimulus that is independent of task relevance [42, 43]. Prefrontal, frontal, and anterior temporal brain regions play an important role in generating P3a, giving it a fronto-central distribution [42]. P3b is mainly generated by posterior temporal, parietal, and posterior cingulate mechanisms, and mostly distributed over the centro–parietal region. Moreover, P3a has a shorter latency and more rapid habituation than P3b.

By visual inspection of the components it is seen that they overlap in both time and the space of electrodes. Therefore, it is not easy to employ conventional source separation systems to separate them. In an attempt by Jarchi et al. [44] a method for separating correlated components such as **X** has been developed. The problem may be expressed in the following simple constrained optimization form:

$$\min \left\| \mathbf{w}_1^T \mathbf{X} - \mathbf{r}_1 \right\|_2^2 \quad \text{subject to} \quad \mathbf{w}_1^T \mathbf{a}_2 = 0 \tag{13.12}$$

where $\mathbf{w}$ is the separating vector, $\mathbf{r}_1$ is a template for the first component (e.g. P3a) and $\mathbf{a}_2$ is the second component (e.g. P3b). $\mathbf{r}_1$ is modelled as a gamma function and the above constrained problem is changed to an unconstrained one using a Lagrange multiplier. The new cost function, then, has the form:

$$J = \left\| \mathbf{w}_1^T \mathbf{X} - \mathbf{r}_1 \right\|_2^2 + \mathbf{w}_1^T \mathbf{a}_2 q = 0 \tag{13.13}$$

where q is the Lagrange multiplier. This parameter has also been accurately estimated as [44]:

$$q = \frac{2\mathbf{r}_1\mathbf{X}^T\mathbf{C}_x^{-1}\mathbf{a}_2}{\mathbf{a}_2^T\mathbf{C}_x^{-1}\mathbf{a}_2} \tag{13.14}$$

where $\mathbf{C}_x^{-1}$ is the inverse covariance matrix of $\mathbf{X}$. Suitable choices for $\mathbf{a}_1$ and $\mathbf{a}_2$ have been considered as the projections of $\mathbf{r}_1$ and $\mathbf{r}_2$ respectively onto $\mathbf{X}$ as $\mathbf{r}_1\mathbf{X}$ and $\mathbf{r}_2\mathbf{X}$. Minimizing J with respect to $\mathbf{w}_1$ and $\mathbf{w}_2$ (as for the second source, P3b, in a similar fashion) using the defined parameters for finding P3a and P3b respectively results in [44]:

$$\mathbf{w}_1 = \mathbf{C}_x^{-1}\mathbf{X}\mathbf{r}_1^T - 0.5q\mathbf{C}_x^{-1}\mathbf{X}\mathbf{r}_2^T \tag{13.15}$$

$$\mathbf{w}_2 = \mathbf{C}_x^{-1}\mathbf{X}\mathbf{r}_2^T - 0.5q\mathbf{C}_x^{-1}\mathbf{X}\mathbf{r}_1^T \tag{13.16}$$

Also, the gamma functions for both $\mathbf{r}_1$ and $\mathbf{r}_2$ have the following general form:

$$r(t) = \beta t^{k-1} exp\left(\frac{-t}{\theta}\right) \tag{13.17}$$

where $k > 0$ is the shape parameter, $\theta > 0$ is a scale parameter, and β is a normalizing constant.

To examine the method two components have been generated in two different hypothetical brain locations using gamma functions with different correlation levels. Using these sources, a 20-channel dataset has been produced. Each channel included 40 trials. In all of the trials, the latency of P3a was fixed at 150 ms and the latency of P3b was fixed for 200 ms. The amplitudes of P3a and P3b change in different trials. The noise variance was fixed at all the trials. In the first attempt the method has been applied to the simulated data considering the reference signal as the actual synthetic source. This has been called *exact match*. For another trial the reference signals for P3a and P3b were not exactly the same as the actual synthetic sources. These waves for P3a and P3b have been considered as gamma waves with different parameters [44]. The method considering these references is called *mismatch*. Since in this work the temporal correlation is high, it is reasonable to compare the results with spatial PCA rather than temporal PCA [45, 46]. Therefore, spatial PCA (from the ERP PCA Toolkit [47]) has been used in which Infomax [48], as the rotation algorithm for whitening the signal mixtures has been employed. It has been shown that the above method performs better than spatial PCA. Conversely the *mismatch* system performs very closely to the *exact match* system where the source model is identical to the template. The results have also been compared in terms of the correlations between the original and estimated sources for 0 and −5 dB signal-to-noise ratios (SNRs) in Figure 13.4.

In an application EEG data were recorded. The stimuli were presented through ear plugs inserted in the ear. Tones of 1 KHz were randomly and infrequently played for 40 times amongst tones of 2 KHz which appeared more frequently for 160 times. Their intensity was 65 dB with 10- and 50-ms duration for rare and frequent tones, respectively. The subjects were asked to press a button as soon as they heard a low tone (1 kHz). ERP components measured in this task included N100, P200, N200, P3a, and P3b. Forty trials related to the infrequent (rare) tones were selected and the above method for estimation of latency, amplitude, and scalp projections of P3a and P3b was employed. These trials and their average from channel Fz is shown in Figure 13.5(a). The estimated amplitudes are also depicted in Figure 13.5(b).

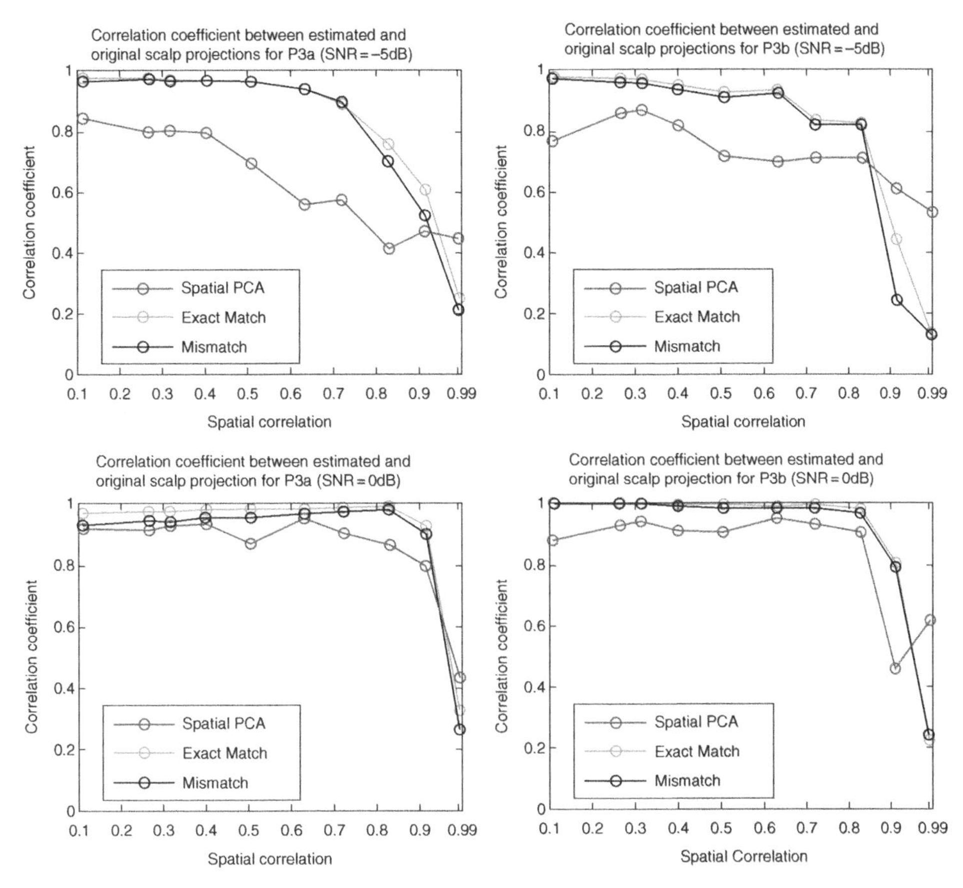

Figure 13.4 Comparison of three methods (spatial PCA, exact match and mismatch) versus the correlation coefficients between the original and estimated scalp projections of P3a and P3b in different spatial correlations and two SNR levels of 0 and –5 dB.

For selection of reference signals, the averaged P300, as shown in Figure 13.6, has been used. In the top row, the selected reference signals have little overlap with the average P300. The normalized estimated scalp projection is depicted for both P3a and P3b. It can be seen that the scalp projection values for P3a are negative or very close to zero. By increasing the correlation, the estimated scalp projections change. In the top row, the reference signal for P3a seems to have no correlation (or very little correlation) with actual P3a so its estimated scalp projection has negative or very close to zero values in all entries; however, it seems that there are some correlations between the reference signal and the actual signal for P3b. In the bottom row, the reference signal for P3a seems to have overlap with the actual P3b; therefore, in the first subcomponent estimations the P3b component has appeared over the scalp projections. The reference signal for P3b seems to have no correlation with actual P3b and the estimated scalp projection has negative values in all entries. So, it is logical to use the reference signals in the middle row of Figure 13.6 for P3a and P3b. Figure 13.7 shows the scalp projections of P3a and P3b in four selected progressive trials.

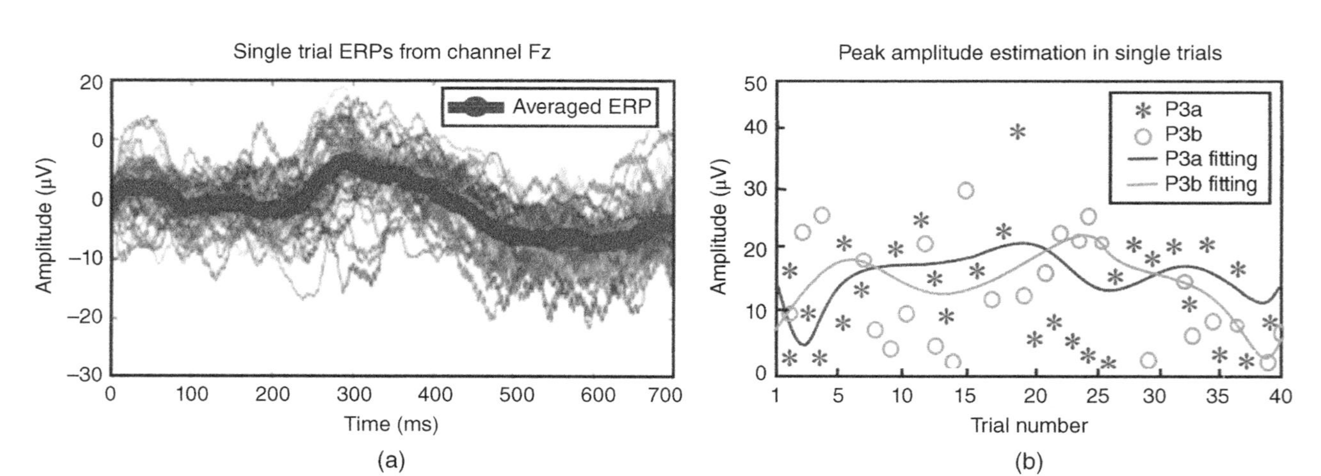

Figure 13.5 (a) Single-trial ERPs (40 trials related to the infrequent tones) and their average from channel Fz, and (b) estimated amplitudes for P3a and P3b in different trials.

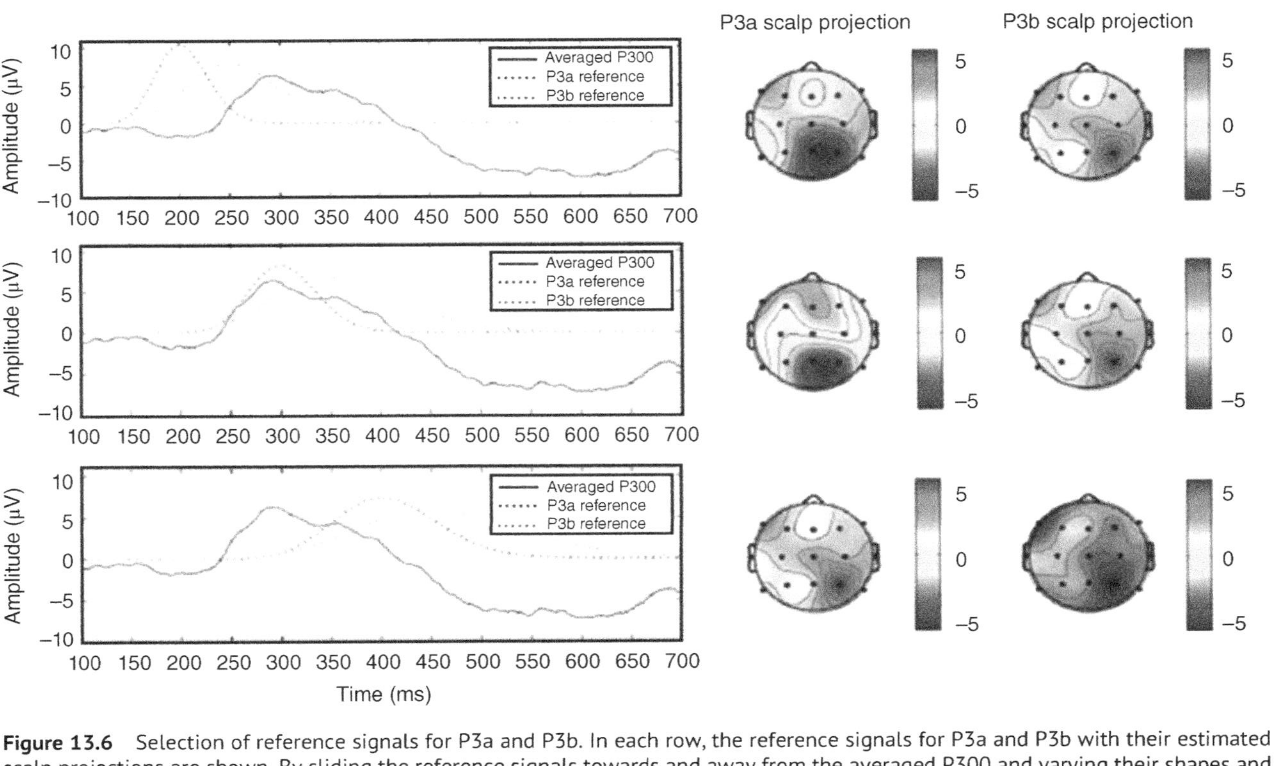

Figure 13.6 Selection of reference signals for P3a and P3b. In each row, the reference signals for P3a and P3b with their estimated scalp projections are shown. By sliding the reference signals towards and away from the averaged P300 and varying their shapes and observing the changes in their estimated scalp projections, an appropriate reference can be selected. The reference signals that are shown in the middle row, and have high correlation with the averaged P300, are used as good candidates for approximating the actual P3a and P3b.

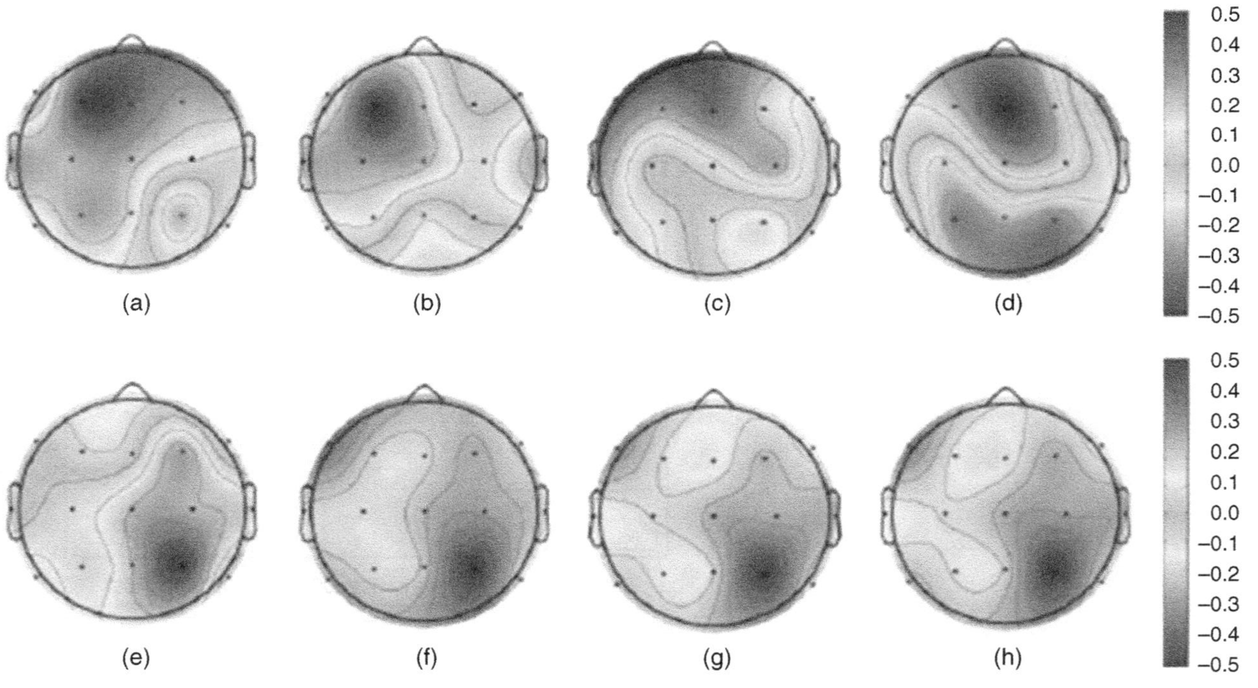

Figure 13.7 Scalp projections of P3a (top row) and P3b (bottom row) in four selected progressive trials.

The average ERP often lacks the early ERPs such as N100 and P200; though these ERPs can be detected by using the average of trials related to the frequent tones.

The proposed method can be applied in order to investigate mental fatigue based on trial-to-trial amplitude and latency variations of the P300 subcomponents and the relative changes of P300 subcomponent variations. In addition, estimation of the scalp projections can be useful for detecting the changes in the locations of P300 subcomponents for schizophrenic patients.

13.6 A Hybrid EEG–ERP-Based Method for Fatigue Analysis Using an Auditory Paradigm

It seems to be favourable to introduce a unified method using both background EEG and ERP data to more efficiently check the state of mental fatigue. An auditory-based paradigm may be implemented when EEG data are recorded in two states of alert and fatigue.

The data are then examined by both approaches (i.e. EEG and ERP-based) simultaneously. Such a recording paradigm allows for sufficient time interval between the stimulations so the state of the brain and its response to the stimulus can be evaluated.

For this test the experiment was run in a quiet room that is illuminated normally. The subject was sited comfortably in an armchair and the EEG data were recorded using a 32-channel QuickAmp amplifier and Ag/AgCl electrodes positioned according to the international 10–20 system and re-referenced to linked ears. In addition, vertical (VEOG) and horizontal (HEOG) electro-oculographic signals were recorded bipolarly using electrodes above and below the left eye and from the outer canthi. The EEG data were recorded in direct current (DC) mode at 1000 Hz with respect to an average reference. The EEG signal was recorded at the start of the experiment during the auditory oddball task. The subject heard 180 tones, 40 of them were infrequent tones while 140 of them were frequent tones. The subject was asked to respond to the infrequent tones by pressing a soft push button. The trial duration was set to four seconds. The subject was instructed to perform a simple arithmetic task for approximately two hours. After that the EEG data of the subject were recorded using the same auditory oddball task.

Elicitation of P300 is expected to be better when the infrequent tones are used. The spatiotemporal filtering method was applied to the 40 trials of infrequent tones. The average of these trials (considering 600 ms) from the Cz channel were used in order to select appropriate reference signals. Forty single-trial ERPs from the Cz channel and their average before and during fatigue are shown in Figures 13.8 and 13.9 respectively.

The two averaged ERPs are shown separately in Figure 13.10 for the two states of before and during fatigue. These average ERPs are used in order to select appropriate reference signals for P3a and P3b.

From the averaged ERPs in the fatigue state there is a reduced amplitude and increased latency in the P300 wave. However, the reduction in amplitude is very trivial and may not be considered as the sign of fatigue. In addition, there is not sufficient consistency in the increase in latency due to fatigue across trials. Therefore, there is a need for single-trial estimation of the ERPs and for that the spatiotemporal filtering method proposed in the

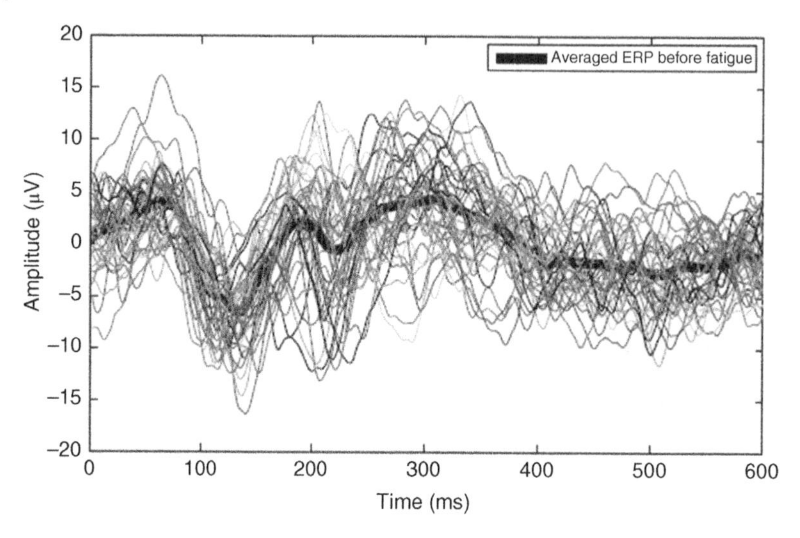

Figure 13.8 Forty single-trial ERPs and their average from the Cz channel before fatigue.

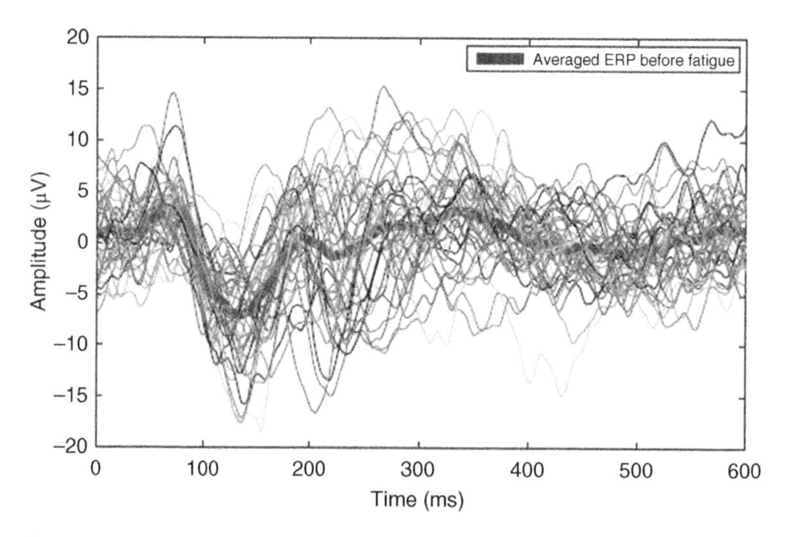

Figure 13.9 Forty single-trial ERPs and their average from the Cz channel during the fatigue state.

previous section is applied to estimate the P300 subcomponent descriptors (latency, amplitude, and scalp projections). The mean latency of P3a and P3b before fatigue was obtained as 279.6 and 335.5 ms respectively. This shows that the latencies of P3a and P3b are increased by increasing mental fatigue as expected. The mean latency of P3a and P3b during fatigue state was obtained as 325.5 and 372.2 ms respectively.

The scalp projections of P3a and P3b for five selected trials are shown in Figures 13.11 and 13.12 for before and during fatigue state respectively.

Although there is not a significant difference in the estimated scalp projections for P3a and P3b before and during fatigue state, the important issue is the separation of these

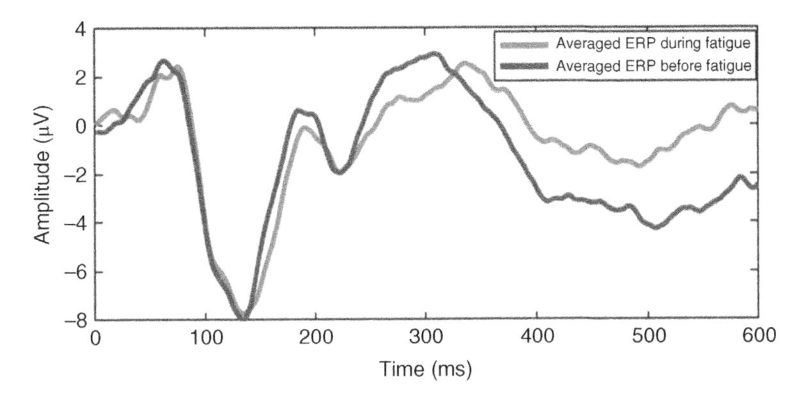

Figure 13.10 The ERP achieved by averaging 40 EEG trials before and during the fatigue state from the Cz channel.

subcomponents and estimation of other parameters. The results obtained by considering one subject confirm that the suggested paradigm is a good option for designing a mental fatigue detection system and the estimation of P300 subcomponent parameters can be good features for discriminating the fatigue state. Other important features can be obtained using phase synchronization measures.

From what has been discussed previously, the phase synchronization of different EEG rhythms especially alpha rhythm can be used as good features for recognition of the fatigue state. To investigate this, three seconds of the data segment has been considered and the EMD applied to decompose the EEG signal into its constituent oscillations. For the beta rhythm, the ALE is applied to the resulting IMF. One second of data segment (1000 samples) before stimulus onset and one second after stimulus onset are considered for measuring the phase synchronization. Beta and theta rhythms are extracted from frontal electrodes (F3 and F4 channels) and the alpha rhythm is extracted from central electrodes (C3 and C4 channels). The phase synchronization is calculated for five trials for one second before and after stimulus onset. The calculated phase synchronization for theta, alpha, and beta rhythms can be seen in Figures 13.13, 13.14, and 13.15 respectively. In these figures the average phase synchronization is also depicted by the thick line. From these figures the changes in phase synchronization can be clearly seen for alpha and theta rhythms before stimulus onset. Therefore, using the recorded EEG signal the EEG phase synchronization can be considered as a good feature for discrimination of the fatigue state.

13.7 Assessing Mental Fatigue by Measuring Functional Connectivity

Similar to many other applications, the brain connectivity is expected to vary during the alert-to-fatigue transitions. One early attempt in this study was in [49] where they estimated the multivariate autoregressive (MVAR) model coefficients from multichannel EEG signals and classified them using different classifiers such as a multiclass SVM. MVAR parameters

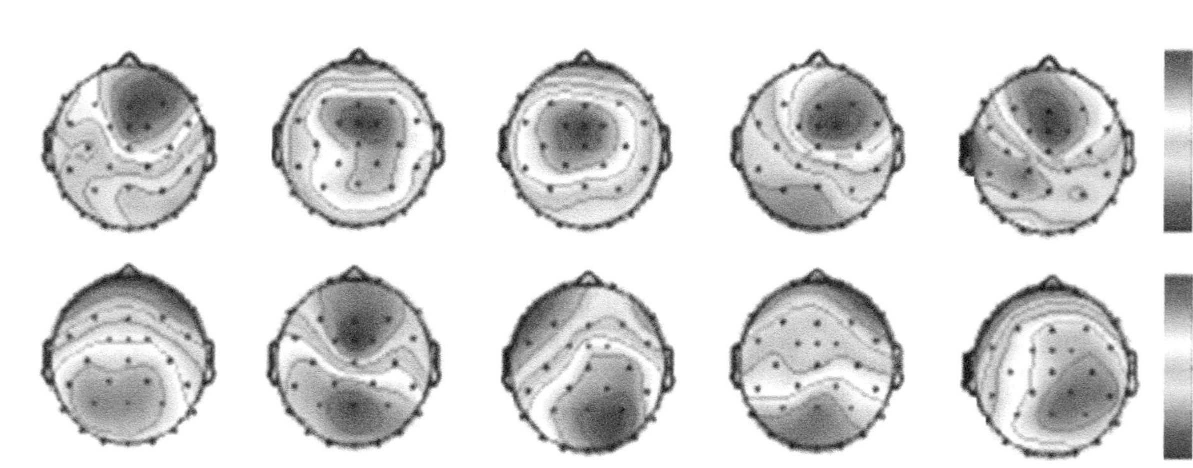

Figure 13.11 The estimated scalp projections of P3a (top row) and P3b (bottom row) before fatigue.

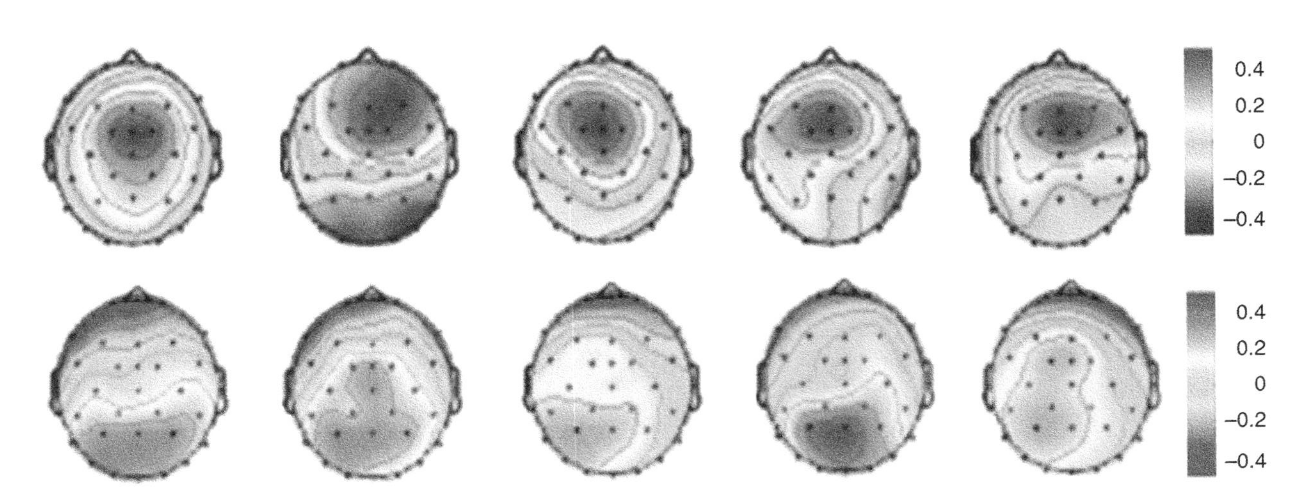

Figure 13.12 The estimated scalp projections of P3a (top row) and P3b (bottom row) during the fatigue state.

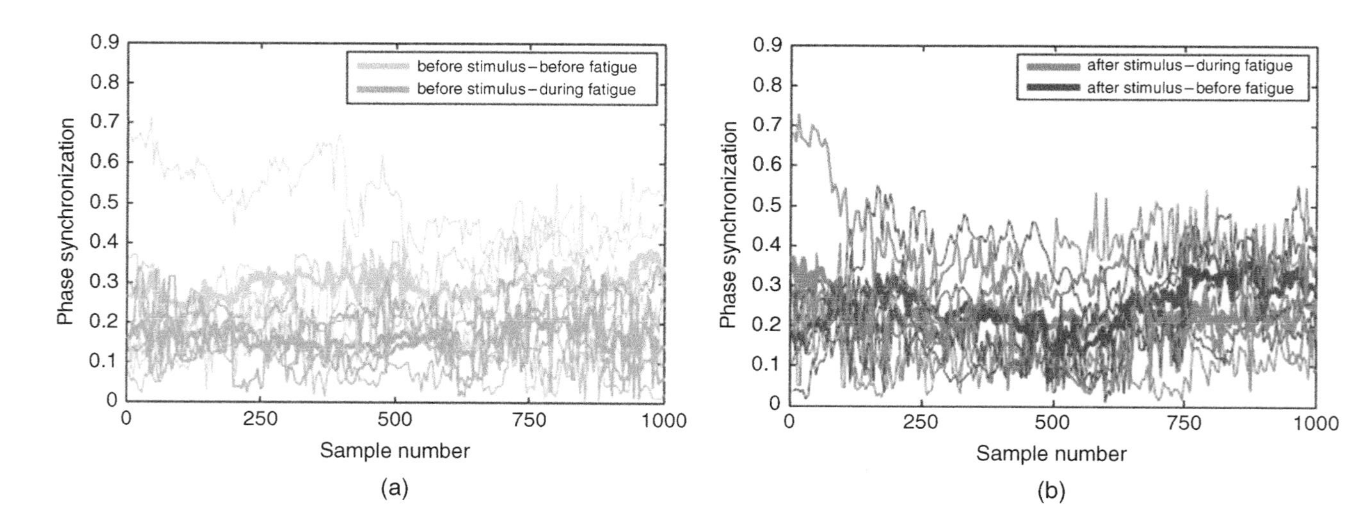

Figure 13.13 Theta phase synchronization of F3–F4: (a) before stimulus and (b) after stimulus.

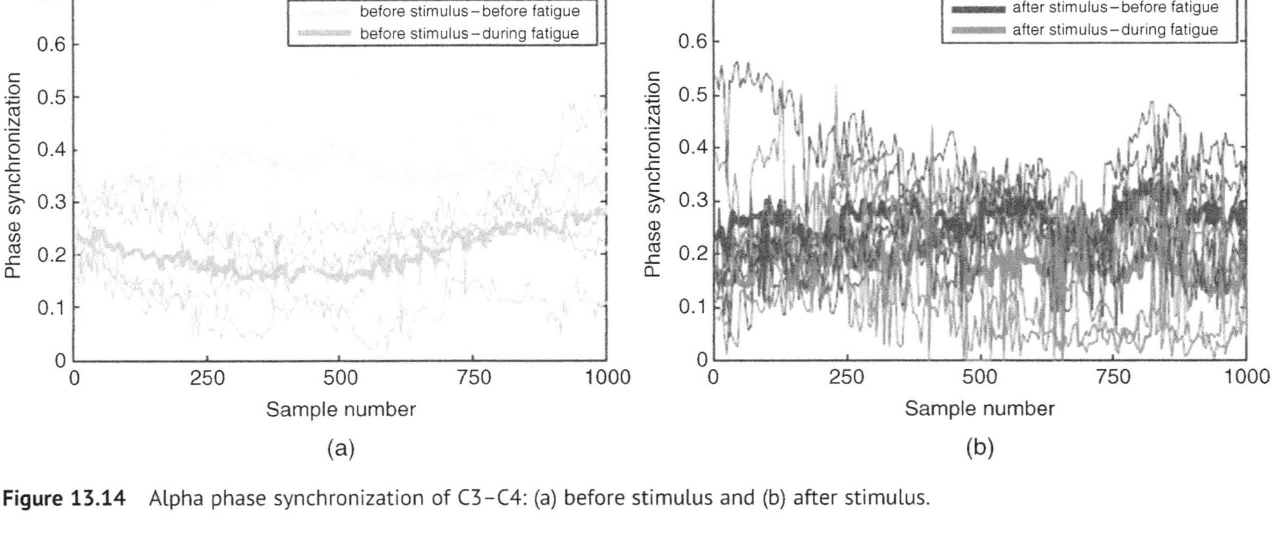

Figure 13.14 Alpha phase synchronization of C3–C4: (a) before stimulus and (b) after stimulus.

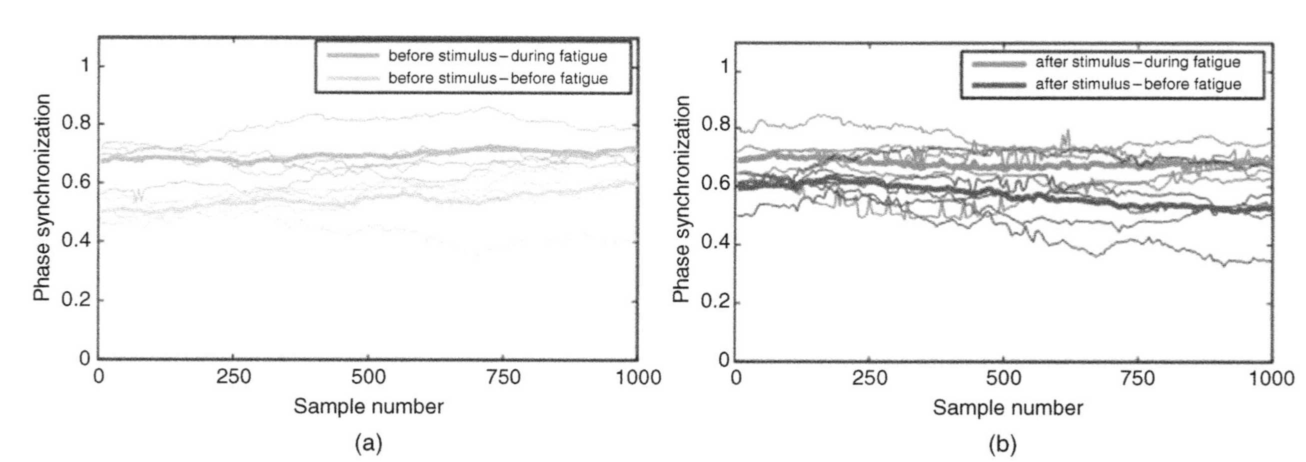

Figure 13.15 Beta phase synchronization of F3–F4: (a) before stimulus and (b) after stimulus.

indeed represent the brain connectivity, as they are estimated not only based on the individual channel history but also the correlations with other channels. They also applied KPCA in order to reduce the number of features before applying the classifier.

In another attempt, Qi et al. [50] used fMRI and introduced an empirical method to examine the reorganization of dynamic functional connectivity in a two-session experiment where one session including a mid-task break (Rest) was compared to a successive task design in the other session (No-rest). In their method, the functional connectivity has been estimated over a sliding window to construct the temporal brain network. The networks were estimated from 20 participants and the spatiotemporal architecture was examined by temporal efficiency analysis.

Task breaks would allow for retrieval of the mental resources, as well as return to the initial task goals and the adjustment of one's effort allocation strategy, which in turn may relieve mental fatigue and improve behavioural performance. Moreover, varying the break length affects the recovery from fatigue.

They demonstrated that taking a mid-task break leads to a restorative effect towards the end of the experiment instead of an immediate post-rest behaviour recovery. Also, the reduced spatiotemporal global integrity of the temporal brain network in a No-rest session was significantly improved with the break opportunity in the last task block of the Rest session. Hence, they provided evidence to support the beneficial effect of rest breaks in both behaviour performance and brain function. Moreover, these findings extended prior static functional connectivity studies of mental fatigue and highlight that altered dynamic connectivity may underlie cognitive fatigue [50].

In [51] the EEGs have been analyzed using functional connectivity estimated by Granger causality. The authors concluded that the topology of the brain networks and the brain's ability to integrate information changed in transition from alert to fatigue. They also observed a significant difference in terms of strength of Granger causality in the frequency domain and the properties of the brain effective network, i.e. causal flow, global efficiency, and characteristic path length between such conditions. Larger variations were observed over the frontal brain lobes for the alpha frequency band.

The use of brain connectivity for analysis of mental fatigue and drowsiness has been reported in a number of other publications including [52, 53]. In [52] the 32-channel EEGs are first restored from eye-blink artefact using second-order blind identification (SOBI) ICA method. After ICA the sources with maximum correlation with EOG are removed as eye-blinks. Then, the differences between the covariances of six regions have been used as the indices for classification. These regions are centro-frontal, centro-posterior, left and right fronto-laterals, and left and right posterio-lateral brain zones.

The authors of [53] estimated the brain functional networks in the alpha band using three different methods, partial directed coherence (PDC), direct transfer function (DTF) and phase lag index (PLI). Then, they used functional connectivity (the values of 22 discriminative connections) as the feature to discriminate between alertness and fatigue states. They reported up to 84.7 accuracy. Based on their achievements, the selected features revealed alterations of the functional network due to mental fatigue and specifically reduction of information flow among the brain zones. They also established a feature ranking strategy to minimize the number of electrodes for fatigue monitoring applications.

13.8 Deep Learning Approaches for Fatigue Evaluation

Machine learning approaches have been recently shifted towards deep learning networks as they can perform feature learning and classification simultaneously while using the wealth of memory and processing power.

In [54] a novel feature extraction strategy based on a deep learning for classification driving fatigue from EEG signals recorded from six healthy volunteers in a simulated driving experiment has been proposed. PCA and deep learning have been integrated into a model called PCA network (PCANet) [55]. The PCANet consists of two PCA-based filtering layers and an output layer that includes processing of binary hashing and block-wise histogram. The PCA has been used for dimensionality reduction, making the approach feasible, and the deep network for feature learning. The features have later been classified using SVM and k-nearest neighbour (KNN) to identify the brain state.

PCANet is widely used in 2-D image processing, such as face recognition [55]. In this study, each optimized multichannel EEG segment was treated as a 2-D data matrix (32×20) and fed into the PCANet for feature extraction. The PCANet block diagram is depicted in Figure 13.16.

Using PCANet, a classification accuracy up to 95% has been claimed [54]. During the analysis, the authors have identified that the parietal and occipital lobes of the brain are strongly associated with the fatigue.

Recurrent neural networks may be used for fatigue or drowsiness prediction from the EEG signals to avoid accidents for those under heavy mental workload such as drivers, pilots, or militaries. Liu et al. [56] recorded both EEG and EOG signals, extracted the spectral features, and used recurrent fuzzy neural network (RFNN) architecture to increase adaptability in realistic EEG applications while classifying and predicting driver drowsiness. They developed a complex network to enable the small changes in power spectrum due to fatigue to be captured.

The authors in [57] combined boosting strategy and transfer learning method to establish a model for classification of driving drowsiness and alertness states based on the power spectral density of different EEG frequency bands. The model was trained using previously recorded data tuned by a small portion of the currently recorded data (as a common practise in transfer learning). They claimed that the proposed boosting transfer learning method significantly outperformed the SVM and Adaboost classifiers. The process has been suggested as a suitable method for alert-fatigue classification.

In a more recent research by Jeong et al. [58] classification of two mental states (i.e. alert and drowsy states) and five drowsiness levels from 32-channel EEG and 5-channel EOG signals has been reported. However, they use EEG only to classify drowsiness levels and EOGs as reference for eye movement-related base line removal. They also used ICA for eye-blink artefact removal. For feature learning and classification, they used a deep spatiotemporal convolutional bidirectional long short-term memory network (DSTCLN) model. The Bi-LSTM network has been used to exploit the long-term dependency problem for time-series data. They constructed a hierarchical convolutional neural network (CNN) architecture for extracting high-level spatiotemporal features and applied the Bi-LSTM network to exploit the EEG time-series characteristics.

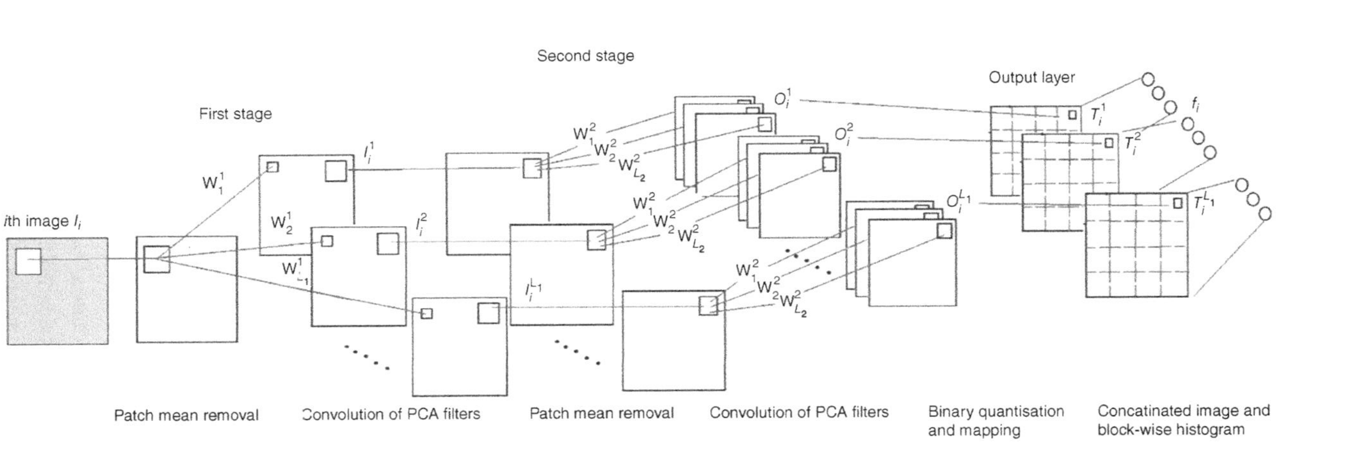

Figure 13.16 Two-stage PCANet block diagram proposed in [55].

They evaluated the classification performance using KSS values [28] and achieved classification accuracies of 0.87 (±0.01) and 0.69 (±0.02) for discrimination of the two mental states and five drowsiness levels, respectively.

Researchers in [59] studied the effect of daily physical activities on mental fatigue. During the evaluation sessions of the recorded EEG signals (before and after the fatigue-inducing sessions), the states of the participants were assessed by evaluating the EEG parameters. WPE, spectral coherence value (SCV), defined as:

$$\mathrm{SCV}_f = \frac{|S_{xy}(f)|^2}{S_{xx}(f)S_{yy}(f)} \tag{13.18}$$

and Lempel-Ziv complexity (LZC) [60] were used to indicate mental fatigue from the perspectives of activation, functional connectivity, and complexity of the brain. The indices are the beta band power P_{beta}, the power ratio P_{alpha}/P_{beta}, inter-hemispheric SCV of beta band SCV_β and LZC. The statistical analysis shows that by using these statistics mental fatigue is detected as the consequence of physical–mental task and the mental fatigue increased faster during physical–mental task.

13.9 Summary

Normal brain rhythms and the brain responses to various stimuli are both affected by the changes in the brain state due to mental fatigue. In addition, synchrony, coherency, and the connectivity of brain lobes are subject to change during the transition of brain state from alert to fatigue and vice versa. In this chapter, these concepts have been investigated in detail and the methods for extracting, tracking, and classification of the mental fatigue indicators from the EEG recordings have been explained in detail. More investigations and experiments may be carried out to find more about mutual effects of memory and fatigue for different aged subjects and those suffering from mental or physical abnormalities.

References

1 Maarten, A.S., Boksem, A.S., Meijman, T.F., and Lorist, M.M. (2005). Effects of mental fatigue on attention: an ERP study. *Cognitive Brain Research* 25: 107–116.

2 Yeo, M.V., Li, X., and Wilder-Smith, E.P. (2007). Characteristic EEG differences between voluntary recumbent sleep onset in bed and involuntary sleep onset in a driving simulator. *Clinical Neurophysiology* 118: 1315–1323.

3 Yeo, M.V.M., Li, X.P., Shen, K.Q. et al. (2004). EEG spatial characterization for intentional & unintentional sleep onset. *Journal of Clinical Neuroscience* 11 (sup. 1): 70.

4 Lorist, M.M., Boksem, M.,.A.S., and Ridderinkhof, K.R. (2005). Impaired cognitive control and reduced cingulate activity during mental fatigue. *Cognitive Brain Research* 24 (2): 199–205.

5 Grandjean, E. (1979). Fatigue in industry. *Occupational and Environmental Medicine* 36 (3): 175–186.

6 Lal, S.K. and Craig, A. (2001). A critical review of the psychophysiology of driver fatigue. *Biological Psychology* 55 (3): 173–194.
7 Fletcher, L., Apostoloff, N., Petersson, L., and Zelinsky, A. (May 2003). Vision in and out of vehicles. *IEEE Intelligent Systems* 18 (3): 12–17.
8 Craig, A., Tran, Y., Wijesuriya, N., and Nguyen, H. (2012). Regional brain wave activity changes associated with fatigue. *Psychophysiology* 49 (4): 574–582.
9 Lal, S.K., Craig, A., Boord, P. et al. (2003). Development of an algorithm for an EEG-based driver fatigue countermeasure. *Journal of Safety Research* 34 (3): 321–328.
10 Craig, A., Tran, Y., Wijesuriya, N., and Boord, P. (2006). A controlled investigation into the psychological determinants of fatigue. *Biological Psychology* 72: 78–87.
11 Borghini, G., Astolfi, L., Vecchiato, G. et al. (Jul. 2014). Measuring neurophysiological signals in aircraft pilots and car drivers for the assessment of mental workload, fatigue and drowsiness. *Neuroscience and Biobehavioral Reviews* 44: 58–75.
12 Huo, X.-Q., Zheng, W.-L., and Lu, B.-L. (2016). Driving fatigue detection with fusion of EEG and forehead EOG. *Proceedings of the International Joint Conference on Neural Networks*. pp. 897–904. IJCNN.
13 Breiman, L. (2001). Random forests. *Machine Learning* 45: 5–32.
14 Shen, K.Q., Li, X.P., Ong, C.J. et al. (Jul 2008). EEG-based mental fatigue measurement using multi-class support vector machines with confidence estimate. *Clinical Neurophysiology* 119: 1524–1533.
15 Shen, K.Q., Ong, C.-J., Li, X.-P. et al. (2007). A feature selection method for multilevel mental fatigue EEG classification. *IEEE Transactions on Biomedical Engineering* 54: 1231–1237.
16 Yeo, M.V.M., Li, X.-P., Shen, K.-Q., and Wilder-Smith, E.P.V. (2009). Can SVM be used for automatic EEG detection of drowsiness during car driving? *Safety Science* 47 (1): 115–124, Elsevier.
17 Zhang, C., Zheng, C., Yu, X., and Ouyang, Y. Estimating VDT mental fatigue using multichannel linear descriptors and KPCA-HMM. *EURASIP Journal on Advances in Signal Processing* 2008: 185638. https://doi.org/10.1155/2008/185638.
18 Liu, J., Zhang, C., and Zheng, C. (2010). EEG-based estimation of mental fatigue by using KPCA–HMM and complexity parameters. *Biomedical Signal Processing and Control* 5: 124–130.
19 Zhang, L.Y., Zheng, C.X., Li, X.P., and Shen, K.Q. (2005). Feasibility study of mental fatigue grade based on Kolmogorov entropy. *Space Medicine & Medical Engineering* 18 (5): 375–380.
20 Fu, R., Wang, H., and Zhao, W. (Nov. 2016). Dynamic driver fatigue detection using hidden Markov model in real driving condition. *Expert Systems with Applications* 63: 397–411.
21 Rabiner, L. and Juang, B.H. (1993). *Fundamentals of Speech Recognition*. Englewood Cliffs, NJ: Prentice-Hall.
22 Lin, C.-T., Wu, R.-C., Jung, T.-P. et al. (2005). Estimating driving performance based on EEG Spectrum analysis. *EURASIP Journal on Advances in Signal Processing* 2005: 521368.
23 Wu, D., Lawhern, V.J., Gordon, S. et al. (Dec. 2017). Driver drowsiness estimation from EEG signals using online weighted adaptation regularization for regression (OwARR). *IEEE Transactions on Fuzzy Systems* 25 (6): 1522–1535.
24 Chai, R., Naik, G.R., Nguyen, T.N. et al. (May 2017). Driver fatigue classification with independent component by entropy rate bound minimization analysis in an EEG-based system. *IEEE Journal of Biomedical and Health Informatics* 21 (3): 715–724.

25 Li, X.-L. and Adali, T. (2010). Blind spatiotemporal separation of second and/or higher-order correlated sources by entropy rate minimization. *Proceedings of the. IEEE International Conference on Acoustics, Speech, and Signaling Process*, 1934–1937.

26 Zhao, C., Zheng, C., and Liu, J. (2010). Physiological assessment of driving mental fatigue using wavelet packet energy and random forests. *American Journal of Biomedical Sciences* 2 (3): 262–274.

27 Putilov, A.A. and Donskaya, O.G. (2013). Construction and validation of the EEG analogues of the Karolinska sleepiness scale based on the Karolinska drowsiness test. *Clinical Neurophysiology* 124 (7): 1346–1352.

28 Akerstedt, T. and Gillberg, M. (1990). Subjective and objective sleepiness in the active individual. *International Journal of Neuroscience* 52: 29–37.

29 Quiroga, R.Q., Kraskov, A., Kreuz, T., and Grassberger, P. (2002). Performance of different synchronization measures in real data: a case study on electroencephalographic signals. *Physical Review E* 65.

30 Haykin, S. (2002). *Adaptive Filter Theory*. New Jersey: Prentice-Hall.

31 Mormann, F., Lehnertz, K., David, P., and Elger, C. (2000). Mean phase coherence as a measure for phase synchronization and its application to the EEG of epilepsy patients. *Physica D* 144: 358–369.

32 Jarchi, D., Makkiabadi, B., and Sanei, S. (2010). Mental fatigue analysis by measuring synchronization of brain rhythms incorporating empirical mode decomposition. In: *IEEE CIP Conference, Italy*.

33 Huang, N.E., Shen, Z., Long, S.R. et al. (1998). The empirical mode decomposition and Hilbert spectrum for nonlinear and non-stationary time series analysis. *Proceedings of the Royal Society A. Mathematical, Physical and Engineering Sciences,* 454: 903–995.

34 Widrow, B. (1975). Adaptive noise cancellation: principles and applications. *Proceedings of the IEEE* 63: 1692–1716.

35 Kalman, R.E. (1960). A new approach to linear filtering and prediction problems. *Journal of Basic Engineering* 82 (1): 35–45.

36 Kong, W., Zhou, Z., Jiang, B. et al. (Jan. 2017). Assessment of driving fatigue based on intra/inter-region phase synchronization. *Neurocomputing* 219: 474–482.

37 Wiener, N. (1949). *Extrapolation, Interpolation, and Smoothing of Stationary Time Series*. New York: Wiley.

38 Mohseni, H.R., Nazarpour, K., Wilding, E., and Sanei, S. (2009). Application of particle filters/ in single-trial event related potential estimation. *Physiological Measurements* 30 (10): 1101–1116.

39 Murata, A., Uetake, A., and Takasawa, Y. (2005). Evaluation of mental fatiguenext term using feature parameter extracted from event-related potential. *International Journal of Industrial Ergonomics* 35 (8): 761–770.

40 Ullsperger, P., Metz, A.M., Yu, X., and Gille, H.G. (1988). The P300 component of the event-related brain potential and mental effort. *Ergonomics* 31: 1127–1137.

41 Jarchi, D., Sanei, S., Mohseni, H.R., and Lorist, M.M. (2011). Coupled particle filtering: a new approach for P300-based analysis of mental fatigue. *Journal of Biomedical Signal Processing and Control* 6 (2): 175–185.

42 Friedman, D., Cycowicz, Y.M., and Gaeta, H. (2001). The novelty p3: an event related brain potential (ERP) sign of the brain's evaluation of novelty. *Neuroscience and Biobehavioral Reviews* 25 (4): 355–373.

43 Comerchero, M.D. and Polich, J. (1999). P3a and P3b from typical auditory and visual stimuli. *Clinical Neurophysiology* 110 (1): 24–30.

44 Jarchi, D., Sanei, S., Makkiabadi, B., and Principe, J. (2011). A new spatiotemporal filtering method for single-trial estimation of correlated ERP subcomponents. *IEEE Transactions on Biomedical Engineering*.

45 Dien, J. (1998). Addressing misallocation of variance in principal components analysis of event-related potentials. *Brain Topography* 11 (1): 43–55.

46 Dien, J., Khoe, W., and Mangun, G.R. (2007). Evaluation of PCA and ICA of simulated ERPs: Promax vs. infomax rotations. *Human Brain Mapping* 28 (8): 742–763.

47 Dien, J. (2010). The ERP PCA toolkit: an open source program for advanced statistical analysis of event-related potential data. *Journal of Neuroscience Methods* 187 (1): 138–145.

48 Delorme, A. and Makeig, S. (2004). EEGLAB: an open source toolbox for analysis of single-trial EEG dynamics including independent component analysis. *Journal of Neuroscience Methods* 134 (1): 9–21.

49 Zhao, C., Zheng, C., Zhao, M., and Liu, J. (March 2011). Multivariate autoregressive models and kernel learning algorithms for classifying driving mental fatigue based on electroencephalographic. *Expert Systems with Applications* 38 (3): 1859–1865.

50 Qi, P., Gao, L., Meng, J. et al. (Jan. 2020). Effects of rest-break on mental fatigue recovery determined by a novel temporal brain network analysis of dynamic functional connectivity. *IEEE Transactions on Neural Systems and Rehabilitation Engineering* 28 (1): 62–71.

51 Kong, W., Lin, W., Babiloni, F. et al. (2015). Investigating driver fatigue versus alertness using the granger causality network. *Sensors* 15 (8): 19181–19198.

52 Charbonnier, S., Roy, R.N., Bonnet, S., and Campagne, A. (Jun. 2016). EEG index for control operators' mental fatigue monitoring using interactions between brain regions. *Expert Systems with Applications* 52: 91–98.

53 Dimitrakopoulos, G.N., Kakkos, I., Vrahatis, A.G. et al. (2017). Driving mental fatigue classification based on brain functional connectivity. *Proceedings of the International Conference on Engineering Applications in Neural Networks*, 465–474.

54 Ma, Y., Chen, B., Li, R. et al. (2019). Driving fatigue detection from EEG using a modified PCANet method. *Hindawi Journal of Computational Intelligence and Neuroscience* 2019: 4721863, 9 pages.

55 Chan, T.-H., Jia, K., Gao, S. et al. (2015). PCANet: a simple deep learning baseline for image classification? *IEEE Transactions on Image Processing* 24 (12): 5017–5032.

56 Liu, Y.-T., Lin, Y.-Y., Wu, S.-L. et al. (Feb. 2016). Brain dynamics in predicting driving fatigue using a recurrent self-evolving fuzzy neural network. *IEEE Transactions on Neural Networks and Learning Systems* 27 (2): 347–360.

57 He, J., Zhou, G., Wang, H. et al. (2018). Boosting transfer learning improves performance of driving drowsiness classification using EEG. *Proceedings of the IEEE International Workshop on Pattern Recognition and Neuroimaging (PRNI)*, 1–4.

58 Jeong, J.-H., Yu, B.-W., Lee, D.-H., and Lee, S.-W. (2019). Classification of drowsiness levels based on a deep spatio-temporal convolutional bidirectional LSTM network using

electroencephalography signals. *Brain Sciences* 9: 348. 18 pages, doi:https://doi.org/10.3390/brainsci9120348.

59 Xu, R., Zhang, C., He, F. et al. (2018). How physical activities affect mental fatigue based on EEG energy, connectivity, and complexity. *Frontiers in Neurology* 9: 915, 13 pages.

60 Lempel, A. and Ziv, J. (January 1976). On the complexity of finite sequences. *IEEE Transactions on Information Theory* 22 (1): 75–81.

14

EEG-Based Emotion Recognition and Classification

14.1 Introduction

No matter how basic or complex, an emotion is a mental and brain-related physiological state associated with a wide variety of feelings, thoughts, and behaviour. Emotions are subjective experiences often associated with mood temperament, personality, and disposition.

The word 'emotion' originates from the French word '*émouvoir*' which is based on the Latin '*emovere*', where *e-* (variant of *ex-*) means 'out' and '*movere*' means 'move'. Theories about emotions stretch back at least as far as the ancient Greek Stoics, as well as Plato and Aristotle. Numerous articles refer to physiological and biological aspects of emotions and their correlation with other states of body and brain. Most of these theories go far beyond the scope of this chapter.

As many as 55 different, some related, emotions can be named as admiration, affection, amusement, annoyance, anxiety, approval, boredom, calm, cold anger, coldness, confidence, contentment, contempt, cruelty, despair, determination, disagreeableness, disappointment, disapproval, disgust, distraction, effervescent, embarrassment, excitement, fear, friendliness, greed, guilt, happiness, hopeful, hot anger, hurt, impatience, indifference, interest, jealousy, mockery, nervousness, neutrality, panic, pleasure, pride, relaxation, relief, resentment, sadness, satisfaction, serenity, shame, shock, stress, surprise, sympathy, wariness, weariness, and worry.

Emotion is central to human daily experience, influencing cognition, perception, and everyday tasks such as learning, communication, and even rational decision-making. However, the large number of emotion states and the overlaps between the corresponding brain regions make analysis of emotion very challenging for technologists and neuroscience researchers.

Symptoms such as body movement, facial expression, change of body temperature, heart rate variability, breathing, blood pressure, muscle contraction, and particularly variation in the brain rhythms may be effectively quantified and used to accurately discriminate between aforementioned emotion types. Often it is difficult and less accurate to perform such classification using a single modality biometric.

Several theories, at least for some particular emotions, have been proposed by neuroscientists for a better understanding of how the brain acts or reacts to emotions. These theories also identify the particular brain regions which might be activated under each emotion.

EEG Signal Processing and Machine Learning, Second Edition. Saeid Sanei and Jonathon A. Chambers.

14.1.1 Theories and Emotion Classification

Some theories are somatic and define emotions such as human expression and body movements [1]. William James and Carl Lang developed the James-Lange theory, a hypothesis on the origin and nature of emotions. It states that the autonomic nervous system of a human being provokes physiological events such as muscular tension, rise in heart rate, perspiration, and dryness of mouth in response to the environment's events. Emotions, hence, are feelings which come about as a result of physiological changes, rather than being their cause.

Other theories based on neurobiological changes describe emotion as a pleasant or unpleasant mental state organized in the limbic system of the mammalian brain. Emotions would then be mammalian elaborations of general vertebrate arousal patterns, in which neurochemicals (e.g. dopamine, noradrenaline, and serotonin) step-up or step-down the brain's activity level, as visible in body movements, gestures, and postures. In mammals, primates, and human beings, feelings are displayed as emotion cues.

Emotion is believed to be related to the limbic system [2]. The limbic system is a set of primitive brain structures located on top of the brainstem on both sides of the thalamus, just under the cerebrum and is involved in many of our emotions and motivations, especially those related to survival. Fear, anger, and emotions related to sexual behaviour are those originated from this area of the brain. It includes the hypothalamus, hippocampus, amygdala (also called amygdale) and some other brain regions. The amygdala is an almond shaped mass of nuclei located deep within the temporal lobe of the brain. It is a limbic system involved in emotions, motivations, and memory. Figure 14.1 illustrates the main regions of the limbic system.

Although in some previous research it has been claimed that the entire brain limbic system is involved in development of emotions, recent research findings have shown that some of these limbic structures are not as directly related to emotion as others and that some non-limbic structures have been found to be of greater emotional relevance [2].

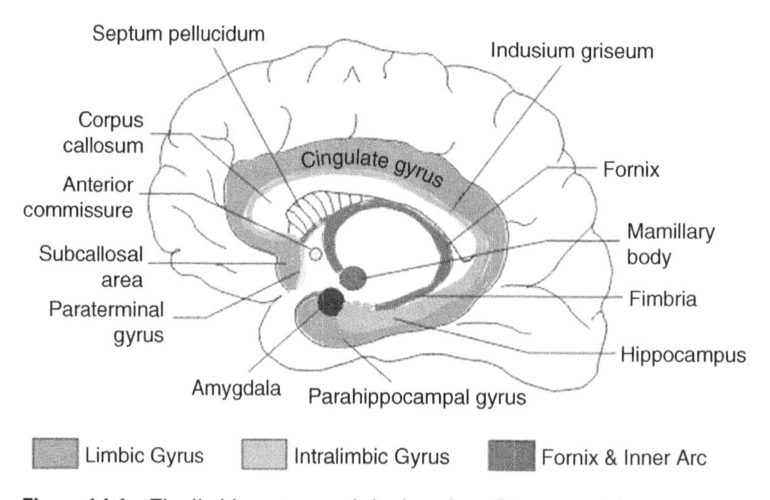

Figure 14.1 The limbic system and the location of the amygdala.

Often, terms such as *affect, emotion,* and *mood* are used interchangeably [3]. Generally, *affect* may be used to refer to both emotions and moods. *Emotion* often has an identifiable cause – a stimulus or antecedent thought, usually a spasmodic, intense experience of short duration and the person is well aware of it (i.e. emotions typically have high cognitive involvement and elaborate content). Conversely, a *mood* tends to be more subtle, longer lasting, less intense, more in the background, often like a frame of mind, casting a positive glow or negative shadow over experiences. Moods for healthy subjects are often nonspecific (e.g. pleasant/unpleasant; energetic/lethargic; anxious/relaxed) compared to emotions, which are usually specific, linked to clear-cut consciously available cognitive representations about their antecedents and are therefore typically focused on an identifiable person, object, or event. In contrast, people may not be aware of their mood unless their attention is drawn to it. Mood variations may be caused by subtle factors such as diurnal fluctuations in brain neurotransmitters, sleep/waking biorhythms, sore or tense muscles, visiting or losing a friend, sunny or rainy weather, and an accumulation of pleasant or unpleasant events [3].

Based on neurobiological theories, models for emotion have been presented with contradictory observation deductions in some cases. Among them are two neurobiological models of emotion making opposing predictions. The valence model which predicts that anger, a negative emotion, would activate the right prefrontal cortex (PFC). In contrast to the valence model, the Direction Model suggests that anger, an approach emotion, would activate the left PFC. The second model was supported by further experiments [4]. It however remains questionable whether the opposite of approach in the PFC is better described as moving away (Direction Model), unmoving but with strength and resistance (Movement Model) or unmoving with passive yielding (Action Tendency Model). Support for the Action Tendency Model (passivity related to right prefrontal activity) comes from research on shyness [5] and on behavioural inhibition research [6]. Some research examined the competing hypotheses generated by all four models and supported the Action Tendency Model [7, 8].

Another neurological approach [9] distinguishes between two classes of emotion: first, 'classical emotions' including lust, anger, and fear, generally evoked by *environmental stimuli*, which motivate us (to copulate/fight/flee respectively); second 'homeostatic emotions' including feelings evoked by *internal body states*, which modulate our behaviour. Thirst, hunger, feeling hot or cold (core temperature), feeling sleep deprived, salt hunger and air hunger are all examples of homeostatic emotion. Homeostatic emotion onset occurs when an imbalance arises in any one of these systems hence prompting us to react and restore the balance to the system. Pain is a homeostatic emotion alerting us of an abnormal state/condition [9].

Another argument on emotions is based on the theory that cognitive activity in the form of judgements, evaluations, or thoughts is necessary for an emotion to occur. As argued by Richard Lazarus, emotions are about something or have intentionality, as such cognitive activity may be conscious or unconscious and may or may not take the form of conceptual processing. It has also been suggested that emotions are often used as shortcuts to process information and influence behaviour [10].

A hybrid of the somatic and cognitive theories of emotion is the perceptual theory. Based on so-called neo-Jamesian theory, bodily responses are central to emotions. In this respect, emotions are held to be analogous to faculties such as vision or touch providing information about the relation between the subject and the world in various ways.

Another theory called Affective Event Theory is a communication-based theory [11] which looks at the causes, structures, and consequences of emotional experience. This theory suggests that emotions are influenced and caused by events which in turn influence attitudes and behaviours. It also emphasizes on so-called emotion episodes. This theory has been utilized by numerous researchers to better understand emotion from a communicative perspective and was reviewed further in [12].

Singer-Schachter theory is another cognitive theory. This is based on experiments purportedly showing that subjects can have different emotional reactions despite being placed into the same physiological state with an injection of adrenaline. Subjects were observed to express anger or amusement if another person in such situation displayed that emotion.

A recent model called the Component Process model considers emotions as synchronization of many different bodily and cognitive components. Therefore, symptoms and physiological signatures may be used to evaluate emotions. Emotions are identified with an overall process where low-level cognitive appraisals, particularly the processing of relevance, trigger bodily reactions, behaviours, feelings, and actions occur. Based on this model, it is clear that a thorough evaluation of emotion requires assessment of many symptoms, facial and body movement features, together with physiological signals.

More common and popular emotions such as fear, anger, disgust, happiness, sadness, surprise, interest, shame, contempt, suffering, love, tension, and mirth have been investigated by researchers in emotion understanding, control, and regulation mainly by analyzing the brain signals and images.

Traditionally, few neuroimaging, imaging, signal processing, and data analysis techniques have been used for detection and recognition of emotions. These modalities include facial pattern and gestures, respiration, blood pressure, skin impedance, body temperature, muscle activity, heart rate, and brain activity [13].

14.1.2 The Physiological Effects of Emotions

Respiration as a peripheral body process changes with emotion. There are significant differences between the states of respiration in emotions such as calmness and fear. Slow and uniform respiration occurs in the calm state as opposed to fast and abrupt respiration when in fear [14, 15]. A respiration belt may be used for measuring this biometric. Generally, breathing rhythm significantly changes with emotion. Frustration causes an increase in the breathing frequency and vice versa. Therefore, although respiration is primarily regulated for metabolic and homeostatic changes in the brainstem, it is studied for emotion recognition. [16]. It has also been deduced that final respiratory output is influenced by a complex interaction between the brainstem and higher centres, including the limbic system and cortical structures. The important and interesting conclusion is the coexistence of emotion and respiration, which is important in maintaining physiological homeostasis [15]. Relationships between emotions and respiration have shown more rapid breathing during an arousal state. It has been shown that the breathing frequency corresponds approximately linearly to trait anxiety [15]. They have gone even further to investigate olfactory function and the piriform–amygdala in relation to respiration including oscillations of piriform–amygdala complex activity and respiratory rhythm [17]. Final respiratory output involves a complex interaction between the brainstem and higher centres, including the limbic system and cortical structures.

Using both electroencephalography (EEG) and functional magnetic resonance imaging (fMRI) it has been observed that the generators of respiration-related anxiety potentials are in the right temporal pole in subjects with low anxiety and for the most anxious subjects in the temporal pole and amygdala, as depicted in Figure 14.2 [18].

The temporal pole located within paralimbic regions is involved in the evaluation of environmental uncertainty or danger [16]. If the respiratory rate is increased by anxiety, these regions may be activated before the onset of inspiration. It is assumed that an increase in respiratory rate is caused by unconscious evaluation in the amygdala and that these two activities occur in parallel. Therefore, stimulation of the amygdala results in a rapid increase in respiratory rate followed by a feeling of fear and anxiety [19].

Blood pressure is another human factor affected by various changes in the body metabolism such as those caused by emotion changes. Blood pressure increases with emotions such as anger or fear and decreases with happiness or calmness. Plethysmograph is used to measure this quantity for investigation of different physiological abnormalities. Also, it is a well-known observation that blood pressure is directly related to stress.

Another biometric, which changes with emotion, is skin impedance. Skin impedance can be checked using a galvanic skin response (GSR) sensor. Emotional state of human beings is a physiological mechanism which affects the body and manifests itself by changing the face impression. Stress and other emotional states are controlled by the hypothalamus, a region in the brain and the hormones released by the adrenal gland situated on the kidney. The sympathetic and para-sympathetic nervous systems are also involved. While adrenaline is responsible for creating stress, emotional feelings such as fear and anxiety stimulate the hypothalamus which in turn increases adrenaline secretion. Adrenaline causes an increase in heartbeat, blood pressure, breathing rate, sweating, and fainting in an extreme case. These events prepare and alert the body to face the situation. One important effect of adrenaline is to increase blood flow to the skin to remove excess heat from the body through sweating. That is why the feeling of burning occurs in shock. During stress, blood flow to the skin increases, vessels become porous, and sweating occurs due to water secretion. This in effect removes heat from the body through evaporation of sweat. This lowers the skin resistance to facilitate removal of water more easily. A moist skin also increases electrical activity. Hence one can conclude that the skin resistance is directly proportional to the emotional state. Consequently, body temperature is subjected to changes when the subject is consistently involved in emotion. This can be due to changes in other physiological and metabolic factors such as heart rate, blood pressure, or muscle activities.

Electromyography (EMG) is used to measure the muscle activities of the face, neck, and shoulder which are very likely to change due to certain emotions. Correlation between facial expression and facial muscle activities has been investigated and verified by many researchers; see for example [20–23]. Facial pattern analysis and recognition can be performed using frontal or lateral view images taken by still or video cameras. In this direction, there has been much work done giving rise to numerous algorithms and solutions. Many databases such as IAPS (international affective picture system) [24] have also been provided emotional facial images.

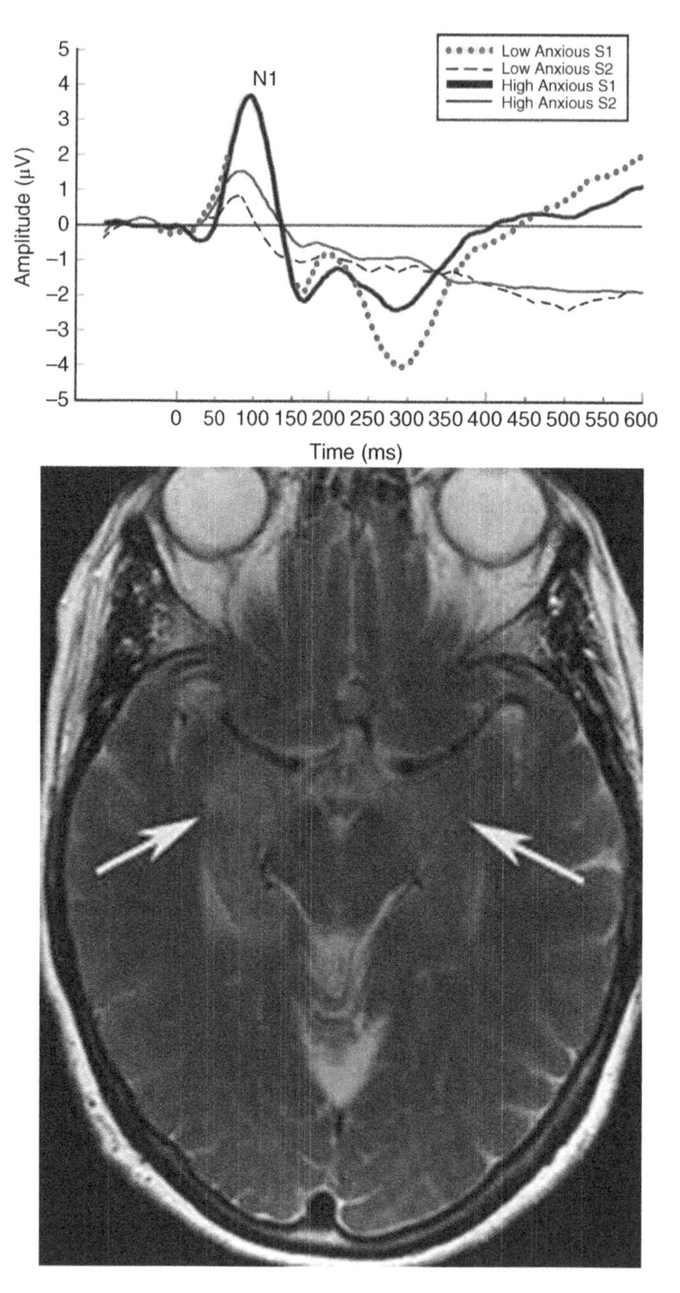

Figure 14.2 The generators of respiration-related anxiety potentials are in the right temporal pole in subjects with low anxiety and in the temporal pole and amygdala in the very anxious subjects (Source: taken from [18].)

Security, superior automatic speaker recognition (ASR) system performance, life-like agents, language etiquette, Student modelling, and many other applications have motivated the study of emotion-influenced language recognition. Fluctuation in voice characteristics such as energy, speaking rate, formant frequencies, and formant bandwidths triggered by the changes in emotions have been studied by many researchers. As an example, Oudeyer [25] developed algorithms that allow a robot to express its emotions by modulating the intonation of its voice. In a different approach emotion recognition in spontaneous speech has been addressed [26].

Later research has revealed the relations between light and emotions, colour and emotions, and more importantly music and emotions [27]. Light is used to treat mood disorders but the logic behind it is not well understood. While rod and cone eye photoreceptors process visible light, a third type of photoreceptor, particularly sensitive to blue light, mediates non-visual responses such as sleep cycles and alertness. So, light may make us feel better because it helps regulate circadian rhythms.

To find out how this pathway directly affects our emotional state, Vandewalle and his colleagues at the University of Liège, Belgium scanned the brains of volunteers exposed to green or blue light while a neutral or angry voice recited meaningless words. As expected, brain areas responsible for processing emotion responded more strongly to the angry voice, and this effect was amplified further by blue light [26]. Vandewalle suggests blue light is likely to amplify emotions in both directions.

The main focus of this chapter is to study the brain response for different emotions. Although some studies have used functional near-infrared spectroscopy (fNIRS) to capture brain activity mainly from the PFC, the most important modality for this purpose would be the EEG. fINRS is a low-cost, user-friendly, and practical measurement modality. It detects the infrared light (photon count) travelling through the cortex and is used to monitor the hemodynamic changes during cognitive and/or emotional activity. A big advantage of this modality over EEG is that fINRS is not affected by the facial muscles EMG during motion expression. However, its application is limited to the '-to-cortex' area and having a small number of sensors. Therefore, it is less practical when study of various emotions is undertaken.

Study of emotion using EEG has become more attractive due to many algorithms developed by the signal processing community. In particular, localization of brain segments involved in particular emotion, connectivity of those regions for expression, control and regulation of emotions, and synchronization of brain lobes to evaluate the extent the brain is influenced and stimulated. Generally speaking, there is no doubt that the brain is involved in or affected by various physical or physiological changes in the human body.

14.1.3 Psychology and Psychophysiology of Emotion

Undoubtedly, emotions pervade our daily life. They can help or prevent us guide our choices, avoid dangers, and also play a key role in nonverbal communication.

Cornelius the [28] introduced three emotion viewpoints: Darwinian, cognitive, and Jamesian. Based on Darwinian viewpoint, emotions are selected by nature in terms of their survival value, e.g. fear exists because it helps avoiding anger. Cognitive theory considers the brain as the centre of emotions, particularly, concentrating on direct and non-reflective processes called appraisal, by which the brain judges a situation or event as good or bad. Both of these few points refer to the psychophysiology of emotion. Finally, the Jamesian theory suggests that emotions are only peripheral (bodily) perception changes such as heart rate or dermal responses. For example, I am afraid because I shiver. This example refers to the role of physiological responses in the study of emotions.

Before any quantification or assessment of emotions three principles or goals in analysis of emotions have to be recognized as emotion awareness, emotional arousal and expression, and emotion regulation. Emotion regulation skills – including those which identify and label emotions, allow and tolerate emotions, establish a working distance, increase positive emotions, reduce vulnerability to negative emotions, self-sooth, breath, and distract – also found to help when in high distress [29].

Quantitatively, emotions are often presented in a two-dimensional space of valence–arousal [30]. Valence represents the way one judges a situation, from unpleasant to pleasant; arousal expresses degree of excitement felt by people, from calmness to agitation. In a two-dimensional space of valence–arousal (such as that in Figure 14.3), it is possible to map a point to a categorical feeling label, although verbal description of the emotion state cannot be achieved.

Alexithymia and Anhedonia – *Alexithymia* is an impaired ability to experience and express emotions. It is a prominent feature among neurological patients with hemispheric commissurotomy [31]. Thus, the division in awareness affecting such patients includes an inability to communicate emotion arising from centres in the right hemisphere, through the language centres of the left hemisphere. In other words, the left hemisphere might not be aware and unable to communicate emotions affecting the right hemisphere – emotion awareness that might be revealed if the right hemisphere possessed the same language skills as the left. In any event, the alexithymic patient's inability to discriminate between such feelings as anger and sadness suggests a rather marked deficit in explicit emotion. The question, then, is whether one can find evidence of *implicit* emotion in these patients, in terms of

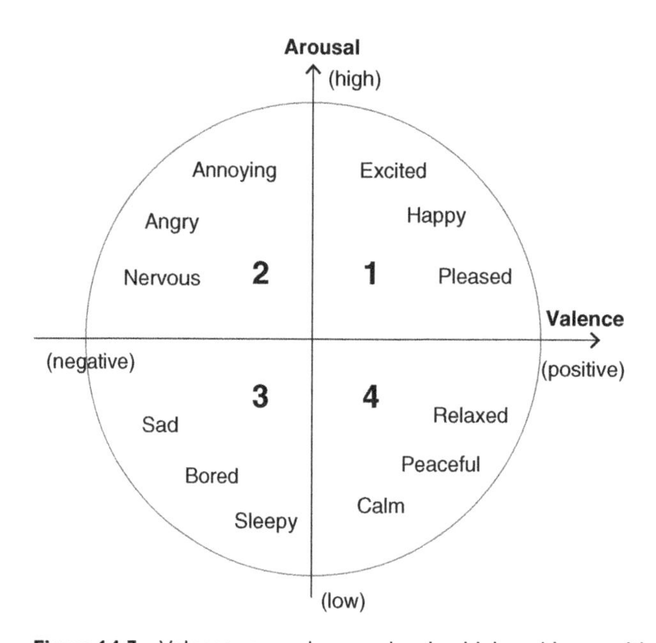

Figure 14.3 Valence–arousal space showing high and low positive and negative emotions.

behavioural or physiological indices. This disorder may lead to psychophysiological and somatoform disorders if not controlled or regulated. Alexithymia can be distinguished from *anhedonia,* an inability to experience positive emotions [31]. However, the alexithymic inability to communicate emotions to others is correlated with social anhedonia, or a preference for solitary rather than social activities [32]. Physical anhedonia affects explicit (subjective) components of positive emotion, leaving implicit (behavioural and physiological) components of positive emotion intact. Alexithymia is known to be associated with hypertension.

Alexithymia and related emotion-processing variables have been examined as a function of hypertension type [33]. The hypothesis here is that if dysregulated emotional processes play a key neurobiological role in essential hypertension, they would be less present in hypertension due to specific medical causes or secondary hypertension. The results achieved are consistent with a contribution of an emotional or psychosomatic component in essential hypertension and may have practical implications for non-pharmacological management of hypertension. Such results also demonstrate the usefulness of complementary measures for emotion processing in medically ill patients.

14.1.4 Emotion Regulation

Emotional self-regulation, also known as emotion regulation, refers to a state where one is able to properly regulate one's emotions. It is a complex process which involves the initiation, inhibition, or modulation of the following functions [34]:

- internal feeling states (i.e. the subjective experience of emotion)
- emotion-related cognitions (e.g. thought reactions to a situation)
- emotion-related physiological processes (e.g. heart rate, hormonal, or other physiological reactions)
- emotion-related behaviour (e.g. reactions or facial expressions related to emotion).

As stated above, in the human brain, emotion is normally regulated by a complex physiological circuit consisting of the orbitofrontal cortex (OFC), amygdala, anterior cingulated cortex (ACC) and several other interconnected regions. There are both genetic and environmental contributions to the structure and function of this circuitry. Diverse regions of PFC, amygdala, hippocampus, hypothalamus, ACC, insular cortex, ventral striatum and other interconnected structures are involved in this complex circuitry as depicted in Figure 14.4. Each of these interconnected structures plays a role in different aspects of emotion regulation. Abnormalities in one or more of these regions or in the interconnections among them are associated with failures of emotion regulation, increased propensity for impulsive aggression and violence [36].

Emotion regulation involves processes that amplify, attenuate, or maintain an emotion. Here our focus is on the associated effective phenomena of anger, general negative effect, and impulsive aggression.

Amygdala is involved in the learning process. It associates stimuli with primary events that are intrinsically punishing or rewarding [37, 38]. In human neuroimaging studies, the amygdala is activated in response to cues that connote threat (such as facial signs of fear) [39, 40], as well as during induced fear (e.g. fear conditioning) [41, 42] and generalized

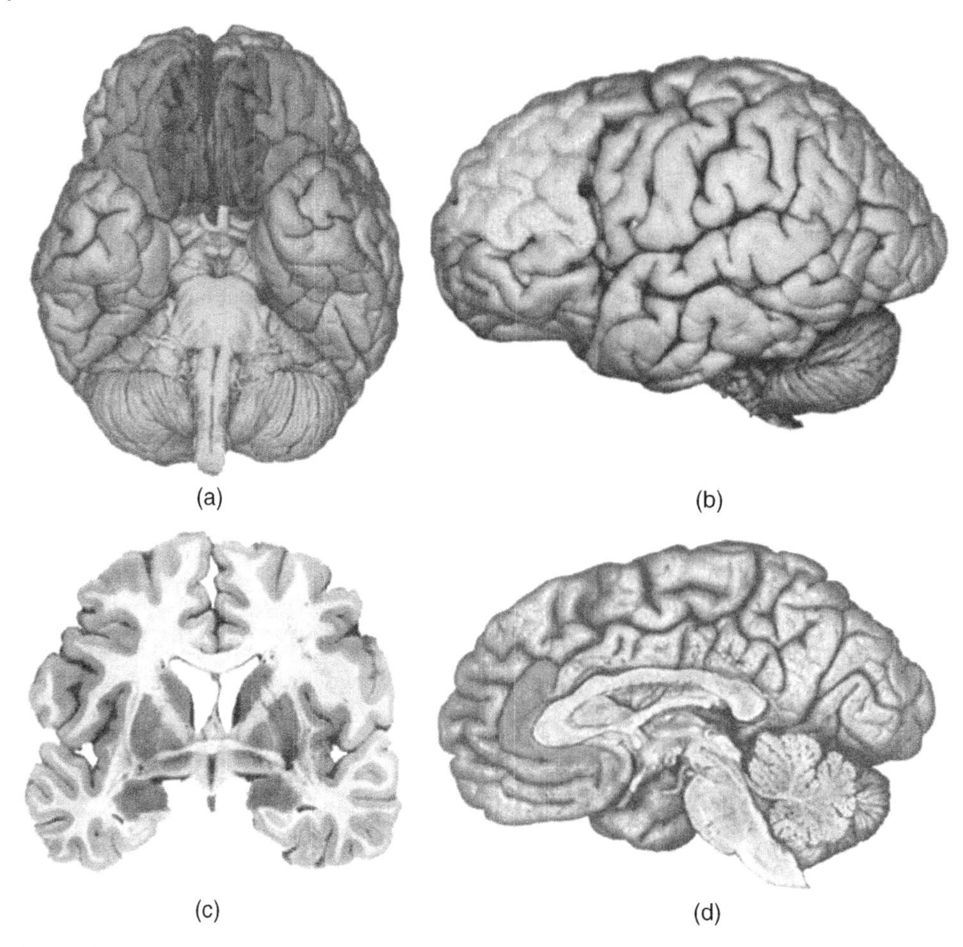

Figure 14.4 Emotion neural circuitry regions involved in emotion regulation. (a) Orbital prefrontal cortex (green regions) and the ventromedial prefrontal cortex (red regions). (b) Dorsolateral prefrontal cortex. (c) Amygdala and (d) anterior cingulate cortex. (Source: adapted from [35].)

negative affect (for example, the negative effect provoked by watching unpleasant pictures) [43].

The pathways of the brain to the amygdala are depicted in Figure 14.5 [45]. An emotional stimulus, something that looks like a snake, is first processed in the brain of a hiker by the thalamus. The thalamus passes on an immediate, but crude representation of the stimulus to the amygdala. This transmission allows the brain to start to respond to a potentially dangerous object, which could be a snake. Therefore, the hiker prepares for a dangerous situation, before he even knows what the stimulus is. Meanwhile, the thalamus also sends information to the cortex for more detailed examination. The cortex then creates a more accurate impression of the stimulus. The hiker realizes that he mistook a root for a snake. This outcome is fed to the amygdala as well and is transmitted to the body.

Patients with selective bilateral damage to the amygdala have a specific impairment in recognizing fearful facial expressions [46]. Amygdala is more strongly activated by facial

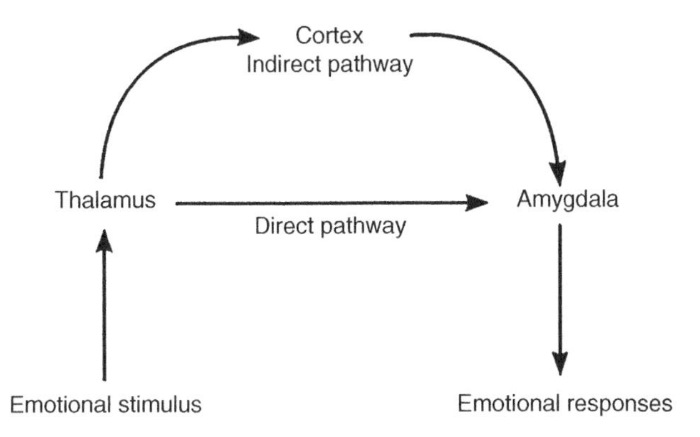

Figure 14.5 Direct and indirect pathways to the amygdala. (Source: taken from [44].)

expressions of fear than it is by other facial expressions, including anger. For example, an increase in intensity of fearful facial expressions is associated with activation of the amygdala. In contrast, the increase in intensity of angry facial expressions is associated with increased activation of the OFC and the ACC [47]. In some neuroimaging studies undertaken to induce anger specifically those presented in [48, 49], normal subjects showed increased activation in the OFC and ACC.

It is expected that in individuals prone to aggression and violence, the increase in OFC and ACC activation, usually observed in such conditions, is attenuated [36].

In humans, when the subjects view unpleasant pictures, there is an increase in the magnitude of the eye-blink reflex (measured from surface EMG recording of the orbicularis oculi muscle) in response to a brief burst of noise. Moreover, the magnitude of the eye-blink reflex to the same stimulus during viewing of pleasant pictures is smaller than that during the viewing of neutral pictures [24, 50].

By triggering a startle response at different times during affective processing, information about the time course of emotion can be gleaned [51–54].

In one experiment [53], normal subjects viewed unpleasant or neutral pictures for 8 seconds. Four seconds after the picture appeared, a digitized human voice instructed the subject to regulate the emotion they were experiencing in response to the picture. For unpleasant pictures, the subjects were asked to suppress, enhance, or maintain their emotional response. Subjects were instructed to continue voluntary regulation of their emotional response even after the picture disappeared. Most importantly, the subjects varied considerably in their skills at suppressing negative emotion. By inserting startle stimuli before presenting the instruction, it was confirmed that this variation in the ability to regulate emotions could not be accounted for by differences in the initial reaction to negative stimuli. In a more recent work [55], it has been found that the baseline levels of regional brain activation inferred from high-density scalp-recorded EEG [55] predicted the ability of subjects to suppress emotions. Those subjects with greater relative left-sided activation in prefrontal scalp regions presented greater startle attenuation in response to suppression instruction.

Suppressing negative effect in response to a stimulus that previously aroused such emotion can be conceptualized as a form of reversal learning. Patients with lesions in the OFC and those vulnerable to impulsive aggression should be particularly deficient in such a task, although they would still show basic enhancement of startle magnitude in response to negative stimuli. In [52] it has been proposed that the mechanism underlying suppression of negative emotion is via an inhibitory connection from regions of the PFC [56].

Based on these experiments a number of important conclusions were drawn [36] first, individual differences in the capacity to regulate emotion are objectively measurable. Second, individual differences in patterns of prefrontal activation predict the ability to perform this task and thus reflect differences in aspects of emotion regulation and third, individual differences in emotion regulation skills, particularly as they apply to suppression of negative effect, may especially be important in determining vulnerability to aggression and violence.

A greater left prefrontal activation, compared to the right one, is associated with increased reactivity to positively valence emotional stimuli [57], increased ability to recover from negative affective challenge [58], better voluntary suppression of negative affect [55], and higher scores on scales measuring psychological well-being [59]. Davidson has interpreted these findings as the left and right sides of PFC play differential roles in emotional processing. Moreover, there are hemispheric differences in goal-directed tendencies (approach versus avoidance) beyond those captured by positive or negative effects [60]. It is proposed that the left PFC is active in response to stimuli evoking the experience of positive affect. That is because these stimuli induce a fundamental tendency to approach the source of stimulation.

Research on the neural mechanism regarding the frontal asymmetry [61] and others on the neural mechanism of executive function, suggests that the right PFC involves in monitoring and checking the environment, while the left PFC is primarily engaged in generating strategies for action. Based on this assumption, Petrantonakis et al. [62] developed a method to evaluate this asymmetricity between the two frontal brain lobes. This is explained later in this chapter. Effective interaction with the environment which is likely to result from initiating appropriate activity may well be associated with increased positive effect. Conversely, vigilance required for monitoring and checking the environment may be associated with a negative effect such as anxiety.

In terms of the regions involved in different emotions there are hypothesis or experimental results which suggest the followings [63]:

14.1.4.1 Agency and Intentionality

Humans often attribute agency and intentionality to others. This is a pervasive cognitive mechanism that allows one to predict the actions of others based on spatiotemporal configurations and mechanical inferences, and on their putative internal states, goals, and motives. Despite a high degree of overlap that may exist between attributing agency and intentionality, they are not the same thing.

Implicated brain regions in agency and intentionality include the parietal cortex, insula, motor cortex, medial PFC and the superior temporal sulcus (STS) region [64, 65].

14.1.4.2 Norm Violation

Norms are abstract concepts firmly encoded in the human mind. They differ from most kinds of abstractions because the code for behavioural standards and expectations are often associated with emotional reactions when violated [66]. This effect, which marks a violation of an arbitrary norm, has been demonstrated to elicit brain responses in regions linked to conflict monitoring, behavioural flexibility, and social response reversals such as the ACC, anterior insula, the lateral OFC [67]).

14.1.4.3 Guilt

Guilt emerges prototypically from: (i) recognizing or envisioning a bad outcome to another person, (ii) attributing the agency of such an outcome to oneself, and (iii) being attached to the damaged person (or abstract value). Recent neuroimaging data showed the involvement of the anterior PFC, anterior temporal cortex, insula, ACC, and STS region in guilt experience [68]).

14.1.4.4 Shame

While there are no available brain imaging studies on shame, brain regions similar to those demonstrated for embarrassment should be involved. The ventral region of the ACC, also known as subgenual area, has been associated with depressive symptoms, [69]). Hence, such brain region might also play a more specific role in neural representation of shame.

14.1.4.5 Embarrassment

Embarrassment has traditionally been viewed as a variant of shame [70]. The related neural structures include the medial PFC, anterior temporal cortex, STS region, and lateral division of OFC.

14.1.4.6 Pride

The polar opposite of shame and embarrassment is pride. So far, there is no clear evidence for neural representation of pride, although it has been shown that patients with OFC lesions may experience this emotion inappropriately [71]. We hypothesize that brain regions involved with mapping the intentions of other persons (e.g. the medial PFC and the STS region) and regions involved in reward responses (OFC, hypothalamus, septal nuclei, ventral striatum) may play a role in this emotion.

14.1.4.7 Indignation and Anger

Indignation and anger are elicited by: (i) observing a norm violation in which (ii) another person is the agent, especially if (iii) the agent acted intentionally. Indignation relies on (iv) engagement of aggressiveness, following an observation of (v) bad outcomes to self or a third party. We and others have shown that indignation evokes activation of the OFC (especially its lateral division), anterior PFC, anterior insula and ACC [72].

14.1.4.8 Contempt

Contempt has been considered a blend of anger and disgust, and sometimes is considered as a more subtle form of interpersonal disgust [73]. For this reason here it is considered together with disgust. Neural representations of disgust have been shown to include the anterior insula, the anterior cingulate, and temporal cortices, basal ganglia, amygdala, and OFC [72, 74, 75].

14.1.4.9 Pity and Compassion

These feelings are described as 'being moved by another's suffering' [76, 77]. Preliminary functional imaging results in normal subjects point to the involvement of anterior PFC, dorsolateral PFC, OFC, the anterior insula and anterior temporal cortex in pity or compassion [78]. Further studies with more advanced imaging techniques may test if certain limbic regions, such as hypothalamus, septal nuclei, and ventral striatum could also be involved.

14.1.4.10 Awe and Elevation

Awe and elevation are self-transcendent emotions, possibly triggered by witnessing acts of human moral beauty or virtue often giving people desire to improve themselves and work towards the greater good. Awe and elevation are poorly understood emotions that have received more attention recently [79]. While the neuroanatomy of awe is still obscure, it is likely to involve limbic regions associated with reward mechanisms, including the hypothalamus, ventral striatum, medial OFC, and cortical regions linked to perspective taking and perceiving social cues, such as the anterior PFC and the STS region [78].

14.1.4.11 Gratitude

This emotion is elicited by detecting a good outcome to oneself, attributed to the agency of another person, who acted in an intentional manner to achieve the outcome. Gratitude is associated with a feeling of attachment to other agents and often promotes the reciprocation of favours. Recent studies using game-theoretic methods show that the activated brain regions encompass ventral striatum, OFC, and ACC [80, 81].

For a more natural human machine interaction emotions have to be registered, conveyed, and emotional relevance of events has to be well understood.

14.1.5 Emotion-Provoking Stimuli

From the above hypothesis, experiments, facts, assessing emotions, their impacts, and the level of regulation by each subject is essential to the understanding and control of human behaviour.

Based on the results of positron emission tomography (PET) it has been revealed and confirmed that transient sadness significantly stimulates bilateral limbic and paralimbic structures, brain stem, thalamus, and caudate/putamen, while transient happiness is associated with marked, widespread reductions in cortical cerebral blood flow (CBF) in the right prefrontal and bilateral temporal–parietal regions [82]. CBF increases during sad mood induction. The frontal pole is the only region differentiating sad states from happy states. Emotion expression in infants is also known to be associated with frontal activity asymmetries in which reflect specific regulation strategies [82]. Moreover, the left frontal region corresponds to regulation involving actions that try to maintain continuity and stability.

Four models for brain lateralization of emotional processing have been proposed by Demaree et al. [83]; the right-hemisphere, valence, approach–withdrawal, and behavioural inhibition system–behavioural activation system models.

Findings from the first model indicate that wide regions of occipital and temporal cortices, particularly within the right hemisphere, are subject to increased activity during dynamical facial expressions viewing. Following this model, it can be deduced that facial expression represents another lateralized motoric function like handedness and footedness, which might be controlled by dominant cerebral hemisphere. Thus, facial expression would be right-sided for right-handers and left-sided for left-handers. The second assumption stems from emotional processing literature available in early 1970s ([83] and references herein) and proposed that facial expression of emotion might be mediated by the right hemisphere. Thus, facial expression would be left-sided in right-handers, but not necessarily predictable in left-handers.

Based on the 'valence model' the right hemisphere is specialized for negative emotion and the left hemisphere is specialized for positive emotion [84, 85]. In this model, as confirmed by other experimental achievements, a variant of the valence hypothesis contends that differential specialization exists for expression and experience of emotion as a function of valence, whereas *percepual* processing of both positive and negative affective stimuli is a right cerebral function [85]. Correspondingly, this variant of valence hypothesis indicates that left and right-anterior regions are specialized for the expression and experience of positive and negative valence, respectively. Conversely, right posterior parietal, temporal, and occipital regions are dominant for the perception of emotion [86, 87].

The valence hypothesis was largely subsumed by the 'approach–withdrawal model' of emotion processing, which establishes that emotions associated with approach and withdrawal behaviours are processed within the left- and right-anterior brain regions, respectively. The overlap between the valence and approach–withdrawal models is extensive, with most negative emotions (e.g. fear, disgust) eliciting withdrawal and most positive emotions (e.g. happiness, amusement) eliciting approach behaviours.

Two anatomical pathways underlying emotional/motivational systems have been termed the behavioural activation system (BAS) and behavioural inhibition system (BIS). BAS appears to activate behaviour in response to conditioned, rewarding stimuli and in relieving non-punishment. Thus, this system is responsible for both approach and active avoidance behaviours. Also, emotions associated with these behaviours are generally positive in nature [88].

Numerous theories about the relationship between the strengths of BIS and BAS and affective disorders have been published. For example, although the BIS and BAS are thought to be rather stable over time, the existing variability around a strong BAS produces mania while variability around a weak BAS may yield depression.

The above lateralization models of emotional processing appear to have certain strengths. The right-hemisphere model, for example, emphasizes emotional perception at least as much as expression/experience. The valence model, as experienced, largely gave way to the approach–withdrawal model, which appears to be a better fit for the majority of data (e.g. anger). The BIS/BAS model, however, is essentially identical to the approach–withdrawal model but focuses on relatively stable emotional traits (BIS and BAS strengths) instead of states (primary focus of the approach–withdrawal model).

Cognitive neuroscience aspects of emotion have been therefore under development and grown significantly [89]. Perception and evaluation of emotional stimuli (emotional processing) and the effects of emotion on memory formation (emotional memory) have possibly been more researched.

Electrophysiological and functional neuroimaging study achievements support the role of PFC and amygdala in evaluating the emotional content of stimuli although the roles of other structures, such as ventral striatum, anterior, cingulate, posterior parietal, and insula regions, should not be ignored.

The amygdala is strongly associated with emotional processing by both lesion and functional neuroimaging studies. As an example, lesion studies show that patients with bilateral amygdala damage are impaired in detecting emotional facial expression and imaging studies on neurologically intact people have reported amygdala activations associated with processing of both pleasant and unpleasant stimuli.

14.2 Effect of Emotion on the Brain

14.2.1 ERP Change Due to Emotion

As for many other brain-related studies, event-related potentials (ERPs) have become an indicator of emotional effects. Consistent and robust modulation of specific ERP components can be achieved by viewing emotional images or listening to emotional vocals [90].

Correct assessment of emotions using ERP components requires accurate detection and estimation of amplitudes, latencies, and source locations of these components preferably on single-trial basis (rather than by averaging over epochs). Popular ERP detection and tracking approaches are presented in Chapter 9 of this book.

These modulations represent different stages of stimulus processing including perceptual encoding, stimulus representation in working memory, and elaborate stimulus evaluation. Selective processing of emotional cues by the brain is a function of stimulus novelty, emotional prime pictures, learned stimulus significance, and the explicit context of the attention tasks. Therefore, ERP measures are useful to assess the emotion–attention interface at the level of distinct processing stages.

The first ERP component reflecting differential processing of emotional compared to neutral stimuli is the early posterior negativity (EPN). A pronounced ERP difference for processing of emotionally arousing and neutral pictures developed around 150 ms maximally pronounced around 250–300 ms. This differential ERP appeared as negative deflection over temporo-occipital sensor sites and a corresponding polarity reversal over frontocentral regions. Despite differences in the overall topography, research presenting pictures discretely with longer presentation times (1.5 seconds and 1.5 seconds inter-trial interval) demonstrated a more pronounced negative potential for emotional pictures in the same latency range over temporo-occipital sensors [90]. In addition, from the biphasic view of emotion the differential processing of pleasant and unpleasant cues varies as a function of emotional arousal. Correspondingly, EPN changes with arousal level of emotional pictures. Particularly, highly arousing picture contents of erotic scenes and mutilations cause a more pronounced posterior negativity compared to less arousing categories of the same valence. Figure 14.6 illustrates the time course of an EPN and the corresponding topography image [90].

Instead of modulation during perceptual encoding, it is observed that emotional (pleasant and unpleasant) visual stimulations increases the amplitude of late positive potentials (LPPs) over centroparietal regions, most apparent around 400–600 ms post-stimulus, namely a P3b wave. [91–93]. This LPP modulation appears sizeable and can be observed

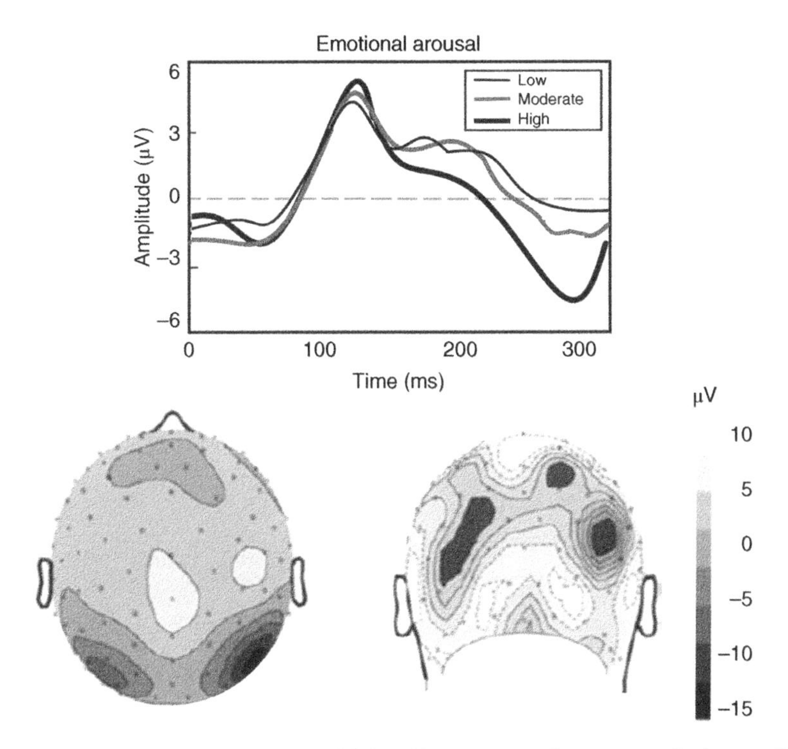

Figure 14.6 Time course of an EPN and its corresponding topography images [90].

in almost every individual when calculating simple difference scores (emotional–neutral). In addition, LPP is also specifically enhanced for more emotionally intense pictures (i.e. described by viewers as more arousing, indicating a heightened skin conductance response). It has been found that picture contents of high evolutionary significance such as pictures of erotica and sexual contents and contents of threat and mutilations further enhance the LPP amplitudes [91]. The effect of stimulation repetition has not shown any significant effect on the results [90]. Figure 14.7 shows the brain's time course during LPP. Topoplots of pleasant, unpleasant, and neutral visual stimuli do not present noticeable difference.

In addition to LPP, in a sustained stimulation scenario, the LPP is followed by a positive slow wave (SW). Therefore, the positive SW reflects sustained attention to visual emotional stimuli.

Generally, ERP modulations, induced by emotional cues, are reliable and consistently observed early in the processing stream. Additionally, emotional attentions of motivated and instructed attentions look alike.

An emotion effect has been found in different ERP components, including P300 component (e.g. [89, 94]), N300 component [95]; and the SW component [89, 96]), but most consistently, an emotion effect is expressed in a P300–SW complex as discussed above. More research has shown that the P300 component may be more sensitive to emotion under intentional emotional processing (e.g. [94, 97]), whereas the N300 component appears to be more sensitive during incidental emotional processing but the evidence is not conclusive [94]. Other studies discuss variations in P3a and P3b for arousal and valence states [98].

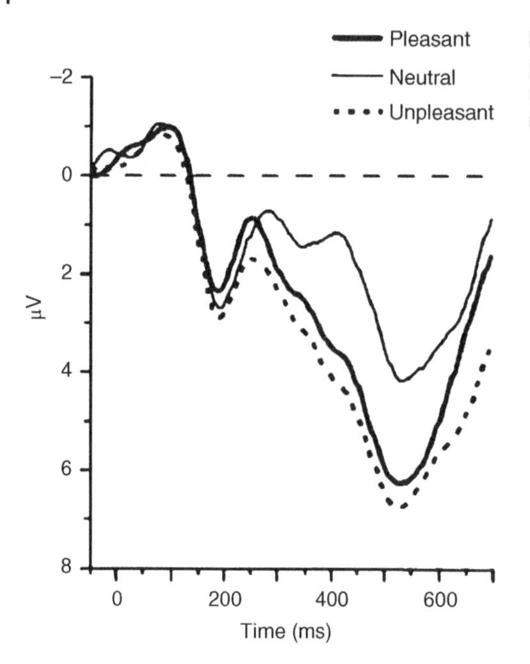

Figure 14.7 Time course of the late positive potential; grand-averaged ERP waveforms of a right parietal sensor while viewing pleasant, neutral, and unpleasant pictures [90].

It has been found that P100 is significantly affected by facial emotions (i.e. the fearful facial expressions elicited higher P100 than happy and neutral ones) [99]. Also, for N100 considered as a sensory component, there is a significant emotion modulation, in which fearful faces elicited larger N100 amplitudes than happy and neutral faces. Conversely, happy vocal stimuli significantly reduced the N100 latencies as compared to sad and anger stimuli but no significant main effect of emotion on N100 amplitudes was observed. These results may show the difference in the effect of different stimuli (audio and visual) on N100.

As another ERP component widely used for EEG-based emotion recognition, N170 is a negative-going ERP component detected at the lateral occipitotemporal electrodes, which peaks around 170 ms after stimulus onset. This component has been found to be sensitive to face stimuli rather than nonface stimuli. A significant emotional modulation of N170, in which happy and fearful facial expressions elicited larger N170 amplitudes and shorter latencies than the neutral faces, has been reported. Nevertheless, this conclusion has not been agreed with by some other researchers [99].

Another ERP component, namely vertex positive peak (VPP), is a positive component with a peak latency similar to that of N170 and is detected at the frontocentral electrodes. Significant emotional effect on VPP amplitudes as well as on VPP latencies, in which happy and fearful faces elicited larger VPP amplitudes than neutral facial expressions, have been experienced. It has also been found that shorter latencies are elicited by fearful facial expressions than by happy and neutral faces. As another example, the disgust facial expressions evoke larger VPP amplitudes than the happy faces.

EPN is another ERP component which peaks between 210 and 350 ms with a topographical distribution over occipitotemporal sites. The EPN effect for emotionality has been normally reported in studies using emotional faces and words to elicit emotion response/

experience. The EPN amplitudes for angry expressions are larger than those elicited by other facial expressions in the right hemisphere and are larger than those for neutral faces in the left hemisphere. Also, there are significantly larger EPN amplitudes for positive and negative words than for neutral words. However, based on some other researches the difference in the amplitude between the two emotional states is not significant [99].

In a study by Luo et al. [100], the P300 amplitudes for facial expressions showed a difference between the fearful and the happy as well as the emotional faces and neutral faces, which to some degree extends the previous findings that angry faces elicited larger P300 than happy and neutral faces.

The P300 is the sum of P3a and P3b subcomponents which have different temporal and spatial distributions over the scalp. There are opposite patterns towards emotionality, in which angry elicited smaller P3a amplitudes than happy and larger P3b amplitudes for angry than for happy. This may be due to the less attention orienting towards happy than angry [99]. Briefly speaking, P300 can be a valuable tool to explore emotion processing, especially for relationship between emotion and attention.

LPP, explained above, also known as LPC, is evident at central and frontal midline sites as compared to the earlier parietal positivities. It has been shown that larger LPP amplitudes are elicited by positive words than for ones by negative or neutral words. Using dynamical emotional faces which constituted emotional clips, significant LPP difference among fear, anger, surprise, disgust, sad, and neutral has been found [99].

There are still two main issues concerning the emotion ERP effect which have not been completely solved. First, it is unclear whether the emotion effect is sensitive to arousal only (emotional vs. neutral) or to both arousal and valence (pleasant vs. unpleasant). Most studies have only found differences owing to arousal. Few recent studies have reported differences that could attribute to emotional valence [89, 96]). Many of these studies have used small number of electrodes and therefore, due to lack of accurate localization, no significant differences could be noted between topographies of pleasant and unpleasant stimulations.

14.2.2 Changes of Normal Brain Rhythms with Emotion

In terms of changes in the normal brain rhythms, the correlation between neural activity and emotional changes have been supported by lesion [101], electrophysical [102], and functional neuroimaging [103, 104]. The same techniques support the role of PFC and amygdala in the evaluation of emotional content of the stimuli. In the study of emotions with respect to pleasant, unpleasant, and neutral pictures the role of PFC regions in emotional processes has been examined [104] and it was deducted that right hemisphere is specialized for perception, expression, and experience of emotion. According to definition of valence and based on electrophysiological and functional neuroimaging evidences, again it has been concluded that the left hemisphere is primarily associated with processing of pleasant emotions, while the right hemisphere with processing of unpleasant emotions [103].

Zheng et al. [105] have attempted to identify the stable patterns over time for EEG-based emotion recognition. Their experimental results indicated that the lateral temporal areas consistently activate more for positive emotion than for negative emotion in beta and gamma bands; the neural patterns of neutral emotion have higher alpha responses at parietal and occipital sites; and for negative emotion, the neural patterns have significant higher delta responses at parietal and occipital sites and higher gamma responses at prefrontal sites.

14.2.3 Emotion and Lateral Brain Engagement

As concluded from the above sections, the brain lateral asymmetricity, coherency, and connectivity change during opposite emotions. Various measures of the above effects have been proposed by different researchers. Generally speaking, the mutual information (MI) between the EEGs recorded over the electrodes on two sides of the brain midline in different EEG conventional frequency bands changes significantly.

14.2.4 Perception of Odours and Emotion: Why Are They Related?

Odours may be considered as human perception after smelling. There are various odours with different strengths and humans may have different impressions about the odours. For perceptions of sensations influencing or producing emotion, the perception of odours is dependent on respiration; our sense of smell is enhanced by inhalation or inspiration. Olfactory cells have cilia (dendrites) extending from the cell body into the nasal mucosa. The axons carry impulses from activated olfactory receptors to the olfactory bulb [15]. The sensors within the olfactory bulb (Figure 14.8) send signals to the prepiriform and piriform cortex, which include the primary olfactory cortex, anterior olfactory nucleus, amygdaloid nucleus, olfactory tubercle, and entorhinal cortex.

Olfactory information ascends directly to the limbic areas and is not relayed through the thalamus, so that each breath activates these areas directly. Direct stimulation of olfactory limbic areas, unconsciously, alters the respiratory pattern. Unpleasant odours increase respiratory rate and induce shallow breathing while pleasant odours induce deep breathing.

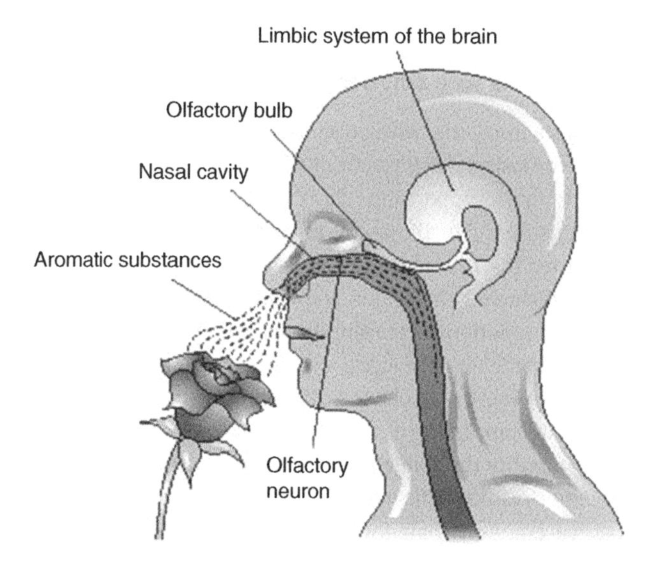

Figure 14.8 Olfactory bulb.

It is interesting to note that the respiratory rate increases even before subjects discern whether a smell is unpleasant. It is likely that physiological outputs respond more rapidly than cognition [106]. PET and fMRI studies in humans have linked olfactory-related brain regions that are related to higher-order olfactory processing, such as discrimination [107] and emotion [108]. EEG studies have also shown three to four negative and positive waves present during inspiration and the waves are phase-locked to inspiration. These waveforms, referred to as inspiration phase-locked α-band oscillation (I-α), are not observed during the expiratory phase or during breathing of normal air. The EEG dipole tracing method has estimated the location of I-α source generators to the entorhinal cortex, amygdala, hippocampus, and orbitofrontal cortex. Over 300–400 ms, the source dipole converges most in the orbitofrontal cortex. During such an event, recognition and identification of odour occurs after evaluation in the olfactory limbic areas [15].

Experiments have shown odour molecules reach olfactory receptors by inspiration, and odour information is directly projected to limbic structures but not through the thalamus. Odour stimuli thus cause respiratory changes, and simultaneously induce emotion and memory recognition via stimulation of olfactory-related limbic structures [109].

Relations between respiration and olfaction have been reviewed by Masaoka and Homma. The same authors have also reviewed the relation between brain rhythm and respiration for both normal subjects and those suffering from Parkinson's disease (PD) with olfactory impairment [109].

14.3 Emotion-Related Brain Signal Processing and Machine Learning

Research in computer science, engineering, psychology, and neuroscience has aimed at developing devices that recognize human affect, monitor, and model emotions. In computer science, affective computing (AfC) is a branch of the study developing artificial intelligence that deals with the design of systems and devices for recognition, interpretation, and processing of human emotions. While the origins of the field may be traced back to early philosophical enquiries into emotion the more modern branch of computer science originated with Rosalind Picard in 1995 in her paper on AfC [110].

It might be easy to discriminate between two contradictory types of emotions (such as sadness and happiness), but a learning system capable of classifying all or a large collection of emotions remains far from reality. This is mainly because a wide range of emotions is tied to certain brain regions and unified by a certain structure.

Detection of emotional information, as for almost all types of physiological signals and data, incorporates use of passive sensors to capture the user's physical state or behaviour without interpreting the input. The collected data are analogous to the cues humans use to perceive emotions in others. The more accurate the analysis of these data, the better the corresponding emotional state is recognized.

There are often debates on the differences between brain dynamic activity and emotion. Although both refer to neuron activations and neural processes within the brain the way brain engagements of the brain might be different [111].

Another facet of AfC is the design of computational devices proposed to exhibit either innate emotional capabilities or that are capable of simulating emotions convincingly. Therefore, creating new techniques to indirectly assess frustration, stress, and mood, through natural interaction and conversation or making the computers more emotionally intelligent, especially responding to a person's frustration in a way that reduces negative feelings, can be a desired research direction in emotion recognition and regulation.

EEG has traditionally been used for investigation and study of cortical activity and deep brain signal sources have not been considered widely. Advances in source separation, however, facilitate tracking and localization of different brain activities.

These signals may be corrupted by and are sensitive to electrical signals emanating from facial muscles while emotions are expressed. Therefore, it is logical to fuse these two modalities in the study of emotion.

14.3.1 Evaluation of Emotion Based on the Changes in Brain Rhythms

In the majority of emotion studies with EEG, changes in the background activity of the brain have been observed or assessed. Frequency domain or time–frequency domain methods (such as by the use of the wavelet transform [112]) aim at exposing the imbalance in the power between the two lateral lobes of the brain for two opposite emotions. However, more detailed processing techniques are necessary to include multichannel EEG to enable detection and scoring a wider range of emotions. As simple approach for analysis of three emotion states of pleasant, unpleasant, and neutral in different brain regions, based on classification of brain responses to visual stimuli (by looking at IAPS images [24]) for different frequency bands within 4–45 Hz frequency interval, has been given. The procedures, selected frequency bands, and selected electrodes are similar to those in [113]. Six features, sub-bands frequency powers from 12 sites (six for each side of the brain) have been measured and used for classification. In another attempt [114] the researchers use the power spectrum, wavelet features, and approximate entropy of the EEG signals to evaluate emotions.

To reduce the redundancy within EEG signals and choose informative EEG features, in [115] an EEG feature selection technique, termed as feature selection with orthogonal regression (FSOR) has been introduced which can employ orthogonal regression to exploit more discriminative information in the classification process.

The selected input features from the EEG channels are linear features including mean absolute amplitude, variance, kurtosis, skewness, Hjorth parameters (activity; $A_x = \frac{1}{T}\sum_{t=1}^{T}(x(t)-\mu_x)^2$, mobility; $M_x = \sqrt{var(\dot{x}(t))/var(x(t))}$, complexity; $C_x = M_{\dot{x}}/M_x$

, where the dot on top of a letter denotes differentiation), absolute power, relative power, maximum power spectral density, frequency at maximum power spectral density, the absolute power ratio of beta band to theta band, and the asymmetry of the two alpha and beta bands, and nonlinear features including Shannon entropy, Kolmogorov entropy, C0-complexity [116] and spectral entropy. Given the feature input matrix as $\mathbf{X} \in \Re^{d\times n}$, a bias $b \in \Re^{k\times 1}$, and the output labels $\mathbf{Y} \in \Re^{k\times n}$, during FSOR they find the regressor (discriminator hyperplane – regression matrix) $\mathbf{W} \in \Re^{d\times k}$ through solving the following regularized optimization problem [115]:

$$\min_{\mathbf{W}, b, \Gamma} \|\mathbf{W}^T \Gamma \mathbf{X} + b\mathbf{1}_n^T - \mathbf{Y}\|_F^2 \quad s.t. \quad \mathbf{W}^T\mathbf{W} = \mathbf{I}_k, \quad \boldsymbol{\gamma}^T \mathbf{I}_d = 1, \quad \gamma \geq 0 \tag{14.1}$$

where $\Gamma \in \mathfrak{R}^{d \times d}$ is a diagonal matrix with $\boldsymbol{\gamma}^T\mathbf{I}_d = 1$, and $\mathbf{I}_n = [1\ 1...1]^T \in \mathfrak{R}^{n \times 1}$. γ_l in $\boldsymbol{\gamma}$ denotes the importance of the lth feature. By increasing the number of features from 30 to 390 the classification accuracy increases less than 4% (from approximately 72–76%).

They also ranked the features using FSOR and empirically verified that the absolute power ratio of beta wave to theta wave is the most discriminative feature, and the beta band is the critical band for emotion recognition as the asymmetry in the beta band is another significant feature.

14.3.2 Brain Asymmetricity and Connectivity for Emotion Evaluation

In the above sections the asymmetricity of the frontal brain lobes during emotion has been addressed. Exploiting the fact that the asymmetry between the left and right brain hemispheres forms the most prominent expression of emotion in brain signals, in [62] EEG has been used and an index called asymmetry index (AsI) introduced for emotion evaluation. This is accomplished by a multidimensional directed information (MDI) analysis between different EEG sites from the two opposite brain hemispheres from the EEG recorded over Fp1, Fp2, and F3/F4 cites in a conventional 10/20 electrode setup. The simple index used by Davidson is $DI = (L - R)/(L + R)$, where L and R are, respectively, the powers of EEG in a specific band in the left and right brain hemispheres.

Estimation of brain lobes connectivity has been shown to be an effective means of assessing various brain abnormalities and functionalities. Although the estimates for connectivity represent communications between cortical sources, additionally, they can show how sensitive the brain can be to emotional stimuli. As an example, insula is believed to process convergent information from several sensory modalities monitoring the state of the body and related to pain and other basic emotions experiences including disgust, anger, and fear [117, 118]. Some researchers have even suggested that its role in mapping visceral states and emotions could be the basis for conscious feelings [119]. In that sense, high levels of connectivity found here could support an integrative role for this brain structure. Similarly, the amygdala, which is linked to emotions, has role in learning modulation [120, 121] and its enhanced coherence is probably related to that in the insula functionally.

Different connectivity measures such as MI [122] have been used. For a clearer assessment of brain connectivity, the connectivity measure is often taken over different conventional EEG frequency bands.

Similar to other brain connectivity measures explained in Chapter 8, to define MDI, consider the simple case of two stationary time series X and Y of length N divided into n epochs of length $L = N/n$; each epoch of length $L = P + 1 + M$ is written as a sequence of two sections of length P and M before and after the samples x_k and y_k of time series X and Y at time k, respectively, i.e. [62]:

$$X = x_{k-P}...x_{k-1}x_kx_{k+1}..x_{k+M} \triangleq X^Px_kX^M \tag{14.2}$$

$$Y = y_{k-P}...y_{k-1}y_ky_{k+1}..y_{k+M} \triangleq Y^Py_kY^M \tag{14.3}$$

Then, the MI between the two time series is defined as:

$$I(X,Y) = \sum_{k} I_k(X,Y) \tag{14.4}$$

where $I_k(X, Y)$ are the directed MI defined as:

$$I_k(X,Y) = I\left(x_k, Y^M \mid X^P Y^P y_k\right) + I\left(y_k, X^M \mid X^P Y^P x_k\right) + I\left(x_k, y_k \mid X^P Y^P\right) \tag{14.5}$$

Obviously, although $I(X, Y) = I(Y, X)$, the directed terms are not necessarily symmetric. To better understand this, you may write, for example, the first term as:

$$I\left(x_k, Y^M \mid X^P Y^P y_k\right) = I\left(x_k \to Y^M \mid X^P Y^P y_k\right) = \sum_{m=1}^{M} I\left(x_k \to y_{k+m} \mid X^P Y^P y_k\right) \tag{14.6}$$

where each term on the right-hand side of (14.6) can be interpreted as information that is first generated in X at time k and propagated with a time delay of m to Y, and can be calculated through the conditional MI as a sum of joint entropy functions:

$$\begin{aligned} I\left(x_k \to y_{k+m} \mid X^P Y^P y_k\right) &= H\left(X^P Y^P x_k y_k\right) + H\left(X^P Y^P y_k y_{k+m}\right) - H\left(X^P Y^P y_k\right) \\ &\quad - H\left(X^P Y^P x_k y_k y_{k+m}\right) \end{aligned} \tag{14.7}$$

If the variables are Gaussian distributed, then the joint entropy of n variables $r_1,\ldots, r_n$ can be calculated using the covariance matrix $R(r_1,\ldots, r_n)$ in the following form:

$$H(r_1,\ldots, r_n) = \frac{1}{2} \log\left[(2\pi e)^n |R(r_1,\ldots, r_n)|\right] \tag{14.8}$$

where $|\cdot|$ denotes the determinant. By using (14.8), (14.7) can be written as:

$$I\left(x_k \to y_{k+m} \mid X^P Y^P y_k\right) = \frac{1}{2} \log \frac{\left|R\left(X^P Y^P x_k y_k\right)\right| \left|R\left(X^P Y^P y_k y_{k+m}\right)\right|}{\left|R\left(X^P Y^P y_k\right)\right| \left|R\left(X^P Y^P x_k y_k y_{k+m}\right)\right|} \tag{14.9}$$

This can be extended to calculation of MDI where the number of variables is more than two [62, 123]. For three sequences of X, Y, and Z, the directed information is then [29]:

$$I\left(x_k \to y_{k+m} \mid X^P Y^P Z^P y_k z_k\right) = \frac{1}{2} \log \frac{\left|R\left(X^P Y^P Z^P x_k y_k z_k\right)\right| \left|R\left(X^P Y^P Z^P y_k z_k y_{k+m}\right)\right|}{\left|R\left(X^P Y^P Z^P y_k z_k\right)\right| \left|R\left(X^P Y^P Z^P x_k y_k z_k y_{k+m}\right)\right|} \tag{14.10}$$

Using (14.10) and (14.6), the total amount of information, namely S, that is first generated in X and propagated to Y, taking into account the existence of Z, across the time delay range is:

$$S^{XY} : I\left(x_k \to Y \mid X^P Y^P Z^P y_k z_k\right) = \frac{1}{2} \sum_{m=1}^{M} \log \frac{\left|R\left(X^P Y^P Z^P x_k y_k z_k\right)\right| \left|R\left(X^P Y^P Z^P y_k z_k y_{k+m}\right)\right|}{\left|R\left(X^P Y^P Z^P y_k z_k\right)\right| \left|R\left(X^P Y^P Z^P x_k y_k z_k y_{k+m}\right)\right|} \tag{14.11}$$

This is used to consolidate the AsI measure by estimating the MI shared between the left and right brain hemisphere, thereby exploiting that way the frontal brain asymmetry concept.

According to the frontal brain asymmetry concept, the experience of negative emotions is related with an increased right frontal and prefrontal hemisphere activity, whereas positive emotions evoke an enhanced left-hemisphere activity. Assuming the EEG signals recorded at FP1, FP2, and F3-F4 (linked together) represent respectively the signals X, Y, and Z and exploiting the asymmetry concept, a measure to evaluate this asymmetry information in signals X and Y, taking into account the information propagated by signal Z to both of them, would introduce an index of how effectively an emotion has been elicited. Towards this, it is assumed that the total amount of information S, as in (14.11), hidden in the EEG signals and shared between the left and right hemisphere becomes maximum when the subject is calm (referring to information symmetry), whereas S reaches minimum when the subject is emotionally aroused (information asymmetry). Following the MDI concept, two variables denoted S_r and S_p have been introduced [62]. S_r refers to bidirectional information sharing between X and Y taking into account Z when the subject is not under any emotional state, i.e. is relaxed meaning that:

$$S_r = S_r^{XY} + S_r^{YX} \tag{14.12}$$

whereas S_p is the same sharing information during the period where she/he is supposed to feel an emotion, i.e.:

$$S_p = S_p^{XY} + S_p^{YX} \tag{14.13}$$

According to what has already been discussed, Sp will presumably be smaller than Sr if the asymmetry concept holds. Finally, in order to directly define a measure for emotion experience, the AsI is introduced which is defined as the distance of the (S_p, S_r) point, corresponding to a specific picture, from the line $S_p = S_r$, i.e.:

$$\text{AsI} = \frac{\sqrt{2}}{2}(S^r - S^p) \tag{14.14}$$

AsI serves as an index for the efficiency of emotion elicitation as used as a metric for emotion arousal through an extensive classification setup. Through experiments the efficiency of AsI for emotion arousal classification using support vector machines (SVM) has been investigated [62]. The experiments involve 5 seconds relaxation followed by another 5 seconds countdown duration before a 5 seconds video clip is presented and finally a 20 seconds self-assessment period is followed. However, the outcomes for the two cases of user-independent and user-dependent are significantly different.

Given a pathophysiological theory of a specific disease, connectivity models might allow one to define an *endophenotype* of that disease, i.e. a biological marker at intermediate levels between genome and behaviour, which enables a more precise and physiologically motivated categorization of patients [124]. Such an approach has received particular attention in the field of schizophrenia research where a recent focus has been on abnormal synaptic plasticity leading to dysconnectivity in neural systems concerned with emotional and perceptual learning [125, 126]. A major challenge will be to establish sufficiently sensitive neural systems models and using their connectivity parameters reliably for diagnostic classification and treatment response prediction of individual patients. Ideally, such models should be used in conjunction with paradigms, minimally dependent on patient compliance and not confounded by factors like attention or performance. Given established validity and sufficient sensitivity and specificity of such a model, one could use its analogy to biochemical tests in internal medicine, i.e. compare a particular model parameter (or combinations thereof) against a reference distribution derived from a healthy population [126]. Such procedures could help to decompose current psychiatric entities like schizophrenia into more well-defined subgroups characterized by common pathophysiological mechanisms which may facilitate the search for genetic underpinnings.

Brain region connectivity may also be influenced during sleep as a result of previous emotional events. It has been demonstrated that rapid eye movement (REM) sleep physiology is associated with an overnight dissipation of amygdala activity in response to previous emotional experiences [61]. This alters functional connectivity and reduces next-day subjective emotionality.

14.3.3 Changes in ERPs for Emotion Recognition

In many applications such as for dementia, Alzheimer's and depression the way the brain responds to various stimuli changes dramatically. Therefore, the study of brain event responses for evaluation of emotion (vice versa) in these subjects attracts psychologists and bioengineering researchers. In a study by Champagne et al. [127] healthy and schizophrenic subjects have been studied and their emotion against pleasant and unpleasant stimuli compared. The authors studied the ERPs P200, N200, and P300 for both groups. They reported that emotional valence generally affects early components (N200), which reflect early process of selective attention, whereas emotional arousal and valence both influence the P300 component, which is related to memory context updating, and stimulus categorization.

Their results show that, in the control group, the N200 amplitude was significantly more lateralized over the right hemisphere, while no such lateralization can be found in the schizophrenic patients. In the schizophrenic patients, significantly smaller anterior P300 amplitude can be observed for the unpleasant stimuli compared with the pleasant ones. This can be viewed in Figure 14.9 [127].

In another study 64 channel wireless EEG and an eye tracker (to mark the time instants in which the pictures first occurred in the visual field) have been employed to study the ERPs in subjects against viewing different type (pleasant and unpleasant) pictures in a gallery setup [128]. They performed both ERP detection and ERP dipole source localization and discovered the changes in the early ERPs (up to 210 ms) due to the emotion level.

14.3.4 Combined Features for Emotion Analysis

In a comprehensive feature-based emotion classification time domain, frequency domain, time–frequency domain, and connectivity-based features have been taken into consideration [129]. In this study the ERP features are left out mainly because the exact onset of stimulus and various ERP timings were difficult to identify in their study.

Among time-domain features, to measure the nonstationary index (NSI), The normalized signal is divided into small parts and the average of each segment is computed. The NSI is then estimated as the standard deviation of all means, where higher index values indicate more inconsistent local averages.

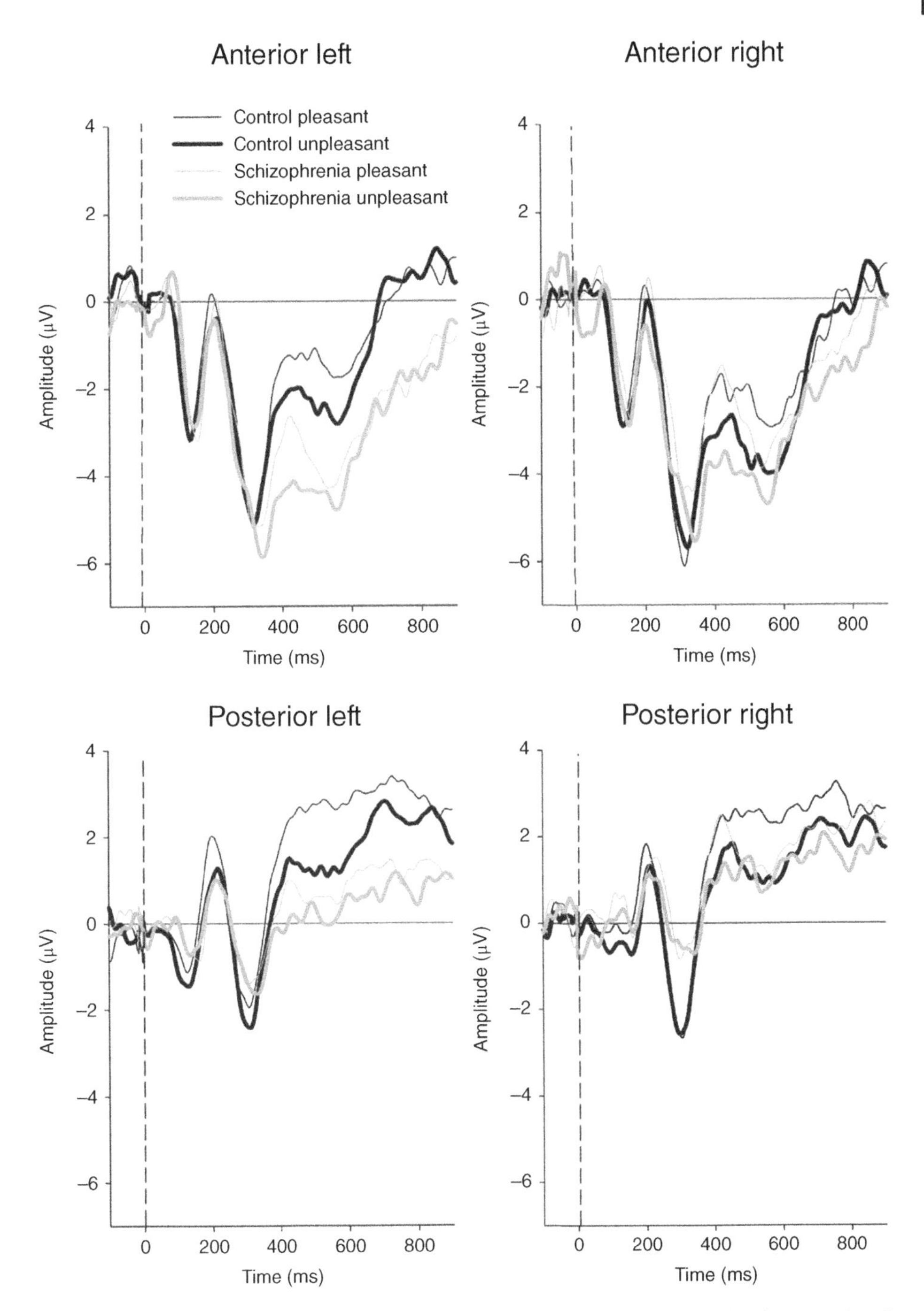

Figure 14.9 Group comparison between anterior and posterior P300 amplitudes for control and schizophrenic subjects. P300 amplitude is generally smaller in schizophrenic patients compared with the control group in all conditions, but the differences between pleasant and unpleasant are larger over the anterior region in this group [127].

Another time-domain feature is fractal dimension (FD) which has been estimated using the Higuchi algorithm in this work. Following this approach, for a time series $x(t)$ of length T, the time series can be written segment-wise as:

$$\left\{x(m), x(m+k), \ldots, x\left(m+\left\lfloor\frac{T-m}{k}\right\rfloor k\right)\right\}$$

where $m = 1,2, \ldots, k$, is the initial time and $\lfloor\bullet\rfloor$ refers to the floor operation. Then, k sets are calculated through:

$$L_m(k) = \frac{T-1}{\left\lfloor\frac{T-m}{k}\right\rfloor k^2} \sum_{i=1}^{\left\lfloor\frac{T-m}{k}\right\rfloor} |x(m+ik) - x(m+(i-1)k| \tag{14.15}$$

The average value over k sets of $L_m(k)$, denoted as $\langle L_m(k)\rangle$, has the following relationship with the fractal dimension FD_x:

$$\langle L_m(k)\rangle \propto k^{FD_x} \tag{14.16}$$

Therefore, FD_x can be obtained by the negative slope of the log–log plot of $\langle L_m(k)\rangle$ against k.

They also use the number of zero-crossings of the filtered data as another time-domain feature.

Among the frequency domain features they use bispectrum, defined as:

$$Bis(f_1, f_2) = E[X(f_1), X(f_2), X^*(f_1 + f_2)] \tag{14.17}$$

and bicoherence defined as:

$$Bic(f_1, f_2) = \frac{Bis(f_1, f_2)}{\sqrt{P(f_1)\bullet P(f_2)\bullet P(f_1 + f_2)}} \tag{14.18}$$

where $P(f) = E[X(f)X^*(f)]$ is the power spectrum. Conversely, to estimate the time–frequency features empirical mode decomposition (EMD) is performed and the Hilbert-Huang transform of each intrinsic mode function (IMF) calculated. As features, amplitude and instantaneous phase for each IMF transform are used [129].

Features representing the relationship between the EEG channels include magnitude squared coherence estimate using the cross-power spectral density between two signals x and y, P_{xy} and individual signal powers P_x and P_y:

$$C_{xy}(f) = \frac{|P_{xy}(f)|^2}{P_x(f)P_y(f)} \tag{14.19}$$

As well as differential and rational asymmetries [129], MI is another measure representing the relationship between the EEG channels.

From the above approaches, multivariate measures, brain connectivity estimates, and the ratio between the lateral powers particularly within the beta band, seem to be the most important indicators of opposite emotions.

Some of the above features have also been used in the EEG-based classification of music like/dislike [130]. The authors have concluded that the bilateral average activity from beta and gamma bands, when referred to the resting period, leads to the best discrimination between liking and disliking judgements.

14.4 Other Physiological Measurement Modalities Used for Emotion Study

The advent of fMRI and other imaging technology has spawned a deluge of research examining how brain functions at the macroscopic level. This work, which treats each voxel as a basic unit of analysis, has yielded tremendous insights as to how cortical region networks cooperate to produce emotion [131].

fMRI has been reported as a method for determination of regions of activity within the brain during emotion processing [132]. However, spatial sampling of the signals from amygdala is a challenging problem. It most probably depends on the emotion intensity. It has been claimed that an unjustified application of proportional global signal scaling (PGSS) leads to an attenuated effect of emotional activation in structures with a positive correlation between local and global blood oxygen level dependence (BOLD) signal [132].

In a research by Eryilmaz et al. [133] a wavelet correlation approach has been developed to investigate the impact of transient emotions on functional connectivity during subsequent resting and that how emotionally positive and negative information influences later brain activity differentially at rest.

The fMRI connectivity analysis has been estimated using temporal correlations between regional brain activities in predefined wavelet sub-bands. In this application the orthogonal cubic B-spline wavelet transform in the temporal domain [134, 135] has been used. The correlations between wavelet coefficients of a given sub-band for all regions of interest (ROIs) identified in general linear model (GLM) contrasts have been measured. Also, cross-correlation matrices between these regions for each condition and for four different wavelet sub-bands including the typical resting-state frequency range (ca. 0.01–0.1 Hz) have been obtained [133].

To compute the normalized correlation between any pair of regions in fMRI, the time courses corresponding to all blocks of a specific condition (e.g. fearful movies) are extracted for both regions. Then, by taking the discrete wavelet transform (DWT) the signal is separated into four different frequency bands: (i) 0.03–0.06 Hz, (ii) 0.06–0.11 Hz, (iii) 0.11–0.23 Hz, and (iv) 0.23–0.45 Hz. The first two frequency intervals are known to correspond to typical default mode (DM) bands, constituted our main 'bands of interest', and therefore, the first two bands are considered here. The cross-correlations between wavelet coefficients of all possible pairs of regions are calculated and the correlation matrices for each block are constructed.

In order to statistically test the differences between two resting conditions (e.g. rest post-fearful > rest post-neutral), corresponding correlation matrices to blocks pertaining to two conditions (12 matrices per condition for each participant) are compared by nonparametric permutation testing [66].

A threshold was defined for each condition and each sub-band and applied to wavelet correlation matrices, using false discovery rate at 0.2. Several node measures based on small-world models were computed after applying a threshold to the matrices. In a so-called small-world network model, each region was considered as a 'node' and, after applying an appropriate threshold, various measures, such as clustering coefficient, mean minimum path length, and degree, can be computed for each node [136].

Clustering the wavelet coefficients led to grouping of the existing links connecting a number of neighbouring nodes. Minimum path length is the number of connections that forms the shortest path between two nodes, whereas mean minimum path length for a specific node is the average of all minimum path lengths between that node and all other nodes. The degree of a node was defined as the number of connections it makes with other network nodes. Based on these measures, network hubs were defined as nodes with largest degree. Eventually, these parameters can be averaged over different participants to obtain a single value per condition and sub-band.

In an experiment two movies, one joyful and one fearful were presented to the participants. As shown in Figure 14.10, different activity profiles can be observed among different brain regions. The posterior cingulate cortices (PCCs) show no significant variation in activity between the neutral and emotional rest conditions as can be seen in Figure 14.10a. This region shows similar increases during rest regardless of the preceding movie type. Inferior parietal lobule (IPL) bilaterally shows strong increases during resting state, although return to resting activity appeared slightly less complete or less rapid after emotional rather than neutral movies as in Figure 14.10b. By contrast, a much stronger emotional effect is observed within the level and time course of activity during rest for insula and ACCs. Both ACC and insula bilaterally show a typical DM pattern with a strong deactivation during both movies as shown in Figures 14.10c and d. However, activation during resting periods clearly differs depending on whether the preceding movie is emotional or not. After neutral movies, resting-state activity recovers immediately to a stable level, whereas a more gradual restoration after both joyful and fearful movies is exhibited. After emotional movies, ventral–medial prefrontal cortices (vMPFC) and precuneus exhibit a marked initial reduction during rest. Then, they recover gradually during the remaining period of rest as illustrated in Figure 14.10e and f. These two regions show no reduction following neutral movies. All regions present symmetric pattern in both hemispheres, but only one side is illustrated for left IPL and right insula while other activation clusters spanned across the midline [133].

As a result of this investigation, resting activity in ACC and insula as well as their coupling are strongly enhanced by preceding emotions, while coupling between ventral–medial PFC and amygdala was selectively reduced. These effects were more pronounced for higher frequency bands after watching fearful rather than joyful movies. Initial suppression of resting activity in ACC and insula after emotional stimuli was followed by a gradual restoration over time. Emotions did not affect IPL average activity but increased its connectivity with other regions.

The variabilities in EMG, electrocardiogram (ECG), skin conductivity, and respiration changes have been evaluated in [137] to recognize the emotion changes in music listeners. The authors have classified a wide range of physiological features from various analysis domains, including time/frequency, entropy, geometric analysis, sub-band spectra, and multiscale entropy.

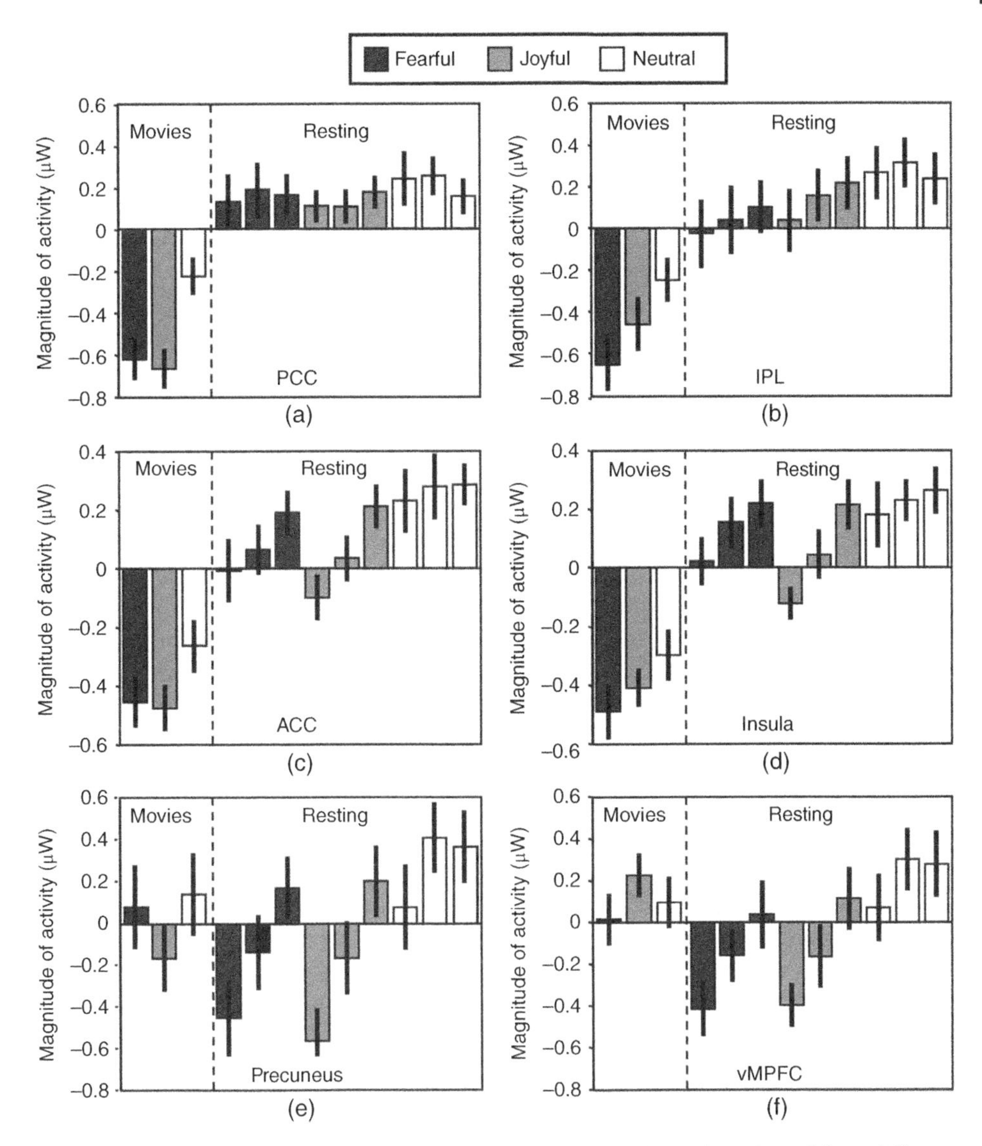

Figure 14.10 Negative and positive magnitudes of activity for the main regions differentially activated during resting periods, across three successive 30-second periods and three emotion conditions during film watching, with different emotion types, and during resting periods, with each successive time bins of 30 seconds shown for periods following fearful (three left-most black bars), joyful (three central grey bars), and neutral movies (three right-most white bars). These three successive time bins in each emotion condition correspond to 0–30, 30–60 and 60–90 seconds of rest following the end of movie. Activity in (a) posterior cingulate cortex and (b) IPL show global increases during rest but regardless of preceding emotion condition. Activity in (c) ACC and (d) right insula, show marked deactivations during movies, with a steep increase in activity during rest after neutral movies but more gradual restoration of rest activity after joyful and fearful movies. Unlike typical DM regions, activity in (e) left precuneus and (f) vMPFC is not reduced during movies, but during the first part of resting periods following emotional movies then progressively increased to a complete restoration level (Source: adapted from [133].)

14.5 Applications

Emotion assessment is a rapidly growing research field especially in the human–computer interface community where assessing the emotional state of a user can greatly improve interaction quality by bringing it closer to human-to-human communication.

Changes in emotions can influence brain activities during awake to sleep transition. It can also affect the states of mental fatigue and many other task or target related brain activities.

Another important impact is on memory formation. According to [89] two conclusions have been revealed. First, there is dissociation between arousal and valence. Results of emotion effect illustrate that processing of emotional content of pictures is sensitive to both arousal and valence and that their effects can be dissociated over the scalp: ERPs from parietal areas were sensitive to arousal, whereas ERPs from frontocentral areas were sensitive to both arousal and valence. Second, this is the first study in which the subsequent memory effect of emotional pictorial stimuli has been investigated. The *subsequent memory effect* results show an early occurrence of this effect for pleasant and unpleasant pictures than for neutral pictures and suggest that emotional stimuli have privileged access to processing resources, possibly resulting in better encoding. These results are compatible with previous evidence regarding the role of different brain areas in emotional information processing and with evidence about beneficial effect of emotion on memory formation. This study emphasizes on the fact that there is a link between two seemingly unrelated ERP phenomena: the emotion effect and the subsequent memory effect.

There are plenty of external effects and applications in the areas of health and economy mainly.

Basal ganglia role, in facial emotions recognition of PD has been investigated [138]. This work was to determine the extent to which visual processing of emotions and objects differ in PD and assess the impact of cognitive load on processing of these types of information.

People with verbal communication difficulty (such as autism) sometimes send nonverbal messages that do not match with what is happening inside them. For example, a child might appear calm and receptive to learning – but have a heart rate over 120 bpm and being about to collapse. This mismatch can lead to misunderstandings such as ‘he became aggressive for no reason’. In order to address this fundamental communication problem, advance technologies have to be developed to enable analysis of emotion-related physiological signals.

New tools and techniques in physiological arousal measurement in children with sensory challenges such as autism spectrum disorder (ASD) and attention deficit hyperactivity disorder (ADHD), can help them understand and control what makes them overexcited.

14.6 Pain Assessment Using EEG

Pain is an integrative phenomenon that results from dynamic interactions between sensory and contextual (i.e. cognitive, emotional, and motivational) processes. It is an unpleasant sensory and emotional experience that signals threat and promotes behaviour to protect the individual. Pain is associated with the activation of an extended network of brain areas including the somatosensory, insular, cingulate, and prefrontal cortices, the thalamus, subcortical areas, and the brainstem [139]. These areas do not constitute a dedicated pain system but belong to different functional brain zones that are transiently orchestrated in the processing of pain. In the brain the experience of pain is associated with neuronal oscillations and synchrony at different frequencies [139]. However, a well-defined framework for the significance of oscillations for pain remains lacking. Study of oscillations at different frequencies and their relation to pain, however, has been an agenda for research. Pain is associated not only with a spatially extended network of dynamically recruited brain areas but also with complex temporal–spectral patterns of brain activity. In particular, pain-related neuronal oscillations at all the conventional bands including gamma band. The pain is generally categorized into phasic, tonic, and chronic.

For phasic pain, EEG and magnetoencephalography (MEG) studies have shown that brief noxious stimuli induce a complex spectral–temporal– spatial pattern of neuronal responses with at least three different changes in the oscillatory rhythms and ERPs. First, pain stimuli evoke increases neural activity at frequencies below 10 Hz. These increases occur between 150 and 400 ms after stimulus application originating from sensorimotor area and correspond to pain-related evoked potentials. Second, phasic pain stimuli transiently suppress oscillations at alpha and beta frequencies. These suppressions are observed at latencies between about 300 and 1000 ms in the sensorimotor cortex and occipital areas. Third, phasic pain stimuli induce oscillations at gamma frequencies over the sensorimotor cortex at latencies of between 150 and 350 ms [139].

Tonic pains are longer lasting pains which can last few minutes. Based on some studies, the tonic pain is associated with suppression of oscillations at alpha frequencies. Some others believe that the suppression of alpha and beta oscillations during tonic pain is more closely related to stimulus intensity indicating that these suppressions reflect stimulus processing rather than perception. In addition, several studies have recorded gamma oscillations during tonic pain but in contrast to phasic pain, they were not recorded over sensorimotor areas but over the medial PFC [139].

Looking at the EEG activities in chronic pain, the most-noticed abnormality is an increase of theta oscillations. At the cortical level, the abnormal theta oscillations are expected to reduce lateral inhibition, which might result in abnormal gamma oscillations too. Therefore, it is concluded that chronic pain mostly changes the theta and beta-gamma oscillations in chronic pain, the latter particularly in frontal brain areas.

In a brief report, [140] the authors used fMRI to show that the intrinsic brain connectivity is decreased in association with reduced clinical pain in fibromyalgia. In another fMRI study during resting state [141] the functional connectivity within and between the brain default mode network (DMN) and salience network (SN) has been examined for 20 patients with chronic pain due to ankylosing spondylitis and 20 healthy controls. A whole-network analysis revealed that compared to healthy controls, patients exhibited less anticorrelated functional connectivity between the SN and DMN, and the degree of cross-network abnormality tracked pain and disease-related symptoms. The authors concluded that the cross-network functional connectivity is a metric of functional brain abnormality in chronic pain.

In a recent joint EEG–fMRI recording approach the functional brain activity and connectivity have been characterized and quantified using EEG and fMRI in patients suffering from the pain associated with sickle cell disease (SCD) [142]. SCD is a red blood cell disorder that causes many complications including lifelong pain. Treatment of pain remains challenging due to a poor understanding of the mechanisms and limitations in characterization and quantification of the pain. They performed independent component analysis (ICA) and seed-based connectivity on the fMRI data and power and microstate analysis on the joint EEG–fMRI data.

In their EEG analysis, they removed the gradient artefact using a principal component analysis (PCA)-based optimal basis set (OBS) algorithm [143]. The ballistocardiogram artefact was also removed using an ECG reference signals. The further denoised the signals using a combination of ICA, OBS, and an information-theoretic rejection criterion [144]. The remaining components are then divided into epochs around each heartbeat and an OBS is obtained across all epochs to fit and remove the artefacts.

Based on this comprehensive study, they concluded that patients have reduced activity of DMN and increased activity in pain processing regions during rest [142].

Decoding of pain perception is also useful in brain–computer interfacing (BCI) studies. The authors in [145], while they believe that the pain perception dynamics and its decoding using effective biomarkers, are still not fully understood, they use EEG to identify and validate the spatiotemporal signature of brain activity during innocuous, moderately more intense, and noxious stimulation of an amputee's phantom limb using transcutaneous electrical nerve stimulation (TENS). Based on these features, they classified three different stimulation conditions with a high accuracy. They have also reported that the noxious stimulation activates the pre-motor cortex with the highest activation shown in the central cortex (Cz electrode) between 450 and 750 ms post-stimulation, whereas the highest activation for the moderately intense stimulation was found in the parietal lobe (P2, P4, and P6 electrodes). Moreover, by localizing the cortical sources they observed an early strong activation of the ACC corresponding to the noxious stimulus condition. They also reported the activation of the PCC during the noxious sensation. As an overall conclusion, more work may be required to understand and quantify the brain dynamics in response to the general types of phasic and chronic pains.

14.7 Emotion Elicitation and Induction through Virtual Reality

In today's life, AfC has emerged as an important field of study in the development of systems that can automatically recognize, model and express emotions [146]. Proposed by Rosalind Picard in 1997, it is an interdisciplinary field based on psychology, computer science, and biomedical engineering. Stimulated by the fact that emotions are involved in many background processes (such as perception, decision-making, creativity, memory, and social interaction), several methods, some described above, have been proposed using signal processing and machine learning algorithms. Architecture is a field where AfC has been infrequently applied, despite its obvious potential; the physical environment has a great impact, on a daily basis, on human emotional states in general, and on well-being in particular. AfC could contribute to improve building design to better satisfy human emotional demands. Irrespective of its application, AfC involves both emotional classification and emotional elicitation. Regarding emotional classification, two approaches have commonly been proposed: discrete and dimensional models. On the one hand, the former posits the existence of a small set of basic emotions, on the basis that complex emotions result from a combination of these basic emotions. For example, Ekman [147] proposed six basic emotions: anger, disgust, fear, joy, sadness, and surprise. Dimensional models, conversely, consider a multidimensional space where each dimension represents a fundamental property common to all emotions. For example, as discussed before in this chapter, the 'Circumplex Model of Affects' (CMA) [148] uses a Cartesian system of axes, with two dimensions, proposed by Russell and Mehrabian [149] as valence and arousal.

Regarding emotional elicitation, the ability to reliably and ethically elicit affective states in the laboratory is a critical challenge in the process of the development of systems that can detect, interpret, and adapt to human affect. Many methods of eliciting emotions have been developed to evoke emotional responses. Based on the nature of the stimuli, two types of method are distinguished, the active and the passive. Active methods can involve behavioural manipulation, social psychological methods with social interaction and dyadic interaction. Conversely, passive methods usually present images, sounds, or films.

Even when elicitation is carried out with a non-immersive stimulus, it has been shown that these passive methods have significant limitations due to the importance of immersion for eliciting emotions through the simulation of real experiences [146]. At present, virtual reality (VR) represents a novel and powerful tool for behavioural research in psychological assessment. It provides simulated experiences that create the sensation of being in the real world [150, 151]. Thus, VR makes it possible to simulate and evaluate spatial environments under controlled laboratory conditions [151, 152], allowing the isolation and modification of variables in a cost and time effective manner, something which is unfeasible in real space. During the last two decades VR has usually been displayed using desktop PCs or semi-immersive systems such as CAVEs or Powerwalls [153]. Today, the use of head-mounted displays (HMD) provides fully-immersive systems that isolate the user from external world stimuli. These provide a high degree of immersion, evoking a greater sense of presence, understood as the perceptual illusion of non-mediation and a sense of 'being-there' [142]. Moreover, the ability of VR to induce emotions has been analyzed in studies which demonstrate that virtual environments do evoke emotions in the user [154]. Other works confirm that the immersive virtual environments (IVEs) can be used as emotional induction tools to create states of relaxation or anxiety [143], basic emotions [144, 145], and to study the influence of the users cultural and technological background on emotional responses in VR [155]. In addition, some works show that emotional content increases sense of presence in an IVE [156] and that, faced with the same content, self-reported intensity of emotion is significantly greater in immersive than in non-immersive environments [157]. Thus, IVEs, showing 360° panoramas or 3D scenarios through an HMD, are powerful tools for psychological research.

14.8 Summary

For psychologists and neuroscientists, it is important to quantify and mark the level of impression and self-regulation of a subject against various emotions as the sense of emotion changes for many abnormal brain states as well as generative and degenerative brain diseases. Different recording modalities such as facial impressions, body movement, eye movement, body temperature, heart rate, and breathing may be carried out. EEG however, allows localization, connectivity, and synchronization of the brain for different emotions. It is then expected to extract and process highly valuable emotion-related information from EEGs. The role of particular brain region such as amygdala in emotion, evaluation of brain asymmetricity for positive and negative emotions and connectivity of the brain regions due to emotion types and levels are probably the most important topics in emotion-based brain signal processing. It is important to note that the sources within limbic brain structure are considered as deep sources. Another problematic issue is the overlap between emotion and cognition. The current view of brain organization verifies a considerable degree of functional specialization and that many regions can be conceptualized as either 'affective' or 'cognitive'; amygdala is involved with emotion and the lateral PFC is the area of neurogenerators for cognition. This prevalent view is the source of many problems in emotion signal processing and machine learning. Complex cognitive–emotional behaviours have their basis in dynamic coalitions of networks of brain areas, none of which should be conceptualized as specifically affective or cognitive. Central to cognitive–emotional interactions are brain areas with a high degree of connectivity, called hubs; these are critical for regulating the flow and integration of information between regions [44].

Conclusively, direct detection and classification of emotion signal sources are generally complex and advanced methods in signal processing are needed. The high impact and enormous applications in this direction are well acknowledged.

References

1 James, W. (1884). What is emotion. *Mind* 9: 188–205.

2 Ten Donkelaar, H.J. (2011). *Clinical Neuroanatomy, Brain Circuitry and its Disorders*. Springer-Verlag.

3 Eich, E., Kihlstrom, J.F., Bower, G.H. et al. (2000). *Cognition and Emotion*, 259. New York: Oxford University Press. ISBN 0-19-511334-9.

4 Harmon-Jones, E., Vaughn-Scott, K., Mohr, S. et al. (2004). The effect of manipulated sympathy and anger on left and right frontal cortical activity. *Emotion* 4: 95–101.

5 Scmidt, L.A. (1999). Frontal brain electrical activity in shyness and sociability. *Psychological Science* 10: 316–320.

6 Garavan, H., Ross, T.J., and Stein, E.A. (1999). Right hemispheric dominance of inhibitory control: an event-related functional MRI study. *Proceedings of the National Academy of Sciences of the United States of America* 96: 8301–8306.

7 Drake, R.A. and Myers, L.R. (2006). Visual attention, emotion, and action tendency: feeling active or passive. *Cognition and Emotion* 20: 608–622.

8 Wacker, J., Chavanon, M.-L., Leue, A., and Stemmler, G. (2008). Is running away right? The behavioral activation–behavioral inhibition model of anterior asymmetry. *Emotion* 8: 232–240.

9 Craig, A.D. (June 2003). A new view of pain as a homeostatic emotion. *Trends in Neurosciences* 26 (6): 303–307.

10 Lazarus, R. (1991). *Emotion and Adaptation*. New York: Oxford University Press.

11 Weiss, H.M. and Cropanzano, R. (1996). Affective events theory: a theoretical discussion of the structure, causes and consequences of affective experiences at work. In: *Research in Organizational Behavior: An Annual Series of Analytical Essays and Critical Reviews* (eds. B. M. Staw and L.L. Cummings), 1–74. Greenwich, CT: JAI Press.

12 Weiss, H.M. and Beal, D.J. Reflections on Affective Events Theory. In: *The Effect of Affect in Organizational Settings* (Research on Emotion in Organizations, vol. 1), (eds. N.M. Ashkanasy, W.J. Zerbe and C.E.J. Härtel), 1–21. Bingley: Emerald Group Publishing Limited.

13 Savran, A., Ciftci, K., Chanel, G. et al. (2006). Emotion detection in the loop from brain signals and facial images. *eNTERFACE'06*, Croatia.

14 Haruki, Y., Homma, I., Umezawa, A., and Masaoka, Y. (eds.) (2001). *Respiration and Emotion*. Springer.

15 Homma, I. and Masaoka, Y. (2008). Breathing rhythms and emotions. *Experimental Physiology* 93 (9): 1011–1021.

16 Reiman, E.M., Fusselman, M.J., Fox, P.T., and Raichle, M.E. (1989). Neuroanatomical correlates of anticipatory anxiety. *Science* 243: 1071–1074.

17 Rolls, E.T. (2001). The rules of formation of the olfactory representations found in the orbitofrontal cortex olfactory areas in primates. *Chemical Senses* 26: 595–604.

18 Masaoka, Y. and Homma, I. (2000). The source generator of respiratory-related anxiety potential in the human brain. *Neuroscience Letters* 283: 21–24.

19 Masaoka, Y. and Homma, I. (2005). Amygdala and emotional breathing in human. *Post-Genomic Perspectives in Modeling and Control of Breathing, Jean Champagnat Monique Denavit-Saubié Gilles Fortin Arthur S. Foutz Muriel Thoby-Brisson* 551: 9–14.

20 Larsen, J.T., Norris, C.J., and Cacioppo, J.T. (2003). Effects of positive and negative affect on electromyographic activity over zygomaticus major and corrugator supercilii. *Psychophysiology* 40 (5): 776–785.

21 Sato, W., Fujimura, T., and Suzuki, N. (2008). Enhanced facial EMG activity in response to dynamic facial expressions. *International Journal of Psychophysiology* 70 (1): 70–74.

22 Dimberg, U. (1990). Facial electromyography and emotional reactions. *Psychophysiology* 27 (5): 481–494.

23 Oberman, L.M., Winkielman, P., and Ramachandran, V.S. (2009). Slow echo: facial EMG evidence for the delay of spontaneous, but not voluntary, emotional mimicry in children with autism spectrum disorders. *Developmental Science* 12 (4): 510–520.

24 Lang, P.J., Bradly, M.M., and Cuthbert, B.N. (2005). International affective picture system (IAPS): digitized photographs, instruction manual and affective ratings. *Technical Report A-6*, Ginesville, FL: University of Florida.

25 Oudeyer, P.-Y. (2003). The production and recognition of emotions in speech: features and algorithms. *International Journal of Human Computer Studies* 59: 157–183.

26 Neiberg, D., Elenius, K., Karlsson, I., and Laskowski, K. (2006). Emotion Recognition in Spontaneous Speech, Lund University, Centre for Languages & Literature. *Department of Linguistics & Phonetics Working Papers* 52: 101–104.

27 LeDoux, J.E. (2011). Music and the brain, literally. *Frontiers in Human Neuroscience* 5: 54.

28 Cornelius, R.R. (2000). Theoretical approaches to emotion. *Proceedings of the ISCA Workshop on Speech and Emotion, Belfast.*

29 Lewis, M., Haviland-Jones, J.M., and Feldman, L. (eds.) (2008). *Handbook of Emotion*. The Guildford Press.

30 Chanel, G., Kronegg, J., Grandjean, D., and Pun, T. (2005). Emotion assessment: arousal evaluation using EEG's and peripheral physiological signals. *Technical Report*, Switzerland: Computer Vision Group, University of Geneva.

31 Chapman, L.J., Chapman, J.P., Raulin, M.L., and Edell, W.S. (1978). Schizotypy and thought disorder as a high risk approach to schizophrenia. In: *Cognitive Defects in the Development of Mental Illness* (ed. G. Serban), 351–360. Brunner/Mazel.

32 Prince, J.D. and Berenbaum, H. (1993). Alexithymia and hedonic capacity. *Journal of Research in Personality* 27: 15–22.

33 Consoli, S.M., Lemogne, C., Roch, B. et al. (2010). Differences in emotion processing in patients with essential and secondary hypertension. *American Journal of Hypertension* 23 (5): 515–521.

34 Siegler, R. (2006). *How Children Develop, Exploring Child Develop Student Media Tool Kit & Scientific American Reader to Accompany How Children Develop*. New York: Worth Publishers.

35 DeArrnond, S.J., Fusco, M.M., and Dewey, M.M. (1989). *Structure of the Human Brain: A Photographic Atlas*, 3e. New York: Oxford University Press.

36 Davidson, R.J., Putnam, K.M., and Larson, C.L. (July 2000). Dysfunction in the neural circuitry of emotion regulation – a possible prelude to violence. *Science* 289 (5479): 591–594.

37 Rolls, E.T. (1999). *The Brain and Emotion*. New York: Oxford University Press.

38 Holland, P.C. and Gallagher, M. (1999). Amygdala circuitry in attentional and representational processes. *Trends in Cognitive Sciences* 3: 65–73.

39 Morris, J.S., Frith, C.D., Perrett, D.I. et al. (1996). A differential neural response in the human amygdala to fearful and happy facial expressions. *Nature* 383: 812–815.

40 Whalen, P.J., Rauch, S.L., Etcoff, N.L. et al. (1998). Masked presentations of emotional facial expressions modulate amygdala activity without explicit knowledge. *The Journal of Neuroscience* 18 (1): 411–418.

41 Buchel, C., Morris, J.S., Dolan, R.J., and Friston, K.J. (1998). Brain systems mediating aversive conditioning: an event-related fMRI study. *Neuron* 20: 947–957.

42 LaBar, K.S., Gatenby, J.C., Gore, J.C. et al. (1998). Human amygdala activation during conditional fear acquisition and extinction: a mixed-trial fMRI study. *Neuron* 20: 937–945.

43 Irwin, W., Davidson, R.J., and Lowe, M.J. (1996). Human amygdala activation detected with echo-planar functional magnetic resonance imaging. *Neuroreport* 7: 1765–1769.

44 Pessoa, L. (2008). On the relationship between emotion and cognition. *Nature Reviews Neuroscience* 9: 148–158.

45 LeDoux, J. (1999). *The Emotional Brain*. London: Phoenix (an Imprint of The Orion Publishing Group).

46 Adolphs, R., Tranel, D., Damasio, H., and Damasio, A. (1994). Impaired recognition of emotion in facial expression following bilateral damage to the human amygdala. *Nature* 372: 662–672.

47 Blair, R.J., Morris, J.S., Frith, C.D. et al. (1999). Dissociable neural responses to facial expressions of sadness and anger. *Brain* 122: 883–893.

48 Dougherty, D.D., Rauch, S.L., and Deckersbach, T. (2004). Ventromedial prefrontal cortex and amygdala dysfunction during an anger induction positron emission tomography study in patients with major depressive disorder with anger attacks. *Archives of General Psychiatry* 61: 795–804.

49 Kimbrell, T.A., George, M.S., and Parekh, P.I. (1999). Regional brain activity during transient self-induced anxiety and anger in healthy adults. *Biological Psychiatry* 46 (4): 454–465.
50 Walla, P., Brenner, G., and Koller, M. (2011). Objective measures of emotion related to brand attitude: a new way to quantify emotion-related aspects relevant to marketing. *PLoS One* 6 (11): e26782.
51 Globisch, J., Hamm, A.O., Esteves, F., and Ohman, A. (1999). Fear appears fast: temporal course of startle reflex potentiation in animal fearful subjects. *Psychophysiology* 36: 66–75.
52 Larson, C.L., Ruffalo, D., Nietert, J.Y., and Davidson, R.J. (2000). Temporal stability of the emotion-modulated startle response. *Psychophysiology* 37: 92–101.
53 Jackson, D.C., Malmstadt, J.R., Larson, C.L., and Davidson, R.J. (2000). Suppression and enhancement of emotional responses to unpleasant pictures. *Psychophysiology* 37: 515–522.
54 Davidson, R.I. (1998). Affective style and affective disorders: perspectives from affective neuroscience. *Cognition and Emotion* 12: 307–330.
55 Jackson, D.C., Burghy, C.A., Hanna, A.J. et al. (2000). Resting frontal and anterior temporal EEG asymmetry predicts ability to regulate negative emotion. *Psychophysiology* 37: S50.
56 Schaefer, S.M., Jackson, D.C., Davidson, R.J. et al. (2002). Modulation of amygdala activity by conscious maintenance of negative emotion. *Journal of Cognitive Neuroscience* 14: 913–921.
57 Heller, A.S., Johnstone, T., Shackman, A.J. et al. (2009). Reduced capacity to sustain positive emotion in major depression reflects diminished maintenance of fronto-striatal brain activation. *Proceedings of the National Academy of Sciences of the United States of America* 106 (52): 22445–22450.
58 Jackson, D.C., Mueller, C.J., Dolski, I. et al. (2003). Now you feel it, now you don't: frontal EEG asymmetry and individual differences in emotion regulation. *Psychological Science* (14): 612–617.
59 Urry, H.L., Nitschke, J.B., Dolski, I. et al. (2004). Making a life worth living: neural correlates of well-being. *Psychological Science* 15: 367–372.
60 Urry, H.L., van Reekum, C.M., Johnstone, T., and Davidson, R.J. (2009). Individual differences in some (but not all) medial prefrontal regions reflect cognitive demand while regulating unpleasant emotion. *NeuroImage* 47: 852–863.
61 van der Helm, E., Yao, J., Dutt, S. et al. (2011). REM sleep depotentiates amygdala activity to previous emotional experiences. *Current Biology* 21 (23): 2029–2032.
62 Petrantonakis, P.C. and Hadjileontiadis, L.J. (September 2011). A novel emotion elicitation index using frontal brain asymmetry for enhanced EEG-based emotion recognition. *IEEE Transactions on Information Technology in Biomedicine* 15 (5): 737–746.
63 Shallice, T. and Cooper, R.P. (2011). *The Organization of Mind*. Oxford University Press.
64 Saxe, R., Xiao, D.-K., Kovacs, G. et al. (2004). A region of right posterior superior temporal sulcus responds to observed intentional actions. *Neuropsychologia* 42: 1435–1446.
65 Sirigu, A., Daprati, E., Ciancia, S. et al. (2004). Altered awareness of voluntary action after damage to the parietal cortex. *Nature Neuroscience* 7: 80–84.
66 Nichols, T.E. and Holmes, A.P. (2002). Nonparametric permutation tests for functional neuroimaging: a primer with examples. *Human Brain Mapping* 15: 1–25.
67 Hornak, J., Bramham, J., Rolls, E.T. et al. (2003). Changes in emotion after circumscribed surgical lesions of the orbitofrontal and cingulate cortices. *Brain* 26: 1691–1712.

68 Takahashi, H., Yahata, N., Koeda, M. et al. (2004). Brain activation associated with evaluative processes of guilt and embarrassment: an fMRI study. *NeuroImage* 23 (3): 967–974.

69 Fu, C.H., Williams, S.C., Cleare, A.J. et al. (2004). Attenuation of the neural response to sad faces in major depression by antidepressant treatment: a prospective, event-related functional magnetic resonance imaging study. *Archives of General Psychiatry* 61 (9): 877–889.

70 Lewis, M. and Steiben, J. (2004). Emotion regulation in the brain: conceptual issues and directions for developmental research. *Child Development* 75: 371–376.

71 Beer, J.S., Heerey, E.H., Keltner, D. et al. (2003). The regulatory function of self-conscious emotion: insights from patients with orbitofrontal damage. *Journal of Personality and Social Psychology* 85: 594–604.

72 Moll, J., de Oliveira-Souza, R., Moll, F. et al. (2005). The moral affiliations of disgust: a functional MRI study. *Cognitive and Behavioral Neurology* 18: 68–78.

73 Haidt, J. (2007). The new synthesis in moral psychology. *Science* 316: 998–1002.

74 Buchanan, T.W., Tranel, D., and Adolphs, R. (2004). Anteromedial temporal lobe damage blocks startle modulation by fear and disgust. *Behavioral Neuroscience* 118: 429–437.

75 Fitzgerald, D.A., Posse, S., Moore, G.J. et al. (2004). Neural correlates of internally-generated disgust via autobiographical recall: a functional magnetic resonance imaging investigation. *Neuroscience Letters* 370: 91–96.

76 Haidt, J. (2003). The moral emotions. In: *Handbook of Affective Sciences* (eds. R.J. Davidson, K.R. Scherer and H.H. Goldsmith), 852–870. Oxford, England: Oxford University Press.

77 Adams, R.E., Boscarino, J.A., and Figley, C.R. (2006). Compassion fatigue and psychological distress among social workers: a validation study. *American Journal of Orthopsychiatry* 76 (1): 103–108.

78 Moll, J. (2003). Morals and the human brain: a working model. *Neuroreport* 14: 299–305.

79 Haidt, J. (2003). Elevation and the positive psychology of morality. In: *Flourishing: Positive Psychology and the Life Well-Lived* (eds. C.L.M. Keyes and J. Haidt), 275–289. Washington DC: American Psychological Association.

80 Rilling, J.K., Gutman, D.A., Zeh, T.R. et al. (2002). A neural basis for social cooperation. *Neuron* 35: 395–405.

81 Singer, B.H., Kiebel, S.J., Winston, J.S. et al. (2004). Brain responses to the acquired moral status of faces. *Neuron* 41 (4): 653–662.

82 Musha, T., Terasaki, Y., Haque, H.A., and Ivanitsky, G.A. (1997). Feature extraction from EEGs associated with emotions. *Artificial Life Robotics* 1: 15–19.

83 Demaree, H.A., Everhart, D.E., Youngstrom, E.A., and Harrison, D.W. (2008). Brain lateralization of emotional processing: historical roots and a future incorporating "dominance". *Behavioral and Cognitive Neuroscience Reviews* 4 (1): 3–20.

84 Ehrlichman, H. (1987). Hemispheric asymmetry and positive-negative affect. In: *Duality and Unity of the Brain* (ed. D. Ottoson), 194–206. Hempshire, Houndmills: Macmillan Press.

85 Papousek, L. (2006). Individual differences in functional asymmetries of the cortical hemisphers. *Cognition, Brain, and Behavior* 2: 269–298.

86 Borod, J.C. (1993). Cerebral mechanisms underlying facial, prosodic, and lexical emotional expression: a review of neuropsychological studies and methodological issues. *Neuropsychology* 7: 445–463.

87 Herrmann, M.J., Huter, T., Plichta, M.M. et al. (2008). Enhancement of activity of the primary visual cortex during processing of emotional stimuli as measured with event-related functional near-infrared spectroscopy and event-related potentials. *Human Brain Mapping* 29 (1): 28–35.

88 Gray, J.M., Young, A.E., and Barker, W.A. (1997). Impaired recognition of disgust in Huntington's disease carriers. *Brain* 120: 2029–2038.

89 Dolcos, F. and Cabeza, R. (2002). Event-related potentials of emotional memory: encoding pleasant, unpleasant, and neutral pictures. *Cognitive, Affective, & Behavioral Neuroscience* 2: 252–263.

90 Schupp, H.T., Flaisch, T., Stockburger, J., and Junghofer, M. (2006). Emotion and attention: event-related brain potential studies. In: *Progress in Brain Research*, vol. 156 (eds. Anders, Ende, Junghofer, et al.), 31–51. Elsevier, Chapter 2.

91 Schupp, H.T., Cuthbert, B.N., Bradley, M.M. et al. (2004). Brain processes in emotional perception: motivated attention. *Cognition and Emotion* 18: 593–611.

92 Schupp, H.T., Cuthbert, B.N., Bradley, M.M. et al. (2004). The selective processing of briefly presented affective pictures: an ERP analysis. *Psychophysiology* 41: 441–449.

93 Amrhein, C., Mühlberger, A., Pauli, P., and Wiedemann, G. (2004). modulation of event-related brain potentials during affective picture processing: a complement to startle reflex and skin conductance response? *International Journal of Psychophysiology* 54: 231–240.

94 Hagemann, D., Naumann, F., Thayer, J.F., and Bartussek, D. (2002). Does resting EEG asymmetry reflect a biological trait? An application of latent state-trait theory. *Journal of Personality and Social Psychology* 82: 619–641.

95 Carretié, L., Iglesias, J., García, T., and Ballesteros, M. (1997). N300, P300 and the emotional processing of visual stimuli. *Electroencephalography and Clinical Neurophysiology* 103 (2): 298–303.

96 Cuthbert, B.N., Schupp, H.T., Bradley, M.M. et al. (2000). Brain potentials in affective picture processing: covariation with autonomic arousal and affective report. *Biological Psychology* 52: 95–111.

97 Johnston, V.S., Miller, D.R., and Burleson, M.H. (1986). Multiple P3s to emotional stimuli and their theoretical significance. *Psychophysiology* 23: 684–694.

98 Delplanque, S., Silvert, L., Hot, P. et al. (2006). Arousal and valence effects on event-related P3a and P3b during emotional categorization. *International Journal of Psychophysiology* 60: 315–322.

99 Ding, R., Li, P., Wang, W., and Luo, W. (2017). Emotion processing by ERP combined with development and plasticity. *Hindawi Neural Plasticity* 2017: 5282670, 15 pages. https://doi.org/10.1155/2017/5282670.

100 Luo, W., Feng, W., He, W. et al. (2010). Three stages of facial expression processing: ERP study with rapid serial visual presentation. *NeuroImage* 49 (2): 1857–1867.

101 Bechara, A., Damasio, H., Damasio, A.R., and Lee, G.P. (1999). Different contributions of the human amygdala and ventromedial prefrontal cortex to decision-making. *Journal of Neuroscience* 19: 5473–5481.

102 Wheeler, R.E., Davidson, R.J., and Tomarken, A.J. (1993). Frontal brain asymmetry and emotional reactivity: a biological substrate of affective style. *Psychophysiology* 30: 82–89.

103 Davidson, R.J. and Irwin, W. (1999). The functional neuroanatomy of emotion and affective style. *Trends in Cognitive Sciences* 3: 11–21.

104 Dolcos, F., Rice, H.J., and Cabeza, R. (2002). Hemispheric asymmetries and aging: right hemisphere decline or asymmetry reduction. *Neuroscience and Biobehavioral Reviews* 26 (7): 819–825.

105 Zheng, W.-L., Zhu, J.-Y., and Lu, B.-L. (2016). Identifying stable patterns over time for emotion recognition from EEG. arXiv:1601.02197v1 [cs.HC].

106 LeDoux, J.E. (2000). Cognition and emotion: listen to the brain. In: *Cognitive neuroscience of emotion* (ed. R. Lane), 129–155. New York: Oxford University Press.

107 Rolls, E.T., Kringelbach, M.L., and de Araujo, I.E. (2003). Different representations of pleasant and unpleasant odours in the human brain. *The European Journal of Neuroscience* 18: 695–703.

108 Royet, J.P., Zald, D., Versace, R. et al. (2000). Emotional responses to pleasant and unpleasant olfactory, visual, and auditory stimuli: a positron emission tomography study. *The Journal of Neuroscience* 20: 7752–7759.

109 Masaoka, Y. and Homma, I. (2010). Respiratory response toward olfactory stimuli might be an index for odor-induced emotion and recognition. *Advances in Experimental Medicine and Biology* 669: 347–352.

110 Diamond, D. (2003). The Love Machine; Building Computers That Care. *Wired*. http://www.wired.com/wired/archive/11.12/love.html (accessed 13 May 2008).

111 Anders, S., Ende, G., Junghofer, M. et al. (2006). *Understanding Emotions*. Elsevier.

112 Murugappan, M., Ramachandran, N., and Sazali, Y. (2010). Classification of human emotion from EEG using discrete wavelet transform. *Journal of Biomedical Science and Engineering* 3: 390–396.

113 Aftanas, L.I., Reva, N.V., Varlamov, A.A. et al. (2004). Analysis of evoked EEG synchronization and desynchronization in conditions of emotional activation in humans: temporal and topographic characteristics. *Neuroscience and Behavioral Physiology*: 859–867.

114 Wang, X.-W., Nie, D., and Lu, B.-L. (2014). Emotional state classification from EEG data using machine learning approach. *Neurocomputing* 129: 94–106.

115 Xu, X., Wei, F., Zhu, Z. et al. 2020. EEG feature selection using orthogonal regression: application to emotion recognition. *Proceedings of IEEE International Conference on Acoustics, Speech, and Signal Processing*. Barcelona, Spain: ICASSP.

116 Chen, F., Gu, F., Xu, J. et al. (1998). A new measurement of complexity for studying EEG mutual information. *5th International Conference on Neural Information Processing: ICONIP '98*, 435–437. Kitakyushu, Japan.

117 Adolphs, R. (2002). Neural systems for recognizing emotion. *Current Opinion in Neurobiology* 12: 169–177.

118 Baliki, M.N., Chialvo, D.R., Geha, P.Y. et al. (2006). Chronic pain and the emotional brain: specific brain activity associated with spontaneous fluctuations of intensity of chronic back pain. *The Journal of Neuroscience* 26: 12165–12173.

119 Damasio, A.R. (1999). *The Feeling of What Happens; Body and Emotion in the Making of Consciousness*. New York: Harcourt Brace.

120 Phelps, E.A. and LeDoux, J.E. (2005). Contributions of the amygdala to emotion processing: from animal models to human behaviour. *Neuron* 48: 175–187.

121 Sigurdsson, T., Doyere, V., Cain, C.K., and LeDoux, J.E. (2007). Long-term potentiation in the amygdala: a cellular mechanism of fear learning and memory. *Neuropharmacology* 52: 215–227.

122 Salvador, R., Martínez, A., Pomarol-Clotet, E. et al. (2008). A simple view of the brain through a frequency-specific functional connectivity measure. *NeuroImage* 39: 279–289.

123 Sakata, O., Shiina, T., and Saito, Y. (2002). Multidimensional directed information and its application. *Electronics and Communications in Japan* 4: 3–85.

124 Irving, I., Gottesman, P.D., Hon, F.R.C. et al. (2003). The Endophenotype concept in psychiatry: etymology and strategic intentions. *The American Journal of Psychiatry* 160: 636–645.

125 Morris, J., Friston, K.J., Buechel, C. et al. (1998). A neuromodulatory role for the human amygdala in processing emotional facial expressions. *Brain* 121: 47–57.

126 Stephan, B.C., Breen, N., and Caine, D. (2006). The recognition of emotional expression in prosopagnosia: decoding whole and part faces. *Journal of the International Neuropsychological Society* 12 (6): 884–895.

127 Champagne, J., Mendrek, A., Germain, M. et al. (June 2014). Event-related brain potentials to emotional images and gonadal steroid hormone levels in patients with schizophrenia and paired controls. *Frontiers in Psychology* 5: 543, 12 pages, doi: https://doi.org/10.3389/fpsyg.2014.00543.

128 Soto, V., Tyson-Carr1, J., Kokmotou, K. et al. (October 2018). Brain responses to emotional faces in natural settings: a wireless mobile EEG recording study. *Frontiers in Psychology* 9: 2003.

129 Jenke, R., Peer, A., and Buss, M. (2014). Feature extraction and selection for emotion recognition from EEG. *IEEE Transactions on Affective Computing* 5 (3): 327–339.

130 Hadjidimitriou, S.K. and Hadjileontiadis, L.J. (Dec. 2012). Toward an EEG-based recognition of music liking using time–frequency analysis. *IEEE Transactions on Biomedical Engineering* 59 (12): 3498–3510.

131 Canli, T., Desmond, J.E., Zhao, Z., and Gabrieli, J.D. (2002). Sex differences in the neural basis of emotional memories. *Proceedings of the National Academy of Sciences of the United States of America* 99: 10789–10794.

132 Junghofer, M., Peyk, P., Flaisch, T., and Schupp, H.T. (2006). Neuroimaging methods in affective neuroscience: selected methodological issues. In: *Understanding Emotions* (ed. S. Anders) Progress in Brain Research, 123–143. Elsevier.

133 Eryilmaz, H., Van De Ville, D., Schwartz, S., and Vuilleumier, P. (2011). Impact of transient emotions on functional connectivity during subsequent resting state: a wavelet correlation approach. *NeuroImage* 54: 2481–2491.

134 Battle, G. (1987). A block spin construction of ondelettes; part I. Lemarié functions. *Communications in Mathematical Physics* 110: 601–615.

135 Mallat, S. (1989). A theory for multiresolution signal decomposition: the wavelet decomposition. *IEEE Transactions on Pattern Analysis and Machine Intelligence* 11: 674–693.

136 Waites, A.B., Stanislavsky, A., Abbott, D.F., and Jackson, G.D. (2005). Effect of prior cognitive state on resting state networks measured with functional connectivity. *Human Brain Mapping* 24: 59–68.

137 Kim, J. and Andre, E. (Dec. 2008). Emotion recognition based on physiological changes in music listening. *IEEE Transactions on Pattern Analysis and Machine Intelligence* 30 (12): 2067–2083.

138 Cohen, H., Gagné, M.-H., Hess, U., and Pourcher, E. (2010). Emotion and object processing in Parkinson's disease. *Journal of Brain and Cognition* 72: 457–463.

139 Ploner, M., Sorg, C., and Gross, J. (February 2017). Brain rhythms of pain. *Trends in Cognitive Sciences* 21 (2): 100–110.

140 Napadow, V., Kim, J., Clauw, D.J., and Harris, R.E. (2012 Jul). Decreased intrinsic brain connectivity is associated with reduced clinical pain in fibromyalgia. *Arthritis and Rheumatism* 64 (7): 2398–2403.

141 Hemington, K.S., Wu1, Q., Kucyi, A. et al. (2016). Abnormal cross-network functional connectivity in chronic pain and its association with clinical symptoms. *Brain Structure & Function* 221: 4203–4219.

142 Slater, M. and Wilbur, S. (1997). A Framework for Immersive Virtual Environments (FIVE): speculations on the role of presence in virtual environments. *Presence Teleoperators and Virtual Environments* 6: 603–616.

143 Riva, G. et al. (2007). Affective interactions using virtual reality: the link between presence and emotions. *Cyber Psychology Behavior* 10: 45–56.

144 Baños, R.M. et al. (2006). Changing induced moods via virtual reality. In: *International Conference on Persuasive Technology*, 7–15. Berlin, H: Springer https://doi.org/10.1007/11755494_3.

145 Baños, R.M. et al. (2012). Positive mood induction procedures for virtual environments designed for elderly people. *Interacting with Computers* 24: 131–138.

146 Marín-Morales, J., Higuera-Trujillo, J.L., Greco, A. et al. (2018). Affective computing in virtual reality: emotion recognition from brain and heartbeat dynamics using wearable sensors. *Scientific Reports* 8: 13657, 15 pages.

147 Ekman, P. (1999). Basic emotions. In: *Handbook of Cognition and Emotion*, 45–60. https://doi.org/10.1017/S0140525X0800349X.

148 Posner, J., Russell, J.A., and Peterson, B.S. (2005). The circumplex model of affect: an integrative approach to affective neuroscience, cognitive development, and psychopathology. *Development and Psychopathology* 17: 715–734.

149 Russell, J.A. and Mehrabian, A. (1977). Evidence for a three-factor theory of emotions. *Journal of Research in Personality* 11: 273–294.

150 Giglioli, I.A.C., Pravettoni, G., Martín, D.L.S. et al. (2017). A novel integrating virtual reality approach for the assessment of the attachment behavioral system. *Frontiers in Psychology* 8: 1–7.

151 Marín-Morales, J., Torrecilla, C., Guixeres, J., and Llinares, C. (2017). Methodological bases for a new platform for the measurement of human behaviour in virtual environments. *DYNA* 92: 34–38.

152 Vince, J. (2004). *Introduction to Virtual Reality*. Media, Springer Science & Business.

153 Vecchiato, Jelic, Tieri, G., A., G. et al. (2015). Neurophysiological correlates of embodiment and motivational factors during the perception of virtual architectural environments. *Cognitive Processing* 16: 425–429.

154 Alcañiz, M., Baños, R., Botella, C., and Te Rey, B. (2003). EMMA Project: emotions as a determinant of presence. *PsychNology Journal* 1: 141–150.

155 Gorini, A. et al. (2009). Emotional response to virtual reality exposure across Diferent cultures: Te role of the attribution process. *Cyber Psychology Behav.* 12: 699–705.

156 Gorini, A., Capideville, C.S., De Leo, G. et al. (2011). Role of immersion and narrative in mediated presence: the virtual hospital experience. *Cyberpsychology, Behavior and Social Networking* 14: 99–105.

157 Chirico, A. et al. (2017). Effectiveness of immersive videos in inducing awe: an experimental study. *Scientific Reports* 7: 1–11.

15

EEG Analysis of Neurodegenerative Diseases

15.1 Introduction

Neurodegenerative diseases are a heterogeneous group of disorders that are characterized by the progressive degeneration of the structure and function of the central nervous system or peripheral nervous system (CNS). It is a collective term used to describe various symptoms of cognitive decline, such as forgetfulness and a symptom of several underlying diseases and brain disorders. This disease umbrella includes major clinicopathological entities such as Alzheimer's disease (AD), Parkinson's disease (PD), and frontotemporal dementia (FTD), as well as less common disease subtypes such as progressive supranuclear palsy (PSP), corticobasal degeneration (CBD), and multiple systems atrophy [1].

Neurodegenerative diseases are challenging to diagnose because there is marked heterogeneity in clinical and pathological biomarkers, some of which are only identifiable by postmortem examination and not under any in vivo assessment [2]. These degenerative diseases are associated with neuronal and synapsis loss, neuroinflammation, gliosis and vascular abnormalities in specific regions of the brain, and with the deposition of certain proteins displaying toxic effects; amyloid β and τ in AD; α-synuclein in PD, Lewy body dementia (LBD) and multiple system atrophy (MSA); τ in PSP and CBD; and transactive response DNA-binding protein 43 kDa (TDP-43) in FTD and amyotrophic lateral sclerosis (ALS). Unfortunately, on neuropathological examination most patients show two, three, or even four of such proteopathy features which makes the diagnosis of these diseases very challenging.

By applying machine learning to semiquantitative pathological features in a large dataset of autopsy cases, Cornblath et al. have reported that six transdiagnostic clusters of neurodegenerative disease summarize relevant aspects of their phenotypes, and that the phenotypes can be reasonably predicted from cognitive scores, biomarkers in cerebrospinal fluid (CSF) or genotype [1]. In their classification they used nine pathological features including amyloid β and τ levels, α-synuclein, TDP-43, ubiquitin, neuritic plaques, angiopathy, gliosis, and neuron loss as the features.

Dementia causes dysfunction relative to a person's previous level of social and occupational functioning. This disease is a very common medical condition that can result from diverse causes. One new case of dementia is registered every four seconds in the world. Dementias are classified based on clinical, genetic, and neuropathological features. The most common cause of dementia is AD. Other major classes of disorders that cause

EEG Signal Processing and Machine Learning, Second Edition. Saeid Sanei and Jonathon A. Chambers.

dementia are vascular dementias, psychiatric diseases, and other neurodegenerative diseases, including FTD and LBD.

Chronic conditions that alter or damage nerve cells and synapses involved in cognition are the biological basis of dementia. Currently, there is no effective treatment to prevent or reverse the underlying disease process for the most common neurodegenerative diseases of ageing that cause dementia. The diagnosis, however, is routine and is often achieved by traditional memory tests (such as Addenbrooke's Cognitive Examination [ACE] [3]). The ACE involves the assessment of attention/orientation, memory, language, verbal fluency, and visuospatial skills.

Electroencephalogram (EEG)-based tests are becoming clinically applicable by evaluating the brain connectivity, mostly explained in Chapter 8 of this book and some in the following sections.

Based on some new hypothesis, the methods to rule out other possibly treatable diseases (such as depression, vitamin B12 deficiency, hydrocephalus, and hypothyroidism) are also used for detecting neurodegenerative diseases such as AD. These methods examine the blood protein and other deficiencies and often benefit from machine learning techniques.

Dementia is commonly seen in older people and characterized by a decline in cognitive performance. There are approximately 47 million people worldwide with dementia and there are 10 million new cases every year. It is a major cause of disability and dependence and impacts on the physical, psychologic, and social well-being of families and carers.

Gait and balance impairments are common in people with dementia and significantly increase the risk of falls, injury, institutionalization, hospitalization, morbidity, and death after a fall particularly in elderlies.

Dementia may be in any of the following four stages:

- Stage 1: *Mild cognitive impairment* (MCI): characterized by general forgetfulness. This affects many people as they age but it only progresses to dementia for some. MCI shares some features with AD. The cognitive deficits of MCI subjects are limited to memory but their activities of daily living are preserved.
- Stage 2: *Mild dementia*: people with mild dementia experience cognitive impairments that occasionally impact their daily life. Symptoms include memory loss, confusion, personality changes, getting lost, and difficulty in planning and performing their daily tasks.
- Stage 3: *Moderate dementia*: in this stage daily life becomes more challenging, and the individual may need more help. The symptoms become more severe compared to the mild case. Individuals may need help getting dressed and combing their hair. They may also show significant changes in personality; for instance, becoming suspicious or agitated for no reason. They may face sleep disturbances too.
- Stage 4: *Severe dementia*: at this stage, the symptoms worsen considerably. There may be a loss of ability to communicate, and the individual might need full-time care. Simple tasks, such as sitting and holding one's head up become impossible and they hardly control their bladders.

Dementias can be caused by brain cell death or a neurodegenerative disease. Hence, as well as progressive brain cell death, like that seen in AD, dementia can be caused by a head injury, a stroke, or a brain tumour, among other causes. The causes include vascular dementia (also called multi-infarct dementia), resulting from brain cell death caused by conditions such as cerebrovascular disease, for example, stroke. This prevents normal blood flow, depriving brain cells of oxygen. It may also result from brain injury. Some types of traumatic brain injury – particularly if repetitive, such as those received by sports players – have been linked to certain dementias appearing later in life. Although with less probability, a single brain injury raises the likelihood of having a degenerative dementia such as AD. Dementia can also be due to prion diseases, for instance, Creutzfeldt–Jakob disease (CJD), human immunodeficiency virus (HIV) infection in which the virus can damage the brain cells, and reversible factors, i.e. those dementias that can be treated by reversing the effects of underlying causes, including medication interactions, depression, vitamin deficiencies, and thyroid abnormalities.

15.2 Alzheimer's Disease

In 1901, Karl Deter, a railway worker, admitted his 51-year old wife, Auguste, to a psychiatric institution with symptoms of memory loss, confusion, violent outbursts, and inability to use language. According to Karl, Auguste's symptoms began to emerge in the late 1890s and were quite uncommon for the person he had come to know. She had become increasingly fearful and anxious and would sometimes scream loudly for several hours. Her condition became so overwhelming and debilitating that her family could no longer manage her care. For the next five years until she passed away, Auguste remained at the institution and was observed by a German psychiatrist, Alois Alzheimer. After her death, Alzheimer performed histological studies on Auguste Deter's brain tissue. In the process, he discovered two abnormalities: large abnormal clumps had formed between neurons, and rope-like tangles had formed inside neurons. Calling these abnormalities 'a peculiar disease of the cortex', Alzheimer presented his findings at a 1906 psychiatric conference in Germany, marking the first documented case of what is now known as AD. Over the next five years, 11 similar cases were reported in medical journals. Today, more than one hundred years later, there are approximately 35.6 million people suffering from AD worldwide. To the shock of many, the 2010 World Alzheimer Report projected that this number will almost double to 65.7 million cases by 2030 and will more than triple to 115.4 million cases by 2050.

AD is categorized as an incurable, degenerative neurological disorder. In popular language, AD is commonly used interchangeably with dementia, but they are not the same. Dementia is the general term used to describe a decline in cognitive abilities (memory, thinking, language, judgement, and behaviour) and can be associated with a number of different neurological or psychiatric diseases. As an important distinction, AD is the most common cause of dementia, accounting for 60–80% of all dementia diagnoses. It is primarily characterized by a loss of short-term memory as well as impairments in other cognitive functions such as language, problem solving, attention, and orientation. As the disease progresses, severe mood swings and unprovoked aggression are common. Eventually, the patient loses almost all language ability and motor function.

For the most definitive diagnosis, scientists use postmortem analysis of brain tissue. The tissue abnormalities first defined by Alzheimer as clumps and tangles among neurons in the cortex are more specifically defined as amyloid plaques and neurofibrillary tangles. Plaques are deposits of a protein called beta-amyloid, which scientists believe is one of the earliest signs of the disease process. Over time, the build-up of this protein in the brains of Alzheimer's patients eventually disrupts effective communication between neurons. Scientists hypothesize that these amyloid accumulations can themselves trigger the formation of neurofibrillary tangles, which are made of a different protein called p-tau and have a distinct string-like appearance. Like plaques, these tangles also interfere with neurons, but they do so by accumulating inside of them in large amounts and eventually killing them.

These tissue pathologies and the behavioural symptoms are not always tightly correlated. However, plaques and tangles may appear in the brain well before symptoms of dementia appear which can be detected by magnetic resonance imaging (MRI).

Although AD is commonly diagnosed in people over the age of 65, early-onset cases – appearing at ages 30–60 and accounting for only 1% of total cases – represent a rare subset in which great advances have been made. This form of the disease can run in families. Studies of these families have identified several genes that, when altered, can cause AD. These genes implicate beta-amyloid as a central player in the disease and formation of plaques in the brain as an early manifestation of a disease process in the brain. Recent studies imaging the brains of older adults with and without dementia suggest that these plaques may first appear as early as 10 years before the onset of clinical symptoms. The growth of this category may be concurrent with a growing recognition of the disease among diagnosticians, but it nevertheless causes great concern and contributes to the expanding effect of the disease.

The cost of care for those with AD can leave supporting families emotionally and financially debilitated. Currently, there are no cures for this disease, but there are two types of drugs available that slow disease progression. Unfortunately, both types provide only marginal improvements in cognitive symptoms. Making matters worse, these drugs tend to work for only about half the patients who take them, and among those patients, the effects are moderate and temporary. Based on the discovery that beta-amyloid plaques appear early on in the disease, even before symptoms appear, many drug companies are trying to inhibit the production of this protein or target its removal as a possible strategy to treat the disease.

As life expectancy is on the rise, increasingly more people are likely to develop AD. Currently, there are more than one billion people over the age of 60. The growing number of potential sufferers and the inevitable increase in care demands will have an enormous economic effect on health care and social services.

15.2.1 Application of Brain Connectivity Estimation to AD and MCI

Brain connectivity analyses show considerable promise for understanding how our neural pathways gradually break down in ageing and AD. Nevertheless, we know very little about how the brain's networks change in AD, and which metrics are best to evaluate these changes. Also, connectivity estimation can be applied to EEG, magnetoencephalogram (MEG), or functional magnetic resonance images (fMRI) [4].

Following the methods described in Chapter 8, rooted in the Granger causality (GC) metric [5], magnitude squared coherence – $c_{x,y}(f)$ – is a widespread connectivity metric applied to EEG and MEG, which provides a measure of the linear dependencies between two signals x and y as a function of frequency [6, 7]. This complex-valued metric, as defined in Chapter 8 of this book, is:

$$c_{x,y}(f) = \frac{S_{xy}^2(f)}{\sqrt{S_{xx}^2(f) \cdot S_{yy}^2(f)}} = Re\left(c_{x,y}(f)\right) + iIm\left(c_{x,y}(f)\right) \tag{15.1}$$

where S_{xy}^2 is cross-power spectrum and S_{xx}^2 and S_{yy}^2 are respectively the power spectra of x and y.

It has been hypothesized that AD decreases $c_{x,y}(f)$ in alpha (α) and beta (β) but there are no conclusive findings about the delta (δ) and theta (θ) bands [8, 9]. Furthermore, AD produces opposite changes in features computed from different spectral bands [10, 11]. Thus, it is advisable to analyse different spectral bands separately [12]. Yet, $c_{x,y}(f)$ has several limitations. Firstly, spurious correlations could appear due to the volume conduction [11, 13, 14]. This effect is due to the fact that nearby channels are likely to record activity from identical currents. This leads to abnormally high correlations in the results that reflect volume conduction rather than actual connectivity. Other alternatives, such as the imaginary part of the coherency [14, 15] or the phase lag index (PLI) [11], have been proposed.

The imaginary part of coherency (iCOH), which is $Im(c_{x,y}(f))$, is used to circumvent the volume-conduction effects in functional connectivity estimation. An improved functional connectivity estimation using the iCOH measure in comparison to coherence analysis has been demonstrated and showed transient interactions between left–right motor cortical signals as a function of time and frequency in a real dataset. However, due to its exclusive dependency on the iCOH, the connectivity estimate based on iCOH becomes negligible in some situations even in the presence of a significant true interaction, e.g. the phase difference between two signals is near zero or kπ. Later some improvements on this limitation were achieved by proposing the PLI [11, 16] and the weighted PLI (wPLI) [17], demonstrated by simulations based on the Kuramoto model as well as with real data. In this latter work the effect of volume conductivity on each coherence estimation technique has been analyzed in detail. It has been shown that iCOH cannot be spuriously increased by volume conduction of independent sources. The PLI and iCOH between the signals from two sensors are non-equiprobable. Compared to iCOH, PLI has the advantage of being less influenced by phase delays. However, sensitivity to volume conduction and noise, and capacity to detect the changes in phase synchronization are hindered by the discontinuity of the PLI, as small perturbations turn phase lags into leads and vice versa. To solve this problem, we introduce a related index, namely the wPLI. Differently from PLI, in wPLI the contribution of the observed phase leads and lags is weighted by the magnitude of the imaginary component of the cross-spectrum [17].

In [15] the authors proposed the envelope of the imaginary coherence (EIC) as another measure for functional connectivity and compared it with lagged coherence, iCOH, PLI, and wPLI, using bivariate autoregressive and stochastic neural mass models.

They provided the definition for modified iCOH as:

$$\text{iCOHm}_{x,y} = \left[\frac{E\text{Im}(X(f)Y(f)^*]}{E[H(\text{Im}(X(f)Y(f)^*)]}\right. \tag{15.2}$$

where $H(\bullet)$ refers to Hilbert transform operation. The phase lock value (PLV) is defined as:

$$\mathrm{PLV}_{x,y}(f) = \left|E\left[e^{\,j[phase\ of\ X[f] - phase\ of\ Y[f]]}\right]\right| \tag{15.3}$$

PLI is a measure of the asymmetry of the distribution of phase differences between two signals. It reflects the consistency with which one signal is phase leading or lagging with respect to another signal which is defined as:

$$\mathrm{PLI}_{x,y}(f) = |E[sgn\ [phase\ of\ X[f] - phase\ of\ Y[f]]]| \tag{15.4}$$

where $sgn(\bullet)$ refers to sign. wPLI is defined as:

$$\mathrm{wPLI}_{x,y}(f) = \left|\frac{E[\mathrm{Im}(X(f)Y(f)^{*}]|}{E[|\ \mathrm{Im}(X(f)Y(f)^{*}\ |]}\right| \tag{15.5}$$

and, lagged coherence, lCOH is defined as:

$$\mathrm{lCOH}_{x,y}(f) = \frac{\mathrm{Im}(c_{X,Y}(f))^{2}}{1 - \mathrm{Re}\,(c_{X,Y}(f))^{2}} \tag{15.6}$$

EIC is simply

$$\mathrm{EIC}_{x,y} = \left|H\left(\mathrm{Im}\left(c_{x,y}(f)\right)\right)\right| \tag{15.7}$$

Also, a modified version of EIC is defined as:

$$\mathrm{EICm}_{x,y} = \left|H\left(\frac{E[\mathrm{Im}(X(f)Y(f)^{*}])}{E[H(\mathrm{Im}(X(f)Y(f)^{*})]|}\right.\right| \tag{15.8}$$

Finally, they conclude and claim that EIC and iCOH demonstrate superior results compared with other functional connectivity estimates.

These measures are usually applied in a channel-wise manner, producing a very high number of variables [13, 18]. All these studies and experiments clearly show that the brain connectivity significantly changes for many of these neurodegenerative diseases particularly AD.

A possible option is to try to estimate the equivalent current dipoles [1, 19, 20]. Once the currents are estimated, the same connectivity metrics used in the channel-wise analyses can be applied to them. Moreover, the volume-conduction effect does not affect these dipole-based connectivity assessments. However, the solution of the inverse problem is not unique and the choice of the reconstruction algorithm and other a priori assumptions influence the dipole estimation leading to potentially misleading results [12, 14].

A research study was carried out by Escudero et al. [21] for differentiation between NC, AD, and MCI using MEG signals. In their research they pre-processed the signals before assessing the coherence between the selected brain zones.

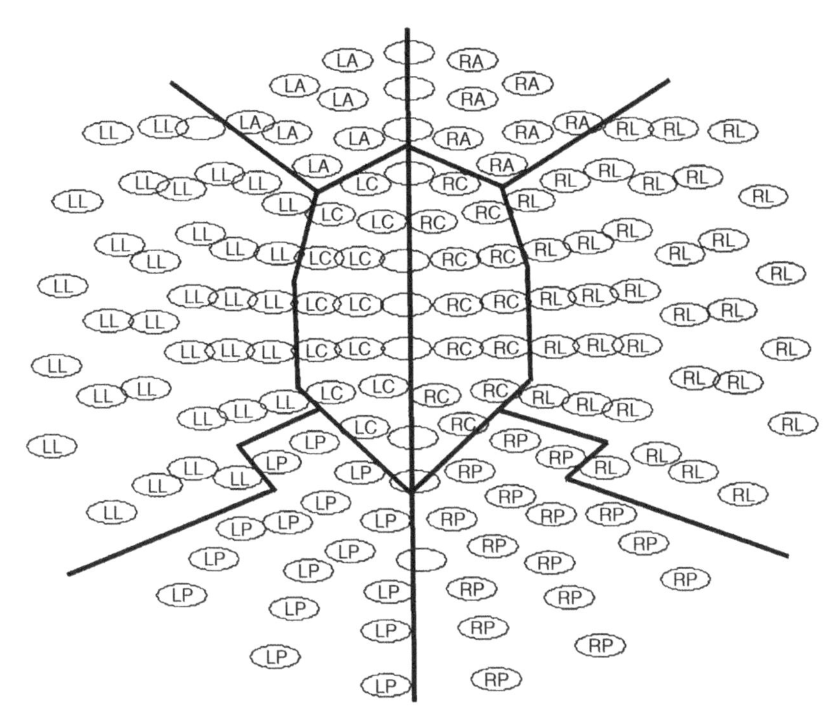

Figure 15.1 Distribution of the MEG sensors into left central (LC), anterior (LA), lateral (LL) and posterior (LP); and right central (RC), anterior (RA), lateral (RL) and posterior (RP) regions. Midline sensors appear empty. The lines depict the region boundaries.

In this work scalp electrode space has been divided into nine regions as illustrated in Figure 15.1. Instead of measuring the connectivity directly, MEG signals have been processed using the empirical mode decomposition (EMD) and blind source separation (BSS).

In the proposed algorithm, the EMD technique was used to decompose each MEG channel into a set of intrinsic mode functions (IMFs). Then, a constraint BSS block extracts the representative delta, theta, alpha, and beta rhythms of each region. In addition to the corresponding spectral coherence, *magnitude squared coherence* (MSC) is used here to measure the brain synchrony. It quantifies linear correlations as a function of frequency. MSC is bounded between 0 and 1. This measure can detect the linear synchronization between two signals, but it does not discriminate the directionality of the coupling [8.50]. A simplified block diagram of the algorithm is presented in Figure 15.2. In Figure 15.3a and b a channel of MEG signal and its corresponding IMFs computed using the EMD algorithm can be viewed.

To select the right IMFs in terms of centre frequency from each region, the IMFs have been clustered based on their amplitudes, frequency, and proximity to any one of the four centre frequencies. As depicted in Figure 15.4 the *k*-means algorithm has been used to select a reference signal frequency for each band. This is necessary since the frequency band centres change from subject to subject.

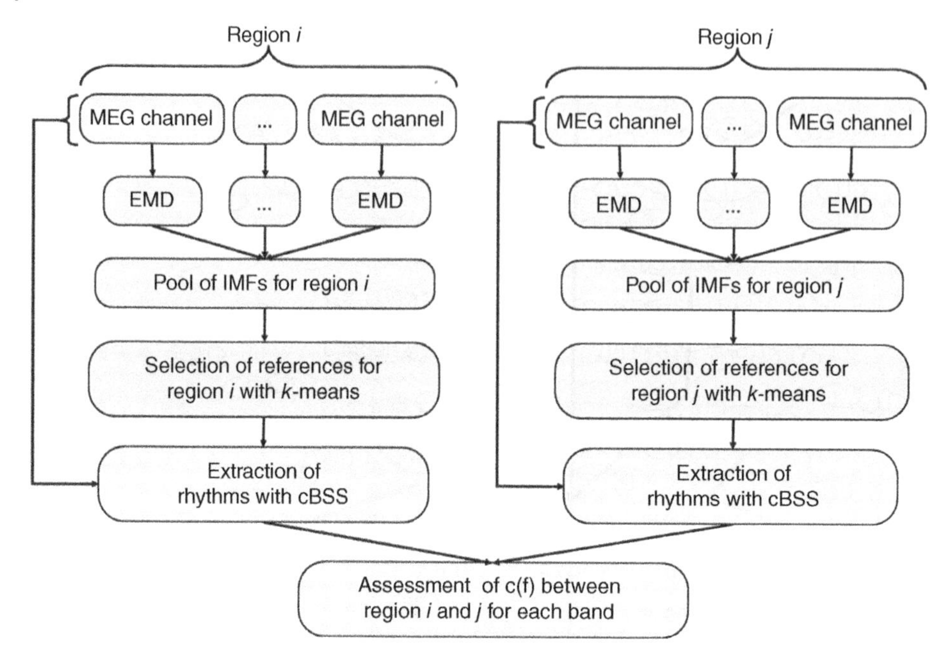

Figure 15.2 Block diagram of the spectral coherency, $c(f)$, measure.

Quantitative measurement of spectral coherency without or with regional extraction of rhythms has led to the charts in Figure 15.5 respectively in the left and right bar graphs.

As a result, raw MEG measurements of subject group pairs produced $c(f)$ accuracy rates of 51.6–61.1%, compared to 66.7–77.3% for the pairs of subject groups when rhythms were extracted.

15.2.2 ERP-Based AD Monitoring

Advanced techniques in signal processing can be employed to detect and reveal the changes in event-related potentials (ERPs), particularly for P300 subcomponents, due to AD. P3b subcomponent elicits as the result of novel or target stimuli in an odd-ball paradigm experiment. For AD patients both amplitude and latency of P3a changes compared to the healthy subjects. For AD the P3b amplitude reduces whereas P3a amplitude does not have any significant change compared with those of healthy subject. Also, the relative (to P3a) latency for P3b significantly increases as the result of AD [22]. Figure 15.6 clearly shows the changes in P3b.

15.2.3 Other Approaches to EEG-Based AD Monitoring

The authors of [11] examined changes in the large-scale structure of resting-state brain networks in patients with AD and compared with non-demented controls, by applying concepts from graph theory to the MEG signals.

For the main frequency bands, synchronization between all pairs of MEG channels was assessed using PLI-weighted connectivity networks and characterized by a mean clustering coefficient and path length.

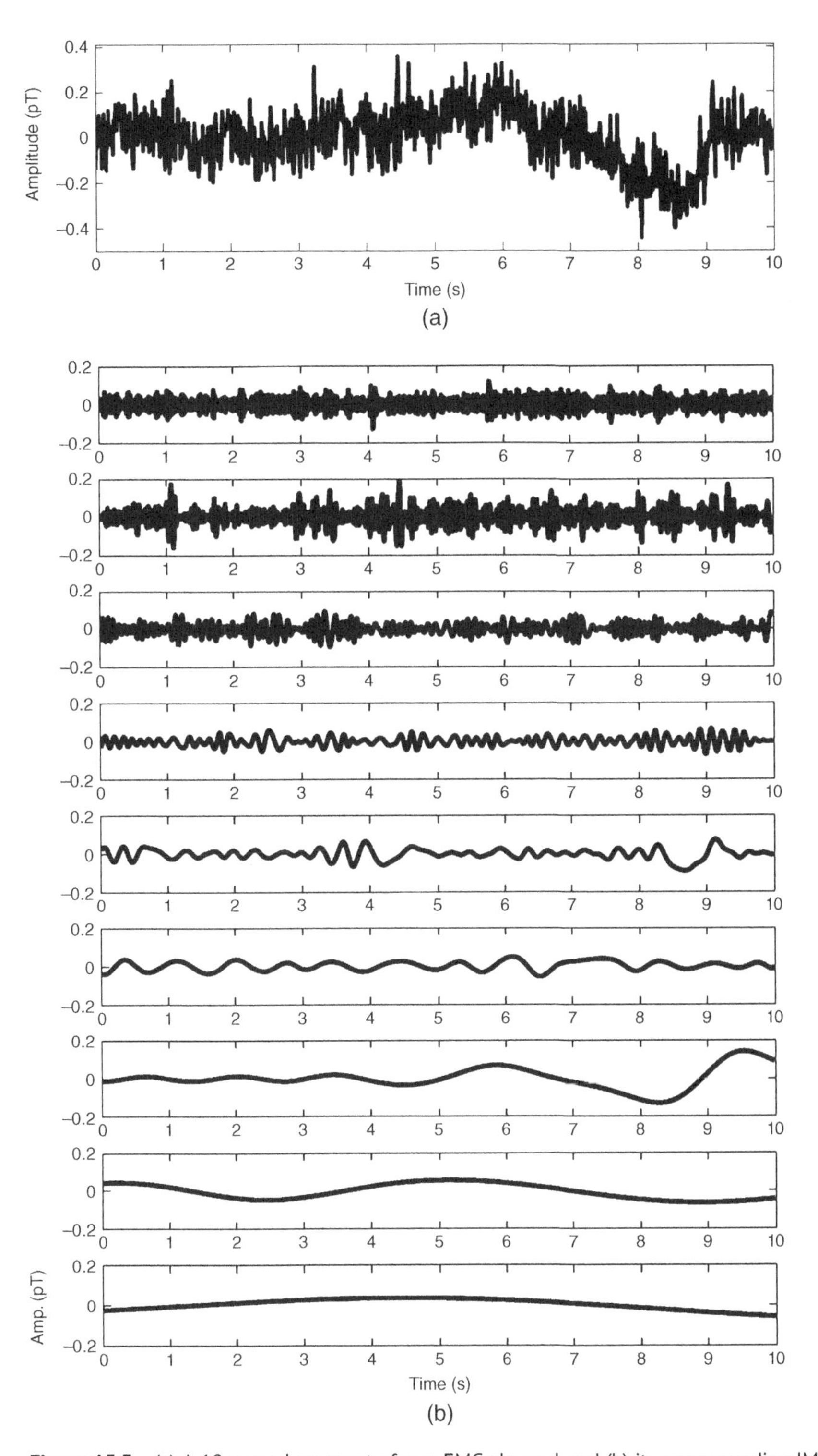

Figure 15.3 (a) A 10 second segment of one EMG channel and (b) its corresponding IMFs.

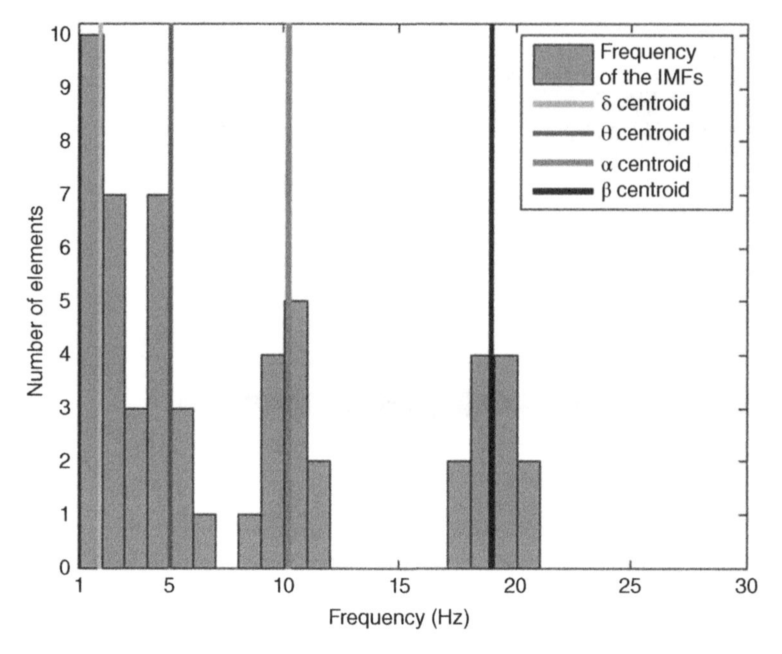

Figure 15.4 Reference selection using the *k*-means algorithm.

Using this method, the AD patients showed a decrease of mean PLI in the lower alpha and beta bands. In the lower alpha band, the clustering coefficient is defined as [11]:

$$C_i = \frac{\sum_{k \neq i} \sum_{\substack{l \neq i \\ l \neq k}} w_{ik} w_{il} w_{kl}}{\sum_{k \neq i} \sum_{\substack{l \neq i \\ l \neq k}} w_{ik} w_{il}} \tag{15.9}$$

where w_{ij} is the weight of the link between vertices i and j and (the average) path length defined as [11]:

$$L_w = \left(\frac{1}{N(N-1)} \sum_{i=1}^{N} \sum_{j \neq i}^{N} w_{ij} \right)^{-1} \tag{15.10}$$

were both decreased in AD patients. The mean values for the above two parameters in different frequency bands are calculated and used in classification between healthy and AD subjects [11]. Network changes in the lower alpha band were better explained by a 'Targeted Attack' model than by a 'Random Failure' model. Thus, the AD patients display a loss of resting-state functional connectivity in lower alpha and beta bands even when a measure insensitive to volume-conduction effects, such as PLI, is used. Moreover, the large-scale structure of lower alpha band functional networks in AD is more random.

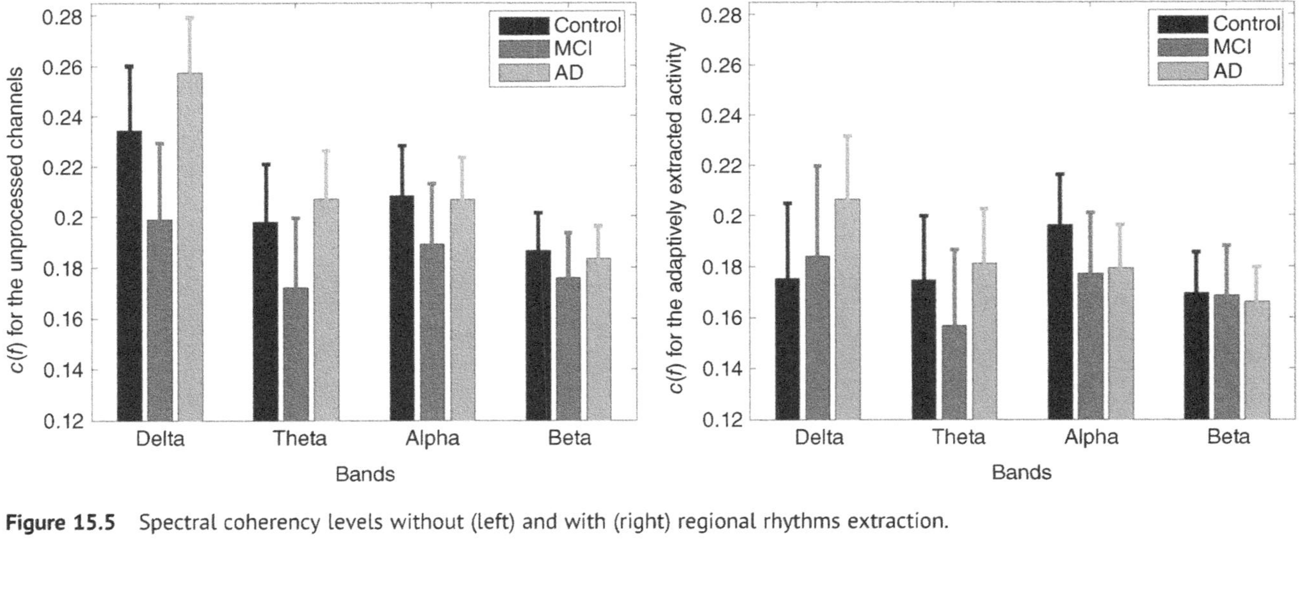

Figure 15.5 Spectral coherency levels without (left) and with (right) regional rhythms extraction.

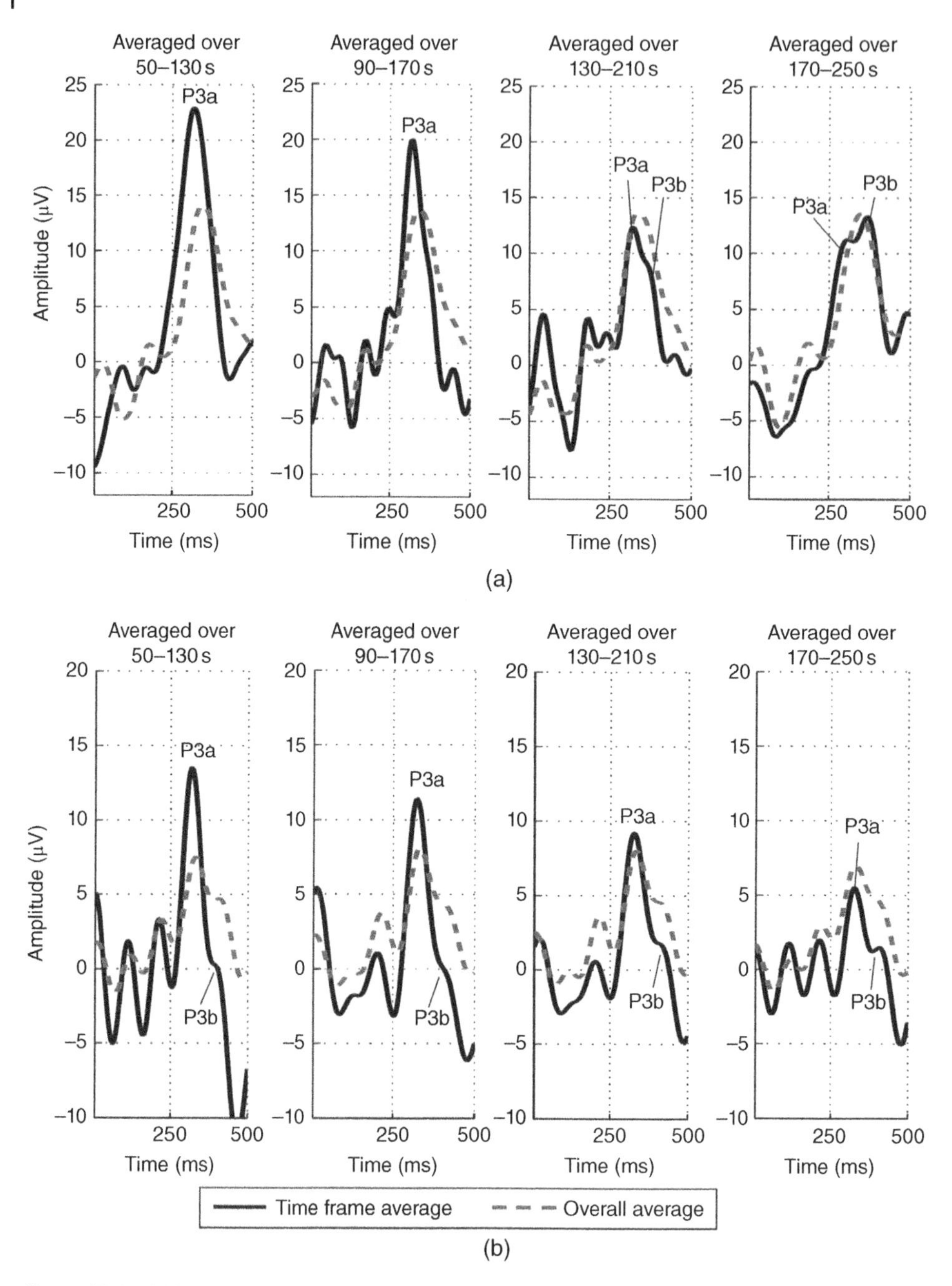

Figure 15.6 ERPs recorded using the Cz electrode for (a) healthy and (b) AD patient. The bold graphs are averaged over a smaller number of 20 second consecutive frames (denoted on the top) and the dashed graphs are the overall average ERPs. There is clear decrease in amplitude and increase in the relative (to P3a) latency of P3b for AD patients.

In another recent research graph theory has been used to discriminate between three states of healthy controls, patients with MCI due to AD (AD-MCI) and AD patients with mild dementia attention-deficit disorder (ADD) to evaluate if the abnormality degree could involve low and/or high degree vertices, the so-called hubs, in both prodromal and over dementia stages [23]. The mean GC magnitudes across subjects for all links in control, AD-MCI and ADD groups can be seen in the vertex-to-vertex graph of Figure 15.7.

The measures are in terms of the number of edges (degree), the number of inward (in-degree), and outgoing edges (out-degree). These parameters were statistically compared to evaluate the degree of abnormality.

15.3 Motor Neuron Disease

Motor neuron disease usually refers to ALS, but it can also refer to other kinds of neurodegenerative disease that affect the motor neurons, such as progressive primary lateral sclerosis, progressive muscular atrophy, and progressive bulbar palsy.

ALS is another heterogeneous neurodegenerative disease characterized primarily by degeneration of upper and lower motor neurons with variable degrees of extra-motor involvement. Clinical manifestations of ALS dichotomize into those associated with apparently pure motor system degeneration involving disruption in motor cortex, corticospinal tracts, and motor networks [24, 25] and degeneration of extra-motor regions, associated with clinical features of cognitive decline, ranging from mild executive impairment to behavioural variant FTD [26]. While there is an urgent need for noninvasive biomarkers that address disease heterogeneity, the majority of imaging [24] and electrophysiological [27] studies to date have focused primarily on quantification of the selective structural degeneration and functional deficiencies of motor pathways. The authors in [26] have indicated that the increasing involvement of broader motor and nonmotor regions and networks requires a more extensive assessment of large-scale brain connectivity.

In [26] the spectral coherency defined in Eq. (15.1), as a brain connectivity measure, followed by a linear discriminant analysis (LDA) classifier has been applied to the EEG data recorded from approximately 100 ALS patients over eight frequency bands.

Based on their study they demonstrated the validity of spectral EEG as a measure of structural degeneration in ALS. They also suggested that the increased connectivity (coherences) reflects network over-activity in affected motor regions (intercortical theta band) and less degenerated regions such as parietal and frontal areas (gamma band), which likely represents an indirect potentially compensatory effect of degeneration [26].

15.4 Parkinson's Disease

PD is the second most common neurodegenerative disorder after AD [28]. It usually affects older people over the age of 50 and results in tremor as its most common symptom. PD is caused by the loss of dopaminergic cells in the substantia nigra which is part of the basal

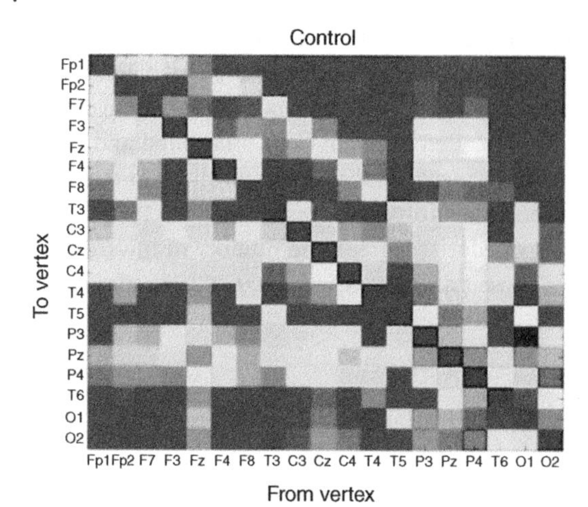

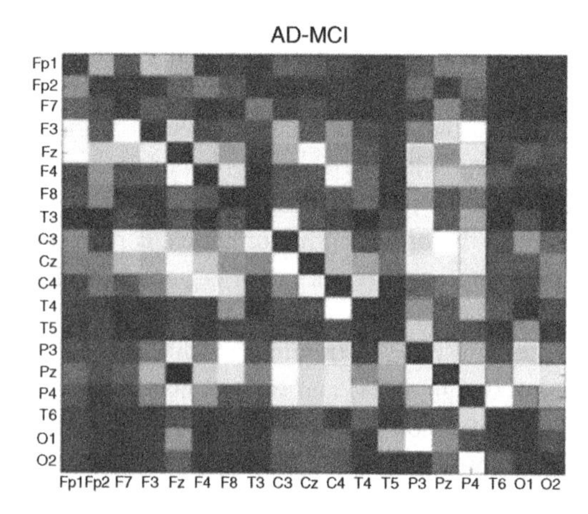

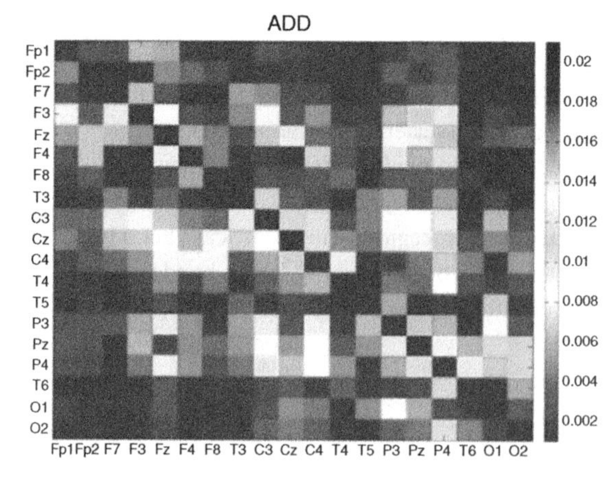

Figure 15.7 Mean GC magnitudes across subjects for all links in control, AD–MCI and ADD groups. In the matrix representation 'from vertex to vertex' indicates the direction of the information transfer between electrode pairs. Electrodes are shown from the left to right side, anteriorly–posteriorly. The colour bar indicates the magnitude of the GC connections [23].

ganglia that is partially responsible for motor actions. Symptoms of the disease are not evident until most of the dopaminergic cells are inactive.

It is characterized by motor symptoms such as tremor, rigidity, slowness of movement, and problems with gait. Motor symptoms are often accompanied with fatigue, depression, pain, and cognitive problems.

Deep brain stimulation has become a common practise in stimulating the dopamine transmitter neurons within the brain to suppress the (mainly resting) tremor.

Monajemi et al. [29] developed a network-based approach for studying the relation between the tremor intensity and the brain connectivity of patients with PD was introduced. They introduced an adaptive multitask diffusion strategy to estimate the underlying model between the gait information and the EEG signals.

In their design they incorporated an S-transform-based connectivity measure [30] in the multitask diffusion strategy to model the relation between tremor and the brain signals as:

$$a_{lk}^{(SC)}(t) = \frac{max\left(\mathrm{Im}\left(C_{lk}^{(ST)}(t,f)\right),0\right)}{\sum\limits_{n \in N_k} max\left(\mathrm{Im}\left(C_{nk}^{(ST)}(t,f)\right),0\right)} \tag{15.11}$$

where $C_{lk}^{(ST)}(t,f)$ represents the combination weights obtained using the imaginary part of S-coherency (ImSCoh), N_k is the neighbourhood of node k and $\mathrm{Im}(\cdot)$ represents the imaginary part of a complex-valued entity.

Although using $a_{lk}^{(SC)}$ as the combination weights incorporates coherency information, it is an over-simplification of the problem. However, since the brain network is a multitask complex network where different regions are associated with different tasks, the authors have proposed a multitask diffusion strategy which incorporates both task-related objectives as well as coherency information. Therefore, to achieve a robust estimation of the combination weights in their diffusion adaptation model and reduce the effect of outliers, they combined the above connectivity values to a set of task-related weights in multitask scenarios defined as:

$$a_{lk}^{(MT)}(t) \approx \begin{cases} \dfrac{\|\mathbf{w}_k\|t-1\| - \psi_l\|t\|\|^2}{\sum_{n \in N_k} \|\mathbf{w}_k\|t-1\| - \psi_n\|t\|\|^2} & l \in N_k \\ 0 & Otherwise \end{cases} \tag{15.12}$$

where $\mathbf{w}_k(t)$ is the adaptive cooperative filter parameter vector and $\psi_l(t)$ is the intermediate vector of parameters, in the following way [5]:

$$a_{lk}(t) \approx \begin{cases} \dfrac{\lambda a_{lk}^{(MT)}(t) + (1-\lambda) a_{lk}^{(SC)}(t)}{\sum_{n \in N_k} \left(\lambda a_{nk}^{(MT)}(t) + (1-\lambda) a_{nk}^{(SC)}(t)\right)} & l \in N_k \\ 0 & Otherwise \end{cases} \tag{15.13}$$

where $0 \leq \lambda \leq 1$ and has to be set empirically. This enhanced brain connectivity measure represents the time–space brain variation relation to the hand tremor.

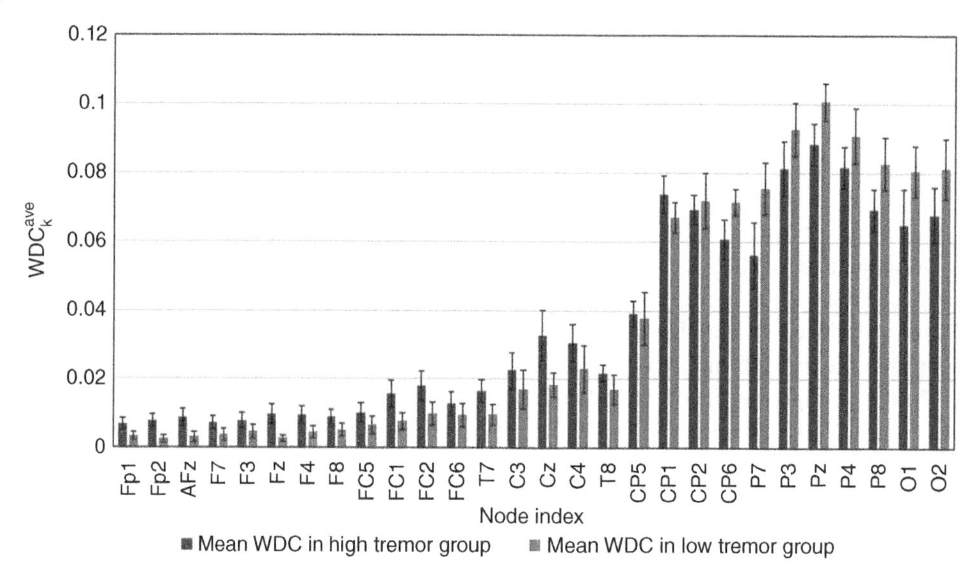

Figure 15.8 Results of the multitask diffusion adaptation method in [5] for high and low Parkinson's disease tremor level groups in terms of averaged WDC for λ = 0.7.

They also obtained the so-called weighted degree centrality (WDC) which is a measure that has been widely used in brain network studies [31–34]. This measure defined as:

$$\mathrm{WDC}_k(t) = \sum_{l=1}^{N} a_{kl}(t) \tag{15.14}$$

evaluates the centrality or importance of a node, k, at time instant t in the network according to its connectivity strength to all the other nodes of the network.

The results show how the differences between the connectivity values of patients with mild and severe hand tremor are most distinguishable from their EEGs when using the proposed method. Figure 15.8 shows the WDC values for different EEG channels (nodes) in low and high tremor patients for an empirically good value of $\lambda = 0.7$.

In another research study, the MEG electrodes were clustered into 10 major cortical regions (as depicted in Figure 15.9) and the data analysis involved calculation of three synchronization likelihood (SL, a general measure of linear and nonlinear temporal correlations between time series) measures which reflect functional connectivity within (local) and between the cortical regions in five frequency bands [35].

In this research it has been concluded an increased resting-state functional connectivity in PD patients compared with the control group. In early-stage PD, the increase is around the lower alpha range, but with disease progression, the functional connectivity progressively increases over a wider range of 4–30 Hz. They also confirm the association between beta band coupling and severity of parkinsonism, in particular bradykinesia, and found some evidence for a similar association between functional connectivity in the theta band and motor symptoms, in particular tremor. As another finding in this study, cognitive perseveration in early-stage PD is positively associated with increased inter-hemispheric functional connectivity in the lower alpha (8–10 Hz) range [35].

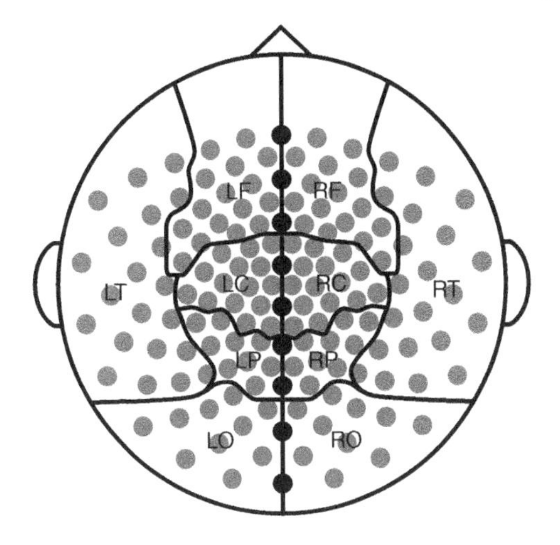

Figure 15.9 Ten brain regions used to estimate the functional connectivity using MEG [35].

15.5 Huntington's Disease

Like PD, Huntington's disease (HD) is also a neurodegenerative disease and affects the basal ganglia. This disease can be passed on from parent to child and leads to severe movement problems.

HD is a result of abnormal protein build-up in the brain, which results in neuronal death. Individuals with this disease show excessive and undesired movements. Gradually, as the disease progresses, people with HD will find it increasingly hard to control their movements and achieve their goals.

Based on a review by Leuchter et al. [36], a major focus in development of novel therapies for HD is identification of treatments that reduce the burden of mutant huntingtin (mHTT) protein in the brain. In order to identify and test the efficacy of such therapies, it is essential to have biomarkers that are sensitive to the effects of mHTT on brain function to determine whether the intervention has been effective at preventing toxicity in target brain systems before onset of the clinical symptoms. Quantitative EEG measures brain oscillatory activity that is regulated by the brain structures that are affected by mHTT in premanifest and early symptom individuals.

Based on this review a number of researchers have examined the changes in human EEG as the result of HD. Lazar et al. [37] used polysomnography (PSG) including EEG and concluded that around the 5–7 Hz range (theta and alpha–theta border) the relative power in pre-HD and early HD is lower during rapid eye movement (REM) compared to controls. Also, the relative power within the 6–7 Hz range (theta and alpha–theta border) is lower in pre-HD and early HD during non-rapid eye movement (NREM) relative to controls.

Piano et al. used EEG source localization using low-resolution brain electromagnetic tomography (LORETA) and discovered that the alpha power is higher in HD during NREM relative to control and alpha power is lower in HD during REM relative to control [38]. They also found that the beta power is lower in HD during NREM relative to control and the theta power is lower in HD during NREM and REM relative to control. Finally, they realized that the delta power is higher in HD during wake compared to control.

Ponomareva et al. [31] utilized quantitative EEG and concluded that within the 8–9 Hz (alpha) range, the relative power in pre-HD is lower relative to control. They also found that in the 7–8 Hz (alpha–theta border) range the relative power in pre-HD is lower relative to control.

EEG-based LORTETA was also used for HD detection by Painold et al. who discovered that compared to a healthy subject, for an HD patient, the alpha, beta, and theta LORETA powers are lower, whereas the delta power is higher [32]. Another quantitative EEG measure also revealed the same results for delta and alpha bands [33].

15.6 Prion Disease

Prion diseases are a group of progressive neurodegenerative diseases that are caused by misfolded proteins, referred to as prions. During post-translational modification of proteins, prions act as a folding template, converting proteins into infectious prion form. The best-known human prion disease is Creutzfeldt–Jakob (mad cow) disease (CJD).

The EEG patterns for CJD are very distinctive and different from the normal EEG patterns during various states of the brain. Figure 15.10 shows a sample of EEG data for a CJD patient.

CJD is a long-latency infection caused by a prion. There are four types of CJD, namely sporadic, variant, familial or inherited, and iatrogenic CJD. Sporadic CJD (also called as classical CJD) is the most popular type of CJD which is shared between human and animals

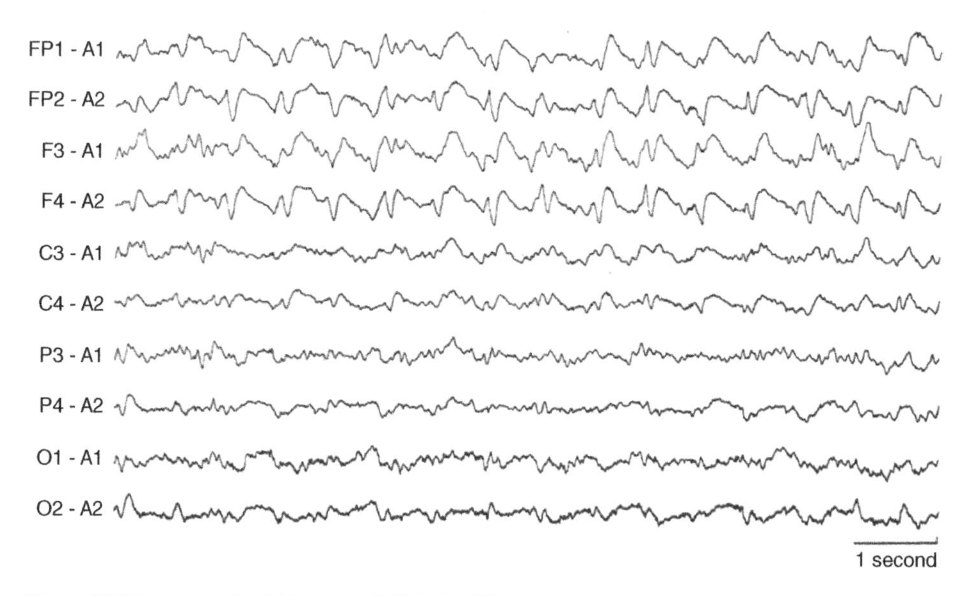

Figure 15.10 A sample of 10-channel EEG of a CJD patient showing clear periodic spikes and diffuse delta (more on the right frontal, i.e. over FP2 and F4). The discharges share some characteristics of triphasic waves and therefore could represent rapidly progressive AD.

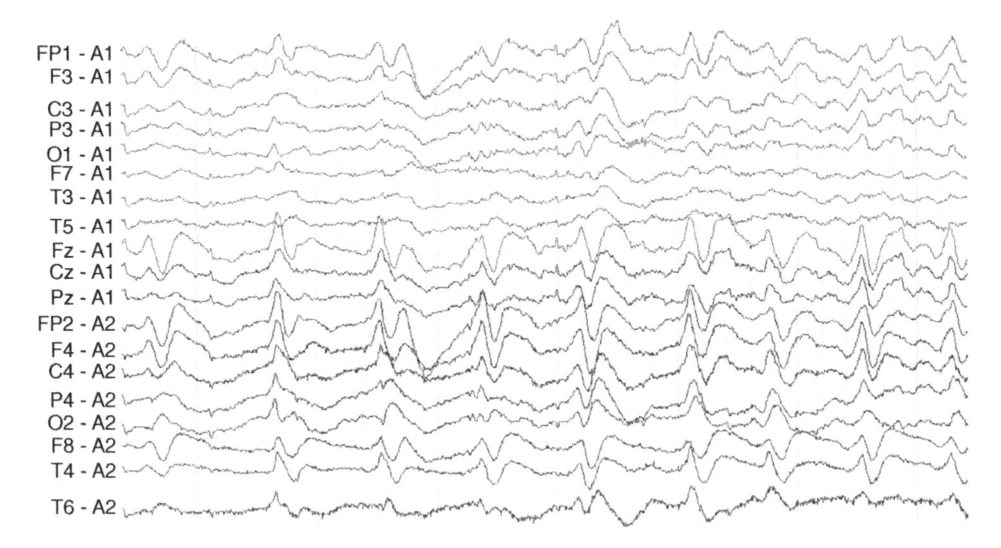

Figure 15.11 EEG in the very early stages of CJD, showing right-lateralised periodic triphasic sharp waves.

and the cause is unknown. Variant CJD which is likely to be caused by consuming meat from a cow that had bovine spongiform encephalopathy (BSE), or 'mad cow' disease, is a similar prion disease to CJD. Familial CJD is a very rare genetic condition where one of the genes a person inherits from their parent (the prion protein gene) carries a mutation that causes prions to form in their brain during adulthood, triggering the symptoms of CJD. Finally, iatrogenic CJD arises from contamination with tissue from an infected person, usually as the result of a medical treatment or surgical operation.

The characteristic EEG shows biphasic or triphasic discharges that are initially sporadic and may even be asymmetric (see Figure 15.11). As the disease advances, the pattern becomes generalized and synchronous with continuous periodic stereotypic 200- to 400-millisecond sharp waves occurring at intervals of 0.5–1.0 seconds.

The investigation in [34] reveals that in sporadic CJD, the EEG exhibits characteristic changes depending on the stage of the disease, ranging from nonspecific findings such as diffuse slowing and frontal rhythmic delta activity (FIRDA) in early stages to disease-typical periodic sharp wave complexes (PSWC) in middle and late stages of coma (around 30–16 days before death) in preterminal EEG recordings. PSWC, either lateralized (in earlier stages) or generalized, occurs in about two thirds of patients with sporadic CJD, with a positive predictive value of 95%. PSWC tend to disappear during sleep and may be attenuated by sedative medication and external stimulation. Conversely, seizures may occur in less than 15% of patients with sporadic CJD. In patients with iatrogenic CJD, PSWC is usually present with more regional EEG.

Similarly, in some recent studies [39] EEG showed slowing and periodic lateralized discharges over the right hemisphere with triphasic morphology, less often involving the left, reflecting clinical asymmetry.

The above waveform patterns are easy to recognize and therefore, classification of CJD from normal and many degenerative brain diseases using EEG signals is often achieved with very high accuracy. This makes the use of EEG for monitoring the disease progress as a diagnostic tool or as the complementary information to behavioural, pathological, and histological findings for CJD favourable.

As an example, in [40] the authors used EEG to differentiate patients with early-stage CJD from rapidly progressive dementia. They used time–frequency representation of the EEG signals through the continuous wavelet transform (CWT) and permutation entropy (PE), as complexity measure. They also used an autoencoder neural network for dimensionality reduction. They also examined support vector machines (SVM) and fully connected multilayer perceptron (MLP-neural networks [NN]) for their supervised classification.

15.7 Behaviour Variant Frontotemporal Dementia

Behaviour variant frontotemporal dementia (bvFTD) is characterized by prominent changes in personality and behaviour that often occurs in people in their 50s and 60s but can develop earlier or later. In behaviour variant FTD, the nerve cell loss is most prominent in areas that control conduct, judgement, empathy, and foresight, among other abilities.

In [41] the authors evaluated the resting-state connectivity in patients with behavioural variant FTD using an SVM. For estimating the brain connectivity, they used a connectivity measure called weighted symbolic mutual information (wSMI) defined in [42] which is basically dependent on the similarity of the wave segments between the nearby channels. This evaluates the extent to which two EEG signals present non-random joint fluctuations, suggesting that they share information and have three main properties. First, it looks for qualitative or 'symbolic' patterns of increase or decrease in the signal, which allows a fast and robust estimation of the signals' entropies. The symbolic transformation depends on the length of the symbols ($k = 3$ in their application) and their temporal separation ($t = 4, 8, 16$, or 32 ms in their application) which follows the PE defined in [43]. Second, wSMI makes few hypotheses on the type of interactions and detects the nonlinear couplings. Third, the wSMI weights discard the spurious correlations between EEG signals arising from common sources and favour nontrivial pairs of symbols.

They found that there was reduced information sharing mainly across mid- and long-range frontotemporal hubs in bvFTD compared to controls. Using these results an enhanced classification of bvFTD patients from controls based on neuropsychological data was achieved. These features offered the best discrimination between bvFTD and AD patients too. These findings suggest that EEG-derived connectivity analyses could offer important contributions to the ongoing quest of massively available, inexpensive biomarkers for bvFTD. Relative to controls, bvFTD patients exhibited hypoconnectivity between frontal and temporal regions of interest (ROIs).

15.8 Lewy Body Dementia

LBD, also called dementia with Lewy bodies (DLB), is the second most common type of progressive dementia after AD. Protein deposits, called Lewy bodies, develop in nerve cells in the brain regions involved in thinking, memory, and movement (motor control). DLB causes a progressive decline in mental abilities. People with DLB may experience visual hallucinations and changes in alertness and attention. Other effects include PD-like signs and symptoms such as rigid muscles, slow movement, and tremors.

Investigations have shown that a combination of quantitative EEG measures (mainly the EEG power in different conventional frequency bands) and the functional connectivity can be used to distinguish between DLB, control, and AD subjects [44].

Also, in [45] the authors combined the spectral domain information with fMRI functional connectivity to compare the resting state in DLB, AD, and control.

The authors in [46] concluded well established slowing of the EEG in the LBD groups compared to healthy controls and AD patients. Although they did not find higher dominant frequency variability (DFV) in DLB patients compared to controls as expected, theta DFV and slow-theta frequency prevalence were positively correlated with cognitive fluctuations (CF) as measured by clinician assessment of fluctuations. This DLB specific correlation suggests that a slower and more temporally variable dominant frequency specifically relates to the CFs seen in DLB and could reveal differential mechanisms underlying CFs in dementia subtypes. Another finding was a significantly higher DFV in AD patients compared to the other groups. Exploratory analysis showed that quantitative EEG measures could predict a DLB versus an AD diagnosis with high accuracy, sensitivity, and specificity.

15.9 Summary

People all over the world are suffering from neurodegenerative diseases, which are illnesses that lead to cell death in the brain. Some neurodegenerative diseases, like PD and HD, affect the basal ganglia and lead to movement difficulties. Other diseases cause more widespread cell death and lead to memory loss, which is seen in AD and LBD. There are also rarer types of neurodegenerative diseases that were not covered in this chapter.

The basis for the assessment of neurodegenerative diseases is to compare the brain network activity of the patient with that of healthy control. This answers a central question in cognitive neuroscience stating how cognitive functions depend upon coordinated and integrated activity of specialized, widely distributed brain regions. There is strong support that a network perspective on the brain is required in order to understand higher brain functioning. Therefore, the brain connectivity estimates, explained before in Chapter 8 and used in this chapter, are very effective tools in the analysis of brain state.

Although normal brain rhythms are less affected by these diseases, the brain response to various stimuli as well as brain connectivity and the transfer of information among various brain zones are seriously affected. Therefore, these parameters are estimated from the EEGs to assess the disease progress.

The use of stem cells to replace the neurons that have died is expected to be a breakthrough in the treatment of neurodegenerative diseases. With this currently in research, hopefully there will soon be help for people with these diseases. Conversely, the collective information from various screening modalities can even help predicting the onset of AD. As an example, recent proposals such as that by Winer et al. [47] suggest that sleep may be a factor associated with accumulation of two core pathological features of AD, i.e. amyloid-β and τ proteins. Then, the combined positron emission tomography (PET) measures of amyloid-β and τ proteins, EEG sleep recordings (for measuring, e.g. time to spindle peak or the NREM slow wave (SW) association), and retrospective sleep evaluations to investigate the potential utility of sleep measures in predicting *in vivo* AD pathology in male and female older adults.

References

1 Cornblath, E.J., Robinson, J.L., Irwin, D.J. et al. (2020). Defining and predicting transdiagnostic categories of neurodegenerative disease. *Nature Biomedical Engineering* 4: 787–800.

2 Villoslada, P., Baeza-Yates, R., and Masdeu, J.C. (2020). Reclassifying neurodegenerative diseases. *Nature Biomedical Engineering* 4: 759–760.

3 Mathuranath, P.S., Nestor, P.J., Berrios, G.E. et al. (Dec. 2000). A brief cognitive test battery to differentiate Alzheimer's disease and frontotemporal dementia. *Neurology* 55 (11): 1613–1620.

4 Daianu, M. et al. (2013). Breakdown of brain connectivity between Normal aging and Alzheimer's disease: a structural k-core network analysis. *Brain Connectivity* 3 (4): 407–422.

5 Granger, C.W.J. (1969). Investigating causal relations by econometric models and cross-spectral methods. *Econometrica* 37 (3): 424–438.

6 Dauwels, J., Vialatte, F., Musha, T., and Cichocki, A. (2010). A comparative study of synchrony measures for the early diagnosis of Alzheimer's disease based on EEG. *NeuroImage* 49: 668–693.

7 Pereda, E., Quiroga, R., and Bhattacharya, J. (2005). Nonlinear multivariate analysis of neurophysiological signals. *Progress in Neurobiology* 77: 1–37.

8 Jeong, J. (2004). EEG dynamics in patients with Alzheimer's disease Clin. *Neurophysiology* 115: 1490–1505.

9 Rossini, P., Rossi, S., Babiloni, C., and Polich, J. (2007). Clinical neurophysiology of aging brain: from normal aging to neurodegeneration. *Progress in Neurobiology* 83: 375–400.

10 Alonso, J., Poza, J., Mañanas, M. et al. (2011). MEG connectivity analysis in patients with Alzheimer's disease using cross mutual information and spectral coherence. *Annals of Biomedical Engineering* 39: 524–536.

11 Stam, C. et al. (2009). Graph theoretical analysis of magnetoencephalographic functional connectivity in Alzheimer's disease. *Brain* 132: 213–224.

12 Stam, C. (2010). Use of magnetoencephalography (MEG) to study functional brain networks in neurodegenerative disorders. *Journal of the Neurological Sciences* 289: 128–134.

13 Gómez, C., Stam, C., Hornero, R. et al. (2009b). Disturbed beta band functional connectivity in patients with mild cognitive impairment: an MEG study. *IEEE Transactions on Biomedical Engineering* 56: 1683–1690.
14 Nolte, G., Bai, O., Wheaton, L. et al. (2004). Identifying true brain interaction from EEG data using the imaginary part of coherency. *Clinical Neurophysiology* 115: 2292–2307.
15 Sanchez Bornot, J.M., Wong-Lin, K., Ahmad, A.L., and Prasad, G. (2018). Robust EEG/MEG based functional connectivity with the envelope of the imaginary coherence: sensor space analysis. *Brain Topography* 31: 895–916.
16 Stam, C.J., Nolte, G., and Daffertshofer, A. (2007). Phase lag index: assessment of functional connectivity from multi channel EEG and MEG with diminished bias from common sources. *Human Brain Mapping* 28: 1178–1193.
17 Vinck, M., Oostenveld, R., van Wingerden, M. et al. (2011). An improved index of phase-synchronization for electrophysiological data in the presence of volume-conduction, noise and sample-size bias. *NeuroImage* 55: 1548–1565. https://doi.org/10.1016/j.neuroimage.2011.01.055.
18 Dauwels, J., Vialatte, F., Musha, T., and Cichocki, A. (2010). A comparative study of synchrony measures for the early diagnosis of Alzheimer's disease based on EEG. *NeuroImage* 49: 668–693.
19 Hoechstetter, K., Bornfleth, H., Weckesser, D. et al. (2004). BESA source coherence: a new method to study cortical oscillatory coupling. *Brain Topography* 16: 233–238.
20 Supp, G., Schlögl, A., Trujillo-Barreto, N. et al. (2007). Directed cortical information flow during human object recognition: analyzing induced EEG gamma-band responses in brain's source space. *PLoS One* 2: e684.
21 Escudero, J., Sanei, S., Jarchi, D. et al. (Aug. 2011). Regional coherence evaluation in mild cognitive impairment and Alzheimer's disease based on adaptively extracted magnetoencephalogram rhythms. *Journal of Physiological Measurements* 32 (8): 1163–1180.
22 Enshaeifar, S., Sanei, S., Cheong-took, C. (2014). Singular spectrum analysis of P300 for classification. *Proceedings of IEEE International Joint Conference on Neural Networks*. IJCNN.
23 Franciotti, R., Falasca, N.W., Arnaldi, D. et al. (2019). Cortical network topology in prodromal and mild dementia due to Alzheimer's disease: graph theory applied to resting state EEG. *Brain Topography* 32: 127–141.
24 Bede, P. and Hardiman, O. (2014). Lessons of ALS imaging: pitfalls and future directions—a critical review. *NeuroImage Clinical* 4: 436–443.
25 Schuster, C., Elamin, M., Hardiman, O., and Bede, P. (2016). The segmental diffusivity profile of amyotrophic lateral sclerosis associated white matter degeneration. *European Journal of Neurology* 23: 1361–1371.
26 Nasseroleslami, B., Dukic, S., and Broderick, M. (January 2019). Characteristic increases in EEG connectivity correlate with changes of structural MRI in amyotrophic lateral sclerosis. *Cerebral Cortex* 29: 27–41. (Oxford Pub.).
27 de Carvalho, M., Chio, A., Dengler, R. et al. (2005). Neurophysiological measures in amyotrophic lateral sclerosis: markers of progression in clinical trials. *Amyotrophic Lateral Sclerosis* 6: 17–28.
28 Dorsey, E.R. et al. (2007). Projected number of people with Parkinson disease in the most populous nations, 2005 through 2030. *Neurology* 68 (5): 384–386.

29 Monajemi, S., Eftaxias, K., Ong, S.-H., and Sanei, S. (2016). An informed multitask diffusion adaptation approach to study tremor in Parkinson's disease. *IEEE Journal of Selected Topics in Signal Processing*; special issue on advanced signal processing in brain networks 10 (7): 1306–1314.

30 Stockwell, R.G., Mansinha, L., and Lowe, R.P. (Apr. 1996). Localization of the complex spectrum: the S-transform. *IEEE Transactions on Signal Processing* 44 (4): 998–1001.

31 Ponomareva, N., Klyushnikov, S., Abramycheva, N. et al. (2014). Alpha-theta border EEG abnormalities in preclinical Huntington's disease. *Journal of the Neurological Sciences* 344: 114–120. https://doi.org/10.1016/j.jns.2014.06.035.

32 Painold, A., Anderer, P., Holl, A.K. et al. (2011). EEG low-resolution brain electromagnetic tomography (LORETA) in Huntington's disease. *Journal of Neurology* 258: 840–854. https://doi.org/10.1007/s00415-010-5852-5.

33 Hunter, A., Bordelon, Y., Cook, I., and Leuchter, A.F. (2010). QEEG measures in Huntington's disease: a pilot study. *PLoS Currents* 2: RRN1192. https://doi.org/10.1371/currents.RRN1192.

34 Wieser, H.G., Schindler, K., and Zumsteg, D. (2006). EEG in Creutzfeldt–Jakob disease. *Clinical Neurophysiology* 117: 935–951.

35 Stoffers, D., Bosboom, J.L.W., Deijen, J.B. et al. (2008). Increased cortico-cortical functional connectivity in early-stage Parkinson's disease: an MEG study. *NeuroImage* 41: 212–222.

36 Leuchter, M.K., Donzis, E.J., Cepeda, C. et al. (30 March 2017). Quantitative electroencephalographic biomarkers in preclinical and human studies of Huntington's disease: are they fit-for-purpose for treatment development? *Frontiers in Neurology* https://doi.org/10.3389/fneur.2017.00091.

37 Lazar, A.S., Panin, F., Goodman, A.O.G. et al. (2015). Sleep deficits but no metabolic deficits in premanifest Huntington's disease. *Annals of Neurology* 78: 630–648. https://doi.org/10.1002/ana.24495.

38 Piano, C., Mazzucchi, E., Bentivoglio, A.R. et al. (2016). Wake and sleep EEG in patients with Huntington disease: an eLORETA study and review of the literature. *Clinical EEG and Neuroscience* 48 (1): 60–71. https://doi.org/10.1177/1550059416632413.

39 Ganesh, A., Hoyte, L.C., Agha-Khani, Y., and Yeung, M.M.C. (April 17, 2018). Teaching NeuroImages: DWI and EEG findings in Creutzfeldt-Jakob disease. *Neurology* 90 (16): e1450–e1451.

40 Morabito, F.C., Campolo, M., Mammone, N. et al. (2017). Deep learning representation from electroencephalography of early-stage Creutzfeldt-Jakob disease and features for differentiation from rapidly progressive dementia. *International Journal of Neural Systems* 27 (02): 1650039. 12 pages.

41 Dottori, M., Sedeño, L., Caro, M.M. et al. Towards affordable biomarkers of frontotemporal dementia: a classification study via network's information sharing. *Nature Scientific Reports* 7: 3822. https://doi.org/10.1038/s41598-017-04204-8.

42 King, J.R., Sitt, J.D., Faugeras, F. et al. (2013). Information sharing in the brain indexes consciousness in noncommunicative patients. *Current Biology: CB* 23: 1914–1919. https://doi.org/10.1016/j.cub.2013.07.075.

43 Bandt, C. and Pompe, B. (2002). Permutation entropy: a natural complexity measure for time series. *Physical Review Letters* 88: 174102.

44 van der Zande, J.J., Gouw, A.A., van Steenoven, I. et al. (2018). EEG characteristics of dementia with lewy bodies, Alzheimer's disease and mixed pathology. *Frontiers in Aging Neuroscience* 10: 190. 10 pages.

45 Schumacher, J., Peraza, L.R., Firbank, M. et al. (2019). Dysfunctional brain dynamics and their origin in Lewy body dementia. *Brain* 142: 1767–1782.

46 Stylianou, M., Murphy, N., Peraza, L.R. et al. (2018). Quantitative electroencephalography as a marker of cognitive fluctuations in dementia with Lewy bodies and an aid to differential diagnosis. *Clinical Neurophysiology* 129: 1209–1220.

47 Winer, J.R., Mander, B.A., Helfrich, R.F. et al. (August 7, 2019). Sleep as a potential biomarker of tau and -amyloid burden in the human brain. *The Journal of Neuroscience* 39 (32): 6315–6324.

16

EEG As A Biomarker for Psychiatric and Neurodevelopmental Disorders

16.1 Introduction

Neurodevelopmental disorders (NDDs) are multifaceted conditions manifested as impairments in cognition, communication, behaviour, and/or motor skills resulting from abnormal brain development. Intellectual disability, communication disorders, autism spectrum disorder (ASD), attention deficit hyperactivity disorder (ADHD) and schizophrenia, among many others, are very popular ones which fall under the umbrella of NDDs [1].

With the help of new diagnostic tools and technologies all these disorders can be diagnosed and some even predicted easily. Various methods and biomarkers are used to discriminate the subjects suffering from psychiatric disorders during different stages of the problem. These include abnormal facial expression, change in voice, deteriorating and slowing gait, and many observable behavioural changes. Nevertheless, from very early medical history, the root of these disorders has been recognized to come from the brain. Therefore, better diagnosis and monitoring of these disorders are expected to be performed by looking at the brain signals.

16.1.1 History

Looking at the Noba website (https://nobaproject.com/modules/history-of-mental-illness), the evolution of mental illness depends on the context surrounding the behaviour and thus changes as a function of a particular time and culture. In the past, uncommon behaviour or behaviour that deviated from the sociocultural norms and expectations of a specific culture and period have been used as reasons and an excuse to silence or control certain individuals or groups. As a result, a less cultural relativist view of abnormal behaviour has focused instead on whether behaviour poses a threat to oneself or others or causes so much pain and suffering that it interferes with one's work responsibilities or with one's relationships with others.

According to the Vandidad, the chapter about social conduct in the Avesta, the holy book of Zoroastrians (*c.* 1000 BC), physicians were divided into three groups: surgeons (*kareto baešaza* in the Avestan language), physicians who worked with herbal medicines (*urvaro baešaza*), and physicians who treated with holy words (*mansrspand baešaza*), which were preferred to other treatments. These physicians were selected from Zoroastrian priests and considered as the first psychiatrists in history [2].

EEG Signal Processing and Machine Learning, Second Edition. Saeid Sanei and Jonathon A. Chambers.

Throughout history there have been three general theories of the aetiology of mental illness: supernatural, somatogenic, and psychogenic. Somatogenic theories identify disturbances in physical functioning resulting from either illness, genetic inheritance, or brain damage or imbalance. Psychogenic theories focus on traumatic or stressful experiences, maladaptive learned associations and cognitions, or distorted perceptions. Etiological theories of mental illness determine the care and treatment mentally ill individuals receive.

Greek physicians rejected supernatural explanations of mental disorders. It was around 400 BC that Hippocrates (460–370 BC) attempted to separate superstition and religion from medicine by systematizing the belief that a deficiency in or especially an excess of one of the four essential bodily fluids (i.e. humours) – blood, yellow bile, black bile, and phlegm – was responsible for physical and mental illness. For example, someone who was too temperamental suffered from too much blood and thus blood letting would be the necessary treatment for him/her.

Hippocrates classified mental illness into one of four categories – epilepsy, mania, melancholia, and brain fever – and like other prominent physicians and philosophers of his time, he did not believe mental illness was shameful or that mentally ill individuals should be held accountable for their behaviour. Mentally ill individuals were cared for at home by family members and the state shared no responsibility for their care. Humourism remained a recurrent somatogenic theory up until the nineteenth century.

Throughout history there have been three general theories of the aetiology of mental illness: supernatural, somatogenic (promoted especially by the father of America psychiatry, Benjamin Rush [1745–1813]), and psychogenic. Supernatural theories attribute mental illness to possession by evil or demonic spirits, displeasure of gods, eclipses, planetary gravitation, curses, and sin. Somatogenic theories identify disturbances in physical functioning resulting from either illness, genetic inheritance, or brain damage or imbalance. Psychogenic theories focus on traumatic or stressful experiences, maladaptive learned associations and cognitions, or distorted perceptions. Etiological theories of mental illness determine the care and treatment mentally ill individuals receive.

Muhammad ibn Zakariya Razi (865–925), known as Rhazes in the west, was an influential Persian physician, philosopher, and scientist, who discovered alcohol, and was among the first in the world to write on mental illness and psychotherapy [3]. During the time he was the chief physician of Baghdad hospital, which was part of the Persian empire at the time, he was also the director of one of the first psychiatric wards in the world. Two of his works in particular, Mansuri Medicine and Hawi, provide descriptions and treatments for mental illnesses [4].

The Persian physician and philosopher Avicenna (aka Ibn Sina, 980–1037 CE) studied the connection between mind and body, an early investigation of psychology in his 'Canon of Medicine.' He also described mental illnesses like depression and anxiety, although in terms like melancholy and fear about death.

European psychiatry in the late eighteenth century and throughout the nineteenth century, however, struggled between somatogenic and psychogenic explanations of mental illness. James Braid (1795–1860) shifted this belief in mesmerism to one in hypnosis, thereby proposing a psychogenic treatment for the removal of symptoms.

Although both somatogenic and psychogenic explanations coexist today, psychoanalysis was the dominant psychogenic treatment for mental illness during the first half of the twentieth century, providing the launch pad for the more than 400 different schools of psychotherapy found today [5]. Most of these schools cluster around broader behavioural, cognitive, cognitive-behavioural, psychodynamic, and client-centred approaches to psychotherapy applied in individual, marital, family, or group formats.

Progress in the treatment of mental illness necessarily implies improvements in the diagnosis of mental illness. A standard diagnostic protocol creates a shared language among mental-health providers and aids in clinical research. While diagnoses were recognized as far back as the Greeks, it was not until 1883 that German psychiatrist Emil Kräpelin (1856–1926) published a comprehensive system of psychological disorders that centred around a pattern of symptoms with the link to being suggestive of an underlying physiological cause.

The use of electroencephalogram (EEG) in ADHD, as an example, began more than 75 years ago with Jasper et al. [6] reporting a slowing of the EEG rhythms at frontocentral sensors, a putative indicator of abnormal brain function in a group of 'behaviour problem children' – described as *hyperactive, impulsive, and highly variable.*

Other clinicians also suggested popular classification systems but the need for a single, shared system paved the way for the American Psychiatric Association's 1952 publication of the first *Diagnostic and Statistical Manual* (DSM). The DSM has undergone various revisions (in 1968, 1980, 1987, 1994, 2000, 2013). The most recent version, DSM-5, reflect an attempt to help clinicians streamline diagnosis and work better with other diagnostic systems such as health diagnoses outlined by the World Health Organization.

16.1.1.1 Different Psychiatric and Neurodevelopmental Disorders

In the most recent DSM factsheet published in 2013 (https://www.psychiatry.org/psychiatrists/practice/dsm/educational-resources/dsm-5-fact-sheets), the following disorders have been listed (alphabetically):

- ADHD
- ASD
- conduct disorder
- disruptive mood dysregulation disorder
- eating disorders
- gender dysphoria
- intellectual disability
- internet gaming disorder
- major depressive disorder and the bereavement exclusion
- mild neurocognitive disorder
- obsessive–compulsive and related disorders
- paraphilic disorders
- personality disorder (including schizotypal disorder)
- post-traumatic stress disorder
- schizophrenia
- sleep–wake disorders (insomnia)
- specific learning disorder
- social communication disorder

- somatic symptom disorder
- substance-related and addictive disorders
- psychiatric disorder due to a terminal illnesses.

In this chapter, there is no intention to cover the details of all the above disorders. Instead, applications of EEG signal processing and machine learning for detection, recognition, or monitoring of some more popular psychiatric disorders for which the EEG signals are used as one of the screening modalities for diagnosis, are explained.

16.1.1.2 NDD Diagnosis

Currently, there is no clear symptomatic biomarker to differentiate between the NDDs. Rather, these disorders are categorized into discrete disease entities, based on clinical presentation. This is mainly because many symptoms are not unique to a single NDD, and several NDDs have clusters of symptoms in common. For example, impaired social cognition is common between ASD and schizophrenia. Also, psychosis is observed not only in schizophrenia but also in those with bipolar disorder (BP) or major depressive disorder. For these reasons the boundaries between various NDDs have been investigated in terms of genomics, interactomics and proteomics in [1] and proposed that identification of unique proteomic signatures that can be strongly associated with patient's risk alleles and proteome-interactome-guided exploration of patient genomes could define biological mechanisms necessary to reformulate disorder definitions.

Conversely, the use of neuroimaging techniques, such as EEG, enables extraction of statistical variables associated with NDD but as we will see in the following sections, many of these measures are in common between the above wide list of disorders. Nevertheless, combined with other historical, behavioural, and peripheral information, these neuroimaging features can indicate the type of NDD with very high accuracy. As soon as the type or at least the category of abnormality is diagnosed some treatments can be prescribed to alleviate their progress or symptoms.

16.2 EEG Analysis for Different NDDs

16.2.1 ADHD

ADHD is one of the most common mental disorders affecting children. ADHD also affects many adults. Symptoms of ADHD include inattention (not being able to keep focus), hyperactivity (excess movement that is not fitting to the setting) and impulsivity (hasty acts that occur in the moment without thought).

An estimated 8.4% of children and 2.5% of adults have ADHD [7, 8]. ADHD is often first identified in school-aged children when it leads to disruption in the classroom or problems with schoolwork. It is more common among boys than girls.

16.2.1.1 ADHD Symptoms and Possible Treatment

Many ADHD symptoms, such as high activity levels, difficulty remaining still for long periods of time and limited attention spans, are common to young children in general. The hyperactivity and inattention are noticeably greater than expected for the ADHD children causing distress or social problems wherever they are.

ADHD can be in any of inattentive, hyperactive/impulsive, or combined types. The diagnosis is based on the symptoms which appeared over six months before being referred to clinic.

No specific causes have been identified for ADHD. There is evidence that genetics contribute to ADHD. Being born prematurely, brain injury, and the mother smoking, using alcohol or having extreme stress during pregnancy might be the other causes.

16.2.1.2 EEG-Based Diagnosis of ADHD

In [9] some early approaches in EEG-based ADHD diagnosis have been reviewed. The emphasis, however, has been on multivariate analyses and the focus towards understanding of the neural generators of EEG. The authors classify the approaches into two main schemes, one by looking at the slow waves initially referred to by Jasper et al. [6] and the changes in event-related potentials (ERPs). Then, the links between EEG features and clinical heterogeneity in ADHD are studied and it is concluded that multivariate analyses and resolution of EEG signals into their neural generators, can put EEG into clinical practise for ADHD.

The most robust EEG feature associated with ADHD, picked up by many researchers, is elevated power of slow-theta waves often together with decreased power of fast beta waves (14–30 Hz) recorded over Cz which can be presented as a univariate theta/beta ratio varying approximately between 0.6 and 3.0 [10–14]. However, although in 2016 the theta/beta ratio measurement was announced as a standard method by the American Academy of Neurology, in a commentary by Nuwer et al. [15] this was criticized mainly because this ratio has a dramatic age effect, decreasing markedly as a child ages from childhood towards adulthood. Certainly, the theta/beta ratio measurement can be compared with an age-dependent baseline and still used in clinical practise. In more recent study [16] by looking at both spectral and spatial footprints of the brain activities, ADHD subjects showed reduced beta activity during the task and higher theta/beta ratio at rest, which are in line with previous research findings. Furthermore, most of the ADHD subjects, during the attention test, presented a beta power falling outside the normality range defined on the normal children group.

As a different direction in using spectral power for ADHD in [17] a penalized logistic regression using elastic net [18] was employed to classify the subjects using the entire EEG spectrum (1–45 Hz) measurements including theta/beta ratio.

Given the data (feature) matrix $\mathbf{X}$ and the vector of labels $\mathbf{y}$ and the vector of parameters (classifier hyperplane) as $\mathbf{w}$, the elastic net based optimization for estimation of $\mathbf{w}$ is very similar to least absolute shrinkage and selection operator (LASSO) regularized optimization. In LASSO the following regularization problem is attempted:

$$\min_{\mathbf{w}} \left(\|\mathbf{y} - \mathbf{X}\mathbf{w}\|^2 + \lambda \|\mathbf{w}\|_1 \right) \tag{16.1}$$

where $\|\bullet\|_1$ refers to l_1 – norm and λ is the regularization parameter. The solution is as follows:

$$\hat{\mathbf{w}}_{LASSO} = \arg\min_{\mathbf{w}} \left(\mathbf{w}^T \left(\mathbf{X}^T\mathbf{X}\right)\mathbf{w} - 2\mathbf{y}^T\mathbf{X}\mathbf{w} + \lambda \|\mathbf{w}\|_1 \right) \tag{16.2}$$

In the elastic net, two regularization parameters are used, namely λ_1 and λ_2, and the solution for $\mathbf{w}$ becomes:

$$\hat{\mathbf{w}}_{Elastic} = \arg\min_{\mathbf{w}} \left(\mathbf{w}^T \left(\frac{\mathbf{X}^T\mathbf{X} + \lambda_2 \mathbf{I}}{1 + \lambda_2} \right) \mathbf{w} - 2\mathbf{y}^T\mathbf{X}\mathbf{w} + \lambda_1 \|\mathbf{w}\|_1 \right) \tag{16.3}$$

Inclusion of λ_2 improves the stability of the results [18]. They concluded that the best predictors of ADHD status could be the increase in power of delta, theta and low-alpha over centroparietal regions, and in frontal low-beta and parietal mid-beta. In addition, the elevated eyes-open power in delta, theta, low-alpha and low-beta could be used to distinguish first-degree relatives from controls. Therefore, it was claimed that, generally, the theta/beta ratio by itself is not an accurate diagnostic marker for ADHD though it works reasonably well for childhood ADHD.

Given that gamma activity is considered to be an index of network functions in the brain that underlie higher-order cognitive processes, in [19] it has been concluded that gamma activity is reduced in adult ADHD and the reduction has a predominantly right centroparietal distribution. This outcome is consistent with childhood ADHD literature with respect to diminished posterior gamma activity in patients, which may reflect altered dorsal attention network functions.

ERPs are the most natural representation of brain functionality and they are expected to be affected in terms of both amplitude and latency by psychiatric diseases including ADHD. In [20] it has been shown that ADHD affects events such as selective attention (P200, P300), response inhibition (N200, P300), error-related negativity (ERN) and error positivity (Pe) and feedback related negativity (FRN) processing, but reported that the results were quite variable, lacking a robust measure for scoring ADHD. Significant changes in the amplitude of P300 due to ADHD in adults has been reported in [21].

In the ERP analysis of ADHD subjects compared with healthy ones in [22], adults with ADHD show attenuated P300, ERN, and Pe, which points to an impairment of attentional resource allocation in demanding conditions, automatic error detection, as well as error awareness. Response-locked activity is generated mainly in the medial frontal cortex and includes the ERN and Pe. The ERN occurs about 50–100 ms after incorrect responses and has a frontocentral scalp distribution. In their experiment [22] the participants were instructed to fixate upon a dot in the centre of the PC screen. The dot was present in a fixed inter-trial interval of 800 ms without jitter. Trials began with six horizontal flanking arrows below the fixation dot that were shown for 100 ms and then followed by the appearance of a centre target arrow that either pointed in the same direction as the flanking arrows (< < < < < < < or > > > > > > >), or in the opposite direction, (< < < > < < < or > > > < > > >), yielding compatible and incompatible trials, respectively. Figure 16.1 illustrates the changes in ERPs (N200, P300, ERP, and Pe) and the comparison between ADHD subject and control [22]. Behavioural performance measures were not significantly different between the groups, while group differences were more readily seen in ERP measures. The authors concluded that the ERP measures are even more sensitive to the underlying liability for ADHD than the overt behavioural task performance.

In a comprehensive and inclusive ERP-based ADHD study, the authors in [23] used an odd-ball paradigm (often used in brain–computer interfacing (BCI)) and implicit navigational images. They analyzed the EEG dynamics with standardized weighted low-resolution

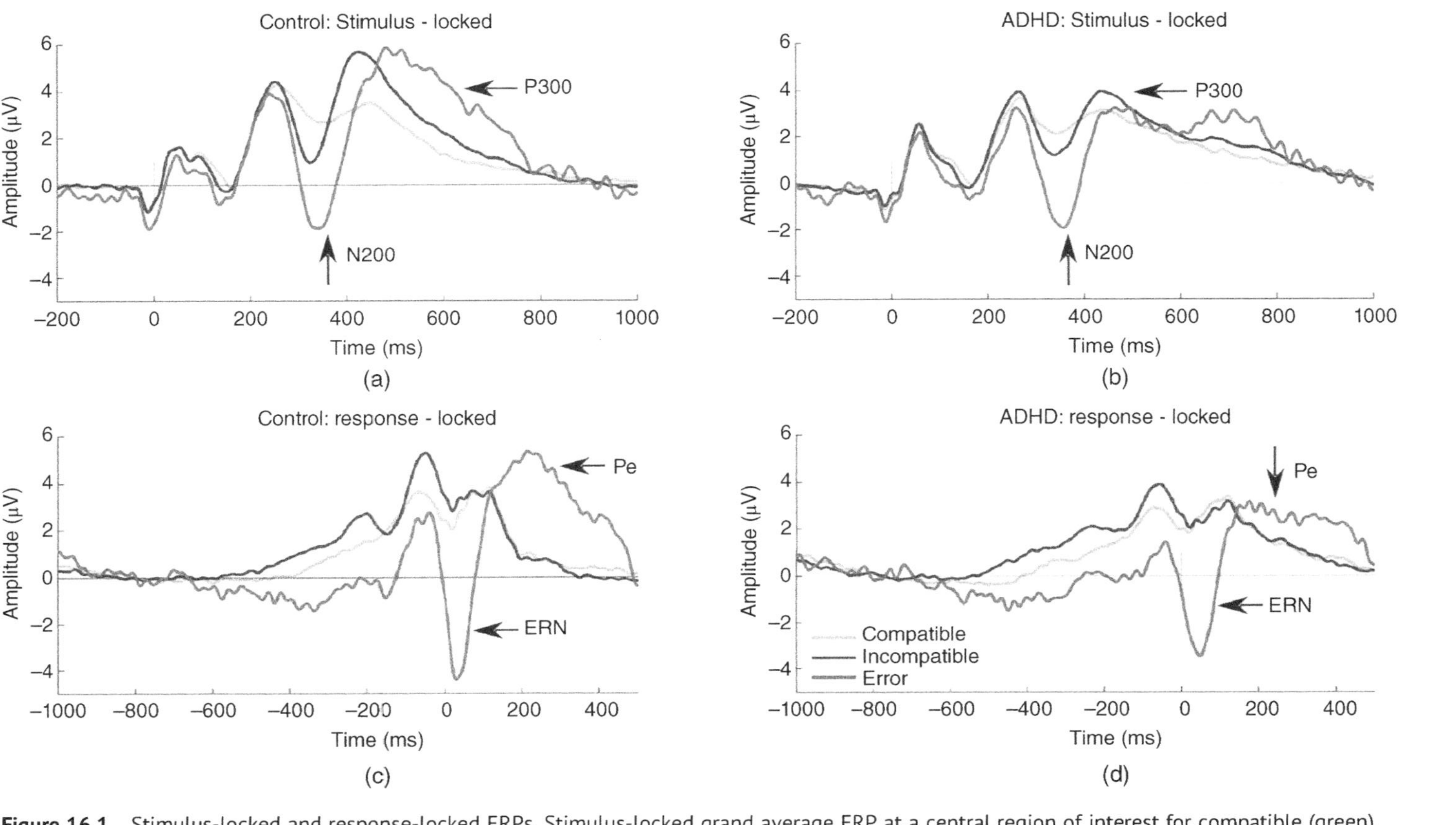

Figure 16.1 Stimulus-locked and response-locked ERPs. Stimulus-locked grand average ERP at a central region of interest for compatible (green), incompatible (blue), and error (red) trials in (a) control participants and (b) ADHD participants. Response-locked grand average ERP for compatible (green), incompatible (blue), and error (red) trials in (c) control participants and (d) ADHD participants [22].

electromagnetic tomography (swLORETA) [24] inverse modelling of the evoked potential generators to study the cortical processing in adults with ADHD and age-matched controls. They demonstrated that P350 amplitude, alpha–beta oscillation event-related synchronization (ERS) anticipation, and beta event-related desynchronization (ERD) were significantly smaller in ADHD. In the active condition, P100 duration was reduced and N140 amplitude increased for both deviant and frequent conditions in ADHD. Alpha ERS and delta-theta ERS were reduced in ADHD in the deviant condition. The left somatosensory area and the right parietal lobe contributed more to the P100 generators in the control than in the ADHD group, while the left frontal lobe contributed more to the P100 generators in the ADHD. The left inferior parietal lobe contributed more to the N140 generators in the control than the ADHD group while the right posterior cingulate contributed more to the N140 generators in ADHD.

The heterogeneity in aetiology, symptoms, and treatment outcomes of the disorder affect the ADHD diagnosis. This has led most theorists to favour multiple pathway models over single-cause explanations of the disorder. Affected neural circuits in ADHD have included nigrostriatal, mesolimbic, and mesocortical dopamine pathways, noradrenergic dysregulation of posterior attentional pathways, delayed development of frontal cortex, and, atypical functioning of the default mode network (DMN) [25]. The plausibility of multiple pathways causing the same set of ADHD symptoms implies that a single EEG measure, like theta/beta ratio, will be effective in predicting ADHD in only a subset of those diagnosed [9].

In a multivariate approach [26] fractal-based features, including Higuchi, Petrosian, and Katz fractal dimensions as well as autoregressive coefficients, band power, and wavelet coefficients have been used to ensure the complexity as well as connectivity have been fully exploited and the spectral features suitably counted. To classify these features they used a number of different classifiers and concluded that a hybrid classification system often outperforms others.

In another approach, [27] independent component analysis (ICA) and time–frequency analysis were used to identify midoccipital alpha (8–12 Hz) to link poor spatial working memory in a group of ADHD children with ineffective stimulus encoding. In their experiments, the children with ADHD showed attenuated alpha band ERD during encoding. This effect was more pronounced when the task difficulty was low (consistent with impaired vigilance) and was predictive of memory task performance and symptom severity. Correlated with alpha ERD in encoding the alpha power increased during the maintenance (when there is no memory activity) period (relative to baseline), suggesting a compensatory effort. Conversely, midfrontal theta power increased during maintenance more strongly in ADHD and in high-load memory conditions. Furthermore, children with ADHD exhibit a maturational lag in the development of posterior alpha power whereas the age-related changes in frontal theta power deviates from the typically developing pattern. Finally, subjects with ADHD showed age-independent attenuation of evoked responses to warning cues, suggesting low vigilance. The combination of these three EEG measures was used to diagnose ADHD with 70% accuracy. They also suggested that the working memory performance deficits in ADHD are at least in part the result of poor vigilance and atypical encoding; possibly from a deficit and/or variability in degree of focus of attention first to the alerting fixation cue and then to the ensuing encoding stimulus.

In [28] the arousal level estimated using VIGALL 2.1 (http://www.uni-leipzig.de/vigall) has been used to recognize the ADHD level in adults. Consecutive one second EEG segments were classified into seven different EEG-vigilance stages (0, A1, A2, A3, B1, B2/3, and C) based on frequency bands and source localization with LORETA. Additionally, horizontal slow eye movements were utilized to separate stages 0 and B1. Alpha frequency and amplitude level were individually adapted. It has been indicated that arousal regulation (i.e. arousal stability score) predicted the retrospectively-assessed severity of childhood ADHD symptoms, supporting the trait aspect of brain arousal regulation.

Although multivariate systems are robust in classification or detection of ADHD, the major problem with these multi-feature approaches is the difficulty in clinical interpretation of the results. It is however expected that with the advances in machine learning approaches and availability of more global (rather than subject-related) trials and features the sensitivity of algorithms to subject and condition variabilities will be decreased.

16.2.2 ASD

ASD is a polygenetic developmental and neurobiological disorder [29] that is characterized by atypical behaviour and lack of social reciprocity. It is defined by a heterogeneous constellation of behavioural symptoms that emerge over the first few years of life. It impacts how a person perceives and socializes with others, causing problems in social interaction and communication. The disorder also includes limited and repetitive patterns of behaviour. The term 'spectrum' in ASD refers to the wide range of symptoms and severity.

ASD includes conditions that were previously considered separate – autism, Asperger's syndrome, childhood disintegrative disorder, and an unspecified form of pervasive developmental disorder. Some people still use the term 'Asperger's syndrome', which is generally thought to be at the mild end of ASD.

16.2.2.1 ASD Symptoms and Possible Treatment

ASD begins in early childhood and eventually causes problems functioning in society, socially in school and at work, for example. Often children show symptoms of autism within the first year. A small number of children appear to develop normally in the first year, and then go through a period of regression between 18 and 24 months of age when they develop autism symptoms. While there is no cure for ASD, intensive, early treatment can make a big difference in the lives of many children.

Some children show signs of ASD in early infancy, such as reduced eye contact, lack of response to their name or indifference to caregivers. Other children may develop normally for the first few months or years of life, but then suddenly become withdrawn or aggressive or lose language skills they have already acquired. Signs usually are seen by age two years.

Each child with ASD is likely to have a unique pattern of behaviour and level of severity.

Some children with ASD have difficulty in learning and some have signs of lower than normal intelligence. Other children with the disorder have normal to high intelligence – they learn quickly, yet have trouble communicating and applying what they know in everyday life and adjusting to social situations. Because of the unique mixture of symptoms in each child, severity can sometimes be difficult to determine. It is generally based on the level of impairments and how they impact the ability to function.

16.2.2.2 EEG-Based Diagnosis of ASD

EEG has been used for ASD research in recent years. Autism may be described as a dynamical disorder and analyzed from the perspective of complex dynamical systems. Measurable changes in cortical excitability may contribute to, or be a manifestation of, connectivity abnormalities. Also, it has been observed that a key feature in autism development is the early brain overgrowth [30] subsequently leading to greater local connectivity and suppressed long-range connectivity. The investigations show evidence of developing shorter range cortico-cortical intra-hemispheric connections with little involvement of connections between hemispheres and cortex and subcortical structures [30]. The behavioural symptoms of autism could be a manifestation of these disrupted neural circuits. In some studies, such as in [31], evidence has been found for supporting the hypothesis of underconnectivity within large-distant networks and also underconnectivity within the local networks. They used the complexity of EEG signals recorded during resting state as a feature to distinguish typically developing children from children with the risk of ASD.

EEG connectivity was attempted for ASD by many researchers such as those in [30]. This was performed on children following 128 electrode EEG recording during face perception task trials to get the individual synchronized state (a.k.a. synchrostates) and their transition sequence. Each synchrostate refers to the time segments where the subjects show a similar emotional face to the face they observe (become happy for observing a happy face, sad for sad face, neutral for neutral face, and fear for an angry face).

They analyzed the EEG from 12 children while processing fearful, happy, and neutral faces. The minimal and maximally occurring synchrostates for each subject are chosen for extraction of brain connectivity features. They use phase synchronization of the continuous wavelet transform (CWT) of EEG signals over the gamma frequency band as the measure of connectivity. These features are then used for classification, using linear discriminant analysis (LDA) and support vector machines (SVM), between these two groups of subjects [30].

The effectiveness of three EEG quantification methods, namely, functional connectivity analysis, spectral power analysis, and information dynamics for diagnosis of ASD have been examined in [32]. However, they concluded that due to high heterogeneity in the results, generalizations could not be inferred and none of the methods alone would be currently useful as a diagnostic tool.

The work in [31] was continued in [33], where the nonlinear (dynamical complexity) features were computed from EEG signals and used as the input to statistical learning systems. In their experiment continuous 30-second segments were selected from the beginning of each of the EEG recordings for each subject when the child was sitting quietly. No splicing of segments or selection based on review was performed. Then, they applied the CWT using Haar wavelet transform to decompose the data into six frequency bands, including the conventional EEG bands. For each band, they computed a broad range of features. These features were sample entropy, detrended fluctuation measure (which is a measure of the 'long-term memory' of a time series), entropy derived from the recurrence plot (associated with Shannon entropy), max line length (related to the largest Lyapunov exponent of a chaotic signal), mean line length (the time that two segments of the recurrence plot trajectory are close to each other), recurrence rate (the probability that a system state recurs in a finite time), determinism (coming from repeating patterns in the system and is an indication of its predictability), laminarity (representing the speed of transitions which instabilities, or the frequency of transitions from one state to another), and trapping time (which is an estimate of the time that a system will remain in a given state) [33]. Finally, they applied an SVM to classify them.

In [34] the authors demonstrated 40 objectively defined EEG coherence factors which reliably separate the neuro-typical controls (healthy subjects) from subjects with autism, and reliable separation of subjects with Asperger's syndrome from all other subjects within the autism spectrum and from neurotypical controls. Each coherence factor represents a particular connectivity pattern. These EEG coherence factors were used prospectively within a large (N = 430) population of subjects with autism in order to determine quantitatively the potential existence of separate clusters within this population.

Their results show that, the control cluster can be discriminated from ASD and Asperger's syndrome subjects reasonably accurately, whereas ASD and Asperger's syndrome clusters have some overlaps. The outcome, however, is very significant.

A step-wise technique for using EEG in classification of ASD has been presented in [35]. In the first stage, they used quantitative EEG (looking at the variations in the spectral components) to select a subset of individuals definitely diagnosed as ASD. In the second stage, the rest were examined by evaluation of their ERPs, where another group was labelled as ASD, and finally, the polysomnography (PSG) recorded (during sleep) from the remaining subjects was analyzed to ensure the maximum discrimination between healthy (control) subjects and those with ASD. Through this analysis the children with seizure and those with ASD were discriminated too, given that there are similar wave patterns for these two brain disorders in children and neonates. For the same reason, the contribution of epilepsy to ASD in developmental neurosciences has grown as an established area of academic and clinical interest [35, 36]. Epilepsy and ASD share several biological pathways that appear to be involved in the disease processes of both. Shared abnormalities are found in gene transcription regulation, cellular growth, synaptic channel function, and maintenance of synaptic structure [37].

The use of more powerful machine learning techniques in EEG recognition is fast growing. In [38] spectral and temporal features have been used by different classifiers including neural networks (NNs) to enhance the accuracy of classification.

In another approach [39] the pre-processed EEG signals were converted to two-dimensional images using higher-order spectra (HOS) in the form of bispectrum. Nonlinear features were extracted thereafter, and then reduced using locality sensitivity discriminant analysis (LSDA) [40]. Significant features were selected from the reduced-dimensionality feature-set using the Student's t-test and were then input to different classifiers. It was found that the probabilistic neural network (PNN) classifier achieves the best classification rate for ASD recognition using only five features [39].

16.2.3 Mood Disorder

Mood disorder often describes all types of depression and BPs and can affect the people of all ages. However, children and teens do not always have the same symptoms as adults. It is harder to diagnose mood disorders in children because they are not always able to express their feelings.

Therapy, antidepressants, and support and self-care can help treat mood disorders. The most popular types of mood disorders are:

- Major depression. Having less interest in usual activities, feeling sad or hopeless, and other symptoms for at least two weeks may indicate depression.
- Dysthymia. This is a chronic, low-grade, depressed, or irritable mood that lasts for at least two years.
- BP. This is a condition in which a person has periods of depression alternating with periods of mania or elevated mood.
- Mood disorder related to another health condition. Many medical illnesses (including cancer, injuries, infections, and chronic illnesses) can trigger symptoms of depression.
- Substance-induced mood disorder. Symptoms of depression that are due to the effects of medicine, drug abuse, alcoholism, exposure to toxins, or other forms of treatment.

Many factors contributing to mood disorders are likely caused by an imbalance of brain chemicals. Life events (such as stressful life changes) may also contribute to a depressed mood. Mood disorders also tend to run in families.

16.2.3.1 EEG for Monitoring Depression

Depression (major depressive disorder) is a common and serious psychiatric disorder that negatively affects person's feeling, thinking, and acting. Depression causes feelings of sadness and/or a loss of interest in activities once enjoyed. It can lead to a variety of emotional and physical problems and can decrease a person's ability to function at work and at home. Depression is considered as a treatable disorder according to the American Psychiatric Association.

Recent findings from EEG analysis have shown that, like many other psychiatric problems, depression affects the normal brain rhythms, brain connectivity, and the brain response to various stimuli. EEG may be used not only for detection or scoring depression but also for monitoring the effect of transcranial magnetic stimulation (TMS) therapy for raising the alpha wave for the patient rehabilitation [41].

In [42] it has been examined whether the complexity of EEG activity, measured by Higuchi's fractal dimension (HFD) and sample entropy (SampEn), differs between healthy subjects, patients in remission, and those in episode phase of recurrent major depression and whether the changes are differentially distributed between the brain hemispheres and the cortical regions.

In their experiments using both male and female patients they measured the HFD and SampEn values for various channels separately across EEG segments and subjects. Then, they applied principal component analysis (PCA) and used only the first three components for classification. From the results, SampEn (when averaged over all 19 electrodes or when only FP1 channel is used) seems to be a better discriminator between depressed and healthy control compared with HFD.

Overall, the authors discovered that depressed patients had higher FD and SampEn complexity compared to healthy subjects. The complexity was higher in patients who were in remission than in those in the acute episode. Altered complexity was present in the frontal and centroparietal regions when compared to the control group. The complexity in frontal and parietal regions differed between the two phases of depressive disorder. This study led to the conclusion that the EEG complexity measure is able to distinguish between the healthy controls, patients in remission and during the depression episodes [42].

In [41] the effect of repetitive TMS (rTMS) on both depression and bipolar subjects has been investigated. It has been shown that HFD, may be a useful marker for evaluation of the rTMS effectiveness and the therapy progress as well as for group differentiation between major depression disorder and bipolar or between responders and non-responders.

In responders, by applying rTMS the HFD for depression reduces for almost the entire frequency range, while for bipolar increases for delta and theta bands. However, for non-responders, FD does not change significantly, whereas it reduces in delta, theta, beta, and gamma bands. Nevertheless, the FD values for responders and non-responders are markedly different for both disorders. Also, the FD values for depression and bipolar cases are significantly different [41], which suggests that the FD can be used for differentiating between depression and BPs.

In addition, the changes of FD under the influence of rTMS allow clinicians to unambiguously conclude whether the effect of stimulation is positive or negative as well as allow therapists to evaluate an optimal time of rTMS.

In a BCI type approach [43] the visual stimuli consisted of both happy-face and sad-face pictures. The male/female and sad/happy population ratios were both 1. Also, the words happy or sad were written on each face centre. In the congruent tasks, the facial expression in the picture was consistent with the meaning of the word (sad or happy). In the incongruent tasks, the facial expression in the face picture was different from the word. One example can be seen in Figure 16.2. The subjects were then asked to determine whether the face was sad or happy and respond by pressing one of the bottoms immediately. The left button is pressed for sad and the right button for happy. The order of the stimuli was pseudo-random, and with equal probabilities of sad and happy.

Following a simple ERP experiment, the parameters of interest were the reaction time (RT) of correctly answered trials and the accuracy of responses (ACR) for both emotional faces and congruent/incongruent states. For both groups (depressed and control), the epochs according to both tasks were extracted and averaged accordingly. The length of each epoch was 1000 ms (200 ms pre-stimulus and 800 ms post-stimulus). N100, P200, N200, P300 and N450 with the expected latency windows of [80, 120], [120, 200], [200, 300], [300, 400] and [400, 600] milliseconds, were respectively selected for analysis and averaged over all the channels. From their resulting ERP waveforms, the decrease in P300 and N450 amplitudes for both incongruent and congruent states is evident.

To estimate the RT and ACR, the authors used the dynamic time warping approach from [44]. The outcome shows approximately 40 ms increase in RT in depressed subjects for both

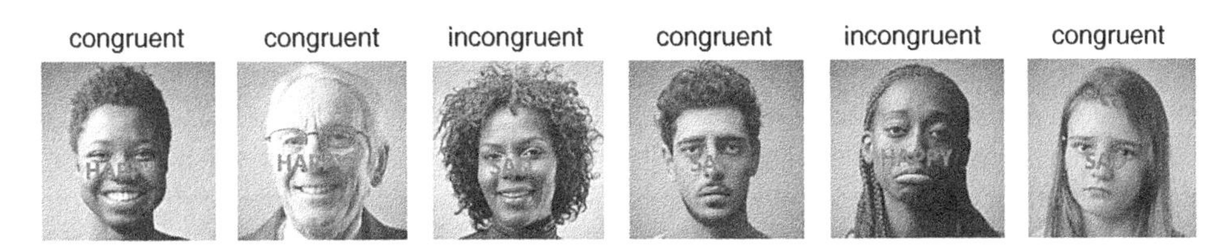

Figure 16.2 Typical faces and labels for congruent and incongruent stimuli used in the stroop experiment for ERP-based depression classification.

congruent and incongruent states. For congruent state about 3% and for noncongruent state approximately 6% decrease in ACR were measured in both cases with considerably higher variance for the depressed subjects.

In a pervasive feature-based approach to EEG-based depression assessment, only three channels of Fp1, Fp2, and Fpz have been used and the earlobe electrode grounded [45]. Six videos including two neutral, two negative, and two positive videos were presented to the subjects each for 90 seconds with 6 seconds breaks between them. A large number of features including linear time and frequency domain features (centroid frequency, relative centroid frequency, absolute centroid frequency, relative power, absolute power, peak, variance, skewness, Kurtosis, and Hjorth) and dynamical (nonlinear) features (power-spectrum entropy, Shannon entropy, correlation dimension, C0-complexity, and Kolmogorov entropy) are computed [45]. For classification of the selected features they tested different classifiers and concluded that the best classification rate (over 76%) could be achieved using k-nearest neighbour (kNN).

It is believed that using various stimuli and applying accurate techniques for single-trial tracking of ERP components can be a very promising approach to monitoring the depression level. Traditionally, the decrease in amplitude of both posterior alpha and parietal mu rhythms is still a robust measure to differentiate between depressed and healthy subjects.

16.2.3.2 EEG for Monitoring Bipolar Disorder

BP, a.k.a. manic-depressive illness, can be seen in one of every 50 adults at some point in their life. BP, including its four types, i.e. classic manic-depressive illness (bipolar I disorder), bipolar II disorder, cyclothymia, and BPs not otherwise classified are common, lifelong, and life-threatening illnesses that are associated with poor functional and clinical outcomes and high suicide rates [46]. It usually starts between the ages of 15 to 25, and rarely after the age of 50. The four types of BP depend on the duration and severity of its episodes, and mood swing [47].

A variety of screening modalities are popular in detection and monitoring of BP such as video/audio based [48], and assessment of social/mood rhythms by the use of self-report instruments [46]. The use of EEG specifically for evaluation of BP and its subtypes has been limited so far.

One example of the application of EEG to BP was addressed in [41], also briefly reviewed in Section 16.2.3.1, where the effect of rTMS on both depression and bipolar subjects was investigated. In their approach they used functional connectivity to distinguish between depression, bipolar, and control.

In another attempt, the researchers attempted classification of the types 1 and 2 of BP [25]. They combined and classified morphological, time, frequency, and time–frequency domain features as well as nonlinear features, such as approximate entropy and mutual information. A reasonable accuracy has been claimed as the result of this research.

Conversely, there are overlaps between BP and schizophrenia symptoms which can cause ambiguity in their diagnosis. In [49] this problem has been considered; a photic stimulation modulated at 16 Hz was presented to two groups of schizophrenic and BD patients for 95 seconds, during which the EEG data were recorded. The steady-state visual evoked potentials (SSVEPs) were evaluated using their statistical measures (mean, skewness, and kurtosis) and classified. The measures from the brain frontal and occipital regions showed a statistically significant difference between the two groups.

16.2.4 Schizophrenia

16.2.4.1 Schizophrenia Symptoms and Management

Schizophrenia is a chronic brain disorder that affects less than 0.8% of the world population. When schizophrenia is active, the patient affected with schizophrenia exhibits a high level of disturbance in thoughts and perceptions and feels extreme difficulty in dealing with relationships. The symptoms can include delusions, hallucinations, disorganized speech, trouble with thinking, and lack of motivation. However, with treatment, most symptoms of schizophrenia improve and the likelihood of a recurrence reduces. While there is no absolute cure for schizophrenia, research is leading to innovative and safer treatments.

Experts try to unravel the disease causes by studying genetics, conducting behavioural research, and using advanced imaging to look at the brain's structure and function. These approaches hold the promise of new, and more effective therapies. While some may argue a diagnostic biomarker for schizophrenia does not yet exist, a company called PsyNova, in collaboration with Rules-Based Medicine, generated a putative blood biomarker assay for the diagnosis of schizophrenia [50]. However, the definition of biomarker changed through time [51]. For example, the Food and Drug Administration (FDA) in collaboration with the National Institute of Health (NIH) Joint Leadership Council convened the FDA-NIH Biomarker Working Group in 2016 which simplified the biomarker definition being considered as 'a defined characteristic that is measured as an indicator of normal biological processes, pathogenic processes, or responses to an exposure or intervention'. In addition, a biomarker should be accessible for its detection and measurement, as would be the case for a plasmatic parameter or a genetic marker, or being detected by histological or image/neuroimaging techniques [51]. With treatment, most people with schizophrenia live with their family, in group homes or on their own. Research has shown that schizophrenia affects men and women fairly equally but may have an earlier onset in males. Rates are similar around the world. People with schizophrenia are more likely to die younger than the general population, mostly because of high rates of co-occurring medical conditions, such as heart disease and diabetes.

When the disease is active, it can be characterized by episodes in which the person is unable to distinguish between real and unreal experiences. As with any illness, the severity, duration, and frequency of symptoms can vary. However, in persons with schizophrenia, the incidence of severe psychotic symptoms often decreases as the person becomes older. Not taking medications as prescribed, the use of alcohol or illicit drugs, and stressful situations tend to increase symptoms. Symptoms fall into three major categories of positive (hallucinations, such as hearing voices or seeing things that do not exit, paranoia and exaggerated or distorted perceptions, beliefs and behaviours), negative (loss or decrease in the ability to initiate plans, speak, express emotion or find pleasure) and disorganized (confused and disordered thinking and speech, trouble with logical thinking and sometimes bizarre behaviour or abnormal movements) symptoms.

Cognition is another area of functioning affected by schizophrenia leading to problems with attention, concentration and memory, and to declining educational performance.

16.2.4.2 EEG as the Biomarker for Schizophrenia

EEG is considerably affected by schizophrenia and therefore, more recently it has been popular in its assessment that. It has been shown that even from the spectral features measured from a single midfrontal EEG channel and applying a suitable time–frequency transform such as Stockwell transform (explained in Chapter 8 of this book), schizophrenic patients can be classified from healthy subjects [52]. A more detailed study of spectral analysis during rest (with both eye open and closed) and task performance can be found in [53]. In this study the delineation of psychotic disorders using delta/alpha frequency activity has been shown which supports the involvement of thalamocortical mechanisms in the psychotic disorders including schizophrenia and BP.

Alterations of EEG gamma activity in schizophrenia have been reported during sensory and cognitive tasks, but it remained unclear whether the changes are present in resting state. In more recent spectral analysis using high-density EEG, the alterations in resting-state gamma activity in patients with schizophrenia have been investigated [54]. Their findings provide support for an increase of gamma power in patients with schizophrenia, and for the association of those increases in several brain regions with more severe psychopathology. Due to the likelihood of the changes in high-frequency oscillations occurring during brain maturation, it is conceivable that neurodevelopmental changes lead to disturbances in gamma activity, which in turn may contribute to the development of schizophrenia. Elevated gamma power may reflect an intrinsic deficit in the temporal coordination of distributed neural activity [54].

In [55] the authors use an evolutionary algorithm optimization approach in the feature optimization and classification of schizophrenia. They use three feature extraction techniques of partial least squares (PLS) nonlinear regression, expectation maximization based principal component analysis (EM-PCA), and isometric mapping (Isomap). The extracted features are further optimized by utilizing four optimization algorithms namely, flower pollination, eagle strategy using different evolution algorithms, backtracking search optimization, and group search optimization. The optimized feature values are then classified using modified Adaboost and Naïve Bayes classifiers. The block diagram of the system can be seen in Figure 16.3.

The individual results show that for normal cases, Isomap features when optimized with the backtracking search optimization algorithm and classified with the modest Adaboost classifier, a classification accuracy as high as 98.77% can be achieved. For the schizophrenia case, the same result can be obtained when Isomap features are optimized with the flower pollination optimization algorithm and classified with the real Adaboost classifier.

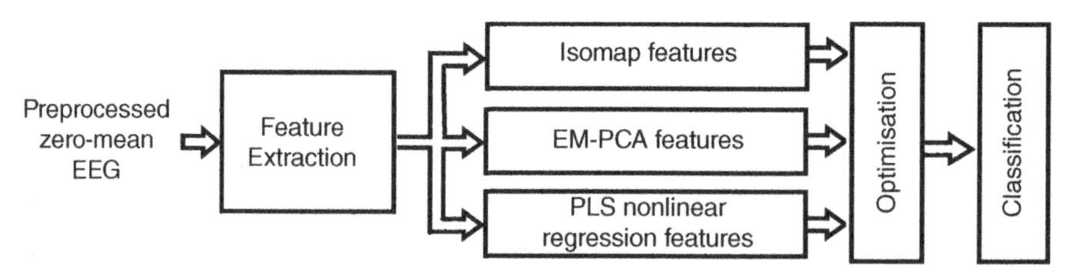

Figure 16.3 Block diagram of the system developed in [55] for classification of schizophrenic and healthy individuals.

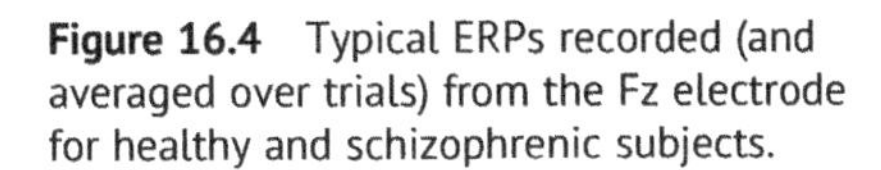

Figure 16.4 Typical ERPs recorded (and averaged over trials) from the Fz electrode for healthy and schizophrenic subjects.

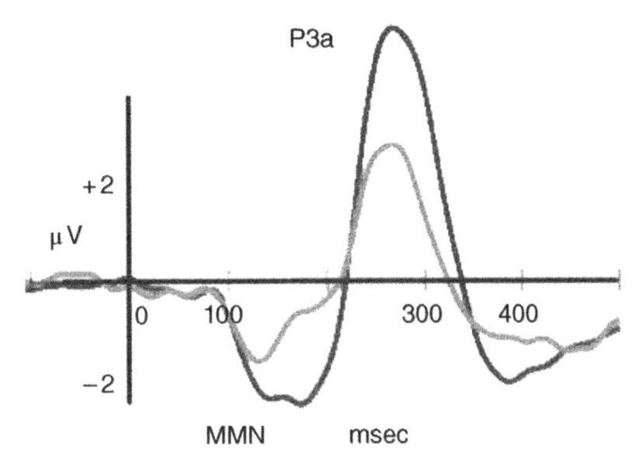

In two almost similar auditory ERP-based studies [56, 57] it has been shown that for schizophrenic subjects the amplitudes of both mismatch negativity (MMN) and P3a subcomponent of P300 ERPs are significantly less than those of healthy subjects. The work in [56] is more in detail as the ERP components have been probed in different brain cortical zones and the components carefully separated using independent component analysis (ICA). It has been shown that the maximum difference between the patient and control can be seen around the dorsal midcingulate of the brain. Typical ERPs recorded (and averaged over trials) from the Fz electrode for healthy and schizophrenic can be seen in Figure 16.4.

Many researchers attempted the use of complexity measures for assessing schizophrenia. Some examples are in [58–62]. Measures of cortical waveform complexity have been used in [59] for EEG channel groups. The right hypothesis in this work is that the difference between complexity levels in healthy and schizophrenic brains changes from the rest state to the state of performing a task. EEGs were analyzed with the classical Lempel-Ziv complexity (LZC) and with the multiscale LZC [63]. Electrodes were grouped in seven regions of interest (ROIs) and it has been concluded that these measures for rest and during task performance are opposite for schizophrenic and control subjects. It has been demonstrated that the LZC values for a schizophrenic are significantly higher during rest and lower during task performance compared with those of healthy subjects. This contrast is better pronounced around frontal, right temporal–parietal, and left temporal–parietal brain regions for the rest state whereas the changes can be seen in all the regions during task performance.

It has been shown that the LZC abnormal temporal and spatial patterns (topographs) are very similar in schizophrenia and depression patients while different from those of healthy subjects [64].

In another recent work researchers examined an auditory task for assessing schizophrenia [65] and came to similar conclusions. In this study, the EEG nonlinear features were estimated under baseline and two paired stimuli conditions using fuzzy entropy (FuzzyEn), focused on the changes in the complexity of sensory gating (SG) between normal controls and schizophrenia patients. By using the SG process the irrelevant stimuli are separated from meaningful ones, which may underlie both sensory overload (which happens when something around us overstimulates one or more of our senses) and the cognitive deficits that are observed in schizophrenic patients.

They found that the FuzzyEn values of schizophrenia patients were higher than those of the controls, in three conditions in the frontal and occipital regions of interest (ROIs) [65].

During the audio task processing the complexities were reduced in the normal controls, but less changes occurred in the patients with schizophrenia. From the perspective of complexity, the suppression ratios of SG in the controls were significantly higher than those in patients with schizophrenia.

The utility of resting EEG measures in characterizing and classification of schizophrenia and BP has been studied in [66]. In this study, the spectral features and coherence are used to study both schizophrenia and BP. The authors concluded that both populations exhibited increased alpha coherence within the hemisphere relative to controls. However, BP was uniquely characterized by increased high-frequency power while schizophrenia was uniquely characterized by increased low frequency connectivity within and across the two brain hemispheres.

In [67] the authors studied the changes in resting-state theta band source distribution and functional connectivity in remitted schizophrenia. They found significantly increased current source density in the dominant anterior cingulate cortex. In addition, they observed increased connectivity between the inferior parietal lobe bilaterally and between the left inferior parietal lobe and right middle frontal gyrus. This might lead to a new conclusion that even during the remission phase the schizophrenia patients have aberrant regional theta band current source density and functional connectivity.

16.2.5 Anxiety (and Panic) Disorder

16.2.5.1 Definition and Symptoms

Anxiety is a normal reaction to stress and can be beneficial in some situations. It can alert us to dangers and help us prepare and pay attention. Anxiety disorders differ from normal feelings of nervousness or anxiousness and involve excessive fear or anxiety. Anxiety disorders are the most common mental disorder and affect nearly 30% of adults at some point in their lives. In addition, anxiety disorders are highly comorbid with one another and with other psychiatric illnesses. Anxiety disorders usually begin during childhood, result in significant suffering and disability, and are associated with a chronic and recurrent lifetime course [68]. According to data from the National Comorbidity Survey Replication study [69] of individuals at least 18 years of age, estimates of the 12-month prevalence of anxiety disorders are 18%, and like mood disorders, there is an approximate two-to-one female prevalence during women's reproductive years. In children and adolescents, the lifetime prevalence of anxiety disorders is estimated to be between 15 and 20% [64].

Within the category of anxiety disorders, DSM-5 includes separation anxiety disorder, selective mutism, social anxiety disorder, panic disorder, agoraphobia, generalized anxiety disorder, substance/medication-induced anxiety disorder, and anxiety disorder due to another medical condition.

Nevertheless, anxiety disorders are treatable and a number of effective treatments are available [64]. Psychotherapy (a.k.a. talk therapy or psychological counselling,) involves working with a therapist to reduce your anxiety symptoms. Cognitive-behavioural therapy is the most effective form of psychotherapy for generalized anxiety disorder. During the treatment the EEG biomarkers can be monitored and the therapy adjusted accordingly.

People with anxiety disorders frequently have intense, excessive, and persistent worry and fear about their daily and future situations. Often, anxiety disorders involve repeated episodes of sudden feelings of intense anxiety and fear or terror that reach a peak within minutes (panic attacks). Many people with anxiety experience symptoms of more than one type of anxiety condition and may experience depression as well. The anxiety types are mainly, generalized anxiety disorder, social anxiety, specific phobias, panic disorder, obsessive–compulsive disorder, and post-traumatic stress disorder.

According to the American Psychiatry Association, although each anxiety disorder has unique characteristics, most respond well to two types of treatment: psychotherapy, or 'talk therapy,' and medications.

16.2.5.2 EEG for Assessing Anxiety

Looking at the EEG spectral power, the EEG gamma band has been assessed for general anxiety disorder (GAD) to differentiate worry from baseline and relaxation [70]. It has been shown that the gamma power is significantly higher than both baseline and relaxed. Moreover, from their results it can be seen that the gamma power reduces to very close to normal after the therapy treatment.

In a hybrid approach the asymmetry between the two brain hemispheres in the frontal and occipital as well as the coherency values between left and right channels for theta, alpha, and beta bands and Hjorth mobility have been measured from the EEG multichannel signals during video watching (causing arousal or valence) [71]. The asymmetry values for frontal, central, and occipital zones are different (see Figure 16.5).

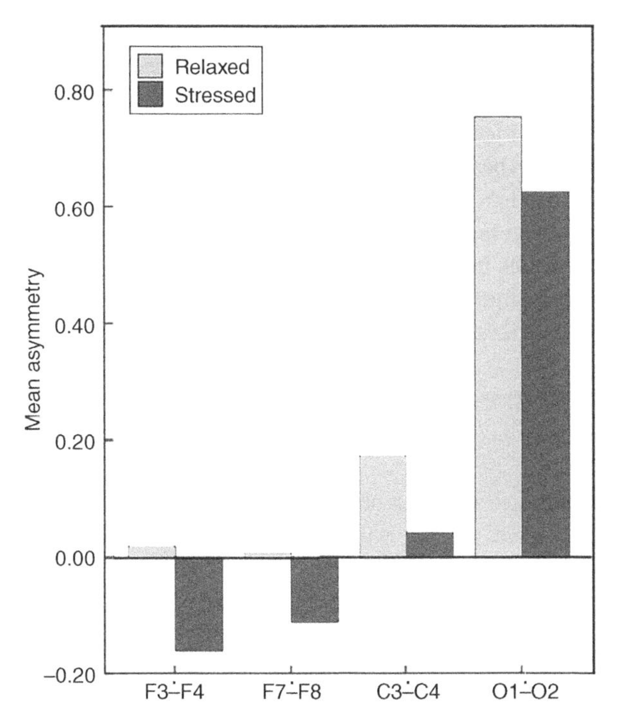

Figure 16.5 The mean asymmetry between the left and right brain hemisphere channels for frontal (F), central (C), and occipital (O) regions. (Source: The figure has been modified from [71]).

In addition, coherence values can be used to discriminate between states mainly between inter-hemispheric locations rather than intra-hemispheric. Hjorth parameters did not show differences during univariate statistical analysis but mobility was selected by a feature selection procedure as a robust feature in the combined data subset [71].

In another trial the frontal asymmetry and beta/alpha ratio were estimated with respect to the valence–arousal domain [72]. Based on this study it was confirmed that the driver's perspective video could induce anxiety in the viewers. The researchers found smaller alpha power reflecting more cognitive control, right dominance of theta and beta oscillations, and greater difference between beta and alpha.

More detailed analysis of linear and nonlinear features has been carried out in [73] to score the level of anxiety which confirms previous achievements.

In [74] the eye movement information has been added to the EEG's for better recognition of anxiety in a population of age between 8 and 17. For fusing these two modalities they used group sparse canonical correlation analysis (GSCCA) to investigate group structure information among the dense EEG and eye movement features. The authors used PCA and the inexact augmented Lagrange multipliers method (ALM) to mitigate the sparse noise from the data. The remaining noisy segments were removed manually. Eye movement data were detected using a dispersion-based and fixed-length moving interval algorithm. The power-spectrum features for the six frequency bands of namely, delta (1–3 Hz), theta (4–7 Hz), slow alpha (8–10 Hz), alpha (8–13 Hz), beta (14–30 Hz), and gamma (31–50 Hz) were derived from the EEGs. They also grouped their participants into different age groups to have more detailed information and results. Finally, they grouped their spectral features according to 13 regions shown in Figure 16.6.

In addition, 21 features were extracted from the eye movement and divided into four groups reflecting blinks, saccades, fixation, and pupil diameter. Saccades and fixation were detected using a dispersion-based and fixed-length moving interval algorithm provided by sample-matrix inversion (SMI). A blink can be regarded as a special case of fixation, where the pupil diameter is either zero or outside a dynamically computed valid pupil, or the horizontal and vertical gaze positions are zero.

For extracting the new combined features from the two modalities using GSCCA, consider $\mathbf{X} \in R^{n \times p}$ and $\mathbf{Y} \in R^{n \times q}$, where n is the number of samples and p and q are respectively the number of EEG and eye movement features, to be the EEG and eye movement feature matrices. Also consider:

$$\mathbf{C}_{\mathbf{X},\mathbf{Y}} = S_{\mathbf{XX}}^{-1/2} S_{\mathbf{XY}} S_{\mathbf{YY}}^{-1/2} = \sum_{i=1}^{k} d_i \mathbf{u}_i \mathbf{v}_i^T \tag{16.4}$$

where $S_{\mathbf{XY}}$, $S_{\mathbf{XY}}$, and $S_{\mathbf{XY}}$ are the covariance matrices of $\mathbf{X}$ and $\mathbf{Y}$, and cross-covariance matrix of $\mathbf{XY}$ respectively. The GSCCA solves the following optimization problem:

$$\min_{\mathbf{u},\mathbf{v}} \left(\|\mathbf{C}_{\mathbf{X},\mathbf{Y}} - d\mathbf{u}\mathbf{v}^T\|_F^2 + \lambda_1 \|\mathbf{u}\|_G + \tau_1 \|\mathbf{u}\|_1 + \lambda_2 \|\mathbf{v}\|_G + \tau_2 \|\mathbf{v}\|_2 \right)$$
$$\text{subject to } \|\mathbf{u}\|_2^2 \leq 1 \text{ and } \|\mathbf{v}\|_2^2 \leq 1 \tag{16.5}$$

Then, the new feature matrices $\mathbf{X}_{new}$ and $\mathbf{Y}_{new}$ are estimated by the following projections:

$$\mathbf{X}_{new} = \mathbf{X}\left(S_{\mathbf{XX}}^{-1/2} \mathbf{u}\right) \tag{16.6a}$$

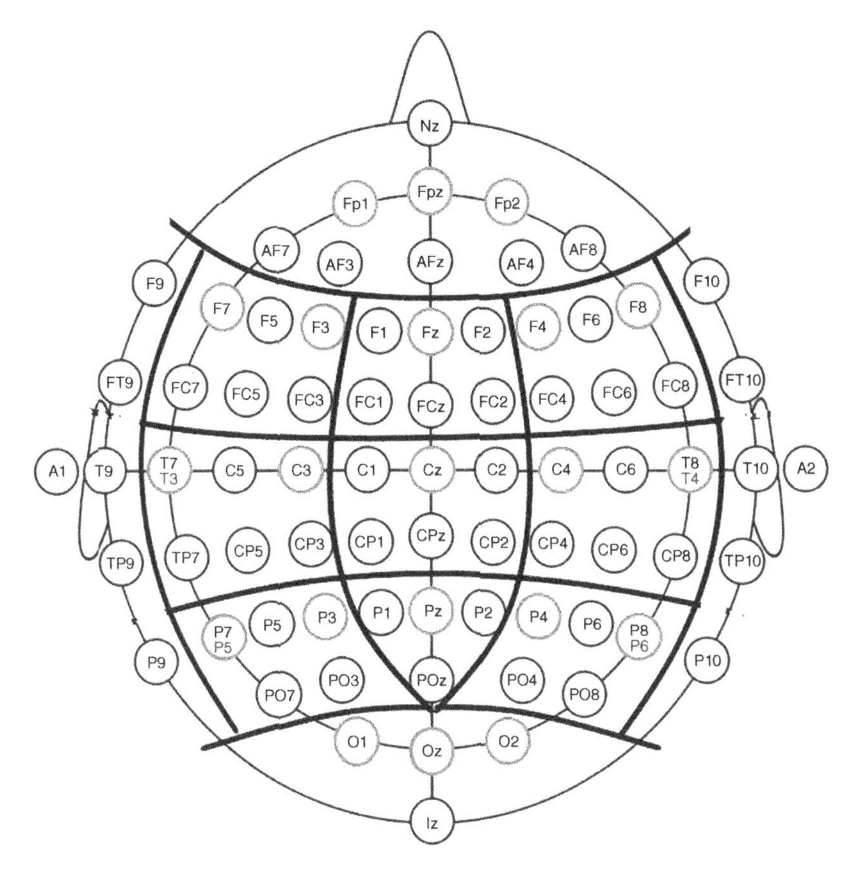

Figure 16.6 The 13 EEG electrode groups used in [74].

$$\mathbf{Y}_{new} = \mathbf{Y}\left(S_{\mathbf{YY}}^{-1/2}\mathbf{v}\right) \tag{16.6b}$$

This will give the canonical-correlation analysis (CCA) feature matrix as $\mathbf{Z} = [\mathbf{X}_{new}; \mathbf{Y}_{new}]$. kNN and SVM classifiers have then been tried to classify the features. They applied the above method to all the frequency bands and across age groups and provided detailed results which show accuracies above 80% for higher frequency bands [74].

Even without having anxiety disorder the human becomes anxious for doing or not doing a task. This anxiety directly affects the brain and therefore by EEG analysis can be scored. An interesting application using convolutional NNs can be seen in [75].

16.2.6 Insomnia

16.2.6.1 Symptoms of Insomnia

This is the only psychiatric disorder which directly affects sleep. According to the National Sleep Foundation (https://www.sleepfoundation.org/insomnia/what-insomnia), insomnia is difficulty falling asleep or staying asleep, even when a person has the chance to do so. People with insomnia can feel dissatisfied with their sleep and usually experience one or more of the following symptoms: fatigue, low energy, difficulty concentrating, mood disturbances, and decreased performance in work or at school. It can be acute or chronic. Acute insomnia is brief and often happens because of life circumstances. Chronic insomnia, conversely, is disrupted sleep that occurs at least three nights per week and lasts at least three months. Chronic insomnia disorders can have many causes. Changes in the environment, unhealthy sleep habits, shift work, other clinical disorders, and certain medications could lead to a long-term pattern of insufficient sleep.

16.2.6.2 EEG for Insomnia Analysis

The study of sleep abnormality using EEG has a long history. One example in the early use of EEG for sleep study can be found in [76]. However, often many other data modalities recorded through PSG are also used together or without EEG for sleep analysis and detection of sleep disorders including insomnia [77]. In [78] actigraphy, which measures the level of physical activity using movement counts per sample time interval, has also been used for insomnia detection.

Single and two-channel signals have been considered by a machine learning system for detection of insomnia from sleep EEGs of 41 controls and 42 primary insomnia patients [79]. The authors used two different approaches, one by using the entire sleep recording irrespective of the sleep stage (*stage-independent classification*), and the second one by considering the EEG data from insomnia-impacted specific sleep stages only (*stage-dependent classification*).

They used statistical measures (mean, variance, skewness, and kurtosis), zero crossing rate, Hjorth parameters (activity, mobility, and complexity), amplitude measures (maximum and mean of the first and second derivate, peak-to-peak amplitude, root mean square, RMS, amplitude, and peak to RMS), and the spectral features for five conventional EEG bands. During their stage-dependent scheme they also involved the estimated sleep stages (through a first-step classifier) into their deep neural network (DNN) classification algorithm [79]. The researcher later used hypnogram to score the sleep stages before applying an SVM classifier for insomnia/control classification [80]. A hypnogram is a form of PSG, a graph that represents the stages of sleep as a function of time from the EEG during sleep.

16.2.7 Schizotypal Personality Disorder

16.2.7.1 What Is Schizotypal Disorder?

Schizotypal personality disorder (a.k.a. schizotypy) is a chronic disorder with manifestations beginning in childhood or adolescence and causes cognitive-perceptual problems (e.g. magical thinking and paranoia), oddness (e.g. odd rapport, affect, and speech), and interpersonal problems (e.g. social anxiety and a lack of close friends) and it is difficult to treat.

The concept of schizotypy has its origins in the early years of the twentieth century when researchers observed that nonpsychotic relatives of schizophrenics showed milder forms of schizophrenic-like symptoms. Today, schizotypy is studied as a multidimensional personality trait on a dimensional continuum with schizophrenia. Individuals with a schizotypal personality are at higher risk for the development of schizophrenia. Although not psychotic, they are considered psychosis-prone [81].

The disorder is associated with significant disability, as well as a wide range of psychiatric comorbidities. Unlike in schizophrenia, the positive symptoms in schizotypy are expressed in forms more commonly experienced in everyday life, such as: 'magical thinking', including a belief that telepathy (mind-to-mind communication without the use of the five senses) is possible; pseudo-hallucinations, such as having had a 'sense of presence' (perhaps evil and foreboding) despite no person being physically present; perceptual aberrations, such as seeing shapes and colours in the dark when nothing is there; and cognitive aberrations, for example, one's thoughts seeming to be so vivid that one can almost hear them (analogous with the most common form of hallucination in diagnoses of schizophrenia – hearing voices). Conversely, the negative symptoms of schizophrenia are characterized by an absence of pleasure and engagement with aspects of everyday life (avolition, anhedonia, alogia). In schizotypy, such experiences include physical anhedonia, gaining little or no pleasure from the senses or from experiences that are usually deemed to be pleasurable (such as enjoying the taste of new foods or dancing), and social anhedonia, preferring solitary activities to spending time with others and not feeling close to other people [81].

This disorder may evolve or lead to other abnormalities such as ADHD or schizophrenia as the age increases. The early diagnosis of this disease was through filling up a questionnaire which was initially in the German language, called schizotypal personality questionnaire (SPQ), provided by Raine [82] and later used by Wolff [83].

16.2.7.2 EEG Manifestation of Schizotypal

The study of schizotypy through brain screening has not become popular yet due to its complexity and the difficulty in managing the patients with schizotypal personality.

A review in [84] has identified a number of perceptual, cognitive, and motor functions that deteriorate in relation to higher schizotypy. However, while there is generally consistent evidence of impairments in certain perceptual-motor abnormalities, there is considerable cross-study variability in higher cognitive deficits such as working memory or top-down attentional control. These evidences will soon lead into more detailed investigation of brain activity using EEG and other functional neuroimaging modalities.

With the help of low-resolution electromagnetic tomography (LORETA), the researchers have investigated the possibility of detecting hypofrontality in schizophrenic patients [85]. They analyzed resting EEGs of unmedicated schizophrenics, normal, schizotypal personality, and depressive subjects. A significant increase of delta activity was found in schizophrenic patients over the whole cortex, most strongly in the anterior cingulate gyrus and temporal lobe (fusiform gyrus). Both schizotypal subjects and depressive subjects showed significantly less delta, theta and beta activity in the anterior cingulum, a decrease of alpha1 activity in the right temporal lobe and a decrease of alpha2 (11–14 Hz) activity in the left temporal lobe. Based on the results, there is evidence for a complex frequency-dependent spatial pattern of hyperactivation in schizotypal subjects and depressive patients.

16.3 Summary

The study of psychiatric disorders is probably the most challenging one due to similarity and overlaps among them and the complexity of brain function in both resting state and during the brain mental or motor activity and workload. Many of the techniques used for the analysis of psychiatric disorders are used for many other brain assessment applications. However, surprisingly, these methods involving machine learning applied to features such as connectivity, complexity, spectral power changes, or variations in the amplitude or latency of ERPs, perform very well for the classification of most psychiatric diseases from healthy individuals. The research in schizotypy is in its enfant stage and is currently under research.

References

1 Mullin, A.P., Gokhale, A., Moreno-De-Luca, A. et al. (2013). Neurodevelopmental disorders: mechanisms and boundary definitions from genomes, interactomes and proteomes. *Translational Psychiatry* 3: e329, 6 pages.

2 Darmesteter, J. (translator) (1898). *Sacred Books of the East* (ed. American). New York, Christian Literature Company. http://www.avesta.org/vendidad/vd7sbe.htm (accessed 1 August 2021).

3 Tubbs, R.S., Shoja, M.M., Loukas, M., and Oakes, W.J. (2007). Abubakr Muhammad Ibn Zakaria Razi, Rhazes (865–925 AD). *Child's Nervous System* 23: 1225–1226.

4 Compier, A.H. (2012). Rhazes in the renaissance of Andreas Vesalius. *Medical History* 56: 3–25.

5 Magnavita, J.J. (2006). In search of the unifying principles of psychotherapy: conceptual, empirical, and clinical convergence. *American Psychologist* 61 (8): 882–892.

6 Jasper, H.H., Solomon, P., and Bradley, C. (1938). Electroencephalographic analyses of behavior problem children. *American Journal of Psychiatry* 95 (3): 641–658.

7 Danielson, M.L., Bitsko, R.H., Ghandour, R.M. et al. (2018). Prevalence of parent-reported ADHD diagnosis and associated treatment among U.S. children and adolescents, 2016. *Journal of Clinical Child & Adolescent Psychology* 47 (2): 199–212.

8 Simon, V., Czobor, P., Bálint, S. et al. (2009). Prevalence and correlates of adult attention-deficit hyperactivity disorder: a meta-analysis. *The British Journal of Psychiatry* 194 (3): 204–211.

9 Lenartowicz, A. and Loo, S.K. (2014 November). Use of EEG to diagnose ADHD. *Current Psychiatry Reports* 16 (11): 498. 19 pages.

10 Barry, R.J., Clarke, A.R., and Johnstone, S.J. (2003). A review of electrophysiology in attention-deficit/hyperactivity disorder: I. Qualitative and quantitative electroencephalography. *Clinical Neurophysiology* 114 (2): 171–183.

11 Lubar, J.F. (1991). Discourse on the development of EEG diagnostics and biofeedback for attention-deficit/hyperactivity disorders. *Biofeedback and Self-Regulation* 16 (3): 201–225.

12 Snyder, S.M. and Hall, J.R. (2006). A meta-analysis of quantitative EEG power associated with attention-deficit hyperactivity disorder. *Journal of Clinical Neurophysiology* 23 (5): 440–455.

13 Boutros, N., Fraenkel, L., and Feingold, A. (2005). A four-step approach for developing diagnostic tests in psychiatry: EEG in ADHD as a test case. *The Journal of Neuropsychiatry and Clinical Neurosciences* 17 (4): 455–464.

14 Arns, M., Conners, C.K., and Kraemer, H.C. (2013). A decade of EEG theta/Beta ratio research in ADHD: a meta-analysis. *Journal of Attention Disorders* 17 (5): 374–383.

15 Nuwer, M.R., Buchhalter, J., and Shepard, K.M. (2016). Quantitative EEG in attention-deficit/hyperactivity disorder. *Neurology Clinical Practice* 6: 543–548.

16 Alex, A., Coelli, S., Bianchi, A.M. et al. (2017). EEG analysis of brain activity. In Attention Deficit Hyperactivity Disorder During an Attention Task. *2017 IEEE 3rd International Forum on Research and Technologies for Society and Industry (RTSI)*, Modena, Italy.

17 Kiiski, H., Bennett, M., Rueda-Delgado, L.M. et al. (2020). EEG spectral power, but not theta/beta ratio, is a neuromarker for adult ADHD. *European Journal of Neuroscience* 51: 2095–2109.

18 Zou, H. and Hastie, T. (2005). Regularization and variable selection via the elastic net. *Journal of the Royal Statistical Society: Series B (Statistical Methodology)* 67: 301–320.

19 Tombor, L., Kakuszi, B., Papp, S. et al. (2018). Decreased resting gamma activity in adult attention deficit/hyperactivity disorder. *The World Journal of Biological Psychiatry* 20 (9): 691–702.

20 Johnstone, S.J., Barry, R.J., and Clarke, A.R. (2013). Ten years on: a follow-up review of ERP research in attention-deficit/hyperactivity disorder. *Clinical Neurophysiology* 124 (4): 644–657.

21 Mueller, A., Candrian, G., Grane, V.A. et al. (2011). Discriminating between ADHD adults and controls using independent ERP components and a support vector machine: a validation study. *Nonlinear Biomedical Physics* 5: 5.

22 Marquardt, L., Eichele, H., Lundervold, A.J. et al. (2018). Event-related-potential (ERP) correlates of performance monitoring in adults with attention-deficit hyperactivity disorder (ADHD). *Frontiers in Psychology* 9: 485.

23 Leroy, A., Petit, G., Zarka, D. et al. (2018). EEG dynamics and neural generators in implicit navigational image processing in adults with ADHD. *Neuroscience* 373: 92–105.

24 Ernesto, P.-S., Kevin, D., Volker, H., and Tass Peter, A. (2007). SwLORETA: a novel approach to robust source localization and synchronization tomography. *Physics in Medicine and Biology* 52 (7): 1783–1800.

25 Khaleghi, A., Sheikhani, A., Mohammadi, M.R. et al. (2015). EEG classification of adolescents with type I and type II of bipolar disorder. *Australasian Physical & Engineering Sciences in Medicine* 38: 551–559.

26 Sadatnezhad, K., Boostani, R., and Ghanizadeh, A. (2011). Classification of BMD and ADHD patients using their EEG signals. *Expert Systems with Applications* 38 (3): 1956–1963.

27 Lenartowicz, A., Delorme, A., Walshaw, P.D. et al. (2014). Electroencephalography correlates of spatial working memory deficits in attention-deficit/hyperactivity disorder: vigilance, encoding, and maintenance. *The Journal of Neuroscience* 34 (4): 1171–1182.

28 Strauß, M., Ulke, C., Paucke, M. et al. (2018). Brain arousal regulation in adults with attention-deficit/hyperactivity disorder (ADHD). *Psychiatry Research* 261: 102–108.

29 Just, M.A., Cherkassky, V.L., Keller, T.A., and Minshew, N.J. (2004). Cortical activation and synchronization during sentence comprehension in high-functioning autism: evidence of underconnectivity. *Brain* 127: 1811–1821.

30 Jamal, W., Das, S., Oprescu, I.-A. et al. (2014). Classification of autism spectrum disorder using supervised learning of brain connectivity measures extracted from synchrostates. *Journal of Neural Engineering* 11: 046019. (19pp).

31 Bosl, W., Tierney, A., Tager-Flusberg, H., and Nelson, C. (2011). EEG complexity as a biomarker for autism spectrum disorder risk. *BMC Medicine* 9: 18.

32 Gurau, O., Bosl, W.J., and Newton, C.R. (2017). How useful is electroencephalography in the diagnosis of autism spectrum disorders and the delineation of subtypes: a systematic review. *Frontiers in Psychiatry* 8: 121.

33 Bosl, W.J., Tager-Flusberg, H., and Nelson, C.A. (2018). EEG analytics for early detection of autism spectrum disorder: a data-driven approach. *Scientific Reports* (8): 6828.

34 Duffy, F.H. and Als, H. (2019). Autism, spectrum or clusters? An EEG coherence study. *BMC Neurology* 19: 27, 13 pages.

35 Swatzyna, R.J., Boutros, N.N., Genovese, A.C. et al. (2019). Electroencephalogram (EEG) for children with autism spectrum disorder: evidential considerations for routine screening. *European Child & Adolescent Psychiatry* 28: 615–624.

36 Trojaborg, W. (1966). Focal spike discharges in children, a longitudinal study. *Acta Paediatrica Scandinavica. Supplement* 55: 1–13.

37 Tuchman, R.F. and Rapin, I. (1997). Regression in pervasive developmental disorders: seizures and epileptiform electroencephalogram correlates. *Pediatrics* 99: 560–566.

38 Sinha, T., Munot, M.V., and Sreemathy, R. (2019). An efficient approach for detection of autism spectrum disorder using electroencephalography signal. *IETE Journal of Research, 2019, 10 pages, ISSN: 0377–2063 (Print) 0974-780X (Online).* Pub Taylor & Francis. http://dx.doi.org/10.1080/03772063.2019.1622462.

39 Pham, T.-H., Vicnesh, J., En Wie, J.K. et al. (2020). Autism spectrum disorder diagnostic system using HOS bispectrum with EEG signals. *International Journal of Environmental Research and Public Health* 17: 971, 15 pages, doi:https://doi.org/10.3390/ijerph17030971.

40 Cai, D., He, X., Zhou, K. et al. (2007). Locality sensitive discriminant analysis. *Proceedings of the 20th International Joint Conference on Artificial Intelligence*, Hyderabad, India (6–12 January 2007), 708–713. IJCAI: San Francisco, CA, USA.

41 Lebiecka, K., Zuchowicz, U., Wozniak-Kwasniewska, A. et al. (2018). Complexity analysis of EEG data in persons with depression subjected to transcranial magnetic stimulation. *Frontiers in Physiology* 9: 1385.

42 Čuki, M., Stoki, M., Radenkovi, S. et al. (2020). Nonlinear analysis of EEG complexity in episode and remission phase of recurrent depression. *Wiley International Journal of Methods in Psychiatric Research* 29: e1816, 11 pages.

43 Guo, Z., Long, H., Yao, L. et al. (2017). Abnormal EEG-based functional connectivity under a face-word stroop task in depression. *Proceedings of the 2017 IEEE International Conference on Bioinformatics and Biomedicine (BIBM).* Kansas City, MO, USA.

44 Gupta, L., Molfese, D.L., Tammana, R., and Simos, P.G. (1996). Nonlinear alignment and averaging for estimating the evoked potential. *IEEE Transactions on Bio-medical Engineering* 43: 348.

45 Cai, H., Han, J., Chen, Y. et al. (2018). A pervasive approach to EEG-based depression detection. *Hindawi Complexity* 2018: 5238028, 13 pages.

46 Matthews, M., Abdullah, S., Murnane, E. et al. (2016). Development and evaluation of a smartphone-based measure of social rhythms for bipolar disorder. *Assessment* 23 (4): 472–483. https://doi.org/10.1177/1073191116656794 (NHS Public Access).

47 World Health Organization (2018). World Health Organization Factsheet – Mental Disorders, http://www.who.int/news-room/fact-sheets/detail/mental-disorders (accessed 19 August 2021).

48 Abaei, N. and Osman, H.A. (2020). A hybrid model for bipolar disorder classification from visual information. *Proceedings of the International Conference on Acoustics, Speech, and Signal Processing (ICASSP)* 2020. Canada.

49 Alimardani, F., Cho, J.-H., Boostani, R., and Hwang, H.-J. (2018). Classification of bipolar disorder and schizophrenia using steady-state visual evoked potential based features. *IEEE Access* 6: 40379.

50 Martins-de-Souza, D. (2012). Translational strategies to schizophrenia from a proteomic perspective. *Translational Neuroscience* 3: 300–302.

51 Garcıa-Gutiérrez, M.S., Navarrete, F., Sala, F. et al. (2020). Biomarkers in psychiatry: concept, definition, types and relevance to the clinical reality. *Frontiers in Psychiatry* 11: 432. https://doi.org/10.3389/fpsyt.2020.00432.

52 Dvey-Aharon, Z., Fogelson, N., Peled, A., and Intrator, N. (2015). Schizophrenia detection and classification by advanced analysis of EEG recordings using a single electrode approach. *PLoS One* 10 (4): e0123033. https://doi.org/10.1371/journal.pone.0123033.

53 Howells, F.M., Temmingh, H.S., Hsieh, J.H. et al. (2018). Electroencephalographic delta/alpha frequency activity differentiates psychotic disorders: a study of schizophrenia, bipolar disorder and methamphetamine-induced psychotic disorder. *Translational Psychiatry* 8: 75. 11 pages.

54 Baradits, M., Kakuszi, B., Bálint, S. et al. (2019). Alterations in resting-state gamma activity in patients with schizophrenia: a high-density EEG study. *European Archives of Psychiatry and Clinical Neuroscience* 269: 429–437.

55 Prabhakar, S.K., Rajaguru, H., and Lee, S.-W. (2020). A framework for schizophrenia EEG signal classification with nature inspired optimization algorithms. *IEEE Access* 8: 39875.

56 Rissling, A.J., Miyakoshi, M., Sugar, C.A. et al. (2014). Cortical substrates and functional correlates of auditory deviance processing deficits in schizophrenia. *Neuroimage. Clinical* 6: 424–437.

57 Light, G.A., Swerdlowa, N.R., Thomas, M.L. et al. (2015). Validation of mismatch negativity and P3a for use in multi-site studies of schizophrenia: characterization of demographic, clinical, cognitive, and functional correlates in COGS-2. *Elsevier Journal of Schizophrenia Research* 163: 63–72.

58 Ibáñez-Molina, A.J., Lozano, V., Soriano, M.F. et al. (2018). EEG multiscale complexity in schizophrenia during picture naming. *Frontiers in Physiology* 9: 1213, 12 pages.

59 Lempel, A. and Ziv, J. (1976). On the complexity of finite sequences. *IEEE Transactions on Information Theory* 22: 75–81. https://doi.org/10.1109/TIT.1976.1055501.

60 Lee, Y.J., Zhu, Y.S., Xu, Y.H. et al. (2001). Detection of non-linearity in the EEG of schizophrenic patients. *Clinical Neurophysiology* 112: 1288–1294.

61 Li, Y., Tong, S., Liu, D. et al. (2008). Abnormal EEG complexity in patients with schizophrenia and depression. *Clinical Neurophysiology* 119: 1232–1241. https://doi.org/10.1016/j.clinph.2008.01.104.

62 Fernández, A., Gómez, C., Homero, R., and López-Ibor, J.J. (2013). Complexity and schizophrenia. *Progress in Neuro-Psychopharmacology & Biological Psychiatry* 45: 267–276. https://doi.org/10.1016/j.pnpbp.2012.03.015.

63 Ibáñez-Molina, A., Iglesias-Parro, S., Soriano, M.F., and Aznarte, J.I. (2015). Multiscale Lempel-ziv complexity for EEG measures. *Clinical Neurophysiology* 126: 541–548. https://doi.org/10.1016/j.clinph.2014.07.012.

64 Lia, Y., Tongb, S., Liu, D. et al. (2008). Abnormal EEG complexity in patients with schizophrenia and depression. *Clinical Neurophysiology* 119: 1232–1241.
65 Xiang, J., Tian, C., Niu, Y. et al. (2019). Abnormal entropy modulation of the EEG signal in patients with schizophrenia during the auditory paired-stimulus paradigm. *Frontiers in Neuroinformatics* 13: 4. https://doi.org/10.3389/fninf.2019.00004.
66 Kam, J.W.Y., Bolbecker, A.R., O'Donnell, B.F. et al. (2013). Resting state EEG power and coherence abnormalities in bipolar disorder and schizophrenia. *Journal of Psychiatric Research* 47 (12): 1893–1901. https://doi.org/10.1016/j.jpsychires.2013.09.009.
67 Umesh, D.S., Tikka, S.K., Goyal, N. et al. (2016). Resting state theta band source distribution and functional connectivity in remitted schizophrenia. *Neuroscience Letters* 630: 199–202.
68 Bandelow, B., Michaelis, S., and Wedekind, D. (2017). Treatment of anxiety disorders. *Dialogues of Clinical Neuroscience* 19 (2): 93–107.
69 Byers, A.L., Yaffe, K., Covinsky, K.E. et al. (2010). High occurrence of mood and anxiety disorders among older adults: the national comorbidity survey replication. *Archives of General Psychiatry* 67 (5): 489–496. https://doi.org/10.1001/archgenpsychiatry.2010.35.
70 Oathes, D.J., Ray, W.J., Yamasaki, A.S. et al. (2008). Worry, generalized anxiety disorder, and emotion: evidence from the EEG gamma band. *Biological Psychology* 79 (2): 165–170. https://doi.org/10.1016/j.biopsycho.2008.04.005.
71 Giannakakis, G., Grigoriadis, D., and Tsiknakis, M. (2015). Detection of stress/anxiety state from EEG features during video watching. *Proceedings of the Engineering in Medicine and Biology Conference*. EMBC.
72 Lee, S., Lee, T., Yang, T. et al. (2018). Neural correlates of anxiety induced by environmental events during driving. *Proceedings of TENCON 2018–2018 IEEE Region 10 Conference*, Jeju, Korea (28–31 October 2018).
73 Li, Z., Wu, X., Xu, X. et al. (2019). The recognition of multiple anxiety levels based on electroencephalograph. *IEEE Transactions on Affective Computing*. Preprint 1–1, 12p. https://www.doi.org/10.1109/TAFFC.2019.2936198.
74 Zhang, X., Pan, J., Shen, J. et al. (2020). Fusing of electroencephalogram and eye movement with group sparse canonical correlation analysis for anxiety detection. *IEEE Transactions on Affective Computing*. 2981440. https://www.doi.org/10.1109/TAFFC.2020.
75 Wang, Y., McCane, B., McNaughton, N. et al. (2019). AnxietyDecoder: an EEG-based Anxiety Predictor using a 3-D Convolutional Neural Network. *International Joint Conference on Neural Networks*, 14–19. Budapest, Hungary: IJCNN.
76 Agnew, H.W., Webb, W.B., and Williams, R.L. (1966). The first night effect: an EEG study of sleep. *Psychophysiology* 2: 263–266.
77 Sanei, S., Jarchi, D., and Constantinides, A.G. (2020). *Body Sensor Networking, Design and Algorithms*. Wiley.
78 Angelova, M., Karmakar, C., ZHU, Y. et al. (2020). Automated method for detecting acute insomnia using multi-night actigraphy data. *IEEE Access* 8: 74413.
79 Shahin, M., Ahmed, B., Tmar-Ben Hamida, S. et al. (2017). Deep learning and insomnia: assisting clinicians with their diagnosis. *IEEE Journal of Biomedical and Health Informatics* 21 (6): 1546–1553.

80 Mulaffer, L., Shahin, M., Glos, M. et al. (2017). Comparing two insomnia detection models of clinical diagnosis techniques. *2017 39th Annual International Conference of the IEEE Engineering in Medicine and Biology Society (EMBC)*, South Korea.

81 Holt, N.J. (2020). Schizotypy. In: *Encyclopedia of Creativity*, 3e (eds. M. Runco, S. Pritzker and R. Reiter-Palmon), 452–459. Academic Press.

82 Raine, A. (1991). The SPQ: a scale for the assessment of schizotypalpersonal ity based on DSM-III-R criteria. *Schizophrenia Bulletin* 17: 555–564.

83 Wolff, M. (1996). *Die Psychometrischen Eigenschaften einer Deutschen Version des Schizotypal Personality Questionary (SPQ)*. Masters Thesis. Humboldt University, Berlin.

84 Ettinger, U., Mohr, C., Gooding, D.C. et al. (2015). Cognition and brain function in schizotypy: a selective review. *Schizophrenia Bulletin* 41 suppl. no. 2: S417–S426.

85 Mientus, S., Gallinat, J., Wuebben, Y. et al. (2002). Cortical hypoactivation during resting EEG in schizophrenics but not in depressives and schizotypal subjects as revealed by low resolution electromagnetic tomography (LORETA). *Psychiatry Research Neuroimaging* 116: 95–111.

17

Brain–Computer Interfacing Using EEG

17.1 Introduction

Often, the effortless way in which our intentions for movement are converted into actions is taken for granted. The loss of motor function can be one of the most devastating consequences of disease or injury to the nervous system. Development of brain–computer interfaces (BCIs) during the last three decades has enabled communications or control over external devices such as computers and artificial prostheses with the electrical activity (e.g. electroencephalogram, EEG) of the human nervous system. However, the performance of these artificial interfaces rarely matches the speed and accuracy of natural limb movements making their clinical applications rather limited. Their practical utility therefore depends not only on the extent to which the patients can learn to operate these devices but also on advancement of the real time, but not necessarily complicated, algorithms that can decode motor intentions efficiently. In a conventional BCI setup, the user is instructed to imagine movement of different body parts (e.g. right-hand or leg movements) and the computer learns to recognize different patterns of the simultaneously recorded EEG activity. In the majority of cases, the signal-processing effort for the progress of EEG-based BCI systems has been focused on extracting more informative features from the EEGs and refinement/reduction of the feature space. More recent developments in machine learning not only enables recognition of the movement but also models the neural pathways from thought to action. Therefore, machine learning has become central to solving many BCI problems. Moreover, the BCI applications are being more extended to human use especially for rehabilitation purposes.

Electromyography (EMG) measurements have been primarily used in both rehabilitation and development of various robotic and haptic tools and machines. EMG, therefore, is still the most popular and effective method for connecting human to machine. New developments in EMG-based interfaces have resulted from more sensitive arrays of surface electrodes as well as availability of neural implants. Nevertheless, there remain major challenges for a prosthesis to replace a fully amputated arm or leg.

BCI (also called brain–machine interfacing [BMI]) is a challenging problem that forms part of a wider area of research, namely human computer interfacing (HCI), which interlinks thought to action. BCI can potentially provide a link between the brain and the physical world without any physical contact. In BCI systems the user messages or commands do not depend on the normal output channels of the brain [1]. As such the main BCI objectives

EEG Signal Processing and Machine Learning, Second Edition. Saeid Sanei and Jonathon A. Chambers.

are to manipulate the electrical signals generated by the brain neurons and generate the necessary signals to control some external systems (devices, prosthesis, or artificial arm/limb). The most important application is to energize the paralyzed organs or bypass the disabled parts of the human body. However, current BCI systems can also help in checking the operation of human cognitive systems such as memory. Conversely, BCI systems may appear as the unique communication mode for people with severe neuromuscular disorders such as spinal cord injury, amyotrophic lateral sclerosis, stroke, and cerebral palsy.

Approximately one hundred years after discovery of the electrical activity of the brain the first BCI research was reported by Vidal [2, 3] during the period 1973–1977. In his research it was shown how the brain signals could be used to build up a mental prosthesis. BCI has moved at a stunning pace since the first experimental demonstration in 1999 which showed that the cortical neurons ensamples could directly control a robotic manipulator [4]. Since then there has been tremendous research in this area [5] while in practise, EMG, has still been more applicable and commercialized.

BCI addresses analyzing, conceptualization, monitoring, measuring, and evaluating the complex neurophysiological behaviours detected and extracted by a set of electrodes over the scalp or from the electrodes implanted over the cortex or inside the brain. Movement-related rhythms for various body parts are captured from the motor cortex. A cross-section of the motor cortex showing the regions related to each organ can be seen in the schematic of Figure 17.1.

A BCI system should be easy to use, effective, efficient in terms of computational time and adaptability, and user friendly. BCI is a multidisciplinary field of research since it deals with cognition, neurophysiology, psychology, sensor design, positioning, circuitry, and communications, machine learning, signal detection, and signal processing.

BCI has now become popular in many institutions and in different countries. However, as the pioneers in BCI, the Berlin BCI (BBCI) group, established in 2000, has followed the objective of transferring the effort of training from the human to machine. The major focus in their work is reducing the inter-subject variability of BCI by minimizing the level of subject training. Some of their works have been reported in [6–8]. The Wadsworth BCI research group has mainly used event-related desynchronization (ERD) of the mu rhythm for EEG classification of real or imagery movements, after training the subject [9, 10]. In Austria, the Graz BCI activity utilizes mu or beta rhythms for training and control. The expert users of their system have been able to control a device based on modulations of the precentral mu or beta rhythms of sensorimotor cortices in a similar way to the Wadsworth BCI approach. However, while Wadsworth BCI directly presents the power modulations to the user, the Graz system for the first time also uses machine adaptation for controlling the BCI. They were also able to allow grasping of the non-functional arm of a disabled patient by functional electrical stimulation (FES) of the arm controlled by EEG signals [11–15]. Conversely, Martigny BCI started with adaptive BCI in parallel with BBCI. The researchers have proposed a neural network (NN) based classifier based on linear discriminant analysis (LDA) for classification of the static features [16]. In their approach three subjects are able to achieve 75% correct classification by imagination of the left- or right-hand movement or by relaxation with closed eyes in an asynchronous environment after a few days of training [17, 18]. Finally, the Thought Translation Device (TTD) which has been developed mainly for locked-in patients, enables the subjects to learn self-regulation of the slow cortical

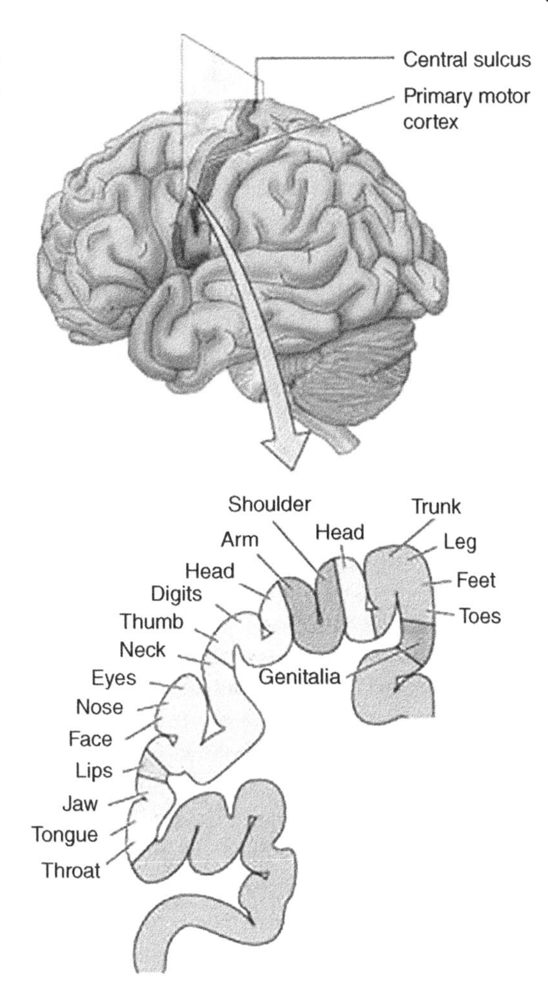

Figure 17.1 A cross-section of the motor cortex and the links to different organs for real or imagery (aka imaginary) movements.

potentials (SCPs) at central scalp positions using EEG or Electrocorticogram (ECoG). The subjects are able to generate binary decisions and hopefully provide a suitable communication channel to the outside world [19].

In recent years the improvements in invasive and noninvasive sensor design, neural implants, and availability of wireless recording systems together with advances in both signal-processing and machine-learning techniques have further enhanced the usage and accuracy of the BCI systems.

Other related technology such as multi-subject recording (also known as EEG hyperscanning) and amazing involvement of virtual and augmented reality have significantly boosted BCI applications for human mental (such as for emotions) and physical (such as for stroke or paralysis) rehabilitations.

In addition to EEG-based BCI, other functional brain-scanning modalities such as fMRI separately or together with EEG started being used. In this chapter the fundamental concepts and the requirement for the BCI design using EEG signals are reviewed and some very recent applications are briefly described.

17.1.1 State of the Art in BCI

The correspondence between EEG patterns and computer actions constitutes a machine-learning problem since the computer should learn how to recognize a given EEG pattern. As for other learning problems, in order to solve this problem, a training phase is necessary, in which the subject is asked to perform prescribed mental activities and a computer algorithm is in charge of extracting the associated EEG patterns. After the training phase is finished the subject should be able to control the computer actions with his thoughts. This is the major goal for a BCI system.

In terms of signal acquisition, BCI systems are classified into invasive (intracranial) and noninvasive. Noninvasive systems primarily exploit EEGs to control a computer cursor or a robotic arm. The techniques in designing such systems have been under development over the years due to their hazardless nature and flexibility [1, 19–27]. However, despite the advantage of not exposing the patient to the risks of brain surgery, EEG-based techniques provide limited information mainly because of the limitations in the existing technology, head volume conduction (which highly distorts the individual neuron cortical activities), and physiological and interfering undesired signals, noise and artefacts. However, despite these shortcomings, EEG-based methods can detect modulations of brain activity that correlate with visual stimuli, gaze angle, voluntary intentions and cognitive states [5]. These advantages led to development of several classes of EEG-based systems, which differ according to the cortical areas recorded, the extracted features of the EEGs, and the sensory modality providing feedback to subjects. In terms of feature detection and classification, common spatial patterns (CSPs), decision trees, Riemannian geometry-based classification [28, 29], and transfer learning using deep neural networks (DNNs) are the most popular techniques.

Among visual, audio, and haptic (tactile or somatosensory), which are often used in a feedback BCI system, the most effective feedbacks are visual and haptic. In a short paper by Kauhanen et al. [30] it has been concluded that both visual and haptic feedbacks have similar effects to the brain. Haptic feedback may be necessary for blind subjects. Less accurate results have been often reported for audio feedback experiments. In an editorial review however, it has been emphasized that haptic feedback provides more complete and immersive sensation than that a purely visual environment [31] can do. Haptic feedback involves tactile and proprioceptive sensory modalities and can thus help increase subjects' attention and motivation during repeated performance.

To detect the start of left or right, hand or leg movements in a noninvasive BCI, analysis of the EEG power around the motor and premotor brain regions is often sufficient. However, for the analysis of multitask movements and recognition of prolonged movement trajectory, a more complex information processing of all the EEG channels and effective estimation of the changes in brain connectivity over time become necessary.

In an EEG-based BCI, detection and separation of the control (movement-related) signals from the raw EEG signals is probably the first objective. The event-related source signals can be effectively clustered or separated if the corresponding control signals are well characterized. Given that these control sources are likely to propagate inside the brain, an exciting research area is also to localize and track these sources in real time. The first step in developing an effective BCI paradigm is therefore determining suitable control signals from the EEG. A suitable control signal has the following attributes: it can be precisely characterized for an individual, readily modulated or translated to express the intention, and consistently and reliably detected and tracked.

17.1.2 BCI Terms and Definitions

There are two main approaches towards BCI: one is based on event-related potentials (ERPs) which can be captured using a small number of electrodes around the Pz and Cz electrode sites and another one is based on the multiple sensor EEG movement-related activities recorded in the course of ordinary brain activity. In many cases however, only a few electrodes are used to detect the changes in the cortical activities around the motor and premotor areas. The latter approach is more comprehensive and does not necessarily require any particular stimulus.

Preparation and execution of movements (or imagination of movement) lead to short-lasting and circumscribed attenuation of the contralateral (to the side of movement) Rolandic mu (μ) (8–13 Hz) and the beta (β) (14–28 Hz) rhythms of the EEGs. This phenomenon is known as ERD. ERD is followed by a rebound amplification phase called event-related synchronization (ERS) in which μ and β rhythms re-emerge. Distinct spatial, temporal, and spectral characteristics of ERD/ERS can only be observed if several trials of EEG are averaged to cancel out the effect of (presumably) zero-mean, Gaussian and spatiotemporally independent recordings of background brain activity and measurement noise or to minimize the effect of volume conduction. Therefore, much effort has been invested in development and data-driven optimization of spatiotemporal filters that can extract the ERD/ERS features from noisy single-trial EEG measurements. Several other works as in [32, 33] have examined the suitability of P300 and steady-state movement-related potentials [34] for BCI.

However, as mentioned before, in many cases such as those where there is a direct connection from the electrodes to the mechanical systems, the number of recording channels, i.e. electrodes, is generally limited. In such cases EEG channel selection across various subjects has become another popular research area within the BCI community [35]. The general idea behind these approaches is based on a recursive channel elimination (RCE) criterion; channels that are well known to be important (from a physiological point of view) are consistently selected whereas task-irrelevant channels are disregarded. The corresponding spatial patterns may be recognized using different techniques. Nonnegative matrix factorization [NMF] has been used to analyze and identify local spatiotemporal patterns of neural activity in the form of sparse basis vectors [36].

17.1.3 Popular BCI Directions

BCI using ERPs: An ERP appears in response to some specific stimulus. The most widely used ERP evoked potential (EP) is the P300 signal, which can be auditory, visual, or somatosensory. It has a latency of approximately 300 ms and is elicited by rare or significant stimuli, and its amplitude is strongly related to the unpredictability of the stimulus; the more unforeseeable the stimulus, the higher the amplitude [37]. Another type of visual EP (VEP) is those for which the ERPs have short latency, representing the exogenous response of the brain to a rapid visual stimulus. They are characterized by a negative peak around 100 ms (N1) followed by a positive peak around 200 ms (P2). The ERPs can provide control when the BCI produces the appropriate stimuli. Therefore, the BCI approach based on ERP detection from the scalp EEG seems to be easy since the cortical activities can be easily measured noninvasively and in real time. Also, an ERP-based BCI needs little training for a new subject to gain control of the system. However, the information achieved through ERP extraction and measurement is not accurate enough for extraction of movement-related features and they have vast variability in different subjects with various brain abnormalities and disabilities. More importantly, the subject has to wait for the relevant stimulus presentation [38, 39].

BCI using SSVER: These are the approaches based on steady-state visual-evoked responses (SSVERs), as the natural responses for visual stimulations at specific frequencies. These responses are elicited by a visual stimulus that is modulated at a fixed frequency. The SSVERs are characterized by an increase in EEG activity around the stimulus frequency. With feedback training, subjects learn to voluntarily control their SSVER amplitude. Changes in the SSVER result in control actions occurring over fixed time intervals [39, 40].

BCI using slow cortical potential shift: Slow cortical potential shifts (SCPSs) are shifts of cortical voltage, lasting from a few hundred milliseconds up to several seconds. Subjects can learn to generate slow cortical amplitude shifts in an electrically positive or negative direction for binary control. This can be achieved if the subjects are provided with feedback on the evolution of their SCP and if they are positively reinforced for correct responses [41]. However, in both of the above methods the subject has to wait for the brain stimulus.

Generally, some BCI approaches rely on the ability of the subjects to develop control of their own brain activity using biofeedback [42–44], whereas others design or utilize classification algorithms that recognize EEG patterns related to particular voluntary intentions [45]. Initial attempts to enable the subjects to use the feedback from their own brain activity started from the 1960s. The system enables the subjects to gain voluntary control over brain rhythms. It has been claimed that after training with EEG biofeedback, humans are able to detect their own alpha [46, 47] and mu rhythms [48]. This has also been tested on cats using their mu rhythms [49] and dogs [50] to control their hippocampal theta rhythm. Classification-based approaches have also been under research recently [45].

17.1.4 Virtual Environment for BCI

BCI has for a long time been considered as a system for pure transduction of brain signals to some effector output. To achieve this goal, the systems needs to be trained, which is often very dull or monotonous, and offering little direct reward or positive reinforcement feedback. In mental rehabilitation, BCI is trained and used to reflect on the brain itself. In order to achieve this one may enhance the reward dimensionality of the training in order to better or effectively enhance neuroplasticity. The linking of BCI and virtual reality (VR) is a logical step in this line of thinking [51]. VR scenes can be made very interactive, rich and complex environments that are effective in demonstrating reward. Although one has to take into account the drawback of potential sources of distraction and attentional drift, which is currently an attractive line of research. People who learn to use BCI to actively and autonomously navigate a VR scene such as move a VR 'phantom' limb or explore emotionally rich learning environments with rewarding feedback, will eventually benefit from this goal-oriented learning process as a consequence of brain plasticity. As such, BCI has evolved from initial 'mechanistic' tools towards plasticity enhancers which hopefully will continue to find their way to the clinical practise of rehabilitation in neurological and psychiatric disorders. BCI–VR paradigms realize a shift of focus from a distal effector perspective to a more proximal point of interest: train your brain and change it.

In a demonstration, subjects navigated through a virtual environment by imagining themselves walking [52]. These works paved the path for more research in this area.

17.1.5 Evolution of BCI Design

Although most of the recent BCI research has been focused upon scalp EEGs, some approaches using invasive EEGs, particularly for rehabilitation of chronic disabilities, have also been reported. Invasive BCI approaches are based on recordings from ensembles of single brain cells or on the activity of multiple neurons using ECoG, or intracranial EEG (iEEG) recordings. They rely on the physiological properties of individual cortical and subcortical neurons or combination of neurons that modulate their movement-related activities. These works started in the 1960s and 1970s through some experiments by Fetz and his co-researchers [53–58]. In these experiments, monkeys learned how to control their cortical neurons activity voluntarily with the help of biofeedback. The information obtained by ECoG is far more useful and meaningful than EEG as the ECoG signals have less noise and the rates and delays of neuron firings can be measured from these signals. A few years later, Schmidt [59] provided that the voluntary motor commands could be extracted from raw cortical neural activity and used them to control a prosthetic device designed to restore motor functions in severely paralyzed patients.

Most of the research on invasive BCI was carried out on monkeys. These works relied on single cortical site recordings either of local field potentials [60–63], or from small samples of neurons or multiple brain zones [64–66]. They are mostly recorded in the primary motor cortex [64, 65], although some work has been undertaken on the signals recorded from the posterior parietal cortex [67].

Normally, subdural electrodes for deep brain recording are not used for BCI. They are either used for seizure or as deep brain stimulators, which have been recently used for monitoring of Parkinson's patients [68].

Kennedy and his colleagues [69] have presented an impressive result from implanted cortical electrodes. In another work [4] a monkey managed to remotely control a robotic arm using implanted cortical electrodes. These works, however, require solutions to possible risk problems, advances in robust and reliable measurement technologies, and clinical competence. Conversely, the disadvantage of using scalp recordings lies in the very low quality of the signals, due to attenuation of the electrical activity signals on their way to the electrodes and the effect of various noises.

Moreover, a diverse range of BCI systems, mostly for rehabilitation purposes, have been developed during the past decade. Manufacturing miniaturized implantable sensors, availability of huge computer clusters, computing power, and highspeed communication networks as well as wider use of virtual [51] or augmented reality have significantly boosted the range of applications and the variety of BCI systems [70].

In the following subsections a number of features used in BCI are explained. Initially, the changes in EEG before, during, and after the externally or internally paced events are observed. These events can be divided into two categories: first, the ERPs including EPs and second, ERD and synchronization (ERS). The main difference between the two types is that the ERP is a stimulus-locked, or more generally, a phase-locked reaction, while the ERD/ERS is a non-phase-locked response. In a finger movement process, for example, often discussed in BCI, pre-movement negativity prominent prior to movement onset (readiness potential [RP]), and post-movement beta oscillations occurring immediately after movement-offset are respectively phase-locked (evoked) and non-phase-locked processes [71].

In terms of detection and classification of brain motor responses (real or imaginary), CSP [72] started being used as an effective and robust technique in feature estimation for movement-related EEG pattern classification. Some variants of this approach have been developed recently which are discussed later in this chapter. Conversely, DNNs using transfer learning revolutionized both feature learning and classification of various brain patterns and waveforms. These networks can be also used in modelling the neural pathways between deep brain sources and scalp electrodes [73] which can be used in future BCI systems.

17.2 BCI-Related EEG Components

17.2.1 Readiness Potential and Its Detection

An early indication of movement can be realized from so-called RP, or *Bereitschaftspotential* (BP), which is the German word for readiness potential, or premotor potential. This is a transient signal hump which appears just before the movement around the brain motor region. This was discovered by Helmut Kornhuber and Lüder Deecke at the University of Freiburg in Germany in 1964. The RP is much (10–100 times) smaller than the EEG alpha rhythm and it can be seen only by averaging, relating the electrical potentials to the onset of the movement. Figure 17.2 illustrates a typical RP together with the onset of voluntary hand (or finger) movement.

A time–frequency-space approach such as that in [34] may be followed to detect and characterize the RP. In this method which uses PARAFAC based tensor factorization [74] a finger can tap the floor with a frequency of at most 2 Hz to avoid the interferences by higher latency ERPs up to 500 ms, to detect the BP and differentiate between left and right-hand (or finger) movements. Detection and tracking of BP over similar trials allow a more accurate estimation of BPs by averaging. It is also useful in monitoring the rehabilitation process in humans. The detected RPs related to left and right finger movements together with their topographic signatures can be seen in Figure 17.3 [34].

In another approach, Ahmadian et al. [75] applied constrained blind source separation to extract the independent component best representing an RP.

17.2.2 ERD and ERS

As explained in Section 17.1, ERD represents a short-lasting and localized amplitude decrease of rhythmic alpha (mu) and lower beta activity just before and during the real or imagery (imaginary) movement. The ERS which follows ERD indicates an amplitude

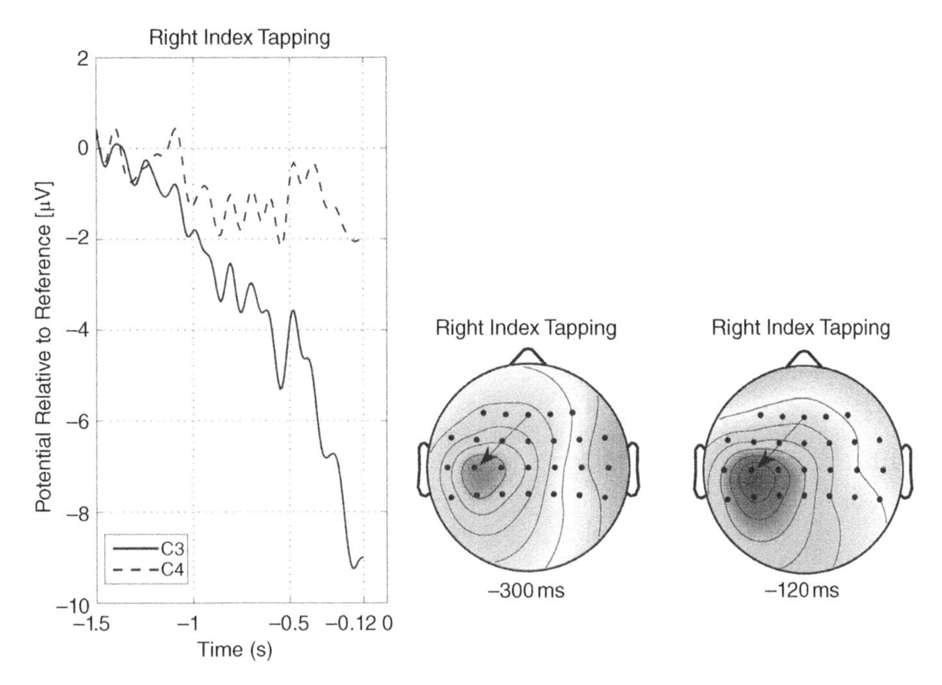

Figure 17.2 Readiness potential elicited around the finger movement time instant 0 (taken from http://neuroskeptic.blogspot.co.uk).

increase. These reactivities are highly frequency-band specific and non-phase locked to the event [76, 77]. A simple measure of ERD is given as:

$$ERD\ Level = \frac{P(f,n) - P_{ref}(f)}{P_{ref}(f)} \tag{17.1}$$

where $P(f, n)$ is the value of a signal power at a given time–frequency point of an average power map, and $P_{ref}(f)$ is an average power during some reference time calculated for frequency f. This represents the level of rhythmic activity within the alpha band just before or during the movement. Any attention dramatically attenuates the alpha rhythms, while an increase of task complexity or attention results in an increased magnitude of ERD.

Increased cellular excitability in thalamocortical systems results in a low amplitude desynchronized EEG. So ERD may be due to electrophysiological correlate of various activated cortical regions involved in processing of sensory or cognitive information or production of motor reaction. Involvement of more neurons increases the ERD magnitude. In the BCI context, explicit learning of a movement sequence, e.g. key pressing with different fingers, is accompanied by an enhancement of the ERD over the contralateral central regions. As the learning progresses and becomes more automatic the ERD decreases.

The cortical mu rhythm, which is in the alpha band and unlike normal occipital alpha has lateral distribution, is of particular interest in BCI applications. This is mainly because it can be modulated/translated through imagery movement and can be monitored via a noninvasive technique.

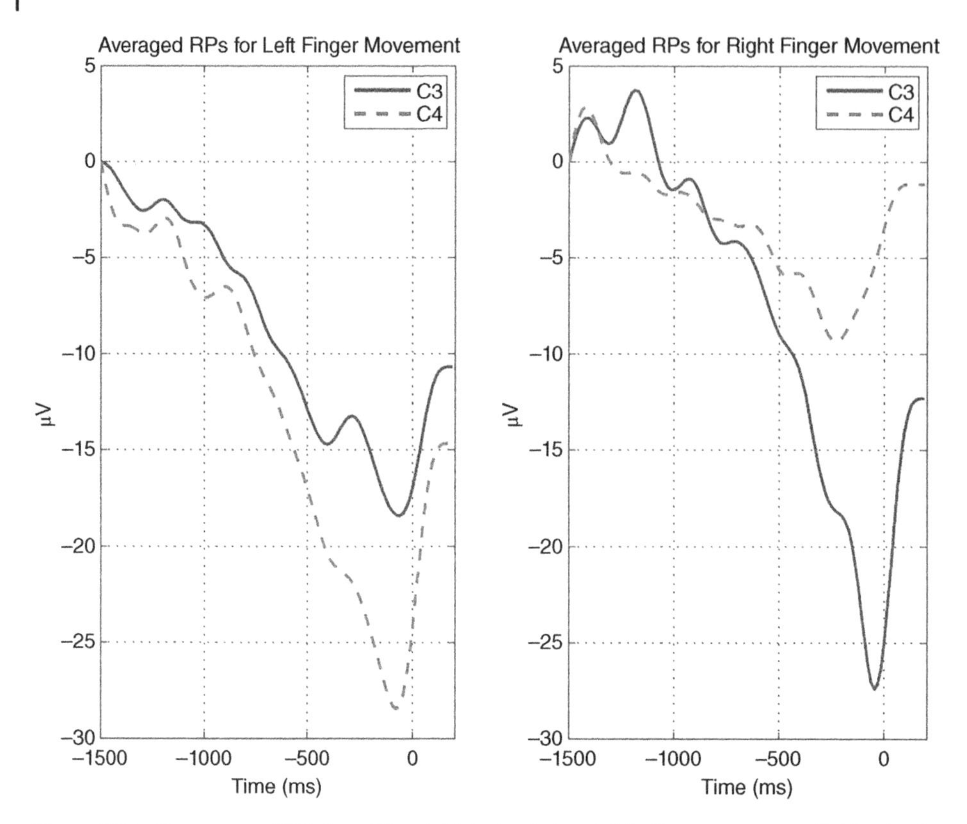

Figure 17.3 The averaged RPs from C3 and C4 during left and right finger movement. Evidently, there exists asymmetries between the averages for C3 and C4 in both cases. In the left finger movement, the voltage at C4 is more negative than that at C3 and vice versa for right finger movement.

The overall alpha band may be divided into lower and higher alphas. Lower alpha (6–10 Hz) is a response to any type of task and topographically is spread over almost all the electrodes. Higher alpha ERD, restricted to parieto-occipital areas, is found during visually presented stimuli.

It is good to mention at this point that the level of ERD is closely linked to semantic memory processes; those with good memory show a larger ERD in the lower alpha band [78].

In an auditory memory task, the absence of an ERD can be explained by the anatomical localization of the auditory cortex below the surface. Detection of the auditory ERD from the EEGs is often difficult.

As related to BCI, voluntary movement also results in a circumscribed desynchronization in the upper alpha and lower beta bands, localized over sensorimotor regions [79]. The ERD starts over the contralateral rolandic region and, during the movement, becomes bilaterally symmetrical with execution of movement. It is of interest that the time course of the contralateral mu desynchronization is almost identical to brisk and slow finger movement, starting about two seconds prior to movement onset. Generally, brisk and slow finger movements have different encoding processes. Brisk movement is pre-programmed and the afferents are delivered to the muscles as bursts. Conversely, slow movement depends on the reafferent input from kinaesthetic receptors evoked by the movement itself.

Finger movement of the dominant hand is accompanied by a pronounced ERD in the ipsilateral side, whereas movement of the non-dominant finger is preceded by a less lateralized ERD [79]. Circumscribed hand area mu ERD can be found in nearly every subject, whereas, a foot area mu ERD is hardly localized close to the primary foot area between both hemispheres.

In another study [80] with cortical electrodes, it was discovered that mu rhythms are not only selectively blocked with arm and leg movements, but also with face movement. The ECoG captures more detailed signals from smaller cortical areas than the conventional EEG-based systems. These signals also contain low amplitude high frequency gamma waves. Consequently, ECoG-based BCIs have higher accuracy and require shorter training time than those of EEGs [81].

In ERS, however, the amplitude enhancement is based on the cooperative or synchronized behaviour of a large number of neurons. In this case, the field potentials can be easily measured even using scalp electrodes. It is also interesting to know that approximately 85% of cortical neurons are excitatory, with the other 15% being inhibitory.

In [77], during an imagery task, the participants imagined an indicated movement for one second (i.e. brief movement imagery) or five seconds (i.e. continuous movement imagery). Based on this study, mu and beta ERD/ERS patterns are elicited during imagined hand movements and the movement duration affects ERS and does not affect ERD patterns, during motor movement imagery. Additionally, brief movement imagery had a greater impact on mu and beta ERD whereas continuous movement imagery had a greater impact on mu and beta ERS. These results can be viewed in Figure 17.4.

The left and right foot representation area is located within the interhemispheric fissure of the sensorimotor cortex and share spatial proximity. This makes it difficult to visualize the cortical lateralization of event-related (de)synchronization (ERD/ERS) during left and right foot motor imageries.

The authors in [82] investigated the possibility of using ERD/ERS in the mu, low beta, and high beta bands, during left and right foot dorsiflexion kinaesthetic motor imageries (KMI), as unilateral control commands for a BCI. EEG was recorded from nine healthy participants during cue-based left–right foot dorsiflexion KMI tasks. The extracted time–frequency features were analyzed for common average and bipolar references. With each reference, mu and beta band-power features were analyzed using time–frequency transforms, scalp topographies, and average time course for ERD/ERS. They confirmed the cortical lateralization of ERD/ERS, during left and right foot KMI. Statistically significant features were classified using LDA, SVM, and *k*-nearest neighbour (KNN) classifiers, and evaluated using the area under receiver operating characteristic (ROC) curves. They also acknowledged an increase in high beta power following the end of KMI for both tasks from the brain right and left hemispheres at the vertex. The selected features can evoke left–right leg differences in single EEG trials. LDA and SVM were examined for classification and good performances were achieved. The outcome certainly benefits clinician and patients working on rehabilitation of the patients suffering from brain damages, such as stroke.

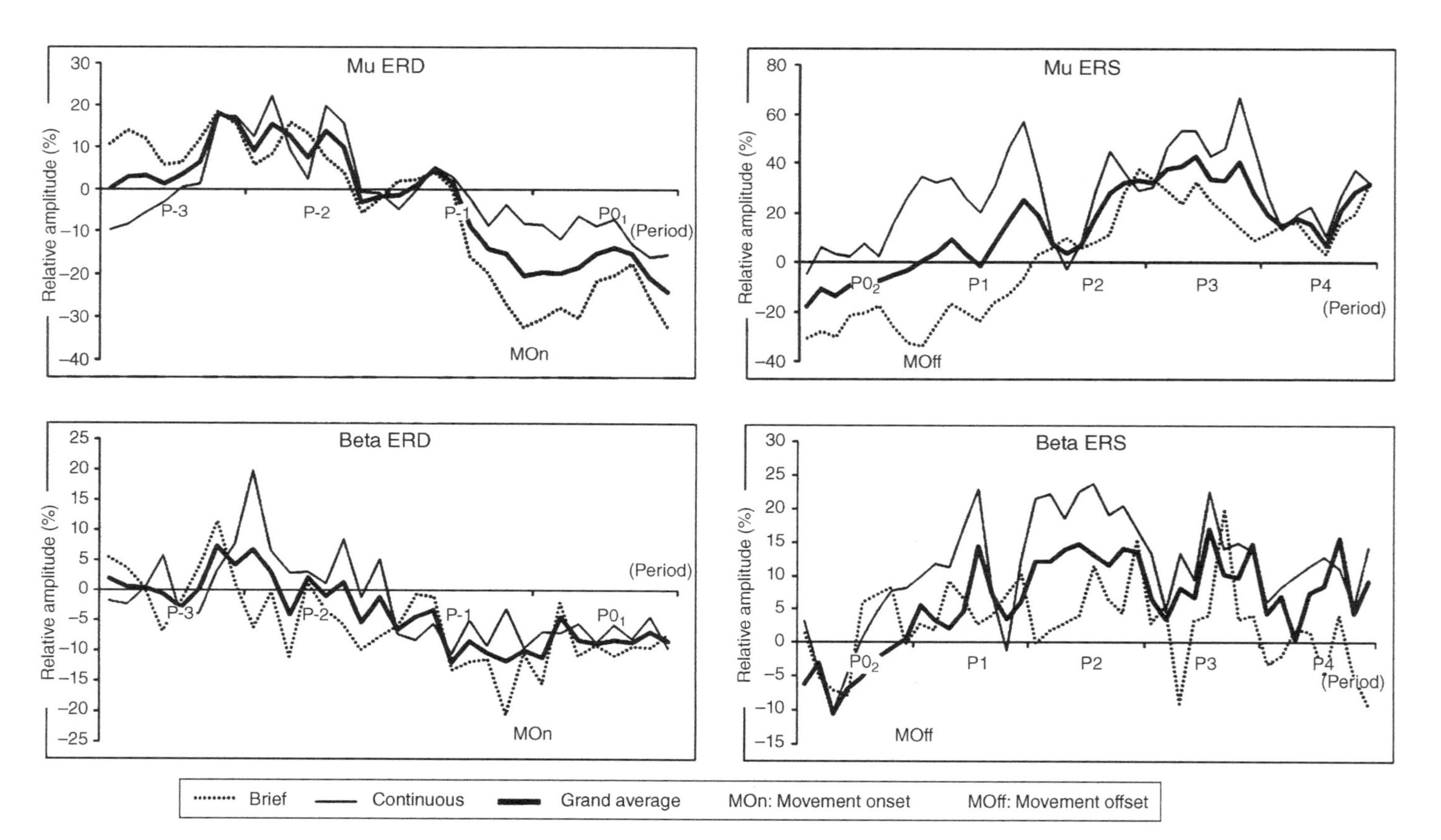

Figure 17.4 ERD/ERS patterns over the central region (*C*3 and *C*4) during imagery movements in mu and beta bands averaged over 32 trials. The trend is in thick (grand average), brief movement imagery in dotted and continuous movement imagery in thin lines [77].

17.2.3 Transient Beta Activity after the Movement

This activity, also called post-movement beta synchronization (PMBS) is another informative and consistent event starting during the movement and continuing for about 600 ms [79]. It is found after finger or foot movement over both hemispheres without any significant bilateral coherence. The frequency band may vary from subject to subject; for finger movement the range is around 16–21 Hz [83] whereas for foot movement it is around 19–26 Hz [84]. The PMBS has similar amplitude for brisk and slow finger movements. This is interesting since brisk and slow movements involve different neural pathways. Moreover, this activity is significantly larger with hand as compared to finger movement [79]. Also, larger beta oscillations with wrist as compared to finger movement can be interpreted as the change of a larger population of motor cortex neurons from an increased neural discharge during the motor act to a state of cortical disfacilitation or cortical idling [79]. This means movement of more fingers results in a larger beta wave. Beta activity is also important in generation of a grasp signal, since it has less overlap with other frequency components [85].

17.2.4 Gamma Band Oscillations

Oscillation of neural activity (ERS) within the gamma band (35–45 Hz) has also been of interest recently. Such activity is very obvious after visual stimuli or just before the movement task. This may act as the carrier for the alpha and lower beta oscillations and relate to binding of sensory information and sensorimotor integration. Gamma, together with other activities in the above bands, can be observed around the same time after performing a movement task. Gamma-ERS manifests itself just before the movement, whereas beta ERS occurs immediately after the event.

Different approaches attempt to detect or highlight the gamma-ERS. As an example, in [86] the authors applied empirical mode decomposition (EMD) to a single (Cz) channel EEG followed by time–frequency analysis and managed to improve the gamma-ERS detection. In a recent approach the power of the filtered and artefact-removed EEG Cz channel signal within 30–60 Hz has been measured during imaginary movement by the same group [87]. In this work they used FieldTrip [88] for undertaking most of the EEG signal processing.

17.2.5 Long Delta Activity

Rather than other known ERS and ERD activities within alpha, beta, and gamma bands a long delta oscillation starts immediately after the finger movement and lasts for a few seconds. Although this has not been widely reported often in the literature, it can be a prominent feature in distinguishing between movement and non-movement states mainly due to its high power and less sensitivity to noise.

The main task in BCI is how to exploit the behaviour of the EEGs in the above frequency bands before, during, and after the imaginary movement, or after certain brain stimulation, in generation of the control signals. The following sections address this problem. A study reported in [89] shows that delta-band EEG signals contain useful information that can be used to infer finger kinematics in decoding the repetitive finger movement. Further, the highest decoding accuracies were characterized by highly correlated delta-band EEG activity mostly localized to the contralateral central areas of the scalp. Delta activity is highly correlated with 3D reach-to-grasp BCI movement too [90]. Generally, finger kinematics can be inferred, to some extent, from the delta-band filtered fluctuations of the amplitude of EEG signals across the scalp using linear decoders with memory [89].

17.2.6 ERPs

ERP variability is widely exploited in the design of many BCI systems. In some important BCI applications such as spelling BCI elicitation of P300 and particularly its P3b subcomponent the response to novel/target stimulus in an odd ball paradigm is examined and quantified. ERPs are used in many other BCI applications too. Beverina et al. [91] have used P300 and steady-state visual-evoked potentials (SSVEPs). They have classified the ERP feature patterns using SVMs. In their SSVEP approach they have used signals from the occipital electrodes (O_z, O_2, PO_8). The stimulations have been considered random in time instants, and a visual feedback [42] has been suggested for training purposes. In normal cases it is possible to amplify the brain waves through the feedback. In another work, SSVEPs have been used in a visually elaborate immersive 3D game [92]. The SSVEP generated in response to phase-reversing checkboard patterns is used for the proposed BCI. A number of BCI applications using ERP, particularly P300, are discussed in the later sections of this chapter.

17.3 Major Problems in BCI

A simple and very popular BCI system set up is illustrated in Figure 17.5. Feature extraction and classification of the features for each particular body movement is the main objective in most of the BCI systems.

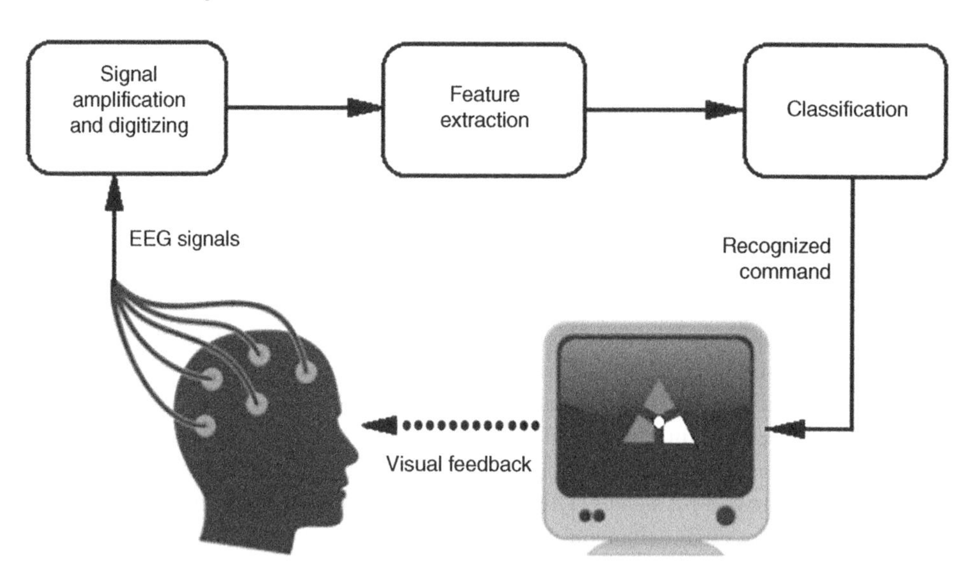

Figure 17.5 A typical BCI system using scalp EEGs when visual feedback is used.

As mentioned previously, the main problem in BCI is separating the control signals from the background EEG. Meanwhile, cortical connectivity, as an interesting identification of various task-related brain activities, has to be studied and exploited. Detection and evaluation of various features in different domains will then provide the control signals. To begin, however, the EEG signals have to be pre-processed since the signals are naturally contaminated by various internal and external interferences. Preprocessing and conditioning the data (including noise and artefact removal) can further enhance the signals. In the case of blind source separation (BSS), for example, pre-whitening may also be necessary before implementation of the source separation algorithms.

17.3.1 Preprocessing of the EEGs

In order to have an artefact-free EEG to extract the control signals, the EEGs have to be restored from the artefacts such as eye blinking, electrocardiograms (ECGs), and any other internal or external disturbing effects.

Eye-blinking artefacts are very clear in both frontal and occipital EEG recordings. ECGs, conversely can be seen more over the occipital electrodes. Many attempts have been made by different researchers to remove these artefacts.

Most of the noise, and artefacts are filtered out by the hardware provided in new EEG machines. As probably the most dominant remaining artefact, interfering eye blinks (ocular artefact; OA) generate a signal within EEGs that is on the order of 10 times larger in amplitude than cortical signals and can last between 200 and 400 ms.

There have been some works by researchers to remove OAs. Certain researchers have tried to estimate the propagation factors, as discussed in [93] based on regression techniques in both the time and frequency domains. In this attempt there is a need for a reference electrooculogram (EOG) channel during the EEG recordings.

Principal component analysis (PCA) and SVMs have also been utilized for this purpose [94]. In these methods the EEGs and OAs are assumed statistically uncorrelated. Adaptive filtering has also been utilized [95]. This approach has considered the EEG signals individually and therefore ignored the mutual information among the EEG channels. Independent component analysis (ICA) has also been used in some approaches. In these works the EEG signals are separated into their constituent independent components (ICs) and the ICs are projected back to the EEGs using the estimated separating matrix after the artefact-related ICs are manually eliminated [96]. In [97] a BSS algorithm based on second order statistics separates the combined EEG and EOG signals into statistically independent sources. The separation is then repeated for a second time with the EOG channels inverted. The estimated ICs in both rounds are compared, and those ICs with different signs are removed. Although, due to the sign ambiguity of the BSS the results cannot be justified it is claimed that using this method the artefacts are considerably mitigated. As noticed, there is also a need to separate EOG channels in this method.

In a robust approach the EEGs are first separated using an iterative SOBI following by SVM to effectively remove the EOG artefacts [98]. The method can also be easily extended to removal of the ECG artefacts. The proposed algorithm consists of BSS, automatic removal of the artefact ICs, and finally reprojection of the ICs to the scalp, providing artefact-free EEGs. This is depicted in Figure 17.6. Iterative SOBI as previously discussed has been

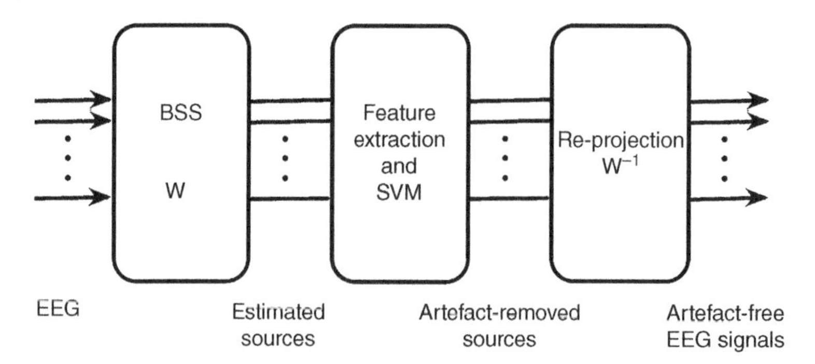

Figure 17.6 A hybrid BSS-SVM system for EEG artefact removal [98].

effectively used to separate the ICs in the first stage. In the second stage only four features were carefully selected and used for classification of the normal brain rhythms from the EOGs. These features are as follows:

Feature I: The large ratio between the peak amplitude and the variance of a signal suggests that there is an unusual value in the data. This is a typical identifier for the eye blink because it causes a large deflection on the EEG trace. This is described mathematically as:

$$f_1 = \frac{max\left(|\mathbf{u}_n|\right)}{\sigma_n^2} \quad \text{for } n = 1, \ldots, N \tag{17.2}$$

where $\mathbf{u}_n$ is one of the N ICs, max(.) is a scalar valued function that returns the maximum element in a vector, σ_n is the standard deviation of $\mathbf{u}_n$, and |•| denotes absolute value. Normal EEG activity is tightly distributed about its mean. Therefore, a low ratio is expected while the eye-blink signals manifest a large value.

Feature II: This is a measure of skewness which is a third order statistic of the data, defined as:

$$f_2 = \left|\frac{E\{\mathbf{u}_n^3\}}{\sigma_n^3}\right| \quad \text{for } n = 1, \ldots, N \tag{17.3}$$

for zero-mean data. The EEG containing eye blink typically has a positive or negative skewness since the eye-blinking signal has considerably larger value for this feature.

Feature III: The correlation between the ICs and the EEG signals from certain electrodes is significantly higher than those of other ICs. The electrodes with most contributed EOG are frontal electrodes FP_1, FP_2, F_3, F_4 and occipital lobe electrodes O_1 and O_2 (in total, six electrode signals) The reference dataset, i.e. the EEG from the aforementioned electrodes, is distinct from the training and test datasets. This will make the classification more robust by introducing a measure of the spatial location of the eye-blinking artefact. Therefore, the third feature can be an average of the correlations between the ICs and the signals from these six electrodes:

$$f_3 = \frac{1}{6}\sum_{i=1}^{6}\left(\left|E\left\{x_i^0(t)u(t+\tau)\right\}\right|\right) \quad \text{for} \quad \forall t \tag{17.4}$$

where $x_i^0(n)$ are eye-blinking reference signals, and i indexes each of the aforementioned electrode locations. The value of this feature will be larger for ICs containing eye-blinking artefact, since they will have a larger correlation for a particular value of τ in contrast to ICs containing normal EEG activity.

Feature IV: The statistical distance between distributions of the ICs and the electrode signals which are more likely to contain EOG is used. This can be measured using Kullback–Leibler (KL) distance defined as:

$$f_4 = \int_{-\infty}^{\infty} p(u(t))\, ln \frac{p(u(t))}{p_{ref}} du(t) \quad \text{for} \quad \forall t \tag{17.5}$$

where $p(.)$ denotes the pdf and p_{ref} is a reference distribution. When the IC contains OA the KL distance between its pdf and the pdf of the reference IC will be approximately zero, whereas the KL distance to the pdf of a normal EEG signal will be larger.

An SVM with an RBF nonlinear kernel is then used to classify the ICs based on the above features. Up to 99% accuracy in detection of the EOG ICs has been reported [98].

After the artefact signals are marked, they will be set to zero. Then, all the estimated sources are re-projected to the scalp electrodes to reconstruct the artefact-free EEG signals.

The same idea has been directly used for extraction of the movement-related features [99] from the EEGs. In this work it is claimed that without any long-term training the decision of whether there is any movement for a certain finger or not, can be achieved by BSS followed by a classifier. A combination of a modified genetic algorithm (GA) and an SVM classifier has been used to condition and classify the selected features.

17.4 Multidimensional EEG Decomposition

All movement-related potentials are limited in duration and in frequency. In addition, each channel contains the spatial information of the EEG data. PCA and ICA have been widely used in decomposition of the EEG multiple sensor recordings. However, an efficient decomposition of the data requires incorporation of the space, time, and frequency dimensions.

Time–frequency (TF) analysis exploits variations in both time and frequency. Most of the brain signals are decomposable in the TF domain. This has been better described as sparsity of the EEG sources in the TF domain. In addition, TF domain features are much more descriptive of the neural activities. In [100] for example, the features from the subject-specific frequency bands have been determined and then classified using LDA.

In a more general approach the spatial information is also taken into account. This is due to the fact that the majority of the events are localized in distinct brain regions. As a favourable approach, joint space–time–frequency classification of the EEGs has been studied for BCI applications [39, 101]. In this approach the EEG signals are measured with reference to *digitally linked ears* (DLE). DLE voltage can be easily found in terms of the left and right earlobes as:

$$V_e^{DLE} = V_e - \frac{1}{2}(V_{A_1} + V_{A_2}) \tag{17.6}$$

where V_{A_1} and V_{A_2} are respectively the left and right earlobe reference voltages. Therefore, the multivariate EEG signals are composed of the DLE signals of each electrode. The signals are multivariate since they are composed of the signals from multiple sources. A decomposition of the multivariate signals into univariate classifications has been carried out after the segments contaminated by eye-blink artefacts are rejected [39].

There are many ways to write the general class of TF distributions for classification purposes [102]. In the above work the characteristic function (CF) $M(\theta, \tau)$ as in:

$$C(t,\omega) = \frac{1}{4\pi^2}\int_{\tau=-\infty}^{\infty}\int_0^{2\pi} M(\theta,\tau)e^{-j\theta t - j\tau\omega}d\theta d\tau \tag{17.7}$$

for a single channel EEG signal, $x(t)$, assumed continuous time, (a discretized version can be used in practise) is defined as:

$$M(\theta,\tau) = \varphi(\theta,\tau)A(\theta,\tau) \tag{17.8}$$

where

$$\begin{aligned} A(\theta,\tau) &= \int_{-\infty}^{\infty} x^*\left(u - \frac{1}{2}\tau\right)x\left(u + \frac{1}{2}\tau\right)e^{j\theta u}du \\ &= \int_0^{2\pi} \hat{X}^*\left(\omega + \frac{1}{2}\theta\right)\hat{X}\left(\omega - \frac{1}{2}\theta\right)e^{j\tau\omega}d\omega \end{aligned} \tag{17.9}$$

and $\hat{X}(\omega)$ is the Fourier transform of $x(t)$, which has been used for classification. This is a representative of the joint time–frequency autocorrelation of $x(t)$. $\varphi(\theta, \tau)$ is a kernel function which acts as a mask to enhance the regions in the TF domain so the signals to be classified are better discriminated. In [39] a binary function has been suggested as the mask.

In the context of EEGs, as multichannel data, a multivariate system can be developed. Accordingly, the multivariate ambiguity function (MAF) of such system is defined as:

$$\boldsymbol{MA}(\theta,\tau) = \int_{-\infty}^{\infty} \boldsymbol{x}\left(t + \frac{\tau}{2}\right)\boldsymbol{x}^H\left(t - \frac{\tau}{2}\right)e^{j\theta t}dt \tag{17.10}$$

where $(.)^H$ denotes conjugate transpose. This ambiguity function can also be written in a matrix form as:

$$\boldsymbol{MA}(\theta,\tau) = \begin{bmatrix} a_{11}\dots & a_{1N} \\ \cdot \quad \cdot & \cdot \\ \cdot \quad \cdot & \cdot \\ a_{N1}\dots & a_{NN} \end{bmatrix} \tag{17.11}$$

where

$$a_{ij} = \int_{-\infty}^{\infty} x_j^*\left(t - \frac{\tau}{2}\right)x_i\left(t + \frac{\tau}{2}\right)e^{j\theta t}dt \tag{17.12}$$

The diagonal terms are called auto-ambiguity functions and the off-diagonal terms are called cross-ambiguity functions. MAF can therefore be an indicator of the multivariate time–frequency–space autocorrelation of the corresponding multivariate system. The space dimension is taken into account by the cross-ambiguity functions.

17.4.1 Space–Time–Frequency Method

It is desirable to exploit the changes in EEG signals in time, frequency, and space (electrodes) at the same time. The disjointedness property of the brain sources may not be achievable in time, frequency or space separately due to the strong overlaps of the sources in each domain. However, in a multidimensional space the sources are more likely to be disjoint. Such a property may be exploited to separate and localize them. An early work in [103] represents this idea. The block diagram in Figure 17.7 illustrates the steps of the approach.

The above concept may be studied in the framework of tensor factorization where the signal variations in all possible dimensions can be considered at the same time. PARAFAC and Tucker methods explained in previous chapters are the two main approaches. Here, application of PARAFAC to BCI particularly for artefact removal is discussed.

Another direction in BCI research is to evaluate the cortical connectivity and phase synchronization for characterization of continuous movement. The work on this area is however limited. Multivariate autoregressive (MVAR) modelling followed by directed transfer functions (DTFs) and evaluation of the diagonal and off-diagonal terms has been the main approach. However, comprehensively discussed in the Chapter 8 of this book, there are other approaches for this evaluation. One application of the MVAR is discussed in a later section of this chapter.

The contrast of MAF is then enhanced using a multidimensional kernel function and the powers of cross-signals (cross-correlation spectra) are used for classification [39]. Using this method (as well as MVAR) the location of the event-related sources can be tracked and effectively used in BCI.

17.4.2 Parallel Factor Analysis

In this approach the events are considered sparse in the space–time–frequency domain and no assumption is made on either independency or uncorrelatedness of the sources. Therefore, the main advantage of PARAFAC over PCA or ICA is that uniqueness is ensured

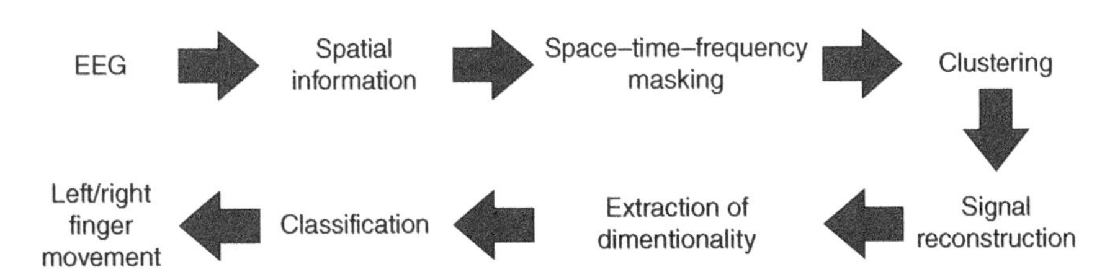

Figure 17.7 Classification of left/right finger movements using space–time–frequency decomposition.

under mild conditions, making it unnecessary to impose orthogonality or statistical independence constraints. Harshman [104] was the first researcher to suggest that PARAFAC be used for EEG decomposition. Carol, and Chang [105] independently proposed PARAFAC in 1970.

Möcks reinvented the model, naming it topographic component analysis, to analyze the ERP of channel × time × subjects [106]. The model was further developed by Field and Graupe [107]. Miwakeichi et al. eventually used PARAFAC to decompose the EEG data to its constituent space–time–frequency components [108]. In [101] PARAFAC has been used to decompose the wavelet transformed event-related EEG given by the inter-trial phase coherence. Figures 17.8 and 17.9 show respectively the space–time–frequency decomposition of the 15 channel EEG signals recorded during left and right index finger movement imaginations. Spectral contents, temporal profiles of the two identified factors, and the topographic mapping of EEG for the two factors are shown in these figures.

Accordingly, space–time–frequency features can be evaluated and used by a suitable classifier to distinguish between the left and right finger movements (or finger movement imagination) [109].

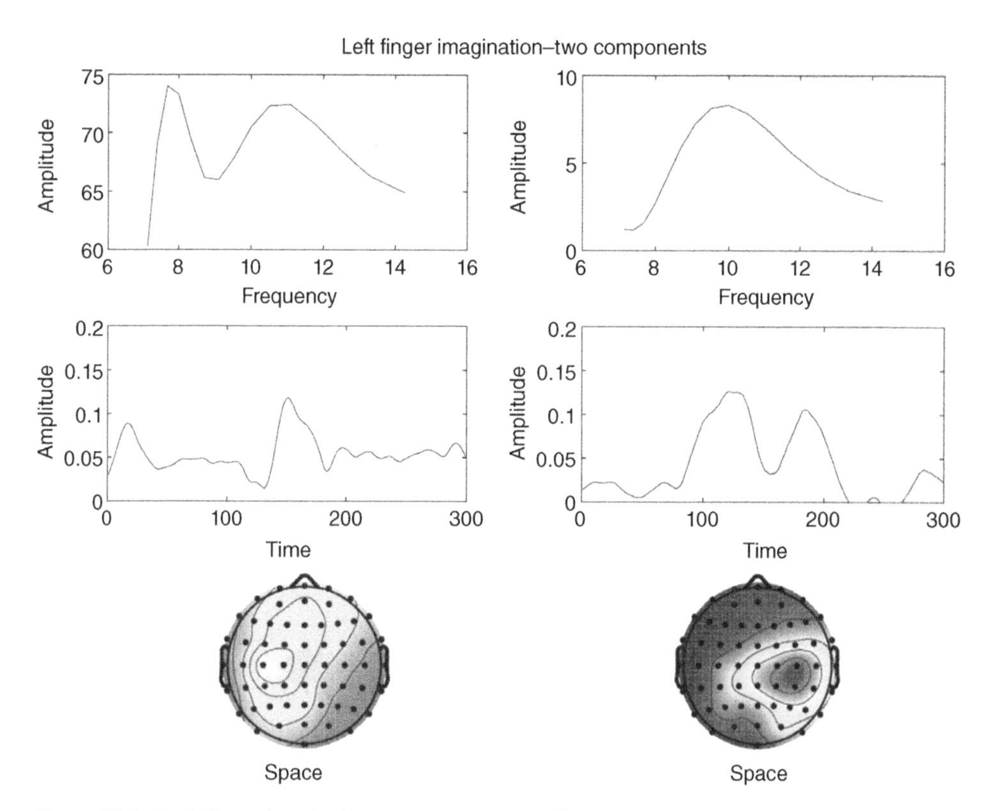

Figure 17.8 Left finger imagination: two components. The upper figure represents the spectral signature of the extracted components in Hz, the middle figure, the temporal signatures in milliseconds and the lower figures demonstrate the spatial distribution. The time onset of finger movement is on 200 ms.

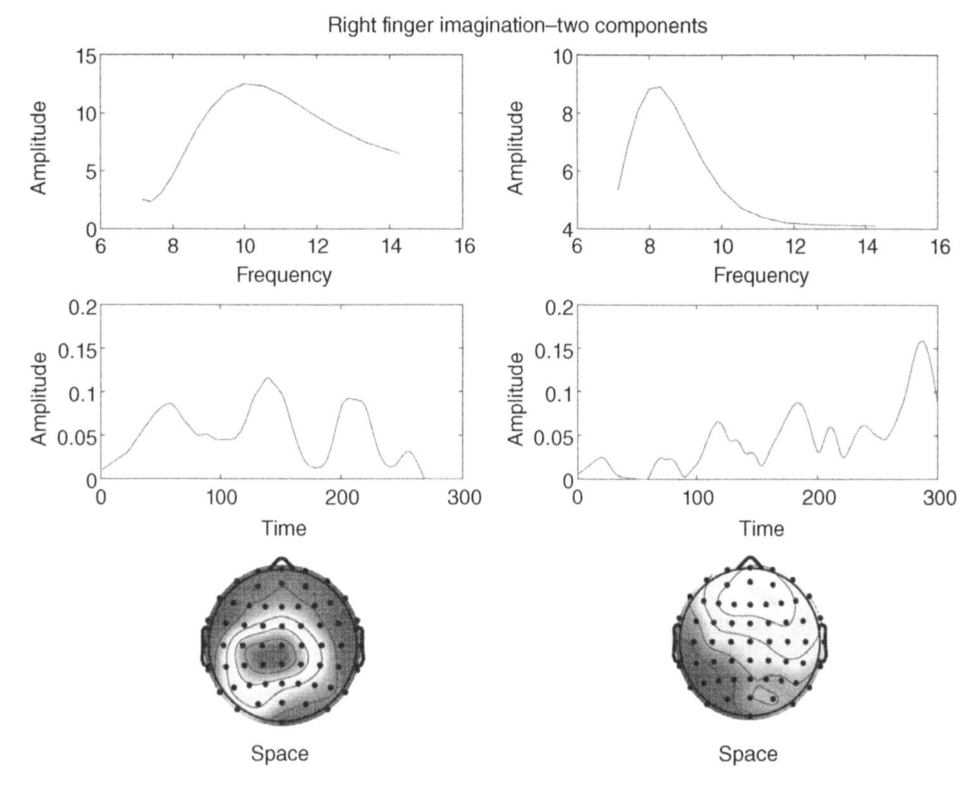

Figure 17.9 Right finger imagination: two components. The upper figures represent the spectral signatures of the extracted components in Hz, the middle figure, the temporal signatures in millisecond, and the lower figures demonstrate the spatial distribution. The time onset of finger movement is on 200 ms.

Combining PARAFAC with other signal-processing modality such as BSS or beamforming can be effectively used for the removal of EEG artefacts. Figure 17.10 represents the results using PARAFAC combined with beamforming [110]. In this approach the solution to the beamforming problem is used as a spatial constraint into the PARAFAC.

17.5 Detection and Separation of ERP Signals

Utilization of the ERP signals provides another approach in BCI design. The ERP-based BCI systems, such as those based on audio or visual EPs, often consider a small number of electrodes to study the movement-related potentials of certain body organs. Although P300 subcomponents, i.e. P3a and P3b can be best detected over F and P electrodes respectively, a single Cz channel recording is used for detection of both subcomponents. However, in recent works multichannel EEGs have been used followed by an efficient means of source separation algorithm in order to exploit the maximum amount of information within the recorded signals. The major problems in this context are related to detection, separation, and classification of the ERP signals which can be studied in Chapter 9. As stated

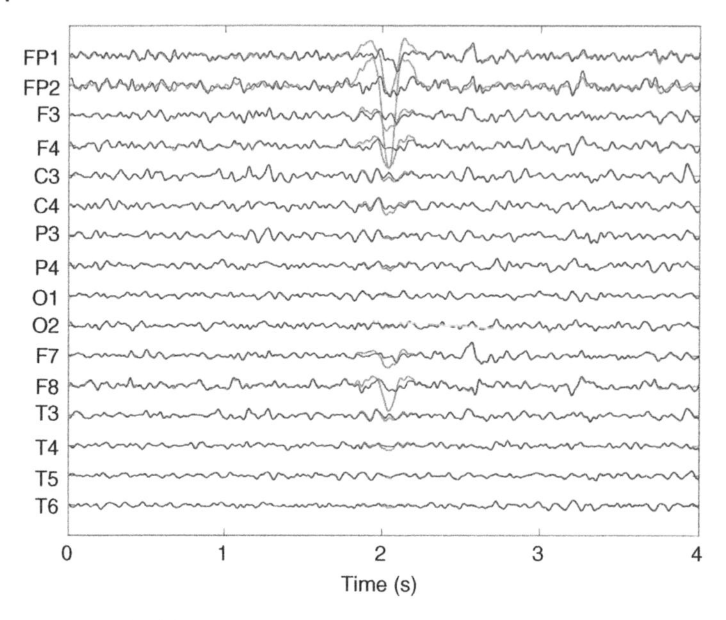

Figure 17.10 The EEG signals before the removal of eye-blinking artefacts in red (pale) and after removing the artefact in blue (dark) colour.

previously, these systems are initiated by introducing certain stimulations of the brain. As soon as the movement-related ERP components are classified the system can be used in the same way as in the previous sections. In single-trial applications the ERP components can be tracked in order to evaluate the state of the brain during BCI.

ERP-based real-time BCI systems require single-trial ERP (particularly P300) detection. Although some simple spatial filters such as xDAWN [111] or Fisher spatial filters [112] for ERP classification based on time point features are popular, accurate estimation of ERP subcomponents is a challenging problem and often complicated signal-processing techniques such as those described in Chapter 9 of this book need to be used for a reliable ERP detection. Moreover, the BCI system should discriminate between routine and novel stimuli. A novel or target stimuli increases the amplitude of P3b whereas the P3a amplitude remains more or less the same for both stimulation types. Therefore, even more challenging solutions have to be provided for P3b detection and tracking. Particle filtering ERP tracking regularized by other available information such as ERP source locations within the brain or the left–right symmetry of the P3a and P3b [113] have been the best solution for this problem.

Conversely, as explained in Chapter 15, the ERPs (particularly the P3b subcomponent of P300) change due to some brain diseases such as Alzheimer's disease (AD). AD reduces the P3b amplitude and increases its latency [114]. Another example can be seen in [113] where a cooperative particle filtering method has been developed to exploit the four-channel (F3, F4, P3, P4) information for estimation of the P3a and P3b parameters, i.e. amplitude, latency, and width. From their results it can be concluded that the P3b amplitude significantly reduces for a Schizophrenic patient whereas its latency increases.

17.6 Estimation of Cortical Connectivity

The planning and the execution of voluntary movements are related to the pre-movement attenuation and post-movement increase in amplitude of alpha and beta rhythms in certain areas of motor and sensory cortex [115, 116]. Also, it has been found that during movement planning, two rhythmical components in the alpha frequency range namely mu1 and mu2 play different functional roles. Differentiation of these two components may be achieved by using the matching pursuit (MP) algorithm [117] based on the signal energy in the two bands [118, 119].

The MP algorithm refers to the decomposition of signals into basic waveforms from a very large and redundant dictionary of functions. MP has been utilized for many applications such as epileptic seizure detection, evaluation, and classification by many researchers.

In modelling the brain activity for performing continuous movements, the changes in brain functional connectivity related to the neural pathways from active potentials and synaptic activities to electromyographic activities and eventually body movements need to be captured and quantified.

Determination of propagation of brain electrical activity, its direction, and the frequency content, is of great importance. DTFs using a MVAR model have been employed for this purpose [120]. In this approach the signals from all EEG channels are treated as realizations of a multivariate stochastic process. A short-time DTF (SDTF) was also developed [119] for estimation and evaluation of auto-regressive (AR) coefficients for short-time epochs of the EEGs.

MP and SDTF have been performed for analysis of the EEG activity during planning of self-paced finger movements, and we will discuss the results with respect to representation of the features of cortical activities during voluntary action.

The MP has been applied to decomposition of the EEGs to obtain reference-free data, and the averaged maps of power were constructed. ERP/ERS were calculated as described in part 17.3.2.

Model order is normally selected at the point where the model error does not decrease considerably. A well known criterion called Akaike information criterion (AIC) [121] has been widely used for this purpose. According to the AIC, the correct model order will be the value of p which minimizes the following criterion:

$$\mathrm{AIC}(p) = 2\,log\,(det(\mathbf{V})) + \frac{2kp}{N} \tag{17.13}$$

where $\mathbf{V}$ is variance matrix of the model noise, N is the data length for each channel, and k is the number of channels. Recently, a more accurate technique for estimation of the model order has been proposed [122]. Using this approach, the information criterion (IC) for the noisy case with unknown noise can be calculated as [122]:

$$IC(\rho) = \left(\rho + 1 - w + \frac{2(n_e-\rho)^2 + n_e - \rho + 2}{6(n_e-\rho)} - \sum\nolimits_{i=1}^{k} \frac{\lambda_{n_e-\rho}^2}{\left(\lambda_i - \lambda_{n_e-\rho}\right)^2}\right) \cdot \log\left(\lambda_{n_e-\rho}^{-(n_e-\rho)} \prod\nolimits_{i=\rho+1}^{n_e} \lambda_i\right) + 2g(\rho, n_e)\beta(w-1) \tag{17.14}$$

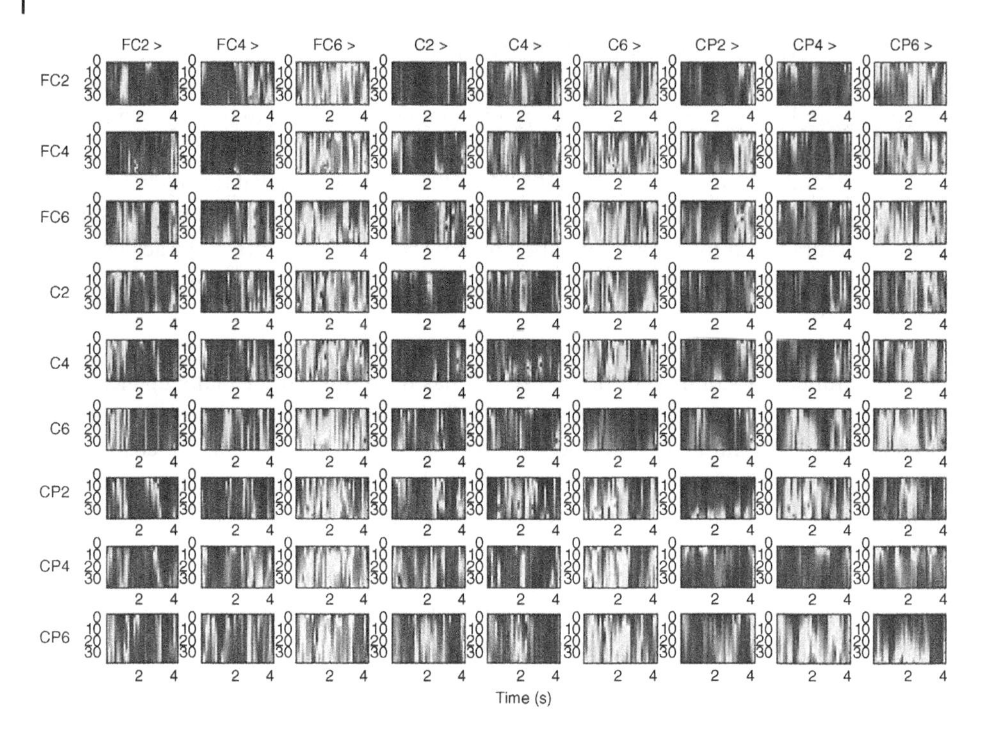

Figure 17.11 Illustration of source propagation from the coherency spectrum for the specified EEG channels for the left hand.

where in this equation $g(\rho, n_e) = k(n_e - \rho + 2)(n_e - \rho + 2)/2$, and $\bar{\lambda}_{n_e-\rho}$ is the average of the $n_e - \rho$ smallest eigenvalues. In this equation n_e is the number of channels, ρ is the order to be estimated, w is the number of time points in the data matrix.

The transitions in the DTF patterns can be illustrated for different EEG channels for left and right hands finger movements as depicted in Figures 17.11 and 17.12 respectively.

The direction of signal source movement is realized from the cross correlations between signals, which are computed for different time shifts in the procedure of correlation $\mathbf{R}(n)$ matrix estimation. These time shifts are translated into phase shifts by transformation to the frequency domain. The phase dependencies between channels are reflected in the transfer matrix. The DTF values express the direction of a signal component in a certain frequency (not the amount of delay) [116]. Analysis of the DTF values, however, will be difficult when the number of channels increases resulting in an increase in the number of MVAR coefficients.

The coherency of brain activities as described in the previous section may be presented from a different perspective namely, brain connectivity. This concept plays a central role in neuroscience. Temporal coherence between the activities of different brain areas are often defined as functional connectivity, whereas the effective connectivity is defined as the simplest brain circuit that would produce the same temporal relationship as observed experimentally between cortical regions [123]. A number of approaches have been proposed to estimate how different brain areas are working together during motor and cognitive tasks from the EEG and fMRI data [124–126].

Figure 17.12 Illustration of source propagation from the coherency spectrum for the specified EEG channels for the right hand.

Structural equation modelling (SEM) [127], explained comprehensively in Chapter 8, has also been used to model such activities from high resolution (both spatial and temporal) EEG data. Anatomical and physiological constraints have been exploited to change an underdetermined set of equations to a determined one.

The imaginary part of the S-transform (or S-coherency) has been often used as an estimation of brain functional connectivity. This technique has been explained in Chapter 8 too. In a pioneering work by Eftaxias et al. [128–130] the imaginary part of the S-transform is used as the functional connectivity measure with parameters $a_{lk}(t, f)$, and are used as the combination weights for a diffusion Kalman adaptative cooperative filter with a parameter vector $\varphi_{k,t}$ to classify clockwise and counter-clockwise hand drawings. A brief block diagram of the method is shown in Figure 17.13.

A similar approach using multitask diffusion adaptation has been introduced by Monajemi et al. [131] to identify the link between brain connectivity and hand tremor in Parkinsonian patients.

Application of EEG-based functional connectivity in BCI can open two new directions in BCI research. One is to model the pathways between thought to action without any need for feedback (visual, haptic, etc.). The second direction is to enable transfer of information from one brain to another in a brain hyperscanning framework. In such a framework multiple brains which perform similar or competitive tasks are EEG scanned. This is perhaps the main foundation of brain-to-brain communications or internet-of-brain (IoB) establishment.

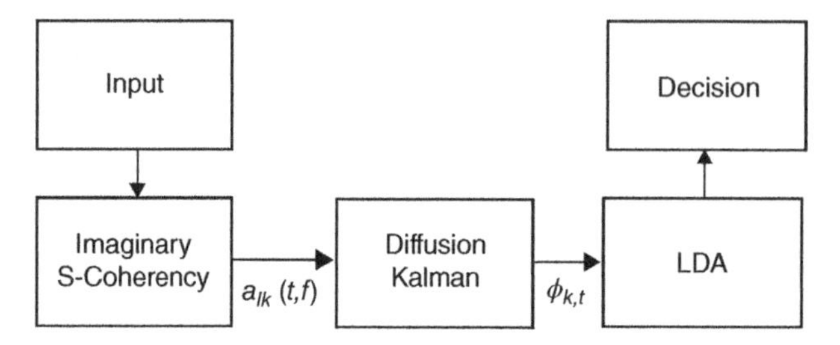

Figure 17.13 A block diagram of the system proposed in [128] for classification of hand movement trajectory.

17.7 Application of Common Spatial Patterns

As stated in Chapter 7, CSPs are probably the most effective feature selection algorithms used in BCI classification. In early 2000 in a two-class BCI setup, Ramoser et al. [72] proposed a CSP that learned to maximize the variance of bandpass filtered EEG signals from one class while minimizing their variance from the other class. Formally, the CSP($\mathbf{w}$) minimizes the Rayleigh quotient of the spatial covariance matrices to achieve the variance imbalance between the two classes of data $\mathbf{X}_1$ and $\mathbf{X}_2$, and is defined using:

$$J(\mathbf{w}) = \frac{\mathbf{w}^T\mathbf{X}_1^T\mathbf{X}_1\mathbf{w}}{\mathbf{w}^T\mathbf{X}_2^T\mathbf{X}_2\mathbf{w}} = \frac{\mathbf{w}^T\mathbf{C}_1\mathbf{w}}{\mathbf{w}^T\mathbf{C}_2\mathbf{w}} \tag{17.15}$$

where T denotes transpose, $\mathbf{X}_i$ is the data matrix for the ith class (with the training samples as rows and the channels as columns) and $\mathbf{C}_i$ is the spatial covariance matrix of the ith class signals, assuming a zero mean for EEG signals. As was stated in Chapter 7, the CSP problem is often solved by the generalized eigenvalue equation:

$$\mathbf{C}_1\mathbf{w} = \lambda\mathbf{C}_2\mathbf{w} \tag{17.16}$$

or $\mathbf{C}_2^{-1}\mathbf{C}_1\mathbf{w} = \lambda\mathbf{w}$. In detection and recognition of two-class patterns in BCI the CSP is widely used as an effective approach. Most of the existing CSP-based methods exploit covariance matrices on a subject-by-subject basis so that inter-subject information is neglected. CSP and its variants have received much attention and have been one of the most efficient feature extraction methods for BCI. However, despite its straightforward mathematics, CSP overfits the data and is highly sensitive to noise. To address these shortcomings, recently it has been proposed to improve the CSP learning process with prior information in terms of regularization terms. Lotte and Guan [132] reviewed, categorized and compared 11 different regularized CSP approaches: from regularization in the estimation of the EEG covariance matrix [133] to several different regularization methods such as composite CSP [134], regularized CSP with generic learning [135], regularized CSP with diagonal loading [134] and invariant CSP [136]. They applied these methods to the EEGs recorded from 17 patients and verified the superiority of CSP with Tikhonov regularization in which the optimization of $J(\mathbf{w})$ is penalized by minimizing $\|\mathbf{w}\|^2$,

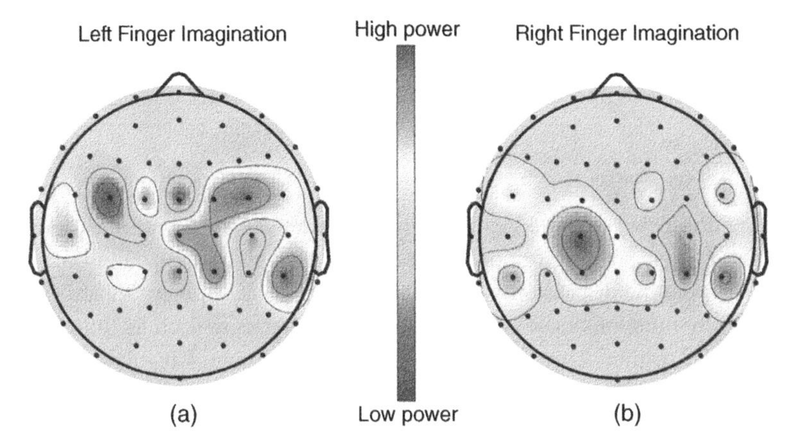

Figure 17.14 The results of applying CSP to classify the cortical activity of the brain for both (a) left-hand and (b) right-hand movements.

hence, minimizing the influence of artefacts and outliers. Regularization of **w** can also be used to reduce the number of EEG channels without compromising the classification score. Farquhar et al. [137] converted CSP into a quadratically constrained quadratic optimization problem with l_1-norm penalty and Arvaneh et al. [138] used l_1/l_2 -norm constraint. Recently, in [133] a computationally expensive quasi l_0-norm based principle has been applied to achieve a sparse solution for **w**.

The topography images in Figure 17.14 demonstrate the effectiveness of applying CSPs to distinguish between left- and right-hand movements from EEG recordings.

In [133] the CSP has been modified for subject-to-subject transfer, where a linear combination of covariance matrices of subjects under consideration has been exploited. In this approach a composite covariance matrix has been used that is a weighted sum of covariance matrices involving subjects, leading to composite CSP.

To alleviate the effect of noise the objective function in (17.24) may be regularized either during the estimation of the covariance matrix for each class or during the minimization process of the CSP cost function by imposing priors to the spatial filters **w** [132].

The proposed method in [135] for estimating the regularized covariance matrix for class i, $\hat{\boldsymbol{C}}_i$, is given as:

$$\hat{\boldsymbol{C}}_i = (1-\gamma)\boldsymbol{P} + \gamma\boldsymbol{I} \tag{17.17}$$

where

$$\boldsymbol{P} = (1-\beta)\boldsymbol{C}_i + \beta\boldsymbol{G}_i \tag{17.18}$$

In these equations $\boldsymbol{C}_i$ is the initial spatial covariance matrix, $\boldsymbol{G}_i$ is so-called generic covariance matrix, and γ and $\beta \in [0, 1]$ are the regularizing parameters. $\boldsymbol{G}_i$ is estimated empirically by averaging the covariance matrices of a number of trials for class i.

In regularizing the CSP objective function, conversely, a regularization term is added to the CSP objective function in order to penalize the resulting spatial filters that do not satisfy a given prior. This results in a slightly different objective function as [132]:

$$J(\mathbf{w}) = \frac{\mathbf{w}^T C_{X_1} \mathbf{w}}{\mathbf{w}^T C_{X_2} \mathbf{w} + \alpha Q(\mathbf{w})} \tag{17.19}$$

where the penalty function $\boldsymbol{Q}(\mathbf{w})$ weighted by the penalty parameter $\alpha \geq 0$ indicates how much the spatial filter $\mathbf{w}$ satisfies a given prior. This term needs to be minimized in order to maximize $J(\mathbf{w})$. Following the discussion in Chapter 7 of this book, different quadratic and non-quadratic penalty functions have been defined in the literature such as those in [132, 137], and [139]. The results achieved in implementing 11 different regularization methods in [132] shows that in places where the data are noisy, the regularization improves the results by approximately 3% in average. However, the best algorithm outperforms CSP by about 3–4% in mean classification accuracy and by almost 10% in median classification accuracy. The regularized methods are also more robust, i.e. they show lower variance across both classes and subjects. Among various approaches the weighted Tikhonov regularized CSP proposed in [132] has been reported to have the best performance.

Often not all the EEG channels are used in CSPs. In practical BCI applications the data recorded from the centro-parietal scalp sites (electrodes over the motor cortex) as highlighted in Figure 17.15, is initially bandpass filtered between 8 and 30 Hz before implementation of the CSP filter.

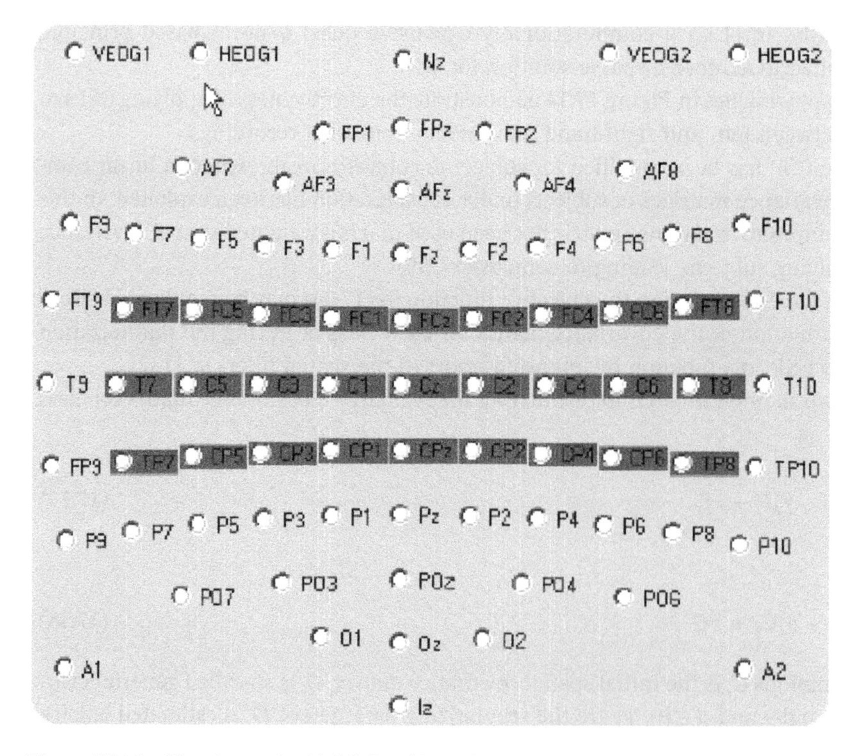

Figure 17.15 The electrodes highlighted in dark grey are those which are over the motor cortex and used in practical CSP applications.

17.8 Multiclass Brain–Computer Interfacing

Often a good model for performing of a full BCI task or detection of various body movements becomes useful and important. This cannot be achieved if a robust multiclass classifier is not in place. Most of the literature refers to two-class problems which can be solved using efficient classifiers such as SVM. For general applications multiclass classifiers such as NNs have always been an option. The efficiency or computational complexity of such algorithms has usually been under question.

Two new classification approaches for BCI have been introduced in [28]. One of these classifiers is based on the distance to Riemannian mean and the other one works in the Riemannian tangent space. Obviously, the popular classification algorithms such as LDA, SVM, and NNs cannot be implemented directly on the Riemannian manifold since they are based on projections into hyperplanes.

In this BCI approach consider short-time segments each including T_s samples of EEG signal or trials in the form of a matrix $\mathbf{X}_i = [\mathbf{x}_{t+T_i} \ldots \mathbf{x}_{t+T_i+T_s-1}] \in \Re^{n \times T_s}$ which corresponds to the ith trial of real or imagined movement started at time T_i. Define $\mathbf{P}_i$ as the $n \times n$ sample covariance matrix as:

$$\mathbf{P}_i = \frac{1}{T_s - 1}\mathbf{X}_i\mathbf{X}_i^T \tag{17.20}$$

and denote the space of all $n \times n$ symmetric matrices in the space of square real matrices by $S(n) = \{\mathbf{S} \in M(n), \mathbf{S}^T = \mathbf{S}\}$ and the set of all $n \times n$ symmetric positive-definite (SPD) matrices by $P(n) = \{\mathbf{P} \in S(n), \boldsymbol{u}^T\mathbf{P}\boldsymbol{u} > 0, \forall \boldsymbol{u} \in \Re^n\}$. Then, for finding the closest class to a new test input $\mathbf{X}$ which is an EEG trial of the unknown class, we use the training set to compute the above sample covariance matrix to find $\mathbf{P}_i$. Then, we apply this to X to calculate the sample covariance matrix $\mathbf{P}$. In the next step we calculate the Riemannian (also called as geometric) mean for all the classes $k = 1, \ldots, K$ as [28]:

$$\vartheta(\mathbf{P}_1, \ldots, \mathbf{P}_I) = arg \min_{\mathbf{P} \in P(n)} \sum\nolimits_{i=1}^{I} \delta_R^2(\mathbf{P}, \mathbf{P}_i) \tag{17.21}$$

where $\delta_R(\mathbf{P}_1\mathbf{P}_2)$ is called the Riemannian geodesic distance between $\mathbf{X}_1$ and $\mathbf{X}_2$ defined as:

$$\delta_R(\mathbf{P}_1\mathbf{P}_2) = \left\| \log\left(\mathbf{P}_1^{-1}\mathbf{P}_2\right) \right\|_F = \left[\sum_{i=1}^{n} \log^2\lambda_i\right]^{1/2} \tag{17.22}$$

where λ_i, $i = 1, \ldots., n$ are the real eigenvalues of $\mathbf{P}_1^{-1}\mathbf{P}_2$. Class $\hat{k}$ for which $\delta_R\left(\mathbf{P}, \mathbf{P}_{\hat{k}}\right)$ has the minimum value will be the corresponding class for $\mathbf{X}$.

The second multiclass classification approach is by using the Riemannian tangent space. In this approach Eq. (17.20) is applied first. Assuming the result of this minimization is denoted as $\mathbf{P}_\vartheta$ the corresponding class to $\mathbf{X}$ is then estimated as [28]:

$$\mathbf{s}_i = upper\left(\mathbf{P}_\vartheta^{-\frac{1}{2}}\mathrm{Log}_{\mathbf{P}_\vartheta}(\mathbf{P}_i)\mathbf{P}_\vartheta^{-\frac{1}{2}}\right) \tag{17.23}$$

Where the *upper*(.) operator keeps the upper triangular part of a symmetric matrix and vectorizing it by applying unity weight for diagonal elements and $\sqrt{2}$ weight for off-diagonal elements.

The above classification methods have been applied to a set of BCI competition IV data [140] from nine subjects who perform four kinds of motor imagery (right hand, left hand, foot, and tongue) movements. The results were compared with those of multiclass CSP (hierarchical CSP) and LDA. The overall performance of the above proposed system was proved to be superior to CSP and LDA.

17.9 Cell-Cultured BCI

Most of the information about this type of BCI comes from the public media and university websites. Researchers have built devices to interface with neural cells and entire NNs in cultures outside animals. Neurochips powered by neuroelectronics have been developed to enable stimulating, sampling and recording from neurons directly [141].

A **neurochip** is a chip (integrated circuit/microprocessor) that is designed for the interaction with neuronal cells. It is made of silicon that is doped in such a way, that it contains EOSFETs (electrolyte–oxide–semiconductor field effect transistors) that can sense the electrical activity of the neurons (action potentials) in the above-standing physiological electrolyte solution. It also contains capacitors for the electrical stimulation of these cells.

The world's first neurochip with 16 neurons connected developed by Pine and Maher in 1997 [142]. Based on another research by Berger in 2003, a neurochip was proposed to function as an artificial or prosthetic hippocampus. The chip was designed to function in rat brains and was intended as a prototype (an ancestor) for the eventual development of higher-brain prosthesis. The hippocampus was chosen because it is the origin of various activations and is the most ordered and structured part of the brain and has been well studied. Its main function is to encode experiences for archiving information as long-term memories in the brain [143].

DeMarse from the University of Florida used a culture of 25 000 neurons taken from a rat's brain to fly an F-22 fighter jet simulator [144]. After collection, the cortical neurons were cultured. The neurons rapidly began to reconnect themselves to form a living NN. The cells were arranged over a grid of 60 electrodes and used to control the pitch and yaw functions of the simulator. This study was mainly to understand and demonstrate how the human brain performs and learns computational tasks at a cellular level.

17.10 Recent BCI Applications

BCI is a direct communication pathway between the brain and the computer without stimulating any muscle activity. The BCI experiments often involved long duration of recordings in different sessions for different subjects. There have been some thorough studies of inter- and intra-subject as well as inter-session variabilities such as [145] and the references herein, to measure and incorporate these variabilities which can be in both event-related and movement-related responses. Nevertheless, EEG signals are highly user-specific, and as such, most current BCI systems are calibrated specifically for each user.

BCI enables physically disabled people, e.g. those suffering from tetraplegia, to communicate with a computer or to control an artificial limb. The BCI applications include, but are

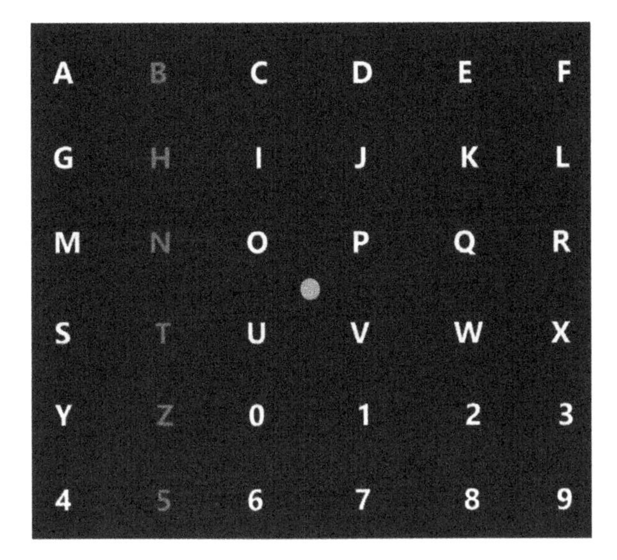

Figure 17.16 A typical stimulus grid for the speller BCI.

not limited to, lie detection [146], brain fingerprinting [147], mood assessment [122] and brain-to-brain interface (BBI) [148].

In the popular speller BCI, often based on SSVEP, a combination of letters and digits is presented visually in a row or column of *N* (such as 6) letters randomly. By detecting the row and the column corresponding to the largest P300s, the letter is determined. Figure 17.16 shows a letter/number board (stimulus grid) for speller BCI. A trial in this task may be defined as a highlight (500 ms) of either a row or a column. At the beginning of each trial block, the subjects become aware of a target letter/number. During subsequent trials, the subjects focus (eye-fix) on the central green dot and see if the target letter/number is shown in the highlighted row/column. Trials could be organized in displaying cycles with each row and column being highlighted, in random order, only once in each cycle.

In a different EEG-based speller BCI paradigm, a new mental spelling system based on SSVEP has been introduced, adopting a QWERTY style layout keyboard with 30 LEDs flickering with different frequencies. The speller system allows the users to spell one target character per target selection, without the need for multiple step selections adopted by conventional SSVEP-based mental spelling systems [149].

Through preliminary offline and online experiments, it has been confirmed that human SSVEPs (recorded over occipital electrodes) elicited by visual flickering stimuli with a frequency resolution of 0.1 Hz could be classified with classification accuracy high enough to be used for a practical BCI system. During the preliminary offline experiments various factors influencing the performance of the mental spelling system, such as distances between adjacent keys, light source arrangements, stimulating frequencies, recording electrodes, and visual angles have been optimized. With additional online experiments the feasibility of the optimized mental spelling system has been verified. They achieved an average typing speed of 9.39 letters per minute (LPM) with an average success rate of 87.58%, corresponding to an average information transfer rate (ITR) of 40.72 bits per minute, demonstrating the good performance of the developed mental spelling system.

In [146] a multichannel ERP-based BCI system for lie detection has been proposed. Bootstrapped geometric difference (BGD) using wavelet features and network analysis using the graph analysis technique addressed in Chapters 8 and 15, were proposed and applied to a feature recognition and classification system. They used visual and auditory stimuli and for all subjects, BGD of the P300 for all the scalp electrodes combined with SVM classifier showed the average rate of recognition accuracy was 84.4 and 82.2% for visual and auditory modality respectively. Statistical analysis of network features indicated the difference in the two groups were significant and the average accuracy rate reached 88.7 and 83.5% respectively. The results suggest the BGD and network analysis-based approaches combined with SVM are effective in lie detection.

In [147] the authors performed a classification concealed information test (CIT) [150] using the 'brain fingerprinting' method of applying P300 in detecting information that is (i) acquired in real life and (ii) unique to US Navy experts in military medicine. The military medicine experts and non-experts were asked to push buttons in response to three types of text stimuli. They used the single channel Pz channel signal filtered within 6–8 Hz. In this work they compared the results of using the P300 alone vs. the P300 plus the late negative (ERP) components. The targets contain known information relevant to military medicine, which are identified to subjects as relevant, and require pushing one button. Subjects are told to push another button to all other stimuli. The probes contain concealed information relevant to military medicine and are not identified to subjects. Those irrelevant to the subject contain equally plausible, but incorrect/irrelevant information.

In [148] a brain-to-brain connection has been proposed which links the EEG measurements recoded during a finger movement from one brain to a transcranial magnetic stimulator (TMS) which induces a magnetic field to another brain. The TMS induction activates the motor cortex for moving the same finger of the second person.

A visuomotor task has been used in which two humans must cooperate through direct brain-to-brain communication to achieve a desired goal in a computer game. The BBI detects motor imagery in EEG signals recorded from one subject (the 'sender') and transmits this information over the Internet to the motor cortex region of a second subject (the 'receiver'). This allows the sender to cause a desired motor response in the receiver (a press on a touchpad) via TMS. This requires decoding the sender's signals, generating a motor response from the receiver upon stimulation, and achieving the overall goal in the cooperative visuomotor task [148].

BCI spellers have rarely been explored for free communication in healthy individuals. True free communication involves translating momentary thoughts to text, continuously and in real time. However, current tests of 'free' spelling often involve users repeating a small number of phrases provided by the experimenter, either from memory or with assistance from salient cues [151]. The same group of researchers in [151, 152] developed such spellers with ITRs of ~267 and ~325 bits per minute (bpm) respectively for cued spelling, both with mean classification accuracies of ~90%. These ITRs are the fastest to date, with other state-of-the-art spellers averaging ~146 bpm. This impressive improvement suggests that BCI spellers may become a viable option for hands-free communication outside of clinical settings. Therefore, in another research, a BCI system has been set up to establish free

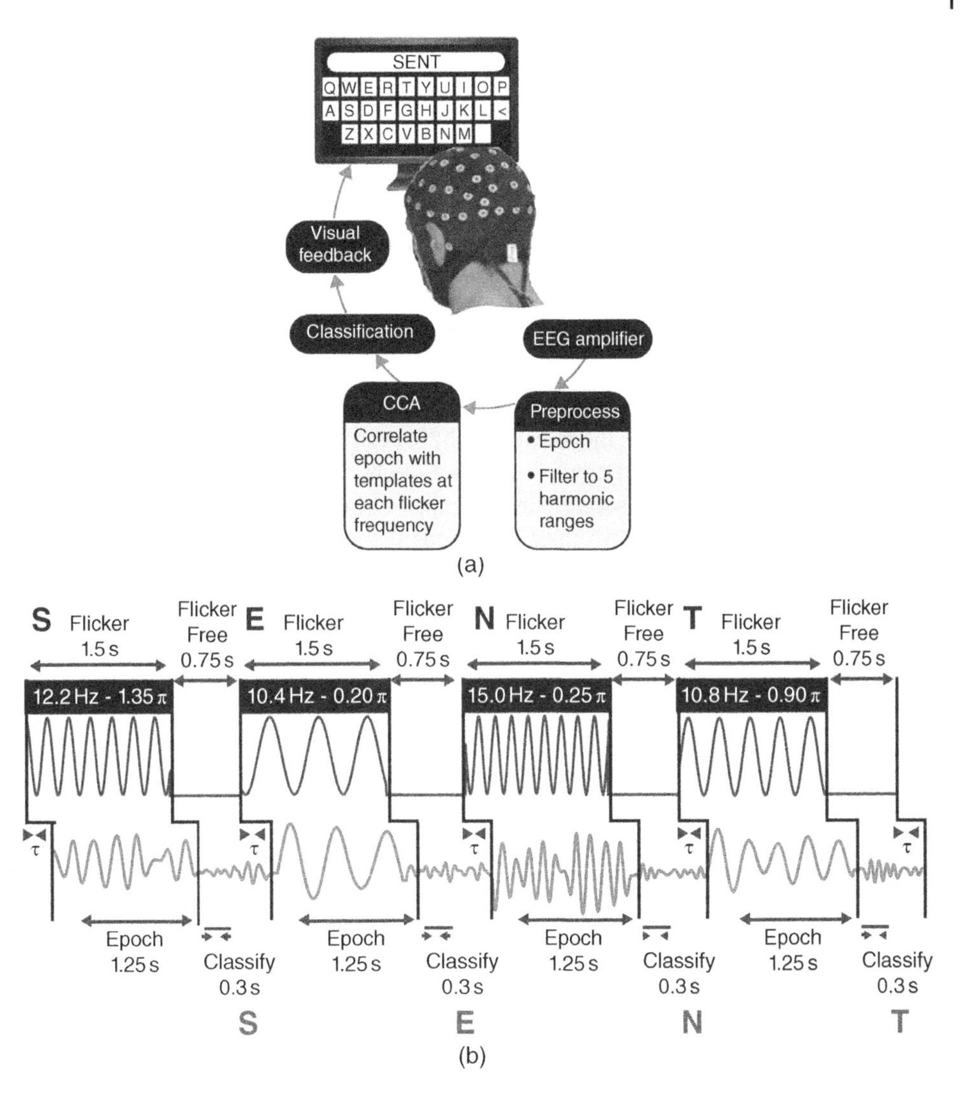

Figure 17.17 (a) The user operates the real-time feedback loop to freely type words and phrases using his/her brain activity alone. (b) An example timeline of visual stimulation and evoked EEG involved in BCI typing of the word 'SENT' can be seen [153].

communications between two brains by assembling a filter-bank canonical correlation analysis (CCA) SSVEP-based speller [153] while overcoming the problem of slow SSVEP-based systems. They used a QWERTY keyboard [154] and tailored the visual flicker frequency to optimize it for each individual user.

In Figure 17.7 (a) the user operates the real-time feedback loop to freely type words and phrases using his/her brain activity alone. The user selects each character in a sequence one-by-one by focusing his/her attention and fixating his/her gaze on sinusoidally flickering keys of a virtual QWERTY keyboard on a computer display, which evokes oscillatory SSVEP responses at the corresponding flicker frequency/ phase in the EEG. EEG time-locked to flicker is extracted, bandpass filtered to five harmonic ranges and then submitted to a filter-bank CCA with respect to a bank of individualized training templates. The classified frequency is the template most highly correlated with the real-time EEG, with the corresponding character displayed as feedback at the top of each key. The user is free to select the next character, or to select the backspace key [153]. In Figure 17.17 (b) an example timeline of visual stimulation and evoked EEG involved in BCI typing of the word 'SENT' can be seen. Each key flickers at a unique frequency/phase for 1.5 seconds, followed by a 0.75 seconds flicker-free period, during which the letter was classified and participants shifted their attention to the next key. Focusing attention on a key which potentiated the corresponding SSVEP response, increases the likelihood that the corresponding letter would be classified. The parameter τ refers to the SSVEP delay relative to flicker onset, calculated separately for each frequency and harmonic. Each key flickers at a unique frequency/phase, ranging from 10 Hz/1.5π – 15.4 Hz/0.95π.

In a project called RoboChair, the researchers have been working on improving the movement of brain-controlled wheelchair since 1996. RobChair is a brain-actuated robotic wheelchair aiming at helping severely motor-impaired people gain some control of their mobility, contributing to improve their life standards, and, ultimately, to increase their social inclusion. It works mainly based on ERPs and neurofeedback [155]. The remaining challenges for this project are (i) sparse and unreliable BCI commands which result in low reliability of EEG decoding, low BCI speed, decay in performance due to time lags, and having sparse and discrete commands which makes the control difficult, (ii) there is need for high mental effort/workload which may not be easily tolerable for all subjects, and (iii) attention shifts and EEG variability leading to mental fatigue and consequently changes in the amplitude and latency of P300 waveform.

17.11 Neurotechnology for BCI

Direct communication with the nervous system is one of the primary goals in neuroengineering. Neurotechnology is a fascinating and, at the same time, controversial field as one of its goals is to directly 'wire up' human brains to machines. It is probably the fastest developing technology related to BCI and brain-related treatments and different groups across the world are active in that. We should indeed expect to encounter such hybrid brain–machine systems more frequently in the near future.

Neuroprosthesis, as the major application of neurotechnology, together with BCI can help restore function for people with neuromuscular disorders such as amyotrophic lateral sclerosis, cerebral palsy, stroke, or spinal cord injury. There is a long history of body prosthesis starting from fifteenth century BCE, when a prosthetic toe was adopted, and evolved until the late 1960s when the first neuroprosthesis was applied to a monkey by Fetz, so the monkey could learn to consciously control the firing rate of its own cortical neurons [156]. Neuroprosthetic interfaces must be biocompatible, so they do not disturb the tissues and their functions as long as the prosthesis is applied.

The most popular and usable public neurotechnology products are flexible and high-density cortical electrode arrays, retinal vision prosthesis, paddle electrodes, cochlear implants, and deep brain stimulators. Figure 17.18 shows the location of these prosthesis and electrodes [157].

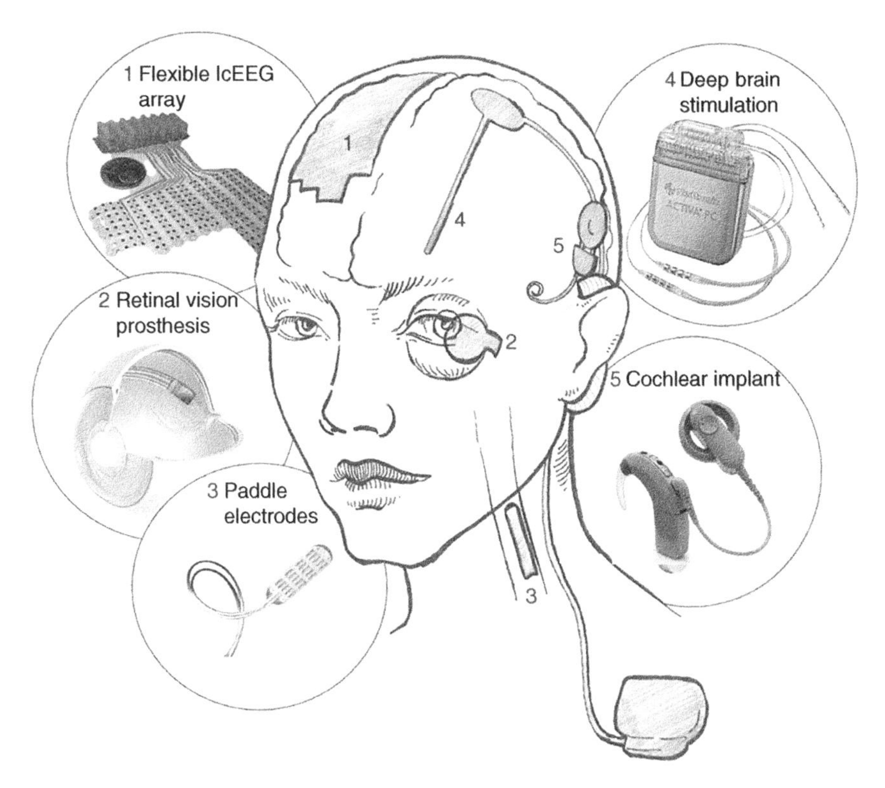

Figure 17.18 The locations of the most popular neurotechnology products [157].

Neuromodulation involves stimulating specific areas of the brain. There are several types of neuromodulatory stimuli, each of which has different properties and uses. More than 150 000 people in the US already have therapeutic brain implants, mainly for treating Parkinson's disease. Neuromodulation therapies offer an alternative to pharmaceuticals for treating chronic conditions. These treatments are generally more targeted, less expensive, and have fewer side effects than pharmaceuticals. One application of this technology is a treatment for insurmountable pain that involves stimulating the spinal and peripheral nervous system in separate parts of the brain.

Deep brain stimulation (DBS) is used to treat motor disorders. A medical device called neurostimulator sends electrical impulses through implanted electrodes to specific targets in the brain for the treatment of movement disorders, including Parkinson's disease, essential tremor, and dystonia. DBS has also been studied in clinical trials as a potential treatment for chronic pain (mostly by spinal cord stimulation) and epilepsy.

Similar to the way a pacemaker corrects an abnormal heartbeat, a neuromodulation device can establish a neurological balance that may help reduce pain. The treatment involves placing electrodes next to a specific spinal area presumed to be the source of pain.

Vagus nerve stimulation involves delivering electrical impulses to the vagus nerve. The relationship between depression, inflammation, metabolic syndrome, and heart disease might be mediated by the vagus nerve. Vagus nerve stimulation is sometimes used for epilepsy and depression.

In Japan, a brain-mapping project named Brain Mapping by Integrated Neurotechnologies for Disease Studies (Brain/MINDS) started in 2014 (http://www.brainminds.jp).

In the USA DARPA has funded and demonstrated advanced neurotechnologies that rely on surgically implanted electrodes to interface with the central or peripheral nervous systems (https://www.darpa.mil/news-events/2019-05-20). The agency has demonstrated achievements such as neural control of prosthetic limbs and restoration of the sense of touch to the users of those limbs, relief of otherwise intractable neuropsychiatric illnesses such as depression, and improvement of memory formation and recall. Due to the inherent risks of surgery, these technologies have so far been limited to use by volunteers with clinical need including military service personals. Currently, there are six teams pursuing a range of approaches that use optics, acoustics, and electromagnetics to record neural activity and/or send signals back to the brain at high speed and resolution. Both invasive and noninvasive approaches are taken. The teams are:

- The Battelle team aiming to develop a minutely invasive interface system that pairs an external transceiver with electromagnetic nanotransducers that are nonsurgically delivered to neurons of interest. The nanotransducers would convert electrical signals from the neurons into magnetic signals that can be recorded and processed by the external transceiver, and vice versa, to enable bidirectional communication.
- The Carnegie Mellon University team, aiming to develop a completely noninvasive device that uses an acousto-optical approach to record from the brain and interfering electrical fields to write to specific neurons. The team will use ultrasound waves to guide light into and out of the brain to detect neural activity. The team's write approach exploits the non-linear response of neurons to electric fields to enable localized stimulation of specific cell types.
- The Johns Hopkins University Applied Physics Laboratory team, aiming to develop a completely noninvasive, coherent optical system for recording from the brain. The system will directly measure optical path-length changes in neural tissue that correlate with neural activity.
- The PARC team, aiming to develop a completely noninvasive acousto-magnetic device for writing to the brain. Their approach pairs ultrasound waves with magnetic fields to generate localized electric currents for neuromodulation. The hybrid approach offers the potential for localized neuromodulation deeper in the brain.
- The Rice University team, aiming to develop a minutely invasive, bidirectional system for recording from and writing to the brain. For the recording function, the interface will use diffuse optical tomography to infer neural activity by measuring light scattering in neural tissue. To enable the write function, the team will use a magneto-genetic approach to make neurons sensitive to magnetic fields.
- The Teledyne team, aiming to develop a completely noninvasive, integrated device that uses micro optically pumped magnetometers to detect small, localized magnetic fields that correlate with neural activity. The team will use focused ultrasound for writing to neurons.

In addition, there are over 21 ongoing companies and startups in the United States involving in brain–machine interfaces, implants, and neuroprosthetics.

17.12 Joint EEG and Other Brain-Scanning Modalities for BCI

Typically, hybrid BCIs such as EEG-functional near-infrared spectroscopy (fNIRS) [158, 159] or EEG–fMRI [160] provide a higher classification accuracy than those of unimodal BCIs [161, 162]. This is mainly due to low spatial resolution of EEG and low temporal resolution of imaging modalities such as fMRI or low depth of imaging of NIRS. Recent MEG technology also allows for joint EEG–MEG due to higher sensitivity of MEG to depth and conductivity.

Joint EEG–fMRI for BCI

Although the use of fMRI solely or joint with EEG is not practical for real-life BCI applications, there have been some experiments to exploit the high spatial resolution of fMRI to boost the BCI performance often achieved using EEG only. In [149] a new tensor factorization approach for transferring the capabilities of fMRI to EEG is proposed which includes simultaneous EEG–fMRI sessions for finding a mapping from EEG to fMRI, followed by a BCI run from only EEG data, but driven by fMRI-like features obtained from the mapping identified previously [163].

In their method, they used orthogonal tensor decomposition using Tucker model as a method for subspace decomposition of EEG and deconvolving fMRI. Then, they designed a higher order partial least squares (HOPLS) model for discovering the relationships between the latent subspace representations of EEG and deconvolved fMRI [149].

Their experiments involve an open loop P300-based speller paradigm wherein the brain activity can be decoded using latent features extracted from simultaneously acquired EEG–fMRI data. Letter decoding accuracy using EEG–fMRI data has been shown to outperform the accuracy obtained from only EEG.

This data-driven method is likely to discover latent linkages between electrical and hemodynamic signatures of neural activity previously unexplored using model-driven methods, and is likely to serve as a template for a multimodal strategy wherein cross-modal EEG–fMRI interactions are exploited for the operation of a unimodal EEG system.

17.12.1 Joint EEG–fNIRS for BCI

NIRS is relatively robust to electrical noise and has good spatial resolution. To alleviate the ITR degradation, the authors in [159] proposed a simultaneous EEG–fNIRS operated by intuitive mental tasks, such as mental arithmetic (MA) and word chain (WC) tasks, performed within a short trial length (5 seconds). The subjects performed MA and WC tasks without preliminary training and remained relaxed (baseline; BL). The ITR (bits per second) is defined as:

$$\mathrm{ITR} = m\bullet\left(\log_2 N + P\log_2 P + (1-P)\log_2\frac{1-P}{N-1}\right) \tag{17.24}$$

where m is the number of trials per minute, N is the available number of commands, and $0 \le P \le 1$ is the classification accuracy. During the MA task the participants were instructed to perform continuous single digit (between 6 and 9) subtraction from a random three digit number (e.g. 567 – 8: 567 – 8 = 559, 559 – 8 = 551, 551 – 8 = 543, etc.). During the WC task, the participants were instructed to continuously come up with a word starting with the last letter of a former word (e.g. in English: B: Boy–year–rabbit–tree, etc.) as fast as possible. The participants were instructed to avoid repeating the same words. Because the first letter changed depending on the word the participants came up with, it was difficult to control the level of task difficulty with the initial letter of the WC task.

Then, they bandpass filtered the EEG data and used ICA to remove eye blink and EOG. Then, they used CSP to extract the features in different frequency bands.

The fNIRS data represent the changes in concentration of deoxyhaemoglobin and oxyhaemoglobin (HbR and HbO) used as the features. Then, they classified the MA- or WC-related brain activations from the BL-related activations using shrinkage LDA (sLDA) [164, 165]. In sLDA the covariance matrix $\mathbf{R}$ is replaced with $(1-\lambda)\mathbf{R} + \lambda\mathbf{I}$, where $\mathbf{I}$ is an identity matrix with the same size as $\mathbf{R}$ and λ is determined based on the Ledoit–Wolf lemma [166]. They achieved average classification accuracies of $90.0 \pm 7.1/85.5 \pm 8.1\%$ and $85.8 \pm 8.6/79.5 \pm 13.4\%$ for MA vs. BL and WC vs. BL, respectively. These are higher than those of the unimodal EEG- or NIRS-BCI in most cases.

17.12.2 Joint EEG–MEG for BCI

In [167] simultaneous EEG and MEG have been used to enhance the accuracy of a motor imagery BCI system. MEG has higher source depth and conductivity sensitivities compared to EEG. These data complement the EEG information. Although generally EEG–MEG is not very practical but the in the light of recent development of portable MEG sensors, based on optically pumped magnetometers such an approach can become more popular. In this particular BCI setup the subjects had to perform a sustained modulation index (MI) (grasping) of the right hand to hit up-targets, while remaining at rest to hit down-targets. Each run consisted of 32 trials with up-targets and down-targets, consisting of a grey vertical bar displayed on the right side of the screen, equally and randomly distributed across trials.

The EEG and MEG data, sampled at 250 Hz, were bandpass filtered to between 4 and 40 Hz and the power spectrum calculated. For each modality, 10 features in each frequency band were calculated and LDA with fivefold cross-validation was used to classify them.

The authors conclude that the integrating information from simultaneous EEG and MEG signals improves BCI performance. E/MEG multimodal BCIs may turn out to be an effective approach to enhance the reliability of brain–machine interactions. However, much of the progress will depend on the miniaturization of MEG scanners, which currently require a magnetic shielding room and sensors cooled via a cryogenic system.

17.13 Performance Measures for BCI Systems

To evaluate BCI performance, different blocks of the BCI system have their own performance measures even though their performance highly depends on the classifiers. Different classifiers used in the BCI systems have been explained in Chapter 7 of this book. For the classifier alone, there are two important measures: one based on the goodness of the design reflected in the cross-validation error and the other one is based on its robustness with respect to different test inputs and the achieved output error. This is valid only if the classes are balanced [168], i.e. with the same number of samples per class and if the classifier is unbiased meaning that it has the same performance for each class [169]. If these conditions are not met, the Kappa metric or the confusion matrix are more informative performance measures [168]. The sensitivity (recall, hit rate, or true positive rate) and specificity (selectivity or true negative rate), or precision, can be computed from the confusion matrix and depends on the number of true positives (TPs), false positives (FPs), true negatives (TNs) and false negatives (FNs):

$$\text{Sensitivity} = \frac{\text{TP}}{\text{TP} + \text{FN}} \tag{17.25}$$

$$\text{Specificity} = \frac{\text{TN}}{\text{FP} + \text{TN}} \tag{17.26}$$

When the classification depends on a continuous parameter (e.g. a threshold), the ROC and the area under the curve (AUC) are often used.

Another important factor is the number of outliers which refers to the number of data samples which are not close to any cluster centres. The classifiers should be capable of rejecting these outliers during the training sessions otherwise, they have to be manually removed from the dataset.

Nevertheless, the contribution of classifier performance to overall BCI performance strongly depends on the contribution of various BCI system subcomponents. The contributions, however, are highly variable given the variety of BCI systems (co-adaptive, hybrid, passive, self- or system-paced).

17.14 Summary

Although EMG is still used widely for both BCI and robotics aiming at rehabilitation, BCI using EEG has been growing during the last three decades. BCI using cortical electrodes is more effective but requires invasive operation. A review of the ongoing research has been provided in this chapter. Static features measured in different EEG conventional frequency bands have been widely used in classification of finger, arm, and leg movements. Dynamic features such as those characterizing the motion of movement-related sources have also been considered recently. Finally, estimation of the cortical connectivity patterns provides a new tool in evaluation of the directivity of brain signals and localization of the movement-related sources.

The advances in signal processing specially in detection, separation, and classification of brain signals have led to very exciting results in BCI, however, as yet, not all physiological, anatomical, and functional constraints have been taken into account. Also, as yet, no gold standard has been defined for any of the approaches in BCI-based applications particularly those designed for paralyzed subjects. The solution effectiveness depends on the type and the level of subject disabilities. Moreover, there has not been any attempt to provide any BCI system for subjects suffering from mental disorders.

Machine-learning systems are central to the BCI design. CSPs and their regularized variants show sufficient robustness and accuracy for the majority of the brain two-class classifications. New multiclass approaches such as in [28] are useful when more than two motor activities are to be separated. In very recent applications Riemannian approaches have been introduced and successfully applied [29]. DNNs are not generally useful due to low speed. Nevertheless, incorporation of transfer learning is paving the way for BCI usage with reasonable speed.

Often, EEG patterns change with time. Visual [42, 43], auditory [44], and other types of feedback BCI systems [170] seem to provide more robust solutions in the presence of these changes. Development of online adaptive BCI systems such as that in [11] enhances the application of BCI in various areas. A complete feedback system, however, requires undertaking more research and investigation.

Generally, to achieve a clinically useful BCI (invasive or noninvasive) system stability, low noise, and long recordings from multiple brain regions/electrodes are necessary. In addition, more computationally efficient algorithms have to be developed in order to cope with the real-time applications.

Combining BCI with VR helps the subjects better learn how to use brain plasticity to incorporate prosthetic devices into the body representation. This will make the prosthetic feel like a natural part of the body of the subject.

Communications between brains by means of simultaneously recordings the EEG signals from multiple brains (named EEG hyperscanning) is part of the plan for next generation BCI systems. To ensure subjects freely moving around new wireless EEG recording devices are filling the market.

Finally, application of BCI for rehabilitation purposes requires a significant improvement in the design of miniaturized implantable neuroprosthesis some equipped with powerful wireless communication systems.

References

1 Wolpaw, J.R., Birbaumer, N., Heetderks, W.J. et al. (2000). Brain-computer interface technology: a review of the first international meeting. *IEEE Transactions on Rehabilitation Engineering* 8 (2): 164–173.

2 Vidal, J.J. (1973). Direct brain-computer communication. *Annual Review of Biophysics and Bioengineering* 2: 157–158.

3 Vidal, J.J. (1977). Real-time detection of brain events in EEG. *Proceedings of the IEEE* 65: 633–664.

4 Chapin, J.K., Moxon, K.A., Markowitz, R.S., and Nicolelis, M.A. (1999). Real-time control of a robot arm using simultaneous recorded neurons in the motor cortex. *Nature Neuroscience* 2: 664–670.

5 Lebedev, M.A. and Nicolelis, M.A.L. (2006). Brain-machine interfaces: past, present and future. *Trends in Neurosciences* 29 (9).

6 Blankertz, B., Dorhege, D., Krauledat, M. et al. (2006). The Berlin brain computer interface: EEG-based communication without subject training. *IEEE Transactions on Neural Systems and Rehabilitation Engineering* 14: 147–152.

7 Blankertz, B., Dornheg, G., Schäfer, C. et al. (2003). Boosting bit rates and error detection for the classification of fast-paced motor commands based on single-trial EEG analysis. *IEEE Transactions on Neural Systems and Rehabilitation Engineering* 11: 127–131.

8 Blankertz, B., Müller, K.-R., Curio, G. et al. (2004). The BCI competition 2003: Progress and perspectives in detection and discrimination of EEG single trials. *IEEE Transactions on Biomedical Engineering* 51: 1044–1051.
9 Wolpaw, J.R., McFarland, D.J., and Vaughan, T.M. (2000). Brain-computer interface research at the Wadsworth Centre. *IEEE Transactions on Neural Systems and Rehabilitation Engineering* 8: 222–226.
10 Wolpaw, J.R. and McFarland, D.J. (2003). Control of two-dimensional movement signal by a non-invasive brain-computer interface in human. *National Academy of Sciences of the United States of America* 101: 17849–17854.
11 Vidaurre, C., Schogl, A., Cabeza, R. et al. (2006). A fully on-line adaptive BCI. *IEEE Transactions on Biomedical Engineering* 53 (6): 1214–1219.
12 Peters, B.O., Pfurtscheller, G., and Flyvbjerg, H. (2001). Automatic differentiation of multichannel EEG signals. *IEEE Transactions on Biomedical Engineering* 48: 111–116.
13 Müller-Putz, G.R., Neuper, C., Rupp, R. et al. (2003). Event-related beta EEG changes during wrist movements induced by functional electrical stimulation of forearm muscles in man. *Neuroscience Letters* 340: 143–147.
14 Müller-Putz, G.R., Scherer, R., Pfurtscheller, G., and Rupp, R. (2005). EEG-based neuroprosthesis control: a step towards clinical practice. *Neuroscience Letters* 382: 169–174.
15 Pfurtscheller, G., Muller, G.R., Pfurtscheller, J. et al. (2003). Thought' - control of functional electrical stimulation to restore hand grasp in a patient with tetraplegia. *Neuroscience Letters* 351: 33–36.
16 Millan, J.D.R., Mourino, J., Franze, M. et al. (2002). A local neural classifier for the recognition of EEG patterns associated to mental tasks. *IEEE Transactions on Neural Networks* 13: 678–686.
17 Millan, J.D.R. and Mourino, J. (2003). Asynchronous BCI and local neural classifiers: an overview of the adaptive brain interface project. *IEEE Transactions on Neural Systems and Rehabilitation Engineering* 11: 1214–1219.
18 Millan, J.D., Renkens, F., Mourino, J., and Gerstner, W. (2004). Noninvasive brain-actuated control of a mobile robot by human EEG. *IEEE Transactions on Biomedical Engineering* 53: 1214–1219.
19 Birbaumer, N., Ghanayim, N., Hinterberger, T. et al. (1999). A spelling device for the paralysed. *Nature* 398: 297–298.
20 Wolpaw, J.R., McFarland, D., and Pfurtscheller, G. (2002). Brain computer interfaces for communication and control. *Clinical Neurophysiology* 113 (6): 767–791.
21 Hinterberger, T., Veit, R., Wilhelm, B. et al. (2005). Neuronal mechanisms underlying control of a brain-computer interface. *The European Journal of Neuroscience* 21: 3169–3181. and N. Birbaumer
22 Kubler, A., Kotchoubey, B., Kaiser, J. et al. (2001). Brain-computer communication: unlocking the locked-in. *Psychological Bulletin* 127: 358–375.
23 Kubler, A., Kotchoubey, B., Kaiser, J. et al. (2001). Brain-computer communication: self regulation of slow cortical potentials for verbal communication. *Archives of Physical Medicine and Rehabilitation* 82: 1533–1539.
24 Obermaier, B., Muller, G.R., and Pfurtscheller, G. (2003). Virtual keyboard controlled by spontaneous EEG activity. *IEEE Transactions on Neural Systems and Rehabilitation Engineering* 11: 422–426.

25 Obermaier, B., Muller, G.R., and Pfurtscheller, G. (2001). Information transfer rate in a five-classes brain-computer interface. *IEEE Transactions on Neural Systems and Rehabilitation Engineering* 9: 283–288.

26 Wolpow, J.R. (2004). Brain computer interfaces (BCIs) for communication and control: a mini review. *Supplements to Clinical Neurophysiology* 57: 607–613.

27 Birbaumer, N., Weber, C., Neuper, C. et al. (2006). Brain-computer interface research: coming of age. *Clinical Neurophysiology* 117: 479–483.

28 Barachant, A., Bonnet, S., Congedo, M., and Jutten, C. (April. 2012). Multiclass brain-computer interface classification by Riemannian Geometry. *IEEE Transactions on Biomedical Engineering* 59 (4): 920–928.

29 Yger, F., Berar, M., and Lotte, F. (Oct. 2017). Riemannian approaches in brain-computer interfaces: a review. *IEEE Transactions on Neural Systems and Rehabilitation Engineering* 25 (10): 1753–1762.

30 Kauhanen, L., Palomaki, T., Jylanki, P. et al. (2006). Haptic feedback compared with visual feedback for BCI. *Proceedings of the 3rd International Brain–Computer Interface Workshop and Training Course 2006*. Graz, Austria.

31 Burdet, E., Sanguineti, V., Heuer, H., and Popovic, D.B. (May 2012). Motor skill learning and neuro-rehabilitation. *Guest Editorial; IEEE Transactions on Neural Systems and Rehabilitation Engineering* 20 (3): 237–238.

32 Salvaris, M., Cinel, C., Citi, L., and Poli, R. (2012). Novel protocols for P300-based brain–computer interfaces. *IEEE Transactions on Neural Systems and Rehabilitation Engineering* 20 (1): 8–17.

33 Li, Y., Long, J., Yu, T. et al. (2010). An EEG-based BCI system for 2-D cursor control by combining Mu/Beta rhythm and P300 potential. *IEEE Transactions on Biomedical Engineering* 57 (10): 2495–2505.

34 Nazarpour, K., Praamstra, P., Miall, R.C., and Sanei, S. (2009). Steady-state movement related potentials for brain–computer interfacing. *IEEE Transactions on Biomedical Engineering* 56 (8): 2104–2113.

35 Schröder, M.I., Lal, T.N., Hinterberger, T. et al. (2005). Robust EEG channel selection across subjects for brain-computer interfaces. *EURASIP Journal on Applied Signal Processing* 2005 (19): 3103–3112.

36 Kim, S.-P., Rao, Y.N., Erdogmus, D. et al. (2005). Determining patterns in neural activity for reaching movements using nonnegative matrix factorization. *EURASIP Journal on Applied Signal Processing* 2005 (19): 3113–3121.

37 Donchin, E., Spencer, K.M., and Wijesinghe, R. (2000). The mental prosthesis: assessing the speed of a P300-based brain-computer interface. *IEEE Transactions on Rehabilitation Engineering* 8: 174–179.

38 Bayliss, J.D. (2001). A Flexible Brain-Computer Interface. PhD thesis. University of Rochester, NY, USA.

39 Molina, G.A., Ebrahimi, T., and Vesin, J.-M. (2003). Joint time-frequency-space classification of EEG in a brain computer interface application. *EURASIP Journal on Applied Signal Processing* 7: 713–729.

40 Middendorf, M., Mc Millan, G., Calhoun, G., and Jones, K.S. (2000). Brain-computer interfaces based on the steady-state visual-evoked response. *IEEE Transactions on Rehabilitation Engineering* 8 (2): 211–214.

41 Kübler, A., Kotchubey, B., and Salzmann, H.P. (1998). Self regulation of slow cortical potentials in completely paralysed human patients. *Neuroscience Letters* 252 (3): 171–174.

42 McFarland, D.J., McCane, L.M., and Wolpaw, J.R. (1998). EEG based communication and control: short-term role of feedback. *IEEE Transactions on Neural Systems and Rehabilitation Engineering* 7 (1): 7–11.

43 Neuper, C., Schlogl, A., and Phertscheller, G. (1999). Enhancement of left-right sensorimotor imagery. *Journal of Clinical Neurophysiology* 4: 373–382.

44 Hinterberger, T., Neumann, N., Pham, M. et al. (2004). A multimodal brain-based feedback and communication system. *Experimental Brain Research* 154 (4): 521–526.

45 Bayliss, J.D. and Ballard, D.H. (2000). A virtual reality testbed for brain brain-computer interface research. *IEEE Transactions on Rehabilitation Engineering* 8: 188–190.

46 Nowlis, D.P. and Kamiya, J. (1970). The control of electroencephalographic alpha rhythms through auditory feedback and the associated mental activity. *Psychophysiology* 6: 476–484.

47 Plotkin, W.B. (1976). On the self-regulation of the occipital alpha rhythm: control strategies, states of consciousness, and the role of physiological feedback. *Journal of Experimental Psychology. General* 105: 66–99.

48 Sterman, M.B., Macdonald, L.R., and Stone, R.K. (1974). Biofeedback training of the sensorimotor electroencephalogram rhythm in man: effects on epilepsy. *Epilepsia* 15: 395–416.

49 Whyricka, W. and Sterman, M. (1968). Instrumental conditioning of sensorimotor cortex spindles in the walking cat. *Physiology and Behavior* 3: 703–707.

50 Black, A.H. (1971). The direct control of neural processes by reward and punishment. *American Scientist* 59: 236–245.

51 Vourvopoulos, A., Pardo, O.M., Lefebvre, S. et al. (June 2019). Effects of a brain-computer interface with virtual reality (VR) neurofeedback: a pilot study in chronic stroke patients. *Frontiers in Human Neuroscience* 13: 210, 17 pages.

52 Pfurtscheller, G., Leeb, R., Keinrath, C. et al. (2006). Walking from thought. *Brain Research* 1071: 145–152.

53 Fetz, E.E. (1969). Operant conditioning of cortical unit activity. *Science* 163: 955–958.

54 Fetz, E.E. (1972). Are movement parameters recognizably coded in activity of single neurons? *The Behavioral and Brain Sciences* 15: 679–690.

55 Fetz, E.E. and Baker, M.A. (1973). Operantly conditioned patterns on precentral unit activity and correlated responses in adjacent cells and contralateral muscles. *Journal of Neurophysiology* 36: 179–204.

56 Fetz, E.E. and Finocchio, D.V. (1971). Operant conditioning of specific patterns of neural and muscular activity. *Science* 174: 431–435.

57 Fetz, E.E. and Finocchio, D.V. (1972). Operant conditioning of isolated activity in specific muscles and precentral cells. *Brain Research* 40: 19–23.

58 Fetz, E.E. (1975). Correlations between activity of motor cortex cells and arm muscles during operantly conditioned response patterns. *Experimental Brain Research* 23: 217–240.

59 Schmidt, E.M. (1980). Single neuron recording from motor cortex as a possible source of signals for control of external devices. *Annals of Biomedical Engineering* 8: 339–349.

60 Mehring, C., Rickert, J., Vaadia, E. et al. (2003). Interface of hand movements from local field potentials in monkey motor cortex. *Nature Neuroscience* 6: 1253–1254.

61 Rickert, J., Oliveira, S.C., Vaadia, E. et al. (2005). Encoding of movement direction in different frequency ranges of motor cortical local field potentials. *The Journal of Neuroscience* 25: 8815–8824.

62 Pezaran, B., Pezaris, J.S., Sahani, M. et al. (2002). Temporal structure in neuronal activity during working memory in macaque parietal cortex. *Nature Neuroscience* 5: 805–811.

63 Scherberger, H., Jarvis, M.R., and Andersen, R.A. (2005). Cortical local field potential encodes movement intentions in the posterior parietal cortex. *Neuron* 46: 347–354.

64 Serruya, M.D., Hatsopoulos, N.G., Paninski, L. et al. (2002). Instant neural control of a movement signal. *Nature* 416: 141–142.

65 Taylor, D.M., Tillery, S.I., and Schwartz, A.B. (2002). Direct cortical control of 3D neuroprosthetic devices. *Science* 296: 1829–1832.

66 Tillery, S.I. and Taylor, D.M. (2004). Signal acquisition and analysis for cortical control of neuroprosthetic. *Current Opinion in Neurobiology* 14: 758–762.

67 Musallam, S., Corneil, B.D., Greger, B. et al. (2004). Cognitive control signals for neural prosthetics. *Science* 305: 258–262.

68 Patil, P.G. et al. (2004). Ensemble recordings of human subcortical neurons as a source of motor control signals for a brain-machine interface. *Neurosurgery* 55: 27–35.

69 Kennedy, P.R., Bakay, R.A.E., Moore, M.M. et al. (2000). Direct control of a computer from the human central nervous system. *IEEE Transactions on Rehanilitation Engineering* 8 (2): 198–202.

70 Si-Mohammed, H., Petit, J., Jeunet, C. et al. (2020). Towards BCI-based interfaces for augmented reality: feasibility, design and evaluation. *IEEE Transactions on Vision and Computer Graphics* 26 (3): 1608–1621.

71 Pfurstcheller, G. Jr., Stancak, A., and Neuper, C. (1996). Post-movement beta synchronization. A correlate of an idling motor area? *Electroencephalography and Clinical Neurophysiology* 98: 281–293.

72 Ramoser, H., Muller-Gerking, J., and Pfurtscheller, G. (2000). Optimal spatial filtering of single trial EEG during imagined hand movement. *IEEE Transactions on Rehabilitation Engineering* 8 (4): 441–446.

73 Antoniades, A., Spyrou, L., Martin-Lopez, D. et al. (2018). Deep neural architectures for mapping scalp to intracranial EEG. *International Journal of Neural Systems* https://doi.org/10.1142/S0129065718500090.

74 Bro, R. (1997). Multi-way analysis in the food industry: Models, algorithms and applications. PhD thesis. University of Amsterdam, and Royal Veterinary and Agricultural University.

75 Ahmadian, P., Sanei, S., Ascari, L. et al. (July 2013). Constrained blind source extraction of readiness potentials from EEG. *IEEE Transactions on Neural Systems and Rehabilitation Engineering* 21 (4): 567–575.

76 Pfurtscheller, G. (2001). Functional brain imaging based on ERD/ERS. *Elsevier Journal of Vision Research* 41: 1257–1260.

77 Jeon, Y., Kim, Y.-J., and Whang, M.C. (Sept. 2011). Event-related (De)synchronization (ERD/ERS) during motor imagery tasks: implications for brain–computer interfaces. *Elsevier International Journal of Industrial Ergonomics* 41 (5): 428–436.

78 Esch, W., Schimke, H., Doppelmayr, M. et al. (1996). Event related desynchronization (ERD) and the Dm effect: does alpha desynchronization during encoding predict later recall performance? *International Journal of Psychophysiology* 24: 47–60.

79 Pfurtscheller, G. (1999). EEG even-related desynchronization (ERD) and event-related synchronization (ERS). In: *Electroencephalography* (eds. E. Niedermeyer and F.L. Da Silva), Chapter 53, 958–966. LW&W.

80 Arroyo, S., Lesser, R.P., Gordon, B. et al. (1993). Functional significance of the mu rhythm of human cortex: an electrophysiological study with subdural electrodes. *Electroencephalography and Clinical Neurophysiology* 87: 76–87.

81 Leuthardt, E.C., Schalk, G., Wolpaw, J.R. et al. (2004). A brain-computer interface using electrocorticographic signals in humans. *Journal of Neural Engineering* 1: 63–71.

82 Tariq, M. Trivailo, P.M. Simic, M. (2020). Mu-Beta event-related (de)synchronization and EEG classification of left–right foot dorsiflexion kinaesthetic motor imagery for BCI. *PLoS ONE*. 20 p. https://doi.org/10.1371/journal.pone.0230184.

83 Pfurtscheller, G., Stancak, A. Jr., and Edlinger, G. (1997). On the existence of different types of central beta rhythms below 30 Hz. *Electroencephalography and Clinical Neurophysiology* 102: 316–325.

84 Neuper, C. and Pfurtscheller, G. (1996). Post movement synchronization of beta rhythms in the EEG over the cortical foot area in man. *Neuroscience Letters* 216: 17–20.

85 Pfurtscheller, G., Müller-Putz, G.R., Pfurtscheller, J., and Rupp, R. (2005). EEG-based asynchronous BCI controls functional electrical stimulation in a tetraplegic patient. *EURASIP Journal on Advances in Signal Processing* 2005 (19): 3152–3155.

86 Usanos, C.A., de Santiago, L., Barea, R. et al. (2017). Analysis of gamma-band activity from human EEG using empirical mode decomposition. *Sensors* 17: 989, 14 pages.

87 Usanos, C.A., Boquete, L., de Santiago, L. et al. (2020). Induced gamma-band activity during actual and imaginary movements: EEG analysis. *Sensors* 20: 1545, 11 pages.

88 Oostenveld, R., Fries, P., Maris, E., and Schoffelen, J.-M. (2011). FieldTrip: open source software for advanced analysis of MEG, EEG, and invasive electrophysiological data. *Computational Intelligence and Neuroscience* 2011: 156869.

89 Paek, A.Y., Agashe, H.A., and Contreras-Vidal, J.L. (2014). Decoding repetitive finger movements with brain activity acquired via non-invasive electroencephalography. *Frontiers in Neuroengineering* 7: 3.

90 Agashe, H. and Contreras-Vidal, J.J.L. (2011). Reconstructing hand kinematics during reach to grasp movements from electroencephalographic signals. *33rd Annual International Conference of the IEEE Engineering in Medicine and Biology Society*, 5444–5447. Boston, MA.

91 Beverina, F., Palmas, G., Silvoni, S. et al. (2003). User adaptive BCIs: SSVEP and P300 based interfaces. *PsychNology Journal* 1 (4): 331–354.

92 Lalor, E.C., Kelly, S.P., Finucane, C. et al. (2005). Steady-state VEP-based brain-computer interface control in an immersive 3D gaming environment. *EURASIP Journal on Advances in Signal Processing* 2005 (19): 3156–3164.

93 Gratton, G. (1969). Dealing with artefacts: the EOG contamination of the event-related brain potentials over the scalp. *Electroencephalography and Clinical Neurophysiology* 27: 546.

94 Lins, O.G., Picton, T.W., Berg, P., and Scherg, M. (1993). Ocular artefacts in EEG and event-related potentials, i: scalp topography. *Brain Topography* 6: 51–63.

95 Celka, P., Boshash, B., and Colditz, P. (2001). Preprocessing and time-frequency analysis of new born EEG seizures. *IEEE Engineering in Medicine and Biology Magazine* 20: 30–39.

96 Jung, T.P., Humphies, C., and Lee, T.W. (1998). Extended ICA removes artefacts from electroencephalographic recordings. *Advances in Neural Information Processing Systems* 10: 894–900.

97 Joyce, C.A., Gorodnitsky, I., and Kautas, M. (2004). Automatic removal of eye movement and blink artefacts from EEG data using blind component separation. *Psychophysiology* 41: 313–325.

98 Shoker, L., Sanei, S., and Chambers, J. (Oct. 2005). Artifact removal from electroencephalograms using a hybrid BSS-SVM algorithm. *IEEE Signal Processing Letters* 12 (10).

99 Peterson, D.A., Knight, J.N., Kirby, M.J. et al. (2005). Feature selection and blind source separation in an EEG-based brain-computer interface. *EURASIP Journal on Advances in Signal Processing* 2005 (19): 3128–3140.

100 Coyle, D., Prasad, G., and McGinnity, T.M. (2005). A time-frequency approach to feature extraction for a brain-computer interface with a comparative analysis of performance measures. *EURASIP Journal on Advances in Signal Processing* 2005 (19): 3141–3151.

101 Mørup, M., Hansen, L.K., Herrmann, C.S. et al. (2006). Parallel factor analysis as an exploratory tool for wavelet transformed event-related EEG. *NeuroImage* 29 (3): 938–947.

102 Cohen, L. (1995). *Time Frequency Analysis, Prentice Hall Signal Processing Series.* Upper Saddle River, NJ, USA: Prentice Hall.

103 L. Shoker, K. Nazarpour, S. Sanei, and A. Sumich, "A novel space-time-frequency masking approach for quantification of EEG source propagation, with application to brain computer interfacing," Proceedings of EUSIPCO, Florence, Italy, 2006.

104 Harshman, R.A. (1970). Foundation of the PARAFAC: models and conditions for an 'explanatory' multi-modal factor analysis. *UCLA Working Papers in Phonetics* 16: 1–84.

105 Shoker, L., Nazarpour, K., Sanei, S., and Sumich, A. (2006). A novel space-time-frequency masking approach for quantification of EEG source propagation, with application to brain computer interfacing. *Proceedings of EUSIPCO.* Florence, Italy.

106 Möcks, J. (1988). Decomposing event-related potentials: a new topographic components model. *Biological Psychology* 26: 199–215.

107 Field, A.S. and Graupe, D. (1991). Topographic component (parallel factor) analysis of multichannel evoked potentials: practical issues in trilinear spatiotemporal decomposition. *Brain Topography* 3: 407–423.

108 Miwakeichi, F., Martinez-Montes, E., Valdes-Sosa, P.A. et al. (2004). Decomposing EEG data into space-time-frequency components using parallel factor analysis. *NeuroImage* 22: 1035–1045.

109 Nazarpour, K., Shoker, L., Sanei, S., and Chambers, J. (2006). Parallel space–time–frequency decomposition of EEG signals for brain computer. *Proceedings of the European Signal Processing Conference.* Italy.

110 Nazarpour, K., Wangsawat, Y., Sanei, S. et al. (Sept 2008). Removal of the eye-blink artifacts from EEGs via STF-TS modeling and robust minimum variance beamforming. *IEEE Transactions on Biomedical Engineering* 55 (9): 2221–2231.

111 Rivet, B., Souloumiac, A., Attina, V., and Gibert, G. (2009). xDAWN algorithm to enhance evoked potentials: application to brain computer interface. *IEEE Transactions on Biomedical Engineering* 56 (8): 2035–2043.

112 Hoffmann, U., Vesin, J., and Ebrahimi, T. (2006). Spatial filters for the classification of event-related potentials. *European Symposium on Artificial Neural Networks (ESANN 2006).*

113 Monajemi, S., Jarchi, D., Ong, S.H., and Sanei, S. (2017). Cooperative particle filtering for detection and tracking of ERP subcomponents from multichannel EEG. *Journal of Entropy, Special Issue on Entropy and Electroencephalography.* 19(5): 199; Invited Feature Paper. http://dx.doi.org/10.3390/e19050199.

114 S. Enshaeifar, S. Sanei, C. Cheong-took, Singular Spectrum Analysis of P300 for Classification," Proceedings of IEEE, Joint Conference on Neural Networks, IJCNN, 2014

115 Enshaeifar, S., Sanei, S., Cheong-took, C. (2014). Singular Spectrum Analysis of P300 for Classification. *Proceedings of IEEE, Joint Conference on Neural Networks, IJCNN.*

116 Ginter, J. Jr., Blinowska, K.J., Kaminski, M., and Durka, P.J. (2001). Phase and amplitude analysis in time-frequency space—application to voluntary finger movement. *Journal of Neuroscience Methods* 110: 113–124.

117 Mallat, S. and Zhang, Z. (1993). Matching pursuits with time-frequency dictionaries. *IEEE Transactions on Signal Processing* 41 (12): 3397–3415.

118 Durka, P.J., Ircha, D., and Blinowska, K.J. (2001). Stochastic time-frequency dictionaries for matching pursuit. *IEEE Transactions on Signal Processing* 49 (3): 507–510.

119 Durka, P.J., Ircha, D., Neuper, C., and Pfurtscheller, G. (2001). Time-frequency microstructure of event-related EEG desynchronization and synchronization. *Medical and Biological Engineering & Computing* 39 (3): 315–321.

120 Kaminski, M.J. and Blinowska, K.J. (1991). A new method of the description of the information flow in the structures. *Biological Cybernetics* 65: 203–210.

121 Bozdogan, H. (1987). Model-selection and Akaike's information criterion (AIC): the general theory and its analytical extensions. *Psychometrika* 52: 345–370.

122 Carreon, F.O., Serna, J.G.G., Rendon, A.M. et al. (Feb. 2016). Induction of emotional states in people with disabilities through film clips using brain computer interfaces. *IEEE Latin America Transactions* 14 (2): 563–568.

123 Astolfi, L., Cincotti, F., Babiloni, C. et al. (2005). Estimation of the cortical connectivity by high resolution EEG and structural equation modelling: simulations and application to finger tapping data. *IEEE Transactions on Biomedical Engineering* 52 (5): 757–767.

124 Gerloff, C., Richard, J., Hardley, J. et al. (1998). Functional coupling and regional activation of human cortical motor areas during simple, internally paced and externally paced finger movement. *Brain* 121: 1513–1531.

125 Urbano, A., Babiloni, C., Onorati, P., and Babiloni, F. (1998). Dynamic functional coupling of high resolution EEG potentials related to unilateral internally triggered one-digit movement. *Electroencephalography and Clinical Neurophysiology* 106 (6): 477–487.

126 Jancke, L., Loose, R., Lutz, K. et al. (2000). Cortical activations during paced finger-tapping applying visual and auditory pacing stimuli. *Cognitive Brain Research* 10: 51–56.

127 Bollen, K.A. (1989). *Structural Equations with Latent Variable.* New York: J. Wiley.

128 Eftaxias, K. and Sanei, S. (2014). Discrimination of task-based EEG signals using diffusion adaptation and S-transform coherency. *Proceedings of the IEEE Workshop on Machine Learning for Signal Processing, MLSP 2014.* France.

129 Eftaxias, K., Sanei, S., and Sayed, A.H. (2013). Modelling brain cortical connectivity using diffusion adaptation. *2013 IEEE International Conference on Acoustics, Speech and Signal Processing (ICASSP),* 959–962. IEEE.

130 Eftaxias, K. and Sanei, S., "Diffusion adaptive filtering for modelling brain responses to motor tasks," in 18th International Conference on Digital Signal Processing (DSP). IEEE, 2013, pp. 1–5.

131 Eftaxias, K. and Sanei, S. (2013). Diffusion adaptive filtering for modelling brain responses to motor tasks. *18th International Conference on Digital Signal Processing (DSP),*1 5. IEEE.

132 Lotte, F. and Guan, C.T. (2011). Regularizing common spatial patterns to improve BCI designs: unified theory and new algorithms. *IEEE Transactions on Biomedical Engineering* 58 (2): 355 362.

133 Goksu, F. Ince, N.F., and Tewfik, A.H. (2011). Sparse common spatial patterns in brain computer interface applications. *Proceeding of the IEEE International Conference on Acoustics, Speech, and Signal Processing, ICASSP*, 533–536.

134 Kang, H., Nam, Y., and Choi, S. (2009). Composite common spatial pattern for subject-to-subject transfer. *IEEE Signal Processing Letters* 16 (8): 683–686.

135 Lu, H., Plataniotis, K., and Venetsanopoulos, A. (2009). Regularized common spatial patterns with generic learning for EEG signal classification. *Proceedings of the IEEE Conference EMBC*, 6599–6602.

136 Blankertz, B., Kawanabe, M., Tomioka, R. et al. (2008). Invariant common spatial patterns: Alleviating nonstationarities in brain-computer interfacing. *Proceeding of NIPS 20.*

137 Farquhar, J., Hill, N., Lal, T., and Schölkopf, B. (2006). Regularised CSP for sensor selection in BCI. *Proceedings of the 3rd International BCI Workshop*. Graz, Austria.

138 Arvaneh, M., Guan, C.T., Kai, A.K., and Chai, Q. (2011). Optimizing the channel selection and classification accuracy in EEG-based BCI. *IEEE Transactions on Biomedical Engineering* 58 (6): 1865–1873.

139 Yong, X., Ward, R., and Birch, G. (2008). Sparse spatial filter optimization for EEG channel reduction in brain-computer interface. *Proceedings of the IEEE International Conference on Acoustic, Speech, and Signal Processing, ICASSP*. 417–420. Taiwan.

140 Leeb, R., Brunner, C., Müller-Putz, G.R. et al. (2014). Pfurtscheller. *BCI Competition 2008 - Graz Dataset B," Graz University of Technology*. Austria.

141 Mazzatenta, A., Giugliano, M., Campidelli, S. et al. (2007). Interfacing neurons with carbon nanotubes: electrical signal transfer and synaptic stimulation in cultured brain circuits. *Journal of Neuroscience* 27 (26): 6931–6936.

142 Robert Tindol. Caltech Media Relation. https://www.caltech.edu/about/news/caltech-scientists-devise-first-neurochip-213 (accessed 9 Apr 2021).

143 Chips coming to a brain near you. (2004). *Wired News* (22 October). https://www.wired.com/2004/10/chips-coming-to-a-brain-near-you (accessed 19 August 2021).

144 Thomas DeMarse (2004). 'Brain' in a dish flies flight simulator. *CNN* (4 November). CNN.com.

145 Saha, S., Iftekhar Uddin, K., Ahmed, R. et al. (Feb. 2018). Evidence of variabilities in EEG dynamics during motor imagery-based multiclass brain–computer Interface. *IEEE Transactions on Neural Systems and Rehabilitation Engineering* 26 (2): 371–382.

146 Wang, H., Chang, W., and Zhang, C. (Jul. 2016). Functional brain network and multichannel analysis for the P300-based brain computer interface system of lying detection. *Expert Systems with Applications* 53: 117–128.

147 Farwell, L.A., Richardson, D.C., Richardson, G.M., and Furedy, J.J. (2014). Brain fingerprinting classification concealed information test detects us navy military medical information with P300. *Frontiers in Neuroscience* 8: 108.

148 Rao, Stocco, Bryan, R.P., A., M. et al. (2014). A direct brain-to-brain interface in humans. *PLoS One* 9 (11): e111332.

149 Hwanga, H.-J., Lima, J.-H., Junga, Y.-J. et al. (2012). Development of an SSVEP-based BCI spelling system adopting a QWERTY-style LED keyboard. *Elsevier Journal of Neuroscience Methods* 208: 59–65.

150 Farwell, L.A. and Richardson, D.C. (2013). Brain fingerprinting: let's focus on the science—a reply to Meijer, Ben-Shakhar, Verschuere, and Donchin. *Cognitive Neurodynamics* 7: 159–166. https://doi.org/10.1007/s11571-012-9238-5.

151 Nakanishi, M., Wang, Y., Chen, X. et al. (2018). Enhancing detection of SSVEPs for a high-speed brain speller using task-related component analysis. *IEEE Transactions on Biomedical Engineering* 65: 104–112.

152 Chen, X., Nakanishi, M., Wang, Y. et al. (2015). High-speed spelling with a noninvasive brain–computer interface. *Proceedings of National Academy of Sciences of the United States of America* 112: E6058–E6067.

153 Renton, A.I., Mattingley, J.B., and Painter, D.R. (2019). Optimising non-invasive brain computer interface systems for free communication between naïve human participants. *Nature Research Scientific Reports* 9: 18705. https://doi.org/10.1038/s41598-019-55166-y.

154 Noyes, J. (1983). The QWERTY keyboard: a review. *International Journal of Man-Machine Studies* 18: 265–281.

155 Lopes, A.C., Rodrigues, J., Perdigão, J. et al. (December 2016). A new hybrid motion planner applied in a brain-actuated robotic wheelchair. *IEEE Robotics & Automation Magazine* 23 (4): 82–93.

156 Adewole, D.O., Serruya, M.D., Harris, J.P. et al. (2016). The evolution of Neuroprosthetic interfaces. *Critical Reviews in Biomedical Engineering* 44 (1–2): 123–152. https://doi.org/10.1615/CritRevBiomedEng.2016017198.

157 Erhardt, J.B., Fuhrer, E., Gruschke, O.G. et al. (2018). Should patients with brain implants undergo MRI? *Journal of Neural Engineering* 15: 041002, 26 pages.

158 Fazli, S., Mehnert, J., Steinbrink, J. et al. (2012). Enhanced performance by a hybrid NIRS-EEG brain computer interface. *NeuroImage* 59: 519–529.

159 Shin, J., Kim, D.-W., Müller, K.-R., and Hwan, H.-J. (2018). Improvement of information transfer rates using a hybrid EEG–NIRS brain–computer interface with a short trial length: offline and pseudo-online analyses. *Sensors* 18: 1827. https://doi.org/10.3390/s18061827, 16 pages.

160 Goldman, R.I., Stern, J.M., Engel, J., and Cohen, M.S. (2002). Simultaneous EEG and fMRI of the alpha rhythm. *Neuroreport* 13: 2487–2492.

161 Weiskopf, N., Mathiak, K., Bock, S.W. et al. (2004). Principles of a brain-computer interface (BCI) based on real-time functional magnetic resonance imaging (fMRI). *IEEE Transactions on Biomedical Engineering* 51: 966–970.

162 Sitaram, R., Weiskopf, N., Caria, A. et al. (2008). fMRI brain-computer interfaces. *IEEE Signal Processing Magazine* 25: 95–106.

163 Deshpande, G., Rangaprakash, D., Oeding, L. et al. (2017). A new generation of brain-computer interfaces driven by discovery of latent EEG-fMRI linkages using tensor decomposition. *Frontiers in Neuroscience* 11: 246. 13 pages.

164 Friedman, J.H. (1989). Regularized discriminant analysis. *Journal of the American Statistical Association* 84: 165–175.

165 BBCI Toolbox. https://github.com/bbci/bbci_public (accessed 30 August 2020).

166 Ledoit, O. and Wolf, M. (2004). A well-conditioned estimator for large dimensional covariance matrices. *Journal of Multivariate Analysis* 88 (2): 365–411.

167 Corsi, M.-C., Chavez, M., Khambhati, A.N. et al. (2019). Integrating EEG and MEG Signals to Improve Motor Imagery Classification in Brain–Computer Interface. *International Journal of Neural Systems*. 29. Open Access.

168 Fatourechi, M., Ward, R., and Mason, S. et al. (2008). Comparison of evaluation metrics in classification applications with imbalanced datasets. *International Conference on Machine Learning and Applications (ICMLA)*. 777–782. IEEE.

169 Schlogl, A., Kronegg, J., Huggins, J., and Mason, S.G. (2007). *Towards brain–computer interfacing, Evaluation Criteria in BCI Research*, 327–342. MIT Press.

170 Kauhanen, L., Palomaki, T., Jylamki, P. et al. (2006). Haptic feedback compared with visual feedback for BCI. *Proceedings of the 3rd International BCI Workshop and Training Course*, 66–67.

18

Joint Analysis of EEG and Other Simultaneously Recorded Brain Functional Neuroimaging Modalities

18.1 Introduction

Electroencephalography (EEG) is a functional brain screening modality with good temporal resolution and poor spatial resolution. Conversely, the records of brain activity from over the scalp seriously suffer from nonlinearity of the head tissues in terms of conductivity. A number of neuroimaging tools and systems have been developed over the years to study different aspects of brain function and compensate for the above EEG drawbacks. The information achieved by these brain screening modalities, if added to those of EEGs, can much better characterize and identify the brain state and its response to various stimuli. The three important brain screening modalities used jointly with EEG are functional magnetic resonance imaging (fMRI), functional near-infrared spectroscopy (fNIRS), and magnetoencephalography (MEG).

EEG and fMRI have been probably used more than other methods to study the brain function. The advances of acquisition systems in recent years have enabled simultaneous EEG–fMRI brain recordings. The information achieved from these two data modalities is important in analysis of brain responses to event-related and movement-related stimulations.

Unfortunately, the EEG recorded in an MRI scanner is highly degraded by the magnetic field. Therefore, to ensure that the EEG signal will be useful and informative, a comprehensive preprocessing technique is required.

In this chapter we review the literature and describe the state of the art in new signal-processing methods for fMRI artefact removal from EEG signals and jointly analysis of these two modalities. The concept of fNIRS together with its applications jointly with EEG is briefly explained and some examples of joint EEG–MEG are also presented. This is followed by presenting a number of new approaches in the above bimodality data processing and solving their associated problems.

18.2 Fundamental Concepts

18.2.1 Functional Magnetic Resonance Imaging

fMRI produces a sequence of low resolution MRI which is able to detect brain activity by measuring the associated changes in blood flow. Unlike structural MRI (sMRI), this

EEG Signal Processing and Machine Learning, Second Edition. Saeid Sanei and Jonathon A. Chambers.

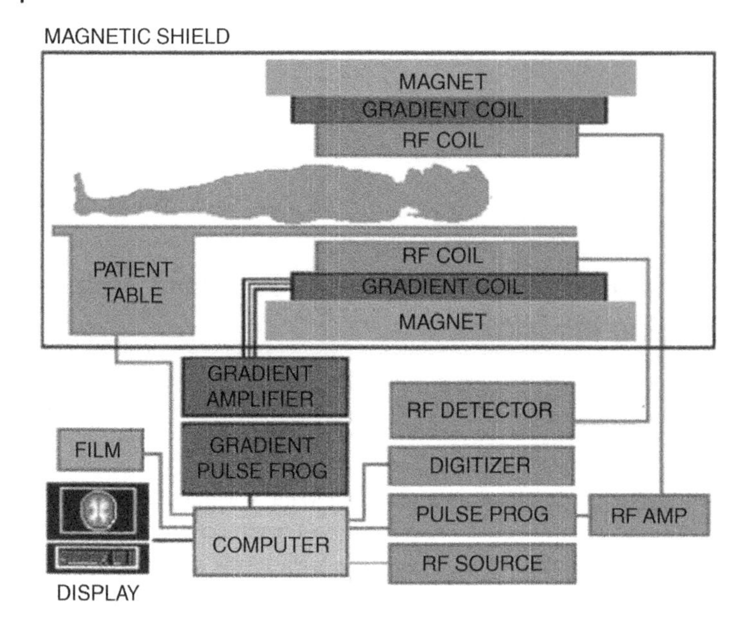

Figure 18.1 Block diagram of an MRI scanner illustrating the components involved in fMRI data recording.

imaging modality has lower contrast. MRI is produced by the nuclear magnetic resonance (NMR) signal from hydrogen nuclei. An MRI session begins when the subject is placed horizontally inside the bore of a large magnetic field. A typical clinical MRI system requires field strength of at least 1.5 tesla (T). The MRI system requires a radio frequency (RF) coil to generate an oscillating magnetic field [1]. Figure 18.1 represents a block diagram of an MRI scanner. The hydrogen nuclei of the water molecules in the subject's body are the source of the NMR signal. The hydrogen nuclei (protons) have an intrinsic property called nuclear spin. The spin of these particles can be considered as a magnetic moment vector which makes the particle behave like a magnet. Inside the magnetic field, a large fraction of magnetic moments aligns parallel to the main magnetic field and forms a net magnetization force which oscillates at the Larmor frequency [2]:

$$f = \gamma B \tag{18.1}$$

where B is the strength of the applied magnetic field and γ is a constant related to the type of nuclear particles. Applying an oscillating magnetic field at a frequency equal to the Larmor frequency causes the spins to absorb energy and be excited. The magnetic field is generated by an RF pulse of magnetic energy. Resonance between the magnetic field excitation and equilibrium states induces a current into a coil placed near the subject. The MRI is then performed by a controlled manipulation of the magnetic field [3].

According to (18.1) a spatially varying magnetic field creates different frequencies in different locations. As a result, the detected signal changes based on varying the magnetic field which leads to a signal containing spectral information. An inverse Fourier transform converts the spectrum into a real signal in the spatial domain as fMRI images. The image produced in each scan is called a volume. Each volume is composed of a number of slices through the body, and each slice has a certain thickness and contains a number of 3D unit elements called voxels.

The fMRI modality extends the use of MRI to find information about biological functions as well as anatomical information. Numerous studies during past decades have shown that neural activity causes change in blood flow and blood oxygenation in the brain. In the early 1990s, Ogawa et al. [4] used the MRI technique to measure the hemodynamic response (HR) (changes in blood flow) in the brain. When a subject performs a particular task during the MRI scanning, the metabolism of neural cells in a brain region responsible for a task is increased and leads to an increase in the need for oxygen. Existing haemoglobin in the red blood cells delivers oxygen through capillaries to the involved neurons. Oxygenated haemoglobin (oxyhaemoglobin) is diamagnetic. In contrast to oxyhaemoglobin, deoxygenated haemoglobins (deoxyhaemoglobin) are paramagnetic [5]. So, the level of oxygenation affects the NMR signal and consequently changes the intensity of the recorded MRI images.

18.2.1.1 Blood Oxygenation Level Dependence

Blood oxygenation level dependence (BOLD) refers to a phenomenon in fMRI which represents the changes in the NMR signal due to the variation of blood deoxyhaemoglobin concentration. There is a biomedical model for dynamic changes in deoxyhaemoglobin content during brain activation. This model is known as the haemodynamic response function (HRF) [6]. HRF presents the temporal properties of brain activation. If the human brain and the fMRI scanner are considered to be linear time invariant systems the variation of the measured intensities of the voxels can be described by a time series $d(n)$ which can be represented as convolution of the stimulus function $h(n)$ and the impulse response $s(n)$. Such a system includes noise $v(n)$ and drift $f(n)$ too:

$$d(n) = h(n)*s(n) + v(n) + f(n) \tag{18.2}$$

where $*$ denotes the convolution operation. The stimulus function $h(n)$ is often called *time course*. This time course should be known to the model-based system for the construction of BOLD. The HRF function is usually modelled by the difference between two gamma functions [7]:

$$s(n) = \left(\frac{n}{k_1}\right)^{\alpha_1} e^{-\frac{n-k_1}{\beta_1}} - c\left(\frac{n}{k_2}\right)^{\alpha_2} e^{-\frac{n-k_2}{\beta_2}} \tag{18.3}$$

where k_j is the time shift of the peak, the constant c determines the contribution of each gamma (often set to satisfy $0.30 < c < 0.5$) and α_1, α_2, β_1, and β_2 are the parameters of the gamma functions (with typical values of 6, 12, 12, 12 respectively). The HR varies for different parts of the brain and different subjects. Figure 18.2 demonstrates the variation of HRF with k and c.

Two common experimental designs are used in fMRI experiments: *block design* and *event-related design* [8]. In block design, the test subjects employ different cognitive processes, alternating between them periodically. This is the most time-efficient approach for comparing brain responses with different tasks. The simplest form of block design experiment includes two states: 'rest' and 'active'. These states are alternated throughout the experiment in order to obtain an optimum experimental design for BOLD detection with sufficient

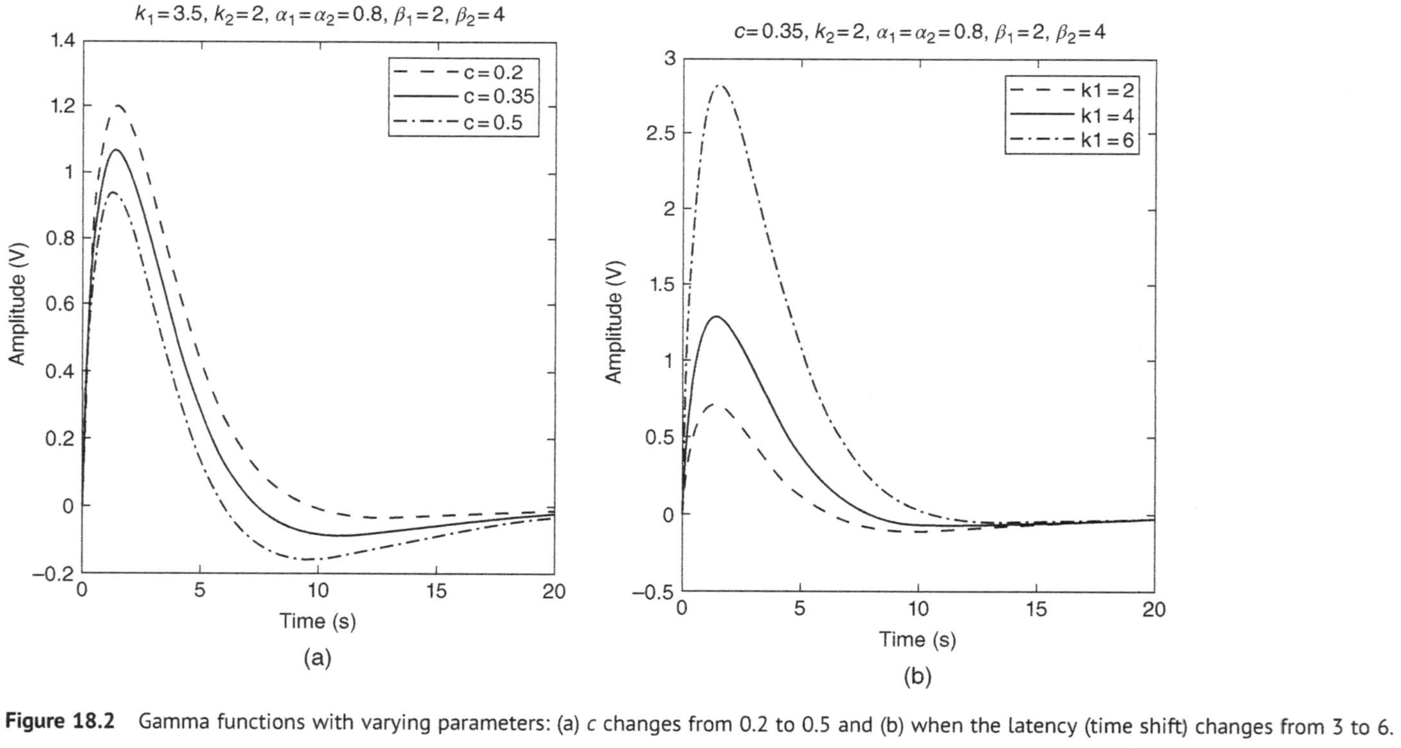

Figure 18.2 Gamma functions with varying parameters: (a) c changes from 0.2 to 0.5 and (b) when the latency (time shift) changes from 3 to 6.

signal-to-noise ratio (SNR). The block design is an ideal choice for many types of experiments because the shape of the brain response is simple, so it can be very useful in an early exploratory stage. Although in block design measuring the temporal response of the brain is hard to achieve limiting the flexibility of block design, the robustness of its results still makes it a valuable method.

Unlike in block design, in event-related design discrete stimuli occur randomly during the scanning. In event-related design, the input is modelled by discrete impulses in the stimulus time instants. Event-related fMRI is a suitable tool to capture temporal properties of the brain response. Moreover, it provides information on HRF occurrence instants [9].

In event-related fMRI experiments, stimuli are separated by inter-stimulus intervals (ISIs).

Large ISI causes the HRFs not to overlap due to sequential stimuli and reveals the transient variation in the brain response. Several types of noises usually appear in fMRI experiments. First, noise related to data acquisition which can be considered due to object variability. Object variability results from quantum thermodynamics and thermal noise. Thermal noise causes white noise with a constant variance in the image domain [10]. Head movement is another source of noise in fMRI data. This noise is considered as physiological noise. Therefore, fMRI needs to be pre-processed before being further analyzed.

18.2.1.2 Popular fMRI Data Formats

The most popular file formats for fMRI are:

- **DICOM** is the standard form of data obtained from the scanner. It is two-dimensional and can be converted to ANALYZE or neuroimaging informatics technology initiative (NIFTI) for further analysis. The statistical parametric mapping (SPM) [11] toolbox enables conversion of DICOM to ANALYZE and NIFTI formats.
- **ANALYZE** is another very commonly used software for fMRI analysis using MATLAB. SPM and FMRIB software library (FSL) use ANALYZE 7.5 format. An ANALYZE 7.5 data format consists of two files, an image file with extension '.img' and header file with extensions '.hdr'. The '.img' file contains the recorded image and the '.hdr' file contains the volume information for the '.img' file, such as voxel size, and the number of pixels in the x, y, and z directions (dimensions).
- **NIFTI** is the most recent standard for fMRI data format. The file extension of NIFTI is '. nii' with arbitrary information. This format can be converted to ANALYZE by the MRIcro toolbox.

18.2.1.3 Preprocessing of fMRI Data

The first stage in analyzing fMRI data is preprocessing with the aim of (i) removing non-task-related variability from the data, (ii) increasing SNR and (iii) preparing data for further analysis in order to detect brain active regions.

Motion correction (realignment), coregistration, segmentation, normalization, and smoothing are important (spatial) preprocessing approaches. Conversely, there are temporal preprocessing methods including slice timing correction which are often required.

- *Realignment*: This is to remove the head movement artefact from fMRI data. Any misalignment due to motion can affect brain source localization.

- *Functional-structural coregistration*: sMRI is a high-resolution image providing static anatomical information. In contrast, fMRI has low resolution and provides dynamic physiological information [12]. Functional images have little structural information, so a high-resolution anatomical image is needed to determine the activated area. Thus, registration between functional and sMRI is required to align images produced by these two processes [1].
- *Normalization*: Recalling that the data are in the form of two-dimensional images making up a volume, intensity normalization refers to rescaling intensities in all volumes such that at the end they have the same average intensity. Spatial normalization is another important step in preprocessing. This resembles coregistration. However, it corrects dissimilarities in shape between volumes.
- *Smoothing*: This is to improve SNR in fMRI. In other words, the high frequency components are removed from fMRI during the smoothing process. Usually a Gaussian kernel with a specified width is used to smooth the fMRI.
- *Slice timing correction*: The differences in image acquisition time between the slices need to be corrected. In fMRI, each slice of a volume is scanned at a time. Since the fMRI slices are acquired separately, the information from different slices come from varying points in time after the task events and need to be aligned by performing this step.

18.2.2 Functional Near-Infrared Spectroscopy

fNIRS measures the HR. More specifically, the neuronal activity is fuelled by glucose metabolism in the presence of oxygen. Increases in neuronal activity set off a series of vascular events that result in the flooding of neuronal tissues with oxygenated haemoglobin (HbO or more precisely HbO_2), the protein molecules that carry oxygen within the blood. During bouts of activity the rate of HbO delivery typically exceeds the rate of oxygen utilization, resulting in a temporary increase in the concentration of HbO and a decrease in the concentration of deoxygenated haemoglobin (HbR). Whereas most biological tissues are transparent to near-infrared (NIR) light, HbO, and HbR are known to absorb and scatter NIR light of slightly different wavelengths in the range of 700–1000 nm as seen in Figure 18.3. Continuous-wave fNIRS capitalizes on this property of HbO and HbR [13]. Light emitters placed on the surface of the scalp (see Figure 18.4) radiate NIR light into the head. Given the differential absorption and backscattering of HbO and HbR, a portion of this NIR light returns to the surface of the scalp, where it is measured with photodetectors. Spectroscopic methods may thus be used to detect changes in the concentrations of HbO and HbR. The most common application of NIR light in the study of the human brain is fNIRS. fNIRS measures changes in the optical properties of brain tissue in the NIR range (650–950 nm) to estimate fluctuations in the concentration of HbO and HbR associated with neural activity. Typical fNIRS sensor pads geometrically position emitters and photoreceptors so that activity at the outer surface of the cortex may be measured with a spatial resolution of the order of square centimetres [13].

Sixteen-channel continuous-wave fNIRS system is popular for the recording of prefrontal cortical activities. The system is composed of a sensor pad with a source–detector separation of 2.50 cm and a data acquisition control box running the software. The sensor pad has

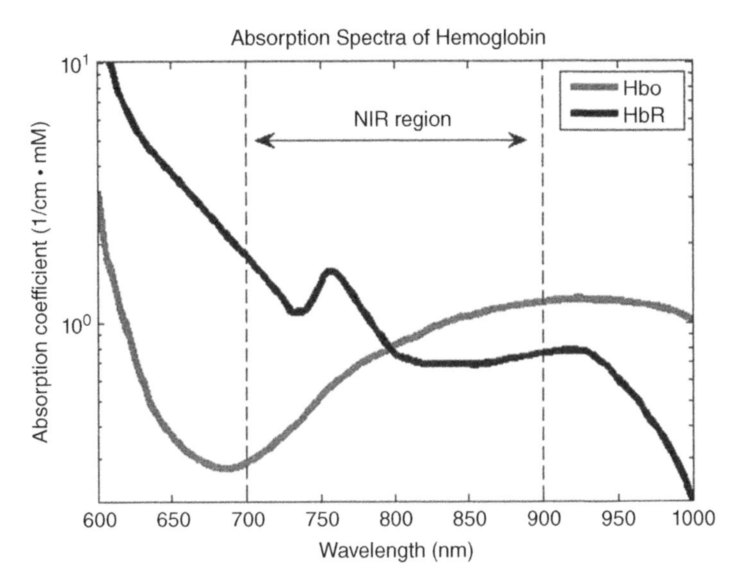

Figure 18.3 The absorption coefficient curves for oxyhaemoglobin and deoxyhaemoglobin versus (near-infrared) light wavelengths.

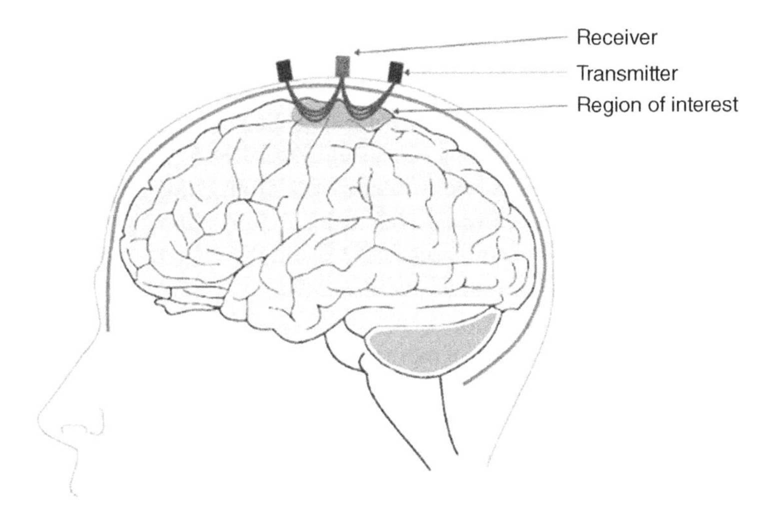

Figure 18.4 The concept of cortex fNIRS imaging.

typical temporal resolution of 500 ms per scan, a penetration depth of 1.25 cm into the prefrontal cortex, and light sources with peak wavelengths at 730 and 850 nm. The sensor pad is secured in alignment with electrode positions F7, FP1, FP2, and F8 based on the international 10/20 EEG system.

To reconstruct the images, the raw fNIRS data are first normalized to unit median and converted from recorded optical intensities to HbO and HbR (deoxy-Hb) concentrations using the modified Beer–Lambert law [14]. This law is used to relate light attenuation to changes in absorption, $\Delta\mu_a$:

$$\varphi_\Delta(t,\lambda) = -\ln\left(\frac{I(t,\lambda)}{I_0(t,\lambda)}\right)$$
$$= d.\mathrm{DPF}(\lambda) \bullet \Delta\mu_a(t,\lambda) \tag{18.4}$$

where $I(t, \lambda)$ describes the measured light intensity at wavelength λ and $I_0(t, \lambda)$ is its baseline measure. This baseline is measured on the same optode pair, but usually a few seconds prior to each activation event. The term 'modified' arises from the inclusion of a differential path-length factor, DPF(λ). The DPF multiplies the source–detector distance d to account for the additional path travelled by light due to scattering events in the diffusive medium [15]. Using two wavelengths of λ_1 and λ_2 respectively sensitive to the concentration of HbO and HbR, the measured absorption (at the above two different wavelengths) can be related to haemoglobin concentrations (C_{HbO} and C_{HbR}) by:

$$\Delta\mu_a(\lambda_1) = \varepsilon_{\mathrm{HbO}}^{\lambda_1} \bullet \Delta C_{\mathrm{HbO}} + \varepsilon_{\mathrm{HbR}}^{\lambda_1} \bullet \Delta C_{\mathrm{HbR}} \tag{18.5a}$$

$$\Delta\mu_a(\lambda_2) = \varepsilon_{\mathrm{HbO}}^{\lambda_2} \bullet \Delta C_{\mathrm{HbO}} + \varepsilon_{\mathrm{HbR}}^{\lambda_2} \bullet \Delta C_{\mathrm{HbR}} \tag{18.5b}$$

where the extinction coefficients ε of haemoglobin species have been measured *in vitro* [15] in the desired wavelengths. Similar to (and as compared with) fMRI, fNIRS can be combined with other neuroimaging modalities as the NIRS optical electrodes (made from glass and plastic optical fibres) are used to deliver light to the head of the subject. It can be used in combination with MEG, EEG, and even MRI without producing magnetic, electric, or susceptibility artefacts. Simultaneous fNIRS and EEG or MEG is appealing because of the ability to explore neurovascular coupling by obtaining concurrent measures of the neuronal and HRs to various stimuli. To gain familiarity with various fNIRS propagation and image reconstruction modalities (continuous-wave, frequency domain, and time domain modalities), the reader may look at the related resources such as [16] and the references herein.

As reported in [17], fNIRS has been applied to study the effects of neurological conditions. Traumatic brain injuries (TBIs) have been investigated using this screening modality. Potential application of fNIRS for early detection of intracranial hematomas, including epidural hematomas, subdural hematomas, and intracerebral hematomas has been investigated. Another application is on patients who are being treated in intensive care units for subarachnoid haemorrhage and TBIs.

Epileptic patients have also been examined during spontaneously occurring complex-partial seizures. During these seizures, extremely large increases in blood volume and oxygenated haemoglobin concentration have been measured, stressing the importance of fNIRS as a useful and simple bedside tool to assess brain function [17].

Continuous brain monitoring enabled by fNIRS allows examination of a disease vulnerable to neurodegeneration or increases in severity over time. fNIRS can provide continuous monitoring of hemodynamic activity over extended periods of time in ecologically valid settings. One important application of fNIRS is for Alzheimer's disease (AD). The changes in HbO in the frontal cortex were monitored while patients with probable AD performed a verbal fluency task. Whereas elderly healthy subjects (as well as patients with major depression, age-associated memory impairments, or vascular dementia) showed increases in the local concentrations of oxy-Hb and total haemoglobin (TotHb), Alzheimer's patients showed significant decreases in oxy-Hb and TotHb (also termed as HbT = HbO + HbR) shortly after beginning the task. This effect was more pronounced in the parietal cortex than in the frontal cortex [17].

In another study fNIRS was used to investigate cerebral blood oxygenation changes in the frontal lobe induced by direct stimulation of the thalamus or globus pallidus in patients with Parkinson's disease (PD) or an essential tremor. Under conditions of neural activation of the frontal lobe, HbO, and HbT increased, while deoxy-Hb decreased in two subjects during globus pallidus stimulation and increased in four subjects during low-frequency stimulation of the thalamus [17].

In addition, fNIRS has been used in the study of psychiatric disorders such as schizophrenia, mood, and anxiety disorders.

Schizophrenia is a complex and enigmatic disorder or group of disorders. The fNIRS-based studies reveal the abnormal patterns of frontal activation in schizophrenia in response to the distinct cognitive demands associated with various tasks. The concordance of fNIRS-based findings with previously established neuroimaging studies has also been concluded, thus providing justification for further work involving noninvasive and continuous explorations of cortical activation patterns in schizophrenia [16, 17].

fNIRS has applications in the study of mood disorders such as depression and bipolar disorder. It has been found that nearly half of patients with major depression show a non-dominant hemisphere response pattern that is not observed in healthy individuals. TotHb increased to a markedly lesser degree in the left hemisphere than in the right hemisphere, based on handedness. The other half of patients with depression showed a bilateral response pattern where, in response to the mirror drawing task, HbT, HbO, and HbR were comparable in both hemispheres. It has also been suggested that there may be some correlations between the course of depression and the response patterns [17]. fNIRS not only permits measurements of course and cerebral activation patterns in depression, but its use is also extendable to treatment modalities such as repetitive transcranial magnetic stimulation (rTMS) and electroconvulsive therapy (ECT) [18].

The studies on anxiety disorder also indicated that left frontal HbO in patients was significantly lower than in control subjects when confronted with anxiety-relevant or anxiety-irrelevant but not emotionally relevant stimuli. This study did not find evidence of frontal brain asymmetry when patients or control subjects observed any of the stimuli [19].

In [20] a multichannel fNIRS system was used to monitor HR in the prefrontal cortex and skin conductance response (SCR) during the video presentation of trauma-related and control stimuli. Their results indicated that both groups of victims, those with and those without post-traumatic stress disorder (PTSD), showed significant elevation of HbO in the prefrontal cortex. However, subjects with PTSD had a smaller increase in HbO and HbT than those without PTSD, along with an enhanced SCR. Because the increase of oxy-Hb lasted at least three minutes after the trauma-related stimuli, this long-lasting psychophysiological response to trauma-related stimuli was consistent with previous studies that had used other measurements such as pulse, blood pressure, SCR, and electromyograms. Therefore, it has been suggested that the measurement of cerebral HR by fNIRS is useful for psychophysiological assessment of PTSD.

Nevertheless, despite these advantages and applications, fNIRS is still an emerging technology that requires more cross-validation efforts with other more established functional neuroimaging modalities. It is, however, a promising neuroimaging technique since, compared to traditional neuroimaging systems such as fMRI, it is much cheaper. A second key advantage of fNIRS is the low monetary cost of running participants, which allows social neuroscientists to enhance the statistical power of their research by recruiting larger samples. Finally, fNIRS imaging systems are fast and relatively insensitive to participant motion and have a portable, compact, and increasingly miniaturized design. This means that fNIRS can be flexibly deployed in naturalistic settings for enhanced ecological validity [13].

18.2.3 Magnetoencephalography

MEG is the measurement of magnetic field generated by the electrical activity of neurons. A common assumption is that each neuronal generator is equivalent to a current dipole. Therefore, the principal of MEG data reconstruction is based on the equivalent current dipole (ECD) model [21]. MEG fields pass through the head without any distortion. This is a significant advantage of MEG over EEG. Due to a relatively large number of electrodes (magnetic coils – 256 and higher) MEG provides a high spatial and temporal resolution. The MEG data look similar to EEGs. As a result of the above reasons MEG has wider applications in the localization of brain sources [22, 23], such as event-related potentials (ERP) and seizure sources, compared with EEG. Nevertheless, MEG has limited use due to its high cost of the system and maintenance [24].

18.3 Joint EEG–fMRI

18.3.1 Relation Between EEG and fMRI

Since the late 1990s, simultaneous EEG–fMRI or more precisely, EEG-correlated fMRI, has emerged as a noninvasive brain imaging technique [25]. Although EEG provides a direct representation of synaptic activities with high temporal resolution, identification of underlying neuronal sources using EEG is incomplete due to low spatial resolution. In contrast, fMRI has high spatial resolution but takes a very long time (on the order of one to two seconds per slice or more than two minutes for the whole head scanning) compared with EEG. Low temporal resolution of fMRI ignores rapid variation of brain responses to stimuli. Hence, EEG and fMRI complement each other.

Haemodynamic brain response, captured by fMRI, reflects the indirect or secondary effect of neuronal activity. In contrast, EEG is a direct measure of synaptic activity. So, deriving the physiological relation between EEG and fMRI depends on understanding how the changes in neuronal activity captured by EEG affect the fMRI BOLD signal. This relationship, however, is not straightforward.

A basic assumption underlying the interpretation of fMRI maps is that an increase of regional 'neuronal activity' (global synaptic product of a local neuronal population in response to inhibitory and excitatory inputs and interneural feed-forward and feedback activity) results in an increase in metabolic demand (of neurons and astrocytes), with

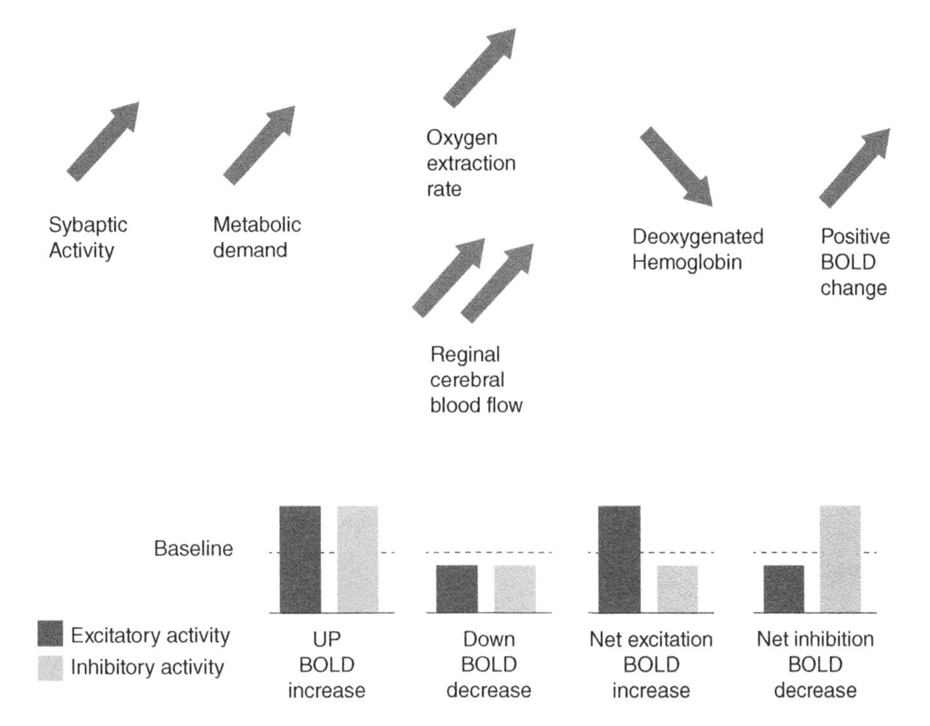

Figure 18.5 The mechanism of inherent link between EEG and fMRI. (Top) An increase in regional 'neuronal activity' (global synaptic product of a local neuronal population in response to inhibitory and excitatory inputs and interneural feed-forward and feedback activity) results in an increase in metabolic demand (of neurons and astrocytes), with increased energy and oxygen consumption. In response to these neurometabolic changes, there is a rise in local brain perfusion that exceeds the metabolic needs. Therefore, although the total oxygen consumption increases, both, part of the oxygen extraction and the percentage of deoxygenated haemoglobin in local venous blood decrease (positive BOLD change as recorded with fMRI). (Bottom) In a neuronal network comprising forward and backward excitatory and inhibitory connections, the changes in BOLD signal reflect changes of the network 'state' when compared to a carefully defined baseline [25, 26].

increased energy and oxygen consumption. In response to these neurometabolic changes, there is a rise in local brain perfusion that exceeds the metabolic needs. Therefore, although the total oxygen consumption increases, the fraction of oxygen extraction and the percentage of deoxygenated haemoglobin in local venous blood decrease (positive BOLD change as recorded with fMRI). In a neuronal network comprising forward and backward excitatory and inhibitory connections, the BOLD signal changes reflect variation of the network 'state' compared to a well and carefully defined baseline. Changes in the local field potential provide the best estimate of the HR [25]. The underlying mechanism is depicted in Figure 18.5.

Among the differences between the two, there is a time lag difference between synaptic responses and haemodynamic responses [27]. The relation is generally nonlinear and the SNR in EEG is significantly higher than the SNR in fMRI [28]. Despite this, the relation between the two depends on the type of neuronal activity. Although the relation between magnitude and spatial scale of the BOLD signal and neuronal physiology is still under debate, some researchers have reported a predominantly linear coupling between BOLD and neuronal activity. For example, a linear relationship between somatosensory-evoked potentials and BOLD has been demonstrated in [29]. Rees et al. [30] and Heeger et al. [31] also declared a linear correlation between averaging spike rate in the cortical area of a monkey's brain and the BOLD signal. This shows EEG and fMRI have an inherent relationship and hence can be analyzed jointly. Simultaneous measurement of EEG and fMRI therefore paves the way for providing more information about the exact relation between the two modalities.

18.3.2 Model-Based Method for BOLD Detection

The general linear model (GLM) is frequently used to analyze fMRI data. The works in [32, 33] are good examples of BOLD detection from fMRI using GLM. Based on this model the stimulus time course is used to construct a HRF to be convolved with the fMRI sequence to highlight the BOLD. The GLM is the most common approach in model-based fMRI analyses. The GLM approach is used to find a good solution for a variety of research questions within an infinite number of different experimental designs. This model can be presented as:

$$\mathbf{x} = \mathbf{Y}\mathbf{b} + \mathbf{e} \tag{18.6}$$

where $\mathbf{x}$ is a vector containing series of multivariate measurements, $\mathbf{Y}$ is called design matrix, $\mathbf{b}$ is a vector containing unknown parameters that are usually to be estimated and $\mathbf{e}$ represents a vector of errors or noise, assumed to have normal distribution.

In GLM, each voxel of the collected fMRI image volume is considered as a linear combination of the haemodynamic responses of stimuli and its corresponding weighted parameters, i.e. each column of matrix $\mathbf{x} \in \Re^{M \times P}$ corresponds to the fMRI time series for each voxel. Hence, the GLM interprets the time-course variation for each voxel. Figure 18.6 shows the time series for one sample voxel in an fMRI experiment. M refers to the number of scans in the experiment and P is the number of all the voxels in one scanned volume. In fMRI analysis using GLM, the design matrix is constructed using a predicted brain response to a given task. Each column of $\mathbf{Y} \in \Re^{M \times N}$ shows one of the responses obtained by convolving the stimulus function and HRF. N refers to the number of available predictors obtained based on some prior knowledge. The stimuli function is determined based on the type of experimental design during the fMRI recording session which can be block design or event-related design. The HRF can be chosen from known functions such as gamma function.

Figure 18.7 represents the schematics of the constructed model for a sample fMRI time series. In this model there are M parameters $\beta_1, \beta_2, \ldots, \beta_M$ to be estimated. These parameters are considered as the coefficients of predefined regressors obtained based on some prior knowledge. After the model is identified the weight parameters, B, should be estimated. This can be obtained by using techniques such as maximum likelihood estimation (MLE) or Bayesian estimation. The parameter values that fit the data and minimize the squared error can be estimated by:

$$\mathbf{b} = \left(\mathbf{Y}^T\mathbf{Y}\right)^{-1}\mathbf{Y}^T\mathbf{x} \tag{18.7}$$

After parameter estimation, the activated regions are detected by evaluating the statistical significance of the whole brain voxels [34]. Common techniques for analyzing fMRI data evaluate the statistical characteristics in each voxel of interest by using the t-statistic

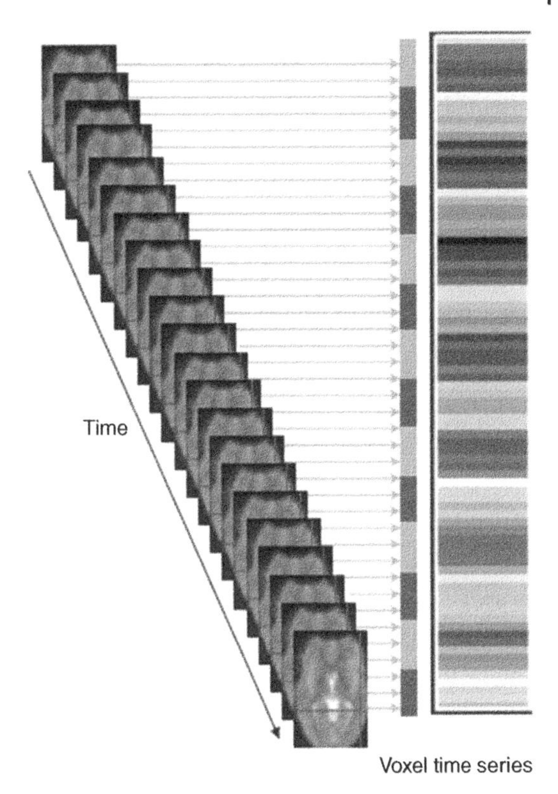

Figure 18.6 fMRI time series for a voxel sample.

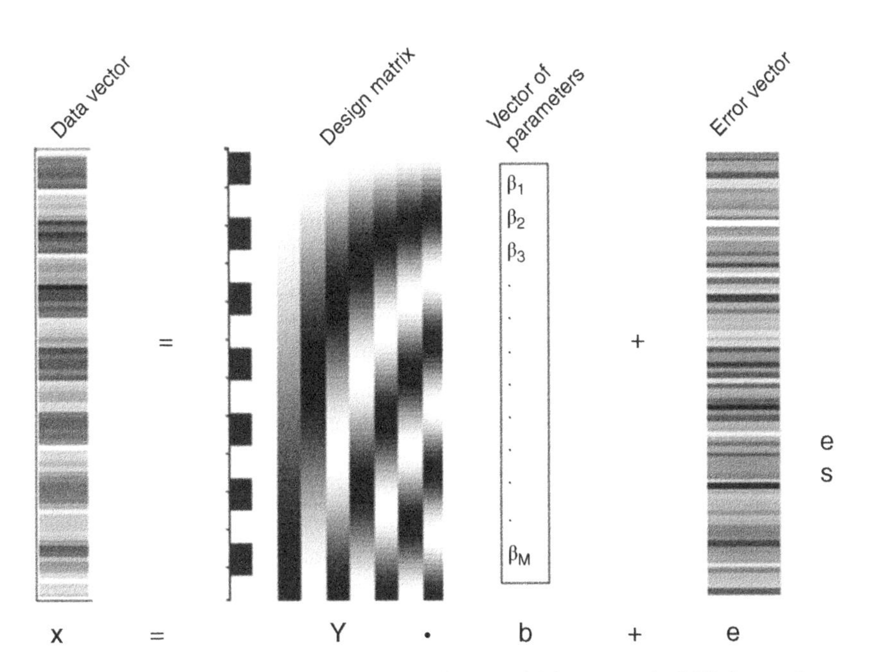

Figure 18.7 A schematic presentation of a GLM model for a sample fMRI time series.

[34]. Then, the activated areas are detected by selecting the voxels with higher statistical significance than a certain threshold level [35]. The SPM [11] and FSL [36] toolboxes are commonly used for analyzing fMRI data modelled using the GLM. The main drawback of the GLM is that the exact time course has to be known.

18.3.3 Simultaneous EEG–fMRI Recording: Artefact Removal from EEG

In simultaneous EEG and fMRI data recordings, the EEG signals are severely affected by magnetic field effects inducing strong artefacts [37]. Without removing the fMRI induced artefacts, the underlying information within the EEG can hardly be realized. EEG signals recorded in the magnetic field suffer from two major artefacts namely gradient and ballistocardiogram (BCG) artefacts. Prior to any joint processing of these data, the artefacts should be effectively removed. In the following sections the main characteristics of these artefacts are summarized and the common and recent approaches for removing them from the EEG signals are provided.

18.3.3.1 Gradient Artefact Removal from EEG

Gradient artefact, or imaging artefact, results from the changes in magnetic field of an MRI scanner during image acquisition. This artefact is characterized by its high frequency and amplitude up to 100 times larger than the EEG average amplitude, and therefore makes any visual data inspection almost impossible [38]. Figure 18.8 presents a segment of the EEG corrupted by gradient artefact. Eleven fMRI slices are recorded during this segment. The enlarged frame on the top right hand of the figure presents the high frequency contents of the gradient artefact.

The easiest way to avoid the gradient artefact is to use interleaved EEG–fMRI protocols such as periodic interleaved scanning [39, 40] or EEG-triggered fMRI [41]. These techniques are not flexible enough for general use and often are less efficient than continuous recording.

The most common technique for gradient removal is average artefact subtraction (AAS). Since this artefact does not show significant variability over time it can be subtracted from the EEG signal using an average template approach. Following this method, first, an artefact template is created by averaging successive artefact cycles in each channel. Then the created template is subtracted from the data.

In an effective work by Allen et al. [42] AAS and an adaptive noise cancellation technique have been combined. In order to generate the artefact template, the EEG signal should be segmented based on certain timing cues or triggers. These triggers are either volume timing triggers or slice timing triggers. These triggers are recorded by the scanner and denote time points when the MRI scanner starts to scan each volume/slice. It is evident obvious that any misalignment between the actual data and the artefact template removes some information within the EEG signals. In order to avoid this problem, slice timing triggers are used. Since these triggers are not recorded by the system due to shortage of memory a correlation-based algorithm is developed to detect the triggers. Figure 18.9 presents a five second segment of data from 13 channels obtained after applying Allen's method for removing the gradient artefact. BCG is evident in these signals.

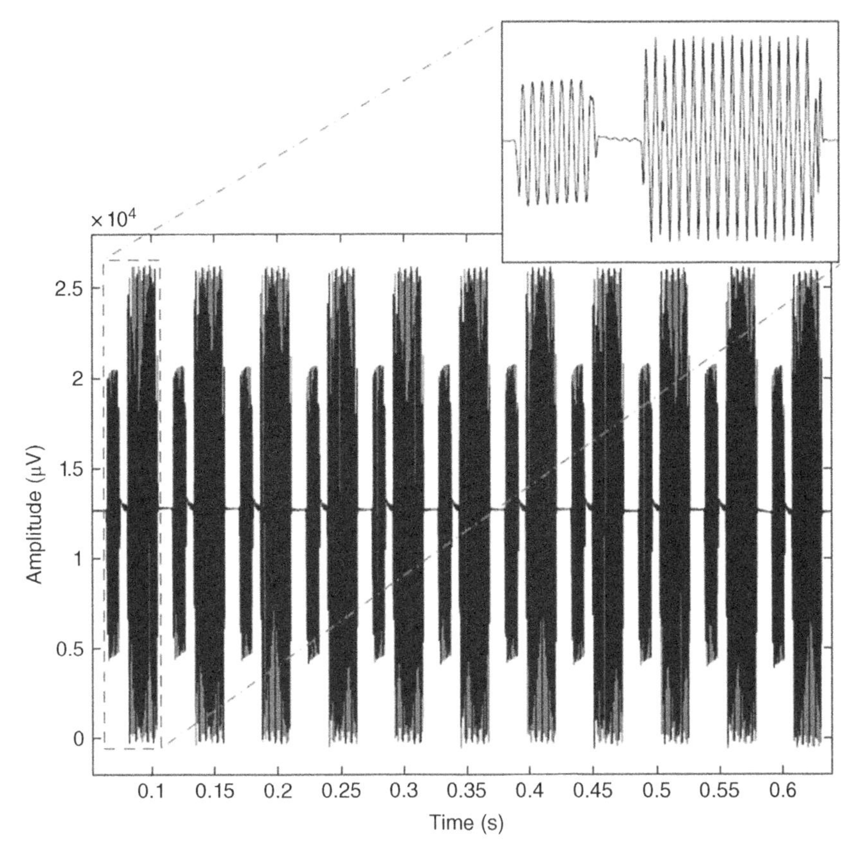

Figure 18.8 Gradient artefact for a sample segments of EEG signal recorded simultaneously with fMRI.

18.3.3.2 Ballistocardiogram Artefact Removal from EEG

Unlike gradient artefact, BCG has a more complex morphology and characterization. This artefact can be seen after the EEG is restored from the gradient artefact. BCG has severe destructive influence on the EEG signals. BCG is caused by movements of EEG electrodes in the magnetic field during the fMRI scans. There is a small movement in each electrode during the cardiac pulsation and as a result, a variable signal is induced into each electrode. The BCG artefact obscures EEG at alpha frequencies (8–13 Hz) and below, with amplitudes around 150 μV inside a magnetic field with the strength of 1.5 T [43]. BCG is quasi periodic with varying morphology. This is because of irregularities in the heartbeat, body movement, and the fact that during cardiac pulsation the electrodes are not affected by the magnetic field uniformly.

There has been intensive research for the removal of this artefact. For example, Allen et al. [38] suggested an efficient technique to reduce BCG by firmly bandaging the electrodes and wires to the subject.

The AAS, more often used for gradient artefact removal, was one of the first algorithms used for cancellation of the BCG artefact [38]. Using this method, BCG is subsequently

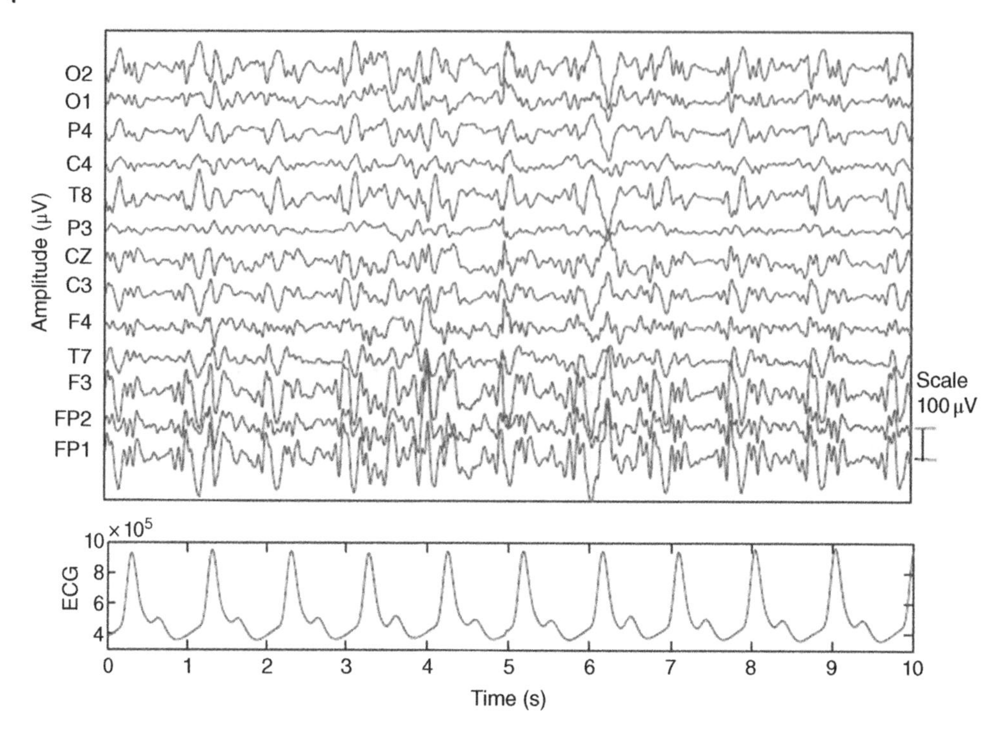

Figure 18.9 A set of 13-channel EEG data covering the entire head, after gradient artefact removal, contaminated with BCG artefact. The electrocardiogram (ECG) trend has also been illustrated at the bottom.

reduced by template subtraction from each trial. One of the major drawbacks of this widely used method is incompatibility with the artefact as it changes over time. Therefore, the assumption of having similar artefact in all trials is not always valid. Moreover, in all the methods based on averaging, the reference ECG channel is needed.

In order to modify the method to deal with heartbeat variation, a dynamic template with adaptive amplitude estimated by sliding and averaging has been proposed [40]. In another research direction, Bonmassar et al. [43] proposed an algorithm based on adaptive filtering to remove both ECG and the motion artefact. Although this method is popular, it requires acquiring another reference signal using a motion sensor attached to the subject. Niazy et al. [44] developed an algorithm using principal component analysis (PCA) known as optimal basis set (OBS). In their method, first a mixture of signals obtained from trials affected by the artefact is formed for each channel. Then, the principal components of this mixture are calculated. In the third step, only a few of the first principal components are selected as the basis set. Finally, a template is created using the selected basis set and subtracted from each BCG trial. Unfortunately, this method does not lead to good results when there are other artefacts in the EEG data which are mainly due to subject movement.

The discrete Hermite transform (DHT) has been proposed for BCG removal [45]. The main objective in this method is to model the BCG artefact using DHT. The shape of the BCG is modelled using Gaussian functions which are initial Hermite functions. These

Gaussian functions are eigenvectors of a centred or shifted Fourier matrix. In [45], the DHT of the EEG signal is obtained by computing the inner product between the signal and the Gaussian functions. This provides a set of transformed values corresponding to a particular shape within the EEG signal. Then, the artefact template is built using some of the transformed values and subtracted from the EEG signal.

Another class of artefact removal methods is based on blind source separation (BSS) mainly using independent component analysis (ICA). Blind approaches often are more useful when no reference signal, such as ECG in his case, is available for template matching. Moreover, they do not consider that BCG is predictable. Methods using ICA assume that the recorded EEG signals can be represented as a linear mixture of independent neural activities inside the brain, and artefacts caused by muscles and noise. Conversely, ICA decomposes the EEG signals into a set of independent components (ICs). Removing the ICs containing the BCG artefact and back-projecting the remaining ICs to the electrode space results in obtaining clean EEG signals. In several studies ICA has been used for removing the BCG artefact [46]. In a research study the performances of different ICA algorithms for removing BCG from EEG data have been evaluated [47]. They also used two different post-processing methods to improve the results of ICA. In a recent method, Ghaderi et al. [48] proposed a blind source extraction technique (BSE) with cyclostationarity constraint to extract the BCG sources. In another attempt, Leclercq et al. [49] proposed a constrained ICA (cICA) to remove the BCG artefact. In their method, first, a template for BCG artefact is estimated for each channel. Then, the artefact-related components based on these constraints are extracted using the cICA algorithm. In the next step, the estimated artefact components are clustered and averaged over each cluster to have better estimation of the components related to BCG artefact. Then, the Gram-Schmidt algorithm is used for orthogonalization and computing the separation matrix. In the last step, clean EEG signals are recovered by deflating (recursively subtracting) the sources selected as artefact and back-projecting the remaining sources to the electrode space.

An important issue in BCG removal using ICA is selection of the correct number of extracted components that should be deflated. Different numbers of ICs, e.g. 3 [47], 3–6 [46], or 1 [50] considered as the BCG artefacts, have been reported by different researchers. Selecting and removing only a small number of sources as BCG may leave some artefacts in the signals out whereas selecting and removing a large number of sources may eliminate useful information from the EEG signals.

In a new approach by Ferdowsi et al. [51] ICA and DHT have been combined. This can be achieved by applying DHT to those ICA sources labelled as BCG artefacts. After detecting the peaks in the BCG by applying a simple peak detection method, each BCG source is divided into time segments centred at the detected peaks. Hence, the segments have one peak in each segment. Then, an adaptive DHT is applied to each segment in order to model the artefact. Finally, the obtained model is subtracted from each segment which gives a clean component with no artefact. All the obtained components based on this procedure will be projected back to the electrode space giving a clean EEG signal.

The main advantage of this approach is its robustness against changes in the shape of artefact over time. The proposed method alleviates the uncertainty in choosing the right number of sources to be deflated in ICA-based methods. Moreover, an adaptive parameter selection strategy is proposed to decrease the sensitivity of DHT-based methods to variations of the model parameters.

Methods solely based on DHT have some drawbacks. For example, a weakness of DHT in BCG removal is its inability to model the whole artefact with minimum number of bases functions. Based on such weaknesses, large numbers of coefficients are needed to reconstruct the BCG template having the same amplitude and duration. Using a large number of basis functions, also, leads to eliminating some details in the EEG data. The hybrid ICA-DHT algorithm may be summarized into the following steps:

1) Apply ICA to the EEG data contaminated by the BCG artefact.
2) Select six ICA components which are more correlated with the ECG channel.
3) Apply DHT on each selected BCG source to find a template for the artefact.
4) Subtract the template from each BCG source and back-project the residuals together with the remaining sources.

In an experiment natural EEG data were recorded from five healthy men. All the subjects were right-handed and their ages ranged from 18 to 50 years. The EEG was acquired using the Neuroscan Maglink RT system (with impedances kept within 10–20 kΩ), providing 64-channel comprising 62 scalp electrodes, one ECG electrode and one electro-oculogram (EOG) electrode.

The sampling rate of raw EEG data was set at 10 kHz. During fMRI acquisition 300 volumes including 38 slices ($3.2969 \times 3.2969 \times 3.3$ mm resolution, repetition time, repeat time (TR) = 2000 ms, and echo time, TE = 25 ms) were acquired.

After preprocessing, Infomax ICA is applied to 10 seconds of EEG data. The number of sources is chosen to be equal to the number of sensors. After applying ICA, the extracted sources are clustered based on their correlations with the ECG channel. DHT is then used for modelling and removing these spikes from the selected sources. For this purpose, these sources are segmented such that the spikes fall in the centre of each segment. In order to have one artefact in each segment, with no overlap, the length of each segment is selected to be 256 samples. This is due to the fact that the period of the BCG artefact is approximately one second, and the sampling rate of data is 250 Hz [51].

The residual contains brain rhythms retrieved by the proposed method. In this work, 15 DHT coefficients (5% of the total coefficients) are selected to model the artefact. The reason for selecting such a small fraction is to avoid losing useful EEG information while removing the BCG. Figure 18.10a shows the results of artefact removal from one 'EEG channel' using ICA and ICA-DHT. The highlighted (circled by dotted lines) areas show some peaks of BCG artefact which have not been removed by ICA. The number of deflated sources in ICA is five, while six sources have been selected to deflate in the ICA-DHT method. Figure 18.10b shows the BCG source which is not deflated in ICA.

Figure 18.11 shows the results of applying different artefact removal methods for a segment of EEG signal labelled as the CZ channel. The threshold and dilation parameter in DHT need to be set appropriately and this is often done using an adaptive approach. In this experiment the averaged dilation parameter and the threshold have been set to 4.25 and 0.001 respectively [51].

In another attempt for BCG removal the structure of BCG has been exploited to develop a source extraction method based on both short-term and long-term prediction [52]. The

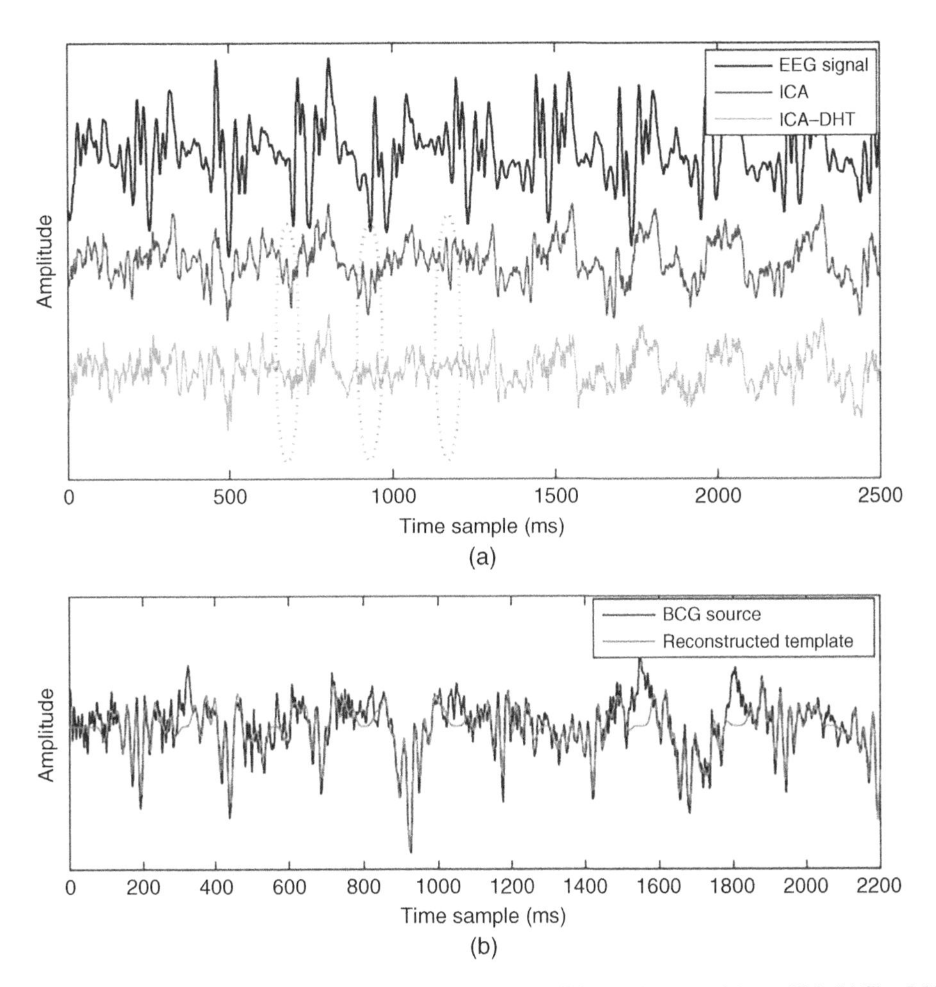

Figure 18.10 (a) Comparison between ICA and ICA-DHT for BCG removal from EEG. (b) The BCG source, not deflated in ICA, together with corresponding model obtained using ICA-DHT.

predictability is therefore used as a criterion for source separation. Looking at Figure 18.12 a sample at time t may be predicted using its previous K_1 samples and K_2 samples of the previous cycle within one cycle interval denoted as τ.

The prediction error resulting from both short- and long-term prediction can be expressed as:

$$e_s(t) = y(t) - \sum_{p=1}^{K_1} b_p y(t-p) \tag{18.8a}$$

$$e_l(t) = y(t) - \sum_{q=1}^{K_2} d_q y(t-\tau-q) \tag{18.8b}$$

where b_p and d_q containing respectively the short- and long-term prediction coefficients, K_1 and K_2 are respectively the short- and long-term prediction orders and $\mathbf{y}(t) = \mathbf{W}\mathbf{x}(t)$ are the separated sources at time instant t. Here, $\mathbf{x}(t) = \mathbf{A}\mathbf{s}(t) + \mathbf{v}(t)$ are the mixtures of EEG signals

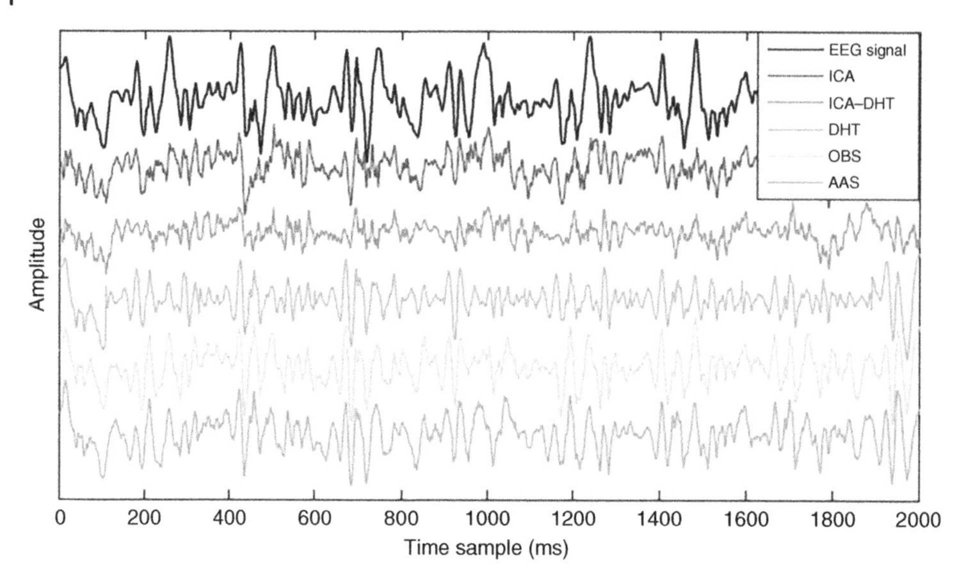

Figure 18.11 Results of artefact removal from CZ channel using ICA (second from top), combined ICA-DHT (third), DHT (fourth), OBS (fifth), and AAS (at the bottom).

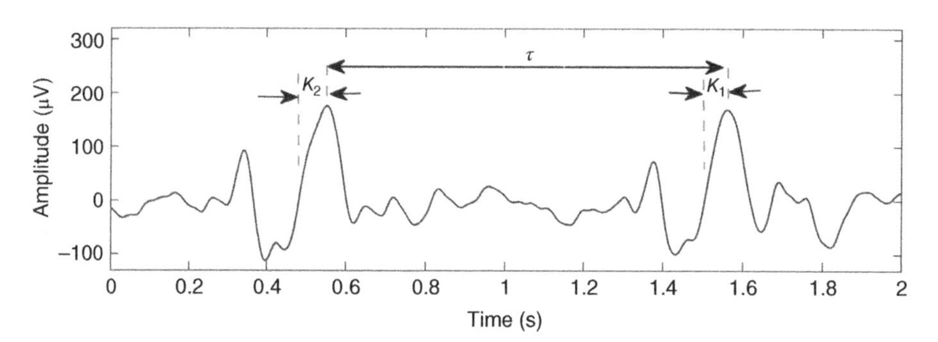

Figure 18.12 A cycle of a BCG artefact in a sample segment of EEG signal.

and artefacts measured at the electrodes and $\mathbf{v}(t)$ is additive noise. Equation (18.6) can be written in matrix/vector form as:

$$e_s(t) = \mathbf{w}^T\mathbf{x}(t) - \mathbf{b}\bar{\mathbf{y}}(t)$$
$$e_l(t) = \mathbf{w}^T\mathbf{x}(t) - \mathbf{d}\bar{\mathbf{y}}_l(t) \tag{18.9}$$

where $\mathbf{y}(t) = [y_1(t), \ldots, y_n(t)]^T$ are the $n \times 1$ estimated source signals at time t, $\mathbf{W}$ is the separating matrix, $\mathbf{w} = [w_1, w_2, \ldots, w_m]^T$, $\mathbf{b} = [b_1, b_2, \ldots, b_{K1}]^T$, $\bar{y}(t) = [y(t-1), y(t-2), \ldots, y(t-K_1)]^T$, K_1 is the short-term prediction order, $\bar{y}_l(t) = [y(t-\tau-1), y(t-\tau-2), \ldots, y(t-\tau-K_1)]^T$ and K_2 is the long-term prediction order, (for notational simplicity we omit index j from $y_j(t)$ and $\mathbf{w}_j$):

$$J(\mathbf{w},\mathbf{b},\mathbf{d}) = \frac{1}{2}E\{e_s^2\} + \frac{1}{2}E\{e_l^2\} = J_s(\mathbf{w},\mathbf{b}) + J_l(\mathbf{w},\mathbf{d}) \tag{18.10}$$

The optimum values for the source extractor filter **w** (a column vector of matrix **W** related to the source of interest) and short-term and long-term coefficient vectors **b** and **d** can be estimated by minimizing the cost function $J(\mathbf{w}, \mathbf{b}, \mathbf{d})$ in (18.8) using alternating least squares (ALS) [53]. The update equations for estimating these parameters can then be derived through some mathematical manipulation and concluded as [52]:

$$\mathbf{w}^{(\kappa+1)} = \mathbf{w}^{(\kappa)} - \eta_{\mathbf{w}}\left(\mathbf{w}^{(\kappa)}\right)^T\left\{2\mathbf{R}_{\tilde{\mathbf{x}}}(0) - 2\sum_{p=1}^{K_1} b_p^{(\kappa)}\mathbf{R}_{\tilde{\mathbf{x}}}(p) - 2\sum_{p=1}^{K_1} d_p^{(\kappa)}\mathbf{R}_{\tilde{\mathbf{x}}}(q+\tau)\right.$$

$$\left. + \sum_{p=1}^{K_1}\sum_{q=1}^{K_1} b_p^{(\kappa)}d_q^{(\kappa)}\mathbf{R}_{\tilde{\mathbf{x}}}(p-q) + \sum_{p=1}^{K_2}\sum_{q=1}^{K_2} b_p^{(\kappa)}d_q^{(\kappa)}\mathbf{R}_{\tilde{\mathbf{x}}}(p-q)\right\} \tag{18.11}$$

$$b_k^{(\kappa+1)} = b_k^{(\kappa)} - \eta_b\left\{-\left(\mathbf{w}^{(\kappa)}\right)^T\mathbf{R}_{\tilde{\mathbf{x}}}(k)\mathbf{w}^{(\kappa)} + \sum_{p=1}^{K_1} b_p^{(\kappa)}\left(\mathbf{w}^{(\kappa)}\right)^T\mathbf{R}_{\tilde{\mathbf{x}}}(p-k)\mathbf{w}^{(\kappa)}\right\} \quad \forall k = 1,\cdots,K_2 \tag{18.12}$$

$$d_k^{(\kappa+1)} = d_k^{(\kappa)} - \eta_d\left\{-\left(\mathbf{w}^{(\kappa)}\right)^T\mathbf{R}_{\tilde{\mathbf{x}}}(k)\mathbf{w}^{(\kappa)} + \sum_{p=1}^{K_2} d_p^{(\kappa)}\left(\mathbf{w}^{(\kappa)}\right)^T\mathbf{R}_{\tilde{\mathbf{x}}}(p-k)\mathbf{w}^{(\kappa)}\right\} \quad \forall k = 1,\cdots,K_1 \tag{18.13}$$

where $\eta_{\mathbf{w}}$, η_b and η_d are step sizes and $\mathbf{w}^{(\kappa+1)}$, $b_k^{(u+1)}$ and $d_k^{(u+1)}$ are the updated parameter values and $\mathbf{R}_{\mathrm{x}}$ is the autocorrelation function of $\tilde{\mathbf{x}}$ (whitened **x**). The step sizes are kept fixed here. The updates continue for all the sources from 1 to n and the values of **w**, **b** and **d** are updated to minimize (18.10). **w** is normalized after each iteration to preserve a unity norm for each column of the separation matrix. In this approach K_1 and K_2 have been selected carefully for optimum solutions [52]. The update process ends when the changes in the parameters remain below a predefined threshold.

To investigate the quality of results one way is to determine how the detected BCG is correlated with the corresponding ECG. Higher correlation values indicate a better separation. Another method is to find out how much of the artefact is still remaining in the original mixtures. This can be achieved by measuring the correlation between the detected artefacts and the mixtures. Better results correspond to smaller correlation values. Figure 18.13 represents the topographic maps illustrating the μ (mu) rhythm after the artefact is removed by employing AAS, OBS, and the prediction-based short-and-long-term linear predictor

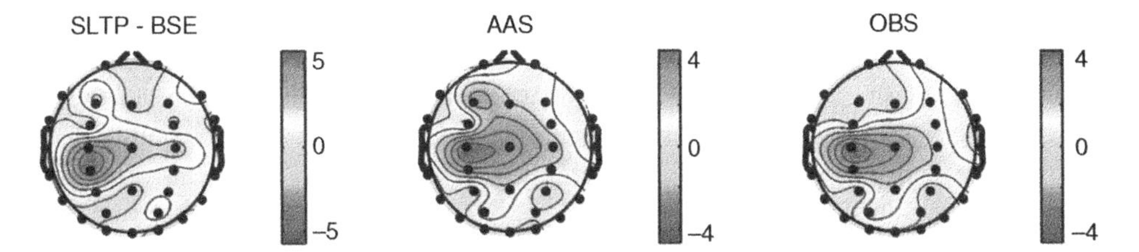

Figure 18.13 Topographic maps illustrating mu rhythm after the artefact is removed by employing AAS, OBS, and the prediction-based BSE (SLTP-BSE) approach.

BSE (SLTP-BSE) approaches [52]. The measure of effectiveness of a method is how well it can localize the mu rhythm. Therefore, the best localization is performed by the proposed SLTP-BSE.

18.3.4 BOLD Detection in fMRI

Detection of BOLD is the prime objective in the analysis of fMRI data. Nonnegative matrix factorization (NMF), as a BSS approach, is able to detect the BOLD signal without relying on any prior knowledge about the instants of stimulus onset. This is in contrast to GLM which is the most widely used technique for fMRI analysis. NMF decomposes a mixture of fMRI data into a set of time courses and their corresponding spatial sources. Extracted time courses represent the brain temporal response to stimuli or artefacts. Each time course is related to a source known as an active map and represents the active area in the brain. Ulfarsson et al. [54] state that a suitable fMRI analysis technique should provide sparse sources. That means having a small number of active (non-zero) voxels in each source. Brain networks of interest such as the motor or visual cortex typically have sparse spatial structure [54]. NMF is able to produce sparse results as a consequence of allowing only non-negative decomposition. These advantages of NMF make it suitable for analysis of fMRI which is inherently positive. To improve the results and ensure uniqueness of the solution, some spatial or temporal constraints are defined and incorporated into the factorization formulation.

The most common model-free methods applied to fMRI data are ICA and NMF. These methods detect BOLD by decomposing a mixture containing measured fMRI data into a set of time courses and their corresponding spatial maps [10, 55]. Generally, separated sources and time courses are divided into two groups: signal of interest and signal not of interest. Signals of interest include task-related, function-related, and transiently task-related ones [10]. BOLD is a task-related signal and is the main outcome of fMRI. If the brain response to a given task dies off before turning off the stimulation or change due to repeated stimuli, it will lead to a transiently task-related signal [10]. Function-related signals reveal the similarities between voxels inside a particular functional domain [56].

Signals not of interest include physiology-related, motion-related, and scanner related signals [10]. Breathing and heartbeat are considered those producing the physiology-related signals. Brain ventricles and areas which contain large blood vessels are the origin of physiology-related signals.

The proposed method by McKeown et al. [57] was the first application of spatial ICA (SICA) for fMRI analysis. In SICA, each row of the mixture matrix refers to the vectorized form of collected fMRI image in one scan. In contrast to SICA, in temporal ICA (TICA) each column of a mixture refers to an fMRI image acquired during one complete scan. In SICA, the algorithm attempts to find a set of spatially ICs and their associated (unconstrained) time courses. However, in TICA, the algorithm attempts to find a set of temporally independent time courses and their associated spatial maps. The decision to choose either spatial or even both depends on the desired criteria [58]. In spite of exploiting both spatial and temporal independence in some research works [58, 59], most approaches still rely on the assumption of spatial independence because of lack of good understanding of the unknown brain activities. Moreover, SICA has lower computational complexity than TICA. This favours SICA for decomposing fMRI data. ICA employs different algorithms to separate independent sources and their associated time courses. Performance evaluation of these algorithms is an important issue which has been studied in some researches [60–62]. Source distribution imposes a suitable criterion to find a suitable ICA algorithm. The Infomax BSS algorithm often leads to reliable separation results for fMRI.

Although ICA has been found very useful and effective in fMRI analysis, it is not able to extract the required sparse sources effectively; mainly due to the number of zero valued components which make higher order averages hard to handle. In [55] constrained NMF has been proposed for detection of active area in the brain. They use sparsity and uncorrelatedness as constraints on decomposed factors and in another research approach they developed their method using the K-means clustering algorithm to improve the initialization of NMF [63]. In these works, spatial NMF is used to decompose a set of fMRI images into a set of nonnegative sources and a set of nonnegative time courses. We also employ spatial NMF throughout this chapter. Although constraints such as sparsity and uncorrelatedness make NMF more reliable, they do not incorporate any information about sources of interest.

18.3.4.1 Implementation of Different NMF Algorithms for BOLD Detection

Squared Euclidean distance and generalized KL divergence are the best known and the most frequently used cost functions for NMF. Csiszar's divergence [64], Bregman divergence [65], generalized divergence measure [66] and α or β divergences [67] are other alternatives. A suitable cost function can be determined based on the assumption about noise distribution. When the noise is normally distributed, the squared Euclidean distance is an optimal choice. In some applications, such as pattern recognition, image processing and statistical learning, the noise is not necessarily Gaussian and the cost functions based on information divergence are often used.

In the following discussion the performances of different NMF algorithms for BOLD detection are compared. These includes the α-divergence [68] algorithm. Given a nonnegative matrix $\mathbf{X} \in \Re^{M \times N}$, containing the input data, two nonnegative matrices $\mathbf{A} \in \Re^{M \times J}$ and $\mathbf{S} \in \Re^{J \times N}$ are estimated such that $\mathbf{X} = \mathbf{AS} + \mathbf{V}$, where $\mathbf{V} \in \Re^{M \times N}$ is the factorization error and $\mathbf{Y} \in \Re^{M \times N}$ denotes rank of factorization and is assumed to be known or estimated by an information-theoretic criterion.

Frobenius norm is the most common NMF cost function:

$$J_F = \|\mathbf{X} - \mathbf{AS}\|_F^2 = \sum_i \sum_t \left(x_{it} - \{\mathbf{AS}\}_{it}\right)^2 \tag{18.14}$$

Multiplicative update rules for the above optimization problem have been derived by Lee and Seung [69]. They proved that the referred cost functions would converge to a local minimum under these update rules. As another algorithm, the basic α -divergence between $\mathbf{X}$ and $\mathbf{AS}$ is defined as:

$$J_\alpha = \frac{1}{\alpha(\alpha - 1)} \sum_i \sum_t \left(x_{it}^{\alpha} \{\mathbf{AS}\}_{it}^{1-\alpha} - \alpha x_{it} + (\alpha - 1)\{\mathbf{AS}\}_{it}\right) \tag{18.15}$$

where $0 \leq \alpha \leq 2$. Special cases for α-divergence algorithm are also defined as follows [68]:

KL I-divergence ($\alpha \to 1$):

$$J_{KL} = \sum_i \sum_t \left(x_{it} \ln \frac{x_{it}}{\{\mathbf{AS}\}_{it}} - x_{it} + \{\mathbf{AS}\}_{it} \right) \tag{18.16}$$

Dual KL I-divergence ($\alpha \to 0$):

$$J_{dKL} = \sum_i \sum_t \left(\{\mathbf{AS}\}_{it} \ln \frac{\{\mathbf{AS}\}_{it}}{x_{it}} + x_{it} - \{\mathbf{AS}\}_{it} \right) \tag{18.17}$$

Squared Hellinger divergence ($\alpha = 0.5$):

$$J_{SH} = \sum_i \sum_t \left(\{\mathbf{AS}\}_{it}^{1/2} - x_{it}^{1/2} \right)^2 \tag{18.18}$$

Pearson divergence ($\alpha = 2$):

$$J_p = \sum_i \sum_t \frac{\left(x_{it} - \{\mathbf{AS}\}_{it}\right)^2}{\{\mathbf{AS}\}_{it}} \tag{18.19}$$

The choice of optimal α depends on the application and the data being analyzed. The following equations present the main learning rules for α -divergence algorithms. These update rules are suitable for large scale NMF [67] and can be expressed as:

$$a_{ij} \leftarrow \left(a_{ij} \left(\sum_t s_{jt} \left(\frac{x_{it}}{\{\mathbf{AS}\}_{it}} \right)^{\alpha} \right)^{\omega/\alpha} \right)^{1-\alpha_{sa}} \tag{18.20}$$

$$s_{ij} \leftarrow \left(s_{ij} \left(\sum_t a_{jt} \left(\frac{x_{it}}{\{\mathbf{AS}\}_{it}} \right)^{\alpha} \right)^{\omega/\alpha} \right)^{1-\alpha_{ss}} \tag{18.21}$$

where ω is the over-relaxation parameter and is typically selected to be between 0.5 and 2. The over-relaxation parameter accelerates the convergence and stabilizes the algorithm. α_{sa} and α_{ss} are small positive parameters which are used to enforce the sparsity constraint on the algorithm.

The α -divergence based NMF algorithm has been used to detect brain activation in a set of synthetic and real fMRI data. In order to find the optimal value of α, the source separation procedure was repeated with different α values. Moreover, the performances of groups of algorithms were compared with those of more common NMF algorithms such as Euclidean distance-based methods.

18.3.4.2 BOLD Detection Experiments

The NMF algorithms have been examined by employing a set of test data [70] and also a set of EEG–fMRI recorded data. The simulated fMRI data included eight sources and their corresponding time courses as depicted in Figure 18.14. $\mathbf{s}_1$ shows the simulated task-related source (BOLD), $\mathbf{s}_2$ and $\mathbf{s}_6$ are transient task-related, and the rest represent the artefact-related sources. Using the ICA and NMF algorithms the task-related source or BOLD is separated from the fMRI sequence.

In order to generate the mixture, a matrix of time courses is multiplied by the matrix of sources. SIR defined as $\text{SIR}_i = \|\hat{\mathbf{s}}_i\|_2 / \|\hat{\mathbf{s}}_i - \mathbf{s}_i\|_2$, for each source i, has been used for evaluation of the algorithm performance. $\|\bullet\|_2$ indicates Euclidean norm. In this equation $\hat{\mathbf{s}}_i$ and $\mathbf{s}_i$ are respectively the extracted and actual task-related (BOLD) sources.

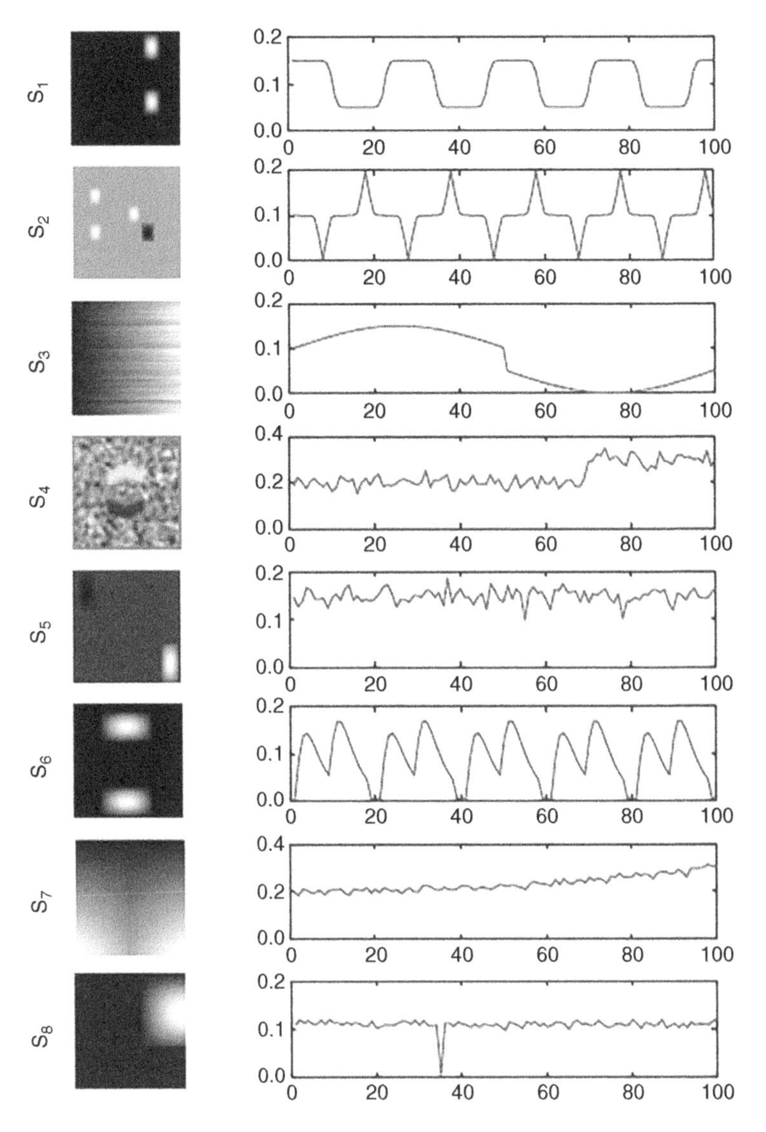

Figure 18.14 Simulated fMRI including the sources and corresponding time courses (using MLSP laboratory tools).

The performances of Euclidean distance, KL I-divergence, dual KL I-divergence, squared Hellinger divergence and Pearson divergence have been evaluated for this data set a number of times and averaged. Our experiments show that the best convergence requires the over-relaxation parameter to be $\omega = 1.9$ and sparsity regularization parameters $\alpha_{sa} = \alpha_{ss} = 0.001$.

Figure 18.15 shows the computed average SIR for the results of Euclidean distance and different α-divergence-based NMF algorithms. It is seen from the figure that the best SIR is related to the Euclidean distance of 29.33 dB. This value is much higher compared to the

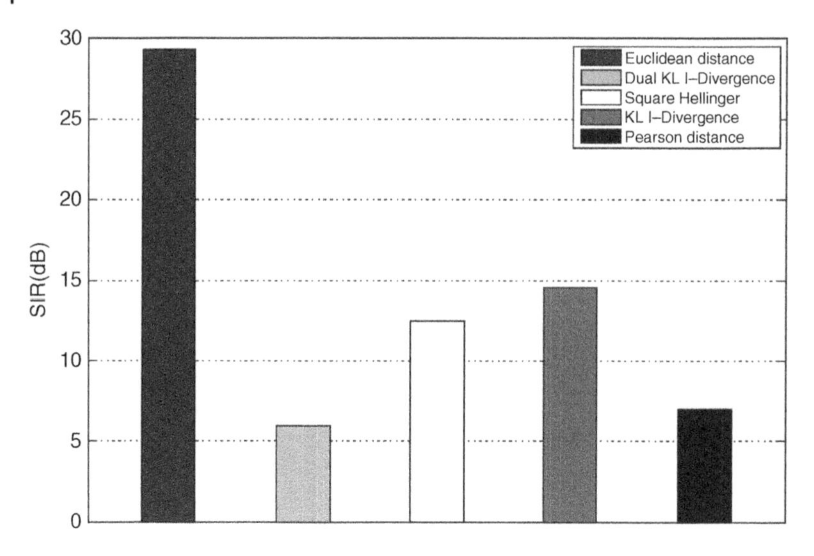

Figure 18.15 Computed SIR of source of interest for different methods.

results of α-divergence-based methods. Among α-divergence based NMF algorithms, KL I-divergence and square Hellinger divergence show higher SIR compared to dual KL I-divergence and Pearson divergence.

It should be noted that the sparsity constraint is used in the Euclidean distance-based algorithm to obtain more accurate results. The sparsity regularization parameter has been set to 0.1 for this experiment.

Real data sets taken from the SPM website [11] are used next. The first data set is auditory fMRI data from a single subject experiment. This dataset has been recorded using a 2 T Siemens MAGNETOM Vision scanner with the scan-to-scan TR of seven seconds. The auditory stimuli are bi-syllabic words presented binaurally at a rate of 60 per minute. The data set contains 96 scans and each scan consists of 64 contiguous slices ($3 \times 3 \times 3$ mm voxel size). The 96 scans include eight blocks of size 12, each of which contains six scans under rest and six scans under auditory stimulations.

The second data set is a visual fMRI data and was collected using a 2 T Siemens MAGNETOM vision system. There are 360 scans during four runs, each consisting of four conditions which are *fixation*, *attention*, *no attention* and *stationary*. In the *attention* condition the subject should detect the changes in scene and during the *no-attention* condition the subject was instructed to just keep their eyes open. During *attention* and *no attention* the subjects fixated centrally, while white dots emerged from the fixation point to the edge of the screen.

The given data sets need preprocessing for subsequent processing for data separation. All the preprocessing steps including realignment, slice timing, coregistration, normalization, and smoothing have been performed using SPM software [11].

In the experiments on real fMRI data, we applied KL I-divergence and square Hellinger to both data sets. Visual comparison of the results for different methods is difficult.

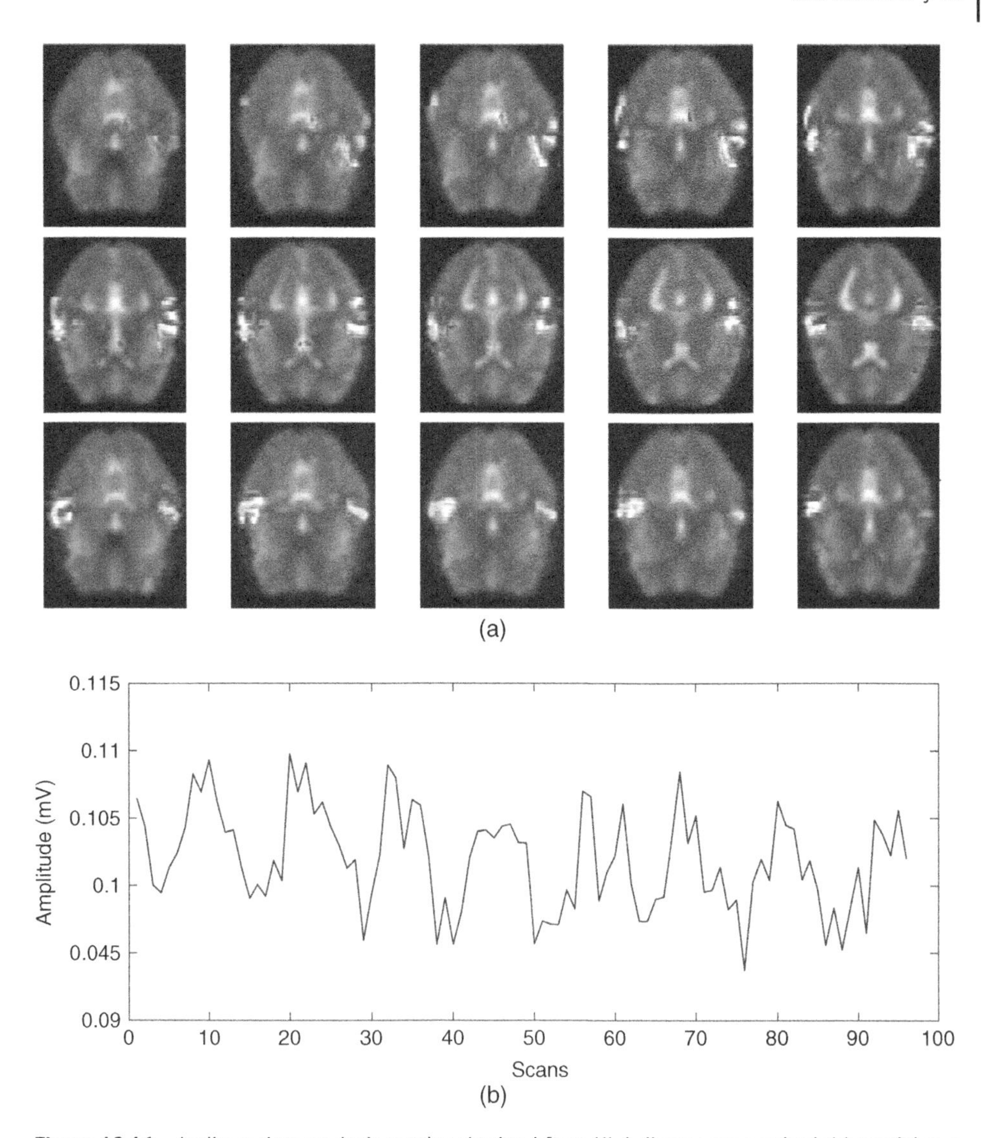

Figure 18.16 Auditory data analysis results obtained from KL I-divergence method: (a) spatial map and (b) corresponding time course.

Figure 18.16 shows the extracted BOLD and its corresponding time course for the first data set. As it is seen, the activated region is correctly detected for different brain slices.

Figure 18.17 presents the results of applying KL I-divergence to the visual fMRI data set. BOLD has been detected in the occipital lobe which is responsible for visual processing tasks. The extracted time course also verifies the temporal behaviour of activation.

In order to compare the results, the normalized correlation between the extracted time course and the predicted temporal response of the brain has been calculated. The brain temporal response to a specific task can be modelled by convolving the task-waveform and the HRF [7].

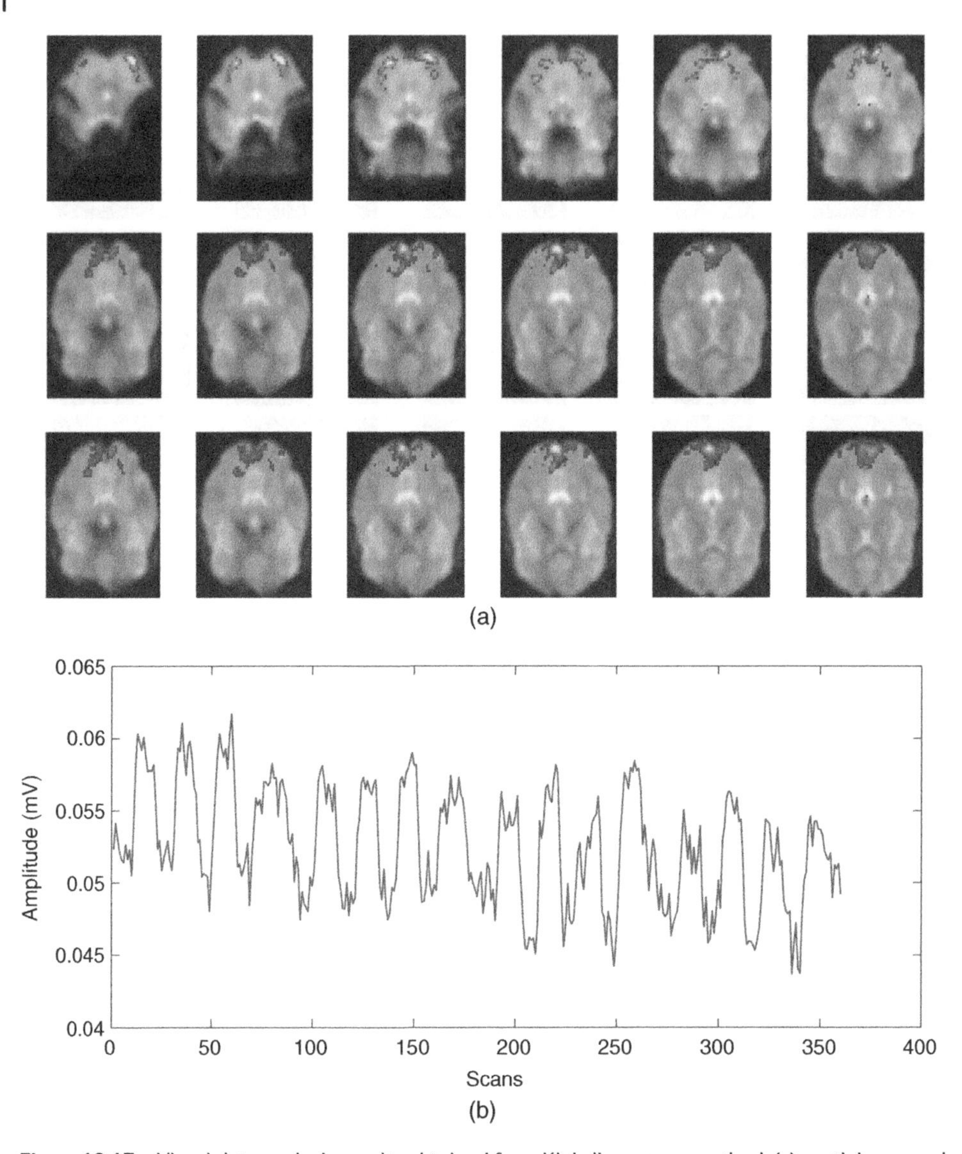

Figure 18.17 Visual data analysis results obtained from KL I-divergence method: (a) spatial map and (b) corresponding time course.

The results show that the normalized correlation between the extracted time course and the predicted temporal response of the brain for the results of Euclidean distance has higher value than those for the two other α -divergence based methods. The numerical results of this comparison for both data sets are given in Table 18.1.

Any temporal or spatial pre-knowledge about the BOLD can enhance the effectiveness of its detection. As an example, auditory BOLD is expected to be seen in lateral lobes, visual BOLD in fronto-temporal lobes, and focal seizure BOLD in frontoparietal lobes. Such information can be incorporated into the separation algorithm to enhance its accuracy.

Table 18.1 Normalized correlation between the extracted BOLD time course and predicted brain temporal response.

Dataset Algorithm	Auditory	Visual
KL I-divergence	0.6950	0.8736
Square Hellinger	0.6371	0.8152
Euclidean distance	0.8689	0.9102

18.3.5 Fusion of EEG and fMRI

18.3.5.1 Extraction of fMRI Time Course from EEG

To construct, enhance, or detect BOLD, GLM is the most robust technique if the time course is available. In places where the recordings are not time-locked, there is no cue, there is voluntary movement, or the BOLD is due to neuro-physiological disorder, such as seizure, the time course is not known. Although blind techniques such as ICA or NMF are able to estimate the spatial and temporal information, they often introduce errors leading to misinterpretation of the results and therefore, estimation of an effective time course allows application of model-based approaches such as GLM which requires accurate construction of HRF.

In the cases of movement-related brain activity, the changes in synchronous activity of the neuronal population cause event-related variations in the amplitude of EEG oscillations [71]. In earlier chapters of this book it has been mentioned that there are two strong event-related changes in brain oscillations: event-related desynchronization (ERD) in alpha rhythm and event-related synchronization (ERS) in the beta rhythm. The act of preparation for a movement in a subject suppresses EEG oscillations in both alpha and beta rhythm over the sensory motor area [72–74]. This is the ERD signal and starts approximately one second before the movement. Following movement termination while the alpha rhythm returns to the baseline, beta power also returns to baseline and exceeds pre-movement level [75, 76]. This sharp increase in the signal power of the beta band is the ERS signal or post-movement beta rebound (PMBR). Using the information obtained from EEG rhythms as a regressor for fMRI analysis enables localization of BOLD which is correlated with a specific neuronal rhythm. Concurrent recording of EEG and fMRI is useful for this purpose.

18.3.5.2 Fusion of EEG and fMRI; Blind Approach

EEG–fMRI integration reveals more complete information about brain functionality which cannot be observed using EEG and fMRI separately. Generally, despite the information provided at the beginning of this chapter the relationship between synchronous neuronal activity and BOLD is not totally clear. Simultaneous recording of EEG and fMRI provides the opportunity to identify different areas of the brain involved during EEG events. Using the extracted time course using the method described in Section 18.3.5, the correlation between PMBR and BOLD can be investigated. The information extracted from EEG analysis is used as a constraint to incorporate information derived from the EEG signal into the fMRI analysis procedure.

Tensor factorization explained in previous chapters of this book may be used for this purpose. In a novel approach PARAFAC2 as a tensor factorization algorithm is selected for EEG–fMRI fusion. The main advantage of using PARAFAC2 is capability to analyze data in multiple modes rather than just two. This feature leads to obtaining valuable information about BOLD and its time course. Moreover, the multi-way data analysis enables processing of the data in a systematic way. This approach can effectively detect the area in the brain which is responsible for PMBR. This can be empirically proved when comparing the results with those obtained from GLM.

Current approaches for EEG–fMRI integration are divided into two main groups: model driven and data driven. The computational biophysical model forms the basis of model-driven approaches [77]. In these methods, the assumptions about the neural activities are used to model the relation between EEG and fMRI. Despite the fact that the model-driven approaches provide deeper understanding about the neuronal mechanisms, they require an explicit description of mutual neuronal substrates to annotate the measured EEG and fMRI data. However, the demand for the application of model-driven methods has decreased due to lack of sufficient knowledge about the neuronal substrates [77].

Data-driven approaches are based on common interactions between EEG and fMRI. These approaches are classified into two main categories: EEG–fMRI fusion based on constraints and EEG–fMRI fusion based on prediction. In methods based on constraint, the fMRI active map obtained by fMRI analysis is used as a priori information for electromagnetic source localization. Since the number of EEG sensors is generally smaller than the number of sources within the brain, often the EEG inverse problem is underdetermined and ill-posed. Therefore, additional constraints or a priori information are needed in order to obtain a unique and stable solution for the location of sources in the EEG.

There are two approaches for EEG–fMRI integration based on the so-called fMRI-constraint dipole fitting and distributed source modelling. Fujimaki et al. [78] proposed a method based on dipole fitting. In their method the neural activity at each fMRI hotspot is modelled as an equivalent regional current dipole. Then, the dipole locations are fixed to fMRI hotspots or the fMRI hotspots are used as seed points for dipole fitting while a maximum distance constraint is applied. After the dipole locations are estimated, the dipole moments can be determined by fitting the ECD model to the EEG data. The temporal dynamics of the regional neural activity is determined after estimating the dipole time courses. Although the methods give unique solutions, they are only able to detect the active dipoles at fMRI activation areas. Some techniques have also been developed based on distributed source modelling [79, 80]. In these works, the geometrical information of BOLD obtained from the fMRI active map is used to derive the covariance prior for source reconstruction. Weighted minimum norm frameworks and Wiener filter are two techniques used for this purpose in [79, 80] respectively. In contrast to methods based on dipole fitting, the fMRI constrained distributed source imaging methods are capable of obtaining dipoles not only at fMRI activation areas but also at areas where fMRI fails to show any activation. However, the performance of these methods is affected by some drawbacks such as identification of fMRI weighting factors and problems which result from the temporal resolution difference between the EEG and fMRI. These problems are mainly how to deal with fMRI extra sources, fMRI invisible sources, and fMRI discrepancy sources [81]. In [82–84], the authors have suggested some empirical values for weighting factors to overcome the problem. In another research, Phillips et al. [85] used expectation maximization (EM) as a data-driven technique to select the fMRI weighting factors.

Integration through prediction refers to incorporation of EEG features as additional regressors for fMRI analysis. The main objective of these techniques is exploring the correlation between the fMRI time series and event-related potentials (ERPs) or EEG rhythm oscillations. The main superiority of these algorithms is the ability to directly use neural responses (measured by EEG) instead of using regressors relying only on the timings of the stimuli or tasks. One of the best examples is in investigating the interictal and ictal epileptic activity to localize epileptic foci and characterize the relationship between epileptic activity and the HR [86, 87]. In another research, Horovitz et al. [88] investigated the neural activations underlying the brain rhythm modulations in the rest state or in pathological brain. Formaggio et al. [89] studied the correlation between the changes of mu rhythm and BOLD signal peak. In both research studies, the regressors are derived from the power of a specific frequency band.

Recently, BSS and its variants have been applied to combine EEG and fMRI. This class of methods works based on measuring mutual dependence between the two modalities. In these approaches, first the original EEG and fMRI data are decomposed into several components. Then, they are cross-matched. Calhoun et al. [90] proposed a multivariate technique using ICA to analyze the features extracted from EEG and fMRI. In another research [91] an algorithm using parallel ICA has been developed for EEG–fMRI fusion. In their method the modalities have been integrated using a pair-wise matching across trial modulation. Joint ICA is another BSS variant proposed by Moosmann et al. [92] to link the components from the two modalities.

Tensor factorization as another variant of BSS may be used for the same purpose. PARAFAC2 has been employed to analyze EEG–fMRI data obtained from a simultaneous recording. Beckmann et al. [93] have used this technique to perform multi-subject and multi-session fMRI analysis. The main advantage of multi-way as compared to two-way data analysis techniques in fMRI applications is to extract meaningful features in more than two modes where the information diversities are more effectively exploited. In this application PARAFAC2 is first used to analyze fMRI with the aim of extracting the signal of interest in temporal, spatial, and slice modes. The main reason of using PARAFAC2 instead of PARAFAC is that we are dealing with a non-trilinear data mixture due to changing the size of the brain area scanned at each slice. Then, a semi-blind method based on PARAFAC2 is proposed to integrate EEG and fMRI. In the proposed method a constrained technique is developed to incorporate the time course obtained from the rolandic beta rhythm into the separation procedure. Imposing such a constraint leads to being able to separate the active area inside the brain which is highly correlated with the time course derived from the EEG signals. The proposed technique can be categorized as a data-driven technique in the family of EEG–fMRI fusion techniques based on prediction.

As part of our work in this area a partially constrained multi-way BSS has been developed as follows. Recalling the PARAFAC2 model, the cost function:

$$J\left(\mathbf{F}_q, \mathbf{A}, \mathbf{D}_1, \ldots, \boldsymbol{D}_q\right) = \sum\nolimits_{q=1}^{Q} \left\| \mathbf{X}_q - \mathbf{F}_q \mathbf{D}_q \mathbf{A}^T \right\|_F^2 \tag{18.22}$$

is subject to the constraint $\mathbf{F}_q^T\mathbf{F}_q = \mathbf{F}_p^T\mathbf{F}_p$ for all pairs $p, q = 1, ..., Q$. Here, $\mathbf{X}_q = \underline{\mathbf{X}}(: , : , q)^T$ is the transposed qth frontal slice of the tensor $\underline{\mathbf{X}}$ for $q = 1, ..., Q$. $\mathbf{A}$ is the component matrix in the first mode which is fixed for all slabs, $\mathbf{F}_q$ is the component matrix in the second mode corresponding to the q th frontal slice of $\underline{\mathbf{X}}$ and $\boldsymbol{D}_q$ is a diagonal matrix holding the q th row of the component matrix $\mathbf{C}$. The above cost function is reformulated as follows to allow us to impose the constraint:

$$J(\mathbf{P}_1, .., \mathbf{P}_q, \mathbf{F}, \mathbf{A}, \mathbf{D}_1, ..., \mathbf{D}_q) = \sum_{q=1}^{Q} \|\mathbf{X}_q - \mathbf{P}_q\mathbf{F}\mathbf{D}_q\mathbf{A}^T\|_F^2 \tag{18.23}$$

subject to $\mathbf{P}_q^T\mathbf{P}_q = \mathbf{I}_R$ and $\mathbf{D}_q$ is diagonal for all q. Using the method proposed by Kiers et al. [94], the PARAFAC2 problem is changed to a PARAFAC problem when $\mathbf{X}_q$ is replace by $\mathbf{P}_q^T\mathbf{X}_q$.

Assuming the data tensor $\underline{\mathbf{X}} \in \Re^{I\times J\times Q}$ contains the recorded fMRI images for one subject. Each fMRI volume recorded in one scan is composed of several slices. In order to arrange the fMRI data in a multi-way tensor, first the slices are converted to vector form. Then, they are inserted as rows of the tensor. Hence, $\underline{\mathbf{X}}(i, : , :)$ holds the recorded volume in the ith scan, $\underline{\mathbf{X}}(:,:, q)$ holds the qth slice of all recorded volumes during all scans and $\underline{\mathbf{X}}(: ,j, :)$ holds the recorded voxel in the jth spatial location. Therefore, matrices $\mathbf{A}$, $\mathbf{F}$, and $\mathbf{C}$ denote the loading factors in the temporal, spatial, and slice domains.

The available information about the loading factor in the temporal domain, extracted from the EEG signals, can then be used as a temporal constraint. In order to incorporate the prior information obtained by analysis of the EEG signals, the following constrained optimization problem is proposed:

$$J_{new}(\mathbf{A}, \mathbf{F}, \mathbf{C}, \mathbf{M}_{uk}, \mathbf{R}) = \|\mathbf{Y} - [\mathbf{A}, \mathbf{F}, \mathbf{C}]\|_F^2 + \lambda\|\mathbf{M} - \mathbf{A}\mathbf{R}^T\|_F^2 \quad \text{subject to } \mathbf{R}^T\mathbf{R} = \mathbf{I} \tag{18.24}$$

where $\boldsymbol{Y} = \boldsymbol{P}_q^T\boldsymbol{X}_q$, $\mathbf{M} \in \Re^{I\times L}$ is a matrix containing the prior information about the temporal signature of $\mathbf{Y}$ including known, $\mathbf{M}_k$, and unknown, $\mathbf{M}_{uk}$, parts, and $\mathbf{R} \in \Re^{L\times L}$ is the permutation matrix. $\|\bullet\|_F$ denotes Frobenius norm and λ is the regularization parameter which stabilizes the trade-off between the main part of the cost function and the constraint.

The constraint matrix $\mathbf{M}$ is designed such that its columns hold the regressors derived as the result of simultaneously recorded EEG analysis. Consider that K out of L columns of $\mathbf{M}$ are known. These columns indicate the available regressors to be used in the fMRI analysis. So, the constraint matrix is defined as $\mathbf{M} = [\mathbf{M}_k \vdots \mathbf{M}_{uk}]$ such that $\mathbf{M}_k \in \Re^{I\times K}$ and $\mathbf{M}_k \in \Re^{I\times (L-K)}$ are the known and unknown sub-matrices respectively. Hence, $\mathbf{M}$ may be written in the following form:

$$\mathbf{M} = \left[\mathbf{M}_k \vdots \mathbf{M}_{uk}\right] = \begin{bmatrix} m_{11}^k \ldots m_{1k}^k \; m_{1K+1}^{uk} \ldots m_{1L}^{uk} \\ \vdots\vdots \\ m_{I1}^k \ldots m_{Ik}^k \; m_{IK+1}^{uk} \ldots m_{IL}^{uk} \end{bmatrix} \tag{18.25}$$

This means each column of $\mathbf{M}$ refers to the time course of a stimulus. This allows detection of BOLDs for a number of stimuli at the same time. In this particular experiment there is only one stimulus and therefore, the first column of $\mathbf{M}$ is known (extracted from the EEG). This regressor is built up by convolving the extracted time course, denoting the onset of the rolandic beta rhythm, and HRF. Matrix $\mathbf{R}$ matches the constraint matrix $\mathbf{M}$ with the estimated factor for temporal mode $\mathbf{A}$. ALS is used to estimate the factors. Following ALS, the gradients of the cost function with respect to all the factors are calculated. For this purpose, the unfolded version of the data array is used. The factors are derived as follows:

$$\mathbf{A} \leftarrow \left(\mathbf{Y}_{(1)}(\mathbf{C} \odot \mathbf{F}) + \lambda \mathbf{M}\mathbf{R}\right)\left((\mathbf{C}^T\mathbf{C}) \otimes (\mathbf{F}^T\mathbf{F}) + \lambda \mathbf{R}^T\mathbf{R}\right)^{\dagger} \tag{18.26}$$

$$\mathbf{F} \leftarrow \mathbf{Y}_{(2)}(\mathbf{C} \odot \mathbf{A})\left((\mathbf{C}^T\mathbf{C}) \otimes (\mathbf{A}^T\mathbf{A})\right)^{\dagger} \tag{18.27}$$

$$\mathbf{C} \leftarrow \mathbf{Y}_{(3)}(\mathbf{F} \odot \mathbf{A})\left((\mathbf{F}^T\mathbf{F}) \otimes (\mathbf{A}^T\mathbf{A})\right)^{\dagger} \tag{18.28}$$

$$\mathbf{M}_{uk} \leftarrow \left(\left[\mathbf{A}\mathbf{R}^{\mathrm{T}}\right] : , K + 1 : L\right) \tag{18.29}$$

where $\odot$ and $\otimes$ are respectively Khatri-Rao and Hadamard products. In order to calculate the permutation matrix, $\mathbf{R}$, the same procedure as what has been used to compute $\mathbf{P}_k$ is performed. Since $\mathbf{R}$ is orthonormal, minimizing (18.22) over $\mathbf{R}$ is reduced to:

$$\mathbf{R} \leftarrow \arg\max_{\mathbf{R}} \left\{tr\left(\mathbf{R}\mathbf{A}^T\mathbf{M}\right)\right\} \tag{18.30}$$

Let $\mathbf{A}^T\mathbf{M} = \mathbf{U}\Sigma\mathbf{V}^T$, so that the unique minimum of (18.28) is obtained as $\mathbf{R} = \mathbf{V}\mathbf{U}^T$ $\mathbf{R} = \mathbf{V}\mathbf{U}^T$. The resulting matrix $\mathbf{R}$ is an orthonormal matrix. The performance improves if λ decreases with the number of iterations.

This method has been applied to the same real data as in the previous section and the results shown in Figure 18.18 present the extracted spatial pattern of BOLD with its

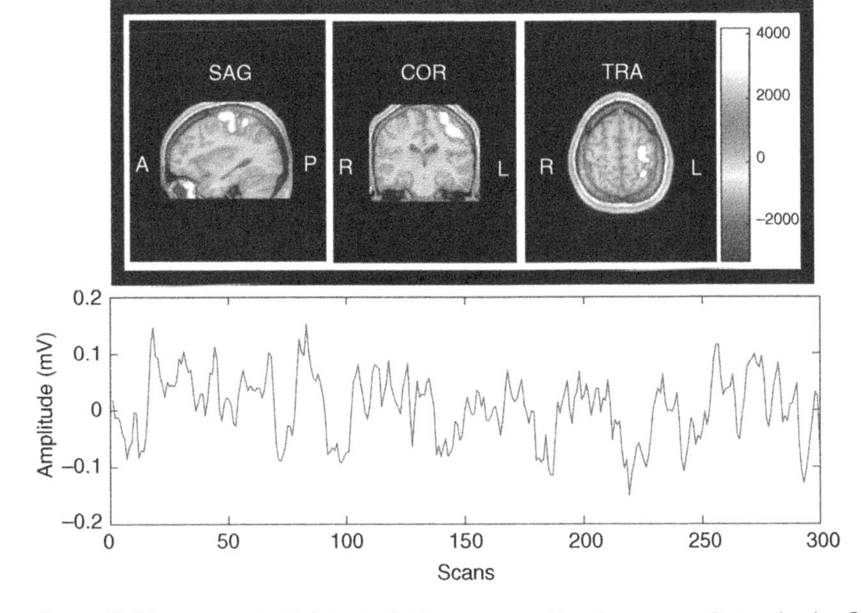

Figure 18.18 Detected BOLD (top) with its corresponding time course (bottom) using PARAFAC2. SAG, COR, and TRA refer respectively to sagittal, coronal, and transverse planes.

corresponding columns in matrix **A** and **C**. The results show activity in the left area of the primary motor cortex. The weights obtained in matrix **C** imply that BOLD has maximum contribution in slice number 62 for the subject. Using GLM on a similar data shows comparable results.

18.3.5.3 Fusion of EEG and fMRI; Model-Based Approach

The correlation between the PMBR and fMRI has been exploited in establishing a data-driven-based EEG–fMRI fusion which uses the extracted information from the EEG signal as a predictor for the BOLD signal. For this purpose, the power time course of the rolandic beta rhythm obtained by the proposed method in the previous section has been utilized. The calculated power time course represents the instantaneous interaction between the EEG activity in the beta band and motor task. The power time course is then convolved with HRF to make the regressor for fMRI analysis. This regressor is used to predict the BOLD response in fMRI data which are collected simultaneously with the EEG. A schematic illustration of data analysis is given in Figure 18.19.

18.3.6 Application to Seizure Detection

Despite limitations due to lack of a suitable time course, in pathological cases such as epilepsy, working under the simplifying assumption of linearity has been a successful strategy to put the high spatial resolution and noninvasiveness of fMRI to good use [95]. Therefore, a growing number of engineering and clinical studies have transformed EEG–fMRI into a powerful tool to investigate the hemodynamic changes associated with spontaneous brain activity in epileptic networks, including subcortical changes, but also to study endogenous brain rhythms of wakefulness and sleep, as well as evoked brain activity [25].

The interactions between neuronal populations and brain regions (macroscale, extrinsic connectivity) form the foundation for dynamic causal modelling (DCM). The EEG signal comes from the instantaneous electrical potentials generated by the pyramidal cells, spread through the head volume. At the cellular level, the fMRI signal is the end result of a metabolic and hemodynamic cascade occurring over a time scale in the order of seconds. Symmetric fusion models rely on finding inverse solutions of these models to estimate the neuronal activity and connectivity of the underlying neuronal population [25].

As a direct imaging tool of neuroelectric activity, the presurgical localization of epilepsy has been validated by several intracranial EEG studies and surgical series. Recent studies also suggest a good concordance between EEG–fMRI findings and similar gold standard localization tools. Studies comparing or combining localization and EEG–fMRI reveals that the combination of these techniques can provide new localizing and temporal information, with potential clinical relevance. The ongoing biophysical modelling and experimental developments attempt to solve the combined 'electrovascular inverse problem' and to uncover the precise nature of microscopic and macroscopic neurovascular coupling. The future translation from group studies of cognitive evoked-related potentials to individual patients with spontaneously occurring epileptic discharges is far from simple but will likely help us to better understand the underlying mechanisms of the generation, propagation, and termination of epileptic discharges. These techniques may also have wider application

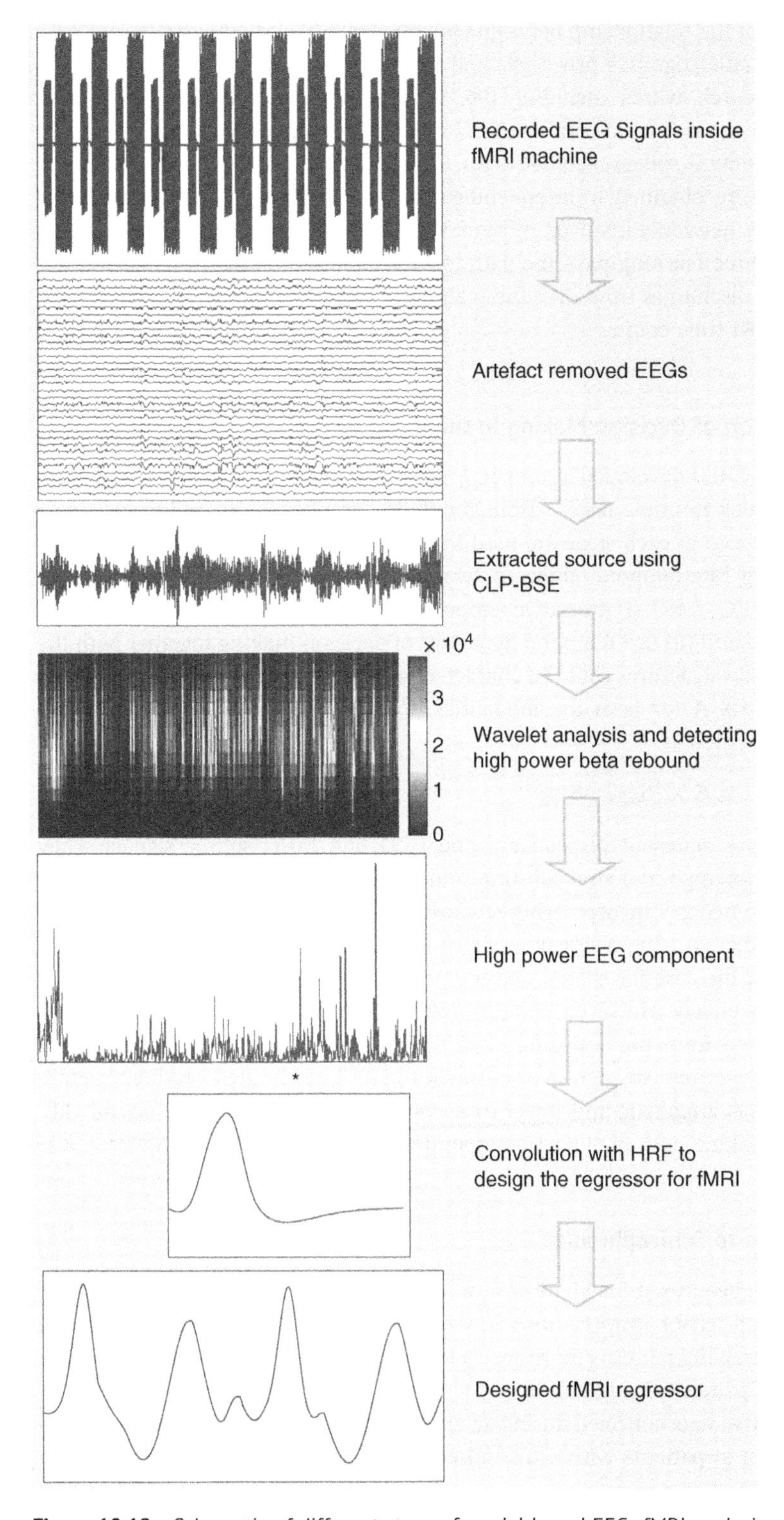

Figure 18.19 Schematic of different steps of model-based EEG–fMRI analysis.

in the investigation of the relationship between epileptic activity and other brain networks involved in resting state, cognitive processes, and sleep [25].

A number of research works including [96, 97] attempted developing new signal-processing tools and algorithms to combine EEG and fMRI in epilepsy studies.

Using marked interictal spikes recorded with EEG to define an activation protocol, a number of results were obtained from epileptogenic zone localization [98] to exploring effective connectivity networks involved in human epilepsy [99, 100]. Advances in signal processing and machine learning pave the path for detection of a larger number of intracranial epileptiform discharges from over the scalp [101–104] which may be directly used for the design of fMRI time course.

18.3.7 Investigation of Decision Making in the Brain

Kyathanahally et al. [105] developed a joint ICA technique for building a layered experimental design in which multiple tasks – from simple to complex – with additional layers of complexity introduced at each stage are used in decision making. This is demonstrated using tasks involving intertemporal choice between immediate and future prospects. In their method using EEG–fMRI, (i) the native temporal/spatial resolutions of either modality are not compromised and (ii) fast temporal dynamics of decision making together with the involved deeper striatal structures could be characterized. The joint ICA algorithm assumes a single mixing matrix **A** for both the modalities. Consequently, the acquired (mixed) signals can be represented as:

$$\left[\mathbf{X}_{EEG} \quad \mathbf{X}_{fMRI}\right] = \mathbf{A} \times \left[\mathbf{S}_{EEG} \quad \mathbf{S}_{fMRI}\right] \tag{18.31}$$

where $\mathbf{S}_{EEG}$ and $\mathbf{S}_{fMRI}$ represent respectively, the EEG and fMRI source signals. They showed that the spatiotemporal neural substrates underlying their proposed complex intertemporal task simultaneously incorporating rewards, costs, and uncertainty of future outcomes could be predicted using a linear model from neural substrates of each of these factors. This was not the case for spatial and temporal features obtained separately from fMRI and EEG, respectively. However, certain prefrontal activations in the complex task could not be predicted from the activations for simpler tasks, indicating that Donder's assumption of pure insertion (or cognitive subtraction) [106] has limited validity. Overall, our approach provides a realistic and novel framework for investigating the neural substrates of decision making with high spatiotemporal resolution.

18.3.8 Application to Schizophrenia

Joint EEG–fMRI has been used for identification of a biomarker for schizophrenia [107]. A coupled matrix and tensor factorizations (CMTF) approach has been followed in this work to capture underlying patterns more accurately without imposing strong constraints on the latent neural patterns, i.e. biomarkers.

EEG, fMRI, and sMRI data collected during an auditory oddball task (AOD) from a group of subjects consisting of patients with schizophrenia and healthy controls, were used in their work.

In their CMTF approach the coupling is along the subject mode. The proposed joint matrix-tensor parameters and latent variables are estimated by minimizing the following cost function through an alternating optimization approach [107]:

$$f(\lambda, \Sigma, \Gamma, \mathbf{A}, \mathbf{B}, \mathbf{C}, \mathbf{D}, \mathbf{E}) = \|\underline{\mathbf{X}} - [\lambda; \mathbf{A}, \mathbf{B}, \mathbf{C}]\|^2 + \left\|\mathbf{Y} - \mathbf{A}\Sigma\mathbf{D}^T\right\|^2 + \left\|\mathbf{Z} - \mathbf{A}\Gamma\mathbf{E}^T\right\|^2 + \beta\left(\|\lambda\|_1 + \|\boldsymbol{\sigma}\|_1 + \|\boldsymbol{\gamma}\|_1\right) \tag{18.32}$$

where $\underline{\mathbf{X}}$ is the fMRI tensor, $\mathbf{Y}$ is the EEG matrix, $\mathbf{Z}$ is the sMRI matrix, and $\boldsymbol{\sigma}$ and $\boldsymbol{\gamma}$ are the vectors constructed respectively from the diagonal elements of Σ and Γ. $\|\bullet\|_1$stands for l_1-norm operation.

Despite the small number of participants, they demonstrated significant and biologically meaningful components in terms of differentiating between patients with schizophrenia and healthy controls while also providing spatial patterns with high resolution and improving the clustering performance compared to the analysis of only the EEG tensor. In addition, they showed that these patterns are reproducible, and tested the reproducibility for different model parameters. In comparison to the jICA data fusion approach, their joint matrix-tensor factorization scheme provides easier interpretation of EEG data by revealing a single summary map of the topography for each component. Furthermore, fusion of sMRI data with EEG and fMRI through this model provides structural patterns.

18.3.9 Other Applications

Brain signalling occurs across a wide range of spatial and temporal scales, and analysis of brain signal variability and synchrony has attracted recent attention as markers of intelligence, cognitive states, and brain disorders. Having limited resolutions either in space or in time does not allow a full capture of spatiotemporal variability, leaving it untested whether temporal variability and spatiotemporal synchrony are a valid and reliable proxy of spatiotemporal variability *in vivo*.

In [105] the authors investigated functional connectivity (FC) and multiscale entropy (MSE) using optical voltage imaging in mice under anaesthesia and wakefulness to monitor cortical voltage activity at both high spatial and temporal resolutions. It was observed that across cortical space, MSE pattern can largely explain regional entropy (RE) pattern at small and large temporal scales with high positive and negative correlation respectively, while FC pattern strongly negatively associated with RE pattern. The time course of FC and small scale MSE tightly followed that of RE, while large scale MSE was more loosely coupled to RE.

The above voltage imaging data were used to simulate the fMRI and EEG data by reducing spatiotemporal resolution of the voltage imaging data or considering haemodynamics yielded MSE and FC measures that still contained information about RE based on the high-resolution voltage imaging data. It was concluded that MSE and FC could still be effective measures to capture spatiotemporal variability under limitation of imaging modalities applicable to human subjects. This verifies FC and MSE as the effective biomarkers for the recognition of brain states.

Ostwald et al. [108] investigated how a core cognitive network of the human brain, the perceptual decision system, can be characterized regarding its spatiotemporal representation of task-relevant information. They used an information-theoretic framework for the analysis of simultaneously acquired EEG–fMRI. They showed how the cognitive domain enables the economic description of neural spatiotemporal information encoding. By performing a visual perceptual decision task and using joint features, they demonstrated how the information-theoretic framework is able to reproduce earlier findings on the neurobiological underpinnings of perceptual decisions from the response signal features' marginal distributions. They tried to deduce an information-theoretic spatiotemporal EEG–fMRI signature by exploiting external state (through external stimulus, etc.), internal state (task related, e.g. attention, motivation, etc.) and behavioural state. Moreover, using the joint EEG–fMRI feature distribution, they provided evidence for a highly distributed and dynamic encoding of task relevant information in the human brain.

A joint common spatial pattern (CSP) approach has been proposed for simultaneous EEG–fMRI recording with the aim of classifying between the eye-closed and eye-open brain states [109]. It has been concluded that the use of joint modalities helps to better separate the CSPs within the brain.

Brookings et al. [110] developed two techniques to solve for the spatiotemporal neural activity patterns using joint EEG–fMRI to transform source localization into an over-constrained problem and produce a solution with the high temporal resolution of EEG and the high spatial resolution of fMRI. Their first method uses fMRI to regularize the EEG solution, while their second method uses ICA and a realistic model of BOLD to relate the EEG and fMRI data. They claimed that both techniques avoid the need for any assumption about the distribution of neural activity. Nevertheless, the solution by the second method is a more accurate inverse solution.

A combination of connectivity measures and joint ICA was used by Hinault et al. [111] to investigate the spatio-temporal neural correlates of cognitive control using simultaneous EEG–fMRI. They studied the spatiotemporal dynamic of the conflict adaptation effects (i.e. reduced interference on items that follow an incongruent stimulus compared to after a congruent stimulus). Joint ICA linked the N200 ERP component to activation of the anterior cingulate cortex (ACC) and the conflict slow potential to widespread activations within the frontoparietal executive control network. They also performed connectivity analyses with psychophysiological interactions and DCM demonstrating the coordinated engagement of the cognitive control network after processing of an incongruent item, which was correlated with better behavioural performance. They concluded that the anterior insula and inferior frontal gyrus are activated when incongruence is detected. These regions then signal the need for higher control to the ACC, which in turn activates the frontoparietal executive control network to improve the performance on the following trial.

18.4 EEG–NIRS Joint Recording and Fusion

Similar to EEG–fMRI systems, simultaneously recorded fNIRS can be fused with EEG and used in various applications [95]. Because of the absence of electro-optical interference, it is quite simple to integrate these two noninvasive recording modalities. fNIRS and EEG are both recorded from over the scalp. A montage for joint EEG–NIRS recording for imaginary movement (left) and Stroop experiments (right) can be seen in Figure 18.20. The Stroop Colour and Word Test (SCWT) is a neuropsychological test extensively used to assess the ability to inhibit cognitive interference that occurs when the processing of a specific stimulus

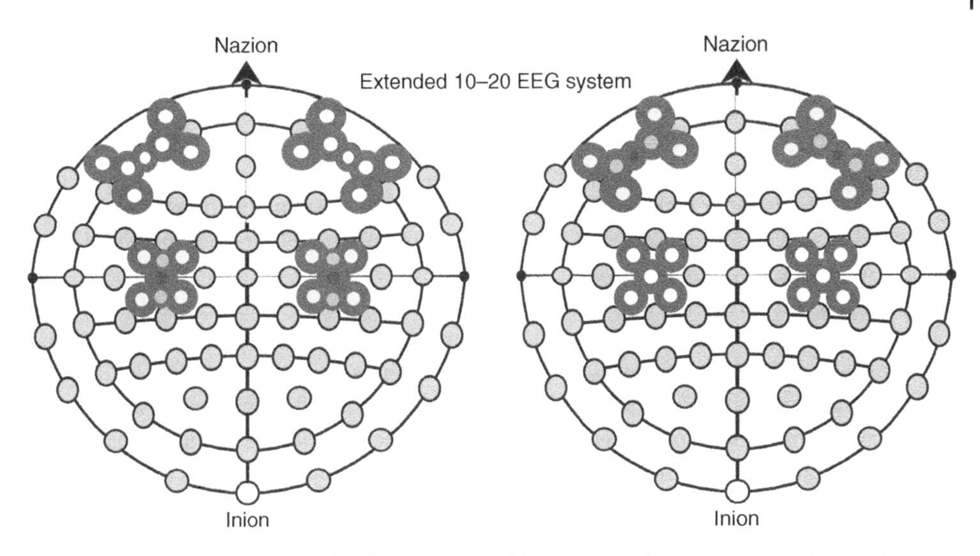

Figure 18.20 EEG electrode and fNIRS optode positions for imaginary movement (left) and Stroop, used for Stroop colour word interference test (right) applications.

feature impedes the simultaneous processing of a second stimulus attribute known as the Stroop Effect.

During an event or movement-related task or whatever changes the brain haemodynamic, there will be a considerable gap between HbO and HbR, with their ratio representing the strength of the task. An example can be seen in Figure 18.21 [15].

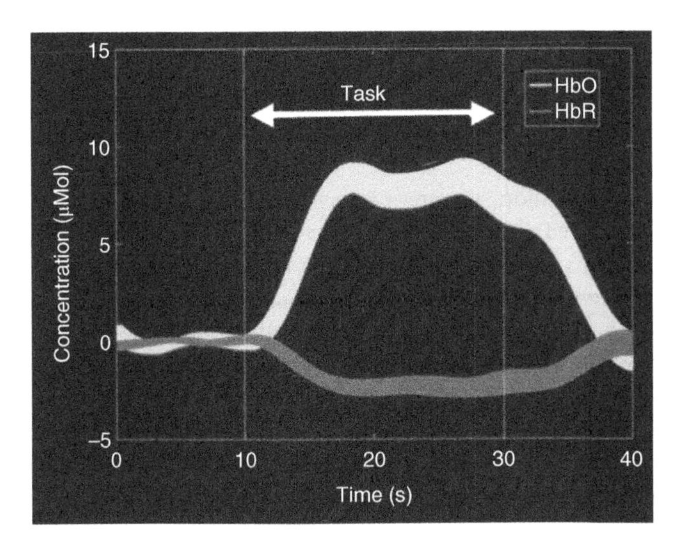

Figure 18.21 Example of a typical 'BOLD' response recorded by fNIRS in a task-activated brain region. Average changes in HbR and HbO concentrations are reported together with their variability for each time instant. The BOLD response in active brain areas is characterized by an over-compensatory supply of HbO with a concurrent wash-out of HbR, typically with a ratio (HbO/HbR) of ∼3 and consisting of a few μmol changes.

As with other imaging modalities, the primary potential use of fNIRS-EEG in epilepsy is in patient assessment for eventual surgical removal of the epileptogenic focus. A standard positive response with fNIRS in a controlled cognitive study shows a functional decrease in HbR with respect to baseline and a larger increase in HbO, resulting in a small increase of HbT which is a proxy for regional cerebral blood volume. However, seizures are infrequent and unpredictable and the question arises whether the same technique could also be applied to detect and characterize the HR to interictal epileptiform discharges (IEDs) which are, conversely, relatively frequent and good marker of activity and location of the epileptic generator though the HR to IEDs are much weaker than the HR to seizures.

In [112] the authors studied the HR to IEDs using simultaneous EEG–fNIRS recordings. For this study, the fNIRS montage was tailored for each patient, the sensitivity to the epileptic generator was optimized. The NIRS optodes were mounted for prolonged acquisition of the signals. The personalization is based on the identification of the epileptic focus and on the design of an optimal fNIRS montage specific for each subject. Without using any HR time course the fNIRS signals were averaged based on the presumed interictal epileptiform discharge (IED) temporal locations read from the concurrent EEG signals. This allowed detection of HR and its spatial and temporal localization.

The combined EEG–fNIRS applications include non-clinical and clinical applications. Brain–computer interfacing (BCI), neurovascular coupling, study of brain functions, emotions, and sleep are among the non-clinical applications. The clinical ones include newborn brain complexity (cerebral hypoxia, activity, and oxygenation immediately after birth, and seizure), brain injury or damage in children, epilepsy as mentioned previously, surgical intraoperative monitoring, psychiatric disorders such as bipolar disorder, schizophrenia and game addiction, and rehabilitation [15].

In a BCI application [113] simultaneously recorded EEG and fNIRS were used to examine inferior parietal lobule (IPL) activity with a high spatiotemporal resolution during single reaching movements. The experimental setup and the movement task can be seen in Figure 18.22. The authors proposed an ERD/ERS-based psychophysiological interaction analysis that estimates fluctuations in NIRS signals associated with the ERD/ERS of EEG signals.

Under a visual feedback-delay condition, gamma-ERS (γ-ERS), i.e. an increase in gamma (31–47 Hz) EEG power occurred during reaching movements. This γ-ERS is considered to reflect processing of information about prediction errors. To integrate this temporal information with spatial information from the fNIRS signals recorded using 16 pairs of optodes (around the motor area), a method was used to enable estimation of the regions that show a HR characterized by EEG fluctuation present in the visual feedback-delay condition.

The analysis involved [113] evaluation of movement trajectory in terms of movement pattern and speed, and the end point accuracy. Motor performance was evaluated according to the trajectory error and the endpoint error. Trajectory error was defined as the degree of deviation from the ideal trajectory drawn in the non-delay condition. Endpoint error was defined as the Euclidean distance from the centre of the goal square to the endpoint square in each trial. For the EEG, after ICA-based artefact removal, the wavelet transform was used to highlight and measure the ERD/ERS power. For the EEG the phase-phase coupling (PPC) between gamma and mu rhythms defined as:

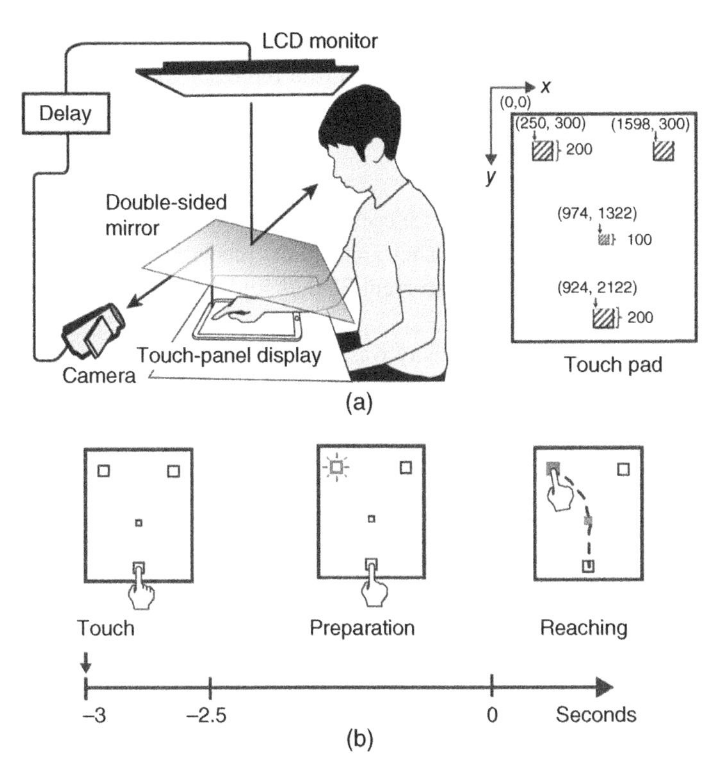

Figure 18.22 (a) Experimental setup and task procedure. The participant performs a reaching task using a tough panel display. He/she observes a moving image of their right hand projected onto a mirror. In the Delay condition, the feedback image is delayed to allow modulating the sense of agency with respect to his/her hand. (b) The participant is instructed to swipe his/her right index finger across the touch panel display from the home position to left/right goal areas via a central square. The movement is expected to be done as quickly and accurately as possible after seeing the 'go' cue. [113].

$$\mathrm{PPC}\left(f_{\mu}, f_{\gamma}\right) = \frac{1}{N}\left|\sum_{t=1}^{N} e^{\,j\left(\varphi\left(t, f_{\mu}\right) - \varphi\left(t, f_{\gamma}\right)\right)}\right| \tag{18.33}$$

has been considered too where t refers to the time point, N is the epoch length, and $\varphi(t, f_{\mu})$ and $\varphi(t, f_{\gamma})$ refer respectively to the mu and gamma phases at time instant t.

They also used a modulation index (MI) defined as [113]:

$$\mathrm{MI} = \frac{1}{N}\sum_{t=1}^{N} A_{\gamma}(t) e^{\,j\varphi_{\mu}(t)} \tag{18.34}$$

where $A_{\gamma}(t)$ and $\varphi_{\mu}(t)$ are respectively the time-varying gamma amplitude and mu phase. In addition, the imaginary part of coherency defined in Chapter 8 of this book was evaluated.

For the fNIRS analysis, the HbO data were filtered between 0.02 and 2 Hz and used. Then, the conventional GLM (also used in fMRI analysis) using a time course and the delayed version of that (as denoted by Delay in the experiment) was employed. To provide the design matrix and the fNIRS regressor the EEG cue and the ERD/ERS power were used. Assuming that the 'Delay' is an approximate time to reach the target, the regressor can be defined as [113]:

$$r_{fNIRS} = \beta_1 \bullet X_{ERD-ERS}^{Delay} + \beta_2 \bullet X_{ERD-ERS}^{non-Delay} + e \tag{18.35}$$

where $X_{ERD-ERS}^{non-Delay}$ and $X_{ERD-ERS}^{Delay}$ are the HR models for each condition represented by a box car function convolved with the Gaussian kernel. They are actually related to peaks in the ERD/ERS power spectrum measured from the concurrent EEG. β_1 and β_2 are the regressor coefficients and e is the error term [113].

As a result, the authors concluded that IPL activity was explained by γ-ERS specific to visual feedback delay during movements. This method is beneficial to investigate whether the IPL responds to the multisensory inconsistency caused by visual feedback delay during controlling a reaching movement. Their results suggest that, although the right IPL receives online prediction error signals, the IPL online response to the prediction error does not directly represent the conscious experience of movement agency.

18.5 MEG–EEG Fusion

Although joint MEG–EEG recording is not popular among researchers but fusion of the information from both modalities has been reported for many applications. Despite separate analysis of EEG and MEG with the help of spectral power, source localization, connectivity, etc., the differences between the two modalities have not yet been exploited. One such approach can be seen in [114] where the authors recorded EEG (64-channel system) and MEG (275 sensor system) simultaneously during open and closed eyes and evaluated spectral power, functional and effective connectivity (using partial directed coherence – PDC), relative power ratio, and spatial resolution were at five different conventional frequency bands. Generally, the parameters achieved from both modalities are very similar. This consistency can be seen in the connectivity patters too. Although in many cases, combining the two modalities provides more consistent results but the experiments also show that EEG outperforms MEG, mainly due to the high noise in MEG. For the parameter coherence EEG outperforms both the combined EEG + MEG approach and MEG alone [114].

18.6 Summary

During the past two decades tremendous amount of research has been carried out to combine and utilize the information achieved by simultaneous EEG and fMRI recordings and some efforts on EEG–NIRS as well as EEG–MEG fusions. After the EEGs are restored from various artefacts, three major lines of research have been developed within the biomedical signal-processing and machine-learning communities. These methods may be referred to as EEG-driven fMRI, fMRI-driven EEG, and joint EEG–fMRI signal processing. The important aspect of using fNIRS, albeit with some shortcomings compared to fMRI–EEG, is fewer artefacts on the EEG signals. Since fNIRS is cheaper to use, its application to BCI neonate brain injury/damage detection, and also a number of psychiatric diseases which affect prefrontal cortex is becoming more popular. Although there are several distinguished findings in these areas, unfortunately, none of these techniques has been widely used for clinical diagnosis and patient treatment. Evoked potentials and their influence on both EEG and fMRI modalities have been the main applications of the algorithms. Conversely, more precise restoration of EEG from BCG artefact needs more research. Simultaneous recording of EEG and MEG has been even less practised mainly because of their similarities and also high cost and less accessibility of MEG.

Despite the above three functional brain screening modalities new techniques such as functional molecular imaging through positron emission tomography (PET) or broadband microwave (ultrawideband) techniques may be able to provide more insight into the brain's structural and functional study.

References

1 Jezzard, P., Matthews, P.M., and Smith, S.M. (2001). Functional Magnetic Resonance Imaging: An Introduction to Methods. Published to Oxford Scholarship. Online: March 2012, doi: https://doi.org/10.1093/acprof:oso/9780192630711.001.0001.

2 Wright, G.A. (1997). Magnetic resonance imaging. *Signal Processing Magazine*. 14:1, 56–66. IEEE.

3 Lauterbur, P.C. (March 1973). Image formation by induced local interactions: examples employing nuclear magnetic resonance. *Nature* 242 (5394): 190–191.

4 Ogawa, S., Lee, T.M., Key, A.R., and Tank, D.W. (1990). Brain magnetic resonance imaging with contrast dependent on blood oxygenation. *Proceedings of the National Academy of Sciences of the United States of America* 87 (24): 9868–9872.

5 Pauling, L. and Coryell, C.D. (1936). The magnetic properties and structure of hemoglobin, oxyhemoglobin and carbonmonoxyhemoglobin. *Proceedings of the National Academy of Sciences of the United States of America* 22 (4): 210–216.

6 Anders, M.D. and Randy, L.B. (1997). Selective averaging of rapidly presented individual trials using fMRI. *Human Brain Mapping* 5 (5): 329–340.

7 Friston, K.J., Fletcher, P., Josephs, O. et al. (1998). Event-related fMRI: characterizing differential responses. *NeuroImage* 7 (1): 30–40.

8 Jezzard, P., Matthews, P.M., Smith, S.M., and Functional, M.R.I. (May 2003). *An Introduction to Methods*. USA: Oxford University Press.

9 Huettel, S.A., Song, A.W., and McCarthy, G. (2004). *Functional Magnetic Resonance Imaging*. Sunderland, MA: Sinauer Associates Inc.

10 Calhoun, V.D., Adali, T., Hansen, L.K. et al. (2003). ICA of functional MRI data: An overview. *Proceedings of the International Workshop on Independent Component Analysis and Blind Signal Separation*, 281–288.

11 Statistical parameter mapping (SPM). http://www.fil.ion.ucl.ac.uk/spm (accessed 19 August 2021).

12 Symms, M., Jager, H., Schmierer, K., and Yousry, T. (2004). A review of structural magnetic resonance neuroimaging. *Journal of Neurology, Neurosurgery, and Psychiatry* 75 (9): 1235–1244.

13 Di Domenico, S.I., Rodrigo, A.H., Dong, M.X. et al. (2019). Functional near-infrared spectroscopy: proof of concept for its application in social neuroscience. *Journal of Neuroergonomics*: 169–173.

14 Delpy, D.T., Cope, M., van der Zee, P. et al. (1988). Estimation of optical pathlength through tissue from direct time of flight measurement. *Physics in Medicine and Biology* 33 (12): 1433–1442.

15 Chiarelli, A.M., Zappasodi, F., Di Pompeo, F., and Merla, A. (2017). Simultaneous functional near-infrared spectroscopy and electroencephalography for monitoring of human brain activity and oxygenation: a review. *Neurophotonics* 4 (4): 041411. https://doi.org/10.1117/1.NPh.4.4.041411.

16 Scholkmann, F., Kleiser, S., Metz, A.J. et al. (2014). A review on continuous wave functional near-infrared spectroscopy and imaging instrumentation and methodology. *NeuroImage* 85: 6–27. https://doi.org/10.1016/j.neuroimage.2013.05.004.

17 Irani, F., Platek, S.M., Bunce, S. et al. (2007). Functional near infrared spectroscopy (fNIRS): an emerging neuroimaging technology with important applications for the study of brain disorders. *The Clinical Neuropsychologist* 21: 9–37. https://doi.org/10.1080/13854040600910018.

18 Eschweiler, G.W., Wegerer, C., Schlotter, W. et al. (2000). Left prefrontal activation predicts therapeutic effects of repetitive transcranial magnetic stimulation (rTMS) in major depression. *Psychiatry Research* 99 (3): 161–172.

19 Akiyoshi, J., Hieda, K., Aoki, Y., and Nagayama, H. (2003). Frontal brain hypoactivity as a biological substrate of anxiety in patients with panic disorders. *Neuropsychobiology* 47 (3): 165–170.

20 Matsuo, K., Kato, T., Taneichi, K. et al. (2003). Activation of the prefrontal cortex to trauma-related stimuli measured by near-infrared spectroscopy in posttraumatic stress disorder due to terrorism. *Psychophysiology* 40 (4): 492–500.

21 Darvas, F., Pantazis, D., Kucukaltun-Yildirim, E., and Leahy, R.M. (2004). Mapping human brain function with MEG and EEG: methods and validation. *NeuroImage* 23 (Suppl 1): S289–S299.

22 Mohseni, H.R., Wilding, E., and Sanei, S. (Oct. 2010). Variational Bayes for spatiotemporal identification of event-related potential subcomponents. *IEEE Transactions on Biomedical Engineering* 57 (10): 2413–2428.

23 Mohseni, H.R. and Sanei, S. (2010). Anew beamforming-based MEG dipole source localization method. *Proceeding of the IEEE International Conference on Acoustics, Speech and Signal Processing, ICASSP*, USA.

24 Singh, S.P. (2014). Magnetoencephalography: basic principles. *Annals of Indian Academy of Neurology* 17 (Suppl 1): S107–S112. https://doi.org/10.4103/0972-2327.128676.

25 Vulliemoz, S., Lemieux, L., Daunizeau, J. et al. (2009). The combination of EEG source imaging and EEG-correlated functional MRI to map epileptic networks. *Epilepsia*: 1–15. https://doi.org/10.1111/j.1528-1167.2009.02342.x.

26 Logothetis, N.K. (June 2008). What we can do and what we cannot do with fMRI. *Nature* 453: 869–878.

27 Shmuel, A. (2010). *Locally Measured Neuronal Correlates of Functional MRI Signals.* Springer Berlin Heidelberg.

28 Logothetis, N.K., Pauls, J., Augath, M. et al. (July 2001). Neurophysiological investigation of the basis of the fMRI signal. *Nature* 412 (6843): 150–157.

29 Ogawa, S., Lee, T.M., Stepnoski, R. et al. (2000). An approach to probe some neural systems interaction by functional MRI at neural time scale down to milliseconds. *Proceedings of the National Academy of Sciences of the United States of America*: 11026–11031.

30 Rees, G., Friston, K., and Koch, C. (2000). A direct quantitative relationship between the functional properties of human and macaque V5. *Nature Neuroscience* 3 (7): 716–723.

31 Heeger, D.J., Huk1, A.C., Geisler, W.S., and Albrech, D.G. (July 2000). Spikes versus BOLD: what does neuroimaging tell us about neuronal activity? *Nature Neuroscience* 3 (7): 631–633.

32 Price, C.J. and Friston, K.J. (1997). Cognitive conjunction: a new approach to brain activation experiments. *NeuroImage* 5 (4): 261–270.

33 Calhoun, V.D., Stevens, M.C., Pearlson, G.D., and Kiehl, K.A. (2004). fMRI analysis with the general linear model: removal of latency-induced amplitude bias by incorporation of hemodynamic derivative terms. *NeuroImage* 22 (1): 252–257.

34 Friston, K.J., Holmes, A.P., Worsley, K.J. et al. (1994). Statistical parametric maps in functional imaging: a general linear approach. *Human Brain Mapping* 2 (4): 189–210.

35 Jing, M. (2008). Predictability of epileptic seizures by fusion of scalp EEG and fMRI. PhD thesis. Cardiff University, UK.

36 http://www.fmrib.ox.ac.uk/fsl (accessed 19 August 2021). FMRIB Software Library v6.0, Created by the Analysis Group under Stephen Smith, Welcome Centre Integrative Neuroimaging, Oxford University, UK. (https://www.win.ox.ac.uk/research/analysis-research/analysis-research), FMRIB, Oxford, UK.

37 Steyrl, D. and Müller-Putz, G.R. (2019). Artifacts in EEG of simultaneous EEG-fMRI: pulse artifact remainders in the gradient artifact template are a source of artefact residuals after average artifact subtraction. *Journal of Neural Engineering* 16: 016011, 11 pages.

38 Allen, P.J., Polizzi, G., Krakow, K. et al. (1998). Identification of EEG events in the MR scanner: the problem of pulse artefact and a method for its subtraction. *NeuroImage* 8 (3): 229–239.

39 Goldman, R.I., Stern, J.M., Engel, J., and Cohen, M.S. (2000). Acquiring simultaneous EEG and functional MRI. *Clinical Neurophysiology* 111: 1974–1980.

40 Kruggel, F., Wiggins, C.J., Herrmann, C.S., and von Cramon, D.Y. (2000). Recording of the event-related potentials during functional MRI at 3.0 tesla field strength. *Magnetic Resonance in Medicine* 44 (2): 277–282.

41 Seeck, M., Lazeyras, F., Michel, C.M. et al. (1998). Non-invasive epileptic focus localization using EEG-triggered functional MRI and electromagnetic tomography. *Electroencephalography and Clinical Neurophysiology* 106 (6): 508–512.

42 Allen, P.J., Josephs, O., and Turner, R. (August 2000). A method for removing imaging artefact from continuous EEG recorded during functional MRI. *NeuroImage* 12 (2): 230–239.

43 Bonmassar, G., Purdon, P.L., Jskelinen, I.P. et al. (2002). Motion and ballistocardiogram artefact removal for interleaved recording of EEG and EPs during MRI. *NeuroImage* 16 (4): 1127–1141.

44 Niazy, R.K., Beckmann, C.F., Iannetti, G.D. et al. (2005). Removal of fMRI environment artefacts from EEG data using optimal basis sets. *NeuroImage* 28 (3): 720–737.

45 Mahadevan, A., Acharya, A., Sheffer, S., and Mugler, D.H. (2008). Ballistocardiogram artefact removal in EEG-fMRI signals using discrete Hermite transforms. *IEEE Journal of Selected Topics in Signal Processing* 2 (6): 839–853.

46 Mantini, D., Perrucci, M.G., Cugini, S. et al. (2007). Complete artefact removal for EEG recorded during continuous fMRI using independent component analysis. *NeuroImage* 34 (2): 698–607.

47 Nakamura, W., Anami, K., Mori, T. et al. (July 2006). Removal of ballistocardiogram artefacts from simultaneously recorded EEG and fMRI data using independent component analysis. *IEEE Transactions on Biomedical Engineering* 53 (7): 1294–1308.

48 Ghaderi, F., Nazarpour, K., McWhirter, J.G., and Sanei, S. (July 2010). Removal of ballistocardiogram artefacts using the cyclostationary source extraction method. *IEEE Transactions on Biomedical Engineering* 57 (11): 2667–2676.

49 Leclercq, Y., Balteau, E., Dang-Vu, T. et al. (2009). Rejection of pulse related artefact (PRA) from continuous electroencephalographic (EEG) time series recorded during functional magnetic resonance imaging (fMRI) using constraint independent component analysis (cICA). *NeuroImage* 44 (3): 679–691.

50 Dyrholm, M., Goldman, R., Sajda, P., and Brown, T.R. (2009). Removal of BCG artefacts using a non-kirchhoffian overcomplete representation. *IEEE Transactions on Biomedical Engineering* 56 (2): 200–204.

51 S. Ferdowsi, S., Sanei, J., Nottage, O., O'Daly et al. (2012). A hybrid ICA-Hermite transform for removal of ballistocardiogram from EEG. *Proceedings of the European Signal Processing Conference, EUSIPCO*. Romania.

52 Ferdowsi, S., Abolghasemi, V., and Sanei, S. (2012). Blind separation of balistocardiogram from EEG via short- and long-term linear prediction filtering. *Proceedings of Machine Learning and Signal Processing, MLSP*. Spain.

53 Haykin, S. (2001). *Adaptive Filter Theory*, 4e. Prentice Hall.

54 Ulfarsson, M.O. and Solo, V. (2007). Sparse variable principal component analysis with application to fMRI. *4th IEEE International Symposium on Biomedical Imaging: From Nano to Macro*, 460–463.

55 Wang, X., Tian, J., Li, X. et al. (2004). Detecting brain activations by constrained non-negative matrix factorization from task-related BOLD fMRI. *Proceeding SPIE 5369, Medical Imaging 2004: Physiology, Function, and Structure from Medical Images*, 675.

56 Biswal, B., Yetkin, F.Z., Haughton, V.M., and Hyde, J.S. (October 1995). Functional connectivity in the motor cortex of resting humasn brain using echo-planar MRI. *Magnetic Resonance in Medicine* 34 (4): 537–541.

57 Mckeown, M.J., Makeig, S., Brown, G.G. et al. (1998). Analysis of fMRI data by blind separation into independent spatial components. *Human Brain Mapping* 6: 160–188.

58 Calhoun, V.D., Adali, T., Pearlson, G.D., and Pekar, J.J. (2001). Spatial and temporal independent component analysis of functional MRI data containing a pair of task-related waveforms. *Human Brain Mapping* 13: 43–53.

59 Stone, J.V., Porrill, J., Porter, N.R., and Wilkinson, I.D. (2002). Spatiotemporal independent component analysis of event-related fMRI data using skewed probability density functions. *NeuroImage* 15 (2): 407–421.

60 Cichocki, A. (2002). ICALAB. http://www.bsp.brain.riken.jp/icalab.

61 Correa, N., Adali, T., Li, Y., and Calhoun, V.D. (2005). Comparison of blind source separation algorithms for fMRI using a new Matlab toolbox: GIFT. *Proceedings of the IEEE International Conference on Acoustics, Speech, and Signal Processing, (ICASSP)*, 5: v/401–v/404.

62 Calhoun, V.D. (2004). Group ICA of fMRI toolbox (GIFT) manual.

63 Wang, X., Tian, J., Yang, L., and Hu, J. (2005). Clustered cNMF for fMRI data analysis. *Medical Images* 5746: 631–638.

64 Cichocki, A., Zdunek, R., and Amari, S.I. (2006). Csiszár's divergences for non-negative matrix factorization: family of new algorithms. LNCS, 32–39. Springer.

65 Dhillon, I.S. and Sra, S. (2005). Generalized nonnegative matrix approximations with Bregman divergences. Neural Information Processing Systems. 283–290.

66 Kompass, R. (2007). A generalized divergence measure for nonnegative matrix factorization. *Neural Computing* 19: 780–791.

67 Cichocki, A., Amari, S.I., Zdunek, R. et al. (2006). Extended smart algorithms for non-negative matrix factorization. In: *Artificial Intelligence and Soft Computing ICAISC 2006*, Lecture Notes in Computer Science, vol. 4029 (eds. L. Rutkowski, R. Tadeusiewicz, A.L. Zadeh and J.M. Żurada), 548–562.

68 Cichocki, A., Lee, H., Kim, Y.D., and Choi, S. (2008). Non-negative matrix factorization with α-divergence. *Pattern Recognition Letters* 29 (9): 1433–1440.

69 Lee, D.D. and Seung, H.S. (2001). Algorithms for non-negative matrix factorization. *Advances in Neural Information Processing Systems*, 13: 556–562. MIT Press.

70 Ferdowsi, S., Abolghasemi, V., and Sanei, S. (2015). A new informed tensor factorization approach to EEG-fMRI fusion. *Journal of Neuroscience Methods*, 254: 27–35.

71 Pfurtscheller, G. and Lopes da Silva, F.H. (1999). Event-related EEG/MEG synchronization and desynchronization: basic principles. *Clinical Neurophysiology* 110 (11): 1842–1857.

72 Jasper, H. and Penfield, W. (1949). Electrocorticograms in man: effect of voluntary movement upon the electrical activity of the precentral gyrus. *European Archives of Psychiatry and Clinical Neuroscience* 183: 163–174.

73 Jong, R., Gladwin, T.E., and Hart, B.M. (2006). Movement-related EEG indices of preparation in task switching and motor control. *Brain Research* 1105 (1): 73–82.

74 Krusienski, D.J., Schalk, G., McFarland, D.J., and Wolpaw, J.R. (2007). A μ-rhythm matched filter for continuous control of a brain-computer interface. *IEEE Transactions on Biomedical Engineering* 54 (2): 273–280.

75 Pfurtscheller, G., Stanck, A. Jr., and Edlinger, G. (1997). On the existence of different types of central beta rhythms below 30 Hz. *Electroencephalography and Clinical Neurophysiology* 102 (4): 316–325.

76 Stevenson, C.M., Brookes, M.J., and Morris, P.G. (2011). β-band correlates of the fMRI bold response. *Human Brain Mapping* 32 (2): 182–197.

77 Valdes-Sosa, P.A., Sanchez-Bornot, J.M., Sotero, R.C. et al. (2009). Model driven EEG/fMRI fusion of brain oscillations. *Human Brain Mapping* 30 (9): 2701–2721.

78 Fujimaki, N., Hayakawa, T., Nielsen, M. et al. (2002). An fMRIConstrained MEG source analysis with procedures for dividing and grouping activation. *NeuroImage* 17 (1): 324–343.

79 Ahlfors, S.P. and Simpson, G.V. (2004). Geometrical interpretation of fMRI-guided MEG/EEG inverse estimates. *NeuroImage* 22 (1): 323–332.

80 Liu, Z. and He, B. (2008). fMRI-EEG integrated cortical source imaging by use of time variant spatial constraints. *NeuroImage* 39 (3): 1198–1214.

81 Liu, Z., Kecman, F., and He, B. (2006). Effects of fMRIEEG mismatches in cortical current density estimation integrating fMRI and EEG: a simulation study. *Clinical Neurophysiology* 117 (7): 1610–1622.

82 Babiloni, F., Babiloni, C., Carducci, F. et al. (2003). Multimodal integration of high-resolution EEG and functional magnetic resonance imaging data: a simulation study. *NeuroImage* 19 (1): 1–15.

83 Wagner, M., Fuchs, M., and Kastner, J. (200). fMRI-constrained dipole fits and current density reconstructions. *Proceedings of the 12th International Conference on Biomaging*, 785–788.

84 Liu, A.K., Belliveau, J.W., and Dale, A.M. (1998). Spatiotemporal imaging of human brain activity using functional MRI constrained magnetoencephalography data: Monte Carlo simulations. *Proceedings of the National Academy of Sciences of the United States of America* 95: 8945–8950.

85 Phillips, C., Mattout, J., Rugg, M.D. et al. (Feb 2005). An empirical Bayesian solution to the source reconstruction problem in EEG. *NeuroImage* 24 (4): 997–1011.

86 Al-Asmi, A., Benar, C.G., Gross, D.W. et al. (2003). fMRI activation in continuous and spike-triggered EEG-fMRI studies of epileptic spikes. *Epilepsia* 44 (10): 1328–1339.

87 Gotman, J., Grova, C., Bagshaw, A. et al. (Oct. 18, 2005). Generalized epileptic discharges show thalamocortical activation and suspension of the default state of the brain. *Proceedings of the National Academy of Sciences of the United States of America* 102 (42): 15236–15240.

88 Horovitz, S.G., Fukunaga, M., de Zwart, J.A. et al. (2008). Low frequency BOLD fluctuations during resting wakefulness and light sleep: a simultaneous EEG-fMRI study. *Human Brain Mapping* 29 (6): 671–682.

89 Formaggio, E., Storti, S., Avesani, M. et al. (2008). EEG and fMRI coregistration to investigate the cortical oscillatory activities during finger movement. *Brain Topography* 21 (2): 100–111.

90 Calhoun, V.D. and Adali, T. (2009). Feature-based fusion of medical imaging data. *IEEE Transactions on Information Technology in Biomedicine* 13 (5): 711–720.

91 Eichele, T., Calhoun, V.D., Moosmann, M. et al. (2008). Unmixing concurrent EEG-fMRI with parallel independent component analysis. *International Journal of Psychophysiology* 67 (3): 222–234.

92 Moosmann, M., Eichele, T., Nordby, H. et al. (2008). Joint independent component analysis for simultaneous EEG-fMRI: principle and simulation. *International Journal of Psychophysiology* 67 (3): 212–221.

93 Beckmann, C.F. and Smith, S.M. (2005). Tensorial extensions of independent component analysis for multisubject fMRI analysis. *NeuroImage* 25 (1): 294–311.

94 Kiers, H.A.L., ten Berge, J.M.F., and Bro, R. (1999). PARAFAC2-part I. a direct fitting algorithm for the PARAFAC2 model. *Journal of Chemometrics* 13 (3–4): 275–294.

95 Pouliot, P., Tremblay, J., Robert, M. et al. (2012). Nonlinear hemodynamic responses in human epilepsy: a multimodal analysis with fNIRS-EEG and fMRI-EEG. *Journal of Neuroscience Methods* 204: 326–340.

96 Jing, M. and Sanei, S. (2009). Simultaneous EEG-fMRI Analysis with Application to Detection and Localizaion of Seizure Signal Sources, Recent Advances in Signal Processing, IN-TECH Pub., (ed. A.A. Zaher) ISBN 978-953-307-002-5.

97 Jing, M., Sanei, S., and Hamandi, K. (2008) A novel ICA approach for separation of seizure BOLD from fMRI, constrained by the simultaneously recorded EEG signals. *Proceedings of the European Signal Processing Conference, EUSIPCO*. Switzerland.

98 Krakow, K., Lemieux, L., Messina, D. et al. (2001). Spatiotemporal imaging of focal interictal epileptiform activity using EEG-triggered functional MRI. *Epileptic Disorders* 3 (2): 67–74.

99 Vaudano, A.E., Laufs, H., Kiebel, S.J. et al. (2009). Causal hierarchy within the thalamo-cortical network in spike and wave discharges. *PLoS One* 4 (8): e6475.

100 Hamandi, K., Powell, H.W.R., Laufs, H. et al. (2008). Combined EEG-fMRI and tractography to visualise propagation of epileptic activity. *Journal of Neurology, Neurosurgery & Psychiatry* 79 (5): 594–597.

101 Antoniades, A., Spyrou, L., Martin-Lopez, D. et al. (2018). Deep neural architectures for mapping scalp to intracranial EEG. *International Journal of Neural Systems* https://doi.org/10.1142/S0129065718500090, online.

102 Antoniades, A., Spyrou, L., Martin-Lopez, D. et al. (2017). Detection of interictal discharges using convolutional neural networks from multichannel intracranial EEG. *IEEE Transactions on Neural Systems and Rehabilitation Engineering* 25 (12): 2285–2294.

103 Spyrou, L., Lopez, D.M., Alarcon, G. et al. (2016). Detection of intracranial signatures of interictal epileptiform discharges from concurrent scalp EEG. *International Journal of Neural Systems* 26 (04) (IF = 7.2), DOI: https://doi.org/10.1142/S0129065716500167.

104 Spyrou, L. and Sanei, S. (2016) Coupled dictionary learning for multimodal data: an application to concurrent intracranial and scalp EEG. *Proceedings of the IEEE International Conference on Acoustics, Speech, and Signal Processing (ICASSP)*. Shanghai, China.

105 Liu, M., Song, C., Liang, Y. et al. (2019). Assessing spatiotemporal variability of brain spontaneous activity by multiscale entropy and functional connectivity. *NeuroImage* 198: 198–220.

106 Ulrich, R., Mattes, S., and Miller, J. (July 1999). Donders's assumption of pure insertion: an evaluation on the basis of response dynamics. *Acta Psychologica* 102 (1): 43–76.

107 Acar, E., Schenker, C., Levin-Schwartz, Y. et al. (May 2019). Unraveling diagnostic biomarkers of schizophrenia through structure-revealing fusion of multi-modal neuroimaging data. *Frontiers in Neuroscience* 13: 416, 16 pages.

108 Ostwald, D., Porcaro, C., Mayhew, S.D., and Bagshaw, A.P. (2012). EEG-fMRI based information theoretic characterization of the human perceptual decision system. *PLoS One* 7 (4): e33896.

109 Tan, A., Fu, Z., Tu, Y. et al. (2015). Joint source separation of simultaneous EEG-fMRI recording in two experimental conditions using common spatial patterns. *Proceedings of the 37th Annual International Conference of the IEEE Engineering in Medicine and Biology Society (EMBC).*

110 Brookings, T., Ortigue, S., Grafton, S., and Carlson, J. (2009). Joint source separation of simultaneous EEG-fMRI recording in two experimental conditions using common spatial patterns. *Journal of Neuroimage* 44: 411–420.

111 Hinault, T., Larcher, K., Zazubovits, N. et al. (2019). Spatio–temporal patterns of cognitive control revealed with simultaneous electroencephalography and functional magnetic resonance imaging. *Human Brain Mapping* 40: 80–97.

112 Pellegrino, G., Machado, A., Ellenrieder, N. et al. (March 2016). Hemodynamic response to interictal epileptiform discharges addressed by personalized EEG-fNIRS recordings. *Frontiers in Neuroscience* 10: 102, 20 pages.

113 Zama, T., Takahashi, Y., and Shimada, S. (2019). Simultaneous EEG-NIRS measurement of the inferior parietal lobule during a reaching task with delayed visual feedback. *Frontiers in Human Neuroscience* 13: 301. https://doi.org/10.3389/fnhum.2019.00301.

114 Muthuraman, M., Moliadze, V., Mideksa, K.G. et al. (2015). EEG-MEG integration enhances the characterization of functional and effective connectivity in the resting state network. *PLoS One* https://doi.org/10.1371/journal.pone.0140832, 23 pages.

Index

b

d

e

g

h

i

j

k

l

m

n

O

p

S

t

u

y

z

Printed and bound by CPI Group (UK) Ltd, Croydon, CR0 4YY

16/04/2025

14658461-0005